Fiesta Owners Workshop Manual

T. H. Robert Jones

Models covered
Ford Fiesta models with petrol engines; Hatchback, Van and Courier, including XR2i, RS Turbo, RS 1800, CTX transmission models & special/limited editions
999 cc, 1118 cc, 1297 cc, 1392 cc, 1596 cc & 1796 cc

Does not cover Diesel engine models

(1595 – 3W2)   ABCD

Haynes Publishing
Sparkford Nr Yeovil
Somerset BA22 7JJ England

Haynes Publications, Inc
861 Lawrence Drive
Newbury Park
California 91320 USA

Acknowledgements

Thanks are due to Champion Spark Plug who supplied the illustrations showing spark plug conditions, to Holt Lloyd Limited who supplied the illustrations showing bodywork repair, and to Duckhams Oils who provided lubrication data. Certain other illustrations are the copyright of the Ford Motor Company, and are used with their permission. Thanks are also due to Sykes-Pickavant, who supplied some of the workshop tools, and to all the staff at Sparkford who assisted in the production of this manual.

© **Haynes Publishing 1994**

A book in the **Haynes Owners Workshop Manual Series**

Printed by J. H. Haynes & Co. Ltd., Sparkford, Nr Yeovil, Somerset BA22 7JJ, England

All rights reserved. No part of this book may be reproduced or transmitted in any form or by any means, electronic or mechanical, including photocopying, recording or by any information storage or retrieval system, without permission in writing from the copyright holder.

ISBN 1 85010 920 6

British Library Cataloguing in Publication Data
A catalogue record for this book is available from the British Library

We take great pride in the accuracy of information given in this manual, but vehicle manufacturers make alterations and design changes during the production run of a particular vehicle of which they do not inform us. No liability can be accepted by the authors or publishers for loss, damage or injury caused by any errors in, or omissions from, the information given.

Restoring and Preserving our Motoring Heritage

Few people can have had the luck to realise their dreams to quite the same extent and in such a remarkable fashion as John Haynes, Founder and Chairman of the Haynes Publishing Group.

Since 1965 his unique approach to workshop manual publishing has proved so successful that millions of Haynes Manuals are now sold every year throughout the world, covering literally thousands of different makes and models of cars, vans and motorcycles.

A continuing passion for cars and motoring led to the founding in 1985 of a Charitable Trust dedicated to the restoration and preservation of our motoring heritage. To inaugurate the new Museum, John Haynes donated virtually his entire private collection of 52 cars.

Now with an unrivalled international collection of over 210 veteran, vintage and classic cars and motorcycles, the Haynes Motor Museum in Somerset is well on the way to becoming one of the most interesting Motor Museums in the world.

A 70 seat video cinema, a cafe and an extensive motoring bookshop, together with a specially constructed one kilometre motor circuit, make a visit to the Haynes Motor Museum a truly unforgettable experience.

Every vehicle in the museum is preserved in as near as possible mint condition and each car is run every six months on the motor circuit.

Enjoy the picnic area set amongst the rolling Somerset hills. Peer through the William Morris workshop windows at cars being restored, and browse through the extensive displays of fascinating motoring memorabilia.

From the 1903 Oldsmobile through such classics as an MG Midget to the mighty 'E' Type Jaguar, Lamborghini, Ferrari Berlinetta Boxer, and Graham Hill's Lola Cosworth, there is something for everyone, young and old alike, at this Somerset Museum.

Haynes Motor Museum

Situated mid-way between London and Penzance, the Haynes Motor Museum is located just off the A303 at Sparkford, Somerset (home of the Haynes Manual) and is open to the public 7 days a week all year round, except Christmas Day and Boxing Day.

Contents

	Page
Acknowledgements	2
About this manual	5
Introduction to the Ford Fiesta	5
General dimensions, weights and capacities *(also see Chapter 13, page 292)*	7
Jacking, towing and wheel changing	8
Buying spare parts and vehicle identification numbers *(also see Chapter 13, page 292)*	10
General repair procedures	12
Tools and working facilities	13
Safety first!	15
Routine maintenance *(also see Chapter 13, page 292)*	16
Recommended lubricants and fluids	23
Conversion factors	24
Fault diagnosis	25
Chapter 1 Engine *(also see Chapter 13, page 292)*	28
Chapter 2 Cooling, heating and ventilation systems *(also see Chapter 13, page 292)*	76
Chapter 3 Fuel, exhaust and emission control systems *(also see Chapter 13, page 292)*	86
Chapter 4 Ignition and engine management systems *(also see Chapter 13, page 292)*	132
Chapter 5 Clutch *(also see Chapter 13, page 292)*	149
Chapter 6 Transmission *(also see Chapter 13, page 292)*	155
Chapter 7 Driveshafts *(also see Chapter 13, page 292)*	184
Chapter 8 Braking system *(also see Chapter 13, page 292)*	190
Chapter 9 Steering *(also see Chapter 13, page 292)*	209
Chapter 10 Suspension *(also see Chapter 13, page 292)*	216
Chapter 11 Bodywork and fittings *(also see Chapter 13, page 292)*	230
Chapter 12 Electrical system *(also see Chapter 13, page 292)*	255
Chapter 13 Supplement: Revisions and information on later models	292
Index	390

Spark plug condition and bodywork repair colour section between pages 32 and 33

Ford Fiesta 1.1 LX

About this manual

Its aim

The aim of this manual is to help you get the best value from your vehicle. It can do so in several ways. It can help you decide what work must be done (even should you choose to get it done by a garage), provide information on routine maintenance and servicing, and give a logical course of action and diagnosis when random faults occur. However, it is hoped that you will use the manual by tackling the work yourself. On simpler jobs it may even be quicker than booking the car into a garage and going there twice, to leave and collect it. Perhaps most important, a lot of money can be saved by avoiding the costs a garage must charge to cover its labour and overheads.

The manual has drawings and descriptions to show the function of the various components so that their layout can be understood. Then the tasks are described and photographed in a step-by-step sequence so that even a novice can do the work.

Its arrangement

The manual is divided into Chapters, each covering a logical sub-division of the vehicle. The Chapters are each divided into Sections, numbered with single figures, eg 5; the Sections are divided into paragraphs, or into sub-sections and paragraphs.

It is freely illustrated, especially in those parts where there is a detailed sequence of operations to be carried out. There are two forms of illustration: figures and photographs. The figures are numbered in sequence with decimal numbers, according to their position in the Chapter – eg Fig. 6.4 is the fourth drawing/illustration in Chapter 6.

Photographs carry the same number (either individually or in related groups) as the Section and paragraph to which they relate.

There is an alphabetical index at the back of the manual as well as a contents list at the front. Each Chapter is also preceded by its own individual contents list.

References to the 'left' or 'right' of the vehicle are in the sense of a person in the driver's seat facing forwards.

Unless otherwise stated, nuts and bolts are removed by turning anti-clockwise, and tightened by turning clockwise.

Vehicle manufacturers continually make changes to specifications and recommendations, and these, when notified, are incorporated into our manuals at the earliest opportunity.

We take great pride in the accuracy of information given in this manual, but vehicle manufacturers make alterations and design changes during the production run of a particular vehicle of which they do not inform us. No liability can be accepted by the authors or publishers for loss, damage or injury caused by any errors in, or omissions from, the information given.

Project vehicles

The main project vehicle used in the preparation of this manual, and appearing in the majority of the photographic sequences, was a 1989 Fiesta 1.1 litre Ghia. Additional work was carried out and photographed on an XR2i model, 1.4 litre CFi model and a 1.1 litre model with CTX automatic transmission.

Introduction to the Ford Fiesta

Introduced early in 1989, the models covered in this manual replace the previous highly successful Fiesta range. A great deal of development has led to the Fiesta being equipped with the standard of refinement more commonly found on executive class cars, yet with emphasis on economical motoring and ease of maintenance.

As on the previous models, front-wheel-drive from a transverse engine/transmission arrangement is retained, whilst the front and rear suspension has been completely redesigned to give a high standard of roadholding and comfort. Key advances over the previous range include the option of an anti-lock braking system on certain models, sophisticated engine management, controlling fuel injected variants, new and modified engines delivering more power and better fuel economy, and the provision of a five-door bodyshell.

The model range is extensive with three-door, five-door and Van bodystyles, overhead valve (HCS) and overhead camshaft (CVH) engines. Four-speed, five-speed and CTX automatic transmissions, as well as an impressive array of trim specification levels are included as extra cost options.

Ford Fiesta 1.1 Ghia

Ford Fiesta XR2i

General dimensions, weights and capacities

For information applicable to later models, see Supplement at end of manual

Dimensions
Overall length:
    XR2i model .... 3801 mm (148.2 in)
    All models except XR2i, with overriders .... 3791 mm (147.8 in)
    All models except XR2i without overriders .... 3743 mm (146.0 in)
Width (excluding mirrors):
    XR2i model .... 1630 mm (63.6 in)
    All other models .... 1605 mm (62.6 in)
Overall height:
    Maximum:
        XR2i model .... 1365 mm (53.2 in)
        All other models .... 1389 mm (54.2 in)
    Minimum:
        XR2i model .... 1363 mm (53.2 in)
        All other models .... 1369 mm (53.4 in)
Wheelbase (all models) .... 2446 mm (95.4 in)
Front track:
    XR2i model .... 1406 mm (54.8 in)
    'S' model .... 1395 mm (54.4 in)
    All other models .... 1392 mm (54.3 in)
Rear track:
    XR2i model .... 1376 mm (53.7 in)
    'S' model .... 1387 mm (54.1 in)
    All other models .... 1384 mm (54.0 in)

Weights
Kerb weight:
    3-door bodystyles:
        1.0 litre models .... 770 kg
        1.1 litre models .... 785 kg
        1.6 litre 'S' model .... 835 kg
        XR2i – 1.6 litre model .... 890 kg
    5-door bodystyles:
        1.0 litre model .... 795 kg
        1.1 litre models .... 810 kg
        1.4 litre models .... 840 kg
    Van (1.0 litre and 1.1 litre models) .... 775 kg
Maximum towing limit (2 up):
    1.0 litre models .... 780 kg
    All other models .... 900 kg
Maximum roof rack load .... 75 kg
Maximum Van payload .... 325 kg

Capacities
Engine oil:
    HCS engine:
        With filter .... 3.25 litres (5.7 pints)
        Without filter .... 2.75 litres (4.8 pints)
    CVH engine:
        With filter .... 3.5 litres (6.2 pints)
        Without filter .... 3.25 litres (5.7 pints)
Transmission:
    4-speed manual .... 2.8 litres (4.9 pints)
    5-speed manual .... 3.1 litres (5.5 pints)
    CTX automatic .... 3.5 litres (6.2 pints)
Cooling system (including the heater)
    HCS engine .... 7.1 litres (12.6 pints)
    CVH engine .... 7.6 litres (13.4 pints)
Windscreen washer reservoir .... 3.5 litres (6.2 pints)
Fuel tank .... 42 litres (9.25 gallons)

Jacking, towing and wheel changing

Jacking and wheel changing

First ensure that the vehicle is on firm, level ground, then apply the handbrake and engage first or reverse gear (on CTX automatic transmission equipped vehicles, engage position 'P'). If parking on a slope is unavoidable, both wheels on the opposite side of the vehicle must be chocked. Suitable load spreading blocks will need to be used if the vehicle is being jacked up on soft ground. **Never** jack under the rear axle twist beam or the front suspension crossmember (XR2i models only) as damage to these components may be incurred. **Never** work under a vehicle raised on a jack without, first securely supporting the vehicle on axle stands.

To change a wheel first remove the jack and wheelbrace assembly from the engine compartment (photo). Remove the spare wheel from its carrier and prise off the plastic wheel trim using the wheelbrace, with its plastic tip. Prior to jacking slacken the relevant roadwheel nuts by half a turn (in a diagonal sequence).

On XR2i models it will be necessary to remove the plastic cover concealing the required jacking point (photo).

Insert the jack into the appropriate jacking point so that its saddle

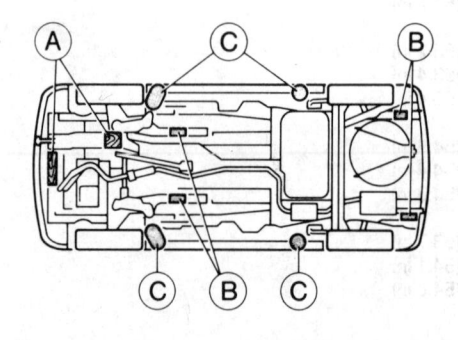

Vehicle jacking positions

A Jacking points for trolley jack (always use a suitable block of wood to protect the vehicle body)
B Axle stand positions
C Jacking points for owner jack and wheel-free hoist

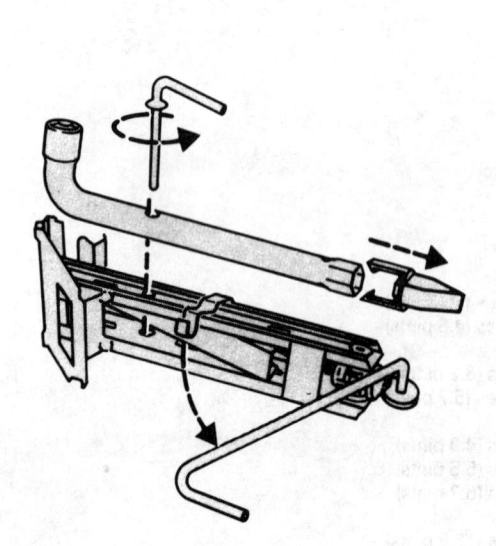

Releasing the spare wheel carrier retaining arrangement

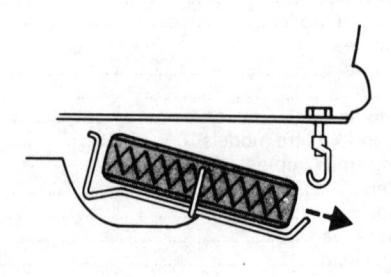

Spare wheel carrier lowered for access to the spare wheel

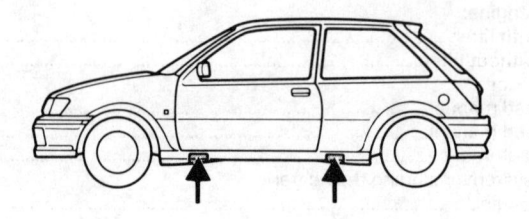

Exploded view of the jack and wheelbrace arrangement

Jacking point cover locations (XR2i model only)

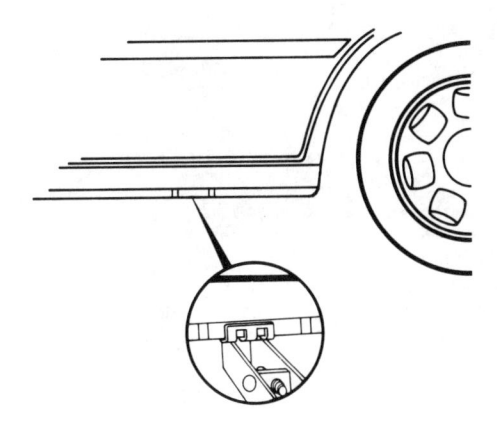

Jacking, towing and wheel changing 9

Lower the vehicle to the ground, remove the jack and tighten the roadwheel nuts progressively (in a diagonal sequence) to their specified torque (photo). If a torque wrench is not available have the tightness of the nuts checked as soon as possible. Check the tyre pressure.

Refit the plastic roadwheel trim (where fitted). On XR2i models, refit the jacking point plastic cover.

Towing

Warning: The ignition key must be set to position II when the vehicle is being towed (with all four wheels on the ground) so that the steering, hazard flasher warning lights and brake lights are fully operational. The brake servo unit only functions when the engine is running, and it must be noted that without the benefit of servo assistance, much greater pedal pressures and stopping distances will be required.

Towing eyes are provided at both the front and rear of the vehicle, for use with a tow rope or an emergency recovery towing bar (photos). For towing of caravans or trailers, a properly fitted towing bracket will be required.

Prior to towing, the **warning** given above must be read and understood.

Before towing, turn the ignition key to position II, release the handbrake and place the gear lever in neutral (or position N on CTX automatic transmission equipped vehicles). Never tow a vehicle with automatic transmission faster than 30 mph (50 km/h) or further than 30 miles (50 km). If it is necessary to tow the vehicle a greater distance, or if the transmission is damaged, the driven wheels must be lifted clear of the ground.

Push or tow starting should not be attempted on vehicles fitted with CTX automatic transmission. Additionally, the only advisable emergency starting procedure on fuel injected vehicles is the use of jump leads.

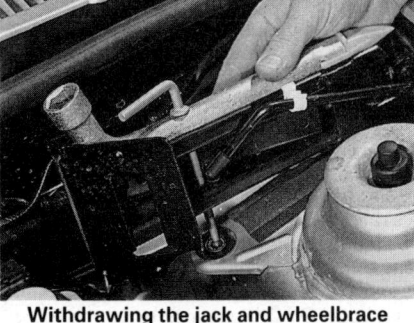

Jack saddle engagement to jacking point

locates positively, then raise the vehicle (photo). When the roadwheel is clear of the ground undo the wheel's nuts, having first marked the roadwheel's relationship with one of the studs. Remove the roadwheel.

When refitting align the marks made on removal (if the same wheel is being refitted) and refit the nuts with their tapered edges against the wheel. A similar procedure must be adopted if fitting the spare wheel but this may need to be balanced afterwards. Note that alloy wheels use special nuts incorporating a washer to prevent damage to a roadwheel. Tighten the nuts as much as possible by hand so that the wheel is correctly positioned.

Withdrawing the jack and wheelbrace assembly

Removing a jacking point cover (XR2i model only)

Jack correctly positioned, with its base making firm level contact with the ground (XR2i model shown)

Tightening the roadwheel nuts to their specified torque

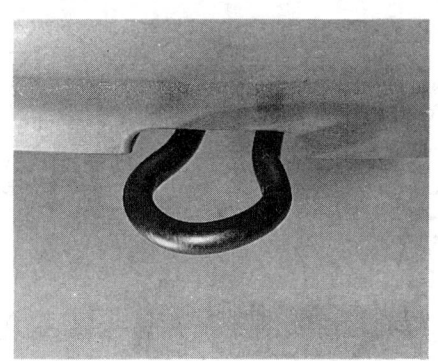

Front towing eye

Rear towing eye

Buying spare parts
and vehicle identification numbers

For information applicable to later models, see Supplement at end of manual

Buying spare parts

Spare parts are available from many sources, for example: Ford garages, other garages and accessory shops, and motor factors. Our advice regarding spare parts is as follows.

Officially appointed Ford garages – This is the best source of parts which are peculiar to your vehicle, and are otherwise not generally available (eg complete cylinder heads, internal gearbox components, badges, interior trim, etc.). It is also the only place at which you should buy parts if the vehicle is still under warranty – non-Ford components may invalidate the warranty. To be sure of obtaining the correct parts, it will be necessary to give the storeman your vehicle's identification number (VIN) and engine details, and if possible, to take the 'old' part along for positive identification. Remember that many parts are available on a factory exchange scheme – any parts returned should always be clean! It obviously makes good sense to go straight to the specialists on your vehicle for this type of part for they are best equipped to supply you.

Other garages and accessory shops – These are often very good places to buy materials and components needed for the maintenance of your vehicle (eg. spark plugs, bulbs, drivebelts, oils and greases, touch-up paint, filters, etc.). They also sell general accessories, usually have convenient opening hours, charge lower prices and can often be found not far from home.

Motor factors – Good factors will stock all of the more important components which wear out relatively quickly (eg. brake cylinders/pipes/hoses/seals/shoes and pads, etc.). Motor factors will often provide new or reconditioned components on a part exchange basis – this can save a considerable amount of money.

Vehicle identification numbers

Modifications are a continued and unpublicised process in vehicle manufacture. Spare parts manuals and listings are compiled upon a numerical basis, the individual vehicle identification numbers being essential to correct identification of the component concerned.

The Vehicle Identification Number (VIN) is located on a plate found under the bonnet (above the radiator). The chassis number (found by lifting the flap in the carpet adjacent to the left-hand front seat) relates directly to the VIN. The plate also carries details concerning paint colour, transmission type, final drive ratio, etc.

The engine details may be found on the engine block, the location of the details being dependent on engine type. On HCS engines, the engine number is found just above the forward of the two engine/transmission flange upper securing bolts. On CVH engines it may be found next to the alternator bracket.

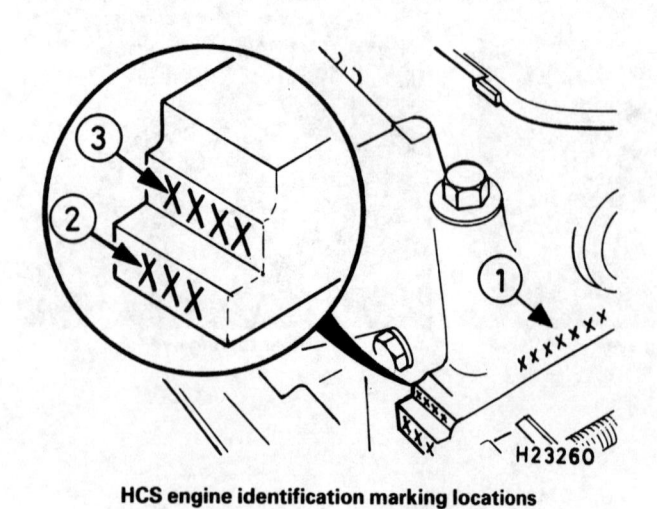

HCS engine identification marking locations

1 Engine number
2 Engine code
3 Engine build date

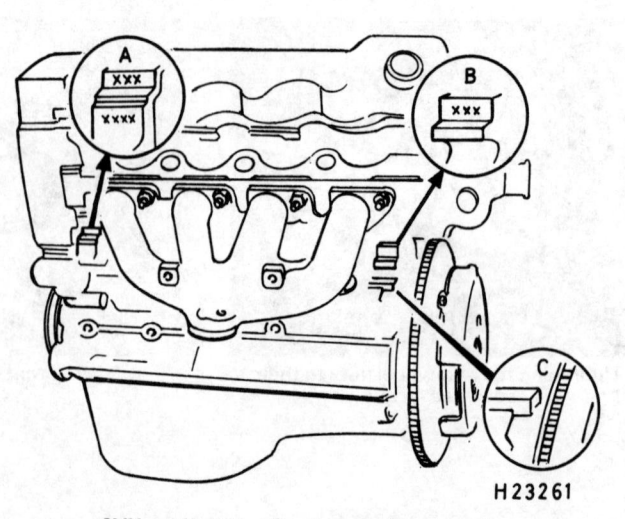

CVH engine identification marking locations

A Engine number
B Engine code
C Engine number in event of repair

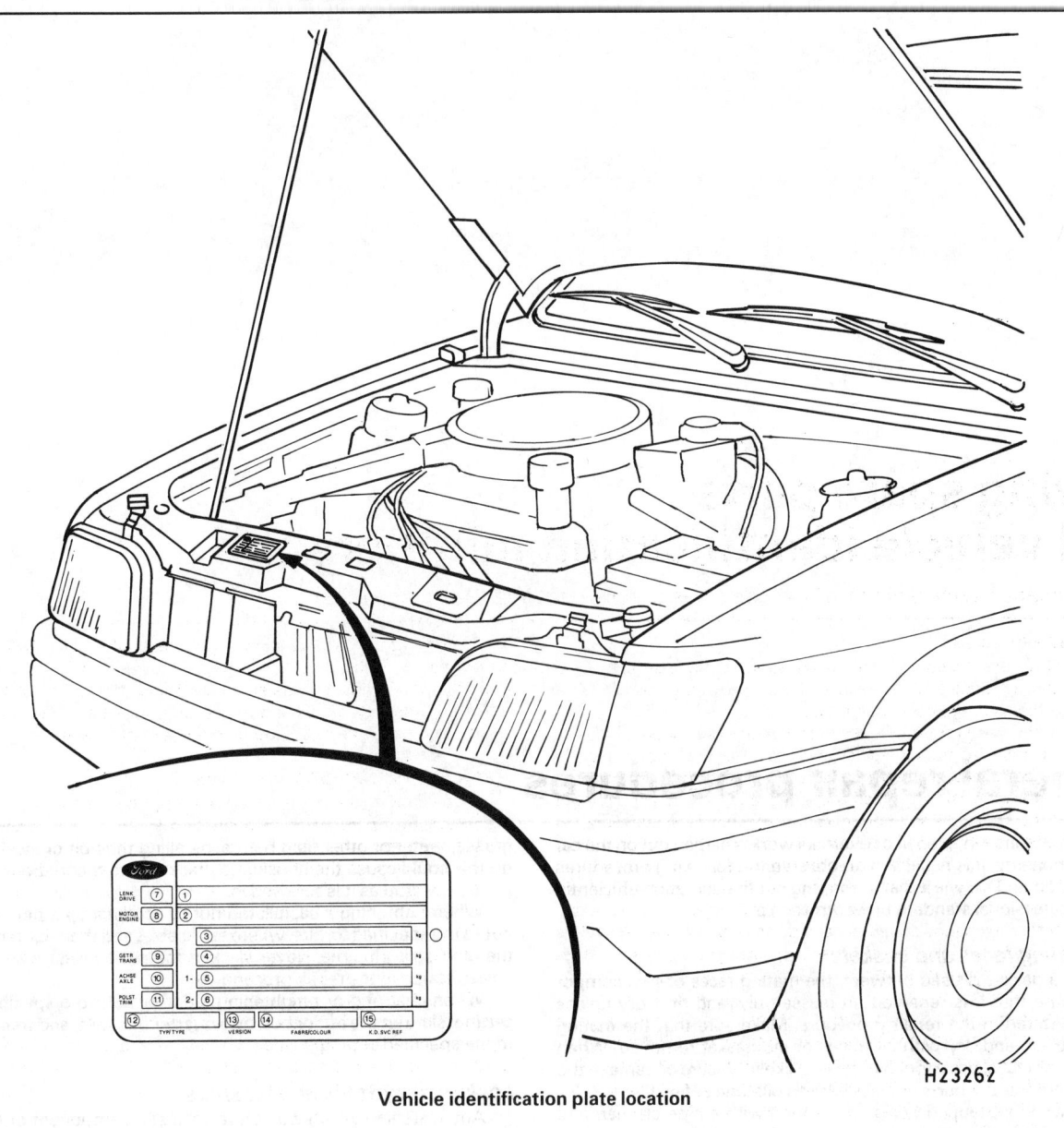

Buying spare parts and vehicle identification numbers

Vehicle identification plate location

1. Type Approval Number
2. Vehicle Identification Number (VIN)
3. Gross vehicle mass
4. Gross train mass
5. Permitted front axle loading
6. Permitted rear axle loading
7. Steering (LHD/RHD)
8. Engine
9. Transmission
10. Axle (final drive ratio)
11. Trim (interior)
12. Body type
13. Special territory version
14. Body colour
15. Emission regulation (see Chapter 13, Section 3)

General repair procedures

Whenever servicing, repair or overhaul work is carried out on the car or its components, it is necessary to observe the following procedures and instructions. This will assist in carrying out the operation efficiently and to a professional standard of workmanship.

Joint mating faces and gaskets

Where a gasket is used between the mating faces of two components, ensure that it is renewed on reassembly, and fit it dry unless otherwise stated in the repair procedure. Make sure that the mating faces are clean and dry with all traces of old gasket removed. When cleaning a joint face, use a tool which is not likely to score or damage the face, and remove any burrs or nicks with an oilstone or fine file.

Make sure that tapped holes are cleaned with a pipe cleaner, and keep them free of jointing compound if this is being used unless specifically instructed otherwise.

Ensure that all orifices, channels or pipes are clear and blow through them, preferably using compressed air.

Oil seals

Whenever an oil seal is removed from its working location, either individually or as part of an assembly, it should be renewed.

The very fine sealing lip of the seal is easily damaged and will not seal if the surface it contacts is not completely clean and free from scratches, nicks or grooves. If the original sealing surface of the component cannot be restored, the component should be renewed.

Protect the lips of the seal from any surface which may damage them in the course of fitting. Use tape or a conical sleeve where possible. Lubricate the seal lips with oil before fitting and, on dual lipped seals, fill the space between the lips with grease.

Unless otherwise stated, oil seals must be fitted with their sealing lips toward the lubricant to be sealed.

Use a tubular drift or block of wood of the appropriate size to install the seal and, if the seal housing is shouldered, drive the seal down to the shoulder. If the seal housing is unshouldered, the seal should be fitted with its face flush with the housing top face.

Screw threads and fastenings

Always ensure that a blind tapped hole is completely free from oil, grease, water or other fluid before installing the bolt or stud. Failure to do this could cause the housing to crack due to the hydraulic action of the bolt or stud as it is screwed in.

When tightening a castellated nut to accept a split pin, tighten the nut to the specified torque, where applicable, and then tighten further to the next split pin hole. Never slacken the nut to align a split pin hole unless stated in the repair procedure.

When checking or retightening a nut or bolt to a specified torque setting, slacken the nut or bolt by a quarter of a turn, and then retighten to the specified setting.

Locknuts, locktabs and washers

Any fastening which will rotate against a component or housing in the course of tightening should always have a washer between it and the relevant component or housing.

Spring or split washers should always be renewed when they are used to lock a critical component such as a big-end bearing retaining nut or bolt.

Locktabs which are folded over to retain a nut or bolt should always be renewed.

Self-locking nuts can be reused in non-critical areas, providing resistance can be felt when the locking portion passes over the bolt or stud thread.

Split pins must always be replaced with new ones of the correct size for the hole.

Special tools

Some repair procedures in this manual entail the use of special tools such as a press, two or three-legged pullers, spring compressors etc. Wherever possible, suitable readily available alternatives to the manufacturer's special tools are described, and are shown in use. In some instances, where no alternative is possible, it has been necessary to resort to the use of a manufacturer's tool and this has been done for reasons of safety as well as the efficient completion of the repair operation. Unless you are highly skilled and have a thorough understanding of the procedure described, never attempt to bypass the use of any special tool when the procedure described specifies its use. Not only is there a very great risk of personal injury, but expensive damage could be caused to the components involved.

Tools and working facilities

Introduction

A selection of good tools is a fundamental requirement for anyone contemplating the maintenance and repair of a motor vehicle. For the owner who does not possess any, their purchase will prove a considerable expense, offsetting some of the savings made by doing-it-yourself. However, provided that the tools purchased meet the relevant national safety standards and are of good quality, they will last for many years and prove an extremely worthwhile investment.

To help the average owner to decide which tools are needed to carry out the various tasks detailed in this manual, we have compiled three lists of tools under the following headings: *Maintenance and minor repair*, *Repair and overhaul*, and *Special*. The newcomer to practical mechanics should start off with the *Maintenance and minor repair* tool kit and confine himself to the simpler jobs around the vehicle. Then, as his confidence and experience grow, he can undertake more difficult tasks, buying extra tools as, and when, they are needed. In this way, a *Maintenance and minor repair* tool kit can be built-up into a *Repair and overhaul* tool kit over a considerable period of time without any major cash outlays. The experienced do-it-yourselfer will have a tool kit good enough for most repair and overhaul procedures and will add tools from the *Special* category when he feels the expense is justified by the amount of use to which these tools will be put.

It is obviously not possible to cover the subject of tools fully here. For those who wish to learn more about tools and their use there is a book entitled *How to Choose and Use Car Tools* available from the publishers of this manual.

Maintenance and minor repair tool kit

The tools given in this list should be considered as a minimum requirement if routine maintenance, servicing and minor repair operations are to be undertaken. We recommend the purchase of combination spanners (ring one end, open-ended the other); although more expensive than open-ended ones, they do give the advantages of both types of spanner.

Combination spanners – 10, 11, 12, 13, 14 & 17 mm
Adjustable spanner – 9 inch
Spark plug spanner (with rubber insert)
Spark plug gap adjustment tool
Set of feeler gauges
Brake bleed nipple spanner
Screwdriver – 4 in long x $\frac{1}{4}$ in dia (flat blade)
Screwdriver – 4 in long x $\frac{1}{4}$ in dia (cross blade)
Combination pliers – 6 inch
Hacksaw (junior)
Tyre pump
Tyre pressure gauge
Oil can
Fine emery cloth (1 sheet)
Wire brush (small)
Funnel (medium size)

Repair and overhaul tool kit

These tools are virtually essential for anyone undertaking any major repairs to a motor vehicle, and are additional to those given in the *Maintenance and minor repair* list. Included in this list is a comprehensive set of sockets. Although these are expensive they will be found invaluable as they are so versatile – particularly if various drives are included in the set. We recommend the $\frac{1}{2}$ in square-drive type, as this can be used with most proprietary torque wrenches. If you cannot afford a socket set, even bought piecemeal, then inexpensive tubular box spanners are a useful alternative.

The tools in this list will occasionally need to be supplemented by tools from the *Special* list.

Sockets (or box spanners) to cover range in previous list
Reversible ratchet drive (for use with sockets)
Extension piece, 10 inch (for use with sockets)
Universal joint (for use with sockets)
Torque wrench (for use with sockets)
Self-locking grips
Ball pein hammer
Soft-faced hammer, plastic or rubber
Screwdriver – 6 in long x $\frac{5}{16}$ in dia (flat blade)
Screwdriver – 2 in long x $\frac{5}{16}$ in square (flat blade)
Screwdriver – 1$\frac{1}{2}$ in long x $\frac{1}{4}$ in dia (cross blade)
Screwdriver – 3 in long x $\frac{1}{8}$ in dia (electricians)
Pliers – electricians side cutters
Pliers – needle nosed
Pliers – circlip (internal and external)
Cold chisel – $\frac{1}{2}$ inch
Scriber
Scraper
Centre punch
Pin punch
Hacksaw
Steel rule/straight-edge
Allen keys (inc. splined/Torx type if necessary)
Selection of files
Wire brush (large)
Axle-stands
Jack (strong trolley or hydraulic type)

Special tools

The tools in this list are those which are not used regularly, are expensive to buy, or which need to be used in accordance with their manufacturers' instructions. Unless relatively difficult mechanical jobs are undertaken frequently, it will not be economic to buy many of these tools. Where this is the case, you could consider clubbing together with friends (or joining a motorists' club) to make a joint purchase, or borrowing the tools against a deposit from a local garage or tool hire specialist.

14 Tools and working facilities

The following list contains only those tools and instruments freely available to the public, and not those special tools produced by the vehicle manufacturer specifically for its dealer network. You will find occasional references to these manufacturers' special tools in the text of this manual. Generally, an alternative method of doing the job without the vehicle manufacturers' special tool is given. However, sometimes, there is no alternative to using them. Where this is the case and the relevant tool cannot be bought or borrowed, you will have to entrust the work to a franchised garage.

> Valve spring compressor (where applicable)
> Coil spring compressors
> Piston ring compressor
> Balljoint separator
> Universal hub/bearing puller
> Impact screwdriver
> Micrometer and/or vernier gauge
> Dial gauge
> Stroboscopic timing light
> Tachometer
> Universal electrical multi-meter
> Cylinder compression gauge
> Lifting tackle
> Trolley jack
> Light with extension lead

Buying tools

For practically all tools, a tool factor is the best source since he will have a very comprehensive range compared with the average garage or accessory shop. Having said that, accessory shops often offer excellent quality tools at discount prices, so it pays to shop around.

There are plenty of good tools around at reasonable prices, but always aim to purchase items which meet the relevant national safety standards. If in doubt, ask the proprietor or manager of the shop for advice before making a purchase.

Care and maintenance of tools

Having purchased a reasonable tool kit, it is necessary to keep the tools in a clean serviceable condition. After use, always wipe off any dirt, grease and metal particles using a clean, dry cloth, before putting the tools away. Never leave them lying around after they have been used. A simple tool rack on the garage or workshop wall, for items such as screwdrivers and pliers is a good idea. Store all normal wrenches and sockets in a metal box. Any measuring instruments, gauges, meters, etc, must be carefully stored where they cannot be damaged or become rusty.

Take a little care when tools are used. Hammer heads inevitably become marked and screwdrivers lose the keen edge on their blades from time to time. A little timely attention with emery cloth or a file will soon restore items like this to a good serviceable finish.

Working facilities

Not to be forgotten when discussing tools, is the workshop itself. If anything more than routine maintenance is to be carried out, some form of suitable working area becomes essential.

It is appreciated that many an owner mechanic is forced by circumstances to remove an engine or similar item, without the benefit of a garage or workshop. Having done this, any repairs should always be done under the cover of a roof.

Wherever possible, any dismantling should be done on a clean, flat workbench or table at a suitable working height.

Any workbench needs a vice: one with a jaw opening of 4 in (100 mm) is suitable for most jobs. As mentioned previously, some clean dry storage space is also required for tools, as well as for lubricants, cleaning fluids, touch-up paints and so on, which become necessary.

Another item which may be required, and which has a much more general usage, is an electric drill with a chuck capacity of at least $\frac{5}{16}$ in (8 mm). This, together with a good range of twist drills, is virtually essential for fitting accessories such as mirrors and reversing lights.

Last, but not least, always keep a supply of old newspapers and clean, lint-free rags available, and try to keep any working area as clean as possible.

Spanner jaw gap comparison table

Jaw gap (in)	Spanner size
0.250	$\frac{1}{4}$ in AF
0.276	7 mm
0.313	$\frac{5}{16}$ in AF
0.315	8 mm
0.344	$\frac{11}{32}$ in AF; $\frac{1}{8}$ in Whitworth
0.354	9 mm
0.375	$\frac{3}{8}$ in AF
0.394	10 mm
0.433	11 mm
0.438	$\frac{7}{16}$ in AF
0.445	$\frac{3}{16}$ in Whitworth; $\frac{1}{4}$ in BSF
0.472	12 mm
0.500	$\frac{1}{2}$ in AF
0.512	13 mm
0.525	$\frac{1}{4}$ in Whitworth; $\frac{5}{16}$ in BSF
0.551	14 mm
0.563	$\frac{9}{16}$ in AF
0.591	15 mm
0.600	$\frac{5}{16}$ in Whitworth; $\frac{3}{8}$ in BSF
0.625	$\frac{5}{8}$ in AF
0.630	16 mm
0.669	17 mm
0.686	$\frac{11}{16}$ in AF
0.709	18 mm
0.710	$\frac{3}{8}$ in Whitworth; $\frac{7}{16}$ in BSF
0.748	19 mm
0.750	$\frac{3}{4}$ in AF
0.813	$\frac{13}{16}$ in AF
0.820	$\frac{7}{16}$ in Whitworth; $\frac{1}{2}$ in BSF
0.866	22 mm
0.875	$\frac{7}{8}$ in AF
0.920	$\frac{1}{2}$ in Whitworth; $\frac{9}{16}$ in BSF
0.938	$\frac{15}{16}$ in AF
0.945	24 mm
1.000	1 in AF
1.010	$\frac{9}{16}$ in Whitworth; $\frac{5}{8}$ in BSF
1.024	26 mm
1.063	$1\frac{1}{16}$ in AF; 27 mm
1.100	$\frac{5}{8}$ in Whitworth; $\frac{11}{16}$ in BSF
1.125	$1\frac{1}{8}$ in AF
1.181	30 mm
1.200	$\frac{11}{16}$ in Whitworth; $\frac{3}{4}$ in BSF
1.250	$1\frac{1}{4}$ in AF
1.260	32 mm
1.300	$\frac{3}{4}$ in Whitworth; $\frac{7}{8}$ in BSF
1.313	$1\frac{5}{16}$ in AF
1.390	$\frac{13}{16}$ in Whitworth; $\frac{15}{16}$ in BSF
1.417	36 mm
1.438	$1\frac{7}{16}$ in AF
1.480	$\frac{7}{8}$ in Whitworth; 1 in BSF
1.500	$1\frac{1}{2}$ in AF
1.575	40 mm; $1\frac{9}{16}$ in Whitworth
1.614	41 mm
1.625	$1\frac{5}{8}$ in AF
1.670	1 in Whitworth; $1\frac{1}{8}$ in BSF
1.688	$1\frac{11}{16}$ in AF
1.811	46 mm
1.813	$1\frac{13}{16}$ in AF
1.860	$1\frac{1}{8}$ in Whitworth; $1\frac{1}{4}$ in BSF
1.875	$1\frac{7}{8}$ in AF
1.969	50 mm
2.000	2 in AF
2.050	$1\frac{1}{4}$ in Whitworth; $1\frac{3}{8}$ in BSF
2.165	55 mm
2.362	60 mm

Safety first!

Professional motor mechanics are trained in safe working procedures. However enthusiastic you may be about getting on with the job in hand, do take the time to ensure that your safety is not put at risk. A moment's lack of attention can result in an accident, as can failure to observe certain elementary precautions.

There will always be new ways of having accidents, and the following points do not pretend to be a comprehensive list of all dangers; they are intended rather to make you aware of the risks and to encourage a safety-conscious approach to all work you carry out on your vehicle.

Essential DOs and DON'Ts

DON'T rely on a single jack when working underneath the vehicle. Always use reliable additional means of support, such as axle stands, securely placed under a part of the vehicle that you know will not give way.

DON'T attempt to loosen or tighten high-torque nuts (e.g. wheel hub nuts) while the vehicle is on a jack; it may be pulled off.

DON'T start the engine without first ascertaining that the transmission is in neutral (or 'Park' where applicable) and the parking brake applied.

DON'T suddenly remove the filler cap from a hot cooling system – cover it with a cloth and release the pressure gradually first, or you may get scalded by escaping coolant.

DON'T attempt to drain oil until you are sure it has cooled sufficiently to avoid scalding you.

DON'T grasp any part of the engine, exhaust or catalytic converter without first ascertaining that it is sufficiently cool to avoid burning you.

DON'T allow brake fluid or antifreeze to contact vehicle paintwork.

DON'T syphon toxic liquids such as fuel, brake fluid or antifreeze by mouth, or allow them to remain on your skin.

DON'T inhale dust – it may be injurious to health (see *Asbestos* below).

DON'T allow any spilt oil or grease to remain on the floor – wipe it up straight away, before someone slips on it.

DON'T use ill-fitting spanners or other tools which may slip and cause injury.

DON'T attempt to lift a heavy component which may be beyond your capability – get assistance.

DON'T rush to finish a job, or take unverified short cuts.

DON'T allow children or animals in or around an unattended vehicle.

DO wear eye protection when using power tools such as drill, sander, bench grinder etc, and when working under the vehicle.

DO use a barrier cream on your hands prior to undertaking dirty jobs – it will protect your skin from infection as well as making the dirt easier to remove afterwards; but make sure your hands aren't left slippery. Note that long-term contact with used engine oil can be a health hazard.

DO keep loose clothing (cuffs, tie etc) and long hair well out of the way of moving mechanical parts.

DO remove rings, wristwatch etc, before working on the vehicle – especially the electrical system.

DO ensure that any lifting tackle used has a safe working load rating adequate for the job.

DO keep your work area tidy – it is only too easy to fall over articles left lying around.

DO get someone to check periodically that all is well, when working alone on the vehicle.

DO carry out work in a logical sequence and check that everything is correctly assembled and tightened afterwards.

DO remember that your vehicle's safety affects that of yourself and others. If in doubt on any point, get specialist advice.

IF, in spite of following these precautions, you are unfortunate enough to injure yourself, seek medical attention as soon as possible.

Asbestos

Certain friction, insulating, sealing, and other products – such as brake linings, brake bands, clutch linings, torque converters, gaskets, etc – contain asbestos. *Extreme care must be taken to avoid inhalation of dust from such products since it is hazardous to health.* If in doubt, assume that they *do* contain asbestos.

Fire

Remember at all times that petrol (gasoline) is highly flammable. Never smoke, or have any kind of naked flame around, when working on the vehicle. But the risk does not end there – a spark caused by an electrical short-circuit, by two metal surfaces contacting each other, by careless use of tools, or even by static electricity built up in your body under certain conditions, can ignite petrol vapour, which in a confined space is highly explosive.

Always disconnect the battery earth (ground) terminal before working on any part of the fuel or electrical system, and never risk spilling fuel on to a hot engine or exhaust.

It is recommended that a fire extinguisher of a type suitable for fuel and electrical fires is kept handy in the garage or workplace at all times. Never try to extinguish a fuel or electrical fire with water.

Note: *Any reference to a 'torch' appearing in this manual should always be taken to mean a hand-held battery-operated electric lamp or flashlight. It does NOT mean a welding/gas torch or blowlamp.*

Fumes

Certain fumes are highly toxic and can quickly cause unconsciousness and even death if inhaled to any extent. Petrol (gasoline) vapour comes into this category, as do the vapours from certain solvents such as trichloroethylene. Any draining or pouring of such volatile fluids should be done in a well ventilated area.

When using cleaning fluids and solvents, read the instructions carefully. Never use materials from unmarked containers – they may give off poisonous vapours.

Never run the engine of a motor vehicle in an enclosed space such as a garage. Exhaust fumes contain carbon monoxide which is extremely poisonous; if you need to run the engine, always do so in the open air or at least have the rear of the vehicle outside the workplace.

If you are fortunate enough to have the use of an inspection pit, never drain or pour petrol, and never run the engine, while the vehicle is standing over it; the fumes, being heavier than air, will concentrate in the pit with possibly lethal results.

The battery

Never cause a spark, or allow a naked light, near the vehicle's battery. It will normally be giving off a certain amount of hydrogen gas, which is highly explosive.

Always disconnect the battery earth (ground) terminal before working on the fuel or electrical systems.

If possible, loosen the filler plugs or cover when charging the battery from an external source. Do not charge at an excessive rate or the battery may burst.

Take care when topping up and when carrying the battery. The acid electrolyte, even when diluted, is very corrosive and should not be allowed to contact the eyes or skin.

If you ever need to prepare electrolyte yourself, always add the acid slowly to the water, and never the other way round. Protect against splashes by wearing rubber gloves and goggles.

When jump starting a car using a booster battery, for negative earth (ground) vehicles, connect the jump leads in the following sequence: First connect one jump lead between the positive (+) terminals of the two batteries. Then connect the other jump lead first to the negative (–) terminal of the booster battery, and then to a good earthing (ground) point on the vehicle to be started, at least 18 in (45 cm) from the battery if possible. Ensure that hands and jump leads are clear of any moving parts, and that the two vehicles do not touch. Disconnect the leads in the reverse order.

Mains electricity and electrical equipment

When using an electric power tool, inspection light etc, always ensure that the appliance is correctly connected to its plug and that, where necessary, it is properly earthed (grounded). Do not use such appliances in damp conditions and, again, beware of creating a spark or applying excessive heat in the vicinity of fuel or fuel vapour. Also ensure that the appliances meet the relevant national safety standards.

Ignition HT voltage

A severe electric shock can result from touching certain parts of the ignition system, such as the HT leads, when the engine is running or being cranked, particularly if components are damp or the insulation is defective. Where an electronic ignition system is fitted, the HT voltage is much higher and could prove fatal.

Routine maintenance

For modifications, and information applicable to later models, see Supplement at end of manual

Maintenance is essential for ensuring safety, and desirable for the purpose of getting the best in terms of performance and economy from your vehicle. Over the years, the need for periodic lubrication – especially oiling, and greasing has been drastically reduced if not totally eliminated. This has unfortunately tended to lead some owners to think that because no such action is required, components either no longer exist, or will last forever. This is certainly not the case. It is essential to carry out regular visual examination as comprehensively as possible in order to spot any possible defects at an early stage before they develop into major expensive repairs.

The following service schedules are a list of maintenance requirements, and the intervals at which they should be carried out, as recommended by the manufacturers. Where applicable, these procedures are covered in greater detail near the beginning of each relevant chapter.

Every 250 miles (400 km) or weekly – whichever occurs first

Engine, cooling system, braking system and general
Check the engine oil level and top up if necessary (photo) (Chapter 1)
Check the coolant level and top up if necessary (photo) (Chapter 2)
Check the brake fluid level and top up if necessary (photo) (Chapter 8)
Visually check for signs of fluid leakage (general)

Lamps, wipers and horn
Check the operation of all interior and exterior lamps, the wipers and washers, and the horn (Chapter 12)
Check the washer fluid level and top up if necessary (Chapter 12), adding a screen wash such as Turtle Wax High Tech Screen Wash

Tyres
Check the tyre pressures (Chapter 10)
Visually examine the tyres for wear or damage (Chapter 10)

Every 6000 miles (10 000 km) or 12 months – whichever occurs first

Engine (Chapter 1)
Change the engine oil and renew the oil filter
Check for oil leaks and rectify as necessary

Cooling system (Chapter 2)
Check the coolant level and top up if necessary
Check for coolant leaks and rectify as necessary (check coolant strength if leaks are evident)
Check all coolant hoses for signs of damage or deterioration
Inspect the radiator matrix for blockage (eg dead insects), and carefully clean it as necessary

Fuel, exhaust and emission control systems (Chapter 3)
Check for fuel leaks and rectify as necessary
Check all fuel pipes and hoses for signs of damage or deterioration
Check and if necessary adjust the idle speed and fuel mixture settings, where applicable
Inspect the crankcase ventilation system for security and condition
Check all vacuum hoses for condition and security

Ignition and engine management systems (Chapter 4)
Check all wiring and vacuum hoses for condition and security

Transmission (Chapter 6)
Check for evidence of fluid leaks and rectify as necessary
Examine the oil cooler pipes for signs of deterioration or damage (CTX automatic transmission models only)

Driveshafts (Chapter 7)
Check the inner and outer CV joint gaiters for signs of deterioration or damage

Braking system (Chapter 8)
Check the brake fluid level and top up if necessary
Check for brake fluid leaks and rectify as necessary
Check the brake hoses and pipes for signs of deterioration or damage
Check the brake pads and shoe friction material for wear, and renew if necessary
Check the operation of the dashboard brake warning lights

Steering (Chapter 9)
Check the steering system components for wear, damage and security (including the steering rack gaiters)

Suspension (Chapter 10)
Check the tightness of the roadwheel nuts (with the vehicle resting on its wheels)

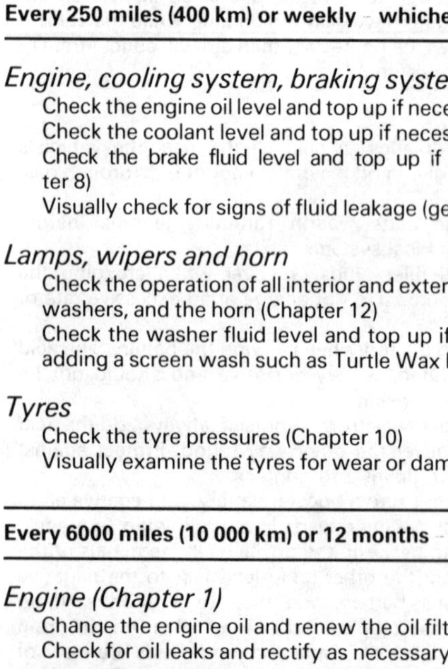

Topping-up the engine oil level (HCS engine shown)

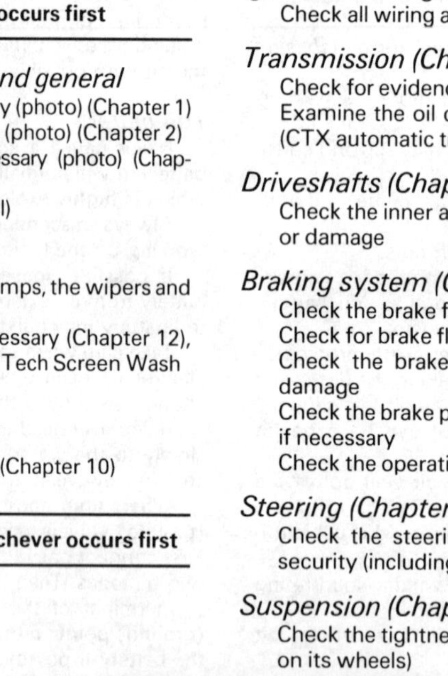

Topping-up the coolant level

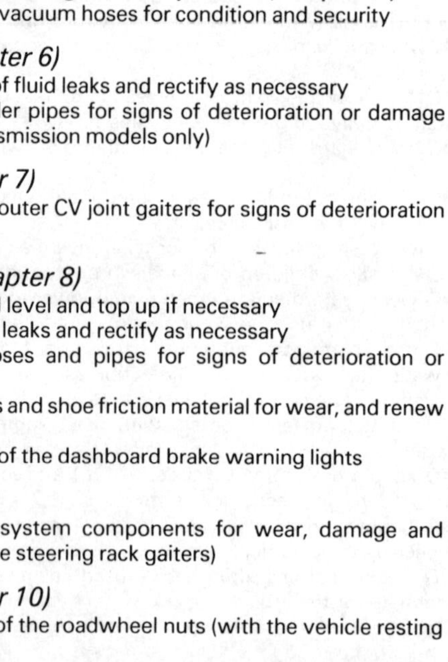

Topping-up the brake fluid level

Engine and underbonnet component locations
1.1 litre model with CTX automatic transmission shown (air filter housing removed for clarity)

1 Engine oil filler cap
2 Engine oil level dipstick
3 Cooling system expansion tank
4 Brake fluid reservoir
5 Windscreen/tailgate washer fluid reservoir cap
6 Battery
7 Vehicle identification plate
8 Thermostat housing
9 Radiator cooling fan thermal switch multi-plug
10 Alternator
11 Starter motor solenoid
12 CTX automatic transmission fluid level dipstick
13 Exhaust heatshield/airbox
14 Brake pressure control valves
15 Top of suspension strut mounting assembly
16 Carburettor
17 Fuel feed hose
18 Anti-dieselling (fuel-cut off) solenoid connection
19 Throttle kicker
20 Throttle kicker control solenoid
21 Ignition module
22 Heater blower motor cover
23 Windscreen wiper motor mounting bracket

**Engine and underbonnet component locations
1.4 litre CFi model shown (air filter housing removed for clarity)**

1. Engine oil filler cap
2. Engine oil level dipstick
3. Cooling system expansion tank
4. Brake fluid reservoir
5. Windscreen/tailgate washer fluid reservoir cap
6. Battery
7. Vehicle identification plate
8. Thermostat housing
9. Pre-heat tube
10. Timing belt (cambelt) cover
11. Distributor
12. Fuel filter
13. Heater blower motor cover
14. Windscreen wiper motor mounting bracket
15. Jack and wheelbrace retaining bolt
16. Top of suspension strut mounting assembly
17. EEC-IV engine management module cover
18. CFi unit
19. Fuel injector
20. Fuel pressure regulator
21. Throttle plate control motor
22. Carbon canister
23. MAP sensor
24. TFI-IV ignition module

**Engine and underbonnet component locations
1.6 litre EFi model (XR2i) shown**

1. Engine oil filler cap
2. Engine oil level dipstick
3. Cooling system expansion tank
4. Brake fluid reservoir
5. Windscreen/tailgate washer fluid reservoir cap
6. Battery
7. Vehicle identification plate
8. Thermostat housing
9. Timing belt (cambelt) cover
10. Top of suspension strut mounting assembly
11. Windscreen wiper motor mounting bracket
12. Jack and wheelbrace retaining bolt
13. Distributorless ignition coil
14. Fuel filter
15. Air filter housing
16. Air filter trunking
17. Idle speed control valve
18. Fuel pressure regulator
19. Throttle plate housing
20. Upper section of inlet manifold
21. Air-charge temperature (ACT) sensor
22. Fuel trap
23. EEC-IV engine management module cover
24. MAP sensor
25. EDIS ignition module

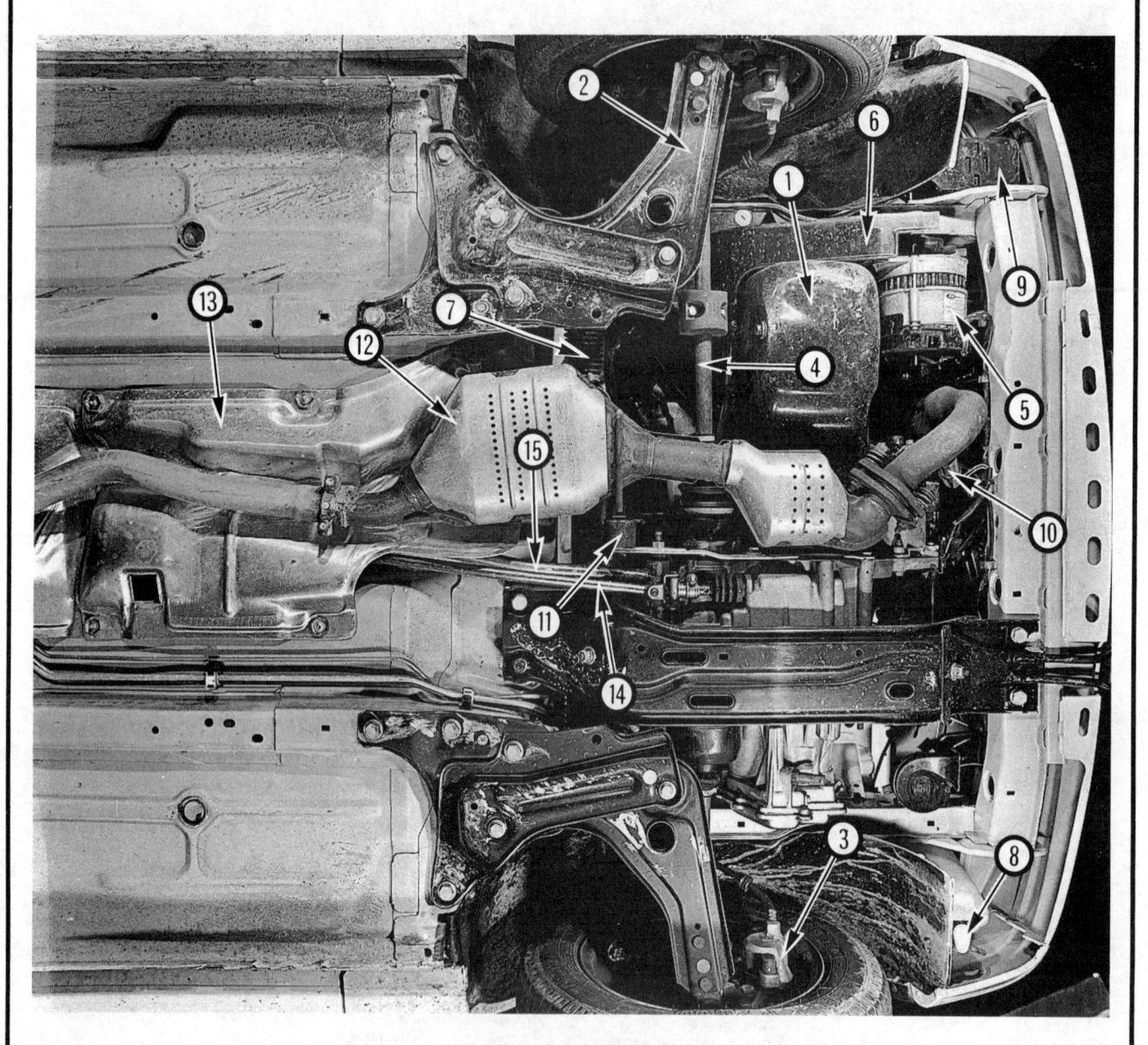

**Front underbody view
1.4 litre CFi model shown**

1 Engine oil sump
2 Front suspension lower arm
3 Brake caliper assembly
4 Driveshaft
5 Alternator
6 Alternator drivebelt shield
7 Steering rack gaiter
8 Windscreen/tailgate washer pump
9 Carbon canister
10 HEGO sensor
11 Catalytic converter (exhaust) rubber insulator mounting
12 Catalytic converter assembly
13 Underbody heatshields
14 Gearchange mechanism shift rod
15 Gearchange mechanism stabiliser bar

**Rear underbody view
1.4 litre CFi model shown**

1 Fuel tank
2 Fuel filler pipe
3 Fuel tank ventilation hose
4 Twist beam rear axle assembly
5 Underbody heatshields
6 Exhaust rear silencer
7 Exhaust rubber insulator mounting
8 Load apportioning valves (on vehicles with the anti-lock braking system)
9 Handbrake cable
10 Rear towing eye
11 Spare wheel carrier hook (on the retaining bolt)

Check and adjust the tyre pressures (including the spare), and visually examine the tyres for wear or damage
Check the suspension components for signs of wear and damage

Bodywork and fittings (Chapter 11)
Check seat belt webbing for damage and cuts
Lubricate door check straps
Lubricate the bonnet lock/safety catch and check its operation
Check the condition of the vehicle underbody and the PVC coated lower panels

Electrical system (Chapter 12)
Check the operation of all interior and exterior lamps and all electrical systems
Check all exposed wiring for condition and security
Check the washer fluid level and top up if necessary
Check for washer fluid leaks and rectify as necessary
Check exposed washer fluid hoses for damage or deterioration

Every 12 000 miles (20 000 km) or 24 months   whichever occurs first

---

In addition to the items in the 12-monthly service, carry out the following:

Engine (Chapter 1)
Check and if necessary adjust the valve clearances (HCS engines only)

Cooling system (Chapter 2)
Check the condition of the water pump/alternator drivebelt (HCS engines only)

Fuel, exhaust and emission control systems (Chapter 3)
Clean the idle speed control valve (XR2i models only)*
Check for evidence of exhaust leaks and rectify as necessary
Check the exhaust system for security and condition

* Applies only to Weber-built valve (mounted on air filter housing) – see also Chapter 13, Section 7.

Ignition and engine management systems (Chapter 4)
Renew the spark plugs

Transmission (Chapter 6)
Check for oil (fluid) leaks and rectify as necessary (on CTX automatic transmission, rectification **must** be carried out by a Ford dealer or other suitably equipped specialist)
Check the transmission fluid level and top up if necessary

Driveshafts (Chapter 7)
Check the driveshaft CV joints for security, condition and leaks

Braking system (Chapter 8)
Check the condition and if necessary adjust the tension of the modulator drivebelts (anti-lock brake equipped vehicles only)
Check the handbrake and footbrake operation

Steering (Chapter 9)
Check the steering system operation

Every 24 000 miles (40 000 km) or 48 months – whichever occurs first

---

In addition to all the items in the 12-monthly and 24-monthly services, carry out the following:

Fuel, exhaust and emission control systems (Chapter 3)
Renew the fuel filter (fuel injected models only)
Check the operation of the air filter intake-air temperature control system, where applicable
Renew the air filter element (required more frequently under dusty or polluted conditions)
Renew the crankcase ventilation adaptor/filter or separate filter, as applicable
Renew the crankcase ventilation pad filter (1.6 litre EFi models)
Visually check the oil filler cap and renew as necessary (HCS engines only)

Ignition and engine management systems (Chapter 4)
Clean and inspect the distributor cap and ignition coil tower, and inspect the distributor and rotor arm for condition (vehicles without the distributorless ignition system)
Clean the HT leads and inspect for condition and connection security

Transmission (Chapter 6)
Renew the transmission fluid (CTX automatic transmission)

Every 36 000 miles (60 000 km) or 24 months – whichever occurs first

---

In addition to the items in the 6, 12 and 24-monthly services, carry out the following:

Cooling system (Chapter 2)
Renew the coolant

Every 36 000 miles (60 000 km) or 36 months – whichever occurs first

---

In addition to the items in the 6, 12 and 24-monthly services, carry out the following:

Braking system (Chapter 8)
Renew the brake fluid
Examine all visible rubber components of the system for deterioration

---

Every 36 000 miles

In addition to the other servicing requirements, carry out the following:

Engine (Chapter 1)
Renew the timing belt (camshaft drive belt) (CVH engines)

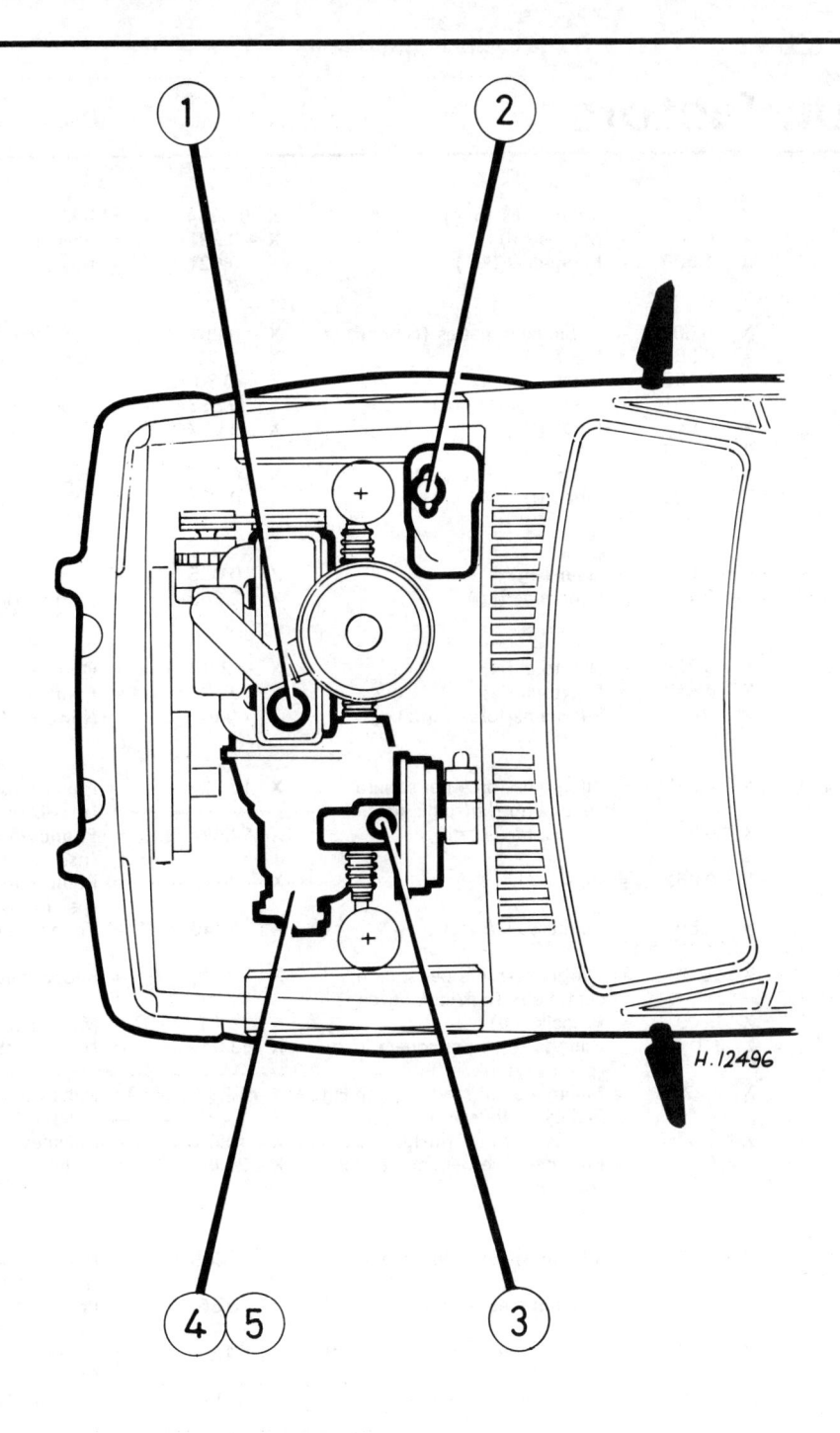

Recommended lubricants and fluids

Component or system	Lubricant type/specification	Duckhams recommendation
Engine (1)	Multigrade engine oil, viscosity range SAE 10W/30 to 20W/50, to API SG	Duckhams QS, QXR, Hypergrade Plus or Hypergrade
Cooling system (2)	Ethylene glycol based antifreeze, to Ford specification SSM-97B-9103-A	Duckhams Universal Antifreeze and Summer Coolant
Braking system (3)	Brake fluid to Ford specification SAM-6C-9103 A	Duckhams Universal Brake and Clutch Fluid
Manual transmission (4)	High pressure gear oil, viscosity SAE 80, to Ford specification SQM 2C-9008A	Duckhams Hypoid 80
CTX automatic transmission (5)	ATF to Ford specification ESP-M2C-166 H	Duckhams Uni-Matic

Conversion factors

Length (distance)

Inches (in)	X	25.4	= Millimetres (mm)	X 0.0394	= Inches (in)
Feet (ft)	X	0.305	= Metres (m)	X 3.281	= Feet (ft)
Miles	X	1.609	= Kilometres (km)	X 0.621	= Miles

Volume (capacity)

Cubic inches (cu in; in³)	X	16.387	= Cubic centimetres (cc; cm³)	X 0.061	= Cubic inches (cu in; in³)
Imperial pints (Imp pt)	X	0.568	= Litres (l)	X 1.76	= Imperial pints (Imp pt)
Imperial quarts (Imp qt)	X	1.137	= Litres (l)	X 0.88	= Imperial quarts (Imp qt)
Imperial quarts (Imp qt)	X	1.201	= US quarts (US qt)	X 0.833	= Imperial quarts (Imp qt)
US quarts (US qt)	X	0.946	= Litres (l)	X 1.057	= US quarts (US qt)
Imperial gallons (Imp gal)	X	4.546	= Litres (l)	X 0.22	= Imperial gallons (Imp gal)
Imperial gallons (Imp gal)	X	1.201	= US gallons (US gal)	X 0.833	= Imperial gallons (Imp gal)
US gallons (US gal)	X	3.785	= Litres (l)	X 0.264	= US gallons (US gal)

Mass (weight)

Ounces (oz)	X	28.35	= Grams (g)	X 0.035	= Ounces (oz)
Pounds (lb)	X	0.454	= Kilograms (kg)	X 2.205	= Pounds (lb)

Force

Ounces-force (ozf; oz)	X	0.278	= Newtons (N)	X 3.6	= Ounces-force (ozf; oz)
Pounds-force (lbf; lb)	X	4.448	= Newtons (N)	X 0.225	= Pounds-force (lbf; lb)
Newtons (N)	X	0.1	= Kilograms-force (kgf; kg)	X 9.81	= Newtons (N)

Pressure

Pounds-force per square inch (psi; lbf/in²; lb/in²)	X	0.070	= Kilograms-force per square centimetre (kgf/cm²; kg/cm²)	X 14.223	= Pounds-force per square inch (psi; lbf/in²; lb/in²)
Pounds-force per square inch (psi; lbf/in²; lb/in²)	X	0.068	= Atmospheres (atm)	X 14.696	= Pounds-force per square inch (psi; lbf/in²; lb/in²)
Pounds-force per square inch (psi; lbf/in²; lb/in²)	X	0.069	= Bars	X 14.5	= Pounds-force per square inch (psi; lbf/in²; lb/in²)
Pounds-force per square inch (psi; lbf/in²; lb/in²)	X	6.895	= Kilopascals (kPa)	X 0.145	= Pounds-force per square inch (psi; lbf/in²; lb/in²)
Kilopascals (kPa)	X	0.01	= Kilograms-force per square centimetre (kgf/cm²; kg/cm²)	X 98.1	= Kilopascals (kPa)
Millibar (mbar)	X	100	= Pascals (Pa)	X 0.01	= Millibar (mbar)
Millibar (mbar)	X	0.0145	= Pounds-force per square inch (psi; lbf/in²; lb/in²)	X 68.947	= Millibar (mbar)
Millibar (mbar)	X	0.75	= Millimetres of mercury (mmHg)	X 1.333	= Millibar (mbar)
Millibar (mbar)	X	0.401	= Inches of water (inH$_2$O)	X 2.491	= Millibar (mbar)
Millimetres of mercury (mmHg)	X	0.535	= Inches of water (inH$_2$O)	X 1.868	= Millimetres of mercury (mmHg)
Inches of water (inH$_2$O)	X	0.036	= Pounds-force per square inch (psi; lbf/in²; lb/in²)	X 27.68	= Inches of water (inH$_2$O)

Torque (moment of force)

Pounds-force inches (lbf in; lb in)	X	1.152	= Kilograms-force centimetre (kgf cm; kg cm)	X 0.868	= Pounds-force inches (lbf in; lb in)
Pounds-force inches (lbf in; lb in)	X	0.113	= Newton metres (Nm)	X 8.85	= Pounds-force inches (lbf in; lb in)
Pounds-force inches (lbf in; lb in)	X	0.083	= Pounds-force feet (lbf ft; lb ft)	X 12	= Pounds-force inches (lbf in; lb in)
Pounds-force feet (lbf ft; lb ft)	X	0.138	= Kilograms-force metres (kgf m; kg m)	X 7.233	= Pounds-force feet (lbf ft; lb ft)
Pounds-force feet (lbf ft; lb ft)	X	1.356	= Newton metres (Nm)	X 0.738	= Pounds-force feet (lbf ft; lb ft)
Newton metres (Nm)	X	0.102	= Kilograms-force metres (kgf m; kg m)	X 9.804	= Newton metres (Nm)

Power

Horsepower (hp)	X	745.7	= Watts (W)	X 0.0013	= Horsepower (hp)

Velocity (speed)

Miles per hour (miles/hr; mph)	X	1.609	= Kilometres per hour (km/hr; kph)	X 0.621	= Miles per hour (miles/hr; mph)

*Fuel consumption**

Miles per gallon, Imperial (mpg)	X	0.354	= Kilometres per litre (km/l)	X 2.825	= Miles per gallon, Imperial (mpg)
Miles per gallon, US (mpg)	X	0.425	= Kilometres per litre (km/l)	X 2.352	= Miles per gallon, US (mpg)

Temperature

Degrees Fahrenheit = (°C x 1.8) + 32   Degrees Celsius (Degrees Centigrade; °C) = (°F - 32) x 0.56

It is common practice to convert from miles per gallon (mpg) to litres/100 kilometres (l/100km), where mpg (Imperial) x l/100 km = 282 and mpg (US) x l/100 km = 235

Fault diagnosis

Introduction

The vehicle owner who does his or her own maintenance according to the recommended schedules should not have to use this section of the manual very often. Modern component reliability is such that, provided those items subject to wear or deterioration are inspected or renewed at the specified intervals, sudden failure is comparatively rare. Faults do not usually just happen as a result of sudden failure, but develop over a period of time. Major mechanical failures in particular are usually preceded by characteristic symptoms over hundreds or even thousands of miles. Those components which do occasionally fail without warning are often small and easily carried in the vehicle.

With any fault finding, the first step is to decide where to begin investigations. Sometimes this is obvious, but on other occasions a little detective work will be necessary. The owner who makes half a dozen haphazard adjustments or replacements may be successful in curing a fault (or its symptoms), but he will be none the wiser if the fault recurs and he may well have spent more time and money than was necessary. A calm and logical approach will be found to be more satisfactory in the long run. Always take into account any warning signs or abnormalities that may have been noticed in the period preceding the fault – power loss, high or low gauge readings, unusual noises or smells, etc – and remember that failure of components such as fuses or spark plugs may only be pointers to some underlying fault.

The pages which follow here are intended to help in cases of failure to start or breakdown on the road. There is also a Fault Diagnosis Section at the end of each Chapter which should be consulted if the preliminary checks prove unfruitful. Whatever the fault, certain basic principles apply. These are as follows:

Verify the fault. This is simply a matter of being sure that you know what the symptoms are before starting work. This is particularly important if you are investigating a fault for someone else who may not have described it very accurately.

Don't overlook the obvious. For example, if the vehicle won't start, is there petrol in the tank? (Don't take anyone else's word on this particular point, and don't trust the fuel gauge either!) If an electrical fault is indicated, look for loose or broken wires before digging out the test gear.

Cure the disease, not the symptom. Substituting a flat battery with a fully charged one will get you off the hard shoulder, but if the underlying cause is not attended to, the new battery will go the same way. Similarly, changing oil-fouled spark plugs for a new set will get you moving again, but remember that the reason for the fouling (if it wasn't simply an incorrect grade of plug) will have to be established and corrected.

Don't take anything for granted. Particularly, don't forget that a 'new' component may itself be defective (especially if it's been rattling round in the boot for months), and don't leave components out of a fault diagnosis sequence just because they are new or recently fitted. When you do finally diagnose a difficult fault, you'll probably realise that all the evidence was there from the start.

Electrical faults

Electrical faults can be more puzzling than straightforward mechanical failures, but they are no less susceptible to logical analysis if the basic principles of operation are understood. Vehicle electrical wiring exists in extremely unfavourable conditions – heat, vibration and chemical attack – and the first things to look for are loose or corroded connections and broken or chafed wires, especially where the wires pass through holes in the bodywork or are subject to vibration.

All metal-bodied vehicles in current production have one pole of the battery 'earthed', ie connected to the vehicle bodywork, and in nearly all modern vehicles it is the negative (–) terminal. The various electrical components – motors, bulb holders etc – are also connected to earth, either by means of a lead or directly by their mountings. Electric current flows through the component and then back to the battery via the bodywork. If the component mounting is loose or corroded, or if a good path back to the battery is not available, the circuit will be incomplete and malfunction will result. The engine and/or gearbox are also earthed by means of flexible metal straps to the body or subframe; if these

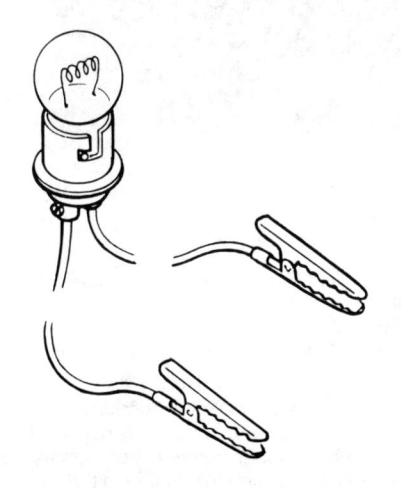

A simple test lamp is useful for tracing electrical faults

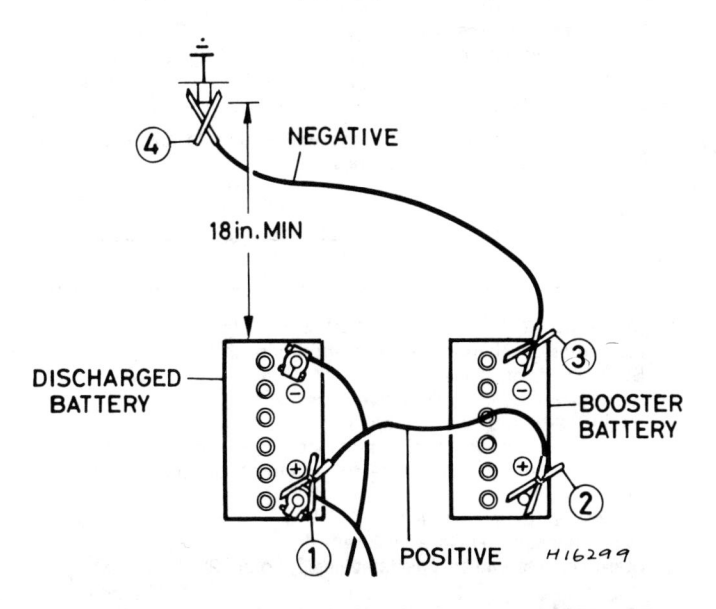

Jump start lead connections for negative earth – connect leads in order shown

Carrying a few spares can save a long walk!

straps are loose or missing, starter motor, alternator and ignition trouble may result.

Assuming the earth return to be satisfactory, electrical faults will be due either to component malfunction or to defects in the current supply. Individual components are dealt with in Chapter 12. If supply wires are broken or cracked internally this results in an open-circuit, and the easiest way to check for this is to bypass the suspect wire temporarily with a length of wire having a crocodile clip or suitable connector at each end. Alternatively, a 12V test lamp can be used to verify the presence of supply voltage at various points along the wire and the break can be thus isolated.

If a bare portion of a live wire touches the bodywork or other earthed metal part, the electricity will take the low-resistance path thus formed back to the battery: this is known as a short-circuit. Hopefully a short-circuit will blow a fuse, but otherwise it may cause burning of the insulation (and possibly further short-circuits) or even a fire. This is why it is inadvisable to bypass persistently blowing fuses with silver foil or wire.

Spares and tool kit

Most vehicles are supplied only with sufficient tools for wheel changing; the *Maintenance and minor repair* tool kit detailed in *Tools and working facilities,* with the addition of a hammer, is probably sufficient for those repairs that most motorists would consider attempting at the roadside. In addition a few items which can be fitted without too much trouble in the event of a breakdown should be carried. Experience and available space will modify the list below, but the following may save having to call on professional assistance:

Spark plugs, clean and correctly gapped
HT lead and plug cap – long enough to reach the plug furthest from the distributor or DIS ignition coil
Distributor rotor (where applicable)
Drivebelt(s) (camshaft and anti-lock brake modulators, as applicable)
Spare fuses
Set of principal light bulbs
Tin of radiator sealer and hose bandage
Exhaust bandage
Roll of insulating tape
Length of soft iron wire
Length of electrical flex
Torch or inspection lamp (can double as test lamp)

Battery jump leads
Tow-rope
Ignition water dispersant aerosol
Litre of engine oil
Sealed can of hydraulic fluid
Emergency windscreen
Worm drive clips

If spare fuel is carried, a can designed for the purpose should be used to minimise risks of leakage and collision damage. A first aid kit and a warning triangle, whilst not at present compulsory in the UK, are obviously sensible items to carry in addition to the above.

When touring abroad it may be advisable to carry additional spares which, even if you cannot fit them yourself, could save having to wait while parts are obtained. The items below may be worth considering:

Clutch and throttle cables
Cylinder head gasket
Alternator brushes
Tyre valve core

One of the motoring organisations will be able to advise on availability of fuel, etc., in foreign countries.

Engine will not start

Engine fails to turn when starter operated

Flat battery (recharge, use jump leads, or push start)
Battery terminals loose or corroded
Battery earth to body defective
Engine earth strap loose or broken
Starter motor (or solenoid) wiring loose or broken
Automatic transmission selector in wrong position, or inhibitor faulty
Ignition/starter switch faulty
Major mechanical failure (seizure)
Starter or solenoid internal fault (see Chapter 12)

Starter motor turns engine slowly

Partially discharged battery (recharge, use jump leads,

Fault diagnosis

or push start)
Battery terminals loose or corroded
Battery earth to body defective
Engine earth strap loose
Starter motor (or solenoid) wiring loose
Starter motor internal fault (see Chapter 12)

Starter motor spins without turning engine
Flat battery
Starter motor pinion sticking on sleeve
Flywheel gear teeth damaged or worn
Starter motor mounting bolts loose

Engine turns normally but fails to start
Damp or dirty HT leads and distributor cap (crank engine and check for spark) - try moisture dispersant such as Holts Wet Start
No fuel in tank (check for delivery)
Excessive choke (hot engine) or insufficient choke (cold engine) (as applicable)
Fouled or incorrectly gapped spark plugs (remove, renew and regap)
Other ignition system fault (see Chapter 4)
Other fuel system fault (see Chapter 3)
Poor compression (see Chapter 1)
Major mechanical failure (eg camshaft drive)

Engine fires but will not run
Insufficient choke (cold engine) (as applicable)
Air leaks at carburettor, inlet manifold or CFi unit, or in EFi idle speed control air bypass hose
Fuel starvation (see Chapter 3)
Ignition fault (see Chapter 4)

Engine cuts out and will not restart

Engine cuts out suddenly - ignition fault
Loose or disconnected LT wires, or loose coil to distributor HT lead (where applicable)
Wet HT leads or distributor cap (after traversing water splash)
Coil failure (check for spark)
Other ignition fault (see Chapter 4)

Engine misfires before cutting out - fuel fault
Fuel tank empty
Fuel pump defective or filter blocked (check for delivery)
Fuel tank filler vent blocked (suction will be evident on releasing cap)
Carburettor needle valve sticking (where applicable)
Carburettor jets or fuel injector(s) blocked (fuel contaminated)
Other fuel system fault (see Chapter 3)

Engine cuts out - other causes
Serious overheating
Major mechanical failure (eg camshaft drive)

Engine overheats

Coolant loss due to internal or external leakage (see Chapter 2)
Thermostat defective

Low oil level
Brakes binding
Radiator clogged externally or internally
Electric cooling fan not operating correctly
Engine waterways clogged
Ignition timing incorrect or automatic advance malfunctioning
Mixture too weak

Note: *Do not add cold water to an overheated engine or damage may result*

Low engine oil pressure

Warning light illuminated with engine running
Oil level low or incorrect grade
Defective sender unit
Wire to sender unit earthed
Engine overheating
Oil filter clogged or bypass valve defective
Oil pressure relief valve defective
Oil pick-up strainer clogged
Oil pump worn or mountings loose
Worn main or big-end bearings

Note: *Low oil pressure in a high-mileage engine at tickover is not necessarily a cause for concern. Sudden pressure loss at speed is far more significant. In any event, check the warning light sender before condemning the engine.*

Engine noises

Pre-ignition (pinking) on acceleration
Incorrect grade of fuel
Ignition timing incorrect
Distributor faulty or worn (where applicable)
Worn or maladjusted carburettor
Excessive carbon build-up in engine

Whistling or wheezing noises
Leaking vacuum hose
Leaking carburettor, CFi unit or manifold gasket
Blowing head gasket

Tapping or rattling
Incorrect valve clearances
Worn valve gear
Worn timing chain or belt
Broken piston ring (ticking noise)
Defective hydraulic tappets

Knocking or thumping
Unintentional mechanical contact (eg fan blades)
Worn drivebelt
Peripheral component fault (alternator, water pump, etc)
Worn big-end bearings (regular heavy knocking, perhaps less under load)
Worn main bearings (rumbling and knocking, perhaps worsening under load)
Piston slap (most noticeable when cold)

Chapter 1 Engine

For modifications, and information applicable to later models, see Supplement at end of manual

Contents

Part A: OHV (HCS) engines

Crankshaft front oil seal – renewal	12
Crankshaft rear oil seal – renewal	16
Cylinder head – removal and refitting	5
Cylinder head and pistons – decarbonising	6
Engine – component examination and renovation	19
Engine – dismantling and reassembly	18
Engine – method of removal/refitting	14
Engine – removal and refitting	15
Engine/transmission mountings – removal and refitting	13
Fault diagnosis	20
General description	1
Lubrication system – description	9
Maintenance and inspection	2
Major operations possible only with engine removed	4
Major operations possible with engine in vehicle	3
Oil pump – removal and refitting	11
Piston/connecting rod assemblies – removal and refitting	17
Rocker shaft assembly – dismantling and reassembly	7
Sump – removal and refitting	10
Valve clearances – adjustment	8

Part B: CVH engines

Camshaft – removal and refitting	27
Camshaft oil seal – renewal	26
Crankshaft front oil seal – renewal	31
Cylinder head (1.4 litre and 1.6 litre carburettor engines including 1.4 litre CFi) – removal and refitting	28
Cylinder head (1.6 litre EFi) – removal and refitting	29
Cylinder head and pistons – decarbonising	30
Engine – component examination and renovation	40
Engine – dismantling and reassembly	39
Engine – method of removal	36
Engine/transmission assembly (1.4 litre and 1.6 litre carburettor engines including 1.4 litre CFi) – removal, separation, reconnection and refitting	37
Engine/transmission assembly (1.6 litre EFi) – removal, separation, reconnection and refitting	38
Engine/transmission mountings – removal and refitting	34
Fault diagnosis	41
General description	21
Lubrication system – description	35
Maintenance and inspection	22
Major operations possible with the engine in the vehicle	23
Major operations requiring engine removal	24
Piston/connecting rod assemblies – removal and refitting	33
Sump – removal and refitting	32
Timing belt – removal, refitting and adjustment	25

Specifications

Part A: OHV (HCS) engines

General

Engine type	Four-cylinder, in-line overhead valve
Capacity:	
1.0 litre	999 cc
1.1 litre	1118 cc
Bore	68.68 mm
Stroke:	
1.0 litre	67.40 mm
1.1 litre	75.48 mm
Compression ratio	9.5 : 1
Firing order	1-2-4-3 (No 1 at timing cover end)

Cylinder block
Material	Cast iron
Number of main bearings	3
Cylinder bore diameter:	
Standard class 1 (or A)	68.680 to 68.690 mm
Standard class 2 (or B)	68.690 to 68.700 mm
Standard class 3 (or C)	68.700 to 68.710 mm
Oversize 0.5 mm	69.200 to 69.210 mm
Oversize 1.0 mm	69.700 to 69.710 mm
Main bearing shell inside diameter (fitted):	
Standard	57.009 to 57.036 mm
0.254 mm undersize	56.755 to 56.782 mm
0.508 mm undersize (service)	56.501 to 56.528 mm
0.762 mm undersize (service)	56.247 to 56.274 mm
Centre main bearing width (excluding semi-circular thrust washers)	22.040 to 22.100 mm
Camshaft bearing bush inside diameter:	
Standard	39.662 to 39.682 mm
Standard and oversize (service)	39.662 to 39.713 mm

Crankshaft
Main bearing journal diameter:	
Standard	56.990 to 57.000 mm
Standard (with yellow line)	56.980 to 56.990 mm
0.254 mm undersize (with green line)	56.726 to 56.746 mm
0.508 mm undersize (service)	56.472 to 56.492 mm
0.762 mm undersize (service)	56.218 to 56.238 mm
Clearance between bearing shell and main bearing journal	0.009 to 0.046 mm
Crankpin (big-end journal) diameter:	
Standard	40.99 to 41.01 mm
0.254 undersize (with green spot)	40.74 to 40.76 mm
0.508 undersize (service)	40.49 to 40.51 mm
0.762 undersize (service)	40.24 to 40.26 mm
Crankshaft endfloat	0.075 to 0.285 mm
Semi-circular thrust washer thickness:	
Standard	2.80 to 2.85 mm
Oversize	2.99 to 3.04 mm

Connecting rod
Bore diameter:	
Big-end	43.99 to 44.01 mm
Small end	17.99 to 18.01 mm
Big-end bearing shell inside diameter (fitted):	
Standard	41.016 to 41.050 mm
0.254 mm undersize	40.766 to 40.800 mm
0.508 mm undersize	40.516 to 40.550 mm
0.762 mm undersize	40.266 to 40.300 mm
1.016 mm undersize	40.016 to 40.050 mm
Clearance between bearing shell and big-end bearing journal	0.006 to 0.060 mm
Endfloat	0.100 to 0.250 mm
Rod weight classifications	A, B, C and D. Do not fit mixed classifications

Piston and piston rings
Piston diameter:	
Standard class 1 (or A)	68.65 to 68.66 mm
Standard class 2 (or B)	68.66 to 68.67 mm
Standard class 3 (or C)	68.67 to 68.68 mm
Standard (service)	68.67 to 68.70 mm
0.5 mm oversize	69.16 to 69.19 mm
1.0 mm oversize	69.66 to 69.69 mm
Piston-to-bore clearance:	
Production	0.020 to 0.040 mm
Service	0.015 to 0.050 mm
Piston ring end-gap (fitted):	
Top and middle (compression)	0.25 to 0.45 mm
Bottom (oil control)	0.20 to 0.40 mm
Ring gap positions:	
Top (compression)	Offset 180° from oil control ring gap
Middle (compression)	Offset 90° from oil control ring gap
Bottom (oil control)	Flush with gudgeon pin

Gudgeon pin
Length	58.6 to 59.4 mm
Diameter:	
White	18.026 to 18.029 mm
Red	18.029 to 18.032 mm
Blue	18.032 to 18.035 mm
Yellow	18.035 to 18.038 mm

Gudgeon pin (continued)
Interference fit in connecting rod at 21°C (70°F) .... 0.016 to 0.048 mm
Clearance in piston at 21°C (70°F) .... 0.008 to 0.014 mm

Camshaft
Number of bearings 3
Drive .... Chain
Thrust plate thickness .... 4.457 to 4.508 mm
Cam lift:
   Inlet valves .... 5.15 mm
   Exhaust valves .... 4.92 mm
Cam length:
   Inlet valve .... 32.036 to 32.264 mm
   Exhaust valve .... 31.806 to 32.034 mm
Camshaft bearing diameter .... 39.615 to 39.635 mm
Camshaft bearing bush inside diameter:
   Standard .... 39.662 to 39.682 mm
   Oversize (service) .... 39.662 to 39.713 mm
Camshaft endfloat .... 0.02 to 0.19 mm

Timing chain
Number of links 46
Length .... 438.15 mm

Cylinder head
Material .... Cast iron
Combustion chamber capacity:
   1.0 litre .... 23.54 to 25.54 cc
   1.1 litre .... 27.24 to 29.24 cc
Maximum permissible cylinder head distortion:
   Measured over a distance of 26 mm .... 0.04 mm
   Measured over a distance of 152 mm .... 0.08 mm
   Measured over entire length .... 0.15 mm
Minimum combustion chamber depth after skimming .... 14.4 ± 0.15 mm
Valve guide bore (inlet and exhaust):
   Standard .... 7.063 to 7.094 mm
   1st oversize (0.2 mm) .... 7.263 to 7.294 mm
   2nd oversize (0.4 mm) .... 7.463 to 7.494 mm

Note: *The cylinder head has valve seat rings on the inlet and exhaust side, and cannot be reworked with conventional tools (see text)*

Valve seat width (inlet/exhaust) .... 1.18 to 1.75 mm
Valve seat angle .... 45°
Seat cutter correction angle:
   Upper .... 30° (see text)
   Lower .... 75° (see text)

Valves – general
Operation .... Cam followers, pushrods and rocker arms
Valve timing:
   Inlet valve opens .... 14° BTDC
   Inlet valve closes .... 46° ABDC
   Exhaust valve opens .... 49° BBDC
   Exhaust valve closes .... 11° ATDC
Valve and guide lubrication (during assembly) .... Hypoid oil to Ford specification SQM-2C-9002-AA (refer to Ford dealer)

Valve clearances (cold):
   Inlet .... 0.20 to 0.25 mm (0.008 to 0.010 in)
   Exhaust .... 0.30 to 0.35 mm (0.012 to 0.014 in)
Cam follower (tappet) diameter .... 13.081 to 13.094 mm
Cam follower (tappet) clearance in bore .... 0.016 to 0.062 mm
Valve spring type .... Single
Valve spring free length .... 41.0 mm
Number of coils 6

Inlet valve
Length .... 103.70 to 104.40 mm
Head diameter .... 32.90 to 33.10 mm
Valve stem diameter:
   Standard .... 7.025 to 7.043 mm
   0.2 mm oversize .... 7.225 to 7.243 mm
   0.4 mm oversize .... 7.425 to 7.443 mm
Valve stem clearance in guide .... 0.020 to 0.069 mm
Valve lift (less clearance) .... 8.450 mm

Exhaust valve
Length... 104.02 to 104.72 mm
Head diameter.. 28.90 to 29.10 mm
Valve stem diameter:
    Standard... 6.999 to 7.017 mm
    0.2 mm oversize... 7.199 to 7.217 mm
    0.4 mm oversize... 7.399 to 7.417 mm
Valve stem clearance in guide... 0.046 to 0.095 mm
Valve lift (less clearance)... 8.070 mm

Lubrication
Oil type/specification... Multigrade engine oil, viscosity range SAE 10W/30 to 20W/50, to API SG (Duckhams QS, QXR, Hypergrade Plus or Hypergrade)
Oil filter... Champion C104
Oil capacity:
    Without filter change... 2.75 litres
    With filter change... 3.25 litres
Initial oil fill capacity (new or rebuilt engine) with filter............ 3.40 litres
Oil pump type.. External, rotor type, driven by gear on camshaft
Minimum oil pressure at 80°C (175°F):
    Engine speed 750 rpm... 0.6 bar
    Engine speed 2000 rpm... 1.5 bar
Oil pressure warning light operates at....................................... 0.32 to 0.53 bar
Pressure relief valve opens at... 2.41 to 2.96 bar
Oil pump clearances:
    Outer rotor to pump body... 0.14 to 0.26 mm
    Inner rotor to outer rotor... 0.051 to 0.127 mm
    Rotor endfloat... 0.025 to 0.060 mm

Torque wrench settings

	Nm	Lbf ft
Main bearing cap bolts	88 to 102	65 to 75
Big-end bearing cap bolts:		
Stage one	4	3
Stage two	Angle-tighten a further 90°	Angle tighten a further 90°
Cylinder head bolts (see text):		
Stage one	30	22
Stage two	Angle-tighten a further 90°	Angle-tighten a further 90°
Stage three	Angle-tighten a further 90°	Angle-tighten a further 90°
Sump bolts (see text):		
Stage one	6 to 8	4 to 6
Stage two	8 to 11	6 to 8
Stage three	8 to 11	6 to 8
Rear oil seal housing	16 to 20	12 to 15
Flywheel	64 to 70	47 to 52
Camshaft sprocket	16 to 20	12 to 15
Camshaft thrust plate	4 to 5	3 to 4
Timing chain tensioner	7 to 9	5 to 7
Timing cover	7 to 10	5 to 7
Crankshaft pulley	110 to 120	81 to 88
Oil pump	16 to 20	12 to 15
Oil pump cover	8 to 12	6 to 9
Oil drain plug	21 to 28	15 to 21
Oil pressure switch	13 to 15	10 to 11
Rocker pedestal bolts	40 to 46	30 to 34
Rocker cover	4 to 5	3 to 4
Lower engine adaptor plate (clutch housing cover):		
Manual transmission	34 to 46	25 to 34
CTX automatic transmission	7.5 to 10	6 to 7
Engine/transmission flange bolts:		
Manual transmission	35 to 45	26 to 33
CTX automatic transmission	27 to 50	20 to 37
Engine mounting (right-hand):		
Bolt to body (in wheel arch)	41 to 58	30 to 43
Nut to body (by suspension strut)	41 to 58	30 to 43
Bracket to cylinder block	54 to 72	40 to 53
Rubber insulator to bracket	70 to 95	52 to 70
Engine/transmission mountings (on transmission bearer)	Refer to Specifications, Chapter 6	

Part B: CVH engines

General
Engine type.. Four-cylinder, in-line overhead camshaft
Capacity:
    1.4 litre... 1392 cc
    1.6 litre... 1596 cc

Bore:
- 1.4 litre .... 77.24 mm
- 1.6 litre .... 79.96 mm

Stroke:
- 1.4 litre .... 74.30 mm
- 1.6 litre .... 79.52 mm

Compression ratio:
- 1.4 litre (not CFi) .... 9.5 : 1
- 1.4 litre (CFi) .... 8.5 : 1
- 1.6 litre (not EFi) .... 9.5 : 1
- 1.6 litre (EFi) .... 9.75 : 1

Firing order .... 1-3-4-2 (No 1 at timing belt end)

Cylinder block

Material .... Cast iron
Number of main bearings 5

Cylinder bore diameter – 1.4 litre (production):
- Standard class 1 .... 77.22 to 77.23 mm
- Standard class 2 .... 77.23 to 77.24 mm
- Standard class 3 .... 77.24 to 77.25 mm
- Standard class 4 .... 77.25 to 77.26 mm
- Oversize class A .... 77.51 to 77.52 mm
- Oversize class B .... 77.52 to 77.53 mm
- Oversize class C .... 77.53 to 77.54 mm

Cylinder bore diameter – 1.4 litre (service):
- Standard .... 77.245 to 77.255 mm
- 0.29 mm oversize .... 77.525 to 77.535 mm
- 0.50 mm oversize .... 77.745 to 77.755 mm

Cylinder bore diameter – 1.6 litre (production):
- Standard class 1 .... 79.94 to 79.95 mm
- Standard class 2 .... 79.95 to 79.96 mm
- Standard class 3 .... 79.96 to 79.97 mm
- Standard class 4 .... 79.97 to 79.98 mm
- Oversize class A .... 80.23 to 80.24 mm
- Oversize class B .... 80.24 to 80.25 mm
- Oversize class C .... 80.25 to 80.26 mm

Cylinder bore diameter – 1.6 litre (service):
- Standard .... 79.965 to 79.975 mm
- 0.29 mm oversize .... 80.245 to 80.255 mm
- 0.50 mm oversize .... 80.465 to 80.475 mm

Main bearing shell inside diameter (fitted):
- Standard .... 58.011 to 58.038 mm
- 0.25 mm undersize .... 57.761 to 57.788 mm
- 0.50 mm undersize .... 57.511 to 57.538 mm
- 0.75 mm undersize .... 57.261 to 57.288 mm

Crankshaft

Main bearing journal diameter:
- Standard .... 57.98 to 58.00 mm
- 0.25 mm undersize .... 57.73 to 57.75 mm
- 0.50 mm undersize .... 57.48 to 57.50 mm
- 0.75 mm undersize .... 57.23 to 57.25 mm

Main bearing journal to bearing shell clearance .... 0.011 to 0.058 mm
Centre main bearing width .... 28.825 to 28.875 mm

Semi-circular thrust washer thickness:
- Standard .... 2.301 to 2.351 mm
- Oversize .... 2.491 to 2.541 mm

Crankshaft endfloat .... 0.09 to 0.30 mm

Crankpin (big-end journal) diameter:
- Standard .... 47.89 to 47.91 mm
- 0.25 mm undersize .... 47.64 to 47.66 mm
- 0.50 mm undersize .... 47.39 to 47.41 mm
- 0.75 mm undersize .... 47.14 to 47.16 mm
- 1.00 mm undersize .... 46.89 to 46.91 mm

Connecting rod

Bore diameter:
- Big-end .... 50.890 to 50.910 mm
- Small end .... 20.589 to 20.609 mm

Big-end bearing shell inside diameter (fitted):
- Standard .... 47.916 to 47.950 mm
- 0.25 mm undersize .... 47.666 to 47.700 mm
- 0.50 mm undersize .... 47.416 to 47.450 mm
- 0.75 mm undersize .... 47.166 to 47.200 mm
- 1.00 mm undersize .... 46.916 to 46.950 mm

Clearance between bearing shell and big-end bearing journal ... 0.006 to 0.060 mm

Are your plugs trying to tell you something?

Normal.
Grey-brown deposits, lightly coated core nose. Plugs ideally suited to engine, and engine in good condition.

Heavy Deposits.
A build up of crusty deposits, light-grey sandy colour in appearance.
Fault: Often caused by worn valve guides, excessive use of upper cylinder lubricant, or idling for long periods.

Lead Glazing.
Plug insulator firing tip appears yellow or green/yellow and shiny in appearance.
Fault: Often caused by incorrect carburation, excessive idling followed by sharp acceleration. Also check ignition timing.

Carbon fouling.
Dry, black, sooty deposits.
Fault: over-rich fuel mixture.
Check: carburettor mixture settings, float level, choke operation, air filter.

Oil fouling.
Wet, oily deposits. Fault: worn bores/piston rings or valve guides; sometimes occurs (temporarily) during running-in period.

Overheating.
Electrodes have glazed appearance, core nose very white – few deposits. Fault: plug overheating. Check: plug value, ignition timing, fuel octane rating (too low) and fuel mixture (too weak).

Electrode damage.
Electrodes burned away; core nose has burned, glazed appearance. Fault: pre-ignition. Check: for correct heat range and as for 'overheating'.

Split core nose.
(May appear initially as a crack). Fault: detonation or wrong gap-setting technique. Check: ignition timing, cooling system, fuel mixture (too weak).

WHY DOUBLE COPPER IS BETTER FOR YOUR ENGINE.

Unique Trapezoidal Copper Cored Earth Electrode — 50% Larger Spark Area — Copper Cored Centre Electrode

Champion Double Copper plugs are the first in the world to have copper core in both centre and earth electrode. This innovative design means that they run cooler by up to 100°C – giving greater efficiency and longer life. These double copper cores transfer heat away from the tip of the plug faster and more efficiently. Therefore, Double Copper runs at cooler temperatures than conventional plugs giving improved acceleration response and high speed performance with no fear of pre-ignition.

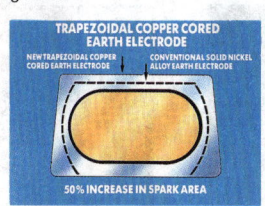

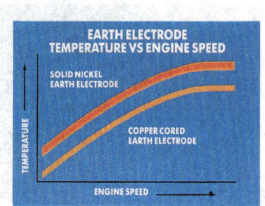

TRAPEZOIDAL COPPER CORED EARTH ELECTRODE
NEW TRAPEZOIDAL COPPER CORED EARTH ELECTRODE | CONVENTIONAL SOLID NICKEL ALLOY EARTH ELECTRODE
50% INCREASE IN SPARK AREA

EARTH ELECTRODE TEMPERATURE VS ENGINE SPEED
SOLID NICKEL EARTH ELECTRODE
COPPER CORED EARTH ELECTRODE

Champion Double Copper plugs also feature a unique trapezoidal earth electrode giving a 50% increase in spark area. This, together with the double copper cores, offers greatly reduced electrode wear, so the spark stays stronger for longer.

- FASTER COLD STARTING
- FOR UNLEADED OR LEADED FUEL
- ELECTRODES UP TO 100°C COOLER
- BETTER ACCELERATION RESPONSE
- LOWER EMISSIONS
- 50% BIGGER SPARK AREA
- THE LONGER LIFE PLUG

Plug Tips/Hot and Cold.
Spark plugs must operate within well-defined temperature limits to avoid cold fouling at one extreme and overheating at the other.
Champion and the car manufacturers work out the best plugs for an engine to give optimum performance under all conditions, from freezing cold starts to sustained high speed motorway cruising.
Plugs are often referred to as hot or cold. With Champion, the higher the number on its body, the hotter the plug, and the lower the number the cooler the plug.

Plug Cleaning
Modern plug design and materials mean that Champion no longer recommends periodic plug cleaning. Certainly don't clean your plugs with a wire brush as this can cause metal conductive paths across the nose of the insulator so impairing its performance and resulting in loss of acceleration and reduced m.p.g.
However, if plugs are removed, always carefully clean the area where the plug seats in the cylinder head as grit and dirt can sometimes cause gas leakage.
Also wipe any traces of oil or grease from plug leads as this may lead to arcing.

CHAMPION

DOUBLE COPPER

1 This photographic sequence shows the steps taken to repair the dent and paintwork damage shown above. In general, the procedure for repairing a hole will be similar; where there are substantial differences, the procedure is clearly described and shown in a separate photograph.

2 First remove any trim around the dent, then hammer out the dent where access is possible. This will minimise filling. Here, after the large dent has been hammered out, the damaged area is being made slightly concave.

3 Next, remove all paint from the damaged area by rubbing with coarse abrasive paper or using a power drill fitted with a wire brush or abrasive pad. 'Feather' the edge of the boundary with good paintwork using a finer grade of abrasive paper.

4 Where there are holes or other damage, the sheet metal should be cut away before proceeding further. The damaged area and any signs of rust should be treated with Turtle Wax Hi-Tech Rust Eater, which will also inhibit further rust formation.

5 *For a large dent or hole* mix Holts Body Plus Resin and Hardener according to the manufacturer's instructions and apply around the edge of the repair. Press Glass Fibre Matting over the repair area and leave for 20-30 minutes to harden. Then ...

5A ... brush more Holts Body Plus Resin and Hardener onto the matting and leave to harden. Repeat the sequence with two or three layers of matting, checking that the final layer is lower than the surrounding area. Apply Holts Body Plus Filler Paste as shown in Step 5B.

5B *For a medium dent*, mix Holts Body Plus Filler Paste and Hardener according to the manufacturer's instructions and apply it with a flexible applicator. Apply thin layers of filler at 20-minute intervals, until the filler surface is slightly proud of the surrounding bodywork.

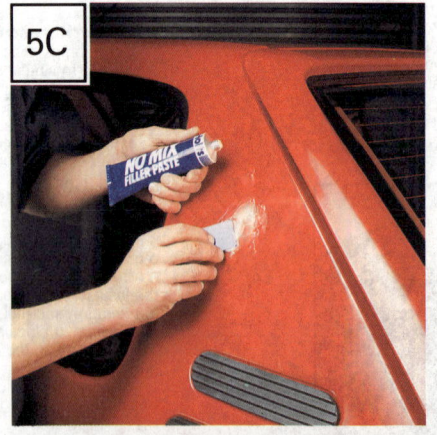

5C *For small dents and scratches* use Holts No Mix Filler Paste straight from the tube. Apply it according to the instructions in thin layers, using the spatula provided. It will harden in minutes if applied outdoors and may then be used as its own knifing putty.

6 Use a plane or file for initial shaping. Then, using progressively finer grades of wet-and-dry paper, wrapped round a sanding block, and copious amounts of clean water, rub down the filler until glass smooth. 'Feather' the edges of adjoining paintwork.

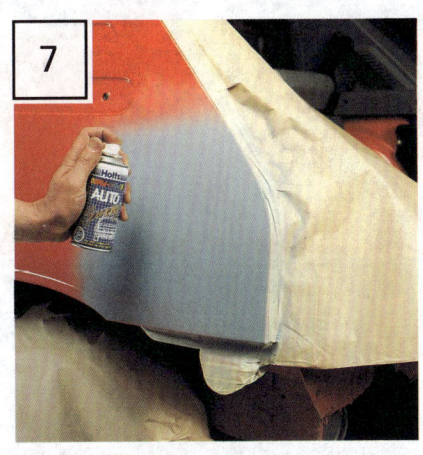

7 Protect adjoining areas before spraying the whole repair area and at least one inch of the surrounding sound paintwork with Holts Dupli-Color primer.

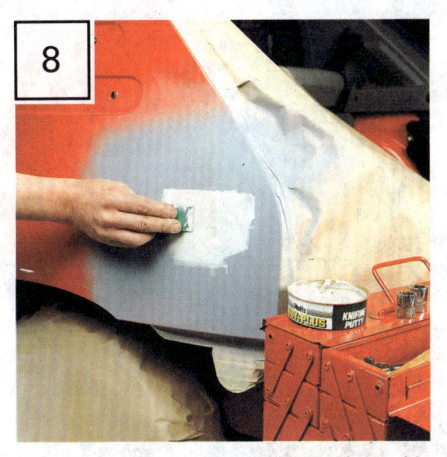

8 Fill any imperfections in the filler surface with a small amount of Holts Body Plus Knifing Putty. Using plenty of clean water, rub down the surface with a fine grade wet-and-dry paper – 400 grade is recommended – until it is really smooth.

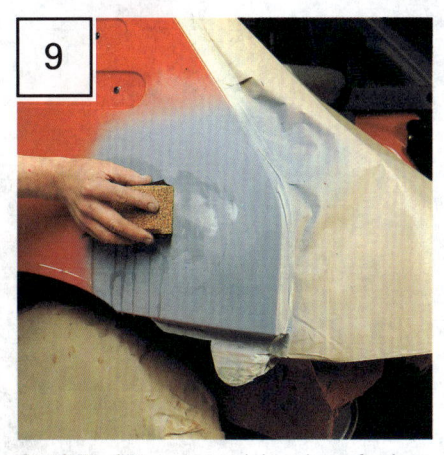

9 Carefully fill any remaining imperfections with knifing putty before applying the last coat of primer. Then rub down the surface with Holts Body Plus Rubbing Compound to ensure a really smooth surface.

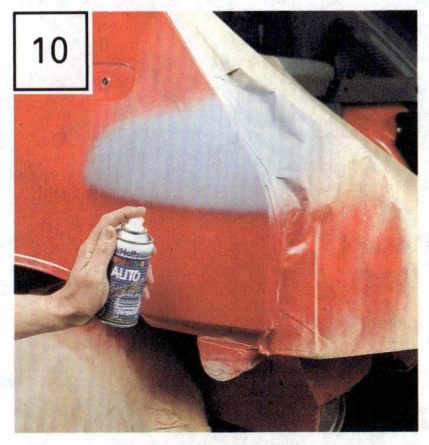

10 Protect surrounding areas from overspray before applying the topcoat in several thin layers. Agitate Holts Dupli-Color aerosol thoroughly. Start at the repair centre, spraying outwards with a side-to-side motion.

10A If the exact colour is not available off the shelf, local Holts Professional Spraymatch Centres will custom fill an aerosol to match perfectly.

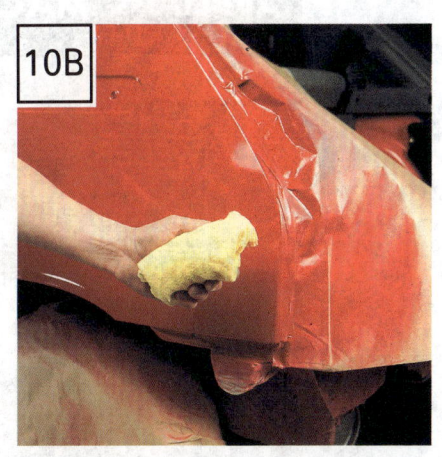

10B To identify whether a lacquer finish is required, rub a painted unrepaired part of the body with wax and a clean cloth.

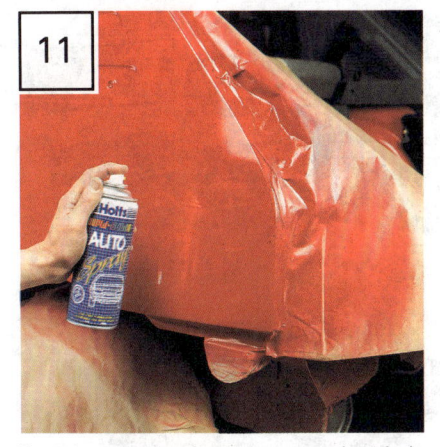

11 If *no* traces of paint appear on the cloth, spray Holts Dupli-Color clear lacquer over the repaired area to achieve the correct gloss level.

12 The paint will take about two weeks to harden fully. After this time it can be 'cut' with a mild cutting compound such as Turtle Wax Minute Cut prior to polishing with a final coating of Turtle Wax Extra.

14 When carrying out bodywork repairs, remember that the quality of the finished job is proportional to the time and effort expended.

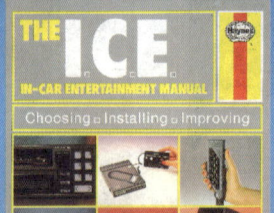

HAYNES No1 for DIY

Haynes publish a wide variety of books besides the world famous range of **Haynes Owners Workshop Manuals**. They cover all sorts of DIY jobs. Specialist books such as the **Improve and Modify** series and the **Purchase and DIY Restoration Guides** give you all the information you require to carry out everything from minor modifications to complete restoration on a number of popular cars. In addition there are the publications dealing with specific tasks, such as the **Car Bodywork Repair Manual** and the **In-Car Entertainment Manual**. The **Household DIY** series gives clear step-by-step instructions on how to repair everyday household objects ranging from toasters to washing machines.

Whether it is under the bonnet or around the home there is a Haynes Manual that can help you save money. Available from motor accessory stores and bookshops or direct from the publisher.

Piston and piston rings

Piston diameter – 1.4 litre (production):
    Standard class 1 .... 77.190 to 77.200 mm
    Standard class 2 .... 77.200 to 77.210 mm
    Standard class 3 .... 77.210 to 77.220 mm
    Standard class 4 .... 77.220 to 77.230 mm
    Oversize class A .... 77.480 to 77.490 mm
    Oversize class B .... 77.490 to 77.500 mm
    Oversize class C .... 77.500 to 77.510 mm
Piston diameter – 1.4 litre (service):
    Standard .... 77.210 to 77.235 mm
    0.29 mm oversize .... 77.490 to 77.515 mm
    0.50 mm oversize .... 77.710 to 77.735 mm
Piston diameter – 1.6 litre (not EFi) (production):
    Standard class 1 .... 79.910 to 79.920 mm
    Standard class 2 .... 79.920 to 79.930 mm
    Standard class 3 .... 79.930 to 79.940 mm
    Standard class 4 .... 79.940 to 79.950 mm
    Oversize class A .... 80.200 to 80.210 mm
    Oversize class B .... 80.210 to 80.220 mm
    Oversize class C .... 80.220 to 80.230 mm
Piston diameter – 1.6 litre (not EFi) (service):
    Standard .... 79.930 to 79.955 mm
    0.29 mm oversize .... 80.210 to 80.235 mm
    0.50 mm oversize .... 80.430 to 80.455 mm
Piston diameter – 1.6 litre EFi (production):
    Standard class 1 .... 79.915 to 79.925 mm
    Standard class 2 .... 79.925 to 79.935 mm
    Standard class 3 .... 79.935 to 79.945 mm
    Standard class 4 .... 79.945 to 79.955 mm
    Oversize class A .... 80.205 to 80.215 mm
    Oversize class B .... 80.215 to 80.225 mm
    Oversize class C .... 80.225 to 80.235 mm
Piston diameter – 1.6 litre EFi (service):
    Standard .... 79.935 to 79.955 mm
    0.29 mm oversize .... 80.215 to 80.235 mm
    0.50 mm oversize .... 80.435 to 80.455 mm
Piston-to-bore clearance (production):
    All CVH except 1.6 litre EFi .... 0.020 to 0.040 mm
    1.6 litre EFi .... 0.015 to 0.035 mm
Piston-to-bore clearance (service):
    All CVH except 1.6 litre EFi .... 0.010 to 0.045 mm
    1.6 litre EFi .... 0.010 to 0.040 mm
Piston ring end-gap (fitted) (all CVH except 1.6 litre EFi):
    Top and middle (compression) .... 0.30 to 0.50 mm
    Bottom (oil control) .... 0.40 to 1.40 mm
Piston ring end-gap (fitted) (1.6 litre EFi):
    Top and middle (compression) .... 0.30 to 0.50 mm
    Bottom (oil control) .... 0.25 to 0.40 mm

Gudgeon pin

Length:
    1.4 litre .... 63.000 to 63.800 mm
    1.6 litre (not EFi) .... 66.200 to 67.000 mm
    1.6 litre (EFi) .... 63.000 to 63.800 mm
Diameter:
    White .... 20.622 to 20.625 mm
    Red .... 20.625 to 20.628 mm
    Blue .... 20.628 to 20.631 mm
    Yellow .... 20.631 to 20.634 mm
Clearance in piston .... 0.005 to 0.011 mm
Interference fit in connecting rod .... 0.013 to 0.045 mm

Camshaft

Number of bearings 5
Drive .... Toothed belt
Thrust plate thickness .... 4.99 to 5.01 mm
Cam lift (inlet and exhaust):
    1.4 litre .... 5.79 mm
    1.6 litre (not EFi) .... 6.09 mm
    1.6 litre EFi .... 6.57 mm
Inlet cam length:
    1.4 litre .... 38.305 mm
    1.6 litre (not EFi) .... 38.606 mm
    1.6 litre EFi .... 37.559 mm

Camshaft (continued)

Exhaust cam length:
    1.4 litre .... 37.289 mm
    1.6 litre (not EFi) .... 37.590 mm
    1.6 litre EFi .... 37.559 mm
Camshaft bearing journal diameter:
    Bearing 1 .... 44.75 mm
    Bearing 2 .... 45.00 mm
    Bearing 3 .... 45.25 mm
    Bearing 4 .... 45.50 mm
    Bearing 5 .... 45.75 mm
Camshaft endfloat .... 0.05 to 0.15 mm
Sealing compound (camshaft sprocket bolt) .... To Ford specification SDM-4G9105-A

Cylinder head

Material .... Light alloy
Combustion chamber volume:
    1.4 litre .... 38.88 to 41.88 cc
    1.6 litre (not EFi) .... 47.36 to 50.36 cc
    1.6 litre EFi .... 53.36 to 55.38 cc
Maximum permissible cylinder head distortion:
    Measured over a distance of 26 mm .... 0.04 mm
    Measured over a distance of 156 mm ...... 0.08 mm
    Measured over entire length .... 0.15 mm
Maximum depth when skimming cylinder head mating face .... 0.30 mm
Minimum combustion chamber depth after skimming:
    1.4 litre .... 17.40 mm
    1.6 litre .... 19.10 mm
Camshaft bearing bore in cylinder head:
    Bearing 1 .... 44.783 to 44.808 mm
    Bearing 2 .... 45.033 to 45.058 mm
    Bearing 3 .... 45.283 to 45.308 mm
    Bearing 4 .... 45.533 to 45.558 mm
    Bearing 5 .... 45.783 to 45.808 mm
Valve tappet bore:
    Standard .... 22.235 to 22.265 mm
    Oversize .... 22.489 to 22.519 mm
Valve guide bore (inlet and exhaust):
    Standard .... 8.063 to 8.094 mm
    0.2 mm oversize .... 8.263 to 8.294 mm
    0.4 mm oversize .... 8.463 to 8.494 mm

Note: *The cylinder head has valve seat rings on the exhaust side. These valve seats cannot be recut with conventional tools (see text)*

Valve seat width .... 1.75 to 2.32 mm
Valve seat angle .... 44° 30′ to 45° 30′
Upper seat cutter correction angle:
    Inlet .... 15°
    Exhaust .... 15° (see text)
Lower seat cutter correction angle:
    Inlet .... 75°
    Exhaust .... 70° (see text)

Valves – general

Operation .... Hydraulic cam followers and rocker arms
Valve timing (1 mm of cam lift):
    1.4 litre:
        Inlet valve opens .... 15° ATDC
        Inlet valve closes .... 30° ABDC
        Exhaust valve opens .... 28° BBDC
        Exhaust valve closes .... 13° BTDC
    1.6 litre (not EFi):
        Inlet valve opens .... 4° ATDC
        Inlet valve closes .... 32° ABDC
        Exhaust valve opens .... 38° BBDC
        Exhaust valve closes .... 10° BTDC
    1.6 litre EFi:
        Inlet valve opens .... 4° ATDC
        Inlet valve closes .... 30° ABDC
        Exhaust valve opens .... 44° BBDC
        Exhaust valve closes .... 10° BTDC
Inlet valve lift:
    1.4 litre .... 9.56 mm
    1.6 litre (not EFi) .... 10.09 mm
    1.6 litre EFi .... 10.80 mm

Valves – general (continued)

Exhaust valve lift:
- 1.4 litre .... 9.52 mm
- 1.6 litre (not EFi) .... 10.06 mm
- 1.6 litre EFi .... 10.80 mm

Valve spring free length (depending on make – different colour code):
- 1.4 litre and 1.6 litre (not EFi) .... 47.20 or 45.40 mm
- 1.6 litre EFi .... 46.90 or 48.30 mm

Fitted valve spring height (valve open):
- 1.4 litre .... 27.700 mm
- 1.6 litre (not EFi) .... 27.000 mm
- 1.6 litre EFi .... 26.284 mm

Fitted valve spring height (valve closed) .... 37.084 mm

Inlet valve

Length:
- 1.4 litre .... 136.29 to 136.75 mm
- 1.6 litre .... 134.54 to 135.00 mm

Head diameter:
- 1.4 litre .... 39.90 to 40.10 mm
- 1.6 litre .... 41.90 to 42.10 (EFi valve marked 'C')

Stem diameter:
- Standard .... 8.025 to 8.043 mm
- 0.2 mm oversize .... 8.225 to 8.243 mm
- 0.4 mm oversize .... 8.425 to 8.443 mm

Valve stem clearance in guide .... 0.020 to 0.063 mm

Exhaust valve

Length:
- 1.4 litre .... 132.97 to 133.43 mm
- 1.6 litre .... 131.57 to 132.03 mm

Head diameter:
- 1.4 litre .... 33.90 to 34.10 mm
- 1.6 litre .... 36.90 to 37.10 mm

Stem diameter:
- Standard .... 7.999 to 8.017 mm
- 0.2 mm oversize .... 8.199 to 8.217 mm
- 0.4 mm oversize .... 8.399 to 8.417 mm

Valve stem clearance in guide .... 0.046 to 0.089 mm

Lubrication

Oil type/specification .... Multigrade engine oil, viscosity range SAE 10W/30 to 20W/50, to API SG (Duckhams QS, QXR, Hypergrade Plus, or Hypergrade)

Oil filter .... Champion C104
Oil pump type .... Rotor type driven by crankshaft

Minimum oil pressure at 80°C (175°F):
- Engine speed 750 rpm .... 1.0 bar
- Engine speed 2000 rpm .... 2.0 bar

Oil pressure warning light operates at .... 0.3 to 0.5 bar
Pressure relief valve opens at .... 4.0 bar

Oil pump clearances:
- Outer rotor to body .... 0.060 to 0.190 mm
- Inner rotor to outer rotor .... 0.05 to 0.18 mm
- Rotor endfloat (relative to mating face) .... 0.014 to 0.100 mm

Torque wrench settings

	Nm	lbf ft
Main bearing caps	90 to 100	66 to 74
Big-end bearing caps	30 to 36	22 to 26
Oil pump	8 to 11	6 to 8
Oil pump cover	8 to 12	6 to 8
Threaded sleeve of oil cooler to cylinder block	55 to 60	40 to 44
Oil pump intake to cylinder block	17 to 23	12 to 17
Oil pump intake to oil pump	8 to 12	6 to 8
Rear oil seal carrier	8 to 11	6 to 8
Sump (with one piece gasket, in two stages)	5 to 8	4 to 6
Lower engine adaptor plate (clutch housing cover):		
Manual transmission	34 to 46	25 to 34
CTX automatic transmission	7.5 to 10	6 to 7
Flywheel	82 to 92	60 to 68
Crankshaft pulley bolt	100 to 115	74 to 85
Cylinder head bolts:		
Stage 1	20 to 40	15 to 30
Stage 2	40 to 60	30 to 44
Stage 3	Angle-tighten a further 90°	Angle-tighten a further 90°
Stage 4	Angle-tighten a further 90°	Angle-tighten a further 90°

Torque wrench settings (continued)

	Nm	lbf ft
Camshaft thrust plate	9 to 13	7 to 10
Camshaft toothed sprocket bolt	54 to 59	40 to 44
Timing belt tensioner	16 to 20	12 to 15
Rocker studs to cylinder head	18 to 23	13 to 17
Rocker arms	25 to 29	18 to 21
Rocker cover	6 to 8	4 to 6
Timing belt cover	9 to 11	7 to 8
Oil pressure switch	18 to 22	13 to 16
Oil drain plug	21 to 28	15 to 21
Engine/transmission flange bolts:		
Manual transmission	35 to 45	26 to 33
CTX automatic transmission	27 to 50	20 to 37
Engine/transmission mountings (on transmission bearer)	Refer to Specifications, Chapter 6	

PART A : OHV (HCS) ENGINES

1 General description

The engine is an overhead valve water-cooled four-cylinder in-line design, designated HCS. It is mounted transversely at the front of the vehicle, together with the transmission, to form a combined power-train.

The crankshaft is supported in three shell type main bearings. The connecting rods are fitted with shell type big-end bearings, and are attached to the pistons by means of interference fit gudgeon pins. The pistons are fitted with three piston rings; two compression and one oil control.

The camshaft, which runs on bearings within the cylinder block, is chain-driven from the crankshaft, and operates the valves via pushrods and rocker arms. The valves are each closed by a single valve spring, and operate in guides integral with the cylinder head. The oil pump is driven by a skew gear on the camshaft, and the fuel pump is driven by an eccentric lobe, also on the camshaft. The oil pump is mounted externally on the cylinder block, and the full-flow oil filter is screwed directly into the oil pump.

2 Maintenance and inspection

1 At the intervals given in *'Routine maintenance'* at the beginning of this manual carry out the following maintenance operations on the engine.

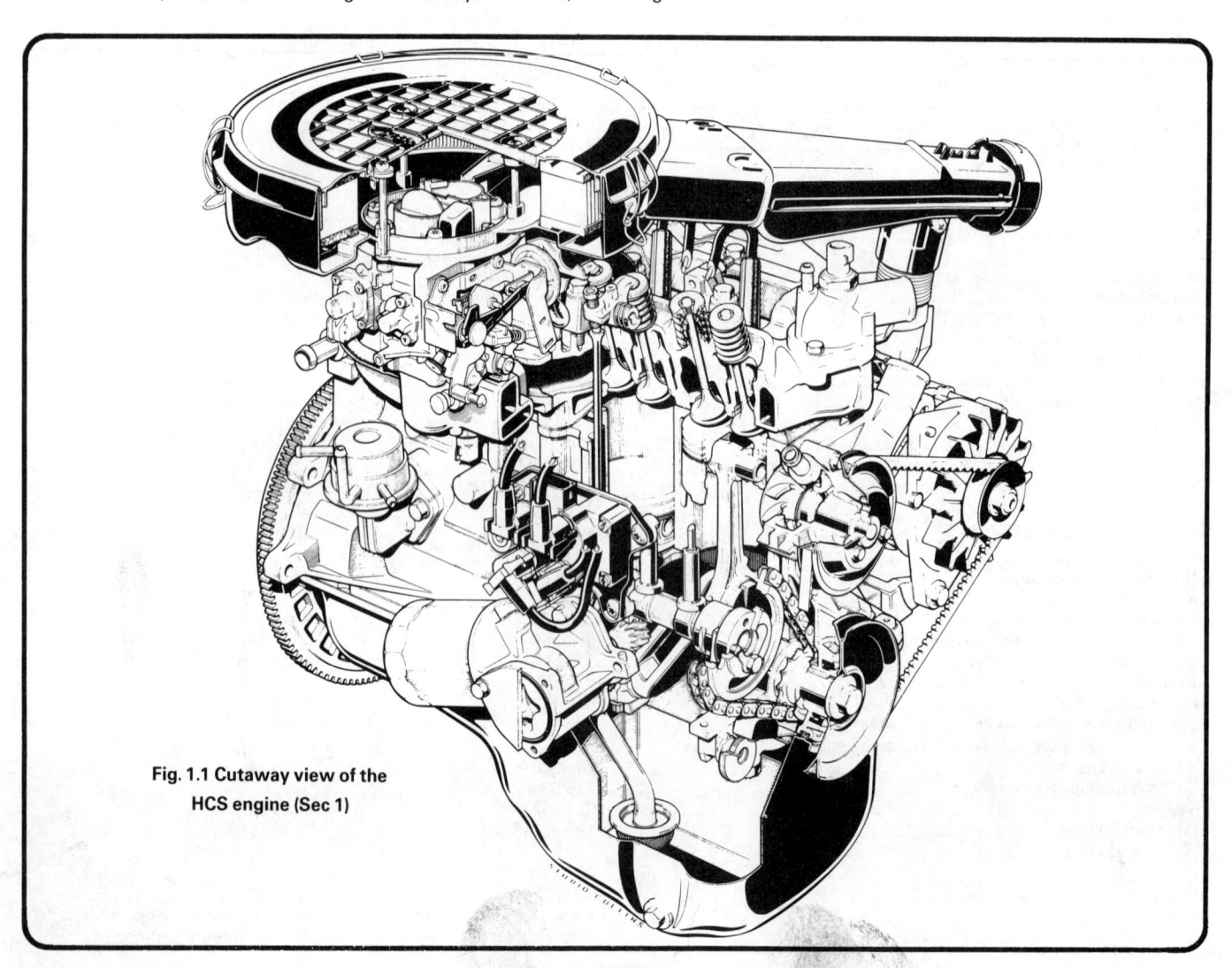

Fig. 1.1 Cutaway view of the HCS engine (Sec 1)

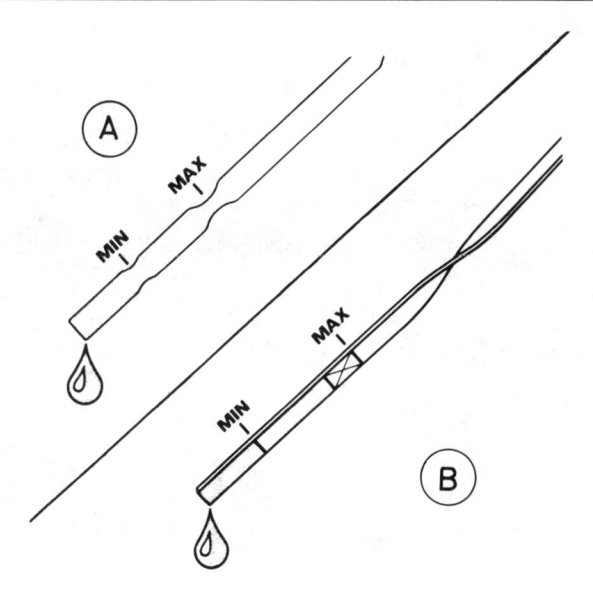

Chapter 1 Engine

11 Lower the vehicle to the ground.
12 Refill the engine with fresh engine oil of the specified type, making reference to paragraph 3. With the engine running, check for leaks around the filter seal.
13 Adjust the valve clearances in accordance with the relevant Section of this Chapter.
14 Carry out an inspection of the crankcase ventilation system (refer to Chapter 3).

3 Major operations possible with engine in vehicle

The following work can be carried out without having to remove the engine:

(a) Cylinder head – removal and refitting
(b) Valve clearances – adjustment
(c) Rocker shaft assembly – overhaul
(d) Crankshaft front oil seal – renewal
(e) Oil filter – renewal
(f) Oil pump – removal and refitting
(g) Sump – removal and refitting
(h) Piston/connecting rod assemblies – removal and refitting
(i) Engine/transmission mountings – removal and refitting

4 Major operations possible only with engine removed

The following work should be carried out only after the engine has been removed from the vehicle:

(a) Crankshaft main bearings – renewal
(b) Crankshaft – removal and refitting
(c) Flywheel – removal and refitting
(d) Crankshaft rear oil seal – renewal
(e) Camshaft – removal and refitting
(f) Timing gears and chain – removal and refitting

5 Cylinder head – removal and refitting

1 If the engine is in the car carry out the preliminary operations described in paragraphs 2 to 12.
2 Disconnect the battery earth terminal.
3 Refer to Chapter 3 and remove the air filter housing.
4 Refer to Chapter 2 and drain the cooling system.
5 Disconnect the hoses from the thermostat housing.
6 Disconnect the heater hose from the inlet manifold (see photo 15.7).
7 Detach the accelerator and choke cables from the carburettor (see Chapter 3).
8 Disconnect the fuel, breather and vacuum hoses from the carburettor and inlet manifold.
9 Disconnect the HT leads from the spark plugs.

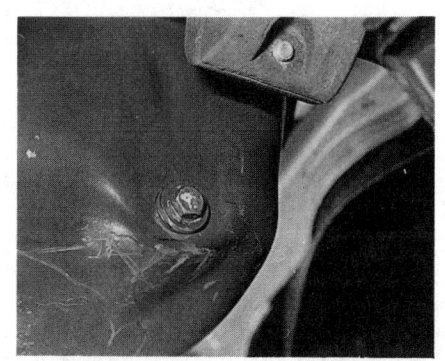

Fig. 1.2 Dipstick markings (Sec 2)

A 1.0 litre and 1.1 litre HCS engines
B 1.4 litre and 1.6 litre CVH engines

2 Visually inspect the engine joint faces, gaskets and seals for any sign of oil or water leaks. Pay particular attention to the areas around the rocker cover, cylinder head, timing cover and sump joint faces. Rectify any leaks by referring to the appropriate Sections of this Chapter.
3 Check the engine oil level with the vehicle parked on level ground, preferably after allowing the engine to cool down. The oil level must be kept between the maximum and minimum markings on the dipstick. Top up the oil level as necessary, using fresh oil of the specified type, through the filler neck on the rocker cover – do **not** overfill. Remember to refit the filler cap.
4 Change the engine oil while the engine is still warm as described in the following eight paragraphs.
5 Apply the handbrake, raise the front of the vehicle and support it securely using axle stands.
6 Place a suitable container beneath the oil drain plug at the rear of the sump (photo). Unscrew the plug using a spanner or socket and allow the oil to drain. Regardless of its condition, the drain plug sealing washer should always be renewed. Refit and tighten the drain plug to its specified torque after draining.
7 Move the container to the rear of the engine, under the oil filter.
8 Using a strap wrench, or filter removal tool, slacken the filter and then unscrew it from its location on the oil pump and discard.
9 Wipe the filter mating face of the oil pump using a clean rag, then lubricate the oil filter seal using clean engine oil (photo).
10 Screw the filter into position, by hand, until its seal makes contact with the oil pump housing (photo). Tighten the filter securely by hand only – do not use any tools.

2.6 Oil drain plug on the rear of the sump

2.9 Lubricating the oil filter seal

2.10 Screw the oil filter onto the oil pump. Oil pressure switch (arrowed)

Chapter 1 Engine

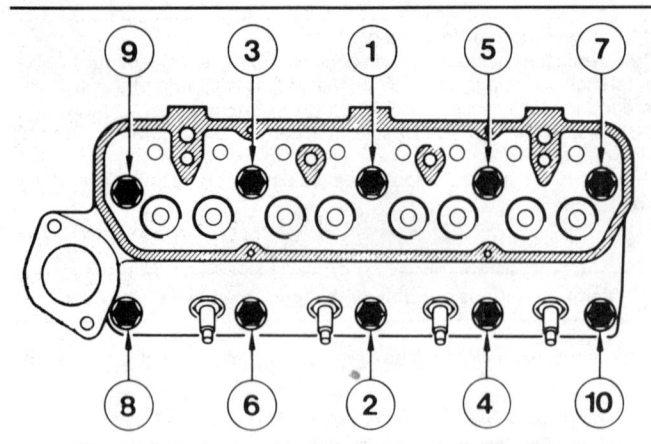

Fig. 1.3 Cylinder head bolt tightening sequence (Sec 5)

10 Disconnect the electrical leads from the temperature gauge sender, radiator cooling fan and its thermal switch, engine coolant temperature sensor on the underside of the inlet manifold, and the anti-run-on solenoid at the carburettor.
11 Raise the front of the vehicle and support it securely using axle stands.
12 Remove the three nuts securing the forward end of the exhaust system to the exhaust manifold, and separate. Suspend the forward end of the system, using a length of strong wire, to prevent it straining. Note that a new gasket will be needed upon reassembly.
13 Remove the oil filler cap (with breather hose).
14 Extract the four bolts securing the rocker cover and remove it.
15 Undo and remove the four rocker shaft securing bolts (photo), then remove the rocker shaft assembly from its cylinder head location.
16 Withdraw the pushrods, keeping them in their original sequence. A simple way to achieve this is to punch holes in a piece of card and number them 1 to 8, from the thermostat housing end of the cylinder head (photos).
17 Remove the spark plugs.

5.15 Rocker shaft securing bolts (arrowed)

18 Unscrew the cylinder head bolts progressively, in the reverse sequence to that given for tightening (Fig. 1.3), then lift off the cylinder head. Note that the cylinder head bolts may be re-used a **maximum** of twice, and should be marked for possible re-use – a single punch mark on each bolt head will suffice. If there is *any* doubt as to how many times the cylinder head bolts have been used, replacement is the **only** safe answer.
19 To dismantle the cylinder head, refer to Section 19.
20 Before refitting the cylinder head, remove every particle of carbon, old gasket and dirt from the mating surfaces of the cylinder head and block. Do not let the removed material drop into the cylinder bores or waterways, if it does, remove it. Normally, when a cylinder head is removed, the head is decarbonised and the valves ground in as described in Section 6 to remove all trace of carbon. Clean the threads of

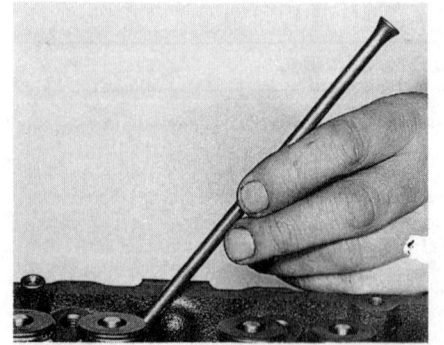

5.16A Withdrawing a pushrod

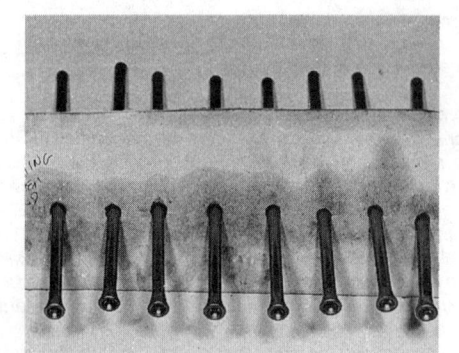

5.16B An easy method of ensuring that the pushrods are kept in their originally fitted order

5.22A Dowel fitment to cylinder block

5.22B Gasket marking

5.22C Locating gasket to cylinder block

5.24A Stage one tightening of the cylinder head bolts

Chapter 1 Engine 39

5.24B Stage two and stage three tightening of the cylinder head bolts – through their specified angle of rotation

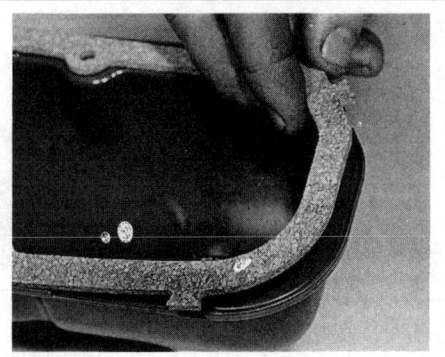

5.30A Engaging the rocker cover gasket tags to the cut-outs in the rocker cover

5.30B Refitting the rocker cover, with a new gasket in position

the cylinder head bolts and mop out oil from the bolt holes in the cylinder block. In extreme cases, screwing a bolt into an oil-filled hole can cause the block to fracture due to hydraulic pressure.

21 If there is any doubt about the condition of the inlet or exhaust gaskets, unbolt the manifolds and fit new ones to perfectly clean mating surfaces.

22 Locate a new cylinder head gasket on the cylinder block, making quite sure that the bolt holes, coolant and lubrication passages are correctly aligned, and that the dowels are located correctly in the cylinder block. The upper surface of the gasket carries a marking. Note that the new cylinder head gasket must be of the same type as the old one (photos).

23 Lower the cylinder head carefully into position on the block, and screw in all the cylinder head bolts to finger tightness only.

24 Tightening of the cylinder head bolts is done in three stages **and in the sequence given** in Fig. 1.3. Stage one involves tightening all the bolts to the torque given in the Specifications. Stage two involves additional tightening of the bolts by turning them through the specified angle of rotation. Stage three is a repeat of Stage two procedure (photos).

25 Refit the pushrods in their original order, having lubricated both ends of each with clean engine oil.

26 Refit the rocker shaft assembly, ensuring that the rocker adjusting screws engage in the sockets at the ends of the push rods.

27 Screw in the rocker pedestal bolts finger tight. At this stage, some of the rocker arms will be applying pressure to the ends of the valve stems and some of the rocker pedestals will not be in contact with the cylinder head. The pedestals will be pulled down however when the bolts are tightened to the specified torque, which should now be done.

28 Adjust the valve clearances as described in Section 8.

29 Refit the spark plugs, tightening to the specified torque (Chapter 4).

30 Refit the rocker cover, tightening bolts to the specified torque. If the gasket is in anything but perfect condition, renew it (photos).

31 Refitting is now a reversal of the removal procedure. Refill the cooling system, making reference to Chapter 2, Sections 3 and 4.

6 Cylinder head and pistons – decarbonising

1 With the cylinder head removed as described in Section 5, the carbon deposits should be removed from the combustion chambers using a scraper and a wire brush fitted into an electric drill. Take care not to damage the valve heads, otherwise no special precautions need be taken as the cylinder head is of cast iron construction.

2 Where a more thorough job is to be carried out, the cylinder head should be dismantled as described in Section 19 so that the valves may be ground in and the ports and combustion chambers cleaned. This is achieved by blowing and brushing out, after the manifolds have been removed.

3 Before grinding-in a valve, remove the carbon and deposits completely from its head and stem. With an inlet valve, this is usually quite easy, simply scraping off the soft carbon with a blunt knife and finishing with a wire brush. With an exhaust valve the deposits are very much harder and those on the head may need a rub on coarse emery cloth to

6.6 Grinding-in a valve

remove them. An old woodworking chisel is a useful tool to remove the worst of the head deposits.

4 Make sure that the valve heads are really clean, otherwise the rubber suction cup of the grinding tool will not stick during the grinding-in operations.

5 Before starting to grind-in a valve, support the cylinder head so that there is sufficient clearance under for the valve stem to project fully without being obstructed.

6 Take the first valve and apply a little coarse grinding paste to the bevelled edge of the valve head. Insert the valve into its guide and apply the suction grinding tool to its head. Rotate the tool between the palms of the hands in a back-and-forth rotary movement until the gritty action of the grinding-in process disappears (photo). Repeat the operation with fine paste and then wipe away all traces of grinding paste and examine the seat and bevelled edge of the valve. A matt silver mating band should be observed on both components, without any sign of black spots. If some spots do remain, repeat the grinding-in-process until they have disappeared. A drop or two of paraffin applied to the contact surfaces will increase the speed of grinding-in, but do **not** allow any paste to run down into the valve guide. On completion, wipe away *every trace* of grinding paste using a paraffin-moistened cloth.

7 Repeat the operations on the remaining valves, taking care not to mix up their originally fitted sequence.

8 The valves are refitted as described in Section 19.

9 An important part of the decarbonising operation is to remove the carbon deposits from the piston crowns. To do this, turn the crankshaft so that two pistons are at the top of their stroke and press some grease between these pistons and the cylinder walls. This will prevent carbon particles falling down into the piston ring grooves. Stuff rags into the other two bores.

10 Cover the oilways and coolant passages with masking tape and then using a blunt scraper remove all the carbon from the piston crowns.

Take care not to score the soft alloy of the crown or the surface of the cylinder bore.
11 Rotate the crankshaft to bring the other two pistons to TDC and repeat the operations.
12 Wipe away the circle of grease and carbon from the cylinder bores.
13 Clean the top surfaces of the cylinder block by careful scraping.

7 Rocker shaft assembly – dismantling and reassembly

Note: *If fitting a new rocker shaft, check that plugs are present in both ends of the shaft before assembling. If plugs are not present, return the shaft and obtain another.*

1 With the rocker assembly removed as described in Section 5, extract the split pin from one end of the rocker shaft (photo).
2 Take off the spring and plain washers from the end of the shaft.
3 Slide off the rocker arms, support pedestals and coil springs, keeping them in their originally fitted order. Clean out the oil holes in the shaft.
4 Apply engine oil to the rocker shaft before reassembling.
5 Make sure that the 'flat' on the end of the rocker shaft is to the same side as the rocker arm adjusting screws (and closest to the thermostat end of the cylinder head when refitted). This is essential for proper lubrication of the components (photo).

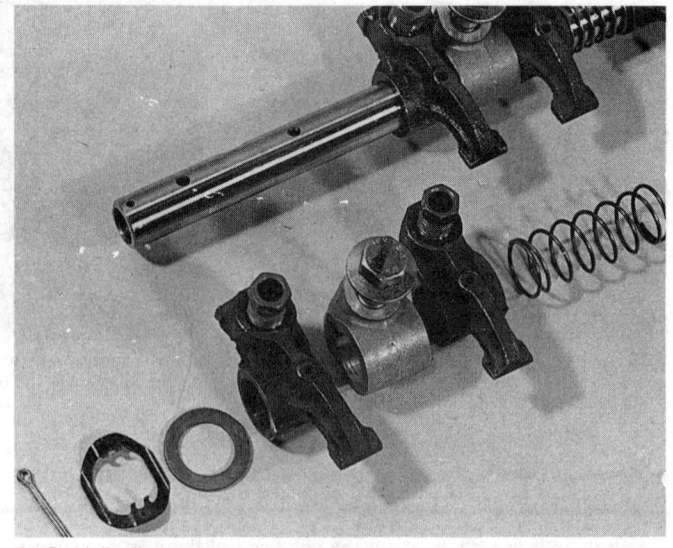

7.1 Partially dismantled rocker shaft

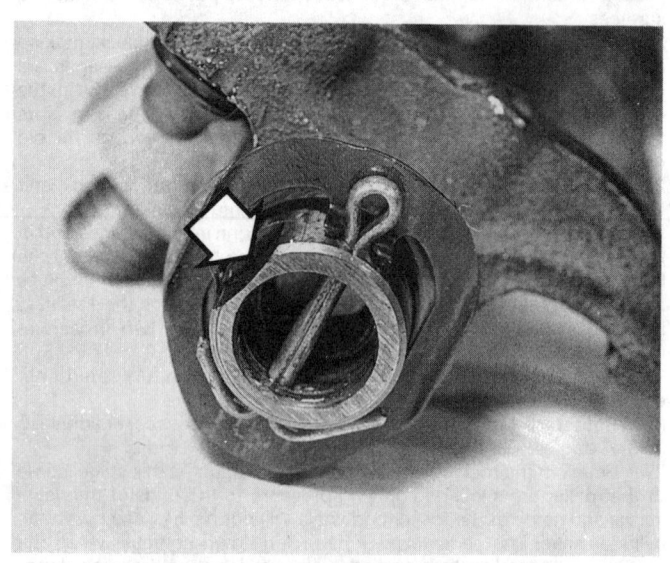

8 Valve clearances – adjustment

1 This operation should be carried out with the engine cold, and the air filter housing and rocker cover removed. The HT leads also need to be disconnected from the spark plugs, noting their order for refitting.
2 Using a ring spanner or socket on the crankshaft pulley bolt, turn the crankshaft in a clockwise direction until No 1 piston is at tdc on its compression stroke. This can be verified by checking that the pulley and timing cover marks are in alignment and that the valves of No 4 cylinder are rocking. When the valves are rocking, this means that the slightest rotation of the crankshaft pulley in either direction will cause one rocker arm to move up and the other to move down.
3 Numbering from the thermostat housing end of the cylinder head, the valves are identified as follows:

Valve No	Cylinder No
1 – Exhaust	1
2 – Inlet	1
3 – Exhaust	2
4 – Inlet	2
5 – Inlet	3
6 – Exhaust	3
7 – Inlet	4
8 – Exhaust	4

7.5 'Flat' on the rocker shaft (arrowed) to the same side as the rocker arm adjusting screws. Note installed split pin

4 Adjust the valve clearances by following the sequence given in the following table. Turn the crankshaft pulley 180° (half a turn) after adjusting each pair:

Valves rocking	Valves to adjust
7 and 8	1 (Exhaust), 2 (Inlet)
5 and 6	3 (Exhaust), 4 (Inlet)
1 and 2	8 (Exhaust), 7 (Inlet)
3 and 4	6 (Exhaust), 5 (Inlet)

5 The clearances for the inlet and exhaust valves are different (see Specifications). Use a feeler gauge of the appropriate thickness to check each clearance between the end of the valve stem and the rocker arm (photo). The gauge should be a stiff sliding fit. If it is not, turn the adjuster bolt with a ring spanner to achieve it. These bolts are of stiff-thread type, and require no locking nut.
6 Refit the rocker cover and tighten its bolts to the specified torque. If the gasket is in anything but perfect condition, renew it.
7 Refit the air filter housing, and reconnect the HT leads to the spark plugs in the correct order.

8.5 Adjusting the valve clearances

Chapter 1 Engine

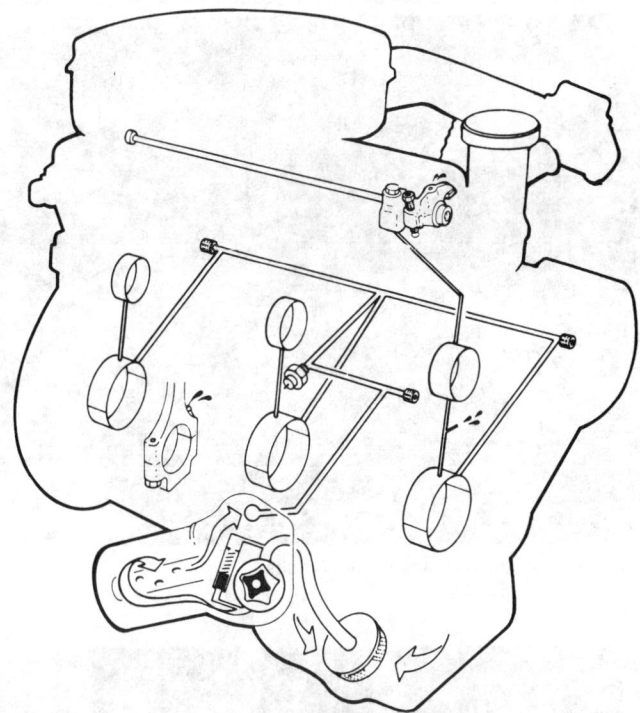

Fig. 1.4 HCS engine lubrication circuit (Sec 9)

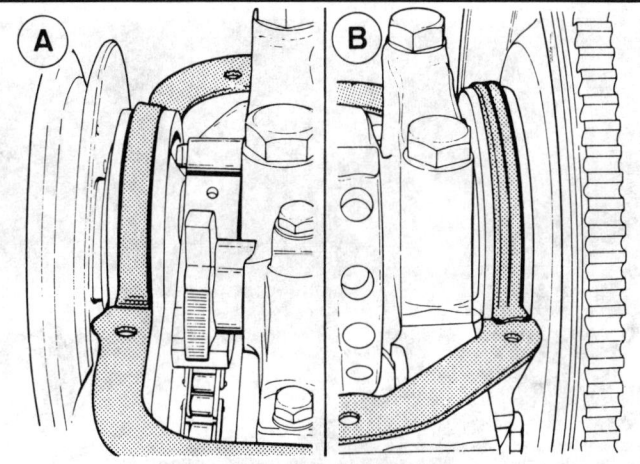

Fig. 1.5 Sump gasket fitting details at timing cover end (A) and flywheel end (B) (Sec 10)

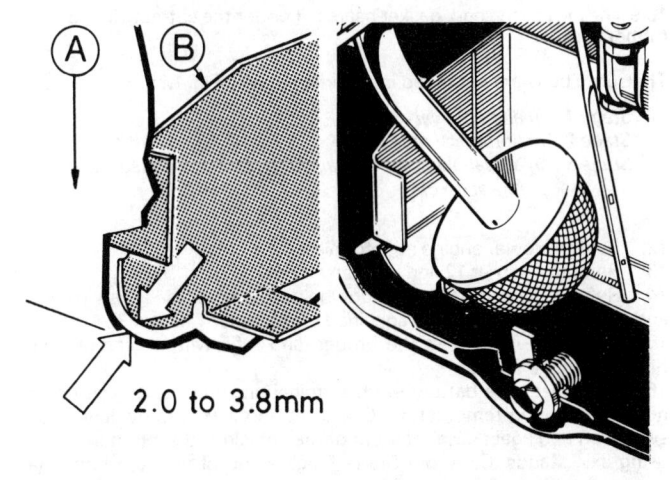

Fig. 1.6 Sump and oil baffle clearance details (Sec 10)

A Sump B Baffle

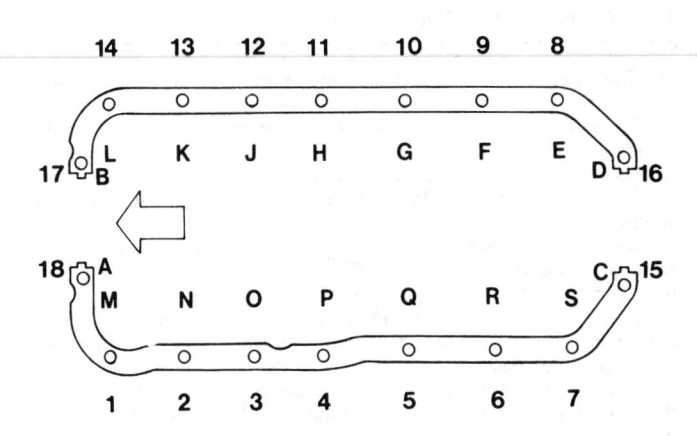

Fig. 1.7 Sump bolt tightening sequence (Sec 10)

9 Lubrication system – description

1 Engine oil contained in the sump is drawn through a strainer and pick-up tube by an externally mounted oil pump of twin rotor design.
2 The oil is then forced through a full-flow, throw-away type oil filter which is screwed onto the oil pump.
3 Oil pressure is regulated by a relief valve integral in the oil pump.
4 The pressurised oil is directed through the various galleries and passages to all bearing surfaces. A drilling in the big-end provides lubrication for the gudgeon pins and cylinder bores. The timing chain and sprockets are lubricated by an oil ejection nozzle.

10 Sump – removal and refitting

1 Disconnect the battery earth terminal.
2 Raise the front of the vehicle and support securely using axle stands.
3 Drain the engine oil (see Section 2, paragraph 6).
4 Remove the forward (downpipe) section of the exhaust system for increased clearance (see Chapter 3). Suspend the free end of the system using a length of strong wire to prevent strain on the mountings. Note that a new gasket will be needed upon reassembly.
5 Refer to Chapter 12 and remove the starter motor assembly.
6 Unbolt and remove the lower engine adaptor plate (see photo 15.14).
7 Extract the sump securing bolts and remove the sump. If it is stuck, prise it gently with a screwdriver but do not use excessive leverage. If it is very tight, cut around the gasket joint using a sharp knife.
8 Prior to refitting, remove the old rubber timing cover and rear oil seal carrier gaskets, and the main cork gasket halves. Clean the mating surfaces of the sump and the cylinder block, and check that the sump mating faces are not distorted. Clean out any sludge present in the sump, and ensure that the oil pick-up strainer is clear.
9 Apply a dab of sealing compound to the cylinder block mating face where the ends of each cork gasket half are to be fitted (Fig. 1.5). Stick new cork gasket halves into position on the cylinder block using thick grease to retain them, then install new rubber gaskets into their slots on the timing cover and the rear oil seal carrier. The lugs of the cork gasket halves fit under the cut-outs in the rubber gaskets (photo).
10 Before offering up the sump, check that the gap between the sump and the oil baffle is between 2.0 and 3.8 mm. Do not use a dented or damaged sump, as the indicated dimension is important for correct engine lubrication. (see Fig. 1.6).
11 With the sump in position and its bolts screwed in finger-tight, make reference to the Specifications for the tightening torque details.

42　　　　　　　　　　　　　　　　　　Chapter 1　Engine

10.9 The lugs of the cork gasket halves fit under the cut-outs in the rubber gaskets

11.4 Undoing the oil pump securing bolts

The sump bolts are tightened in accordance with Fig. 1.7:

> Stage 1 – in alphabetical order.
> Stage 2 – in numerical order.
> Stage 3 – in alphabetical order again (after task completed – see paragraph 16).

12　Refit the lower engine adaptor plate.
13　Refer to Chapter 12 and refit the starter motor.
14　Refit the exhaust forward (downpipe) section. Use a new gasket and tighten the bolts to their specified torques (see Chapter 3).
15　Lower the vehicle to the ground and refill with engine oil (see Section 2, paragraph 3).
16　Reconnect the battery earth terminal and run the engine up to normal operating temperature. Check that no leaks are evident, then switch off and again raise the front of the vehicle and support securely using axle stands. Carry out Stage 3 tightening of the sump bolts (a re-torque to Stage 2 figures), then lower the vehicle to the ground.

11　Oil pump – removal and refitting

1　The oil pump is externally mounted on the rear facing side of the crankcase.
2　Raise the front of the vehicle and support securely using axle stands.
3　Using a suitable removal tool (strap wrench or similar), unscrew and remove the oil filter cartridge and discard it. Catch any spillage in a suitable container.
4　Unscrew its three securing bolts and withdraw the oil pump from the engine (photo).
5　Clean away all traces of old gasket from the mating faces.
6　If a new pump is being fitted, it should be primed with engine oil before installation. Do this by turning its shaft while filling it with clean engine oil, for at least one complete revolution.
7　Locate a new gasket on the pump mounting flange, insert the pump drive and tighten the oil pump securing bolts to the specified torque (photo).
8　Fit a new oil filter to the oil pump assembly in accordance with Section 2, paragraphs 9 and 10.
9　Lower the vehicle to the ground and top up the engine oil to replenish any lost during the operations (refer to Section 2, paragraph 3).

12　Crankshaft front oil seal – renewal

1　Disconnect the battery earth terminal.

11.7 Refitting the oil pump with a new gasket in position

2　Raise the front of the vehicle and support it securely using axle stands.
3　Slacken the alternator pivot and adjuster bolts and after pushing the alternator in towards the engine, slip off the drivebelt.
4　Unscrew and remove the crankshaft pulley bolt. To prevent the crankshaft turning while the bolt is being released, jam the teeth of the starter ring gear on the flywheel after removing the lower engine adaptor plate or starter motor (Chapter 12) for access.
5　Remove the crankshaft pulley. This should come out using the hands but if it is tight, prise it carefully with two levers placed at opposite sides under the pulley flange.
6　Using a suitable claw tool, prise out the defective seal and wipe out the seat. (Fig. 1.8).
7　Lubricate the sealing lip of the new oil seal and the crankshaft stub with clean engine oil.
8　Install the new seal using a suitable distance piece, the pulley and its bolt to draw it into position. If it is tapped into position, the seal may be distorted or the timing cover fractured.
9　When the seal is fully seated, remove the crankshaft pulley bolt, pulley and distance piece. Slide the distance piece off the pulley.
10　Lightly lubricate the seal rubbing surface of the pulley, install the pulley onto the keyed crankshaft stub, and tighten its retaining bolt to the specified torque.

Chapter 1 Engine 43

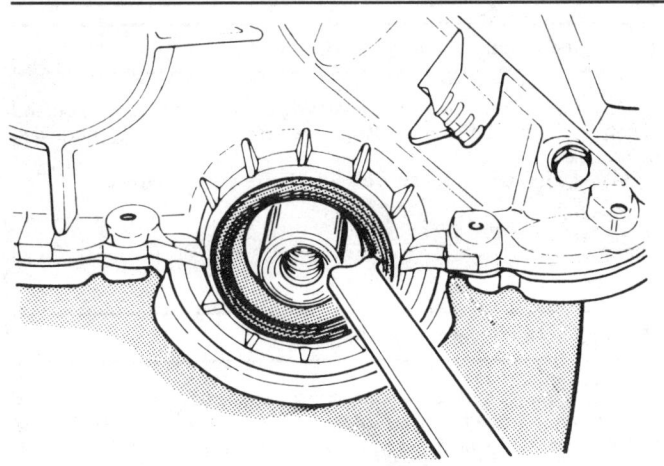

Fig. 1.8 Prise out the crankshaft front oil seal using a suitable claw tool (Ford special tool shown) (Sec 12)

11   Refit the lower engine adaptor plate or starter motor (see Chapter 12), as applicable.
12   Refit the alternator drivebelt, tensioning it as described in Chapter 12, Section 7.
13   Lower the vehicle to the ground and reconnect the battery earth terminal.
14   Check the engine oil level and top up as necessary (see Section 2, paragraph 3).

13   Engine/transmission mountings – removal and refitting

1   The engine mountings can be removed if the weight of the engine/transmission is first taken by one of the three following methods.
2   Either support the assembly from underneath using a jack and a block of wood, or attach a hoist to the engine (refer to Section 15, paragraph 18). A third method is to make up a bar with end pieces which will engage in the water channels at the sides of the bonnet lid aperture. Using an adjustable hook and chain connected to the engine (see above), the weight of the engine can be taken off the mountings.

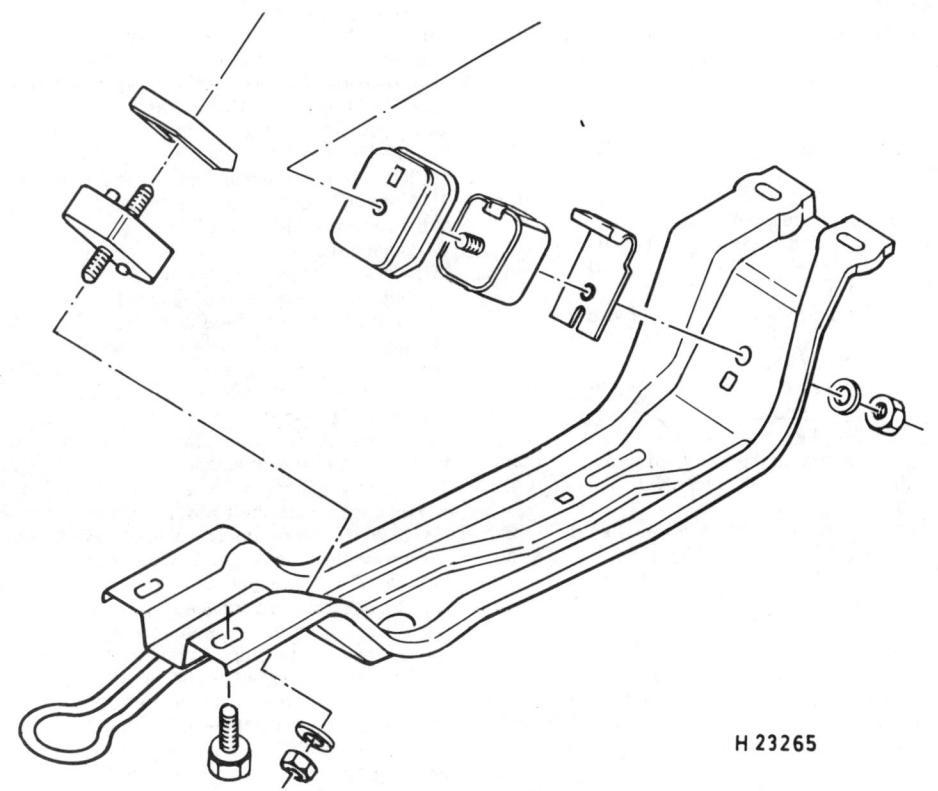

H 23265

Fig. 1.9 Exploded view of transmission bearer mountings (Sec 13)

13.3 Engine right-hand mounting. Bracket side bolt (arrowed) in wheelarch

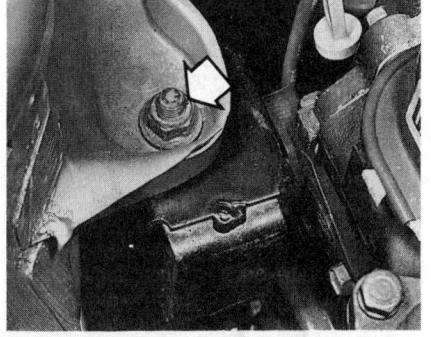

13.4 Engine right-hand mounting. Retaining nut (arrowed) rear suspension strut cup retaining plate

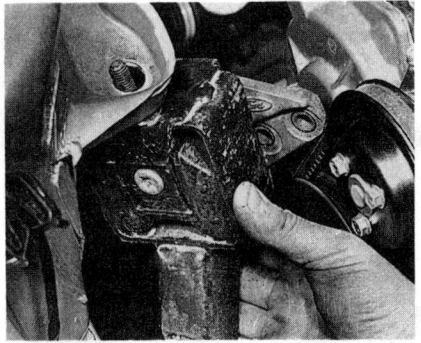

13.5 Removing the engine right-hand mounting

44   Chapter 1 Engine

13.7 Rear view of transmission bearer, showing mounting retaining nut (A) and transmission bearer retaining bolts (B). (Front end has similar layout)

Engine right-hand mounting

3   Unscrew and remove the mounting side bolt from under the right-hand wheel arch (photo).
4   Unscrew and remove the mounting retaining nut and washer from the suspension strut cup retaining plate (photo).
5   Undo the three bolts securing the mounting unit to the cylinder block. The mounting unit and bracket can then be lowered from the engine (photo).
6   Unbolt and remove the mounting from its support bracket.

Transmission bearer and mountings

7   Unscrew and remove the two nuts securing the mountings (front and rear) to the transmission bearer (photo).
8   Support the transmission bearer, then undo and remove the four retaining bolts from the floorpan, two at the front and two at the rear, and lower the transmission bearer from the vehicle. Note plate fitment, as applicable, for reassembly.
9   Unscrew the single nut securing each mounting and its retainer to the transmission support bracket, and remove. The transmission support brackets are fixed externally to the transmission casing and do not need to be removed for this operation.

All mountings

10   Refitting of all mountings is a reversal of removal. Make sure that the original sequence of assembly of washers and plates is maintained.
11   Do not fully tighten any mounting bolts until they are all located. As the mounting bolts and nuts are tightened, check that the mounting rubbers do not twist.

14 Engine – method of removal/refitting

The engine is removed by separating it from the transmission and swinging the flywheel end towards the bulkhead before withdrawing upwards.

Install the engine from above, reversing removal method.

15 Engine – removal and refitting

Note: *On anti-lock brake equipped vehicles, it may be necessary to unbolt the right-hand modulator and tie it securely to the bulkhead (see Chapter 8), then remove the modulator mounting bracket.*

1   Disconnect the battery earth terminal.
2   Remove the bonnet assembly, as described in Chapter 11, Section 6.
3   Drain the cooling system in accordance with Chapter 2, Section 3.
4   Remove the air filter housing by undoing its retaining bolts and lifting it off, disconnecting its hoses as necessary. Certain 1.1 litre models may be fitted with an air charge temperature sensor, located on the underside of the air filter housing, which must be disconnected. The in-line fuel trap attached to the air filter housing should be removed, and its vacuum hoses disconnected from the inlet manifold and the ignition module on the bulkhead panel.
5   Remove the upper and lower radiator hoses (Fig. 1.10).
6   Detach the expansion tank hoses from the thermostat housing and radiator, and tie them out of the way.
7   Detach the heater hoses from the water pump and the inlet manifold (photo), and tie them out of the way.
8   Detach the choke and accelerator cables from their carburettor location, and the fuel feed and return lines from the fuel pump (see Chapter 3). Plug the fuel lines to avoid excessive spillage, and take all normal fuel handling safety precautions (see 'Safety first!').
9   Detach the brake servo vacuum hose from the inlet manifold (see Chapter 8, Section 16).
10   Disconnect the wiring from the radiator cooling fan and its thermal switch (on thermostat housing), temperature gauge sender (just below thermostat housing), oil pressure switch (near oil filter), DIS ignition coil, engine speed/crankshaft position sensor, engine coolant temperature sensor (located on underside of inlet manifold) and anti-run-on solenoid (on carburettor – see Chapter 3) (photos).
11   Disconnect the vacuum pipe from the throttle kicker on the carburettor assembly, as applicable (see Chapter 3).
12   Raise the front of the vehicle and support it securely using axle stands, then drain the engine oil by unscrewing the oil drain plug at the rear of the sump, collecting the oil in a suitable container.
13   Remove the three nuts securing the forward (downpipe) section of the exhaust system to the exhaust manifold, and separate. Suspend the forward end of the system, using a length of strong wire, to prevent it straining. Note that a new gasket will be needed upon reassembly.

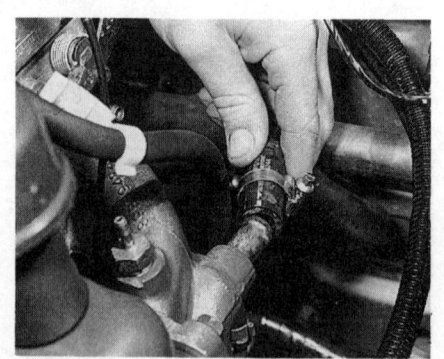

15.7 Detaching the heater hose from the inlet manifold

15.10A Engine speed/crankshaft position sensor and its multi-plug (on forward face of engine – flywheel end)

15.10B Engine coolant temperature sensor (arrowed) on inlet manifold (viewed from underneath vehicle)

Chapter 1 Engine 45

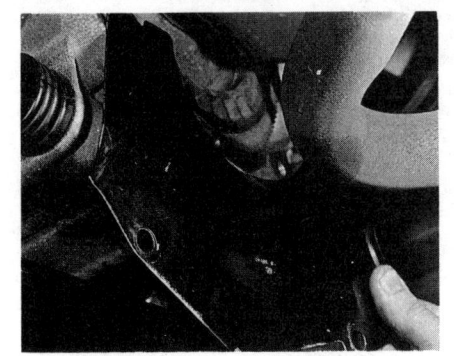

15.14 Remove the lower engine adaptor plate

15.18 Lifting eye attached using an inlet manifold bolt

15.22 Swing the flywheel end towards the bulkhead as the engine is withdrawn

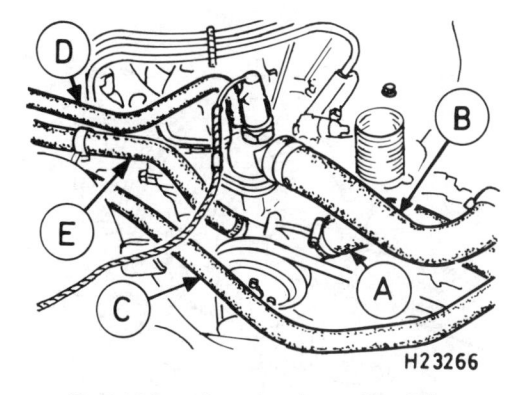

Fig. 1.10 Detach coolant hoses (Sec 15)

A   Lower radiator hose
B   Upper radiator hose
C   Expansion tank/radiator connecting hose
D   Expansion tank/thermostat housing connecting hose
E   Heater hose

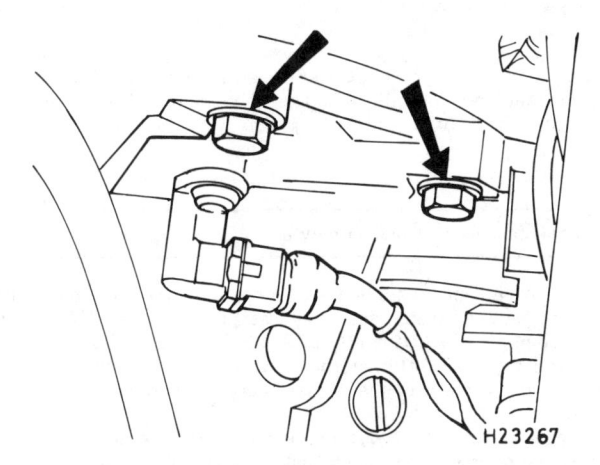

Fig. 1.11 Lower engine/transmission flange bolts (arrowed) (front of engine shown) (Sec 15)

14 Remove the lower engine adaptor plate by undoing its retaining bolts (photo).
15 Remove the alternator and starter motor in accordance with the relevant Sections of Chapter 12.
16 Remove the four lower engine/transmission flange bolts (Fig. 1.11).
17 Carefully lower the vehicle to the ground.
18 Attach a suitable hoist to the engine. It is possible to fabricate lifting eyes to connect the hoist to the engine, but make sure that they are strong enough. The lifting eyes should be held, using an inlet manifold bolt and an exhaust manifold nut, on opposite ends of the engine (photo).
19 Take the weight of the engine by raising the hoist slightly, then detach the right-hand (timing cover end) engine mounting from the vehicle body. Detach and remove the engine mounting from the engine block. Refer to Section 13 of this Chapter.
20 Check to ensure that engine removal will not be impeded by any non-standard installations, taking whatever action necessary.
21 Unscrew the two upper engine/transmission flange bolts, and remove the wiring loom securing brackets (Fig. 1.12). Ensure that all wiring, hoses, etc, are clear.
22 Separate the engine from the transmission by carefully drawing it out, then swing the flywheel end towards the bulkhead and withdraw upwards (photo).
23 Refitting is a reversal of the removal procedure, with a few conditions. Ensure that the clutch driven plate has been centralised if the clutch assembly has been dismantled, and disconnect the clutch cable from its release lever on the transmission casing, before commencing refitting (manual transmission vehicles only – see Chapter 5). Additionally, on manual transmission equipped vehicles **only,** lightly grease the transmission input shaft before refitting the engine (see Chapter 5 for recommended grease). Having ensured that the oil drain plug seal is serviceable, and that the oil drain plug is tightened to its specified torque, refill the lubrication system in accordance with Section 2, paragraph 3 of this Chapter. Ensure that all coolant hoses are properly secured, and that the radiator drain plug is tightened to its specified torque before refilling the cooling system in accordance with Chapter 2, Section 3, paragraph 9. Reconnect the clutch cable to its release lever on the transmission casing (manual transmission vehicles only – see Chapter 5). Refit the right-hand modulator (anti-lock brake equipped vehicles only – see Chapter 8).
24 Refer to the warning in Chapter 2, Section 1, then, having checked the security of all connections, run the engine up to normal operating temperature and check the idle speed and mixture adjustment (refer to Chapter 3). Check that no leaks are evident, then switch the engine off. Check the coolant and oil levels when the engine has cooled down.
25 If a number of new internal components have been installed, or if a new/reconditioned engine has been fitted, run the vehicle at restricted speed and engine load for the first few hundred miles, to allow time for the new components to bed in. In the case of a reconditioned engine, consult your supplier to see if any specific running-in conditions apply

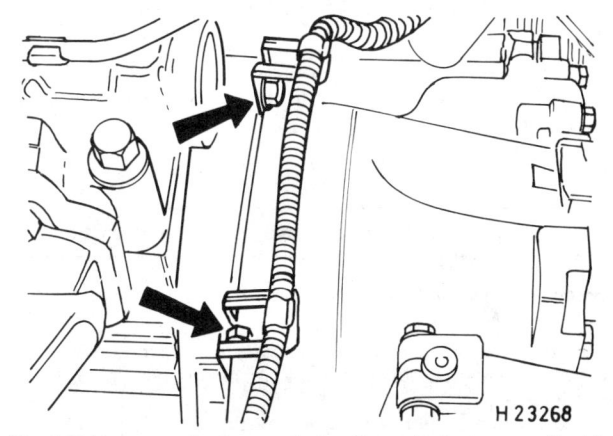

Fig. 1.12 Upper engine/transmission flange bolts (arrowed) with wiring loom securing brackets (Sec 15)

46   Chapter 1   Engine

16.2 Home-made device to jam the starter ring gear teeth and prevent the flywheel rotating

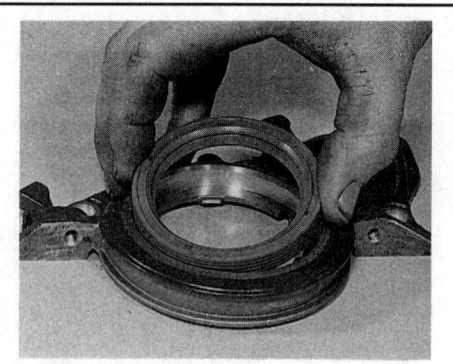

16.6A Crankshaft rear oil seal being positioned in its housing

16.6B Crankshaft rear oil seal being tapped squarely into its housing

(possibly for warranty reasons). It is strongly recommended that with a new, reconditioned or rebuilt engine, the engine oil and filter are changed at the end of the running-in period.

16 Crankshaft rear oil seal – renewal

1 With the engine or transmission removed from the vehicle for access, remove the clutch assembly (see Chapter 5) or the torsional vibration damper (see Chapter 6 – CTX automatic transmission only).
2 Prevent the flywheel from turning as its securing bolts are removed, by jamming the teeth of the starter ring gear (photo). Remove the flywheel.
3 Using a suitable claw tool, prise out the defective seal.
4 Two methods may be used to install a new seal, depending on tool availability.
5 If Service tool 21-059B is available, wipe out the seal housing seat, then lubricate the crankshaft flange and the internal lip of the seal, using clean engine oil. Position the seal fully onto the Service tool and press into its housing (Fig. 1.13).
6 If the Service tool is not available, remove the sump (see Section 10) then release the (Torx head) bolts securing the rear oil seal housing to the cylinder block, and remove the housing. Note that new gaskets for both will be required upon reassembly. Wipe out the seal housing seat then position the seal squarely into the housing (photo). Place a flat piece of wood on top of the seal to avoid distortion, then carefully tap it fully into the housing (photo). Do **not** allow the seal to tilt as it is being installed. Lubricate the crankshaft flange and the internal lip of the seal using clean engine oil then, using a new gasket, refit the rear oil seal housing being careful not to damage the lip of the oil seal as it passes over the crankshaft flange. Tighten the rear oil seal housing securing bolts to the specified torque, having ensured that the seal is centralised. Refit the sump, with reference to Section 10.
7 Clean the crankshaft flange and the flywheel mounting face, then refit the flywheel by reversing the method of removal and tighten its (lightly-oiled) retaining bolts to the specified torque. Note that the ribbed face of the flywheel faces the crankshaft flange.
8 Refit the clutch assembly (see Chapter 5), or on CTX automatic transmission equipped vehicles, refit the torsional vibration damper (see Chapter 6).

17 Piston/connecting rod assemblies – removal and refitting

Note: *Big-end cap securing bolts must **not** be refitted once removed. New bolts should always be used.*

1 Remove the cylinder head and the sump, as described in Sections 5 and 10 respectively. Do not remove the oil pick-up filter or pipe, which is an interference fit.
2 Note the location numbers stamped on the connecting rod big-ends and caps, and to which side they face. No 1 assembly is nearest the timing cover (Fig. 1.14). Keep the components of each piston/connect-

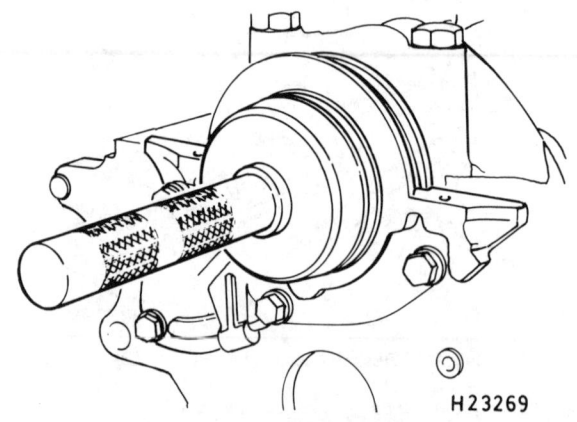

Fig. 1.13 Using Ford Service tool 21-059B to press the new crankshaft rear oil seal into its housing (shown during engine overhaul) (Sec 16)

ing rod assembly grouped together if more than one assembly is to be removed.
3 Turn the crankshaft with the pulley bolt until the big-end cap bolts for No 1 connecting rod are in their most accessible position. Unscrew and remove the (Torx head) bolts and the big-end cap complete with bearing shell. If the cap is difficult to remove, tap it off with a plastic-faced hammer.
4 If the bearing shells are to be used again (refer to Section 19), keep the shell taped to its cap.
5 Feel the top of the cylinder bore for a wear ridge. If one is detected, it should be scraped off before the piston/rod is pushed out of the top of the cylinder block. Take care when doing this not to score the cylinder bore surfaces.
6 Push the piston/connecting rod out of the block, retaining the bearing shell with the rod if it is to be used again.
7 Dismantling the piston/rod is covered in Section 19.
8 Repeat the operations on the remaining piston/rod assemblies.
9 To install a piston/rod assembly, have the piston ring gaps staggered, as shown in the diagram (Fig. 1.16) oil the rings and apply a piston ring compressor. Compress the piston rings.
10 Oil the cylinder bores.
11 Wipe out the bearing shell seat in the connecting rod and insert the shell.
12 Lower the piston/rod assembly into the cylinder bore until the base of the piston ring compressor stands squarely on the top of the block.
13 Check that the directional arrow on the piston crown faces towards the timing cover end of the engine and then apply the wooden handle of a hammer to the piston crown. Strike the head of the hammer sharply to drive the piston into the cylinder bore.
14 Oil the crankpin and draw the connecting rod down to engage with it (photo). Check that the bearing shell is still in position in the connecting rod.
15 Wipe the bearing shell seat in the big-end cap clean and insert the bearing shell (photo).
16 Lubricate the threads of the **new** big-end cap securing bolts, fit the

Chapter 1 Engine 47

17.14 Oil the crankpin (journal) and draw the connecting rod (with its bearing installed) down to engage with it

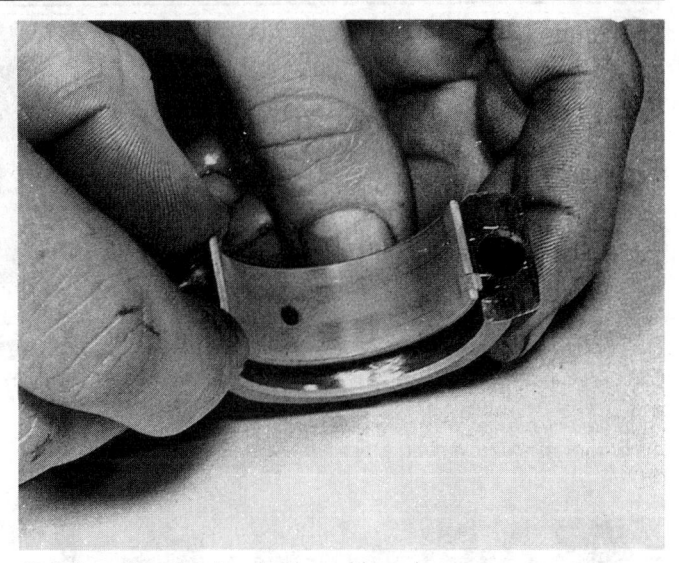

17.15 Inserting a bearing shell into a big-end cap

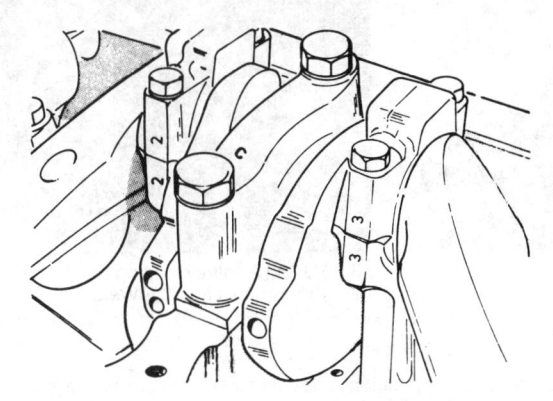

Fig. 1.14 Connecting rod and big-end cap markings (Sec 17)

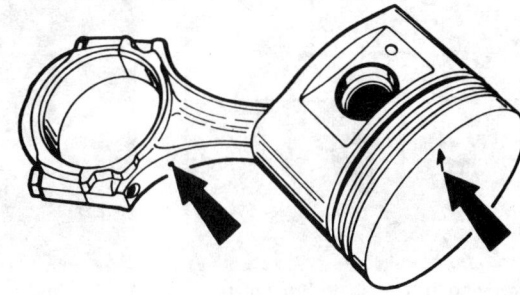

Fig. 1.15 Relative positions of arrow marking on piston crown and connecting rod oil splash bore (arrowed) (Sec 17)

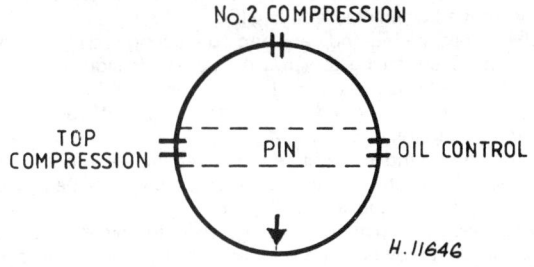

Fig. 1.16 Piston ring end-gap positioning diagram (Sec 17)

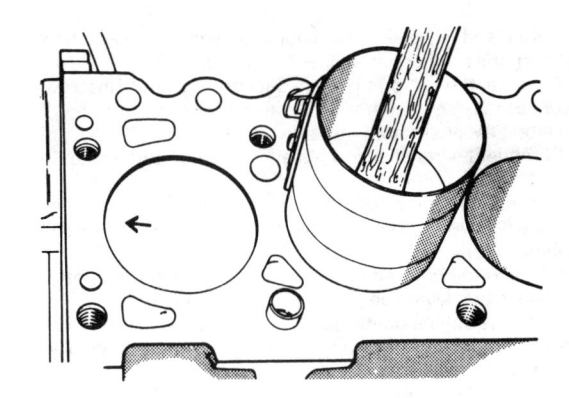

Fig. 1.17 Installing a piston/connecting rod assembly (Sec 17)

cap and tighten the bolts to the specified torque in two stages. Ensure that the endfloat is within the tolerances specified.
17   Repeat the operations on the remaining pistons/connecting rods.
18   Refit the sump (Section 10) and the cylinder head (Section 5). Refill with oil and coolant.

18   Engine – dismantling and reassembly

1   The need for dismantling will have been dictated by wear or noise in most cases. There is no reason why only partial dismantling cannot be carried out to renew such items as the timing chain or crankshaft rear oil seal. However when the main bearings or big-end bearings have been knocking and especially if the vehicle has covered a high mileage, it is recommended that a complete strip down is carried out and every engine component examined as described in Section 19.
2   Remove the engine from the vehicle, as described in Section 15.
3   Position the engine so that it is upright on a bench or other convenient working surface. If the exterior is very dirty it should be cleaned before dismantling using paraffin and a stiff brush or a water-soluble solvent.
4   Remove the dipstick and unscrew and discard the oil filter.
5   Disconnect the HT leads from the spark plugs and the DIS ignition coil, noting fitment.
6   Unscrew and remove the spark plugs.
7   Remove the oil filler cap (with breather hose).
8   Disconnect the fuel and vacuum pipes from the carburettor, and remove the carburettor (see Chapter 3).
9   Unbolt the thermostat housing and remove thermostat (see Chapter 2).
10   Remove the cylinder head (see Section 5 of this Chapter), complete with inlet and exhaust manifolds.
11   Detach the DIS ignition coil from the cylinder block by unscrewing its three (Torx head) bolts (photo).

Chapter 1 Engine

18.11 DIS ignition coil
A Mounting bracket to cylinder block bolts
B DIS ignition coil to mounting bracket screws

18.20 Remove the oil slinger from the front face of the crankshaft sprocket

18.21 Sliding the chain tensioner arm from its pivot pin, having retracted the tensioner cam against its spring pressure. Tensioner assembly retaining bolts (arrowed)

18.25 Removing the camshaft thrust plate

18.27 Withdrawing the camshaft

18.28 Using a valve grinding tool suction cup to withdraw the cam followers

12 Unbolt and remove the fuel pump and gasket/spacer, and the fuel vapour separator where fitted (see Chapter 3).
13 Remove the oil pump (see Section 11), and unscrew the oil pressure switch (see Chapter 12, Section 15).
14 Unbolt the alternator brackets.
15 Remove the water pump pulley and water pump (see Chapter 2, Section 11).
16 Unscrew the crankshaft pulley bolt. To do this, the flywheel starter ring gear will have to be jammed to prevent the crankshaft from turning. (see photo 16.2).
17 Remove the crankshaft pulley. If this does not pull off by hand, carefully use two levers behind it placed at opposite points.
18 Place the engine on its side and remove the sump. Do not invert the engine at this stage, or sludge and swarf may enter the oilways.
19 Turn the engine so that it rests on its cylinder head mating face, on a suitable wooden support.
20 Unbolt and remove the timing chain cover, and remove the oil slinger from the front face of the crankshaft sprocket having noted which way it faces (photo).
21 Slide the chain tensioner arm from its pivot pin on the front main bearing cap (photo).
22 Unbolt and remove the chain tensioner.
23 Bend back the lockplate tabs from the camshaft sprocket bolts and unscrew and remove the bolts.
24 Withdraw the sprocket complete with timing chain.
25 Unbolt and remove the camshaft thrust plate (photo).
26 Rotate the camshaft until each cam follower (tappet) has been pushed fully into its hole by its cam lobe.
27 Withdraw the camshaft, taking care not to damage the camshaft bearings (photo).
28 Withdraw each of the cam followers, keeping them in their originally fitted sequence by marking them with a piece of numbered tape or using a box with divisions (photo).
29 From the front end of the crankshaft, draw off the sprocket using a two-legged extractor.
30 Check that the three main bearing caps have marks to indicate their original fitting position. They also carry on arrow marking which points to the timing cover end of the engine. Note that the caps must be fitted to their original locations (and with the arrows pointing towards the timing cover) when reassembling (Fig. 1.18).
31 Check that the big-end caps and connecting rods have adjacent matching numbers facing towards the camshaft side of the engine. Number 1 assembly is nearest the timing chain end of the engine. If any markings are missing or indistinct, make some of your own with quick-drying paint (refer to Fig. 1.14).
32 Unbolt and remove the big-end bearing caps and note that **new** big-end cap bolts will be needed upon reassembly. If the bearing shell is to be used again, tape the shell to the cap.
33 Now check the top of the cylinder bore for a wear ring. If one can be felt, it should be removed with a scraper before the piston/rod is pushed out of the cylinder.

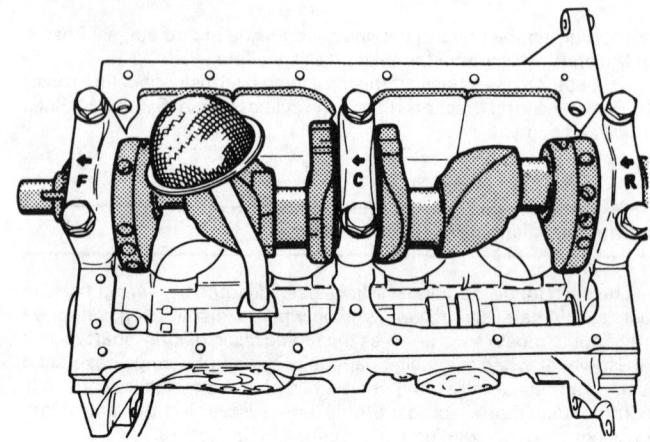

Fig. 1.18 Main bearing cap markings (1.1 litre engine shown) (Sec 18)

Chapter 1 Engine

18.49 Securing the camshaft thrust plate retaining bolts by bending up the locktabs

18.50 Oiling the centre main bearing shell, with shell and thrust washers fitted to crankcase. Thrust washer oil grooves (arrowed)

18.51 Lowering the crankshaft into its bearings

18.54 Using feeler blades to check crankshaft endfloat

18.57A Fit the timing chain and camshaft sprocket ...

18.57B ... so that the timing marks on both sprockets align (with the shaft centres)

34 Remove the piston/rod by pushing it out of the top of the block. Tape the bearing shell to the connecting rod.
35 Remove the remaining three piston/rod assemblies in a similar way.
36 Unbolt the clutch pressure plate cover from the flywheel. Unscrew the bolts evenly and progressively until spring pressure is relieved, before removing the bolts. Be prepared to catch the clutch driven plate as the cover is withdrawn. On CTX automatic transmission equipped vehicles, remove the torsional vibration damper (see Chapter 6).
37 Unbolt and remove the flywheel. It is heavy, but do not drop it. Note that the ribbed face of the flywheel faces the crankshaft flange. If necessary, the starter ring gear can be jammed to prevent the flywheel rotating (see photo 16.2). There is no need to mark the fitted position of the flywheel to its mounting flange as it can only be fitted one way. Remove the upper adaptor plate (engine backplate) from around the periphery of the engine to transmission mating flange.
38 Detach the engine speed/crankshaft position sensor by undoing its single (Torx head) retaining bolt and withdrawing (see photo 15.10A).
39 Unbolt and remove the crankshaft rear oil seal housing (refer to Fig. 1.17).
40 Unbolt the main bearing caps. Note that **new** main bearing cap bolts will be needed upon reassembly. Remove the caps, tapping them off if necessary with a plastic faced hammer. Retain the bearing shells with their respective caps if the shells are to be used again, although unless the engine is of low mileage this is not recommended (see Section 19).
41 Lift the crankshaft from the crankcase and lift out the upper bearing shells, noting the thrust washers either side of the centre bearing. Keep these shells with their respective caps, identifying them for refitting to the crankcase if they are to be used again.
42 Remove the crankshaft oil seals from the timing cover and the rear oil seal housing (see Sections 12 and 16).
43 With the engine now dismantled, each component should be examined as described in the following Section before reassembling.
44 Ensure that everything is clean before commencing reassembly. Dirt and metal particles can quickly destroy bearings and result in major engine damage. Use clean engine oil to lubricate during reassembly.
45 Fit a new oil seal to the rear oil seal housing (see Section 16), and to the timing cover. If Service tool 21-046 is not available, fit the seal to the timing cover using a suitably sized piece of tube, ensuring that it is pressed fully home – avoid tapping it in as timing cover damage may result.
46 Check that the Woodruff key is in position on the front end of the crankshaft then position the crankshaft sprocket over it, with its marking facing outwards (see photo 18.57B) and draw the crankshaft sprocket into position using the crankshaft pulley, bolt and washer.
47 Lubricate the cam follower bores and fully insert the cam followers in their original positions.
48 Lubricate the camshaft bearings, camshaft and thrust plate, then insert the camshaft from the timing cover end of the engine.
49 Fit the thrust plate and tighten the fixing bolts to the specified torque. The endfloat will already have been checked as described in Section 19. Secure the bolts with the locktabs (photo).
50 Wipe clean the main bearing shell seats in the crankcase and fit the shells. Using a little grease, stick the semi-circular thrust washers on either side of the centre bearing so that the oil grooves are visible when the washers are installed (photo).
51 Oil the bearing shells and lower the crankshaft into the crankcase (photo).
52 Wipe clean the seats in the main bearing caps and fit the bearing shells into them (shells without oil groove). Install the caps so that their markings are correctly positioned, as explained during dismantling, in paragraph 30, having first liberally oiled the running faces of the bearings.
53 Screw in the **new** main bearing cap bolts and tighten evenly to the specified torque.
54 Now check the crankshaft endfloat. Ideally a dial gauge should be used, but feeler blades are an alternative if inserted between the face of the thrust washer and the machined surface of the crankshaft balance weight after having prised the crankshaft first in one direction and then the other (photo). Provided the thrust washers at the centre bearing have been renewed, the endfloat should be within the specified tolerance. If it is not, oversize thrust washers are available (see Specifications).
55 Bolt the timing chain tensioner into position and check that the face of the tensioner cam is parallel with the face of the cylinder block,

Chapter 1 Engine

18.58 Secure the camshaft sprocket bolts with the locktabs

18.61 Fitting the timing cover. Crankshaft pulley used as an aid to centring

18.62 Fitting the rear oil seal housing, with a new gasket in position on the cylinder block

ideally using a dial gauge. The maximum permissible error between the two measuring points is 0.2 mm. Release and turn the timing chain tensioner to achieve this, as necessary. Refer to the Specifications for tightening torque details.

56 Rotate the crankshaft so that the timing mark on its sprocket is directly in line with the centre of the crankshaft sprocket mounting flange.

57 Engage the camshaft sprocket within the timing chain and then engage the chain around the teeth of the crankshaft sprocket. Push the camshaft sprocket onto its mounting flange. The camshaft sprocket bolt holes should now be in alignment with the tapped holes in the camshaft flange and both sprocket timing marks in alignment. Turn the camshaft as necessary to achieve this, also withdraw the camshaft sprocket and reposition it within the loop of the chain. This is a 'trial and error' operation which must be continued until exact alignment of bolt holes and timing marks is achieved (photos).

58 Screw in the sprocket bolts to the specified torque and bend up the tabs of a new lockplate (photo).

59 Retract the timing chain tensioner cam then slide the tensioner arm onto its pivot pin. Release the cam so that it bears upon the arm.

60 Fit the oil slinger to the front of the crankshaft sprocket so that its convex side is against the sprocket.

61 Lightly lubricate the front end of the crankshaft and the radial lip of the timing cover oil seal (already installed in timing cover). Using a new gasket, fit the timing cover centring it with the aid of the crankshaft pulley (photo) – lubricate the seal contact surface of the crankshaft pulley beforehand. One timing cover retaining bolt should be left out at this stage as it also holds the water pump. Refer to the Specifications for tightening torque details. Secure the crankshaft pulley loosely at this stage by screwing its bolt in to finger tightness only.

62 Lubricate the crankshaft rear flange and the radial lip of the rear oil seal (already installed in its housing). Using a new gasket, fit the rear oil seal housing being careful not to damage the lip of the oil seal as it passes over the crankshaft flange (photo). Ensuring that the seal is centralised, tighten the rear oil seal housing bolts to the specified torque.

63 Locate the upper adaptor plate (engine backplate) on its dowels, then refit the flywheel. Screw in and tighten the flywheel bolts to the specified torque, preventing the flywheel from rotating in a similar manner to that used during dismantling (photo). Note that the flywheel bolts should be lightly oiled before fitting. With the starter ring gear still locked, tighten the crankshaft pulley bolt to its specified torque.

64 Install and centralise the clutch as described in Chapter 5, or, on CTX automatic transmission equipped vehicles, fit the torsional vibration damper (see Chapter 6).

65 The pistons/connecting rods should now be installed. Although new pistons may have been fitted to the rods by your dealer or supplier (see Section 19), it is worth checking to ensure that with the piston crown arrow pointing to the timing cover end of the engine, the oil hole in the connecting rod is on the left as shown (Fig. 1.19). Oil the cylinder bores. Install the pistons/connecting rods as described in Section 17.

66 Fit the engine speed/crankshaft position sensor and secure with its single Torx head bolt.

67 Fit the sump as described in Section 10.

68 Turn the engine and support it so that it rests securely in an upright position.

18.63 Tightening the flywheel retaining bolts, with the starter ring gear jammed

69 Screw in the oil pressure switch and tighten to the specified torque.

70 Fit the oil pump complete with new gasket and a new oil filter as described in Section 11.

71 Fit the DIS ignition coil and secure with its three Torx head bolts.

72 Fit the fuel pump using a new gasket/spacer. The protrusion on the gasket/spacer must face towards the pump. Also fit the fuel vapour separator, where removed.

73 Fit the alternator brackets.

74 Refit the water pump its gasket being integral with the new timing cover gasket (see Chapter 2, Section 11). Tighten the bolts to the specified torque. Refit the water pump pulley – refer to the Specifications in Chapter 2 for tightening torque details.

75 Fit the cylinder head as described in Section 5.

76 Refit the pushrods in their original sequence and the rocker shaft, also as described in Section 5.

77 Adjust the valve clearances (Section 8) and refit the rocker cover using a new gasket. Refit the oil filler cap (with breather hose).

78 If the inlet or exhaust manifolds have been removed from the cylinder head for whatever reason, refit in accordance with Chapter 3 using new gaskets and tightening according to specification.

79 Fit the thermostat to the cylinder head ensuring correct location. Align its gasket and refit the thermostat housing, tightening the bolts to the specified torque (see Chapter 2).

80 Refit the carburettor and its associated fuel and vacuum pipes. Use a new flange gasket (see Chapter 3).

81 Fit the spark plugs, and the temperature gauge sender (if removed).

82 Connect the HT leads to the DIS ignition coil and the corresponding spark plugs.

Chapter 1 Engine   51

83 Refit the engine oil dipstick, and check that the sump drain plug is tightened to its specified torque. A new seal should be fitted at regular intervals to prevent leakage. Refilling with oil should ideally be left until the engine is installed in the vehicle.

19 Engine – component examination and renovation

1 Clean all components using paraffin and a stiff brush, except the crankshaft, which should be wiped clean and the oil passages cleaned out with a length of wire.

2 Never assume that a component is unworn simply because it looks all right. After all the effort which has gone into dismantling the engine, refitting worn components will make the overhaul a waste of time and money. Depending on the degree of wear, the overhauler's budget and the anticipated life of the vehicle, components which are only slightly worn may be refitted, but if in doubt it is always best to renew.

Crankshaft, main and big-end bearings

3 The need to renew the main bearing shells or to have the crankshaft reground will usually have been determined during the last few miles of operation when perhaps a heavy knocking has developed from within the crankcase or the oil pressure warning lamp has stayed on denoting a low oil pressure probably caused by excessive wear in the bearings.

4 Even without these symptoms, the journals and crankpins on a high mileage engine should be checked for out-of-round (ovality) and taper. For this a micrometer will be needed to check the diameter of the journals and crankpins at several different points around them. A motor factor or engineer can do this for you. If the average of the readings shows that either out-of-round or taper is outside permitted tolerance (see Specifications), then the crankshaft should be reground by your dealer or engine reconditioning company to accept the undersize main and big-end shell bearings which are available. Normally, the company doing the regrinding will supply the necessary undersize shells.

5 If the crankshaft is in good condition, it is wise to renew the bearing shells as it is almost certain that the original ones will have worn. This is often indicated by scoring of the bearing surface or by the top layer of the bearing metal having worn through to expose the metal underneath.

6 When of 'standard' diameter, the crankshaft main bearing journals are classified into two sizes, dependent on tolerances. Crankshafts at the lower end of the tolerance range are marked with a yellow paint line, whilst those more central to the 'standard' classification carry no marking. When they are 0.254 mm undersize, they are marked with a green paint line. This classification marking is carried on the first balance weight at the front of the crankshaft (Fig. 1.19). Refer to the Specifications for further details.

7 When of 'standard' diameter, the crankshaft big-end bearing journals are also unmarked. When the big-end bearing journals are 0.254 mm undersize, the crankshaft is marked with a green paint spot. This classification marking is carried on the first crank throw web next to the big-end bearing journal. Refer to the Specifications for further details.

8 'Standard' main and big-end bearing shells are classified according to tolerance; they either carry no paint mark on the side, or they carry a yellow mark (denoting 0.010 mm thicker) (Fig. 1.20). In the case of production repair sizes, they carry green or black marks. Larger service repair sizes do not have colour markings. Bearing shells are usually marked with the part number and size classification on the back.

9 If in any doubt regarding parts suitability, seek the advice of your parts supplier.

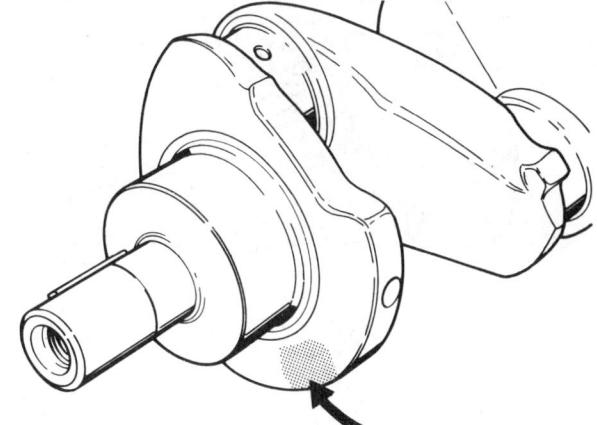

Fig. 1.19 Crankshaft main bearing journal size identification mark on balance weight (Sec 19)

Cylinder bores, pistons, rings and connecting rods

10 Cylinder bore wear will usually have been evident from the smoke emitted from the exhaust during recent operation of the vehicle on the road, coupled with excessive oil consumption and fouling of spark plugs.

11 Engine life can be extended by fitting special oil control rings to the pistons, These are widely advertised and will give many more thousands of useful mileage without the need for a rebore, although this will be inevitable eventually. If this remedy is decided upon, remove the piston/connecting rods as described in Section 17 and fit the proprietary rings in accordance with the manufacturer's instructions.

12 Where a more permanent solution is decided upon, the cylinder block can be rebored by your dealer or engineering works, or by one of the mobile workshops which now undertake such work. The cylinder bore will be measured both for out-of-round and for taper to decide how much the bores should be bored out. A set of matching pistons will be supplied in a suitable oversize to suit the new bores.

13 Due to the need for special heating and installing equipment for removal and refitting of the interference type gudgeon pin, the removal and refitting of pistons to the connecting rods is definitely a specialist job, preferably for your Ford dealer.

14 The removal and refitting of piston rings is however well within the scope of the home mechanic. Do this by sliding two or three old feeler blades round behind the top compression ring so that they are at equidistant points. The ring can now be slid up the blades and removed. Repeat the removal operations on the second compression ring and then the oil control ring. This method will not only prevent the rings dropping onto empty grooves as they are withdrawn, but it will also avoid ring breakage.

15 Examine the pistons for ovality, scoring and scratches. Check the connecting rods for wear and damage. The connecting rods carry a letter indicating their weight category; all the rods fitted to one engine must be of the same weight category.

16 Even when new piston rings have been supplied to match the

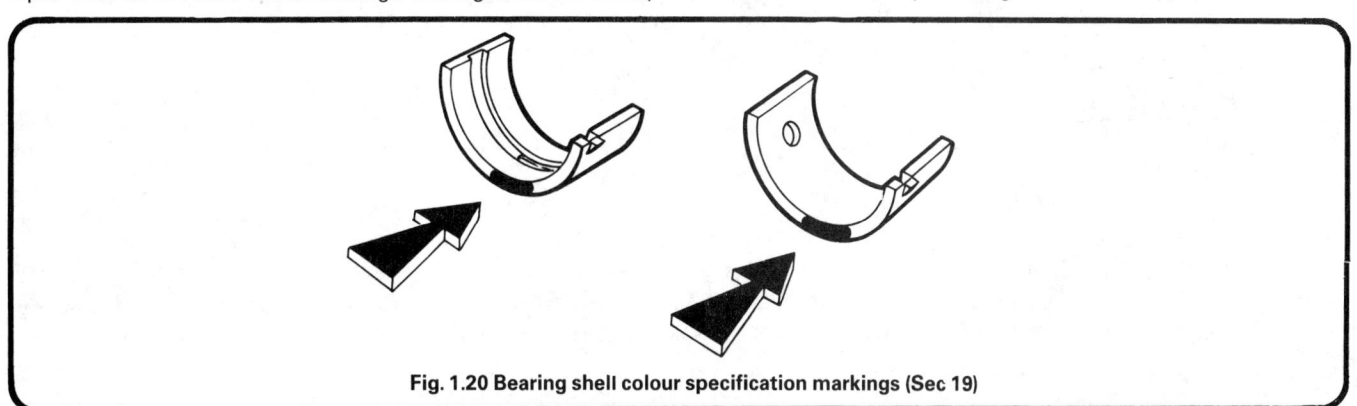

Fig. 1.20 Bearing shell colour specification markings (Sec 19)

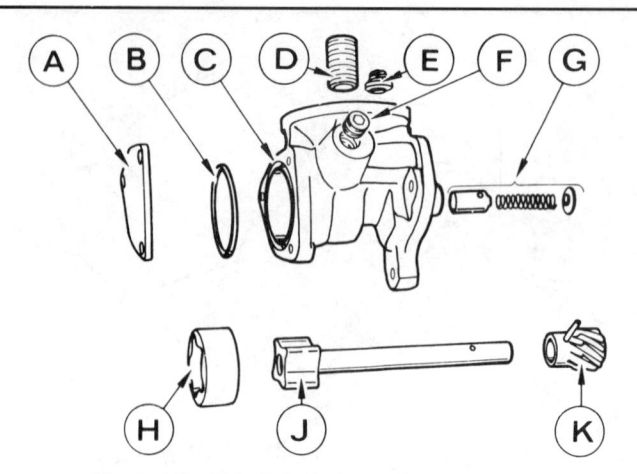

Fig. 1.21 Exploded view of the oil pump (Sec 19)

- A Pump cover
- B O-ring
- C Pump body
- D Oil filter attachment stud
- E Filter relief valve
- F Blind plug
- G Oil pressure relief valve
- H Outer rotor
- J Inner rotor
- K Drive gear

pistons, always check that they are not tight in their grooves and also check their end gaps by pushing them squarely down their particular cylinder bore and measuring with a feeler blade. Adjustment of the end gap can be made by careful grinding to bring it within the specified tolerance (photo).

17 If new rings are being fitted to an old piston, always remove any carbon from the grooves beforehand. The best tool for this job is the end of a broken piston ring. Take care not to cut your fingers, piston rings are sharp. Piston rings must be fitted to the piston with their manufacturer's mark facing upwards. The cylinder bores should be roughened with fine glass paper to assist the bedding-in of the new rings.

Timing sprockets and chain

18 The teeth on the timing sprockets rarely wear, but check for broken or hooked teeth even so.
19 The timing chain should always be renewed at time of major engine overhaul. A worn chain is evident if when supported horizontally at both ends it takes on a deeply bowed appearance.
20 Finally check the rubber cushion on the tensioner spring leaf. If grooved or chewed up, renew it.

Flywheel

21 Inspect the starter ring gear on the flywheel for wear or broken teeth. If evident, the ring gear should be renewed. Due to the method of renewing, you are advised to consult your dealer or local motor engineering works about them performing the task on your behalf.
22 The clutch friction surface on the flywheel should be checked for grooving or tiny hair cracks, the latter being caused by overheating. If these conditions are evident, it may be possible to surface grind the flywheel provided its balance is not upset. Otherwise, a new flywheel will have to be fitted – consult your dealer about this.

Oil pump

23 The oil pump should be checked for wear by unbolting and removing the cover plate and O-ring (photo) and checking the following tolerances:

   (a) Outer rotor to pump body gap
   (b) Inner rotor to outer rotor gap
   (c) Rotor endfloat (use a feeler blade and straight-edge across pump body)

24 Use feeler blades to check the tolerances and if they are outside the specified values, renew the pump (photos).
25 Check also for wear or damage to the pump drive gear and, if present, renew it. If wear or damage is present on the oil pump drive-gear, the camshaft gear may also be defective.

Oil seals and gaskets

26 Renew the oil seals on the timing cover and the crankshaft rear retainer as a matter of routine at time of major overhaul. Oil seals are

19.16 Checking piston ring end gap

19.23 Having removed the oil pump cover, lift the O-ring out of its annual groove

19.24A Checking outer rotor to pump body clearance

19.24B Checking inner rotor to outer rotor clearance

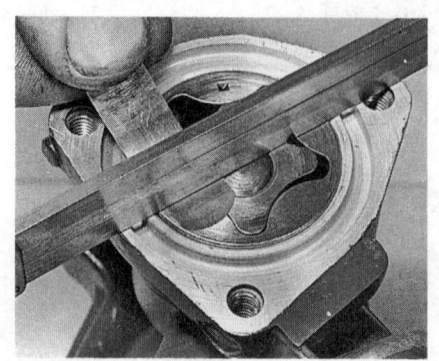

19.24C Checking rotor endfloat

19.33 Checking camshaft endfloat

Chapter 1 Engine

19.37 Valve spring compressor engaged on No 1 valve

19.39A Remove the valve spring retainer, followed by the spring ...

19.39B ... then withdraw the valve from the cylinder head

19.40 Prise up and remove the valve stem oil seal

19.46A Having covered the split collet grooves with adhesive tape, slide the new valve stem oil seal onto the valve stem

19.46B Using a suitably sized socket, ensure that the valve stem oil seal is fully seated. A gentle tap will suffice

cheap, oil is not! Using clean engine oil, lubricate the oil seal lips and check that the small tensioner spring in the oil seal has not been displaced by the vibration caused during fitting of the seal.

27 Renew all the gaskets by purchasing the appropriate 'de-coke', short or full engine set. Oil seals may be included in the gasket sets.

Crankcase

28 Clean out the oilways with a length of wire or by using compressed air. Similarly clean the coolant passages. This is best done by flushing through with a cold water hose. Examine the crankcase and block for stripped threads in bolt holes; if evident, thread inserts can be fitted.

29 Renew any core plugs which appear to be leaking or which are excessively rusty.

30 Cracks in the casting may be rectified by specialist welding, or by one of the cold metal key interlocking processes available.

Camshaft and bearings

31 Examine the camshaft gear and lobes for damage or wear. If evident a new camshaft must be purchased, or one which has been 'built-up' such as are advertised by firms specialising in exchange components.

32 The bearing internal diameters should be checked against the Specifications if a suitable gauge is available; otherwise, check for movement between the camshaft journal and the bearing. Worn bearings should be renewed by your dealer.

33 Check the camshaft endfloat by temporarily refitting the camshaft and the thrust plate. If the endfloat exceeds the specified tolerance, renew the thrust plate (photo).

Cam followers

34 It is seldom that the cam followers wear in their bores, but it is likely that after a high mileage, the cam lobe contact surface will show signs of a depression or grooving.

35 Where this condition is evident, renew the cam followers. Grinding out the wear marks will only reduce the thickness of the hardened metal of the cam follower and accelerate further wear.

Cylinder head and rocker gear

36 The usual reason for dismantling the cylinder head is to decarbonise and to grind in the valves. Reference should therefore be made to Section 6, in addition to the dismantling operations described here. First remove the manifolds (see Chapter 3).

37 Using a standard valve spring compressor, compress the spring on No 1 valve (nearest the timing cover) (photo). Do not overcompress the spring or the valve stem may bend. If it is found that when screwing down the compressor tool, the spring retainer does not release from the collets, remove the compressor and place a piece of tubing on the retainer so that it does not impinge on the collets and strike the end of the tubing with a sharp blow with a hammer. Refit the compressor and compress the spring.

38 Extract the split collets and then gently release the compressor and remove it.

39 Remove the valve spring retainer and the spring, then withdraw the valve from the cylinder head (photos).

40 Prise up and remove the valve stem seals (photo).

41 Repeat the removal operations on the remaining seven valves. Keep the valves in their originally fitted sequence by placing them in a piece of card which has holes punched in it and numbered 1 to 8 (from the timing cover end), and group their springs, retainers and spring collets in a similar manner.

42 Place each valve in turn in its guide so that approximately one third of its length enters the guide. Rock the valve from side to side. If there is any more than an imperceptible movement, the guides will have to be reamed (working from the valve seat end) and oversize stemmed valves fitted. You are advised to entrust this work to your Ford dealer.

43 Examine the valve seats. Normally the seats do not deteriorate but the valve heads are more likely to burn away, in which case, new valves can be ground-in as described in Section 6. If the valve seats are cracked or require re-cutting, consult your Ford dealer; the valve seats cannot be re-cut using conventional tools.

44 If the cylinder head mating surface is suspected of being distorted due to persistent leakage of coolant at the gasket joint, then it can be checked and surface ground to your dealer or motor engineering works. Distortion is unlikely under normal circumstances with a cast iron head.

45 Check the rocker shaft and rocker arms pads which bear on the valve stem end faces for wear or scoring, also for any broken coil springs. Renew components as necessary after dismantling as described in Section 7. If the valve springs have been in use for 50 000 miles (80 000 km) or more, they should be renewed.

46 Reassemble the cylinder head by first lubricating the valves and valve guides (refer to the Specifications for lubricant details). Install No 1 valve fully into its guide, then cover the split collet grooves with tape before sliding a **new** valve stem seal over the top (photo); support the valve to prevent it dropping out. Using a suitable-sized socket, ensure that the valve stem seal is fully seated (photo). Take care not to distort the seal and ensure that its spring is correctly located, otherwise there is no guarantee of effective sealing. Remove the tape from the valve stem, then position the valve spring and retainer over the valve stem and engage the valve spring compressor. Compress the spring and refit the spring collets to their valve stem cut-outs. Hold the split collets in position while the compressor is gently released and removed.

47 Repeat the operations on the remaining valves, making sure that each valve is returned to its original guide or if new valves have been fitted, into the seat into which it was ground.

48 On completion, support the ends of the cylinder head on two wooden blocks and strike the end of the valve stem with a plastic or copper-faced hammer, just a light blow to settle the components.

49 Refit the manifolds (refer to Chapter 3).

20 Fault diagnosis

Symptom	Reason(s)
Engine fails to turn over when starter operated	Discharged or defective battery Loose or corroded battery connections Engine earth lead disconnected Defective starter motor, solenoid or switch Selector lever not in positions N or P (CTX equipped vehicles only)
Engine turns over but fails to start	No petrol in petrol tank Ignition damp or wet DIS ignition coil multi-plug disconnected Ignition HT leads loose (or incorrectly connected at DIS ignition coil after overhaul) Defective ignition switch DIS ignition coil defective Ignition module defective Vapour lock in fuel line (in hot conditions or at high altitude) Blocked float chamber needle valve Blocked fuel pump filter Blocked carburettor jets Fuel pump defective
Engine stalls and refuses to restart	Ignition failure (possible in severe rain or after traversing a water splash) No petrol in petrol tank Petrol tank breathing system blocked Sudden obstruction in carburettor Water in fuel system
Engine misfires or idles unevenly	Battery leads loose on terminals Battery earth lead loose at body attachment point Engine earth lead loose Ignition HT leads loose at DIS ignition coil or spark plugs Dirty or incorrectly gapped spark plugs Defective ignition module Defective DIS ignition coil Air leak in carburettor, inlet manifold-to-carburettor, or inlet manifold-to-cylinder head mating faces Carburettor incorrectly adjusted Incorrect valve clearances Sticking valve Weak or broken valve springs Worn valve guides or stems Worn pistons and piston rings
Lack of power and poor compression	Burnt exhaust valves Sticking or leaking valves Worn valve guides and stems Weak or broken valve springs Incorrect valve clearances Worn or scored cylinder bores Worn pistons and piston rings Blown head gasket (accompanied by increase in noise) Defective ignition module or blocked vacuum advance hose Incorrectly gapped spark plugs Fuel filter partially blocked causing top and fuel starvation Defective fuel pump giving top end fuel starvation Carburettor incorrectly adjusted Air filter choked Brakes binding

Symptom	Reason(s)
Excessive oil consumption	Badly worn, perished or missing valve stem oil seals Excessively worn valve stems and valve guides Worn piston rings Worn pistons and cylinder bores Excessive piston ring gap allowing blow-by Piston oil return holes clogged
Engine oil pressure warning light illuminates with engine running	Oil level low (possibly through leakage) or incorrect grade used Oil filter clogged or relief valve defective Oil pick-up strainer clogged Defective oil pressure switch or wiring Oil pump worn or defective Worn main or big-end bearings
Oil being lost due to leakage	Leaking sump drain plug Leaking gaskets or seals
Unusual noises from engine	Worn valve gear or incorrect valve clearances (noisy tapping from rocker cover) Worn big-end bearings (regular heavy knocking) Worn main bearings (rumbling and vibration) Worn crankshaft (knocking, rumbling and vibration) Unintentional mechanical contact (eg fan blades contacting shroud, for whatever reason)

Note: *In the event that an internal fault is suspected in the ignition module or DIS ignition coil, the vehicle should be taken to a Ford dealer for an accurate diagnosis as the equipment used in the diagnosis is of a specialist nature. Ensure that all connections are secure before taking the above course of action.*

PART B : CVH ENGINES

21 General description

The 1.4 litre and 1.6 litre CVH (Compound Valve angle, Hemispherical combustion chambers) engines are of four-cylinder in-line overhead camshaft type, mounted transversely, together with the transmission, at the front of the vehicle.

The crankshaft is supported in five main bearings within a cast iron crankcase.

The cylinder head is of light alloy construction, supporting the overhead camshaft in five bearings. Camshaft drive is by a toothed composite rubber belt, driven from a sprocket on the crankshaft.

The distributor is driven from the rear (flywheel) end of the camshaft by an offset dog. Where the distributor has been replaced by the E-DIS-4 (distributorless ignition system), drive from the camshaft is no longer required and a plastic blanking cover is fitted.

The cam followers are of hydraulic type, which eliminates the need for valve clearance adjustment. The cam followers operate in the following way. When the valve is closed, pressurised engine oil passes through a port in the body of the cam followers and four grooves in the plunger and into the cylinder feed chamber. From this chamber, oil flows through a ball type non-return valve into the pressure chamber. The tension of the coil spring causes the plunger to press the rocker arm against the valve and to eliminate any free play.

As the cam lifts the cam follower, the oil pressure in the pressure chamber increases and causes the non-return valve to close the port feed chamber. As oil cannot be compressed, it forms a rigid link between the body of the cam follower, the cylinder and the plunger which then rise as one component to open the valve.

The clearance between the body of the cam follower and the cylinder is accurately designed to meter a specific quantity of oil as it escapes from the pressure chamber. Oil will only pass along the cylinder bore when pressure is high during the moment of valve opening. Once the valve has closed, the escape of oil will produce a small amount of free play and no pressure will exist in the pressure chamber. Oil from the feed chamber can then flow through the non-return valve into the pressure chamber so that the cam follower cylinder can be raised by the pressure of the coil spring, thus eliminating any play in the arrangement until the valve is operated again.

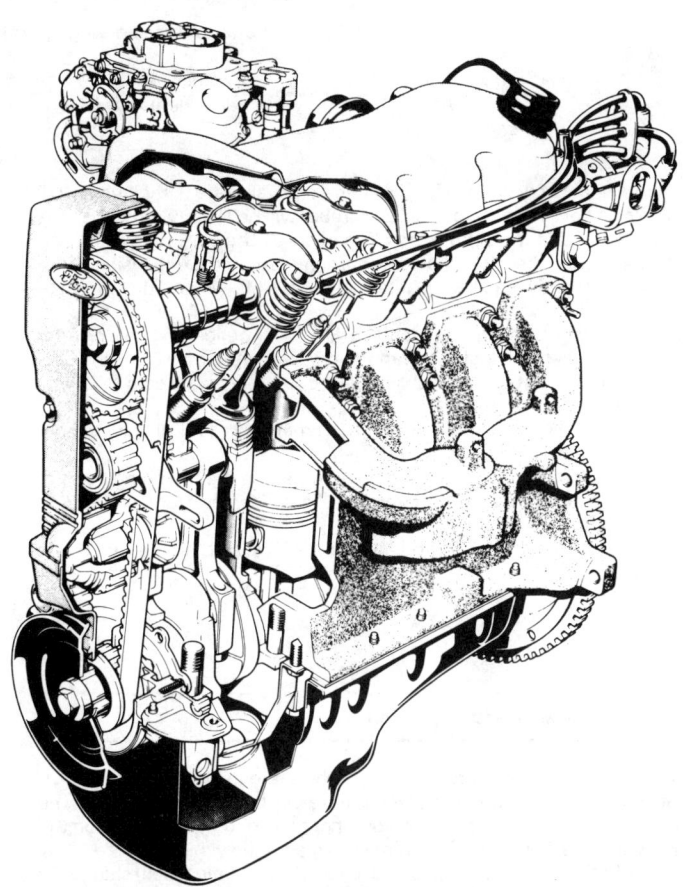

Fig. 1.22 Cutaway view of the CVH engine (Sec 21)

56 Chapter 1 Engine

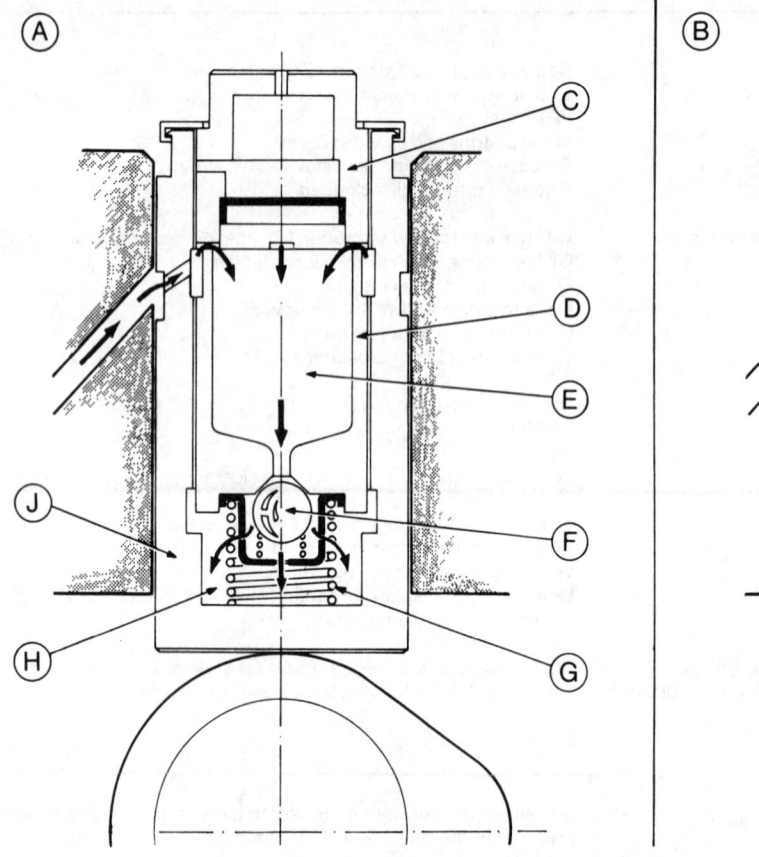

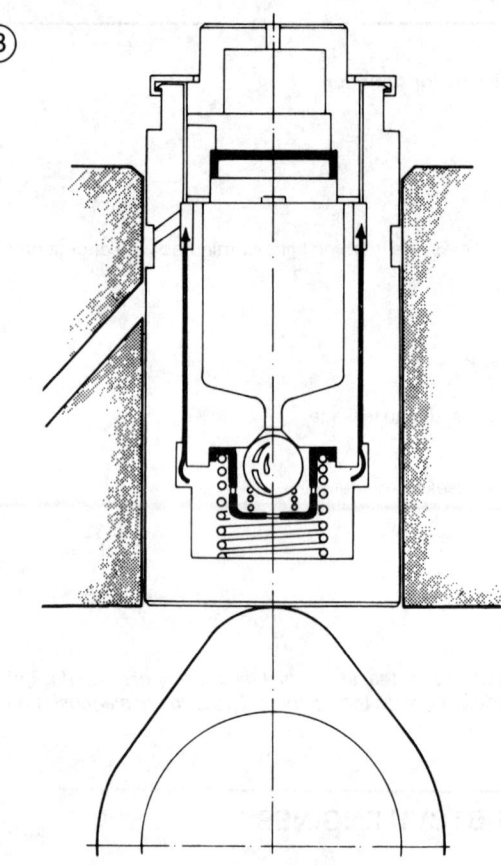

Fig. 1.23 Sectional views showing operation of hydraulic cam followers (Sec 21)

A Valve closed
B Valve open
C Plunger
D Cylinder
E Feed chamber
F Non-return valve
G Coil spring
H Pressure chamber
J Body

As wear occurs between rocker arm and valve stem, the quantity of oil which flows into the pressure chamber will be slightly more than the quantity lost during the expansion cycle of the cam follower. Conversely, when the cam follower is compressed by the expansion of the valve, a slightly smaller quantity of oil will flow into the pressure chamber than was lost.

If the engine has been standing idle for a period of time, or after overhaul, when the engine is started up, valve clatter may be heard. This is a normal condition and will gradually disappear as the cam followers are pressurised with oil.

A water pump is mounted on the timing belt end of the cylinder block and is driven by the toothed belt.

A gear or rotor type oil pump is mounted on the timing belt end of the cylinder block and is driven by a gear on the front end of the crankshaft.

A full-flow oil filter of throw-away type is located on the side of the crankcase.

22 Maintenance and inspection

1 The maintenance procedures are the same as described in Part A, Section 2, but ignore any references to valve clearance adjustment. The oil filter locates to the cylinder block, not to an external pump as on the HCS engines. It is accessible from underneath the vehicle.
2 Additionally, at the specified interval, the timing belt should be renewed as described in Section 25.

23 Major operations possible with the engine in the vehicle

The following work can be carried out without having to remove the engine:

(a) *Timing belt – renewal*
(b) *Camshaft oil seal – renewal*
(c) *Camshaft – removal and refitting*
(d) *Cylinder head – removal and refitting*
(e) *Crankshaft front oil seal – renewal*
(f) *Sump – removal and refitting*
(g) *Piston/connecting rod – removal and refitting*
(h) *Engine/transmission mountings – removal and refitting*

24 Major operations requiring engine removal

The following work can only be carried out after removal of the engine from the car:

(a) *Crankshaft main bearings – renewal*
(b) *Crankshaft – removal and refitting*
(c) *Flywheel – removal and refitting*
(d) *Crankshaft rear oil seal – renewal*
(e) *Oil pump – removal and refitting*

Chapter 1 Engine

Fig. 1.24 Upper timing cover section retaining bolts (arrowed) (Sec 25)

Fig. 1.25 Crankshaft sprocket TDC mark alignment (arrowed) (Sec 25)

25 Timing belt – removal, refitting and adjustment

Note: Accurate adjustment of the timing belt entails the use of Ford special tools. An approximate setting can be achieved using the method described in this Section, but the tension should be checked by a dealer on completion.

1 Disconnect the battery earth terminal.
2 Remove the alternator drivebelt (see Chapter 12, Section 7). Depending on model, remove the air filter housing/trunking for additional working clearance.
3 Unscrew the two bolts securing the upper half of the timing cover and remove it. (Fig. 1.24).
4 Using a spanner on the crankshaft pulley bolt, turn the crankshaft until the TDC pointer on the camshaft sprocket aligns with the TDC mark on the cylinder head (photo). To prevent the crankshaft rotating from this position, remove the starter motor (see Chapter 12, Section 12) and lock the flywheel ring gear with a cold chisel or other suitable tool (photo).
5 Remove the crankshaft pulley bolt and draw the pulley off the crankshaft.
6 Unscrew the two bolts securing the lower half of the timing cover and remove it. Note that the small projection on the crankshaft sprocket is aligned with the TDC mark on the oil pump casing, if all is correct (Fig. 1.25).
7 Slacken the two bolts which secure the timing belt tensioner and, using a large screwdriver, prise the tensioner to one side to relieve the tautness of the belt. If the tensioner is spring loaded, tighten one of the bolts to retain it in the slackened position (photo).
8 If the original belt is to be refitted, mark it for direction of travel and also the exact tooth positions on all three sprockets.
9 Slip the timing belt off the camshaft, water pump and crankshaft sprockets.
10 Before refitting the belt, check that the crankshaft is still at TDC

25.4A TDC pointer on the camshaft sprocket aligned with the TDC mark on the cylinder head (mirror image shown)

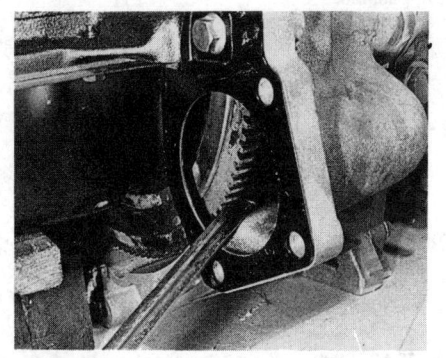

25.4B Using a stout bar to lock the flywheel ring gear

25.7 Timing belt tensioner securing bolts (arrowed)

25.13 Refit the crankshaft pulley and secure with its bolt and washer

25.15 Crankshaft pulley notch aligned with the TDC (0) mark on the lower half of the timing cover

(the small projection on the belt sprocket front flange in line with the TDC mark on the oil pump housing (Fig. 1.25) and that the timing mark on the camshaft sprocket is opposite the TDC mark on the cylinder head (see photo 25.4A). Adjust the position of the sprockets slightly, but avoid any excessive movement of the sprockets while the belt is off, as the piston crowns and valve heads may make contact.

11 Engage the timing belt with the teeth of the crankshaft sprocket and then pull the belt vertically upright on its right-hand run. Keep it taut and engage it with the teeth of the camshaft sprocket. Check that the positions of the sprockets have not altered.
12 Wind the belt round the camshaft sprocket around and under the tensioner and over the water pump sprocket (see photo 25.7).
13 Refit the lower half of the timing cover and tighten its retaining bolts to the specified torque. Refit the crankshaft pulley and tighten its bolt to the specified torque, using the same method as before to stop the crankshaft rotating (photo).
14 To adjust the belt tension, slacken the tensioner and move it towards the front of the car to apply an initial tension to the belt. Secure the tensioner in this position, then remove the flywheel ring gear locking device.
15 Rotate the crankshaft through two complete revolutions, then return to the TDC position (camshaft sprocket to cylinder head). Check that the crankshaft pulley notch is aligned with the TDC (0) mark on the lower half of the timing cover (photo).
16 Grasp the belt between thumb and forefinger at a point midway between the crankshaft and camshaft sprocket on the straight side of the belt. When the tension is correct it should just be possible to twist the belt through 90° at this point. Slacken the tensioner and using a large screwdriver as a lever, move it as necessary until the tension is correct. Tighten the tensioner bolts, rotate the camshaft to settle the belt, then recheck the tension. It will probably take two or three attempts to achieve success.
17 It must be emphasised that this is an approximate setting only and should be rechecked by a Ford dealer at the earliest opportunity.
18 Refit the starter motor (see Chapter 12, Section 12).
19 Refit the upper half of the timing cover and tighten its bolts to the specified torque.
20 Refit the alternator drivebelt and adjust its tension as described in Chapter 12, Section 7. Refit the drivebelt shield, as applicable.
21 Refit the air filter housing/trunking, as applicable, then reconnect the battery earth terminal.

26 Camshaft oil seal – renewal

1 Disconnect the battery earth terminal.
2 Release the timing belt from the camshaft sprocket, as described in the preceding Section.
3 Pass a bar through one of the holes in the camshaft sprocket to anchor the sprocket while the retaining bolt is unscrewed. Remove the sprocket. Note Woodruff key fitted to camshaft.
4 Using a suitable tool, hooked at its end, prise out the oil seal.
5 Apply a little oil to the lips of the new seal and the running face of the camshaft, and draw the seal into position using the sprocket bolt and a suitable distance piece.
6 Refit the sprocket using a **new** retaining bolt, having smeared the threads of the bolt with sealing compound. Tighten the bolt to the specified torque (see Specifications).
7 Refit and tension the timing belt, as described in the preceding Section.
8 Reconnect the battery earth terminal.

27 Camshaft – removal and refitting

1 Disconnect the battery earth terminal.
2 On 1.6 litre EFi models, remove the battery.
3 On carburettor and CFi equipped vehicles, disconnect the HT leads from the spark plugs and remove the distributor cap with HT leads attached. Remove the air filter housing (refer to Chapters 3 and 4).
4 On 1.6 litre EFi models, disconnect the HT leads from the spark plugs and the air filter trunking guides, and unclip them from the rocker

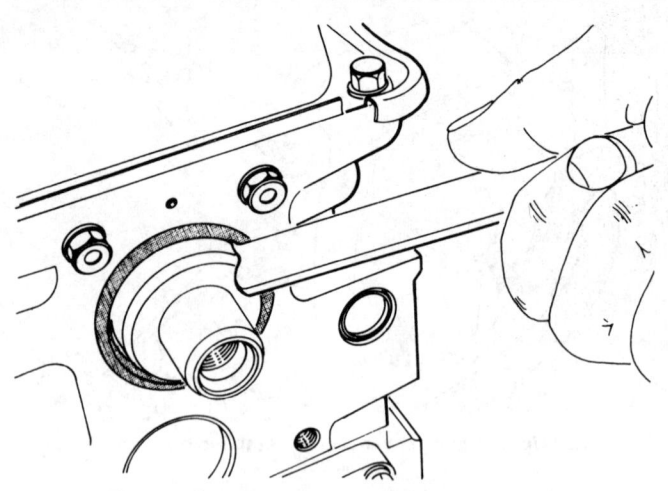

Fig. 1.26 Removing the camshaft oil seal (Sec 26)

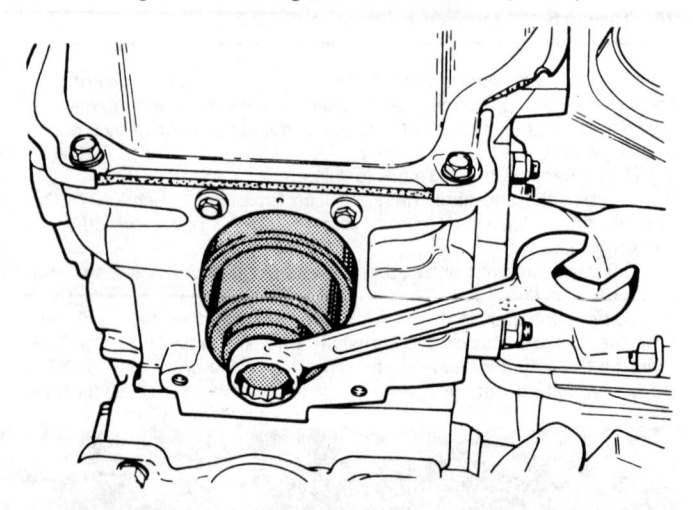

Fig. 1.27 Installing a new camshaft oil seal (Sec 26)

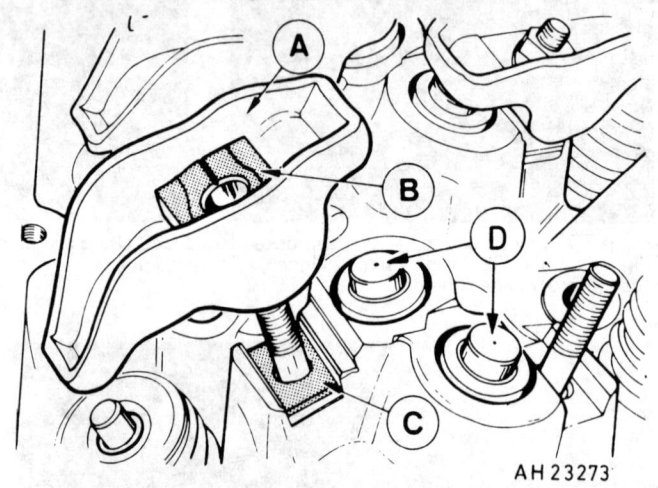

Fig. 1.28 Rocker arm and associated compounds (Sec 27)

A Rocker arm   C Spacer plate
B Rocker guide   D Hydraulic cam followers

cover. Remove the complete air filter housing/trunking, including the idle speed control valve (see Chapter 3).
5 Unscrew the two bolts securing the upper half of the timing cover and remove it.
6 Disconnect the breather hoses from the rocker cover.
7 Remove the windscreen washer reservoir (Chapter 12).

Chapter 1 Engine

27.20A Unscrewing the camshaft sprocket bolt

27.20B Removing the camshaft sprocket

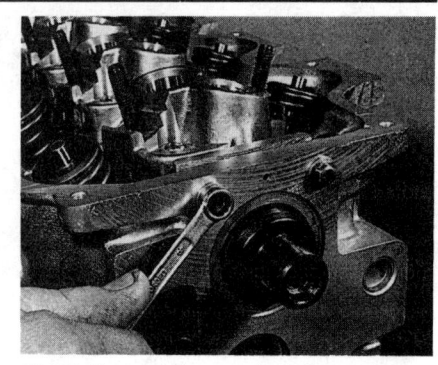

27.21A Unscrew the camshaft thrust plate bolts ...

27.21B ... and withdraw the thrust plate

27.22 Withdrawing the camshaft

8   Refer to Chapter 4 and remove the distributor (not 1.6 litre EFi).
9   On carburettor equipped vehicles, remove the fuel pump, its gasket and plunger (see Chapter 3).
10   Disconnect the choke cable (where applicable) then disconnect the throttle cable and remove it, complete with bracket, from the carburettor (carburettor equipped vehicles only – refer to Chapter 3).
11   On 1.6 litre EFi models, disconnect the throttle cable, then remove the throttle cable bracket from the throttle housing (refer to Chapter 3).
12   On 1.4 litre CFi models, disconnect the throttle cable complete with the bracket and the thrust rod and spring (refer to Chapter 3).
13   On 1.6 litre EFi models, remove the DIS ignition coil and its bracket from the cylinder head (see Chapter 4), then remove the plastic blanking plug underneath.
14   Remove the rocker cover, noting reinforcing plate fitment under bolts.
15   Unscrew the securing nuts and remove the rocker arms and guides. Keep the components in their originally installed sequence by marking them with a piece of numbered tape or by using a suitable sub-divided box.
16   Withdraw the hydraulic cam followers, again keeping them in their originally fitted sequence.
17   Remove the alternator drivebelt (refer to Chapter 12, Section 7).
18   Using a spanner on the crankshaft pulley bolt, turn the crankshaft until the TDC pointer on the camshaft sprocket aligns with the TDC mark on the cylinder head (see photo 25.4A).
19   Slacken the bolts on the timing belt tensioner, lever the tensioner against the tension of its coil spring (if fitted) and retighten the bolts. With the belt now slack, slip it from the camshaft sprocket.
20   Pass a rod or large screwdriver through one of the holes in the camshaft sprocket to lock it and unscrew the sprocket bolt. Remove the sprocket (photos). A new bolt will be needed upon reassembly. Note the Woodruff key location.
21   Extract the two bolts and pull out the camshaft thrustplate (photos).
22   Carefully withdraw the camshaft from the rear (transmission side) of the cylinder head (photo). Prise out the oil seal as described in the previous Section.

23   Refitting the camshaft is a reversal of removal, but observe the following points.
24   Oil the camshaft bearings, camshaft and thrust plate before refitting. As the camshaft is installed, take care not to damage the bearings. Refer to the Specifications for the thrust plate tightening torque.
25   A new oil seal must be fitted after the camshaft has been installed (see the previous Section).
26   Refit the camshaft sprocket using a **new** retaining bolt, having smeared the threads of the bolt with sealing compound. Tighten the bolt to the specified torque (see Specifications).
27   Fit and tension the timing belt, as described in Section 25.
28   Lubricate the hydraulic cam followers with hypoid oil before inserting them into their original bores.
29   Refit the rocker arms and guides in their original sequence, use **new** nuts and tighten to the specified torque. It is essential that before each rocker arm is installed and its nut tightened, the respective cam follower is positioned at its lowest point (in contact with cam base circle). Turn the camshaft (using the crankshaft pulley bolt) as necessary to achieve this.
30   Use a new rocker cover gasket, and to ensure that a good seal is made, check that its location groove is clear of oil, grease and any portions of the old gasket. When in position tighten the cover retaining screws to the specified torque setting, having ensured that the re-inforcing plates are fitted with the grooving facing upwards.
31   Refit the remainder of the components with reference to their relevant Chapters.

28   Cylinder head (1.4 litre/1.6 litre carburettor engines and 1.4 litre CFi) – removal and refitting

1   Disconnect the battery earth terminal.
2   Remove the air filter housing as described in Chapter 3.
3   Drain the cooling system in accordance with Chapter 2.
4   Disconnect the coolant hoses from the water pump, thermostat

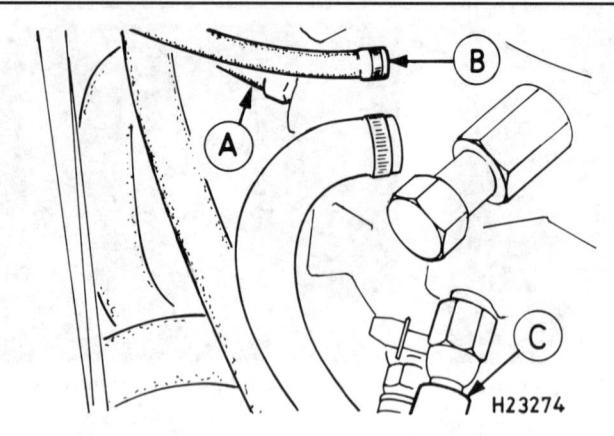

Fig. 1.29 Vacuum hoses/pipes to the rear of the CFi unit (Sec 28)

A   To brake servo
B   To MAP sensor
C   To canister purge solenoid

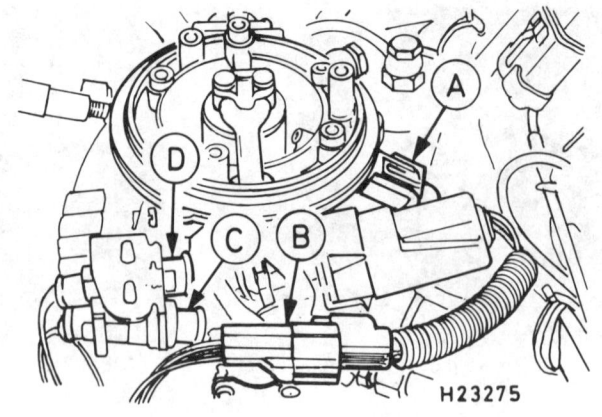

Fig. 1.30 Wiring connections to CFi unit (Sec 28)

A   Air charge temperature sensor
B   Throttle plate control motor
C   Throttle position sensor
D   Injector

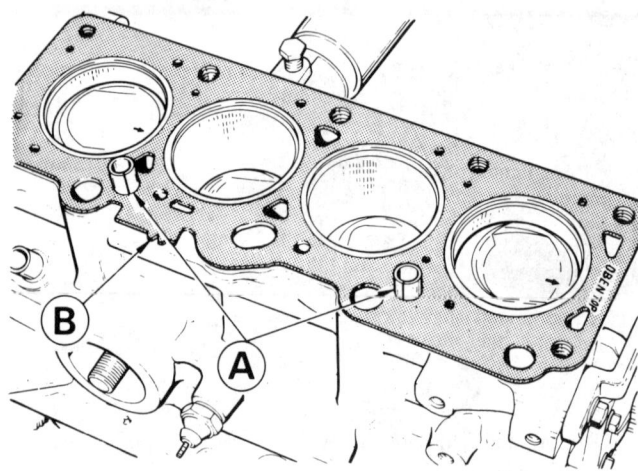

Fig. 1.31 Fit the cylinder head gasket (Sec 28)

A   Locating dowels
B   Gasket identification teeth

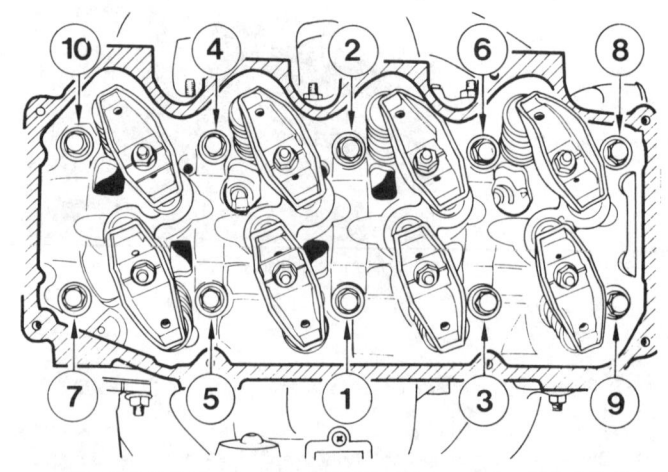

Fig. 1.32 Cylinder head bolt tightening sequence (Sec 28)

housing, automatic choke, inlet manifold and CFi unit as applicable.
5   Disconnect the choke cable (as applicable) then disconnect the throttle cable and remove it, complete with bracket, from the carburettor (carburettor equipped vehicles only – refer to Chapter 3).
6   On 1.4 litre CFi models, disconnect the throttle cable complete with the bracket and the thrust rod and spring (refer to Chapter 3).
7   On carburettor equipped vehicles, refer to Chapter 3 and disconnect the fuel feed pipe from the fuel pump, and the fuel return pipe from the carburettor (where fitted). Plug the fuel pipe(s) to prevent dirt entering and avoid spillage.
8   On 1.4 litre CFi models, disconnect the fuel feed and return pipes from the CFi unit (refer to Chapter 3). Plug the fuel pipes to prevent dirt entering, and avoid spillage (photo).
9   Disconnect the brake servo vacuum hose from the inlet manifold (see Chapter 8, Section 16).
10   Disconnect the remaining vacuum hoses from the inlet manifold, carburettor and CFi unit, as applicable, noting their locations for subsequent refitting.
11   Disconnect the wiring or multi-plugs from the temperature gauge sender unit, radiator cooling fan and its thermal switch, engine coolant temperature sensor (on inlet manifold) and distributor.
12   On carburettor equipped vehicles, disconnect the lead(s) from the carburettor, as applicable.
13   On 1.4 litre CFi models, disconnect the multi-plugs from the road-speed sensor (see Fig. 12.45, in Chapter 12), air-charge temperature sensor (on CFi unit), throttle plate control motor, throttle position sensor and injector (refer to Fig. 1.30).
14   Release the forward (downpipe) section of the exhaust from the exhaust manifold by undoing the three nuts securing it. Note that a **new** gasket will be needed upon reassembly. Suspend the exhaust system

28.8 CFi unit fuel feed pipe (A), and fuel return pipe (B)

using a length of strong wire to prevent excessive strain on its mountings. On 1.4 litre CFi models, it will be necessary to disconnect the Lambda probe multi-plug (see Chapter 3).
15   Remove the alternator drivebelt (see Chapter 12, Section 7).

Chapter 1 Engine 61

16 Unbolt and remove the upper half of the timing cover.
17 Slacken the timing belt tensioner bolts, lever the tensioner to one side against the pressure of the coil spring (if fitted) and retighten the bolts.
18 With the timing belt now slack, slip it from the camshaft sprocket.
19 Disconnect the leads from the spark plugs and unscrew and remove the spark plugs.
20 Disconnect the rocker cover breather hoses then, having noted reinforcing plate fitment, unbolt and remove the rocker cover. Note that a new rocker cover gasket will be needed upon reassembly.
21 Before it is removed, the cylinder head **must** have cooled down to room temperature (approximately 20°C/68°F).
22 Unscrew the cylinder head bolts progressively and in the reverse sequence to that shown in Fig. 1.32. Discard the bolts as new ones **must** be used upon reassembly.
23 Remove the cylinder head complete with manifolds. Use the manifolds if necessary as levers to rock the head from the block. Do not attempt to tap the head sideways off the block as it is located on dowels, and do not attempt to lever between the head and the block or damage will result.
24 To dismantle the cylinder head, refer to Section 40.
25 Before installing the cylinder head, make sure that the mating surfaces of head and block are perfectly clean, with the head locating dowels in position. Clean the bolt holes free from oil. In extreme cases it is possible for oil left in the holes to crack the block due to hydraulic pressure.
26 Turn the crankshaft to position No 1 piston about 20 mm (0.8 in) before it reaches TDC.
27 Place a **new** gasket on the cylinder block and then locate the cylinder head on its dowels. The upper surface of the gasket is marked OBEN-TOP.
28 Install and tighten the **new** cylinder head bolts, tightening them in four stages (see Specifications). After the first two stages, the bolt heads should be marked with a spot of quick-drying paint so that the paint spots all face the same direction. Now tighten the bolts (Stage 3) through 90° (quarter turn) followed by a further 90° (Stage 4). Tighten the bolts at each stage only in the sequence shown in Fig. 1.32 before going on to the next stage. If all the bolts have been tightened equally, the paint spots should now all be pointing in the same direction. The bolts must **not** be retorqued.
29 Turn the camshaft sprocket so that its TDC pointer aligns with the TDC mark on the cylinder head, *then* rotate the crankshaft so that its pulley notch lines up with the TDC (0) mark on the lower half of the timing cover (taking the shortest route) – **not** *vice-versa*. Fit and tension the timing belt (refer to Section 25).
30 Refitting and reconnection of all other components is a reversal of the dismantling, with reference to the relevant Chapter.
31 On completion refill the cooling system as described in Chapter 2.

29 Cylinder head (1.6 litre EFi) – removal and refitting

1 Disconnect the battery earth terminal.
2 Disconnect the HT leads from the spark plugs and the air filter trunking guides, then unclip them from the rocker cover before disconnecting at the DIS ignition coil (note their fitment). Remove the complete air filter housing/trunking, including the idle speed control valve (see Chapter 3) (photo).
3 Drain the cooling system in accordance with Chapter 2.
4 Disconnect the coolant hoses from the thermostat housing, noting their fitted positions.
5 Disconnect the vacuum pipe to the MAP sensor from the upper part of the inlet manifold (see Fig. 1.33).
6 Disconnect the crankcase breather hose and the oil trap hose from the rocker cover, then disconnect the oil trap vacuum hose at the T-piece (refer to Fig. 1.33).
7 Disconnect the brake servo vacuum hose from the inlet manifold (see Chapter 8, Section 16).
8 Raise the vehicle and support it securely using axle stands.
9 Undo the three nuts securing the exhaust forward (downpipe) section to the exhaust manifold. Note that a **new** gasket will be needed upon reassembly. Suspend the exhaust system using a length of strong wire to prevent excessive strain on its mountings.
10 Remove the alternator drivebelt (see Chapter 12, Section 7).

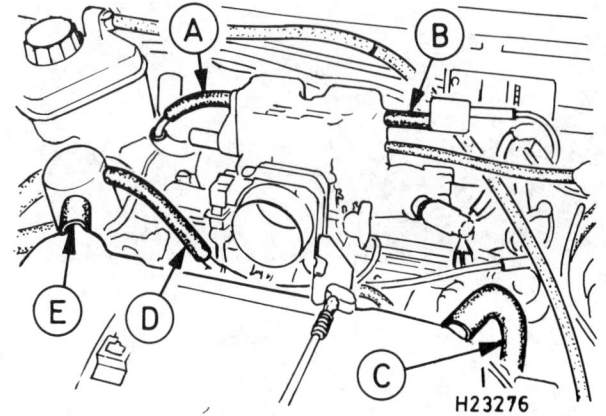

Fig. 1.33 Vacuum/breather hoses (1.6 litre EFi) (Sec 29)

A To canister purge solenoid (vehicles with catalytic converter only)
B To MAP sensor
C Crankcase breather hose
D Oil trap vacuum hose
E Oil trap hose

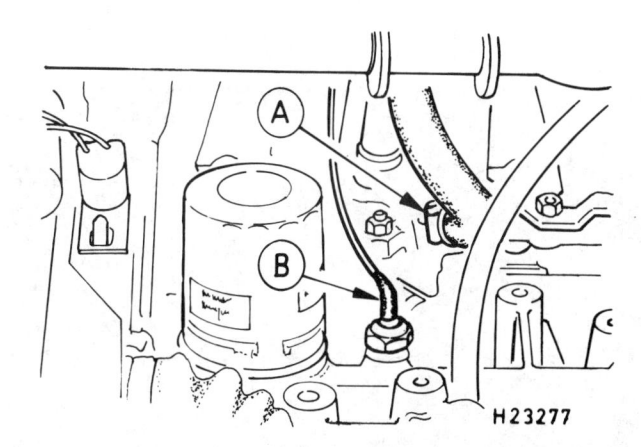

Fig. 1.34 Heater hose attached to lower part of inlet manifold (A) and oil pressure switch connection (B) (Sec 29)

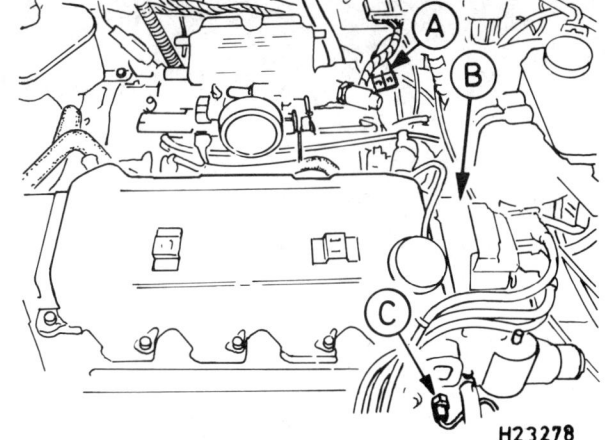

Fig. 1.35 General view of connections (1.6 litre EFi) (Sec 29)

A Main (injector) wiring loom
B DIS ignition coil
C Temperature gauge sender unit

11 Disconnect the heater hoses from the lower part of the inlet manifold (Fig. 1.34).
12 Remove the axle stands and lower the vehicle to the ground.
13 Unclip the throttle cable from its bracket on the throttle housing and disconnect it, then unscrew the bolt and nuts securing the bracket and remove it (photo).
14 Disconnect the multi-plugs of the temperature gauge sender unit,

62   Chapter 1   Engine

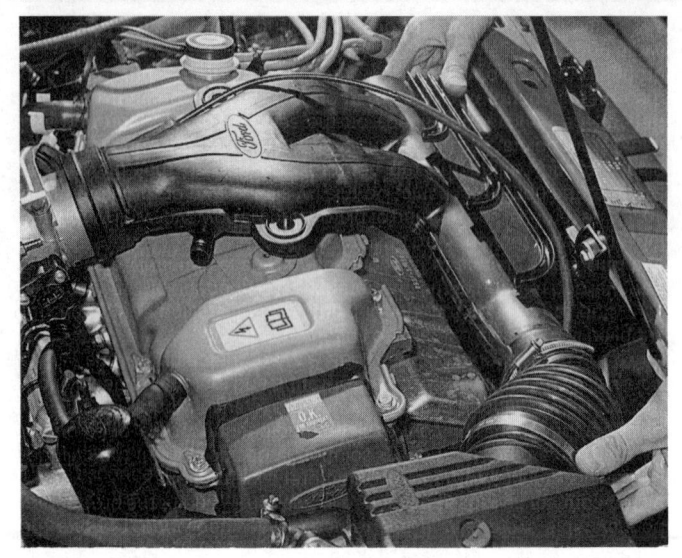

29.2 Removing air filter trunking (XR2i model)

29.13 Disconnect the throttle cable and remove its bracket (XR2i model)

A Throttle cable to bracket securing clip
B Bracket retaining bolt
C Bracket retaining nuts

radiator cooling fan and its thermal switch, main wiring loom and DIS ignition coil (see Fig. 1.35). Additionally, detach the radio suppression earth lead (as applicable).
15 Disconnect the fuel feed pipe from the fuel distributor pipe (fuel rail), and the fuel return pipe from the fuel pressure regulator (see Chapter 3). Note that the fuel system is pressurised.
16 Unbolt and remove the upper half of the timing cover.
17 Slacken the timing belt tensioner bolts, lever the tensioner to one side against the pressure of the coil spring (if fitted) and retighten the bolts.
18 With the timing belt now slack, slip it from the crankshaft sprocket.
19 Remove the spark plugs.
20 Unbolt and remove the rocker cover, noting that the reinforcing plates are fitted with their grooving facing upwards. A **new** rocker cover gasket will be needed upon subsequent reassembly.
21 The remainder of the removal and refitting sequence is the same as described in Section 28, paragraph 21 onwards.

30 Cylinder head and pistons – decarbonising

1 With the cylinder head removed as described in Section 28 to 29 the carbon deposits should be removed from the combustion surfaces using a blunt scraper. Take great care as the head is of light alloy construction and avoid the use of a rotary (power-driven) wire brush.
2 Where a more thorough job is to be carried out, the cylinder head should be dismantled as described in Section 40 so that the valves may be ground-in, and the ports and combustion spaces cleaned and blown out after the manifolds have been removed.
3 Before grinding-in a valve, remove the carbon and deposits completely from its head and stem. With an inlet valve this is usually quite easy, simply a case of scraping off the soft carbon with a blunt knife and finishing with a wire brush. With an exhaust valve, the deposits are very much harder and those on the valve head may need a rub on coarse emery cloth to remove them. An old woodworking chisel is a useful tool to remove the worst of the valve head deposits.
4 Make sure that the valve heads are really clean, otherwise the rubber suction cup grinding tool will not stick during the grinding-in operations.
5 Before starting to grind-in a valve, support the cylinder head so that there is sufficient clearance under it for the valve stem to project fully without being obstructed, otherwise the valve will not seat properly during grinding.
6 Take the first valve and apply a little coarse grinding paste to the bevelled edge of the valve head. Insert the valve into its guide and apply the suction grinding tool to its head. Rotate the tool between the palms of the hands in a back-and-forth rotary movement until the gritty action of the grinding-in process disappears. Repeat the operation with fine paste and then wipe away all trace of grinding paste and examine the seat and bevelled edge of the valve. A matt silver mating band should be observed on both components, without any sign of black spots. If some spots do remain, repeat the grinding-in process until they have disappeared. A drop or two of paraffin applied to the contact surfaces will speed the grinding process, but do not allow any paste to run down into the valve guide. On completion, wipe away every trace of grinding paste using a paraffin-moistened cloth.
7 Repeat the operations on the remaining valves, taking care not to mix up their originally fitted sequence.
8 An important part of the decarbonising operation is to remove the carbon deposits from the piston crowns. To do this (engine in vehicle), turn the crankshaft so that two pistons are at the top of their stroke and press some grease between the pistons and the cylinder walls. This will prevent carbon particles falling down into the piston ring grooves. Plug the other two bores with rag.
9 Cover the oilways and coolant passages with masking tape and then using a blunt scraper, remove all the carbon from the piston crowns. Take great care not to score the soft alloy of the crown or the surface of the cylinder bore.
10 Rotate the crankshaft to bring the other two pistons to TDC and repeat the operations.
11 Wipe away the circles of grease and carbon from the cylinder bores.
12 Clean the top surfaces of the cylinder block by careful scraping.

31 Crankshaft front oil seal – renewal

1 Disconnect the battery earth terminal.
2 On 1.6 litre EFi models, remove the complete air filter housing (see Chapter 3).
3 Slacken the right-hand front wheelnuts then, with the handbrake fully applied, raise the front of the vehicle and support it securely on axle stands. Remove the right-hand front wheel.
4 Remove the right-hand front wheelarch liner (see Chapter 11, Section 65).
5 Remove the timing belt (see Section 25).
6 Withdraw the crankshaft sprocket. If it is tight you will need to use a suitable extractor. Before resorting to this, try levering the sprocket free using screwdrivers.
7 Remove the dished (thrust) washer from the crankshaft, noting that the concave side is against the oil seal.
8 Using a suitably hooked tool, prise out the oil seal from the oil pump housing.

Chapter 1 Engine 63

9 Lightly oil the crankshaft journal and the lips of the new seal, then draw the seal into position using the pulley bolt and a suitable installer. An installer may be fabricated from suitably sized tubular sections. **Never** attempt to tap the seal into position, and take great care to avoid damaging its lip.
10 Fit the thrust washer (concave side to oil seal), then grease the crankshaft journal in preparation for refitting the crankshaft sprocket.
11 Draw the crankshaft sprocket into position using the crankshaft pulley and its bolt, having aligned the Woodruff key locations. Note that the crankshaft sprocket flange faces outwards. Remove the crankshaft pulley.
12 Carry out the operations detailed in Section 25, paragraphs 10 to 19.
13 Refit the right-hand front wheelarch liner (Chapter 11, Section 65).
14 Refit the right-hand front wheel but do not **fully** tighten the wheelnuts until the vehicle is on the ground.
15 Refit the alternator drivebelt and adjust its tension as described in Chapter 12, Section 7. Refit the drivebelt shield, as applicable.
16 Lower the vehicle to the ground and tighten the wheelnuts to their specified torque (refer to the Specifications in Chapter 10).
17 Refit the air filter housing, as applicable.
18 Reconnect the battery earth terminal.

32 Sump – removal and refitting

1 Disconnect the battery earth terminal.
2 Apply the handbrake, raise the front of the vehicle and support it securely on axle stands.
3 On XR2i models, remove the front suspension crossmember (see Chapter 10).
4 Remove the forward (downpipe) section of the exhaust system (all except 1.4 litre CFi). On 1.4 litre CFi models, disconnect the Lambda probe multi-plug then unbolt the exhaust from the manifold, and at the rear of the catalytic converter (see Chapter 3). Suspend the forward end of the remaining system using a length of strong wire to prevent strain on the remaining mountings when the forward mounting bracket is unbolted later. Note that a new gasket will be needed upon reassembly.
5 Drain the engine oil (refer to Part A, Section 2).
6 Remove the alternator drivebelt shield (where fitted) by releasing its three bayonet type fasteners.
7 Refer to Chapter 12, Section 12 and remove the starter motor. Depending on the clearance available, it may also be necessary to remove the alternator (Chapter 12, Section 9).
8 Undo the nut and bolt(s) securing the gearchange mechanism stabiliser bar/exhaust forward mounting bracket (manual transmission vehicles) and ease it out of the way (refer to photo 37.24).
9 Remove the lower engine adaptor plate for access to the rearmost sump bolts.
10 Unscrew the sump bolts progressively and remove them. Remove the sump. If it sticks, prise it gently with a screwdriver but do not use excessive leverage. If it is very tight, cut around the gasket joint using a sharp knife.
11 Prior to refitting, remove the old rubber oil seal carrier and oil pump casing seals and the old main gasket. Clean the mating surfaces of the sump and the cylinder block and check that the mating faces are not distorted. Clean out any sludge present in the sump. Fit the oil drain plug using a **new** seal, and tighten to its specified torque.
12 Smear the cylinder block with sealing compound at the junction of the oil seal carrier and oil pump casing – four locations (Fig. 1.36).
13 Insert a **new** rubber seal into the groove in the rear oil seal carrier and the oil pump casing. Screw ten M6 studs into the cylinder block (represented by circled numbers in Fig. 1.37) as an aid to refitting.
14 Fit a **new** gasket over the studs. Fit the sump (making sure that the spacing pimples sit in the gasket) and insert its bolts into the available holes, tightening them finger-tight only at this stage. Replace the studs with the remaining sump bolts, then tighten all sump bolts to the specified torque in two stages, following the numerical sequence shown in Fig. 1.37.
15 Refit the lower engine adaptor plate, and gearchange mechanism stabiliser bar/exhaust forward mounting bracket.
16 Refit the starter motor and (if removed) and alternator (see Chapter 12).
17 Refit the alternator drivebelt shield (if removed) after the alternator drivebelt is correctly tensioned.

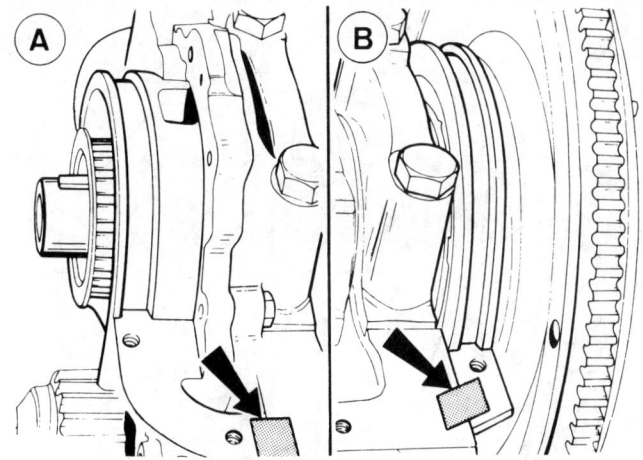

Fig. 1.36 Apply sealing compound before fitting sump gasket (arrowed) (Sec 32)

A At the cylinder block/oil pump casing junction
B At the cylinder block/rear oil seal carrier junction

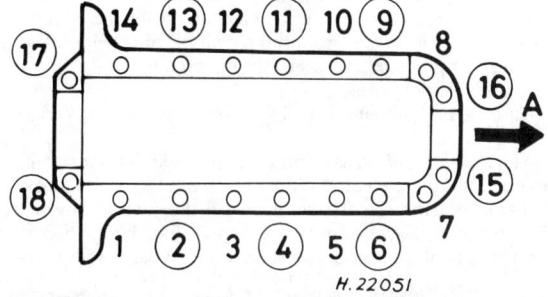

Fig. 1.37 Sump bolt tightening sequence (with one-piece gasket). Front of engine (A) arrowed (Sec 32)

18 Refit and secure the exhaust system by reversing the method of removal and reconnect the Lambda probe multi-plug (1.4 litre CFi). Remove the wire used to suspend the exhaust system. The gasket should be renewed as a matter of course (see Chapter 3).
19 On XR2i models, refit the front suspension crossmember (see Chapter 10, Section 6).
20 Lower the vehicle to the ground and refill with fresh engine oil of the specified type (refer to Part A, Section 2, paragraph 3).
21 Reconnect the battery earth terminal.
22 Run the engine up to normal operating temperature and check that no leaks are evident.

33 Piston/connecting rod assemblies – removal and refitting

1 Remove the sump, as described in the preceding Section, and the cylinder head, as described in Section 28 or 29 as applicable.
2 Check that the connecting rod and cap have adjacent numbers at their big-end to indicate their position in the cylinder block (No 1 nearest timing cover end of engine). (Fig. 1.38).
3 Bring the first piston to the lowest point of its throw by turning the crankshaft pulley bolt and then check if there is a wear ring at the top of the bore. If there is, it should be removed using a scraper, but do not damage the cylinder bore.
4 Unscrew the big-end bolts and remove them.
5 Tap off the big-end cap. If the bearing shell is to be used again, make sure that it is retained with the cap. Note the two cap positioning roll pins.
6 Push the piston/rod out of the top of the block, again keeping the bearing shell with the rod if the shell is to be used again.
7 Repeat the removal operations on the remaining piston/rod assemblies.

Chapter 1 Engine

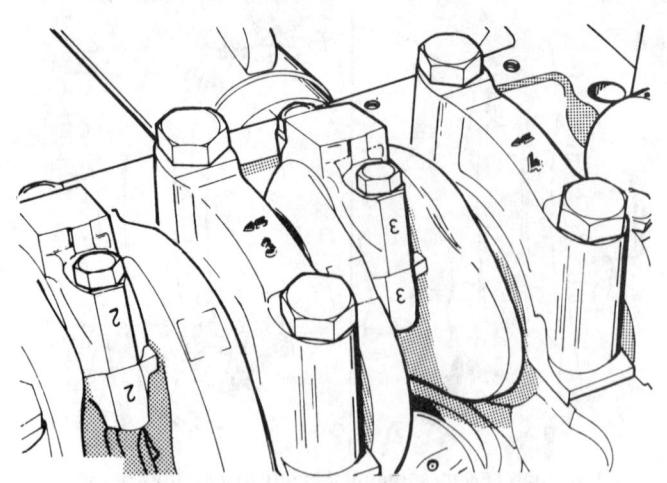

Fig. 1.38 Connecting rod, big-end cap and main bearing cap markings (Sec 33)

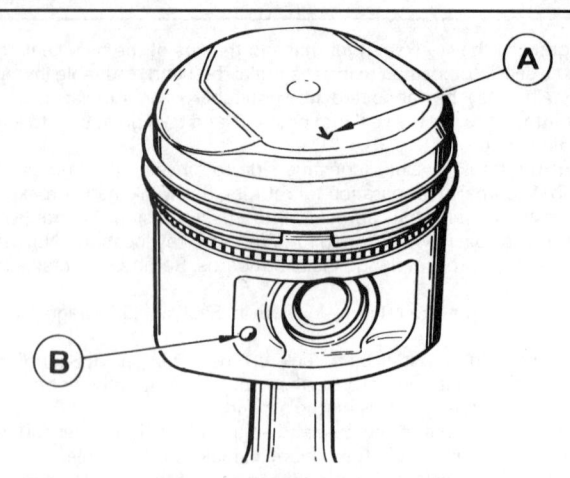

Fig. 1.39 Arrow (A) or cast nipple (B) must face the timing belt end of the engine when installed (Sec 33)

8 Dismantling a piston/connecting rod is covered in Section 40.
9 To refit a piston/rod assembly, have all the piston ring gaps staggered at uniform intervals around the piston. Oil the rings and apply a piston ring compressor. Compress the piston rings.
10 Oil the cylinder bores.
11 Wipe clean the bearing shell seat in the connecting rod and insert the shell (photo).
12 Insert the piston/rod assembly into the cylinder bore until the base piston ring compressor stands squarely on the top of the block.
13 Check that the directional arrow on the piston crown faces towards the timing cover end of the engine (Fig. 1.39), then apply the wooden handle of a hammer to the piston crown. Strike the head of the hammer sharply to drive the piston into the cylinder bore and release the ring compressor (photo).
14 Oil the crankpin and draw the connecting rod down to engage with the crankshaft. Make sure the bearing shell is still in position.
15 Wipe the bearing shell seat in the big-end cap clean and insert the bearing shell (photo).
16 Lubricate the bearing shell with oil then fit the cap, aligning the numbers on the cap and the rod, screw in the bolts and tighten them to the specified torque setting (photos). Check that the connecting rod has sufficient axial play when fitted.
17 Repeat the operations on the remaining pistons/connecting rods.
18 Refit the sump (Section 32) and the cylinder head (Section 28 or 29). Refill the engine with oil and coolant.

33.11 Fitting the bearing shell to the connecting rod

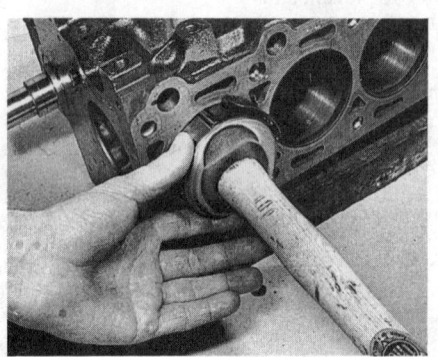

33.13 Fitting a piston/connecting rod assembly

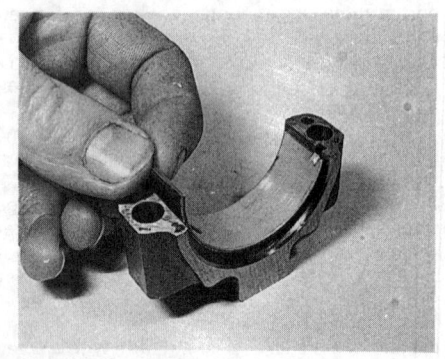

33.15 Fitting the bearing shell to the big-end cap

33.16A Fitting the big-end cap to the connecting rod

33.16B Tightening the big-end cap bolts to the specified torque

Chapter 1 Engine 65

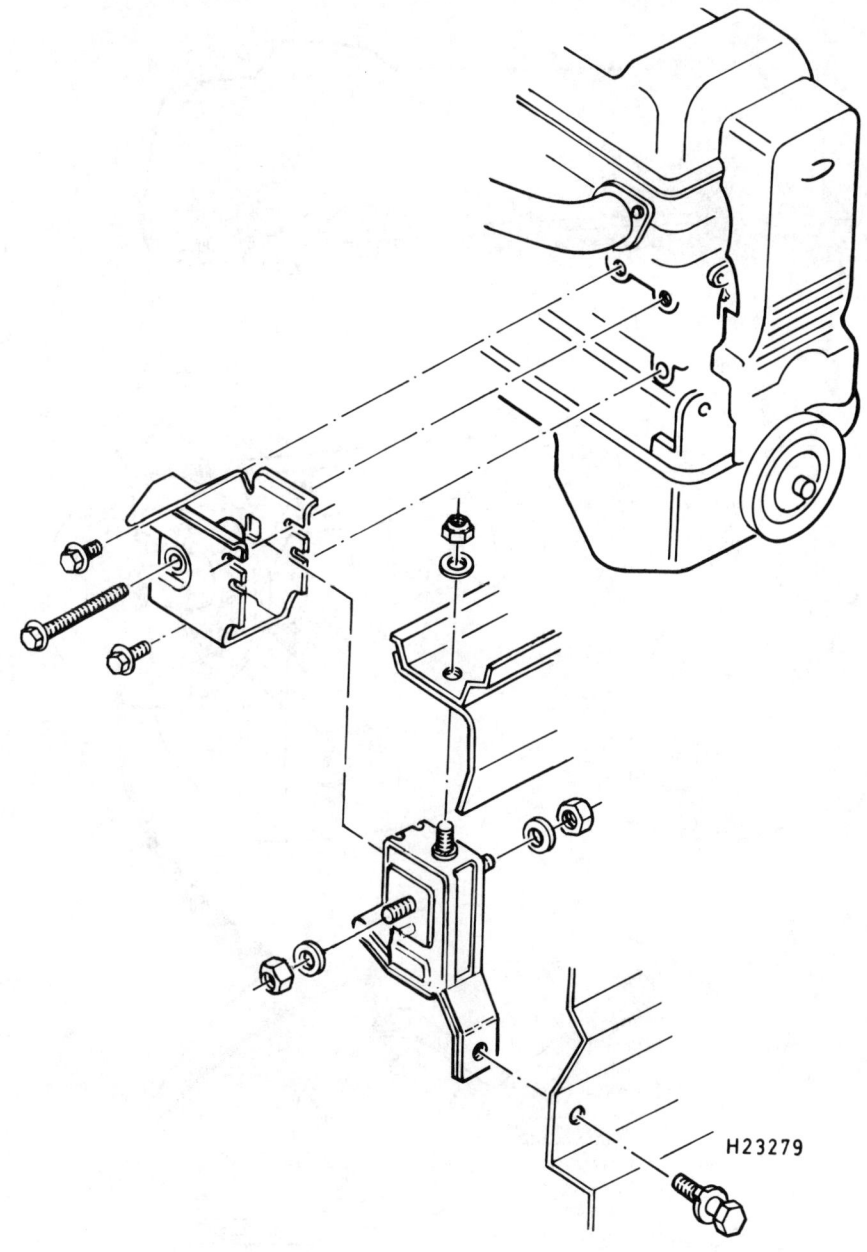

Fig. 1.40 Exploded view of right-hand engine mounting (CVH engine) (Sec 34)

34 Engine/transmission mountings – removal and refitting

1 The removal and refitting method for the transmission bearer and mountings is covered in Part A, Section 13. On XR2i models, it will be necessary to remove the front suspension crossmember first (refer to Chapter 10).
2 The engine right-hand mounting is significantly different to that fitted to HCS engines, in that it is a two-piece bracket, and its removal and refitting procedure is detailed below.
3 First of all support the engine/transmission assembly as described in Part A, Section 13, then unscrew and remove the two nuts securing the two halves of the mounting bracket assembly.
4 Unscrew and remove the three bolts securing the engine bracket section to the cylinder block.
5 Unscrew and remove the mounting retaining nut and washer from their location near the suspension strut (see photo 13.4).
6 Unscrew and remove the mounting side bolt from under the right-hand wheel arch (see photo 13.3).
7 Refitting is a reversal of the removal procedure. Make sure when refitting engine/transmission mountings, that any washers and plates removed during the dismantling process are refitted in their original sequence. Do not fully tighten any mounting bolts until they are all located. As the mounting bolts and nuts are tightened, check that the mounting rubbers do not twist.

35 Lubrication system – description

1 The oil pump draws oil from the sump through a pick-up pipe and then supplies pressurised oil through an oilway on the right-hand side of the engine into a full-flow oil filter. A pressure relief valve is incorporated inside the pump casing.
2 Filtered oil passes out of the filter casing through the central threaded mounting stud into the main oil gallery.
3 Oil from the main gallery lubricates the main bearings, and the big-end bearings are lubricated from oilways in the crankshaft.
4 The connecting rods have an oil hole in the big-end on the side towards the exhaust manifold. Oil is ejected from this hole into the gudgeon pins and cylinder bores.

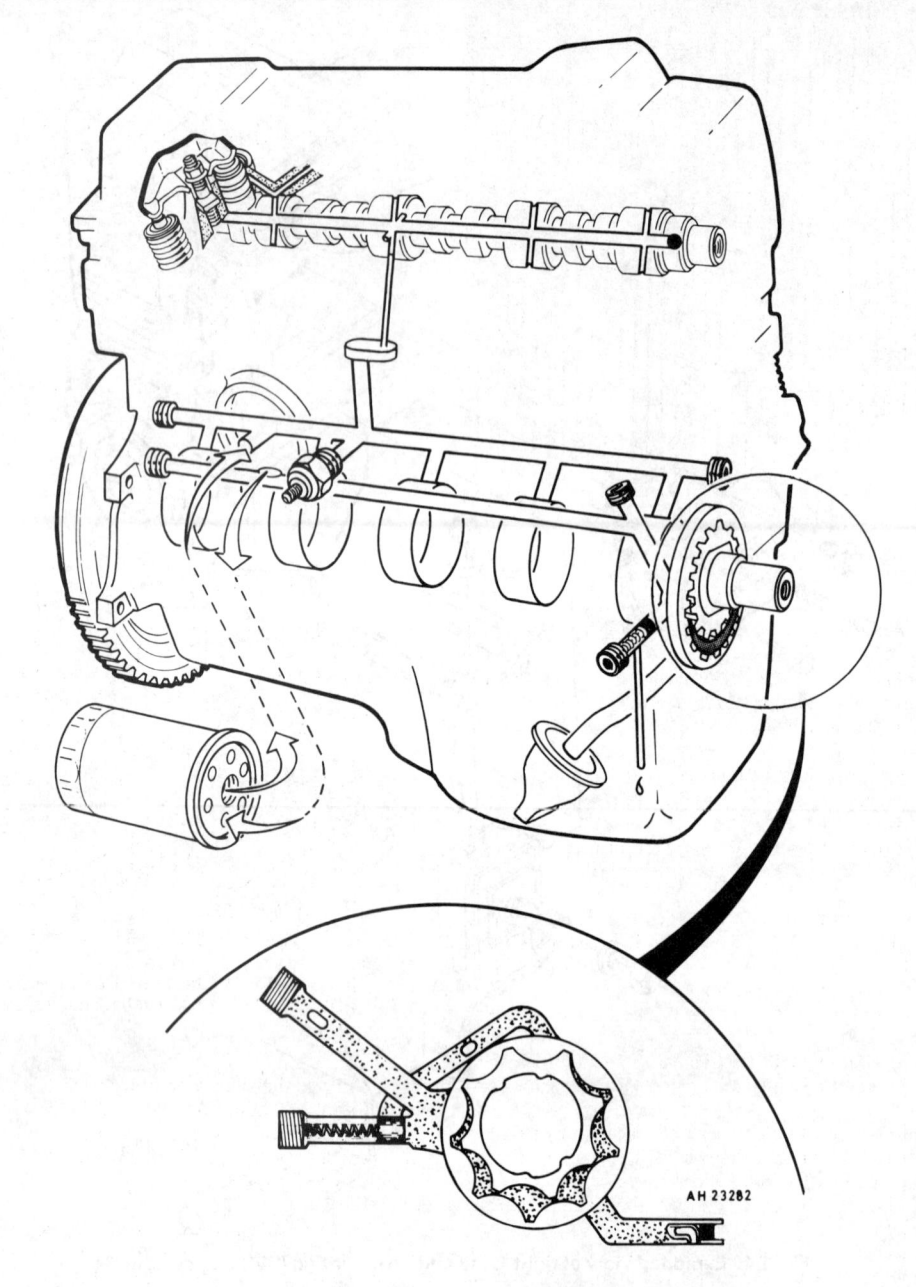

Fig. 1.41 Engine lubrication circuit (Sec 35)

5 The oil pressure warning switch is located next to the oil filter and connected by an internal passage to the main oil gallery. Oil from this passage is supplied to the centre camshaft bearing.
6 Oil is provided to the other camshaft bearings through a longitudinal drilling within the camshaft.
7 The hydraulic cam followers (tappets) are supplied with oil through the grooves in the camshaft bearing journals and oilways in the cylinder head.
8 The contact face of the rocker arm is lubricated from the ports in the tappet guides, while the end faces of the valve stems are splash lubricated.
9 For some markets, an engine oil cooler is used.

36 Engine – method of removal

The engine is removed complete with the transmission in a downward direction, and then withdrawn from under the front of the vehicle.

37 Engine/transmission assembly (1.4 litre/1.6 litre carburettor engines and 1.4 litre CFi) – removal, separation, reconnection and refitting

1 Disconnect the battery earth terminal.
2 Refer to Chapter 11 and remove the bonnet.
3 Engage the transmission in fourth gear on five-speed manual transmissions to ensure correct adjustment of the gear shift linkage upon subsequent refitting. On CTX automatic transmission equipped vehicles, select position 'P'.
4 Refer to Chapter 3 and remove the air filter housing.
5 Refer to Chapter 2 and drain the cooling system.
6 Disconnect the coolant hoses from the water pump, thermostat housing, inlet manifold, automatic choke and CFi unit as applicable. Additionally, on CTX automatic transmission equipped vehicles, disconnect the automatic transmission fluid cooling pipe connections from the transmission, fitting blanking plugs to prevent excessive fluid loss and avoid dirt ingress (photo).

Chapter 1 Engine

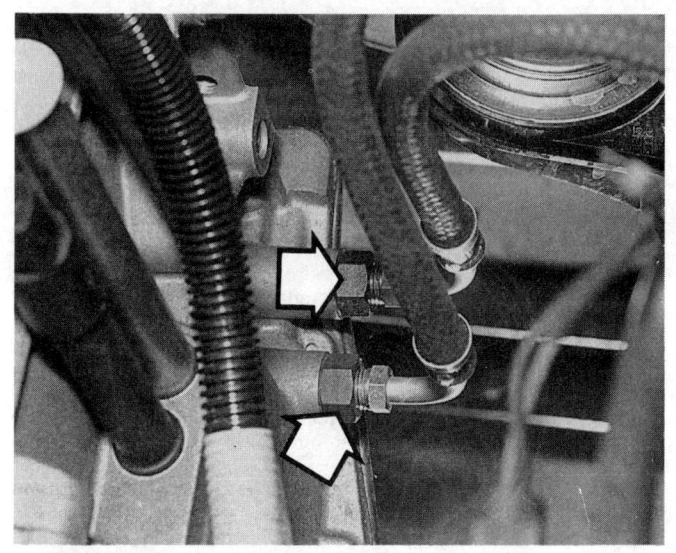

37.6 CTX automatic transmission fluid cooling pipe connections on the transmission casing (arrowed)

7 Disconnect the choke cable (as applicable) then disconnect the throttle cable and remove it, complete with bracket, from the carburettor (refer to Chapter 3). Additionally, on vehicles fitted with CTX automatic transmission, the accelerator pedal cable must be disconnected from the cable bracket assembly on the transmission (Fig 1.42) – tie it out of the way.
8 On 1.4 litre CFi models, disconnect the throttle cable complete with the bracket and the thrust rod and spring (refer to Chapter 3).
9 On carburettor equipped vehicles, refer to Chapter 3 and disconnect the fuel feed pipe from the fuel pump, and the fuel return pipe from the carburettor (where fitted). Plug the fuel pipe(s) to prevent dirt entering and avoid spillage.
10 On 1.4 litre CFi models, disconnect the fuel feed and return pipes from the CFi unit (refer to Chapter 3). Plug the fuel pipes to prevent dirt entering and avoid spillage.
11 Disconnect the brake servo vacuum hose from the inlet manifold (see Chapter 8, Section 16), and the breather hoses from the rocker cover.
12 Disconnect the remaining vacuum hoses from the inlet manifold, carburettor, and CFi unit, as applicable, noting their locations for subsequent refitting.
13 Disconnect the wiring or multi-plugs from the temperature gauge sender unit, radiator cooling fan and its thermal switch, engine coolant temperature sensor (on the inlet manifold) and distributor.
14 On carburettor equipped vehicles, disconnect the lead(s) from the carburettor, as applicable.
15 On 1.4 litre CFi models, disconnect the multi-plugs from the road speed sensor, air charge temperature sensor (on CFi unit), throttle plate control motor, throttle position sensor and injector (refer to Fig. 1.30).
16 Unscrew the union nut at the transmission end of the speedometer drive cable, and withdraw the cable. Tie it out of the way.
17 Disconnect the clutch cable from the clutch release lever on the transmission (manual transmission only), having supported the clutch pedal.
18 On vehicles fitted with the anti-lock braking system, refer to Chapter 8 and release the left-hand modulator from its mounting bracket, *without disconnecting the rigid brake pipes or return hose*. Tie the modulator securely to the bulkhead.
19 Raise the vehicle and support it securely using axle stands.
20 On 1.4 litre CFi models, remove the Lambda probe multi-plug from its bracket and disconnect it. Unscrew the three nuts securing the exhaust forward (downpipe) section to the exhaust manifold and disconnect the exhaust system to the **rear** of the catalytic converter. Release the downpipe/catalytic converter section from its mounting and remove it. Suspend the free end of the remaining system to prevent strain on its remaining mountings. Note that a new gasket will be required upon reassembly.
21 On carburettor equipped vehicles, disconnect the exhaust forward (downpipe) section, release the whole system from its mountings and remove it. Note that a new gasket will be needed upon reassembly.

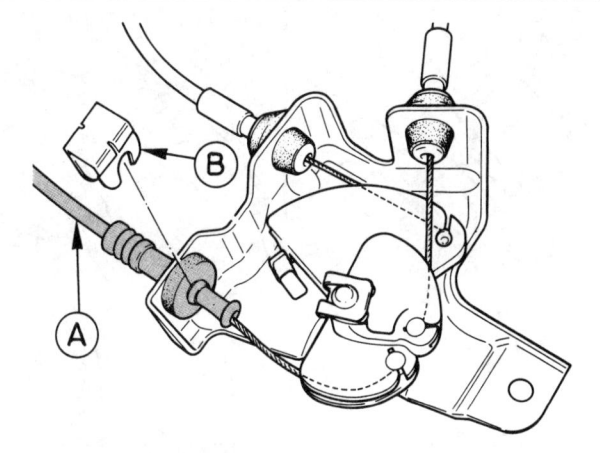

Fig. 1.42 CTX cable bracket assembly (Sec 37)

A Accelerator pedal cable   B Retaining clip

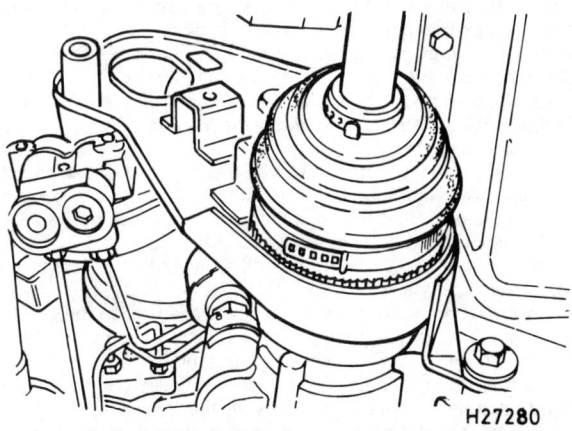

Fig. 1.43 Undo the bolts securing the right-hand modulator mounting bracket (anti-lock braking system equipped vehicles) (Sec 37)

22 Remove the alternator and starter motor and disconnect all earth leads (refer to Chapter 12).
23 Disconnect the multi-plugs from the reversing light switch and the oil pressure switch. Additionally, on CTX automatic transmission equipped vehicles, disconnect the starter inhibitor switch (refer to Chapter 12).
24 On manual transmission vehicles, detach the gearchange mechanism stabiliser bar/exhaust forward mounting bracket, and remove the lower engine adaptor plate located behind it. Disconnect the stabiliser bar and the gearchange mechanism shift rod from the transmission and suspend them out of the way. The shift rod is removed by undoing its clamp bolt and withdrawing. Keep a careful note of any washers removed for subsequent reassembly (photo).
25 On CTX automatic transmission equipped vehicles, refer to Chapter 6 and disconnect the selector cable from the lever on the transmission selector shaft, and the transmission casing. Remove the lower engine adaptor plate (flywheel cover).
26 Refer to Chapter 10 and disconnect the lower suspension arm balljoint from the spindle carrier, and the upper end of the anti-roll bar link from the suspension strut (if fitted), on both sides of the vehicle.
27 Refer to Chapter 9 and disconnect the track rod end from the steering arm on the spindle carrier, on both sides of the vehicle.
28 On vehicles fitted with the anti-lock braking system, refer to Chapter 8 and release the right-hand modulator from its mounting bracket without disconnecting the rigid brake pipes or return hose. Tie the modulator securely to the bulkhead. Additionally, undo the three bolts securing the modulator bracket (Fig. 1.43).
29 Disconnect the left-hand driveshaft assembly from the transmission. Locate a large lever between the transmission casing and the inner CV joint then, with an assistant applying pressure to the roadwheel (away from the centre of the vehicle), strike the lever sharply with

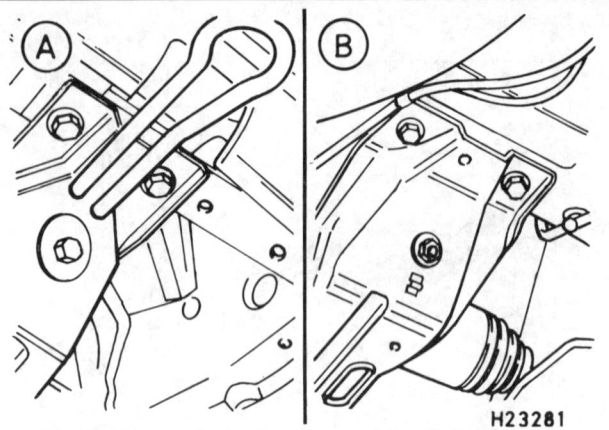

Chapter 1 Engine

Fig. 1.44 Transmission bearing retaining bolts (Sec 37)

A Front
B Rear

hand. Insert a suitable plug (an old inner CV joint is ideal) to prevent excessive transmission oil loss, and to prevent the differential from rotating when the right-hand driveshaft is removed. Suspend the driveshaft assembly to avoid straining the CV joints; the inner joint must not be bent more than 20°, and the outer joint 40°.

30 Repeat the above procedure in paragraphs 28 and 29 on the right hand driveshaft assembly and remove the previously-released modulator mounting bracket.

31 Connect a suitable lifting hoist to the engine using strong chains and the lifting eyes provided.

32 Take the weight of the engine/transmission assembly so that the tension is relieved from the mountings.

33 Refer to Section 34 and detach the right-hand engine mounting from the vehicle body.

34 Undo the four bolts securing the transmission bearer to the underside of the vehicle body (Fig. 1.44). The transmission bearer is removed with the engine/transmission assembly.

35 Carefully lower the engine/transmission assembly, ensuring that nothing becomes snagged, then detach the hoist and withdraw the assembly from under the vehicle. To ease the withdrawal operation, lower the assembly onto a crawler board or a sheet of substantial chipboard placed on rollers or lengths of pipe.

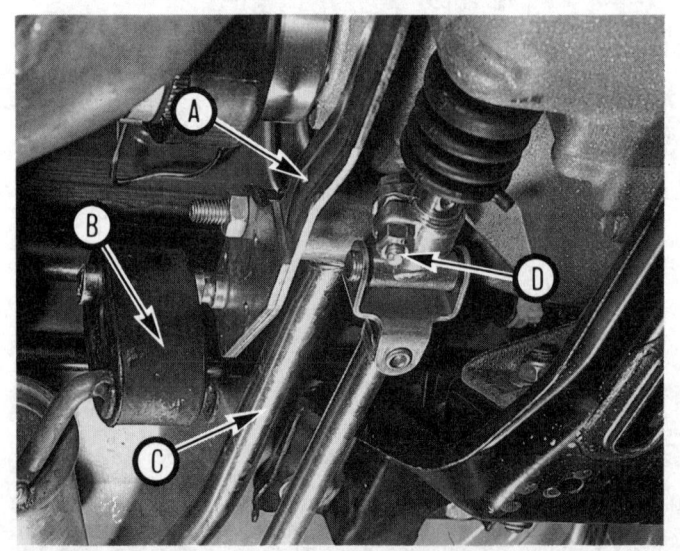

Separation

36 Unscrew and remove the engine/transmission flange bolts. Additionally, on CTX automatic transmission equipped vehicles, disconnect the carburettor cable from the carburettor.

37 Withdraw the transmission from the engine. Support in such a manner that the transmission input shaft bears no weight whilst it is being withdrawn.

Reconnection

38 This is a reversal of the separation procedure given above, with a few considerations.

39 Make sure that the upper engine adaptor plate is correctly located on its dowels.

40 On vehicles with manual transmission, check that the clutch driven plate has been centralised and smear the transmission input shaft splines with a thin film of special grease (see Chapter 5). While supporting the weight of the transmission, connect it to the engine by passing the input shaft through the splined hubs of the clutch plate until the transmission locates onto the dowels. Screw in the flange bolts and tighten to the specified torque. Do **not** attempt to draw the transmission onto the engine by tightening the flange bolts, or damage may result.

41 On CTX automatic transmission equipped vehicles, do **not** lubricate the transmission input shaft as this may contaminate, and affect the operation of, the torsional vibration damper. As on manual transmissions, do **not** attempt to draw the transmission onto the engine by tightening the flange bolts. With engine and transmission mated, screw in the flange bolts and tighten to the specified torque.

Refitting

42 Manoeuvre the engine/transmission assembly under the vehicle and attach the lifting hoist. Raise the engine/transmission assembly carefully until the right-and engine mounting can be loosely fitted to its body locations (refer to Section 34).

43 Align and fit the transmission bearer to the underside of the vehicle body.

44 Lower the hoist to allow the engine/transmission assembly to rest on its mountings, then fit and tighten all mounting nuts and bolts.

45 Disconnect and remove the lifting hoist.

46 On vehicles fitted with the anti-lock braking system, ensure that the belts are located over the driveshafts and that the right-hand modulator mounting bracket is in position before reconnecting the driveshaft assemblies.

47 With a **new** snap ring fitted to both inner CV joint splined shafts, lubricate the splined shafts with transmission oil (see Chapter 7).

48 Remove the plug (or oil inner CV joint) from the transmission casing and insert the right-hand driveshaft inner CV joint through the transmission casing into its splined differential location. Apply pressure at the roadwheel to ensure that the snap ring **fully** locates in the differential.

49 Repeat the above procedure on the left-hand driveshaft assembly.

50 Secure the right hand modulator mounting bracket, as applicable, then refit the right hand modulator and associated components (refer to Chapter 8).

51 Refer to Chapter 9 and reconnect the track rod ends. Note that **new** split pins must be used.

52 Refer to Chapter 10 and reconnect the lower suspension arm balljoints to the spindle carriers on both sides of the vehicle. Note that the bolts must locate to the annular grooves on the balljoint spindles. Also reconnect the anti-roll bar links, as applicable.

53 On manual transmission vehicles, fit the washer between the transmission casing and gearchange mechanism stabiliser bar, then fit and secure the stabiliser bar. Refit the lower engine adaptor plate, followed by the gearchange mechanism stabiliser bar/exhaust forward mounting bracket. Reconnect and adjust the gearchange mechanism shift rod as described in Chapter 6, Section 15.

54 On CTX automatic transmission equipped vehicles, refer to Chapter 6, Section 23 and reconnect and adjust the selector cable. Additionally, refit the lower engine adaptor plate (flywheel cover).

55 The remainder of the refitting procedure is a reversal of that used for removal, with reference to the relevant Chapters.

56 Check and top-up/refill the engine oil level, transmission oil level and coolant (anti-freeze) level, as necessary.

57 On CTX automatic transmission equipped vehicles, note that the accelerator pedal cable adjustment, together with that of the other two cables attached to the cams on the cable bracket assembly, will need to

37.24 Forward end of gearchange mechanism

A Stabiliser bar/exhaust forward mounting bracket
B Exhaust forward mounting rubber
C Gearchange mechanism stabiliser bar
D Gearchange mechanism shift rod clamp bolt

Chapter 1 Engine

be checked by a Ford dealer at the earliest opportunity.
58   Refer to the warning in Chapter 2, Section 1 then, once the engine is running at normal operating temperature, refer to Chapters 3 and 4 and check and adjust the fuel and ignition systems as necessary.
59   If a number of new internal components have been installed, or if a new/reconditioned engine has been fitted, run the vehicle at restricted speed and engine load for the first few hundred miles to allow time for the new components to bed-in. In the case of a reconditioned engine, consult your supplier to see if any specific running-in conditions apply (possibly for warranty reasons). It is strongly recommended that with a new, reconditioned or rebuilt engine, the engine oil and filter are changed at the end of the running-in period.

38   Engine/transmission assembly (1.6 litre EFi) – removal, separation, reconnection and refitting

1   Disconnect the battery earth terminal.
2   Remove the bonnet (refer to Chapter 11).
3   Engage the transmission in fourth gear to ensure correct adjustment of the gearshift linkage upon subsequent refitting.
4   Disconnect the HT leads from the spark plugs and the air filter trunking guides, then unclip them from the rocker cover before disconnecting at the DIS ignition coil (note their fitment). Remove the complete air filter housing/trunking, including the idle speed control valve (see Chapter 3).
5   Disconnect the throttle cable from the throttle housing (see Chapter 3).
6   Drain the cooling system (see Chapter 2).
7   Disconnect the coolant hoses from the thermostat housing and the water pump, noting their fitted positions.
8   Disconnect the vacuum pipe to the MAP sensor from the upper part of the inlet manifold (see Fig. 1.33).
9   Disconnect the crankcase breather hose and the oil trap hose from the rocker cover, then disconnect the oil trap vacuum hose at the T-piece (refer to Fig. 1.33).
10   Disconnect the brake servo vacuum hose from the inlet manifold (see Chapter 8, Section 16).
11   Disconnect the fuel feed pipe from the fuel distributor pipe (fuel rail), and the fuel return pipe from the fuel pressure regulator (see Chapter 3). Note that the fuel system is pressurised.
12   Disconnect the multi-plugs of the main (injector) wiring loom, DIS ignition coil and temperature gauge sender unit (refer to Fig. 1.35), the road speed sensor, the engine speed/crankshaft position sensor (refer to Fig. 1.48), and the radiator cooling fan and its thermal switch (refer to Chapter 2). Additionally, disconnect the earth leads form the engine and transmission.
13   Unscrew the union nut on the transmission casing end of the speedometer drive cable, and withdraw the cable. Tie the cable out of the way.
14   Disconnect the clutch cable from the clutch release lever on the transmission, having first supported the clutch pedal.
15   On vehicles fitted with the anti-lock braking system, refer to Chapter 8 and release the left-hand modulator from its mounting bracket without disconnecting the rigid brake pipes or return hose. Tie the modulator securely to the bulkhead.
16   Raise the vehicle and support it securely on axle stands.
17   Refer to Chapter 10 and remove the front suspension crossmember.
18   Disconnect the exhaust forward (downpipe) section, release the whole system from its mountings and remove it. Note that a new gasket will be needed upon reassembly.
19   Refer to Chapter 12 and remove the alternator and starter motor.
20   Detach the gearchange mechanism stabiliser bar/exhaust forward mounting bracket, and remove the lower engine adaptor plate located behind it. Disconnect the stabiliser bar and the gearchange mechanism shift rod from the transmission, and suspend them out of the way. The shift rod is removed by undoing its clamp bolt and withdrawing it. Keep a careful note of any washers removed for subsequent reassembly (refer to photo 37.24).
21   Disconnect the multi-plugs from the reversing light switch and the oil pressure switch (refer to Chapter 12).
22   Disconnect the heater hose from the inlet manifold.
23   The remainder of the removal operation is as detailed in Section 37, paragraphs 26 to 35 inclusive.

Separation
24   Refer to Section 37.

Reconnection
25   Refer to Section 37, but ignore reference to CTX automatic transmission.

Refitting
26   Refit in accordance with Section 37, paragraphs 42 to 53 inclusive, then complete the refitting operation by reversing the remaining removal operations given in this Section, with reference to the relevant Chapters.
27   Check and top up/refill the engine oil level, transmission oil level and engine coolant level, as necessary.
28   If a number of new internal components have been installed, or if a new/reconditioned engine has been fitted, run the vehicle at restricted speed and engine load for the first few hundred miles to allow time for the new components to bed-in. In the case of a reconditioned engine, consult your supplier to see if any specific running-in conditions apply (possibly for warranty reasons). It is strongly recommended that with a new, reconditioned or rebuilt engine, the engine oil and filter are changed at the end of the running-in period.

39   Engine – dismantling and reassembly

1   The need for dismantling will have been dictated by wear or noise in most cases. Although there is no reason why only partial dismantling cannot be carried out to renew such items as the oil pump or crankshaft rear oil seal, when the main bearings or big-end bearings have been

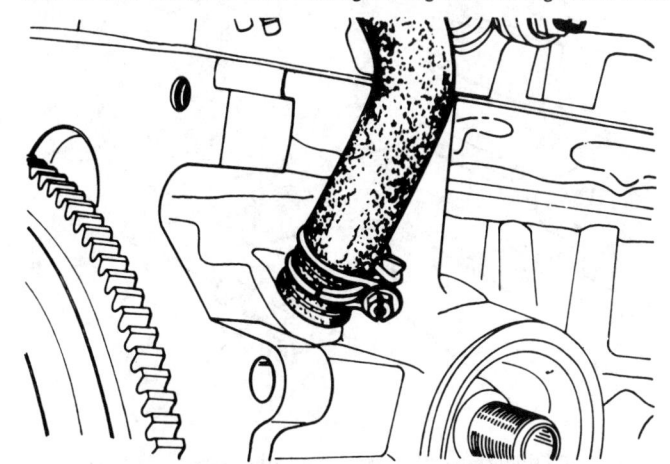

Fig. 1.45 Crankcase breather hose attached to the cylinder block (Sec 39)

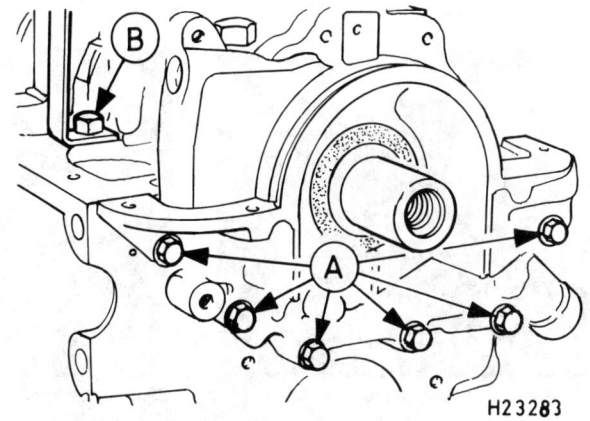

Fig. 1.46 Undo the bolts securing the oil pump (A) and the pick-up tube support (B), and remove as an assembly (Sec 39)

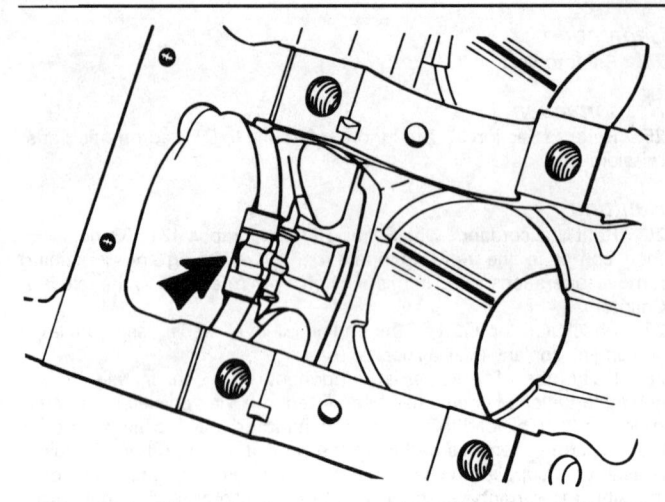

Fig. 1.47 Engine ventilation cap located in cylinder block (Sec 39)

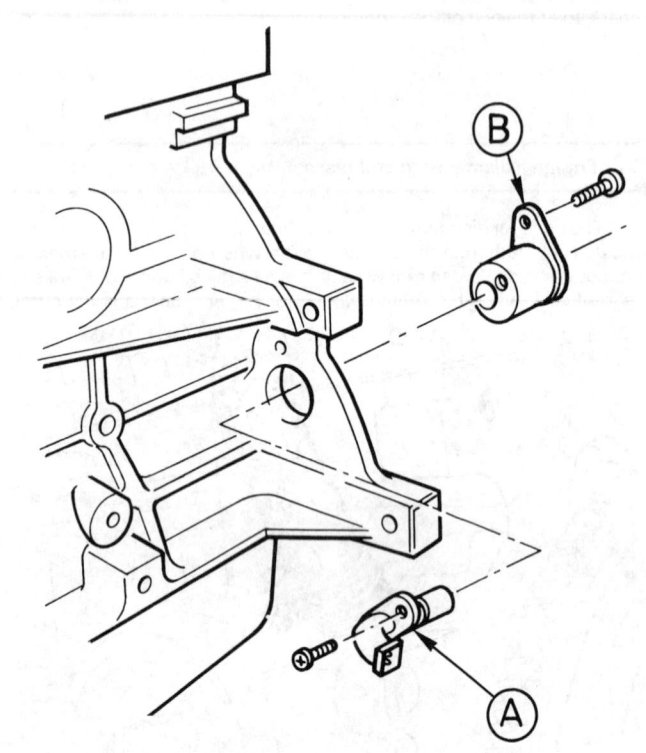

Fig. 1.48 Engine speed/crankshaft position sensor (A) and its bushing (B) (Sec 39)

knocking and especially if the vehicle has covered a high mileage, then it is recommended that a complete strip-down is carried out and every engine component examined as described in Section 40.

2 Position the engine so that it is upright and **safely** chocked on a bench or other convenient working surface. If the exterior of the engine is very dirty it should be cleaned before dismantling, using paraffin and a stiff brush or a water-soluble solvent.

3 Drain the engine oil and remove the dipstick, then unscrew and discard the oil filter.

4 Jam the flywheel starter ring gear to prevent the crankshaft rotating and unscrew the crankshaft pulley bolt. Remove the pulley.

5 Undo the four bolts securing the two-piece timing cover and remove both sections.

6 Slacken the two bolts which secure the timing belt tensioner and, using a large screwdriver, prise the tensioner to one side to relieve the belt tension. If the tensioner is spring-loaded, tighten one of the bolts to retain it in the slackened position.

7 With the belt now slack, note its running direction and mark the mating belt and sprocket teeth with a spot of quick-drying paint. This is not necessary if the belt is being renewed.

8 Remove the timing belt, then unbolt and remove the timing belt tensioner.

9 Disconnect the spark plug leads and remove the distributor cap complete with HT leads, if not already done (where applicable).

10 Unscrew and remove the spark plugs.

11 Disconnect the crankcase breather hose from the cylinder block.

12 Remove the rocker cover.

13 Unscrew the cylinder head bolts in the sequence shown in Fig.1.39, and discard them. **New** bolts must be used at reassembly.

14 Remove the cylinder head complete with manifolds. Dismantling and reassembly is covered in Section 40.

15 Turn the engine on its side. Do not invert it as sludge in the sump may enter the oilways. Remove the sump bolts, withdraw the sump and peel off the gaskets and sealing strips.

16 Remove the bolts form the clutch pressure plate in a progressive manner until the pressure of the assembly is relieved and then remove the cover, taking care not to allow the driven plate to fall to the floor. On CTX automatic transmission equipped vehicles, refer to Chapter 6 and remove the torsional vibration damper.

17 Unbolt and remove the flywheel. The bolt holes are offset so it will only fit one way. Note that **new** bolts will be needed upon reassembly.

18 Remove the engine adaptor plate.

19 Using a hooked implement, remove the crankshaft rear oil seal from the rear oil seal carrier.

20 Unbolt and remove the crankshaft rear oil seal carrier.

21 Unbolt and remove the water pump.

22 Remove the belt sprocket from the crankshaft using the hands or if tight, a puller. Take off the thrust washer.

23 Remove the radial oil seal from the oil pump casing, using a similar method to that used for the crankshaft rear oil seal.

24 Unbolt the oil pump and pick-up tube and remove them as an assembly (Fig. 1.46).

25 Unscrew and remove the oil pressure switch (photo).

26 Turn the crankshaft so that all the pistons are half-way down the bores, and feel if a wear ridge exists at the top of the bores. If so, scrape the ridge away, taking care not to damage the bores.

27 Inspect the big-end and main bearing caps for markings. The main bearings should be marked 1 to 5 with a directional arrow pointing to the

39.25 Unscrew the oil pressure switch

39.42 Inserting a main bearing shell into the cylinder block

39.43 Locating a semi-circular thrust washer to the centre main bearing

Chapter 1 Engine

39.45A Fit the bearing shells to the main bearing caps ...

39.45B ... then fit the caps to their correct numbered locations, with the arrows on the top pointing towards the timing belt end of the engine

39.47 Checking crankshaft endfloat using feeler blades

timing belt end. The big-end caps and connecting rods should have adjacent matching numbers. Number 1 is at the timing belt end of the engine. Make your own marks if necessary (see Fig. 1.38).

28 Unscrew the bolts from the first big-end cap and remove the cap. The cap is located on two roll pins, so if the cap requires tapping off make sure that it is not tapped in a sideways direction.
29 Retain the bearing shell with the cap if the shell is to be used again.
30 Push the piston/connecting rod out of the top of the cylinder block, again retaining the bearing shell with the rod if the shell is to be used again.
31 Remove the remaining pistons/rods in a similar way.
32 Remove the main bearing caps, keeping the shells with their respective caps if the shells are to be used again. Lift out the crankshaft.
33 Take out the bearing shells from the cylinder block noting the semi-circular thrust washers on either side of the centre bearing. Keep the bearing shells and thrust washers identified as to position in the cylinder block if they are to be used again.
34 Remove the engine ventilation cap from the cylinder block, with its retaining spring (Fig. 1.47).
35 Unbolt the alternator and right-hand engine mounting brackets from the cylinder block, as required.
36 Remove the engine speed/crankshaft position sensor and its bushing (1.6 litre EFi only – see Fig. 1.48).
37 With the engine dismantled, each component should be examined as described in Section 40 before reassembling.
38 With everything clean and parts renewed where necessary (see Section 40), prepare to commence reassembly. Remember that dirt and metal particles can quickly destroy bearings and result in major engine damage. Use clean engine oil to lubricate during reassembly.
39 Lift and secure the engine speed/crankshaft position sensor and its bushing (1.6 litre EFi only).
40 Refit and secure the alternator and right-hand engine mounting brackets (refer to the Specifications), if removed.
41 Insert the engine ventilation cap and its retaining spring into the cylinder block. Make sure that the spring arms engage securely.
42 Insert the bearing half shells into their seats in the cylinder block making sure that the seats are perfectly clean.
43 Stick the semi-circular thrust washers on either side of the centre bearing with thick grease. Make sure that the oil channels face outwards.

44 Oil the bearing shells and the semi-circular thrust washers, then carefully lower the crankshaft into position.
45 Insert the bearings shells into the main bearing caps, making sure that their seats are perfectly clean. Oil the bearings and install the caps in their correct numbered location and with the directional arrow pointing towards the timing belt end of the engine (photos).
46 Tighten the main bearing cap bolts to the specified torque.
47 Check the crankshaft endfloat. Ideally a dial gauge should be used, but feeler blades are an alternative if inserted between the face of the thrust washer and the machined surface of the crankshaft balance web, having first prised the crankshaft in one direction and then the other (photo). Provided the thrust washers at the centre bearing have been renewed, the endfloat should be within specified tolerance. If it is not, oversize thrust washers are available (see Specifications).
48 The pistons/connecting rods should now be installed. Although new pistons may have been fitted to the rods by your dealer or supplier due to the special tools needed, it is worth checking to ensure that with the piston crown arrow or cast nipple in the piston oil cut-out pointing towards the timing belt end of the engine, the F mark on the connecting rod or the oil ejection hole in the big-end is as shown (Fig. 1.49).
49 Oil the cylinder bores and the piston rings, then install the piston/connecting rod assemblies as described in Section 33.
50 Fit the oil pressure switch.
51 Refer to Section 40 with regard to seal fitment to the oil pump casing and rear oil seal carrier.
52 Before fitting the oil pump, action must be taken to prevent damage to the pump oil seal from the step on the front end of the crankshaft. First remove the Woodruff key and then build up the front end of the crankshaft using adhesive tape to form a smooth inclined surface to permit the pump seal to slide over the step without its lip turning back or the seal spring being displaced during installation (photo).
53 If the oil pump is new, pour some fresh engine oil into it before installation in order to prime it and rotate its driving gear a few turns. Lightly oil the lip of the seal and the crankshaft running surface.
54 Align the pump gear flats with those on the crankshaft, and install the oil pump complete with the oil pick-up tube, using a new gasket. Refer to the Specifications for tightening torque details.
55 Remove the adhesive tape and tap the Woodruff key into its groove (photo).
56 With a new seal fitted to the rear oil seal carrier (Section 40), oil the lip of the seal and the running surface of the crankshaft. Using a new gasket, align the rear oil seal carrier to the sump mating face and tighten its bolts to the specified torque. Take care to avoid damaging the lip of the seal during fitting.
57 Engage the engine adaptor plate on its locating dowels.
58 Refit the flywheel. It will only go on in one position, as it has offset holes. Note that the ribbed face of the flywheel faces towards the crankshaft flange, as applicable. Insert **new** bolts and tighten to the specified torque; the bolts must be pre-coated with thread sealant.
59 Refer to Chapter 5 and refit the clutch assembly or, on CTX automatic transmission equipped vehicles, refer to Chapter 6 and refit the torsional vibration damper.
60 To the front end of the crankshaft, fit the thrust washer (belt guide) so that its concave side is towards the pump (photo).
61 Fit the crankshaft belt sprocket. If it is tight, jam the flywheel starter ring gear and draw it into position with the aid of the crankshaft pulley

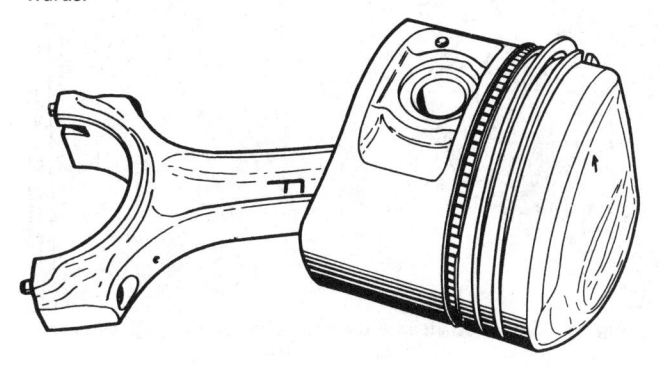

Fig. 1.49 Piston/connecting rod orientation (Sec 39)

72 Chapter 1 Engine

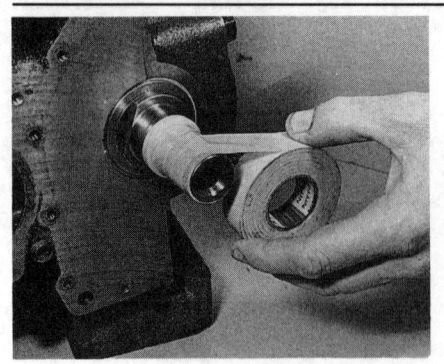

39.52 Using tape to eliminate the step on the crankshaft

39.55 Crankshaft Woodruff key installation

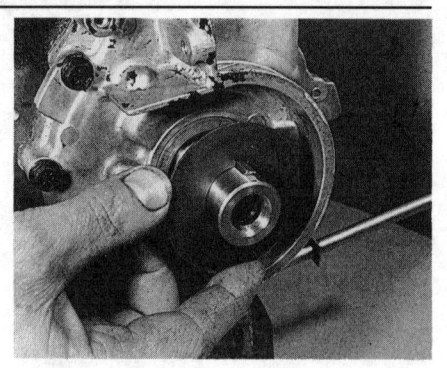

39.60 Fit the thrust washer (belt guide) with its concave side towards the pump

39.61 Fitting the crankshaft sprocket with its belt retaining flange (thrust face) outwards

bolt and a suitable distance piece. Make sure that the belt retaining flange (thrust face) on the sprocket faces outwards, that the nose of the shaft has been smeared with a little grease before fitting, and that the Woodruff key locations are aligned (photo).
62   Install the water pump using a new gasket and tightening the bolts to the specified torque.
63   With the engine resting on its side (not inverted unless you are quite sure that the pistons are not projecting from the block), fit the sump, gaskets and sealing strips as described in Section 32.
64   Loosely refit the timing belt tensioner in its slackened position.
65   The cylinder head must be installed in accordance with Section 28, paragraphs 25 to 29 inclusive. The crankshaft TDC mark alignment referred to in Section 28, paragraph 29 is taken in this case as the small projection on the crankshaft sprocket aligned with the mark on the oil pump housing (see Fig. 1.33). Additional reference must be made to Section 25 for fitting and tensioning of the timing belt.
66   Refit the cylinder head ancillaries, if removed (if not already done); the fuel pump and manifolds (refer to Chapter 3), the thermostat (refer to Chapter 2), the distributor or DIS ignition coil (as applicable – refer to Chapter 4), and the temperature gauge sender unit (Chapter 12, Section 26).
67   Using a new gasket, refit the rocker cover. Fit the reinforcing plates with the knurling facing upwards, and tighten the bolts to the specified torque.
68   Reconnect the crankcase breather hose to the cylinder block.
69   Screw in a new set of spark plugs, correctly gapped, and tighten to the specified torque – this is important. If the specified torque is exceeded, the plugs may be impossible to remove. (See Chapter 4).
70   Fit the timing belt cover sections and tighten to the specified torque.
71   Fit the crankshaft pulley and tighten the bolt to the specified

torque while the flywheel ring gear is locked to prevent it turning.
72   Fit the distributor cap, as applicable, and reconnect the HT leads to the spark plugs (refer to Chapter 4).
73   Lubricate the rubber seal of the new oil filter with fresh engine oil, then screw the filter cartridge into position by hand until the seal touches the casing, then tighten a further three quarters of a turn (again by hand).
74   Fit the sump drain plug using a new seal, and tighten to its specified torque. Insert the dipstick. Refill with oil with the engine installed in the vehicle.

40 Engine – component examination and renovation

Crankshaft, bearings, cylinder bores and pistons
1   Refer to paragraphs 1 to 17 of Section 19. The information applies equally to the CVH engine, except that standard sized crankshafts are unmarked and the following differences in the piston rings should be noted.
2   The upper rings are coated with molybdenum. Avoid damaging the coating when fitting the rings to the pistons.
3   The oil control ring may be a one piece or multi-piece item. With garter spring fitment, make sure that the ends abut and do not overlap. If in any doubt regarding parts suitability, seek the advice of your parts supplier.

Timing sprockets and belt
4   It is very rare for the teeth of the sprockets to wear, but attention should be given to the tensioner idler pulley. It must turn freely and smoothly, be ungrooved and without any shake in its bearing. Otherwise renew it.
5   Check the timing belt carefully for any signs of uneven wear, splitting or oil contamination, and renew it if there is the slightest doubt

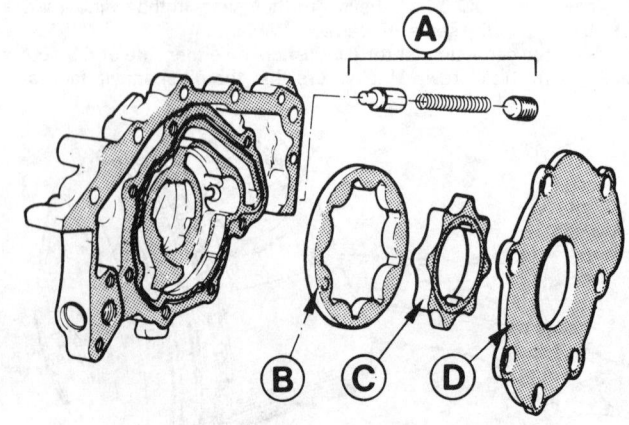

Fig. 1.50 Exploded view of the rotor type oil pump (Sec 40)

A Pressure relief valve   C Inner (driving) rotor
B Outer (driven) rotor   D Cover

Chapter 1 Engine

Fig. 1.51 Compressing a valve spring using a fork type valve spring compressor (Sec 40)

about its condition. As a precaution, the belt must be renewed as a matter of course at the intervals given in *'Routine maintenance'* and Chapter 13 (as applicable); if its history is unknown, the belt should be renewed irrespective of its apparent condition whenever the engine is overhauled.

Flywheel
6 Refer to paragraphs 21 and 22 of Section 19.

Oil pump
7 The oil pump is a low friction rotor type, driven by the crankshaft. If a high mileage engine is being reconditioned it is recommended that a new pump is fitted. Inspection of the pump is of similar method to that described in paragraphs 23 and 24 of Section 19.

Oil seals and gaskets
8 Renew the oil seals in the oil pump and in the crankshaft rear oil seal retainer as a matter of routine at time of major overhaul. It is recommended that new seals are drawn into these components using a nut and bolt and distance pieces. Do **not** tap them into position, or damage will almost certainly result.
9 Renew the camshaft oil seal after the oil seal has been installed.
10 Always smear the lips of a new oil seal with fresh engine oil and check that the small tensioner spring in the oil seal has not been displaced during installation.
11 Renew all gaskets by purchasing the appropriate engine set, which usually includes the necessary oil seals.

Crankcase
12 Refer to paragraphs 28 to 30 in Section 19. Particular attention must be given to the oil drillings as even tiny residues of dirt can impair the operation of the hydraulic cam followers (tappets) and thus lead to failure of the valve gear. Dirt or metal particles can also quickly destroy bearings leading to major engine damage.

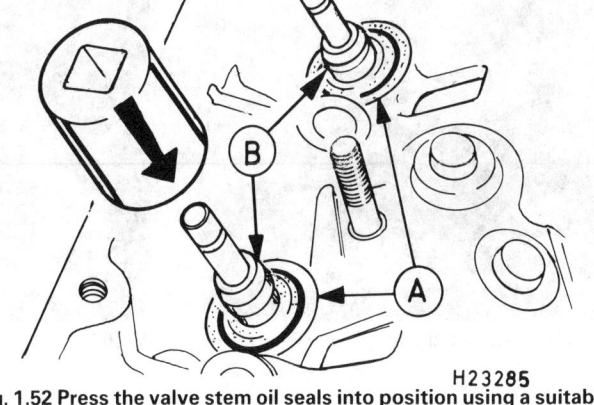

Fig. 1.52 Press the valve stem oil seals into position using a suitably sized socket (Sec 40)

A Lower valve spring retainers   B Valve stem oil seals

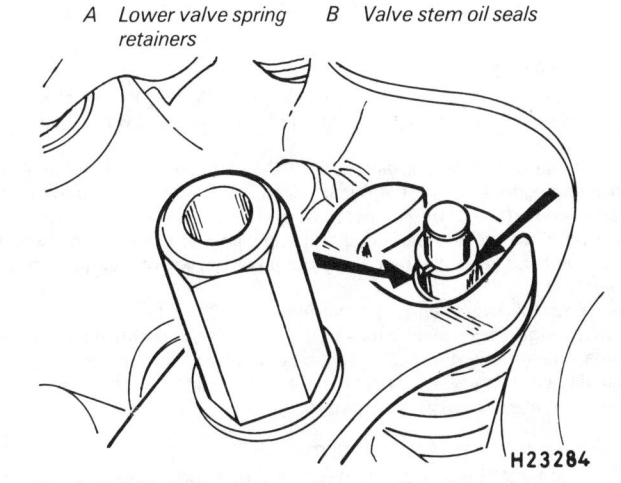

Fig. 1.53 Split collets (arrowed) engaged to retain the valve components (Sec 40)

Camshaft and bearings
13 Examine the camshaft gear and lobes for damage or wear. If evident, a new camshaft must be purchased, or one which has been built-up, such as are advertised by firms specialising in exchange components.
14 Check the camshaft endfloat by temporarily refitting the camshaft and thrust plate. If the endfloat exceeds the specified tolerance, renew the thrust plate.
15 The bearing internal diameters in the cylinder head should be checked against the Specifications if a suitable gauge is available, otherwise check for movement between the camshaft journal and the bearing. If the bearings are proved to be worn, then a new cylinder head is the only answer as the bearings are machined directly in the cylinder head.

40.39 Inserting a cam follower (tappet) to its bore

40.40 Inserting a valve into its guide

40.42A Fit the valve spring ...

Chapter 1 Engine

40.42B ... followed by the upper valve spring retainer

40.46A Fit the rocker arm spacer plate ...

40.46B ... followed by the rocker arm and guide

Cam followers

16 It is seldom that the hydraulic type cam followers (tappets) wear in their cylinder head bores. If the bores are worn then a new cylinder head is called for.
17 If the cam lobe contact surface shows signs of a depression or grooving, grinding out the wear surface will not only remove the hardened surface of the follower but may also reduce its overall length to a point where the self-adjusting capability of the cam follower is exceeded and valve clearances are not taken up, with consequent noisy operation.
18 Cam followers cannot be overhauled so if they become worn after high mileage, they must be renewed. On refitting, it is only necessary to smear the outside surfaces with hypoid oil, as they are self priming and will fill with engine oil once the engine is running, although initial operation may be noisy until primed.

Cylinder head and rocker arms

19 The usual reason for dismantling the cylinder head is to decarbonise and to grind in the valves. Reference should therefore be made to Section 30 in addition to the dismantling operations described here. Support the cylinder head on blocks of wood to avoid valve damage.
20 To completely dismantle, the procedure is as follows. Remove the inlet and exhaust manifolds and their gaskets (Chapter 3), the fuel pump (carburettor equipped vehicles only – Chapter 3), the thermostat housing (Chapter 2), the distributor or DIS ignition coil (as applicable – Chapter 4), and the temperature gauge sender unit (Chapter 12, Section 26).
21 Remove the camshaft sprocket bolt whilst preventing the sprocket from turning by passing a stout bar or screwdriver through one of its holes. Note that a new retaining bolt will be needed upon reassembly. Remove the camshaft sprocket.
22 Unscrew the nuts from the rocker arms and discard the nuts. New ones must be fitted at reassembly.
23 Remove the rocker arms keeping them in their originally fitted sequence. Keep the rocker guide and spacer plates in order (refer to Fig. 1.28).
24 Valve removal should commence with No 1 valve (nearest timing belt end). Keep the valves and their components in their originally installed order by placing them in a piece of card which has holes punched in it and numbered 1 to 8.
25 The No 1 valve spring should now be compressed. A standard type of compressor will normally do the job, but a forked tool (Part No 21-097) can be purchased or made up to engage on the rocker stud using a nut and distance piece to compress it.
26 Compress the valve spring and extract the split collets. Do not overcompress the spring, or the valve stem may bend. If it is found when screwing down the compressor tool that the spring retainer does not release from the collets, remove the compressor and place a piece of tubing on the retainer so that it does not impinge on the collets and place a small block of wood under the head of the valve. With the cylinder head resting flat down on the bench, strike the end of the tubing a sharp blow with a hammer. Refit the compressor and compress the spring.
27 Extract the split collets and then gently release the compressor and remove it.
28 Remove the upper valve spring retainer and the valve spring, then withdraw the valve from the combustion chamber side of the cylinder head and prise up the valve stem oil seal using a screwdriver.
29 Remove the lower valve spring retainer.
30 Repeat the operation on the remaining valves.
31 Withdraw the hydraulic cam followers (tappets) from their bores, keeping them in their originally installed order.
32 If the camshaft is to be renewed, remove the thrust plate and withdraw the camshaft from the rear (thermostat end) of the cylinder head (see Section 27). On 1.6 litre EFi models, it will be necessary to remove the plastic blanking plug from the rear of the cylinder head before withdrawing the camshaft. Remove the old oil seal.
33 To check for wear in the valve guides, place each valve in turn in its guide so that approximately one third of its length enters the guide. Rock the valve from side to side. If any more than the slightest movement is possible, the guides will have to be reamed (working from the valve seat end) and oversize stemmed valves fitted. You are advised to entrust this task to your Ford dealer.
34 Examine the valve seats. Normally the seats to not deteriorate, but the valve heads are more likely to burn away, in which case new valves can be ground in as described in Section 30. If any seat is cracked, or if the exhaust valve seats require recutting, consult your Ford dealer. Exhaust valve seat rings are fitted to the cylinder head to allow the use of unleaded petrol.
35 If the rocker arm studs must be removed for any reason, a special procedure is necessary. Warm the upper ends of the studs with a blow-lamp flame (**not** a welder) before unscrewing them. Clean out the cylinder head threads with an M10 tap and clean the threads of oil or grease. Discard the old studs and fit new ones, which will be coated with adhesive compound on their threaded portion or will have a nylon locking insert. Screw in the studs without pausing, otherwise the adhesive will start to set and prevent the stud seating.
36 If the cylinder head mating surface is suspected of being distorted, it can be checked and surface ground by your dealer or motor engineering works. Distortion is possible with this type of light alloy head if the bolt tightening method is not followed exactly, or if severe overheating has taken place.
37 Check the rocker arm contact surfaces for wear. Renew the valve springs if they have been in service for 80 000 km (50 000 miles) or more.
38 Fit the camshaft (if removed), and a new oil seal, as described in Section 27.
39 Lubricate the hydraulic cam followers (tappets) with hypoid oil and insert them into their original bores (photo).
40 Valve refitting should be carried out in the reverse order to removal. Lubricate No 8 valve with hypoid oil and insert to its guide (photo), then fit the lower valve spring retainer. Support the valve to prevent it from dropping out.
41 Mask the split collet grooves with tape, then lubricate the **new** valve stem oil seal and slide it over the valve. Using a suitably sized socket, ensure that the valve stem oil seal is fully seated; take care not to distort it. Remove the tape from around the split collet grooves.
42 Fit the valve spring and the upper valve spring retainer (photos).
43 Compress the spring and engage the split collets in the cut-out in the valve stem. Hold them in position while the compressor is gently

released and removed (see Fig. 1.53).

44 Repeat the operations on the remaining valves, making sure that each valve is returned to its original guide or, if new valves have been fitted, into the seat into which it was ground.

45 Once all the valves have been fitted, support the ends of the cylinder head on two wooden blocks and strike the end of each valve stem with a plastic or copper-faced hammer, just a light blow to settle the components.

46 Fit the rocker arms with their guides and spacer plates, use **new** nuts and tighten to the specified torque. It is important that each rocker arm is installed **only** when its particular cam follower is at its lowest point (in contact with the cam base circle) (photos). Make sure that the rocker guides are seated correctly.

47 Refit the camshaft sprocket using a **new** retaining bolt, having smeared the threads of the belt with sealing compound. Tighten the bolt to the specified torque.

48 Refit the components removed in paragraph 20, with reference to the appropriate text, as applicable.

41 Fault diagnosis

Refer to Section 20, but also consider (on vehicles not fitted with distributorless ignition systems – 'DIS') the possibility that the ignition timing may be incorrect (see Chapter 4). The 'DIS' function does away with the need for a separate coil and distributor so, wherever reference is made to the 'DIS' ignition coil or its connections in Section 20, these separate components or associated connections should be substituted (1.4 litre/1.6 litre carburettor engines and 1.4 litre CFi).

The advice given in the note regarding suspected internal faults, at the end of Section 20, is extended here to cover engine management modules, as applicable.

On fuel-injected engines (1.4 litre CFi and 1.6 litre EFi), any reference to carburettor jets should be substituted by fuel-injector nozzle(s).

Rough engine idling or misfiring may also be caused on CVH engines by a slack or worn timing belt which has jumped a sprocket tooth.

A certain amount of 'chatter' is normal when starting an engine with hydraulic cam followers (tappets), but the noise should disappear as they fill with oil.

Chapter 2
Cooling, heating and ventilation systems

For modifications, and information applicable to later models, see Supplement at end of manual

Contents

Antifreeze (coolant) mixture – general	4	Heater radiator (matrix) – removal and refitting	17
Cooling system – draining, flushing and refilling	3	Maintenance and inspection	2
Expansion tank – removal and refitting	13	Radiator cooling fan and motor – removal and refitting	5
Fault diagnosis – cooling system	19	Radiator cooling fan thermal switch – removal and refitting	7
Fault diagnosis – heating and ventilation system	20	Radiator – removal, cleaning and refitting	6
General description – cooling system	1	Thermostat – removal and refitting	8
General description – heating and ventilation system	14	Thermostat – testing	9
Heater casing assembly – removal and refitting	18	Water pump (CVH engines) – removal and refitting	12
Heater controls – removal and refitting	15	Water pump (HCS engines) – removal and refitting	11
Heater fan motor and resistor assembly – removal and refitting	16	Water pump drivebelt – general	10

Specifications

System type .... Pressurised, pump-assisted. Front mounted crossflow radiator with electric cooling fan and expansion tank

Pressure cap rating .... 1.0 to 1.2 bar (14.5 to 17.0 lbf/in^2)

Thermostat
Type .... Wax
Initial opening temperature .... 85° to 89°C (185° to 192°F)
Fully open temperature .... 102°C (216°F) (± 3°C/5°F for used thermostats)

Antifreeze
Type/specification .... Ethylene glycol based antifreeze, to Ford specification SSM-97B-9103-A (Duckhams Universal Antifreeze and Summer Coolant)
Recommended concentration .... 40% by volume

Total cooling system capacity
HCS engine .... 7.1 litres (12.5 pints)
CVH engine .... 7.6 litres (13.4 pints)

Torque wrench settings

	Nm	lbf ft
Radiator securing bolts	20 to 27	14.8 to 20
Thermostat housing retaining bolts:		
CVH	9 to 12	6.6 to 8.9
HCS	17 to 21	12.5 to 15.5
Water pump retaining bolts:		
CVH	6.8 to 9.5	5 to 7
HCS	7.0 to 10	5.2 to 7.4
Water pump pulley retaining bolts (HCS only)	8.5 to 10.6	6.3 to 7.8
Temperature gauge sender:		
HCS engine	4 to 8	2.9 to 5.9
CVH engine	5 to 7	3.7 to 5.2
Radiator cooling fan shroud retaining bolt	3 to 5	2.2 to 3.7
Radiator cooling fan motor to shroud nuts	8.6 to 12	6.3 to 8.8
Radiator drain plug	1.2 to 1.5	0.9 to 11
Automatic transmission fluid cooling pipe connections to radiator (CTX equipped vehicles only)	18 to 22	13.2 to 16.2

Chapter 2 Cooling, heating and ventilation systems

1 General description – cooling system

Warning: *Take particular care when working under the bonnet with the engine running, or ignition switched on, as the coolant temperature controlled radiator cooling fan may suddenly actuate without warning. Remember that the coolant temperature will continue to rise for a short time after the engine is switched off. Ensure that ties, clothing, hair and hands are kept away from the fan.*

*When removing the expansion tank pressure cap while the coolant is hot, first relieve the system pressure by **slowly** unscrewing the cap, holding a thick cloth over it. Failure to observe this safety precaution may result in scalding.*

The cooling system consists of a radiator, water pump, thermostat, electric cooling fan, expansion tank and associated hoses.

The system is pressurised, and functions as follows. When the coolant is cold the thermostat is closed, restricting coolant flow. As the coolant temperature increases, the thermostat opens allowing initially partial and then full coolant circulation through the radiator. The coolant temperature is reduced as it passes through the radiator by the inrush of air when the vehicle is moving forwards, supplemented by the operation of the radiator cooling fan. The water pump circulates the coolant around the system from the base of the radiator through the cylinder head and then back into the radiator.

When the engine is at normal operating temperature, the coolant expands, and some of it is displaced into the expansion tank. The coolant collects in the tank and returns to the radiator when the system cools.

The radiator cooling fan is mounted behind the radiator and is controlled by a thermal switch in the thermostat housing. When the coolant reaches a predetermined temperature the switch contacts close, thus actuating the fan.

On vehicles equipped with the CTX automatic transmission option, facility is provided in the cooling system radiator for cooling the transmission fluid.

2 Maintenance and inspection

1 At the intervals specified in the Routine maintenance Section at the beginning of this manual, check the coolant level in the expansion tank **when the engine is cold**. The level should be between the 'MAX' and 'MIN' marks on the side of the tank, visible from the outside.

2 If the level is low, unscrew the pressure cap on the expansion tank (refer to warning in previous Section). Top up the expansion tank with a water and antifreeze mix (see Section 4) until the level is up to the 'MAX' mark then refit the cap.

3 With a sealed type cooling system, topping-up should only be necessary at *very* infrequent intervals. If this is not the case, it is likely that there is a leak in the system. Check all hoses (heater hoses included) and joint faces for staining or actual wetness, and rectify as necessary. If no leaks can be found it is advisable to have the system pressure tested as the leak could be internal. It is a good idea to keep a check on the engine oil level as a serious internal leak can often cause the level in the sump to rise, thus confirming suspicions.

4 Regularly inspect the hose clips for security, the hoses for signs of cracking, leakage or deterioration, and the visible joint gaskets of the system for leakage. Renew any suspect items.

5 At the specified intervals, check the water pump and alternator drivebelt tension (HCS engines only) – see Chapter 12, Section 7; on CVH engines the water pump drive should need no attention as, on this engine type, the water pump is driven by the timing belt.

6 At the less frequent specified intervals, flush and refill the system using fresh antifreeze (see Sections 3 and 4).

3 Cooling system – draining, flushing and refilling

1 It is preferable to drain the system when the coolant is cold. If it must be drained when hot, refer to the warning in Section 1.

2 Set the heater control to the maximum heat position.

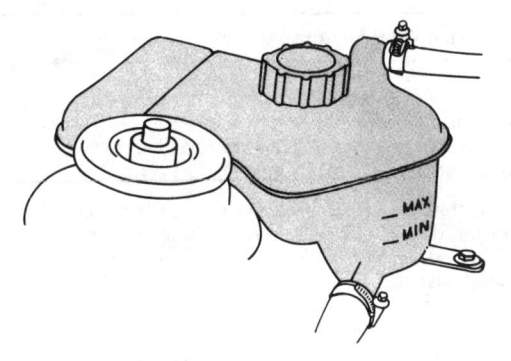

Fig. 2.1 Expansion tank showing coolant level markings (Sec 2)

3 Position a suitable container beneath the radiator to catch escaping coolant, then unscrew the radiator drain plug (photo) and allow the coolant to drain.

4 Provided that the correct mixture of antifreeze and water has previously been maintained in the system, then flushing should not be necessary and the system can be refilled immediately, as described below.

5 Where the system has been neglected however, and rust or sludge is evident at draining, then the system should be flushed through. Release the radiator bottom hose retaining clip and disconnect the hose, then remove the cylinder block drain plug (located on the forward side of the cylinder block, towards the flywheel end). Insert a garden hosepipe into the thermostat housing (thermostat removed – see Section 8). Flush through until clear water emerges.

6 If the radiator is suspected of being clogged, remove it and reverse flush it, as described in Section 6.

7 When the cooling system is being drained/flushed, it is recommended that attention is also given to the expansion tank. Ensure that no old or dirty coolant remains in it or its pipes. Examine the interior of the tank. If it is dirty, remove it (see Section 13) and clean thoroughly – the presence of oil may indicate a leaking cylinder head gasket. Refit the expansion tank once it is clean.

8 Refit and secure the hoses and refit the thermostat (see Section 8) if removed. Refit the radiator drain plug, tightening to its specified torque, and the cylinder block drain plug (if removed).

9 Using the correct antifreeze mixture (see Section 4), refill the system via the expansion tank until it reaches the 'MAX' mark (see Fig. 2.1). Pour the coolant in slowly, and allow any trapped air to escape. Refit the pressure cap.

10 Start the engine and allow it to warm up to normal operating temperature, check that no leaks are evident, then switch off. Once it has cooled, check the level in the expansion tank, topping up as necessary.

3.3 Drain plug on base of radiator

Chapter 2 Cooling, heating and ventilation systems

4 Antifreeze (coolant) mixture – general

Warning: *Antifreeze is poisonous. Keep it out of reach of children and pets. Precautions should be taken to avoid skin contact. Wash splashes off skin with plenty of water. If clothing is splashed it should be washed before being worn again to avoid prolonged skin contact.*

1 Never operate the vehicle with plain water in the cooling system. Apart from the danger of freezing during winter conditions, an important secondary purpose of antifreeze is to inhibit the formation of rust and reduce corrosion.
2 The coolant **must** be renewed at the intervals specified in 'Routine maintenance' at the beginning of this manual, as although the antifreeze will protect against frost damage indefinitely, its corrosion inhibitors lose their effectiveness after prolonged use.
3 A solution of 40% antifreeze (see Specifications) is recommended, to be maintained in the system all year round as adequate protection against frost, rust and corrosion.
4 After filling with coolant, note the date somewhere prominent (on the inside cover of this manual perhaps) so that checks may be made as to when it next needs to be renewed.
5 Do not use engine antifreeze in the windscreen/tailgate washer reservoir, as it will cause damage to the vehicle paintwork. A screen-wash additive as recommended in *'Routine maintenance'* should be used.

5 Radiator cooling fan and motor – removal and refitting

1 Disconnect the battery earth terminal.
2 Disconnect the multi-plug from the rear of the radiator cooling fan motor (photo), and release all wiring from the clips on the fan shroud.
3 Remove the bolt securing the fan shroud (Fig. 2.2), noting the insulating arrangement, then lift the fan shroud and motor assembly from the vehicle.
4 To separate the fan from the motor shaft, first remove its retaining clip and washer, then withdraw (see Fig. 2.3). A **new** clip will be needed upon reassembly.
5 Remove the three nuts securing the motor to the shroud and separate.
6 Refitting is a reversal of the removal procedure, ensuring that the locating tags on the base of the shroud locate correctly to their slots in the body crossmember. Make reference to the Specifications for details of tightening torque (photo).

Note: To test the operation of the radiator cooling fan motor, bridge out the two connections in its thermal switch multi-plug (see Section 7) and switch the ignition on – the motor should operate. Switch the ignition off, remove the bridging wire and reconnect the multi-plug to complete. This test does **not** *determine the correct functioning of the thermal switch.*

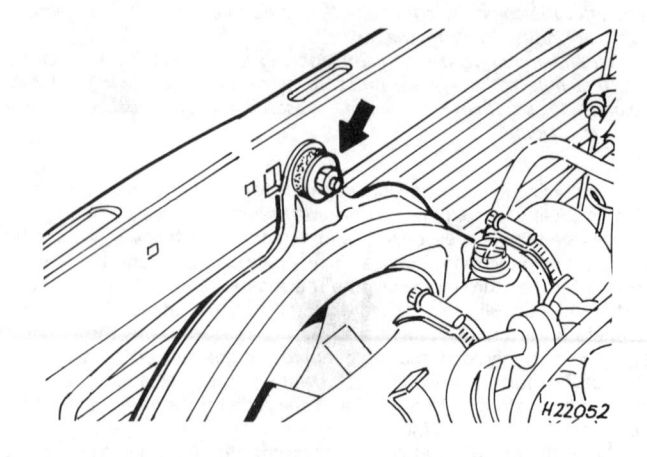

Fig. 2.2 Radiator cooling fan shroud securing bolt (arrowed) (Sec 5)

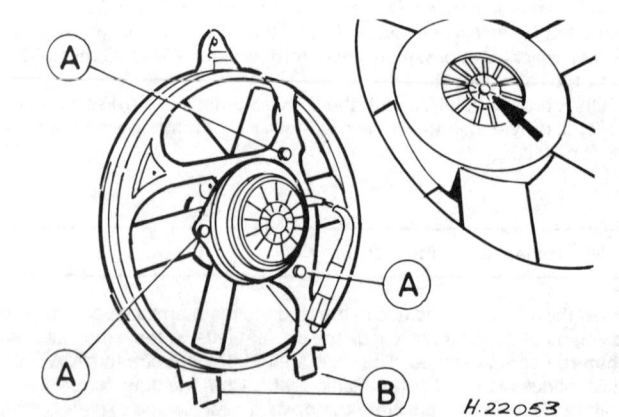

Fig. 2.3 Nuts securing fan motor to shroud (A), and shroud to body crossmember locating tags (B). Inset shows fan to motor shaft retaining clip (arrowed) (Sec 5)

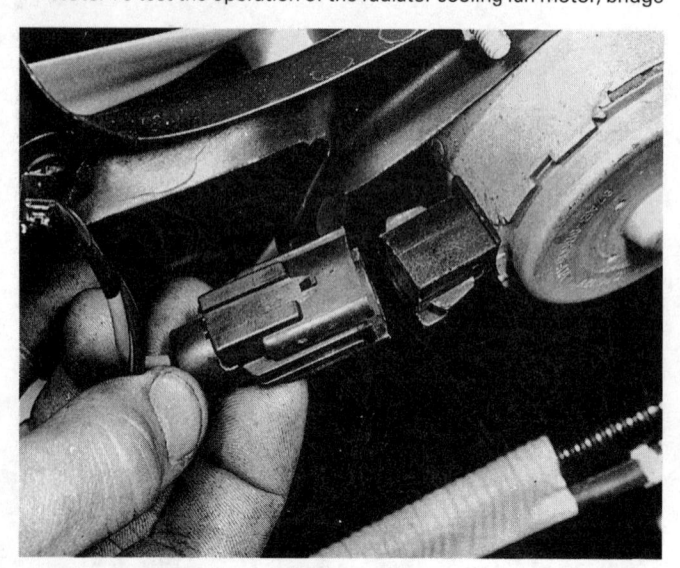

5.2 Disconnecting the multi-plug from the radiator cooling fan motor

5.6 Fan shroud locating slot (arrowed)

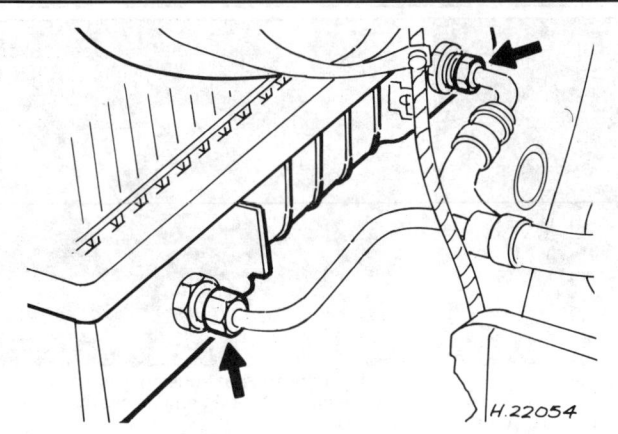

Chapter 2 Cooling, heating and ventilation systems

Fig. 2.4 CTX automatic transmission cooling pipe connections (arrowed) at radiator (Sec 6)

6 Radiator – removal, cleaning and refitting

1 Disconnect the battery earth terminal.
2 Drain the cooling system, as described in Section 3.
3 Release the hose clips and disconnect the hoses from the radiator. Additionally, on CTX automatic transmission equipped vehicles, disconnect the automatic transmission fluid cooling pipe connections (Fig. 2.4) fitting blanking plugs to prevent excessive fluid loss.
4 Remove the radiator cooling fan shroud assembly (see Section 5).
5 Remove the radiator securing bolts and lift the radiator out of its locating slots in the body crossmember. Note rubber insulators fitted to the locating lugs on the base of the radiator (photos).
6 If the purpose of removal was to clean the radiator thoroughly, first reverse-flush it using a garden hosepipe. The normal direction of flow is from the thermostat to the top of the radiator.
7 If the radiator fins are clogged with flies or dirt, remove them with a soft brush or blow compressed air from the rear face of the radiator. In the absence of a compressed air line, a strong jet from a water hose may suffice.
8 If a radiator is leaking, it is recommended that a reconditioned or new one is obtained from specialists. In an emergency, minor leaks from the radiator may be cured by using a radiator sealant such as Holts Radweld. If due to neglect the radiator requires the application of chemical cleaners, then these are best used when the engine is hot with the radiator *in situ*. Holts Radflush or Holts Speedflush are also extremely effective products but in each case it is important that the manufacturer's instructions are followed precisely.
9 Refit by reversing the removal operations, making reference to the Specifications for details of tightening torques. Refill the cooling system (refer to Sections 3 and 4). Check the automatic transmission fluid level on completion (where applicable), topping up as necessary with fresh fluid of the specified type (see Chapter 6).

7 Radiator cooling fan thermal switch – removal and refitting

1 The thermal switch which controls the actuation of the radiator cooling fan is located on the thermostat housing (photo).
2 To renew the thermal switch, wait until the engine is cold, then partially drain the cooling system in accordance with Section 3.
3 Disconnect the battery earth terminal, followed by the thermal switch multi-plug and unscrew the switch from its location.
4 Refit the switch, using a new sealing washer as applicable, and tighten securely. Reconnect the thermal switch multi-plug followed by the battery earth terminal.
5 Refill the cooling system (refer to Sections 3 and 4).

Note: *When testing the operation of the radiator cooling fan motor, the multi-plug should be bridged*

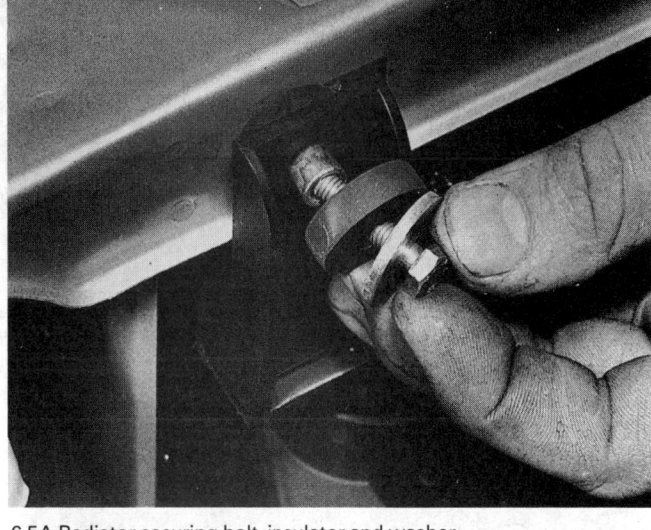

6.5A Radiator securing bolt, insulator and washer

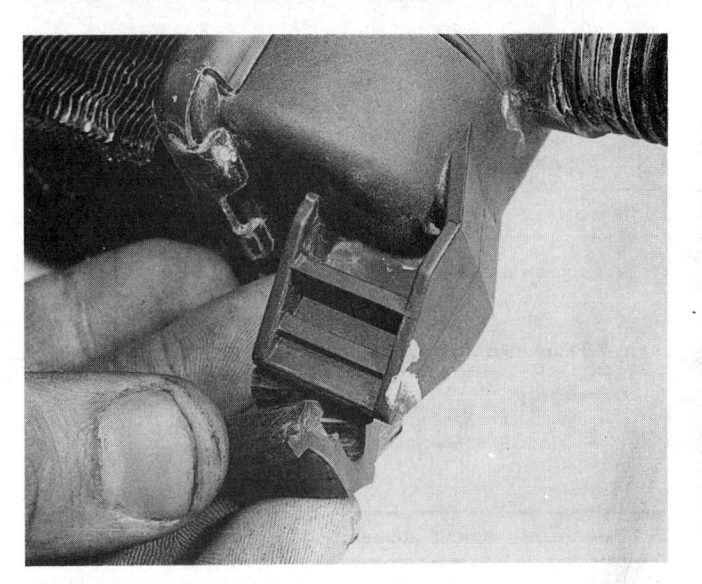

6.5B Rubber insulator on radiator locating plug

7.1 Radiator cooling fan thermal switch location on thermostat housing (CVH engine)

Chapter 2 Cooling, heating and ventilation systems

8.3A Thermostat housing (CVH engine)

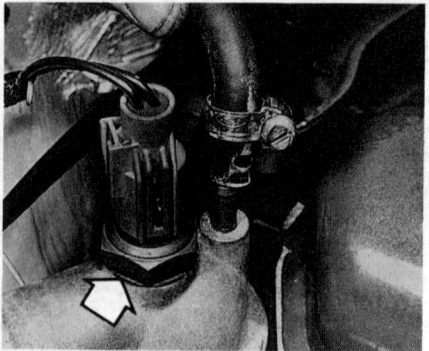

8.3B Disconnecting the expansion tank top hose from the thermostat housing (HCS engine). Radiator cooling fan thermal switch (arrowed)

8.7 Refit thermostat housing using a new gasket (shown during engine overhaul)

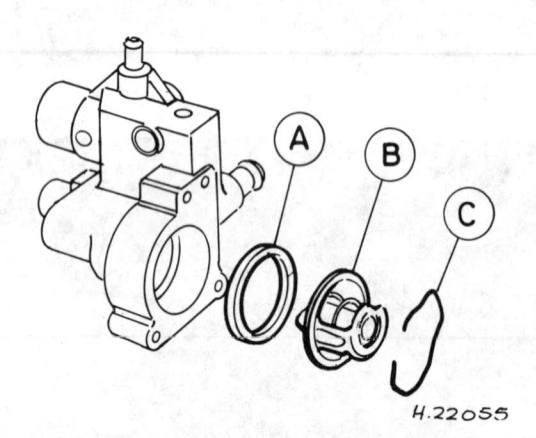

Fig. 2.5 Exploded view of thermostat and housing (CVH engines) (Sec 8)

A   Sealing ring
B   Thermostat
C   Retaining clip

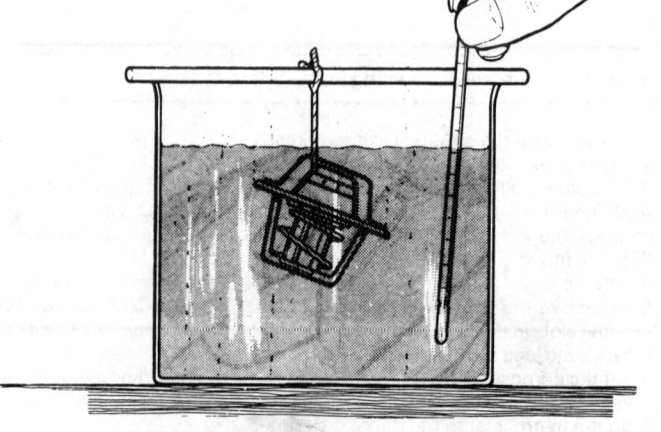

Fig. 2.6 Testing the thermostat (Sec 9)

8 Thermostat – removal and refitting

1   Disconnect the battery earth terminal.
2   Partially drain the cooling system in accordance with Section 3.
3   Undo the hose clips and disconnect the hoses and the radiator cooling fan thermal switch multi-plug from the thermostat housing. The thermostat fitted to CVH engines is located on the right-hand side of the cylinder head (looking into the engine compartment from the front of the vehicle), whilst that fitted to HCS engines is on the left-hand side (looking from the same viewpoint). (Photos).
4   Remove the two bolts retaining the HCS type thermostat housing, or the three bolts retaining the CVH type, as applicable Remove the thermostat housing – if it is stuck, tap it gently using a soft-faced mallet. Remove the thermostat retaining clip, thermostat and sealing ring, as applicable (Fig. 2.5), noting position for subsequent reassembly.
5   Test the thermostat in accordance with the following Section.
6   Prior to refitting, scrape away all traces of old gasket from the cylinder head and thermostat housing and ensure that the mating faces are clean and dry.
7   Refitting is a reversal of the removal procedure, but use a new sealing ring, as applicable, and a new gasket lightly smeared with jointing compound (photo). Tighten the retaining bolts to their specified torque. Refill the cooling system (refer to Sections 3 and 4).

9 Thermostat – testing

1   To test the thermostat, first check that in a cold condition its valve plate is closed. Suspend it on a string in a pan of cold water together with a thermometer. Neither should be allowed to touch the sides of the pan (see Fig. 2.6).
2   Heat the water and check that the thermostat starts to open at the temperature given in the Specifications. It is not possible to check the fully open condition in this manner, as the temperature required to fully open the thermostat is above the boiling point of water at normal atmospheric pressure.
3   Remove the thermostat from the water and check that it returns to a closed condition as it cools.
4   If the thermostat does not function satisfactorily it must be renewed.

10 Water pump drivebelt – general

1   There are two water pump drivebelt arrangements applicable, dependent on engine type.
2   HCS engines have a water pump which is driven the by the same drivebelt as the alternator. Drivebelt removal, refitting and adjustment procedures are covered in Chapter 12, Section 7.
3   CVH engines have a water pump which is driven by the timing belt. Refer to Section 12 of this Chapter.

11 Water pump (HCS engines) – removal and refitting

1   Disconnect the battery earth terminal.
2   Drain the cooling system in accordance with Section 3.
3   Slacken the water pump pulley retaining bolts with the drivebelt still attached (Fig. 2.7).
4   Remove the drivebelt in accordance with Chapter 12, Section 7.
5   Release the retaining clips and disconnect the hoses from the water pump (photos).

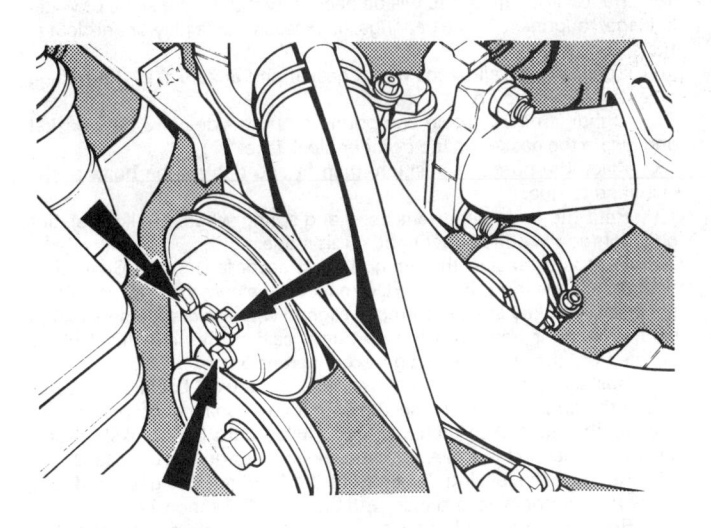

Chapter 2 Cooling, heating and ventilation systems   81

6 Remove the previously-slackened water pump pulley retaining bolts, and remove the pulley.
7 Remove the water pump retaining bolts, and remove the water pump from the cylinder block (see photos 11.10A and 11.10B).
8 Clean all traces of old gasket from the cylinder block and, if it is to be refitted, the water pump. Ensure that mating faces are clean and dry. Note that the water pump gasket fitted during production is integral with the timing cover gasket, and should be removed by cutting with a sharp knife, keeping as close to the timing cover as possible.
9 No provision is made for repair of the water pump, and if the unit is leaking, noisy, or in any way unserviceable, renewal will be necessary.
10 Refitting is a reversal of the removal procedure. Use a new gasket lightly smeared with jointing compound, and tighten all retaining bolts to their specified torques (photos). Adjust drivebelt tension (with the pulley fully secured) in accordance with Chapter 12, Section 7, and refill the cooling system, making reference to Sections 3 and 4 of this Chapter.

12 Water pump (CVH engine) – removal and refitting

1 Disconnect the battery earth terminal.
2 Drain the cooling system in accordance with Section 3 of this Chapter.

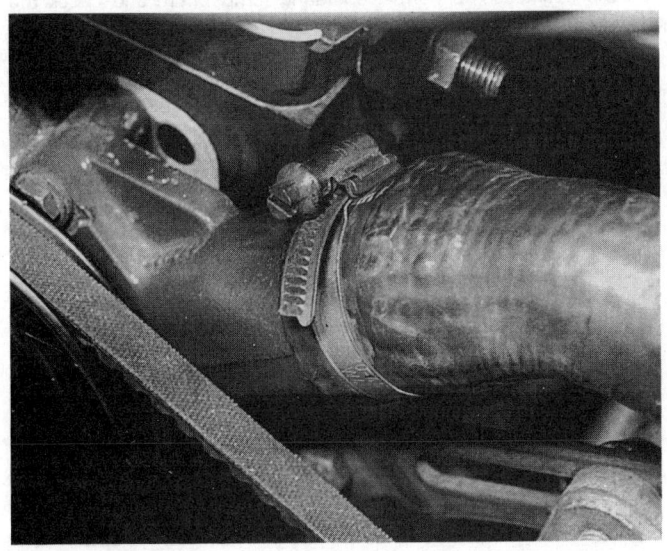

Fig. 2.7 HCS engine water pump pulley retaining bolts (arrowed) (Sec 11)

11.5A Radiator bottom hose to water pump

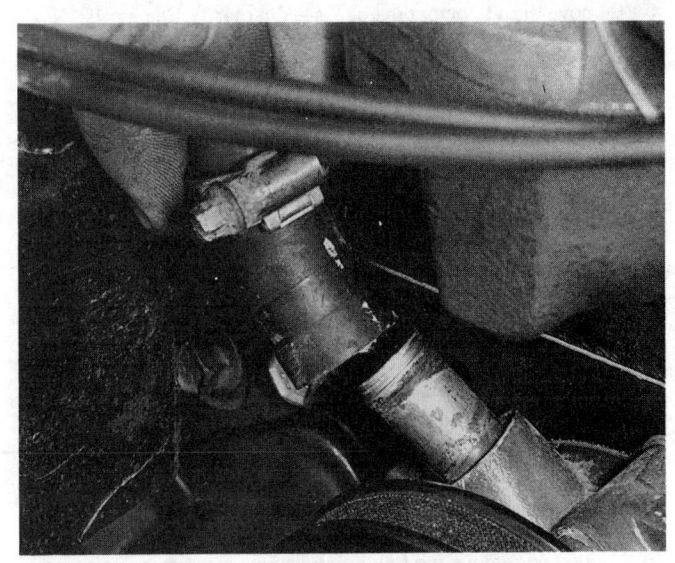

11.5B Disconnecting the heater hose from the water pump

11.10A Refitting water pump with new gasket in position on cylinder block

11.10B Tightening the water pump retaining bolts (arrowed) to their specified torque (shown during engine overhaul)

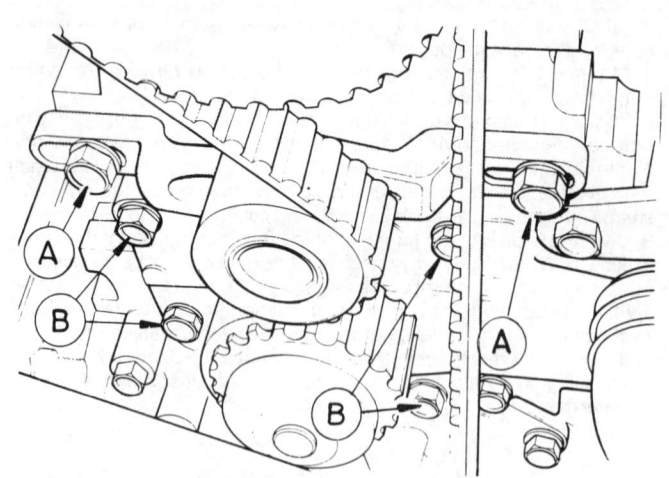

82   Chapter 2 Cooling, heating and ventilation systems

Fig. 2.8 Timing belt tensioner retaining bolts (A) and water pump securing bolts (B) (Sec 12)

3 Depending on model, remove the air filter housing for increased working clearance (see Chapter 3).
4 Remove the alternator drivebelt (see Chapter 12, Section 7).
5 Unscrew the two bolts securing the upper half of the timing cover and remove it.
6 Using a spanner on the crankshaft pulley bolt, turn the crankshaft until the TDC pointer on the camshaft sprocket aligns with the TDC mark on the cylinder head. To prevent the crankshaft rotating from this position, remove the starter motor (see Chapter 12, Section 12), and lock the flywheel ring gear with a cold chisel or other suitable tool.
7 Remove the crankshaft pulley bolt and draw the pulley off the crankshaft.
8 Unscrew the two bolts securing the lower half of the timing cover and remove it.
9 Using a dab of quick drying paint, mark the teeth of the timing belt and their notches on the sprockets so that the belt can be engaged in its original position on reassembly.
10 Slacken the two timing belt tensioner retaining bolts and slide the tensioner sideways to relieve the tautness of the belt. If the tensioner is spring-loaded, tighten one of the bolts to retain it in the slackened position (see Fig. 2.8).
11 Slip the timing belt off its sprockets and remove.
12 Slacken the water pump hose clip and remove the hose.
13 Remove the timing belt tensioner retaining bolts and remove the tensioner assembly.
14 Undo and remove the four bolts securing the water pump to the cylinder block, and withdraw the water pump from the vehicle.

15 Renewal of the pump will be necessary if there are signs of water leakage, roughness of the bearings, or excessive side play or endfloat at the sprocket.
16 Scrape away all traces of old gasket and ensure that the mating faces are clean and dry.
17 Lightly smear jointing compound on both sides of a new gasket and locate the gasket on the cylinder block face.
18 Place the pump in position, then fit and tighten the bolts to the specified torque.
19 Refit the timing belt tensioner (and spring where applicable), but only tighten the bolts finger tight at this stage.
20 Refit and tension the timing belt in accordance with Chapter 1, Section 25, ensuring that the TDC mark on the camshaft sprocket and the crankshaft key are still correctly aligned with their respective marks on the cylinder head and the oil pump casing. Remove the locking device from the flywheel ring gear during tensioning.
21 Reinsert the flywheel ring gear locking device. Locate the lower half of the timing cover and secure with its two bolts, tightening to the specified torque. Slide the crankshaft pulley onto the crankshaft and secure with its bolt, tightening to the specified torque (in Chapter 1).
22 Remove the locking device from the flywheel ring gear, and refit the starter motor in accordance with Chapter 12, Section 12.
23 Refit the upper half of the timing cover, tightening its bolts to the specified torque (in Chapter 1).
24 Refit and secure the water pump hose.
25 Reverse the removal operations given in paragraphs 1 to 4, then run the engine up to normal operating temperature and check the system for leaks.

13 Expansion tank – removal and refitting

1 Disconnect the battery earth terminal.
2 Partially drain the cooling system, in accordance with Section 3, to allow the expansion tank hoses to be disconnected from the expansion tank.
3 Release the hose clips and disconnect the hoses from the expansion tank. Secure the bottom hose with its disconnected end raised to avoid spillage.
4 Remove the single bolt retaining the expansion tank, and slide the other side of the tank free from its retaining bracket (photos).
5 Refitting is a reversal of the removal procedure. Refill the cooling system, making reference to Sections 3 and 4.

14 General description – heating and ventilation system

The heater is of the type which utilizes waste heat from the engine

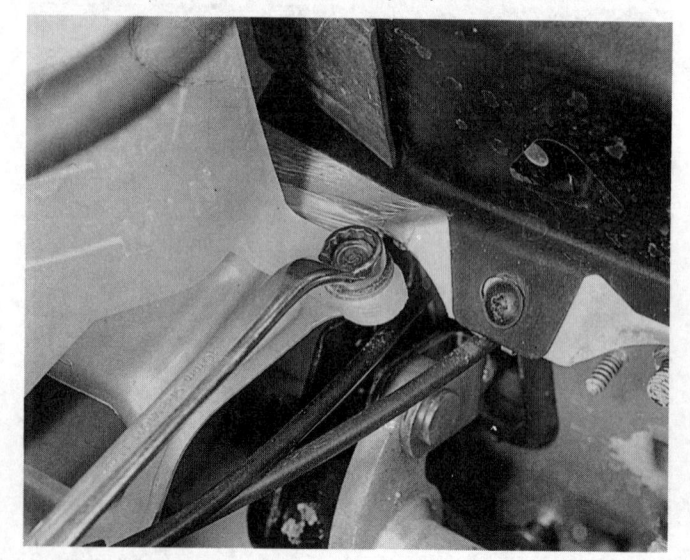

13.4A Remove the single bolt retaining the expansion tank ...

13.4B ... and release the expansion tank from its retaining bracket

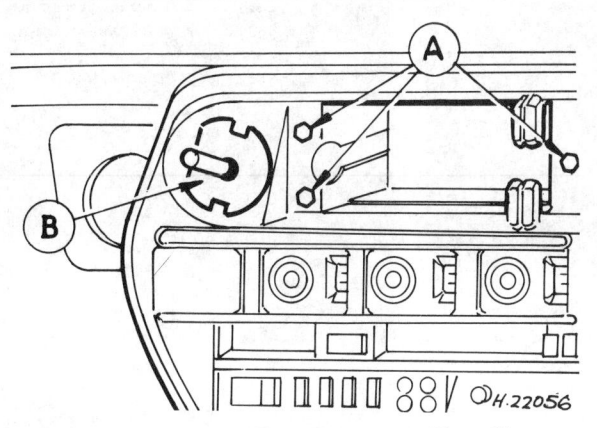

Chapter 2 Cooling, heating and ventilation systems

Fig. 2.9 Heater control panel securing screws (A), and fan motor control switch (B) (Sec 15)

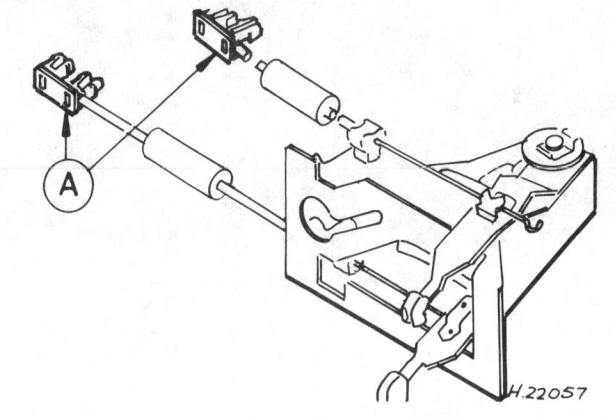

Fig. 2.10 Heater control panel. Assembly aids (A) fitted to heater casing flap valve end of cables (Sec 15)

coolant. The coolant is pumped through a radiator (matrix) in the heater casing assembly, in a similar manner to a domestic central heating system. Air is forced through this radiator by a three-speed radial fan, dispersing the heat into the vehicle interior.

Fresh air enters the vehicle through the grille slats between the windscreen and the rear edge of the bonnet, and passes through to the heater casing. Depending on the position of the heater slide controls, which actuate cable-controlled flap valves within the heater casing, the air is distributed, either heated or unheated, via the ducting to outlet vents. The main outlet vents in the facia are adjustable. The airflow passes through the passenger compartment to exit at the rear of the vehicle.

15 Heater controls – removal and refitting

1 Disconnect the battery earth terminal, and remove the radio (see Chapter 12).
2 Remove the heater slide facia and fan motor control switch, as described in Chapter 12, Section 15, paragraphs 11 and 12.
3 Disconnect the heater control cables from the heater casing assembly by releasing the outer cable abutments and disengaging the cable inner cores from their flap operating mechanisms (photo).
4 Undo the three heater control panel securing screws (Fig. 2.9), and remove the control panel (with its cables attached) from behind the facia.
5 Disconnect the heater control cables from their control panel levers, as required, by releasing their outer cable clamping covers and inner cable core securing clips (photos).
6 If renewing a heater control panel, note that the new unit, is supplied with control cables and assembly aids fitted (Fig. 2.10). The assembly aids ensure correct heater control adjustments during fitting, and must be removed thereafter.

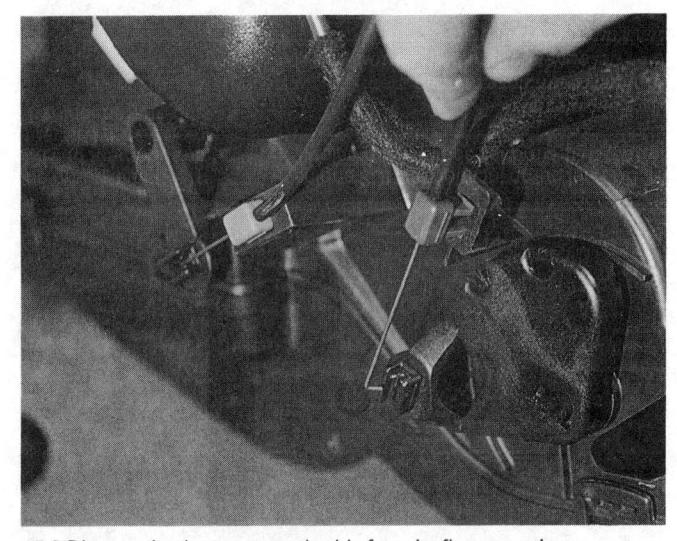

15.3 Disengaging heater control cable from its flap operating mechanism on heater casing

7 Refitting is a reversal of the removal procedure, adjusting the heater control cables to complete. The adjustment is made automatically by moving the heater slide control levers from their left-hand stop to their right-hand stop. When moving the control levers, a considerable amount of resistance may be encountered, which must be overcome.

16 Heater fan motor and resistor assembly – removal and refitting

1 Refer to Chapter 12, Section 49.

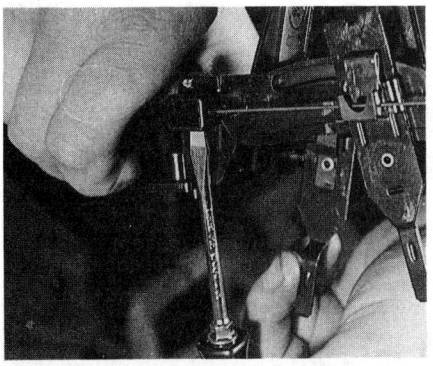

15.5A Releasing outer cable clamping cover ...

15.5B ... and removing it

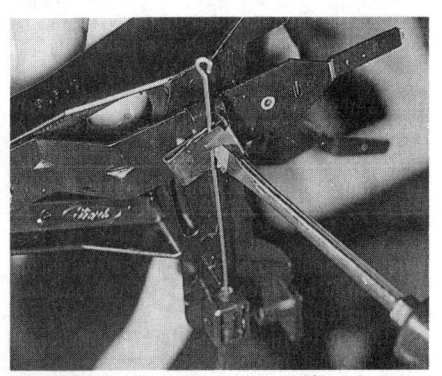

15.5C Releasing inner cable core from heater control lever

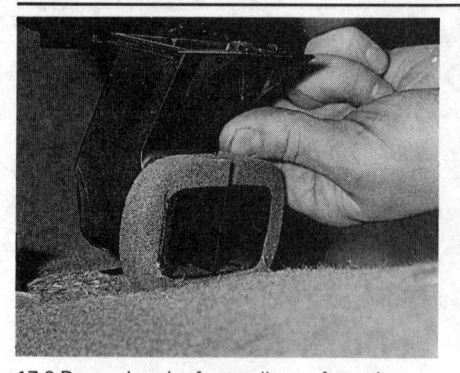

84   Chapter 2   Cooling, heating and ventilation systems

17.8 Removing the footwell vent from the base of the heater casing

17.9A Unclipping the lower heater casing cover

17.9B Withdrawing the heater matrix

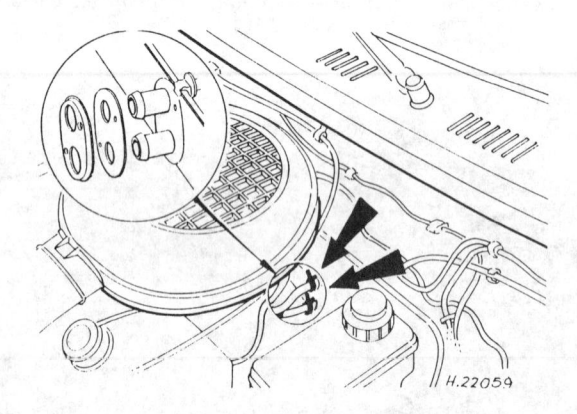

Fig. 2.11 Heater hoses (arrowed) on the dual heater matrix connector, located on the bulkhead (Sec 17)

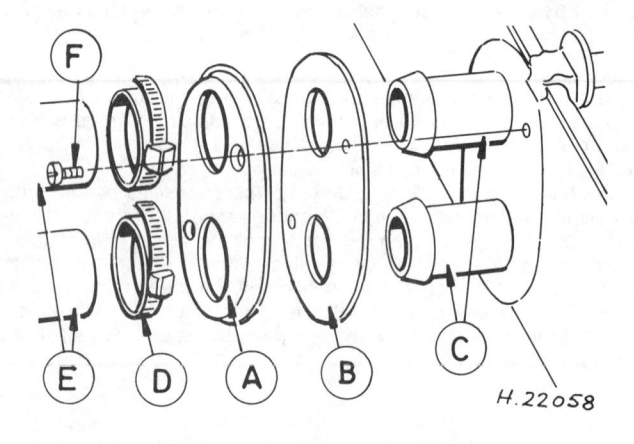

Fig. 2.12 Exploded view of dual heater matrix connector (Sec 17)

A   Cover plate
B   Gasket
C   Heater matrix connector pipe
D   Hose clips
E   Heater hoses
F   Screw

17 Heater radiator (matrix) – removal and refitting

1   It is advisable to perform this operation when the coolant is cold.
2   Disconnect the battery earth terminal.
3   Remove the cooling system pressure cap (refer to the warning in Section 1) to release any residual pressure.
4   Release the hose clips and disconnect the two heater hoses from their dual heater matrix connector on the bulkhead, at the rear of the engine compartment (Fig. 2.11). Plug the hoses to prevent excessive coolant loss, collecting spillage and coolant from the heater matrix in a suitable container.
5   If a compressed air line is advisable, blow through the heater matrix to remove any residual coolant, thus avoiding the possibility of spillage when the matrix is removed through the passenger compartment.
6   Undo the screws securing the heater matrix connector to the bulkhead, and detach its cover plate and gasket (Fig. 2.12).
7   Refer to Chapter 11, Section 56 and remove the centre console, if fitted.
8   Unclip and detach the footwell vent from the base of the heater casing (photo).
9   Unclip the lower cover from the heater casing and remove the heater matrix, being careful to avoid any residual coolant spillage in the passenger compartment (photos).
10 Refitting is a reversal of the removal procedure. Refill the cooling system, making reference to Sections 3 and 4.

18 Heater casing assembly – removal and refitting

1   Remove the facia, as described in Chapter 11, Section 58.
2   Remove the heater radiator (matrix), as described in the previous Section.
3   Disconnect the heater control cables from the heater casing assembly (see Section 15).
4   Disconnect the side outlet vent ducting from the heater casing. The side outlet ducts can be removed by undoing the screws securing them to the bulkhead, as required (photo).
5   Undo the two nuts securing the heater casing (Fig. 2.13), detach and remove it from the vehicle.
6   Refitting is a reversal of the removal procedure, making reference to the above mentioned text (photo).

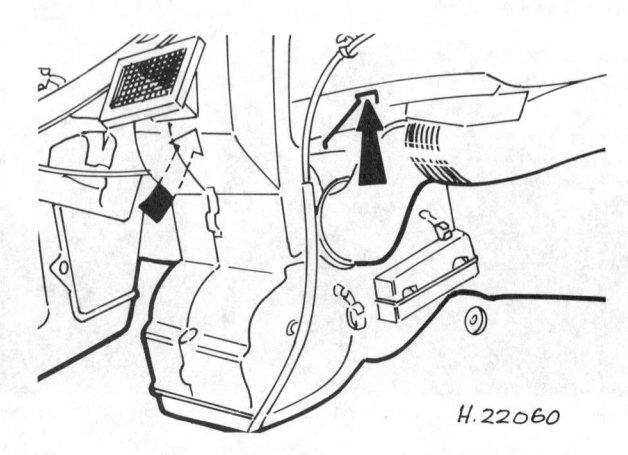

Fig. 2.13 Heater casing securing nuts (arrowed) (Sec 18)

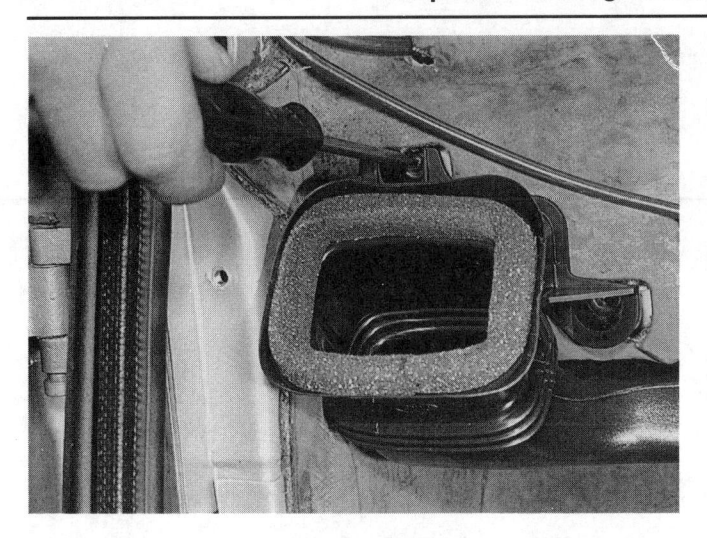

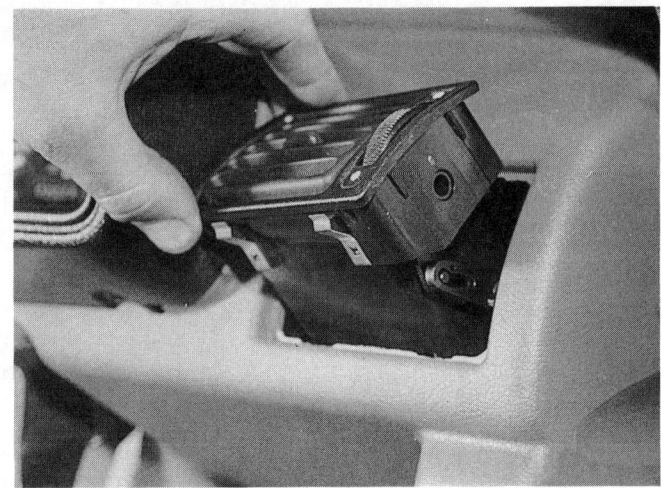

Chapter 2 Cooling, heating and ventilation systems

18.4 Detaching a side outlet duct (facia removed)

18.6 Refitting an adjustable side vent

19 Fault diagnosis – cooling system

Symptom	Reason(s)
Overheating	Low coolant level (this may be the result of overheating for other reasons) Drivebelt slipping (sometimes accompanied by a shrieking noise under a sharp rise in engine speed) – OHV (HCS) engine only Pressure cap defective Thermostat defective Radiator core blockage, or airflow impeded Kinked or internally collapsed hose impeding coolant flow Ignition timing incorrect Radiator cooling fan motor or thermal switch defective Cylinder head gasket blown Water pump defective Oil level in sump too low Brakes binding
Overcooling	Thermostat missing, defective or incorrect heat range Radiator cooling fan thermal switch faulty (fan runs continuously)
Coolant loss	Perished, cracked or damaged hoses Loose hose clips Radiator core, expansion tank or heater matrix leaking Pressure cap defective Boiling due to overheating Localised leakage at water pump or thermostat Cylinder head gasket blown Cylinder head or block cracked
Corrosion	Infrequent draining and flushing Incorrect antifreeze mixture strength or incorrect type

Note: *Never add cold water to an overheated engine, as severe internal damage may result*

20 Fault diagnosis – heating and ventilation system

Symptom	Reason(s)
Poor heat output	Thermostat missing, defective or incorrect heat range Heater matrix blocked Kinked or internally collapsed heater hose impeding coolant flow Radiator cooling fan thermal switch faulty (fan runs continuously)
Poor demisting/defrosting and/or air distribution capability	Poor heat output (see above) Incorrectly adjusted or defective controls Heater fan motor inoperative Air ducting blocked or disconnected

Chapter 3
Fuel, exhaust and emission control systems

For modifications, and information applicable to later models, see Supplement at end of manual

Contents

Part A: Carburettor fuel system
Accelerator cable (CTX automatic transmission) – general	8
Accelerator cable (manual transmission) – removal, refitting and adjustment	7
Accelerator pedal – removal and refitting	9
Accelerator pump diaphragm (Weber TLM carburettor) – removal and refitting	21
Air filter element – renewal	4
Air filter housing – removal and refitting	5
Air filter intake-air temperature control – description and testing	6
Automatic choke (Weber TLD carburettor) – removal, overhaul and refitting	36
Automatic choke and fast-idle speed (Weber TLD carburettor) – adjustment	35
Carburettor (Weber DFTM) – description	28
Carburettor (Weber DFTM) – dismantling, cleaning, inspection and reassembly	32
Carburettor (Weber DFTM) – idle speed, fuel mixture and fast-idle adjustment	29
Carburettor (Weber DFTM) – removal and refitting	31
Carburettor (Weber TLD) – description	33
Carburettor (Weber TLD) – dismantling, cleaning, inspection and reassembly	38
Carburettor (Weber TLD) – idle speed and fuel mixture adjustment	34
Carburettor (Weber TLD) – removal and refitting	37
Carburettor (Weber TLDM) – description	22
Carburettor (Weber TLDM) – dismantling, cleaning, inspection and reassembly	27
Carburettor (Weber TLDM) – idle speed, fuel mixture and fast-idle adjustment	23
Carburettor (Weber TLDM) – removal and refitting	26
Carburettor (Weber TLM) – cleaning	19
Carburettor (Weber TLM) – description	16
Carburettor (Weber TLM) – idle speed, fuel mixture and fast-idle adjustment	17
Carburettor (Weber TLM) – removal and refitting	20
Choke control cable – removal, refitting and adjustment	10
Fuel pump (mechanical) – testing, removal and refitting	15
Fuel system – precautions	2
Fuel tank – removal and refitting	12
Fuel tank filler pipe – removal and refitting	13
Fuel tank sender unit – removal and refitting	14
Fuel tank ventilation tube – description, removal and refitting	11
General description	1
Maintenance and inspection	3
Needle valve and float (Weber TLM carburettor) – removal, refitting and adjustment	18
Throttle kicker (Weber DFTM) carburettor – checking and adjustment	30
Throttle kicker (Weber TLDM) – checking and adjustment	24
Throttle kicker control solenoid (Weber TLDM/DFTM carburettor) – removal and refitting	25

Part B: Fuel injection system
Accelerator cable – removal, refitting and adjustment	44
Accelerator pedal – removal and refitting	45
Air filter element – renewal	42
Air filter housing – removal and refitting	43
Base idle speed (1.6 litre EFi) – checking and adjustment	61
Fuel filter – removal and refitting	52
Fuel injection unit (1.4 litre CFi) – removal and refitting	57
Fuel injector (1.4 litre CFi) – removal and refitting	54
Fuel injectors (1.6 litre EFi) – removal and refitting	59
Fuel mixture/CO content (1.6 litre EFi) – checking and adjustment	60
Fuel pressure regulator (1.4 litre CFi) – removal and refitting	53
Fuel pressure regulator (1.6 litre EFi) – removal and refitting	58
Fuel pump – testing	49
Fuel pump/sender unit – removal and refitting	50
Fuel system – precautions	40
Fuel tank – removal and refitting	47
Fuel tank filler pipe – removal and refitting	48
Fuel tank ventilation tube – general	46
General description	39
Idle speed control valve (1.6 litre EFi) – removal, cleaning and refitting	62
Inertia switch (fuel cut-off) – removal and refitting	51
Maintenance and inspection	41
Throttle-plate control motor (1.4 litre CFi) – removal and refitting	55
Throttle position sensor (1.4 litre CFi) – removal and refitting	56
Throttle position sensor (1.6 litre EFi) – removal and refitting	63

Part C: Manifolds, exhaust and emission control systems
Emission control systems – general description	69
Emission control system components (including vacuum valves and fuel traps) – removal and refitting	70
Exhaust manifold – general	67
Exhaust system – renewal	68
General description	64
Inlet manifold – general	66
Maintenance and inspection	65

Part D: Fault diagnosis
Fault diagnosis – carburettor fuel systems	71
Fault diagnosis – fuel injection systems	72

Chapter 3 Fuel, exhaust and emission control systems

Specifications

Part A: Carburettor fuel system

General

Fuel tank capacity	42 litres (9.25 gals)
Fuel octane rating:	
Leaded	97 RON (4-star)
Unleaded	95 RON (Premium)
Carburettor application and type:	
1.0 litre HCS engines	Weber (1V) TLM
1.1 litre HCS engines	Weber (2V) TLDM
1.4 litre CVH engines	Weber (2V) DFTM
1.6 litre CVH engines	Weber (2V) TLD
Air cleaner element	Champion W153 (1.0, 1.1) or Champion W226 (1.4, 1.6)

Fuel pump

Type	Mechanical, operated by an eccentric lobe on the camshaft
Delivery pressure	0.24 to 0.38 bar (3.5 to 5.5 lbf/in^2)

Weber (1V) TLM carburettor

Idle speed	750 ± 50 rpm (radiator cooling fan on)
Idle mixture CO content	1.0 ± 0.5%
Fast idle speed	3400 ± 100 rpm
Float height	26.0 ± 1.0 mm
Venturi diameter	23 mm
Main jet	110
Air correction jet	220

Weber (2V) TLDM carburettor

Idle speed	750 ± 50 rpm (radiator cooling fan on)	
Idle mixture CO content	1.0 ± 0.5%	
Fast idle speed:		
Manual transmission	2800 rpm	
CTX automatic transmission	2600 rpm	
Float height	29.0 ± 1.0 mm	
Throttle kicker speed (CTX automatic transmission only)	1050 to 1150 rpm	
	Primary	**Secondary**
Venturi diameter	26 mm	28 mm
Main jet:		
Manual transmission	92	122
CTX automatic transmission	92	112
Emulsion tube	F113	F75
Air correction jet	195	155

Weber (2V) DFTM carburettor

Idle speed:		
Manual transmission	800 ± 50 rpm (radiator cooling fan on)	
CTX automatic transmission	850 ± 50 rpm (radiator cooling fan on)	
Idle mixture CO content	1.5 ± 0.25%	
Fast idle speed	2800 ± 100 rpm	
Choke pull-down	2.7 to 3.2 mm	
Float height	8.0 ± 0.5 mm	
Throttle kicker speed:		
Manual transmission	1300 ± 50 rpm	
CTX automatic transmission	1100 ± 50 rpm (in Neutral)	
	Primary	**Secondary**
Venturi diameter	21 mm	23 mm
Main jet	100	125
Air correction jet	210	155
Emulsion tube	F22	F60
Idle jet	42	60

Weber (2V) TLD carburettor

Idle speed	800 ± 50 rpm (radiator cooling fan on)	
Idle mixture CO content	1.5 ± 0.5%	
Fast idle speed	1800 ± 50 rpm (refer to text)	
Choke pull-down	4.7 ± 0.5 mm	
Float height	29.0 ± 0.5 mm	
	Primary	**Secondary**
Venturi diameter	21	23
Main jet	117	127
Emulsion tube	F105	F71
Air correction jet	185	125

Torque wrench settings

	Nm	lbf ft
Fuel pump bolts (mechanical):		
HCS engines	16 to 20	12 to 15
CVH engines	14 to 18	10.5 to 13
Inlet manifold nuts	16 to 20	12 to 15
Carburettor:		
HCS engines	17 to 21	12.5 to 16
CVH engines	12 to 21	9 to 16
Exhaust manifold:		
HCS engines	21 to 25	16 to 18
CVH engines	14 to 17	10.5 to 12.5
Exhaust manifold-to-downpipe bolts	35 to 40	26 to 29.5
Exhaust downpipe-to-exhaust system bolts	48 to 64	35 to 47
Exhaust U-bolt clamps	35 to 40	26 to 29.5

Part B: Fuel injection system

General

Fuel tank capacity	42 litres (9.25 gals)
Fuel octane rating:	
Leaded (**not** 1.4 litre (CFi)	97 RON (4-star)
Unleaded	95 RON (Premium)
Fuel pump	In-tank electric fuel pump, integral with fuel level sender unit
Fuel filter element	Champion L204 (XR2i only)
System application:	
1.4 litre CFi	Single-point fuel injection system, controlled by EEC-IV engine management module. Catalytic converter reducing exhaust emissions
1.6 litre EFi	Multi-point fuel injection system, controlled by EEC-IV engine management module
Air cleaner element	Champion W226 (1.4) or Champion U557 (1.6)

1.4 litre CFi system

Idle speed	900 ± 50 rpm (refer to Ford dealer for adjustment - see text Section 55)
Fuel mixture	Closed loop control (not adjustable)
Regulated fuel pressure	1 bar (fuel pump running continuously)
Lubricant for injector seals	Clean engine oil

1.6 litre EFi system

Idle speed (operational)	900 ± 50 rpm
Idle mixture CO content	0.8 ± 0.25% (radiator cooling fan on)
Base idle speed	750 ± 50 rpm (refer to text)
Regulated fuel pressure:	
Engine running	2.3 to 2.5 bar
Engine not running	3.0 ± 0.1 bar
Lubricant for pressure regulator and injector seals	Clean engine oil

Torque wrench settings

	Nm	lbf ft
Inlet manifold	16 to 20	12 to 15
Coolant hose connector to lower section of inlet manifold (1.6 litre EFi)	13 to 17	10 to 12.5
CFi unit to inlet manifold	9 to 11	6.5 to 8
Exhaust manifold	14 to 17	10.5 to 12.5
Exhaust manifold-to-downpipe bolts	35 to 40	26 to 29.5
Exhaust downpipe-to-exhaust system bolts	48 to 64	35 to 47
Exhaust U-bolt clamps	35 to 40	26 to 29.5
Fuel pressure regulator bolts (1.6 litre EFi)	8 to 12	6 to 9
Fuel distributor pipe (rail) (1.6 litre EFi)	20 to 25.5	15 to 19
Fuel filter unions	14 to 20	10.5 to 15
HEGO sensor (1.4 litre CFi)	50 to 70	37 to 51

PART A: CARBURETTOR FUEL SYSTEM

1 General description

The fuel system on all models with carburettor induction is composed of a rear mounted fuel tank, a carburettor and an air filter.

The fuel tank is mounted under the floor pan, behind the rear seats. The tank has a 'ventilation to atmosphere' system through a combined roll-over/anti-trickle-fill valve assembly mounted in the left hand rear wheelarch. A conventional fuel level sender unit is mounted through the tank upper surface. A filler neck sensing pipe, integral with the fuel tank filler pipe arrangement, will shut-off the petrol pump filler gun when the pre-determined level of fuel in the tank has been reached.

Two types of mechanical fuel pump are featured, dependent on engine type. The pump fitted to HCS engines is operated by a pivoting rocker arm; one end rests on an eccentric lobe on the engine camshaft, and the other end is attached to the fuel pump diaphragm. The pump fitted to CVH engines is operated by a separate independent pushrod; one end rests on an eccentric lobe on the engine camshaft, and the other end rests on the pump actuating centre rod which moves the diaphragm. Both types of mechanical fuel pump incorporate a nylon mesh filter, located in the pump top housing, and also feature a fuel return line sending surplus fuel back to the fuel tank. The pumps are sealed units and cannot be overhauled.

Chapter 3 Fuel, exhaust and emission control systems

Four different types of Weber carburettor are featured in the range (see Specifications). Further description of the individual carburettors may be found later in this Chapter.

The air filter housing has a waxstat-controlled air intake, supplying either hot air from a heat box mounted around the exhaust manifold or cold air from the front of the engine compartment. Further description and testing may be found later in this Chapter.

2 Fuel system – precautions

Warning: *Many of the procedures in this Chapter entail the removal of fuel pipes and connections which may result in some fuel spillage. Before carrying out any operation on the fuel system refer to the precautions given in Safety First! at the beginning of this manual and follow them implicitly. Petrol is a highly dangerous and volatile liquid and the precautions necessary when handling it cannot be overstressed*

Reference must also be made to Chapter 4, Section 2 or 12, as applicable, and any further safety-related referred text contained within the appropriate Section, before working on the vehicle. When disconnecting the automatic choke or other coolant hoses, ensure that the cooling system is not pressurised (refer to Chapter 2). Do not work on or near a hot exhaust system.

Certain adjustment points in the fuel system are protected by tamperproof caps, plugs or seals. In some territories, it is an offence to drive a vehicle with broken or missing tamperproof seals. Before disturbing a tamperproof seal, check that no local or national laws will be broken by doing so, and fit a new tamperproof seal after adjustment is complete, where required by law. **Do not** break tamperproof seals on a vehicle while it is still under warranty.

When working on fuel system components, scrupulous cleanliness must be observed, and care must be taken not to introduce any foreign matter into fuel lines or components. Carburettors in particular are delicate instruments, and care should be taken not to disturb any components unnecessarily. Before attempting work on a carburettor, ensure that the relevant spares are available; note that a complete stripdown of a carburettor is unlikely to cure a fault which is not immediately obvious, without introducing new problems. If persistent problems are encountered, it is recommended that the advice of a Ford dealer or carburettor specialist is sought. Most dealers will be able to provide carburettor re-jetting and servicing facilities, and if necessary it should be possible to purchase the relevant reconditioned carburettor.

3 Maintenance and inspection

Note: *Refer to Section 2 before proceeding.*

1 At the intervals specified in the *'Routine maintenance'* Section at the beginning of this manual, carry out the following operations.
2 With the vehicle raised on a hoist or supported securely on axle stands, carefully inspect the fuel pipes, hoses and unions for chafing, leaks and corrosion. Renew any pipes that are severely pitted with corrosion or in any way damaged. Renew any hoses that show signs of cracking or other deterioration/damage.

4.1 Undoing the air filter housing securing screws (1.1 litre HCS model shown)

3 Examine the fuel tank for signs of leakage, corrosion and damage.
4 From within the engine compartment, check the security of all fuel and vacuum hose attachments, and inspect the hoses for kinks, chafing, leaks and any damage or deterioration.
5 Check the inlet manifold nuts/bolts for correct tightness.
6 Check the operation of the throttle linkage and lubricate the linkage and the cable with a few drops of clean engine oil. Where fitted, the choke cable should be similarly attended to.
7 Check that the air filter intake air temperature control is functioning correctly (as applicable), and also renew the air filter element(s).
8 On CTX automatic transmission equipped vehicles, check that the throttle kicker operation is within specification; adjust the throttle kicker as necessary, making reference to the appropriate text.
9 Check and if necessary adjust the carburettor idle speed and fuel mixture settings, making reference to the appropriate text.
10 Refer to Section 65 for details of maintenance and inspection procedures on the exhaust and emission control systems.

4 Air filter element – renewal

1 Remove the air filter housing securing screws on the housing top face (photo), then release the housing lid retaining clips. Detach the lid.
2 On some models the air filter element may be lifted out at this stage (photo), while on others, two further nuts have to be removed. Discard the old element.
3 Wipe the inside of the air filter housing using a clean lint-free cloth, and fit a new element.

4.2 Lifting out the air filter element (1.1 litre HCS model shown)

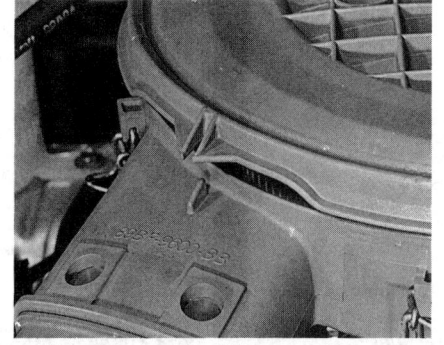

4.4A Air filter housing lid locating lug

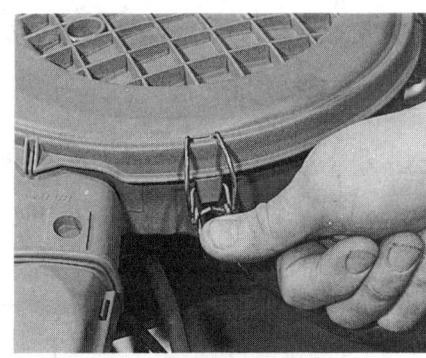

4.4B 'Locking' an air filter housing lid retaining clip

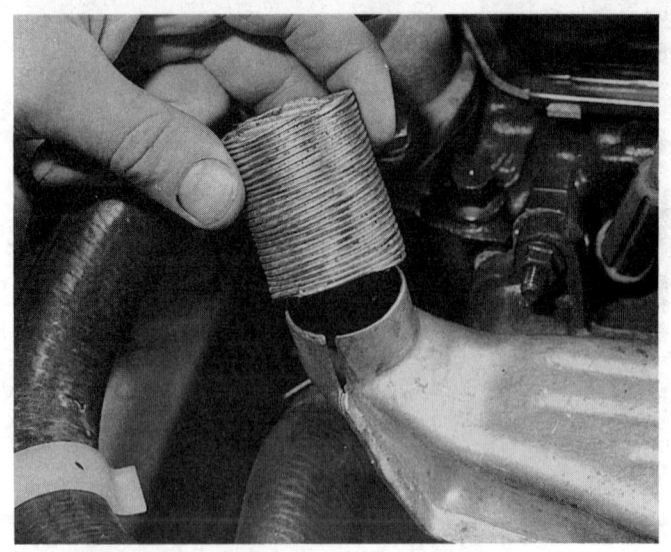

Chapter 3 Fuel, exhaust and emission control systems

the underside of the air filter housing. Disconnect all vacuum hoses, having identified them for subsequent refitting (as applicable).
8 Remove the three securing screws on the housing top face, and detach the assembly.
9 Refer to paragraph 6 of this Section.

6 Air filter intake-air temperature control – description and testing

1 The air filter housing features a waxstat intake air temperature control system which functions automatically to meet engine requirement and operates in the following manner. The air filter housing has two sources of air supply – cold air from the front of the engine compartment and hot air from a heat box mounted around the exhaust manifold. A flap valve in the air filter housing intake spout regulates the flow from each air source to achieve the required temperature. The flap valve is operated by a rod and spring mechanism that is controlled by a wax pellet, located in the intake spout. As the wax pellet starts to melt (at a temperature of 15°C) it expands and closes the flap valve, thus shutting off the hot air to the engine. The flap valve is fully closed at a temperature of 35°C.

Testing
2 With the engine cold, remove the air filter housing (refer to Section 5) and check that the flap valve is open to allow hot air to enter.
3 Refit the air filter housing, reconnect the battery earth terminal and warm the engine to normal operating temperature.
4 Switch off the engine and again disconnect the battery earth terminal and remove the air filter housing. If the intake air temperature control system is functioning correctly, the flap valve should be closed to allow only cold air to enter. If the system is not functioning correctly, the air filter housing should be renewed.
5 Refit the air filter housing making reference to Section 5, then reconnect the battery earth terminal to complete.

5.6 Refitting the pre-heat tube to the exhaust manifold heat box

4 Refit the housing lid and secure with its retaining clips (photos).
5 Insert and tighten the air filter housing securing screws.

5 Air filter housing – removal and refitting

1 Disconnect the battery earth terminal.

1.0 litre and 1.0 litre HCS engines (with carburettor)
2 Disconnect the crankcase ventilation hose or filter/adaptor, as applicable, from the air filter housing.
3 Disconnect all vacuum hoses, having identified them for subsequent refitting (as applicable), and unclip the fuel trap located to the rear of the air filter housing.
4 Where fitted, disconnect the multi-plug from the air-charge temperature (ACT) sensor located on the underside of the air filter housing.
5 Remove the securing screws on the housing top face, and detach the assembly.
6 Refitting is a reversal of the removal procedure, having ensured that the pre-heat tube between the exhaust manifold heat box and the air filter housing is in good condition and is correctly located (photo).

1.4 litre and 1.6 litre CVH engines (with carburettor)
7 Release and pull out the crankcase ventilation filter/adaptor from

7 Accelerator cable (manual transmission) – removal, refitting and adjustment

1 Disconnect the battery earth terminal.
2 Disconnect the cable from the top of the accelerator pedal, release the grommet from the bulkhead panel and lift the grommet clear of the pedal. An assistant may be required to push the grommet out from the driver's footwell whilst you withdraw the cable into the engine compartment.
3 Disconnect any cable retaining clips/ties in the engine compartment as applicable.
4 Refer to Section 5 and remove the air filter housing.

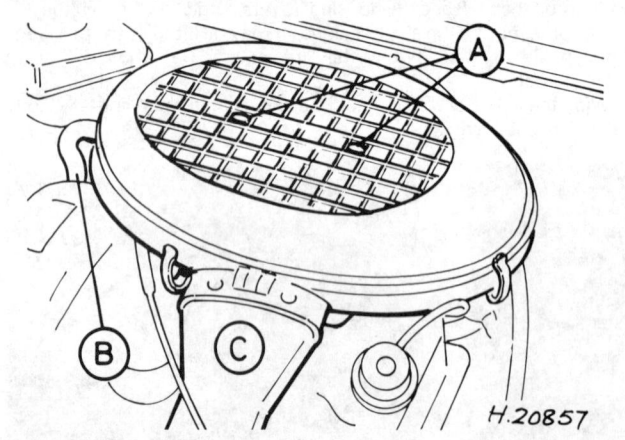

Fig. 3.1 Air filter housing (1.4 and 1.6 litre carburettor models) (Sec 5)

A Securing screws
B Crankcase ventilation hose leading to the filter/adaptor
C Main air intake

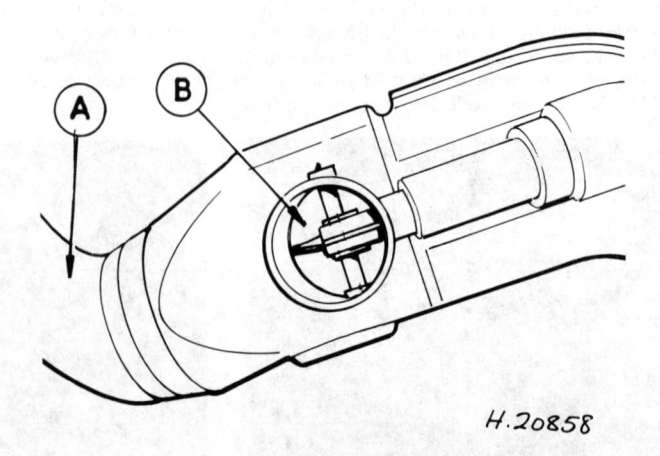

Fig. 3.2 Intake-air temperature control flap valve (underside view of air filter housing) (Sec 6)

A Main air intake
B Flap valve open

Chapter 3 Fuel, exhaust and emission control systems 91

7.5A Remove the accelerator outer cable securing clip ...

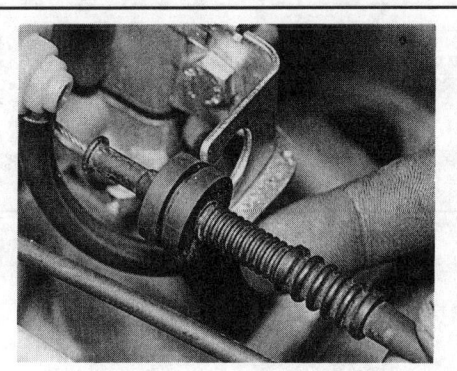

7.5B ... then release from its abutment bracket

7.5C Disconnecting the accelerator inner cable from its throttle plate linkage (cable slot on linkage arrowed)

5 Remove the cable securing clip(s) and detach the outer cable from its abutment bracket, and the inner cable from the throttle-plate linkage (photos). Remove the cable from the vehicle.
6 To install, route the cable through the bulkhead panel and reconnect it to the accelerator pedal (note that a plastic sleeve is supplied with new cables for the purpose of routing through the bulkhead panel (Fig. 3.3). Secure the cable grommet to the bulkhead panel.
7 Lightly lubricate the grommet on the cable abutment bracket with a soap solution, and reconnect the inner cable to the carburettor throttle-plate linkage (through the grommet).
8 Fully depress the accelerator pedal and release it – the outer cable should move into the grommet. Refit the securing clip to the abutment bracket, then have an assistant fully depress the accelerator pedal to enable a check that the throttle-plate is fully open. Repeat the adjustment operation if necessary.
9 Refit the air filter housing in accordance with Section 5.
10 Reconnect the battery earth terminal.

Adjustment only operation
11 Disconnect the battery earth terminal, and remove the air filter housing in accordance with Section 5.
12 Remove the clip securing the outer cable to its abutment bracket, and partially withdraw the outer cable from the grommet. Lightly lubricate the grommet with a soap solution.
13 Proceed as described in paragraphs 8 to 10 inclusive.

8 Accelerator cable (CTX automatic transmission) – general

1 The system for operating the throttle-plate in the carburettor on CTX automatic transmission equipped vehicles is completely different to that employed on manual transmission vehicles. The cable from the accelerator pedal leads to a linkage mechanism which is bolted to the transmission housing. Two further cables lead from this linkage mechanism (refer to Chapter 6), one of which operates the throttle-plate in the carburettor.
2 As all three cables have to be adjusted at the same time, and access to a Ford special tool is required, it is recommended that a Ford dealer be entrusted with cable adjustments, or renewal.

9 Accelerator pedal – removal and refitting

1 Pull back the carpet and insulation in the area of the accelerator pedal, as necessary.
2 Disconnect the accelerator cable from the top of the pedal.
3 Remove the circlip securing the accelerator pedal to its shaft and slide the pedal off.
4 Refitting is a reversal of the removal procedure. Check that the throttle-plate is fully open when the accelerator pedal is fully depressed; if not, adjust the accelerator cable in accordance with Section 7 or Section 8 (as applicable).

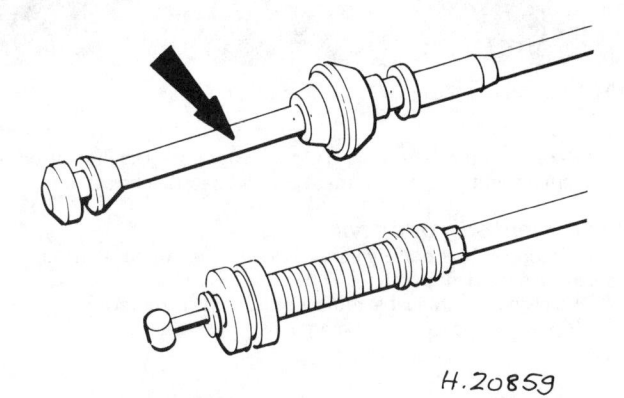

Fig. 3.3 Accelerator cable end fittings, with plastic sleeve

10 Choke control cable – removal, refitting and adjustment

1 Disconnect the battery earth terminal.
2 Refer to Section 5 and remove the air filter housing.
3 Remove the choke control knob on the side of the steering column shroud by pushing in the pin located on the side of the knob and withdrawing.
4 Detach the steering column lower shroud, disconnect the multi-plug from the choke warning light switch/pull control assembly, and unscrew the collar securing the switch/pull control assembly to the shroud (refer to Chapter 9).
5 Disconnect the choke inner cable from its location on the carburettor choke linkage, then release the outer cable retaining clip (photos).
6 Push the bulkhead panel grommet through the bulkhead and remove the cable.
7 To install, first route the cable through the bulkhead then secure the bulkhead panel grommet to its location.
8 Reconnect the choke inner cable to its location on the carburettor choke linkage.
9 Refit the choke warning light switch/pull control assembly to the lower steering column shroud and secure with the collar, then reconnect the multi-plug.
10 Refit and secure the lower steering column shroud, then refit the choke control knob.
11 Pull the choke control knob to the fully 'on' position. Hold the choke in the fully 'on' position at the carburettor, then secure the outer cable with its retaining clip.
12 Re-check the choke cable adjustment. Ensure that the control knob is pulled fully out and that the full-choke stop on the carburettor is reached; if this is not the case, re-adjust as necessary.
13 Check the operation of the choke, and ensure that the choke linkage returns to its 'off' position with the choke plate at 90° to the carburettor venturi.
14 Refit the air filter housing in accordance with Section 5.

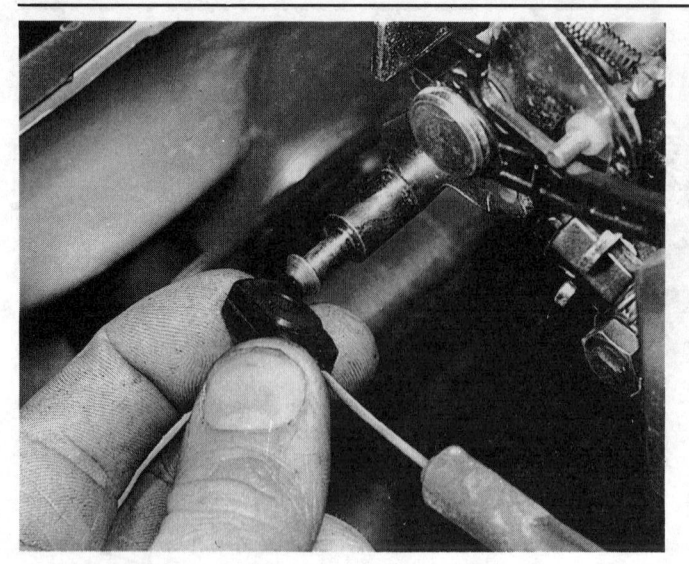

92   Chapter 3   Fuel, exhaust and emission control systems

10.5A Disconnect the inner cable from the choke linkage

10.5B Choke outer cable retaining clip released

15   Reconnect the battery earth terminal, and establish that the choke 'on' warning light on the instrument cluster operates correctly.

Adjustment only operation
16   Disconnect the battery earth terminal, then remove the air filter housing in accordance with Section 5.
17   Release the outer cable retaining clip at the carburettor.
18   Proceed as described in paragraphs 11 to 15 inclusive.

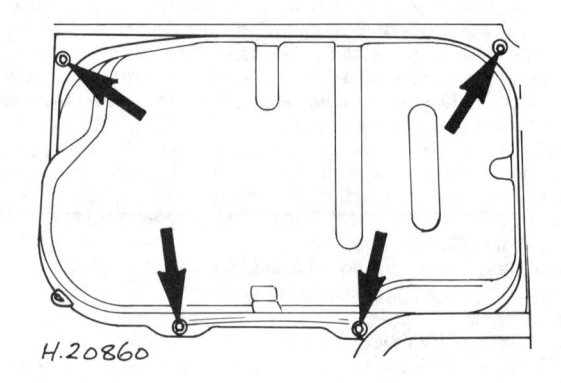

Fig. 3.4 Fuel tank securing bolts (arrowed) (Sec 11)

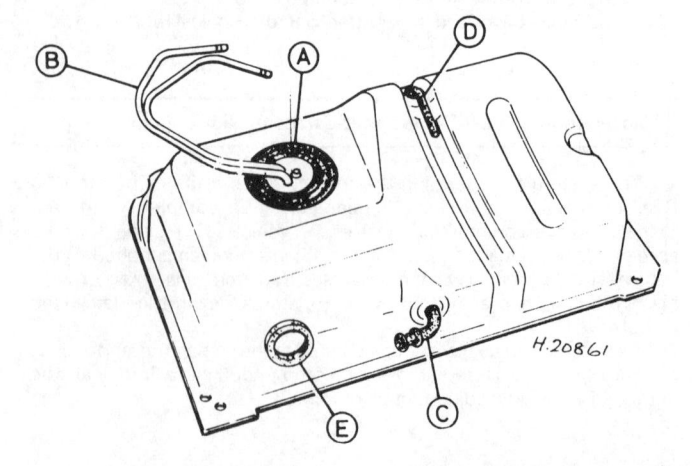

11   Fuel tank ventilation tube – description, removal and refitting

Note: *Refer to Section 2 before proceeding.*

1   The fuel tank ventilation tube runs from the top surface of the fuel tank to the combined roll-over/anti-trickle-fill valve assembly mounted in the left-hand rear wheelarch (photo). *Its purpose is to eliminate any possibility of vacuum or pressure build-up in the fuel tank.*
2   To remove, first disconnect the battery earth terminal.
3   Slacken the left-hand rear roadwheel nuts, raise the rear of the vehicle and support securely using axle stands. Remove the left-hand rear roadwheel.
4   Support the fuel tank from underneath on a suitable jack, using a large thick sheet of board to spread the weight, then undo and remove the four fuel tank securing bolts (Fig. 3.4).
5   Lower the fuel tank slightly in such a manner so as to allow access to disconnect the ventilation tube from the tank top surface. Ensure that the fuel tank does not foul or strain any adjacent components as it is lowered; take appropriate action, as necessary.
6   Disconnect the ventilation tube from the combined roll-over/anti-trickle-fill valve, release the tube from its retaining clips and remove.
7   Refitting is a reversal of the removal procedure, ensuring that the fuel tank filler pipe is located correctly with the tank.

12   Fuel tank – removal and refitting

Note: *Refer to Section 2 before proceeding*

1   Run the fuel level as low as possible before removing the tank.
2   Disconnect the battery earth terminal.
3   Remove the tank filler cap, then syphon (**not** by mouth) or pump out the tank contents (there is no drain plug). The fuel should ideally be pumped directly into a suitable sealed container to minimise risks.
4   Raise the rear of the vehicle and support securely using axle stands.
5   Unclip and disconnect the fuel feed and return hoses located in front of the fuel tank (Fig. 3.6), and allow any residual fuel to drain into a container which can be sealed.

Fig. 3.5 General view of the fuel tank assembly (Sec 11)

A   Sender unit/combined pump and sender unit (as applicable)
B   Fuel feed and return pipes
C   Fuel tank ventilation tube
D   Filler neck sensing pipe connection
E   Fuel filler pipe seal

6   Disconnect the filler neck sensing pipe connection from the rear of the tank (photo).
7   Support beneath the tank to hold it in position and remove its four securing bolts (see Fig. 3.4).

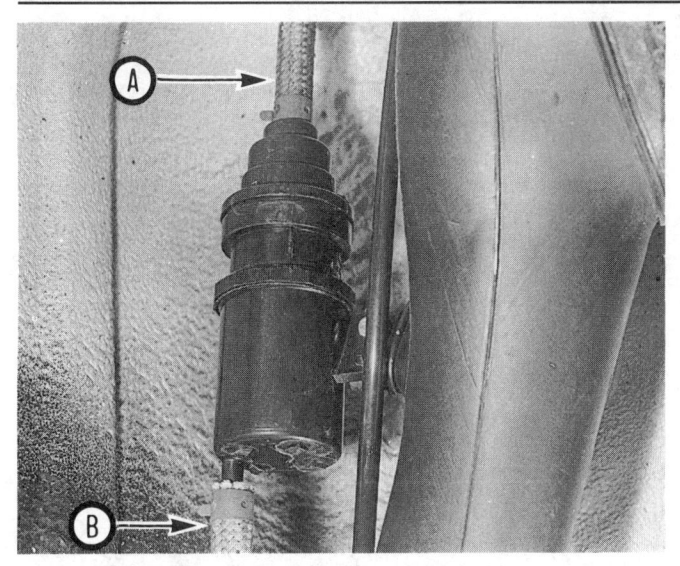

Chapter 3 Fuel, exhaust and emission control systems   93

11.1 Combined roll-over anti-trickle-fill valve assembly

A Tube ventilating to atmosphere
B Ventilation tube from fuel tank

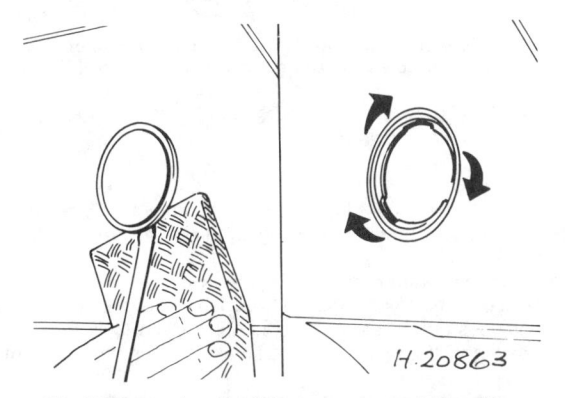

Fig. 3.7 Removing the filler cap surround (Sec 13)

8 Partially lower the fuel tank and disconnect the ventilation tube from the tank top surface and also disconnect the sender unit multi-plug (photo). The filler pipe should release from its fuel tank seal location as the tank is withdrawn.

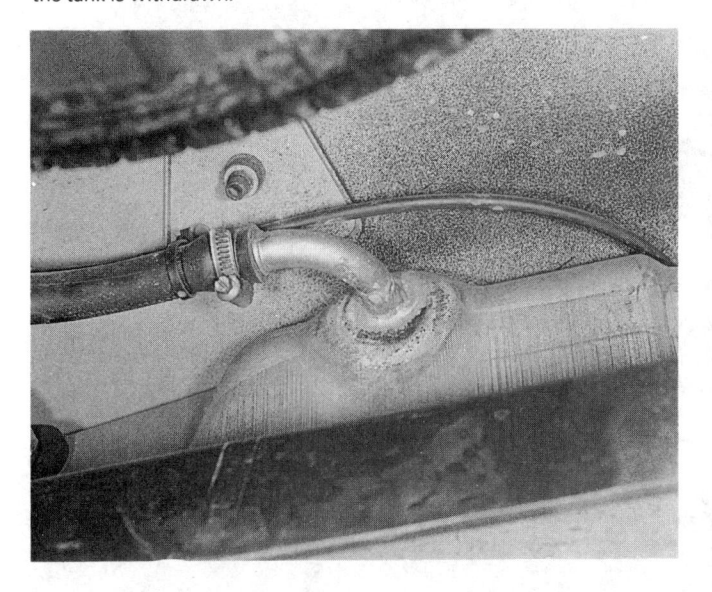

12.6 Filler neck sensing hose connection at the rear of the fuel tank

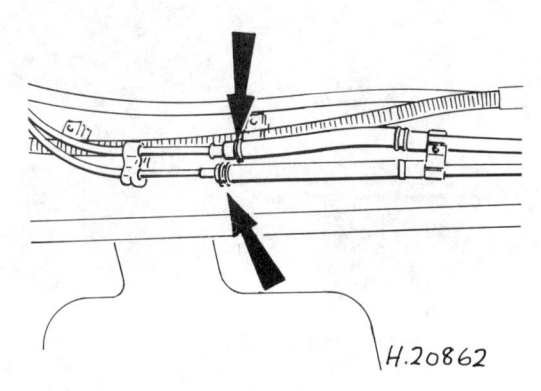

Fig. 3.6 Fuel feed and return pipe connections (arrowed) (Sec 12)

9 Prior to refitting, check the fuel tank for any damage or corrosion. If required, the sender unit may be withdrawn for internal examination of the tank. If the tank is contaminated with sediment or water, swill it out with clean petrol. Do **not**, *under any circumstances*, attempt to repair a leaking or damaged petrol tank – this **must** be left to a specialist. The only alternative is renewal of the tank.
10 Check the condition of the filler pipe seal in the fuel tank and renew if necessary.
11 Refitting is a reversal of the removal procedure, but apply a light smear of grease to the filler pipe seal to aid filler pipe entry. Ensure that all connections are securely fitted. If evidence of contamination has been found do not return any previously removed fuel to the tank. Disposal services are widely available.

13 Fuel tank filler pipe – removal and refitting

Note: *Refer to Section 2 before proceeding*

1 Proceed as described in paragraphs 1 to 3 (inclusive) of the previous Section.
2 Remove the filler cap surround (Fig. 3.7).
3 Raise the rear of the vehicle and support securely using axle stands.
4 Disconnect the ventilation tube from the combined roll-over/anti-trickle-fill valve, release the ventilation tube from its retaining clips and detach the valve from the vehicle (refer to photo 11.1).
5 Remove the fuel tank, as described in Section 12.

12.8 Fuel tank sender unit multi-plug (viewed from inside vehicle, with rear seat cushion removed)

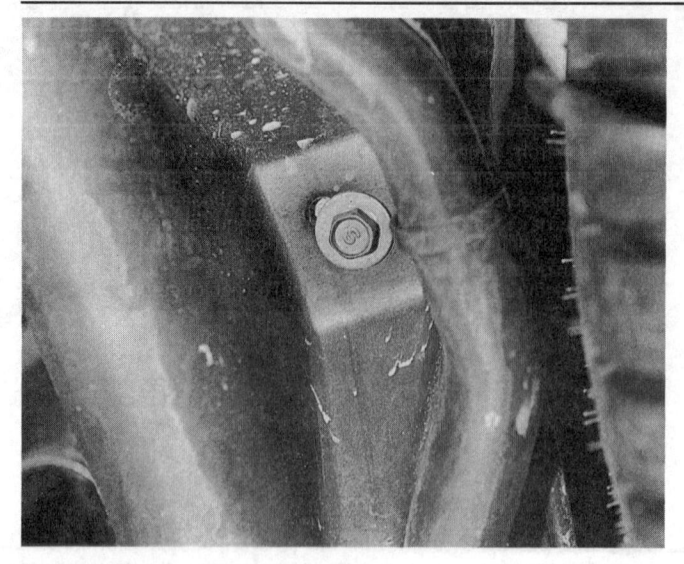

94 Chapter 3 Fuel, exhaust and emission control systems

13.6 Fuel filler pipe securing bolt

6 Remove the filler pipe securing bolt (photo), then twist and withdraw the filler pipe unit.
7 Prior to refitting, check the condition of the filler pipe seal in the fuel tank and renew if necessary.
8 Refitting is a reversal of the removal procedure, but apply a light smear of grease to the filler pipe seal to aid filler pipe entry.

14 Fuel tank sender unit – removal and refitting

Note: *Ford Special tool 23-014 is required to remove and refit the sender unit. If this is not available, or cannot be suitably fabricated, entrust this task to your Ford dealer*

1 With the fuel tank removed from the vehicle, access is gained to remove the sender unit from the top surface of the tank.
2 Engage the tool, then carefully release the sender unit before withdrawing from the fuel tank.
3 Refitting is a reversal of the removal procedure, but use a **new** seal and ensure that any locating lugs on the sender unit (where applicable) align with their cut-outs in the fuel tank.

15 Fuel pump (mechanical) – testing, removal and refitting

Note: *Refer to Section 2 before proceeding*

1 The fuel pump may be tested by disconnecting the fuel feed pipe from the carburettor and placing its open end in a container.

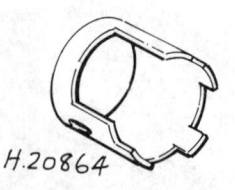

Fig. 3.8 Ford Special tool 23-014, required for sender unit removal and refitting (not fuel injected models) (Sec 14)

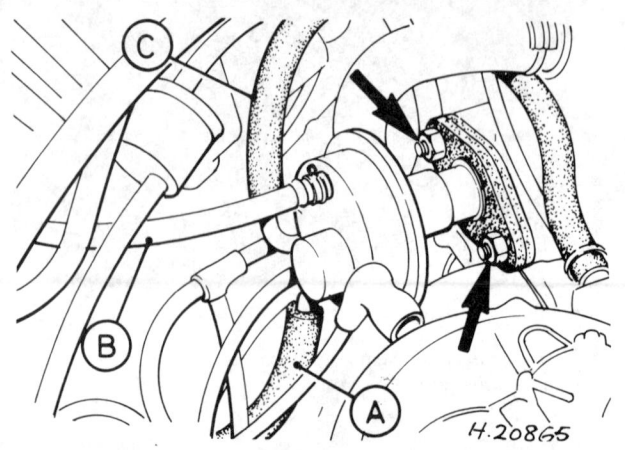

Fig. 3.9 Fuel pump assembly fitted to carburettor equipped CVH engines (securing nuts arrowed) (Sec 15)

A Fuel feed from tank   C Fuel feed to carburettor
B Fuel return to tank

2 Disconnect the multi-plug from the DIS ignition coil, or the LT lead from the negative terminal of the ignition coil, to prevent the engine from firing.
3 Actuate the starter motor. Regular well-defined spurts of fuel should be seen being ejected from the open end of the fuel feed pipe.
4 If this does not occur, and there is fuel in the tank, the pump will need renewing. The pump is a sealed unit and cannot be dismantled or repaired.
5 Two types of mechanical fuel pump are fitted, the application dependent on engine type (refer to Section 1). Certain models may also be fitted with a fuel vapour separator. If this is removed label its pipes as an aid to refitting (photo).
6 To remove a fuel pump, first disconnect the battery earth terminal.
7 On some models, access is improved by removing the air filter housing.
8 Disconnect the fuel pipes from the fuel pump, having noted their locations for subsequent refitting, and plug them.
9 Remove the bolts or nuts (as applicable) securing the pump to its

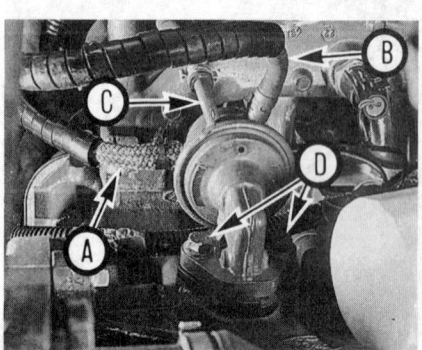

15.5A Fuel pump location on HCS engine type (1.1 litre model, shown from below)

A Fuel inlet hose
B Fuel return hose to tank
C Fuel outlet hose to carburettor
D Pump securing bolts

15.5B Fuel pump and fuel vapour separator arrangement on HCS engine type (1.1 litre model with CTX automatic transmission, shown from below)

15.10 Gasket/spacer fitment on HCS engine type. Note position of the lug (arrowed)

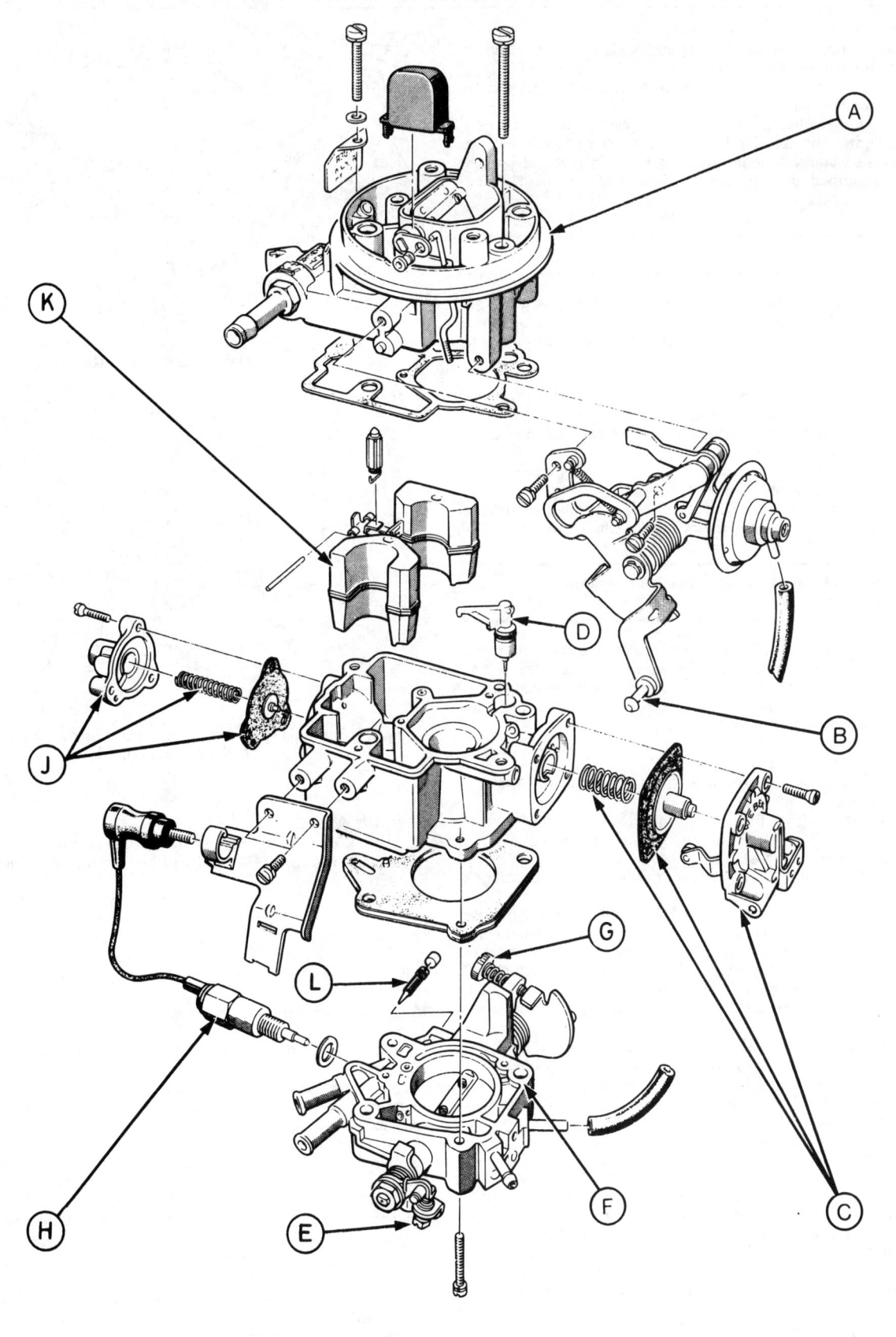

Fig. 3.10 Exploded view of the carburettor (Weber TLM) (Sec 16)

A   Upper body
B   Choke mechanism
C   Accelerator pump assembly
D   Accelerator pump discharge tube
E   Fast-idle speed adjusting screw
F   Throttle housing
G   Idle speed adjusting screw
H   Anti-dieselling (fuel cut-off) solenoid
J   Power valve assembly
K   Float
L   Fuel mixture adjusting screw

location, and withdraw it from the vehicle.
10 Recover the gasket/spacer (photo) and, if desired, withdraw the operating pushrod (CVH engines only).
11 Thoroughly clean the gasket mating surfaces of the pump and the cylinder head or cylinder block.
12 Refitting is a reversal of the removal procedure, but use a new gasket and tighten the securing bolts to the specified torque. Ensure that the hoses are correctly connected, and if the hoses were originally secured with crimped type clips, discard chapter, and fit nut and screws type clips.

16 Carburettor (Weber TLM) – description

The carburettor is of the single (fixed) venturi downdraught type, featuring a fixed size main jet system with a mechanically-operated accelerator pump and vacuum-operated power valve to provide optimum fuelling.

A manually-operated choke system is fitted, featuring a vacuum-operated pull-down mechanism which brings the choke partially off during conditions of high manifold vacuum.

Provision is made for idle speed, fuel mixture and fast-idle speed adjustments. Adjustment procedures are given in the following Section, and it should be understood that *accurate* adjustments can **only** be made using the necessary equipment.

An anti-dieselling (fuel cut-off) solenoid (where fitted) prevents the possibility of engine run-on when the ignition is switched off.

17 Carburettor (Weber TLM) – idle speed, fuel mixture and fast-idle adjustment

Note: *Refer to Section 2 before proceeding. Before carrying out any carburettor adjustments, ensure that the spark plug gaps are set as specified and that all electrical and vacuum connections are secure. To carry out the adjustments, an accurate tachometer and an exhaust gas analyser (CO meter) will be required*

Idle speed and fuel mixture

1 Warm the engine up to normal operating temperature (the radiator cooling fan will cut in).
2 Switch off the engine, then bridge the contacts in the radiator cooling fan thermal switch multi-plug (using a suitable piece of wire) to allow the fan to run continuously.
3 Connect a tachometer and an exhaust gas analyser in accordance with the manufacturer's instructions.
4 Start the engine and run it at 3000 rpm for approximately 30 seconds, ensuring that all electrical loads (headlamps, heater fan motor, etc.) are all switched off, then allow it to idle. Wait for the meters to stabilise (a period of between 5 and 25 seconds is normally sufficient), then record the idle speed and CO content readings.
5 If necessary, adjust the idle speed screw to achieve the specified idle speed.
6 Checking and adjustment should be completed within 30 seconds of the meters stabilising. If this has not been possible, repeat paragraphs 4 and 5, ignoring the reference to starting the engine.
7 Adjustment of the fuel mixture is not normally required during routine maintenance, but if the reading noted in paragraph 4 is not as specified, proceed as follows.
8 Stop the engine and remove the air filter housing then, using a short spike, prise out the tamperproof seal covering the mixture screw.
9 Loosely refit the air filter housing ensuring that all hoses (as applicable), including the crankcase ventilation hose are reconnected.
10 Repeat the procedure given in paragraph 4, then adjust the mixture screw to give the specified CO content.
11 Checking and adjustment should be completed with 30 seconds after meters stabilising. If this has not been possible, repeat paragraph 10.
12 Recheck the idle speed and CO content.
13 On satisfactory completion of the adjustments, stop the engine and disconnect the tachometer and exhaust gas analyser. Remove the bridging wire from the radiator cooling fan thermal switch multi-plug, and reconnect the multi-plug to the thermal switch. Fit a new tamperproof seal to the mixture screw then fully refit and secure the air filter housing.

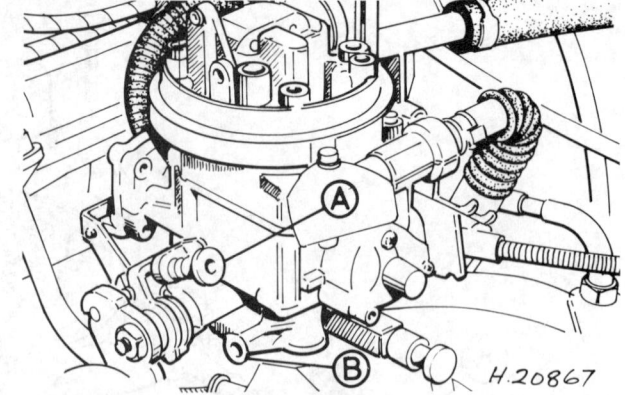

Fig. 3.11 Idle speed adjusting screw (A) and fuel mixture adjusting screw (B) (Weber TLM carburettor) (Sec 17)

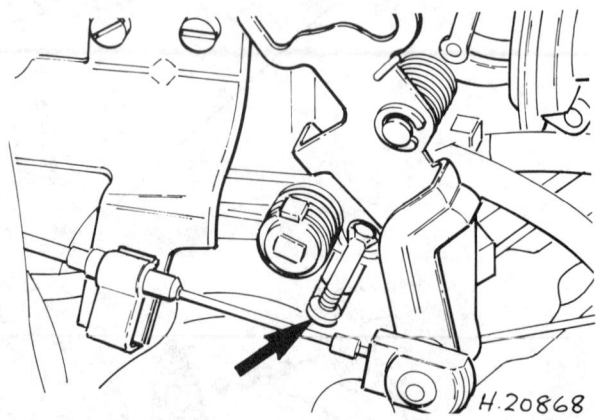

Fig. 3.12 Fast-idle speed adjusting screw (arrowed) (Weber TLM carburettor) (Sec 17)

Fast-idle speed

14 Check the idle speed and fuel mixture settings, and adjust as necessary. These **must** be correct before attempting to check or adjust the fast-idle speed.
15 With the engine at its normal operating temperature, and a tachometer connected in accordance with the manufacturer's instructions, remove the air filter housing (if not already done). Ensure that the ignition module vacuum hose is connected. Bridge the contacts in the radiator cooling fan thermal switch multi-plug, so that the fan runs continuously.
16 Repeat the procedure given in paragraph 4, but ignore the meter reading.
17 Actuate the choke by pulling its control knob fully out.
18 Hold the choke plate open using a 5 mm twist drill held between the plate and the venturi, and record the fast-idle speed achieved. If adjustment is necessary, turn the fast-idle adjusting screw until the specified speed is obtained.
19 Re-check the fast-idle and basic idle speeds.
20 On satisfactory completion of the adjustment, stop the engine and disconnect the tachometer. Fully refit and secure the air filter housing. Remove the bridging wire from the radiator cooling fan thermal switch multi-plug, and reconnect the multi-plug to the thermal switch.

18 Needle valve and float (Weber TLM carburettor) – removal, refitting and adjustment

Note: *Refer to Section 2 before proceeding. Note that new gaskets and a new washer (seal) will be required upon reassembly/refitting. A tachometer and an exhaust gas analyser will be required to check the idle speed and fuel mixture settings on completion*

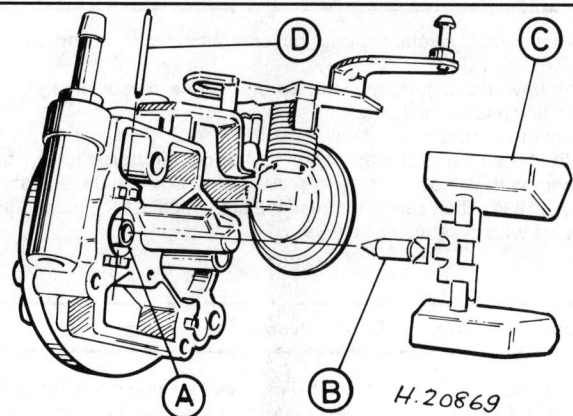

Chapter 3 Fuel, exhaust and emission control systems

Fig. 3.13 Float assembly dismantled (Weber TLM carburettor) (Sec 18)

A Needle valve housing   C Float
B Needle valve   D Float retaining pin

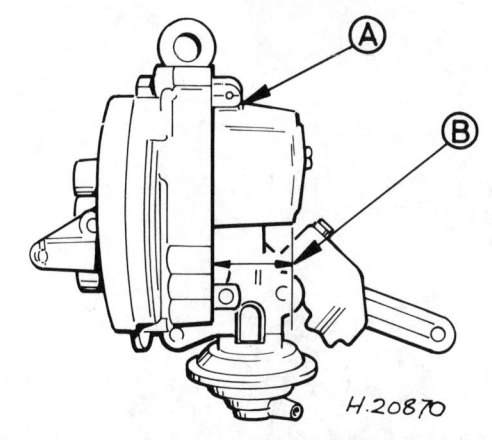

Fig. 3.14 Float level adjustment (Weber TLM carburettor) (Sec 18)

A Adjusting tag   B Float level setting dimension

Removal and refitting

1 Disconnect the battery earth terminal.
2 Remove the air filter housing.
3 Clean the exterior of the carburettor, then disconnect the fuel feed hose.
4 Disconnect the choke cable and the choke vacuum hose.
5 Remove the four screws securing the carburettor upper body (two of these screws are Torx head type), and detach it. Note that the carburettor lower body is now loose on the inlet manifold.
6 Tap out the float retaining pin, remove the float and withdraw the needle valve. Unscrew the needle valve housing, as required, noting washer fitment (refer to Fig. to 3.13).
7 Inspect the components for damage and renew as necessary. Check the needle valve for wear, and check the float assembly for leaks by shaking it to see if it contains petrol. Whilst accessible, clean the float chamber and jets (refer to the following Section).
8 Using a **new** washer, refit the needle valve housing.
9 Refit the needle valve, float and retaining pin, ensuring that the tag on the float engages between the ball and clip on the needle valve.
10 Before refitting the carburettor upper body, check and if necessary adjust the float level as described in paragraph 16 onwards. Also check the float and needle valve for full and free movement.
11 Clean the gasket contact faces (including the inlet manifold) then, using **new** gaskets for the carburettor upper body and the inlet manifold faces, refit the carburettor upper body and secure the carburettor assembly to the inlet manifold (refer to the Specifications for tightening torque details).
12 Reconnect the choke vacuum hose. If the fuel feed hose was originally secured with a crimped type clip, discard this and secure the fuel feed hose with a nut and screw type clip.

13 Reconnect and adjust the choke cable, then refit the air filter housing.
14 Reconnect the battery earth terminal.
15 Check the idle speed and fuel mixture settings, and adjust as necessary (refer to the previous Section).

Float level adjustment

16 With the carburettor upper body removed as described in paragraphs 1 to 5 inclusive, proceed as follows.
17 Hold the carburettor upper body in the position shown in Fig. 3.14, ensuring that the needle valve is shut off. Fit the new upper body gasket to the carburettor upper body, then measure the distance between the gasket and the step on the float.
18 If the measurement is not as specified, adjust by bending the tag on the float, then re-check.
19 Refitting should be carried out in accordance with paragraphs 11 to 15 inclusive.

19 Carburettor (Weber TLM) – cleaning

Note: *Refer to Section 2 before proceeding. Note that new gaskets will be required upon reassembly. A tachometer and an exhaust gas analyser will be required to check the idle speed and fuel mixture settings upon completion*

1 Proceed as described in paragraphs 1 to 5 (inclusive) of the previous Section.
2 Soak out the fuel in the float chamber using a clean rag; this must be safely disposed of. Clean the float chamber and jets. The careful use of an air line (or footpump) is ideal to 'blow out' the upper and lower bodies, but do **not** direct full air pressure into the bleed location or accelerator pump diaphragm damage will occur.
3 Proceed to completion, as described in paragraphs 10 to 15 (inclusive) of the previous Section.

20 Carburettor (Weber TLM) – removal and refitting

Note: *Refer to Section 2 before proceeding. Note that a new gasket will be required upon refitting, and that a tachometer and an exhaust gas analyser will also be needed to check the idle speed and fuel mixture, upon completion*

1 Disconnect the battery earth terminal.
2 Remove the air filter housing.
3 Disconnect the accelerator inner and outer cable from the carburettor.
4 Disconnect the choke inner and outer cable from the carburettor.
5 Disconnect the fuel feed hose from the carburettor, and plug its end to avoid spillage and prevent dirt ingress. If a crimped type hose clip is fitted, cut this free taking care to avoid damage to the hose.
6 Where applicable, disconnect the electrical lead from the anti-dieselling (fuel cut-off) solenoid, and all relevant carburettor vacuum pipes (having labelled them for correct subsequent refitting).
7 Remove the two Torx head screws securing the carburettor to the

Fig. 3.15 Removing one of the two Torx head screws (arrowed) securing the carburettor to the inlet manifold (Weber TLM carburettor) (Sec 20)

98   Chapter 3   Fuel, exhaust and emission control systems

inlet manifold (Fig. 3.15), then withdraw it from the vehicle.
8 Clean the inlet manifold and carburettor gasket mating faces.
9 Refitting is a reversal of the removal procedure, bearing in mind the following points. A new gasket **must** be used, and the Torx head carburettor securing screws must be tightened to the specified torque. When reconnecting the fuel feed hose, use a nut and screw type clip to secure. After refitting the accelerator and choke cables, check their adjustment and re-adjust as necessary.
10 Upon completion, check the idle speed and fuel mixture settings, and adjust as necessary.

21 Accelerator pump diaphragm (Weber TLM carburettor) – removal and refitting

1 With the carburettor removed from the vehicle, place it on a clean flat work surface.

2 Remove the accelerator pump cover retaining screws and detach the cover.
3 Withdraw the diaphragm and spring. Check the diaphragm for damage, and renew if evident.
4 Clean the carburettor and cover mating faces.
5 Refit the spring and diaphragm to the carburettor, aligning the diaphragm with its cover retaining screw holes. Position the actuating lever on its cam, then carefully press the cover against the diaphragm and secure with its retaining screws.

22 Carburettor (Weber TLDM) – description

The carburettor is of the dual (fixed) venturi downdraught type, featuring a fixed size main jet system, adjustable idle system, a mechanically-operated accelerator pump and a vacuum-operated power valve. A manually-operated choke is fitted and, on CTX automatic transmission versions, a throttle kicker is utilised.

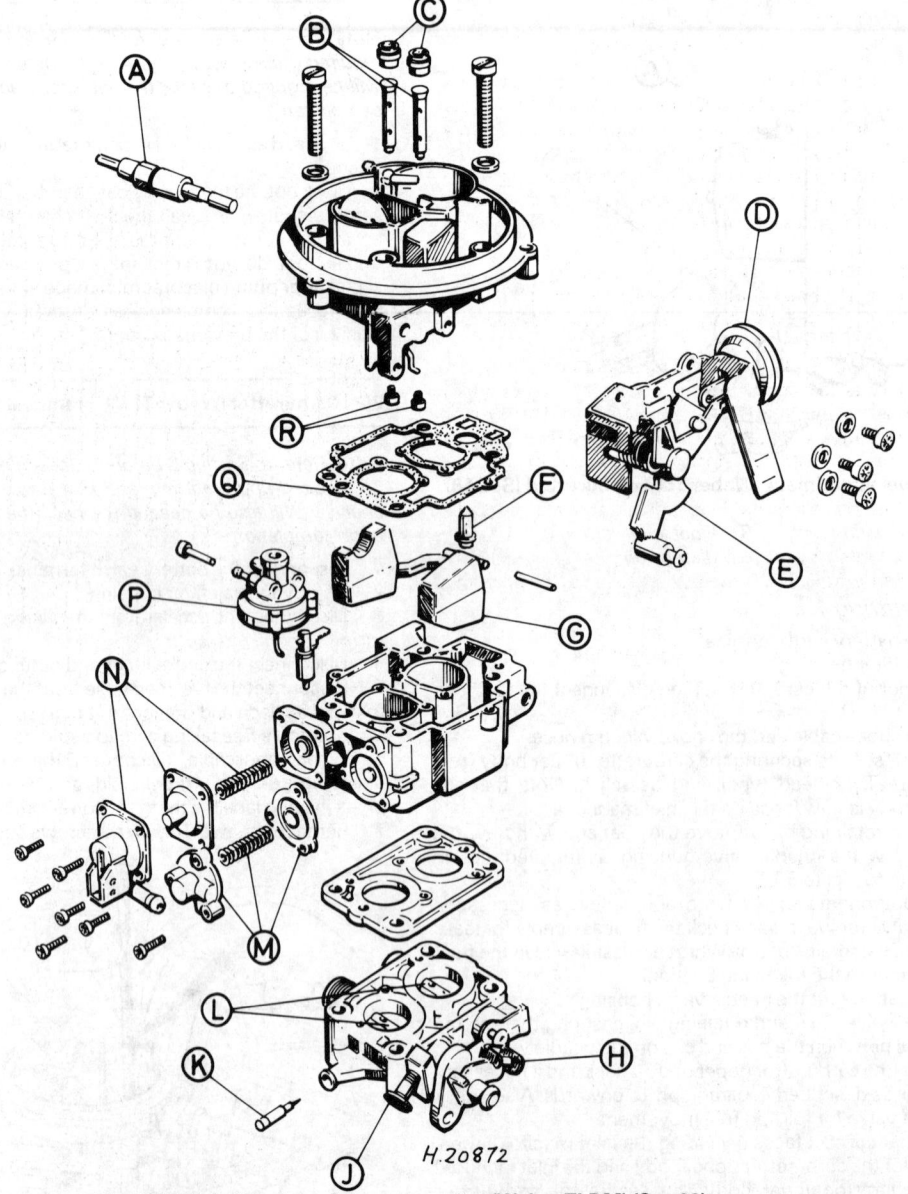

Fig. 3.16 Exploded view of the carburettor (Weber TLDM) (Sec 22)

A Anti-dieselling (fuel cut-off) solenoid	E Manual choke linkage	J Idle speed adjusting screw	N Accelerator pump assembly
B Emulsion tubes	F Needle valve	K Fuel mixture adjusting screw	P Throttle kicker (if fitted)
C Air-correction jets	G Float	L Throttle plates	Q Upper body gasket
D Choke pull-down diaphragm assembly	H Fast-idle adjusting screw	M Power valve assembly	R Main jets

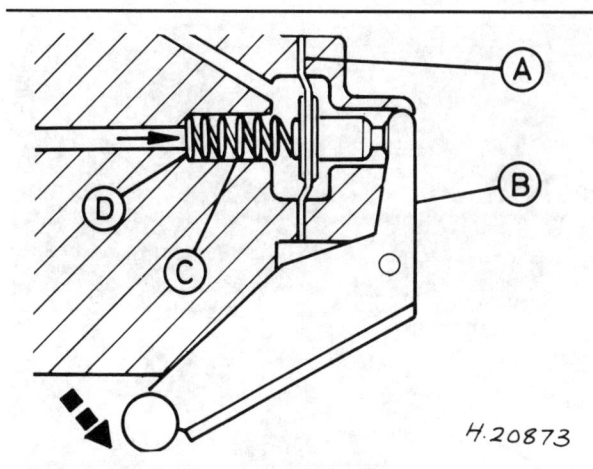

Chapter 3 Fuel, exhaust and emission control systems

Fig. 3.17 Sectional view of the accelerator pump assembly (Sec 22)

A Diaphragm C Return spring
B Lever D One-way valve

In order to meet emission regulations and maintain good fuel consumption, the main jets are calibrated to suit only the quarter-to-three-quarter throttle range. The power valve is therefore only used to supply additional fuel during full throttle conditions, to enrich the fuel mixture.

The accelerator pump is fitted to ensure a smooth transition from the idle circuit to the main jet system. As the accelerator pedal is depressed, a linkage moves the diaphragm within the accelerator pump and a small quantity of fuel is injected into the venturi, thus avoiding a momentary weak fuel mixture and resultant engine hesitation.

The manually-operated choke features a vacuum-operated pulldown mechanism which controls the single choke plate under certain vacuum conditions.

The throttle kicker (fitted on UK specification, CTX automatic transmission equipped vehicles) acts as an idle speed compensator for when the transmission shift lever positions R, D or L are selected. The throttle kicker is operated by vacuum supplied from the inlet manifold. When the appropriate transmission shift lever position is selected, the throttle kicker control solenoid allows the vacuum to pass to the throttle kicker which maintains the idle speed by employing a diaphragm and mechanical linkage.

Adjustment procedures are given in later text, and it should be understood that *accurate* adjustments can *only* be made using the necessary equipment.

An anti-dieselling (fuel cut-off) solenoid is fitted to prevent the possibility of engine run-on when the ignition is switched off.

23 Carburettor (Weber TLDM) – idle speed, fuel mixture and fast-idle adjustment

Idle speed and fuel mixture

1 Refer to Section 17 of this Chapter, but note the following points.
2 The throttle kicker vacuum pipe must be disconnected from the throttle kicker (where fitted) when adjusting the idle speed and fuel mixture, and reconnected upon completion of the adjustments.
3 The idle speed and fuel mixture screw locations are as shown (photo).

Fast-idle speed

4 Refer to Section 17 of this Chapter, but note the following points.
5 Ignore the instruction in Section 17, paragraph 16, starting the engine immediately after fully actuating the choke and holding the choke plate open. Record the fast-idle speed achieved after the tachometer has stabilised, and adjust until the specified speed is obtained.
6 The fast-idle adjusting screw is shown (photo).

24 Throttle kicker (Weber TLDM carburettor) – checking and adjustment

Note: *Refer to Section 2 before proceeding. To carry out the checking and adjustment procedure, an accurate tachometer and an exhaust gas analyser (CO meter) will be required*

23.3 Idle speed adjustment screw (A), and fuel mixture adjustment screw (B) (Weber TLDM carburettor)

23.6 Fast-idle speed adjustment screw (arrowed) (Weber TLDM carburettor)

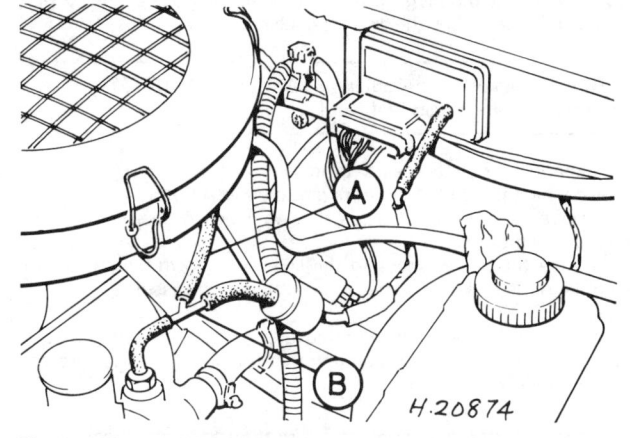

Fig. 3.18 Throttle kicker vacuum test connections (Weber TLDM carburettor) (Sec 24)

A Vacuum pipe to throttle kicker
B T-piece

Chapter 3 Fuel, exhaust and emission control systems

24.7 General view of throttle kicker arrangement (Weber TLDM carburettor)

A Tamperproof plug covering adjusting point
B Throttle kicker securing screws
C Vacuum supply pipe

25.2 Throttle kicker control solenoid

A Multi-plug
B Vacuum pipes
C Securing screw

1   The throttle kicker is only likely to require adjustment after removal and refitting of the unit.
2   Check the idle speed and fuel mixture settings, and adjust as necessary. These **must** be correct before attempting to check or adjust the throttle kicker. Do not disconnect the test meters or temporary wiring.
3   Remove the air filter housing, and plugs its vacuum supply at the inlet manifold.
4   Disconnect the throttle kicker vacuum supply pipe (if not already done).
5   Using a new length of vacuum pipe and a suitable T-piece, connect the throttle kicker into the ignition module vacuum supply pipe, as shown in Fig. 3.18.
6   Start the engine and record the engine-speed attained. Check this against the speed quoted in the Specifications.
7   If the engine speed recorded is outside specification, remove the tamperproof seal in the top of the throttle kicker unit and adjust accordingly (photo).
8   Upon satisfactory completion of adjustment, switch off the engine and fit a new tamperproof seal to the throttle kicker unit. Remove the test length of vacuum pipe and the T-piece, and reconnect the throttle kicker and ignition module vacuum supply pipes to their original operational locations. Remove the temporary bridging wire from the radiator cooling fan thermal switch multi-plug, and reconnect the multi-plug to the thermal switch. Disconnect the tachometer and the exhaust gas analyser. Remove the plug from the air filter housing vacuum supply, and fully refit and secure the air filter housing.

25 Throttle kicker control solenoid (Weber TLDM/DFTM carburettor) – removal and refitting

1   Disconnect the battery earth terminal.
2   Disconnect the multi-plug from the solenoid (photo).
3   Remove both vacuum pipes, having labelled them for correct subsequent refitting.
4   Remove the screw securing the solenoid and mounting bracket assembly to the bulkhead panel, then withdraw the assembly from the vehicle.
5   Refitting is a reversal of the removal procedure, ensuring that the locating lug 'snaps' into position and that the vacuum pipes are connected to their correct terminals.

26 Carburettor (Weber TLDM) – removal and refitting

1   Refer to Section 20 of this Chapter, but note that the carburettor is secured to the inlet manifold by four Torx head screws (photo).

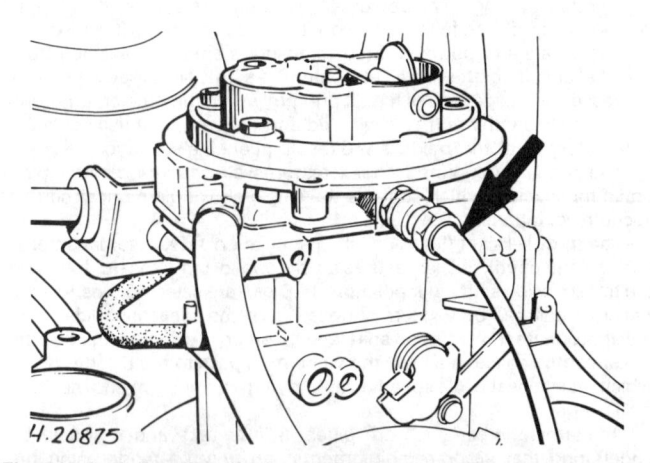

Fig. 3.19 Anti-dieselling (fuel cut-off) solenoid connection (arrowed) (Weber TLDM carburettor) (Sec 26)

26.1 Refitting the carburettor to the inlet manifold. Note new gasket in position (Weber TLDM carburettor shown)

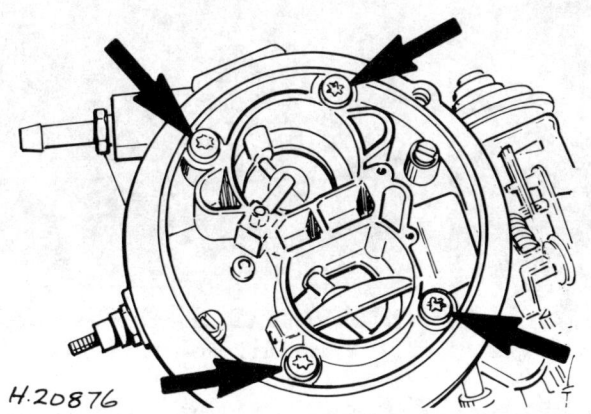

Chapter 3 Fuel, exhaust and emission control systems

Fig. 3.20 Four Torx head screws (arrowed) secure the carburettor to the inlet manifold (Weber TLDM carburettor) (Sec 26)

27 Carburettor (Weber TLDM) – dismantling, cleaning, inspection and reassembly

Note: *Check parts availability before proceeding. Where possible, obtain an overhaul kit containing all the relevant gaskets, seals, etc., required for reassembly*

1 With the carburettor removed from the vehicle, prepare a clean flat work surface prior to commencing dismantling. The procedure given here may be used for partial dismantling, or complete dismantling within practical limits, as required.
2 Clean the exterior of the carburettor, and disconnect the choke pulldown vacuum pipe.
3 Detach the carburettor upper body assembly after removing its securing screws (photos).
4 Dismantle the upper body assembly, as described in the following paragraphs. First, slide out the float retaining pin, then detach the float and needle valve. Remove the needle valve housing and its washer (photos).
5 Remove the anti-dieselling (fuel cut-off) solenoid and ensure that its sealing washer comes away with it (photo).
6 Undo the three screws securing the choke mechanism and detach it (photos).
7 Unscrew and remove both the air correction jets, having noted the size and fitted location of each one to ensure correct subsequent reassembly.
8 Invert the upper body and remove the emulsion tubes, again having noted the size and fitted location of each one. The emulsion tubes are located beneath the air correction jets, and should fall out as the upper body is inverted.
9 Unscrew and remove both the main jets, again having noted the size and fitted location of each one.
10 Dismantle the carburettor main body, as described in the following paragraphs.
11 Carefully prise out the accelerator pump discharge tube assembly (photo).
12 Undo the four screws securing the accelerator pump assembly, and remove the cover followed by the diaphragm and return spring. The valve assembly should come out on the end of the return spring (photos). Check that the valve is complete, and with its O-ring (where applicable).
13 Undo the three screws securing the power valve assembly, and withdraw the cover and return spring followed by the diaphragm (photos).
14 Where fitted, undo the two screws securing the throttle kicker unit and detach it from the carburettor main body (refer to photo 24.7). This may be dismantled, as required, after removing four further screws (Fig. 3.22).
15 Remove the tamperproof seal covering the fuel mixture screw, then undo and remove the mixture screw (photo).
16 Remove the retaining screws and separate the throttle housing from the carburettor main body (photo).
17 Clean out the carburettor drillings, jets and passages using an air-line (or compressed air from a footpump), but do **not** apply high

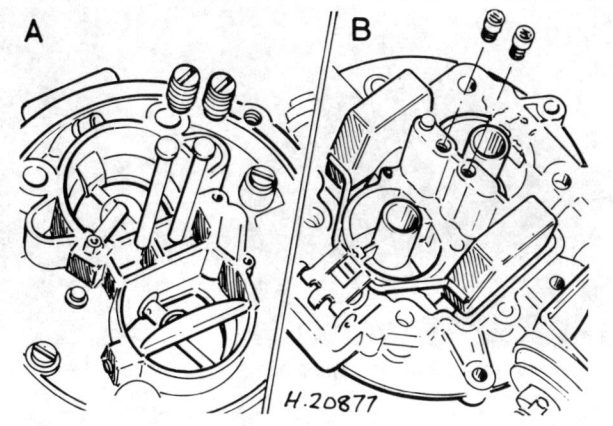

Fig. 3.21 Upper body assembly (Weber TLDM carburettor) (Sec 27)

A Air correction jets and emulsion tubes
B Main jets

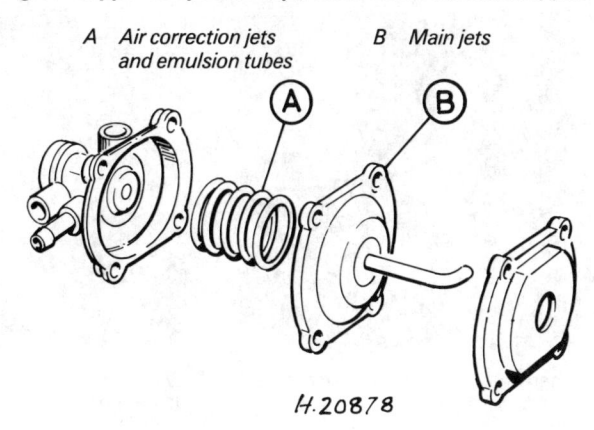

Fig. 3.22 Exploded view of the throttle kicker unit (Sec 27)

A Return spring
B Diaphragm

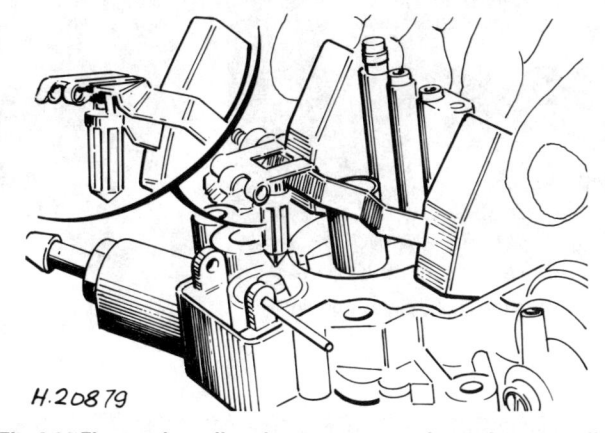

Fig. 3.23 Float and needle valve arrangement. Inset shows needle valve engaged on the float tag (Weber TLDM carburettor) (Sec 27)

pressure air to the accelerator pump discharge assembly, or the pump supply valve, as these two components have a rubber Vernay valve which can be easily damaged. **Never** use a length of wire for cleaning purposes, as this may cause unseen damage.
18 Examine **all** carburettor components for wear or damage, paying particular attention to the diaphragms, throttle spindle and plates, needle valve and mixture screw; the power valve jet is adjacent to the primary main jet. Renew all diaphragms, sealing washers and gaskets as a matter of course.
19 Refit the throttle housing to the carburettor main body (using a **new** gasket) and secure with its screws, then reassemble the remainder of the main body components as follows.
20 Refit the fuel mixture screw and initially set it; fully wind the screw

102　　　　　　　　　　　　**Chapter 3　Fuel, exhaust and emission control systems**

27.3A Remove the carburettor upper body securing screws ...

27.3B ... then detach the carburettor upper body

27.4A Slide out the float retaining pin ...

27.4B ... then detach the float and needle valve

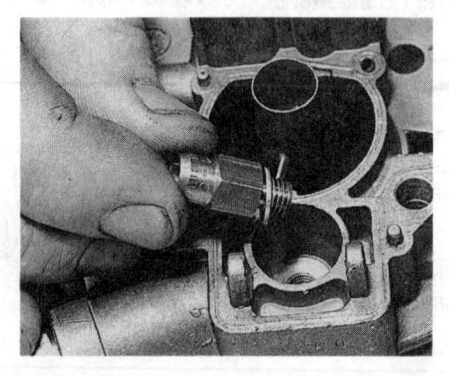

27.4C Remove the needle valve housing and its washer

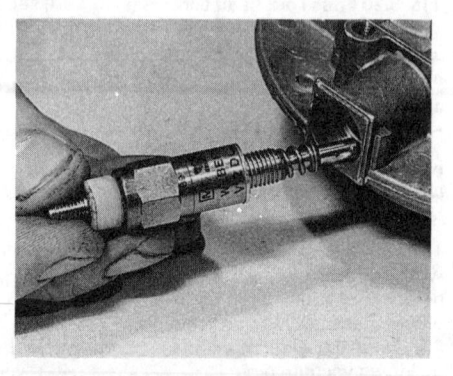

27.5 Withdrawing the anti-dieselling (fuel cut-off) solenoid

27.6A Detach the choke plate operating link ...

27.6B ... and undo the three screws securing the choke mechanism (arrowed). Detach the mechanism

27.11 Carefully prising out the accelerator pump discharge tube assembly

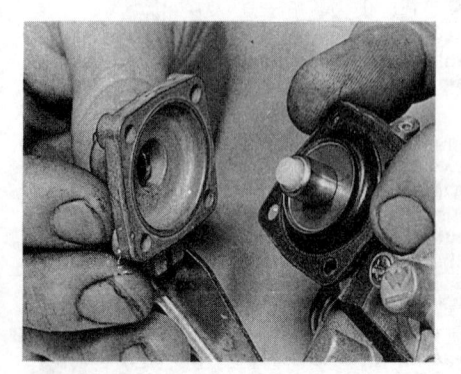

27.12A Remove the accelerator pump cover ...

27.12B ... followed by its diaphragm ...

27.12C ... and the return spring and valve assembly

Chapter 3 Fuel, exhaust and emission control systems

27.13A Undo the power valve assembly securing screws ...

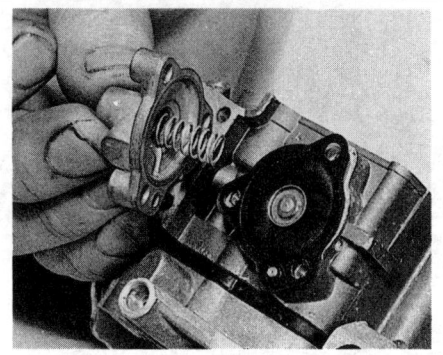

27.13B ... then detach the cover and return spring to expose the diaphragm

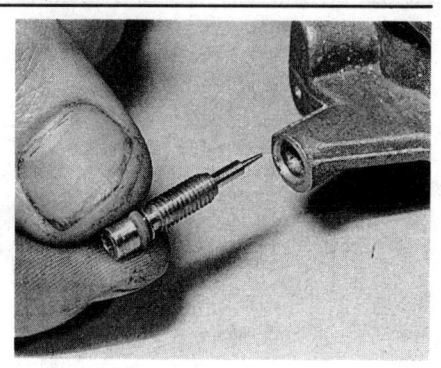

27.15 Unscrew and withdraw the fuel mixture screw

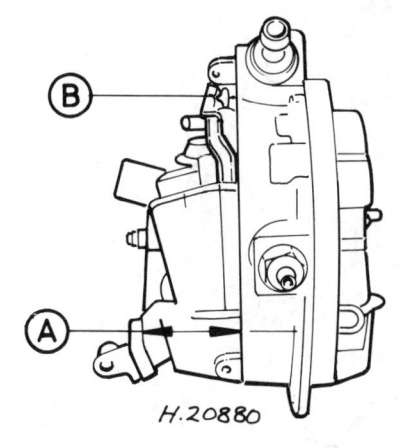

Fig. 3.24 Float level adjustment (Weber TLDM carburettor) (Sec 27)

A Float level setting dimension
B Adjusting tag

27.16 Separating the throttle housing from the carburettor main body

in, then unwind three turns to give an approximate setting.
21   Where fitted, reassemble the throttle kicker (as applicable) ensuring that its diaphragm lies flat, and that the relative position of the operating link to the throttle kicker cover is as shown in Fig. 3.22.
22   Refit the power valve assembly ensuring that the diaphragm lies flat, and that the vacuum gallery lines up correctly with the diaphragm and housing.
23   Refit the accelerator pump assembly ensuring that the valve assembly is not damaged as it is inserted, and that the diaphragm lies flat.
24   Refit the accelerator pump discharge tube assembly, being careful not to damage it or its sealing washer during insertion, and ensuring correct orientation.
25   Reassemble the upper body assembly as described in the following paragraphs.
26   Refit the emulsion tubes and air correction jets, referring to the notes made during dismantling to ensure correct fitted locations.
27   Refit the main jets, again referring to the notes made during dismantling to ensure correct fitted locations.
28   Refit the anti-dieselling (fuel cut-off) solenoid, ensuring that its washer is fitted and that the solenoid is not overtightened.
29   Refit the needle valve housing, using a new washer.
30   Refit the needle valve and float assembly, having ensured that the float tag hooks under the spring clip on the needle valve (Fig. 3.23). Insert the float retaining pin to secure.
31   Refit the choke mechanism and secure with its three screws.
32   Adjust the float level as follows. Ensuring that a **new** gasket is fitted to the carburettor upper body, hold the upper body in the vertical position; the needle valve **must** be shut off. Measure the distance shown in Fig. 3.24, and adjust by bending the float tag if the measurement is outside the specification. Recheck the float level adjustment after bending the float tag, as necessary.
33   Refit the carburettor upper body to the carburettor main body,

ensuring that its new gasket seats correctly and that the float does not foul during assembly. Insert and tighten the securing screws.
34   Reconnect the choke pulldown vacuum pipe.
35   Once the carburettor has been refitted to the vehicle, check and adjust the idle speed, fuel mixture, fast-idle speed and throttle kicker setting (as applicable), making reference to the appropriate text. Renew any tamperproof seals after the successful completion of adjustments.

28 Carburettor (Weber DFTM) – description

Note: *The main jets are a combined assembly incorporating air correction jets and emulsion tubes*

The carburettor operates in essentially the same manner as that described in Section 22, but the following features should be noted.
A throttle kicker is fitted to both CTX automatic *and* manual transmission models. The operation of this unit may be found in the above referred text for CTX automatic transmission equipped vehicles. On manual transmission equipped vehicles, the throttle kicker acts as a damper by slowing down the closing action of the throttle plate. Under deceleration, this maintains the combustion of the air/fuel mixture entering the cylinders, thus improving the exhaust emission levels. A vacuum sustain valve controls the carburettor-sourced vacuum applied to the throttle kicker unit; this allows the vacuum to decay slowly, allowing normal engine idling speed to be achieved.
The secondary venturi (barrel) is vacuum-operated on manual transmission equipped vehicles. On CTX automatic transmission equipped models it is operated sequentially.

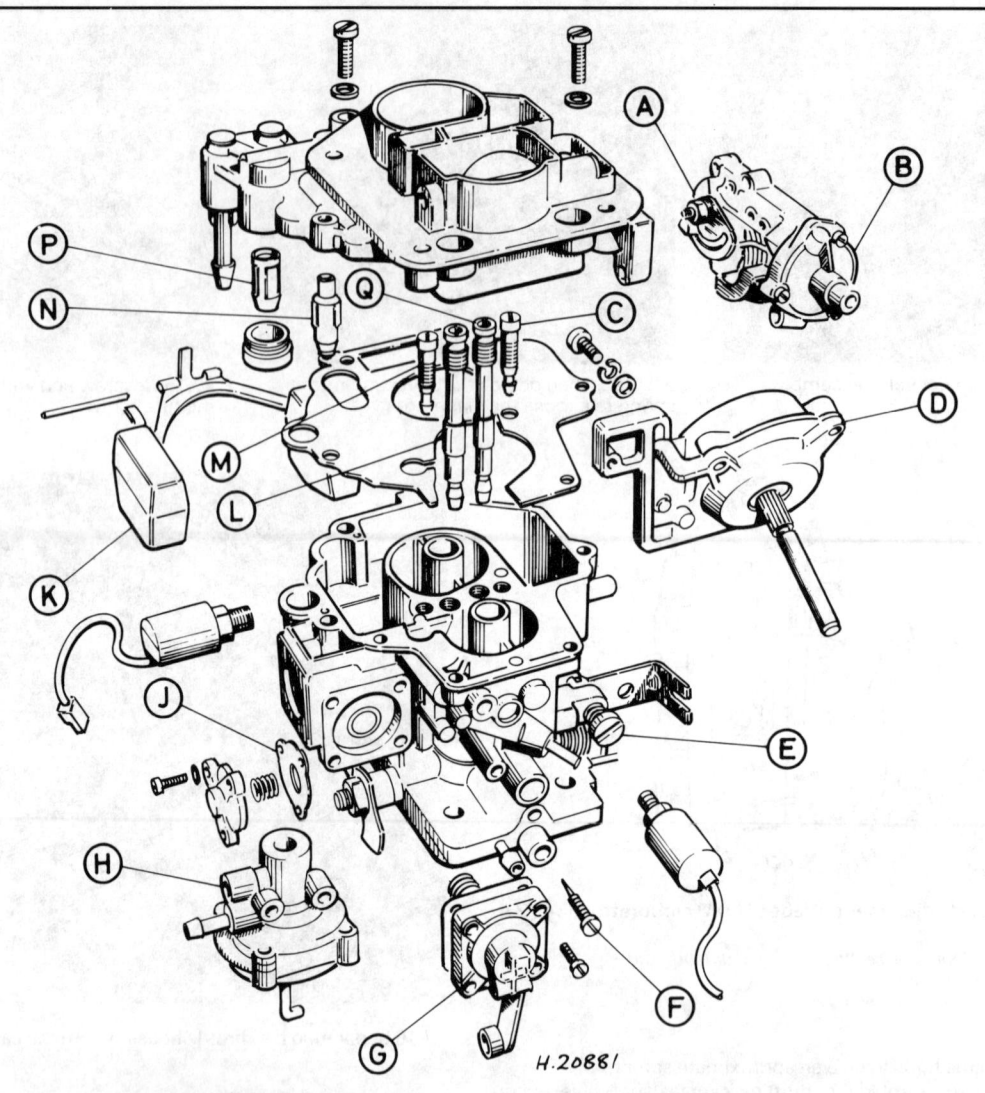

104 Chapter 3 Fuel, exhaust and emission control systems

Fig. 3.25 Exploded view of the carburettor (Weber DFTM) (Sec 28)

- A Manual choke assembly
- B Choke vacuum pull-down
- C Secondary idle jet
- D Secondary barrel diaphragm assembly
- E Idle speed adjusting screw
- F Fuel mixture adjusting screw
- G Accelerator pump assembly
- H Throttle kicker
- J Power valve diaphragm
- K Float
- L Primary main jet combined assembly
- M Primary idle jet
- N Needle valve
- P Fuel feed filter
- Q Secondary main jet combined assembly

A bleed back solenoid (if fitted) is used to control the amount of fuel being delivered to the venturi by the action of the accelerator pump.

Adjustment procedures are given in later text, and it should be understood that *accurate* adjustments can *only* be made using the necessary equipment.

29 Carburettor (Weber DFTM) – idle speed, fuel mixture and fast-idle adjustment

Idle speed and fuel mixture

1 Refer to Section 17 of this Chapter, but note that the idle speed and fuel mixture screw locations are as shown in Fig. 3.27.

Fast-idle speed

2 Refer to Section 17 of this Chapter, but note the following points.
3 Ignore the reference to the ignition module in paragraph 15.
4 When detaching the air filter housing, do not disconnect the vacuum supply or crankcase ventilation hoses; the air filter housing should be positioned clear of the carburettor assembly.

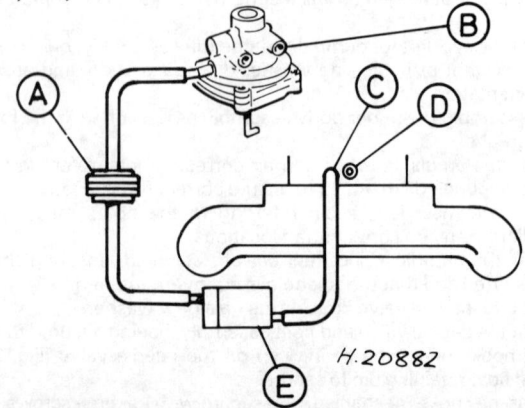

Fig. 3.26 Schematic view of the throttle kicker system fitted to 1.4 litre manual transmission equipped models (Sec 28)

- A Vacuum sustain valve
- B Throttle kicker
- C Vacuum connection at the carburettor
- D Fuel mixture adjusting screw
- E Fuel trap

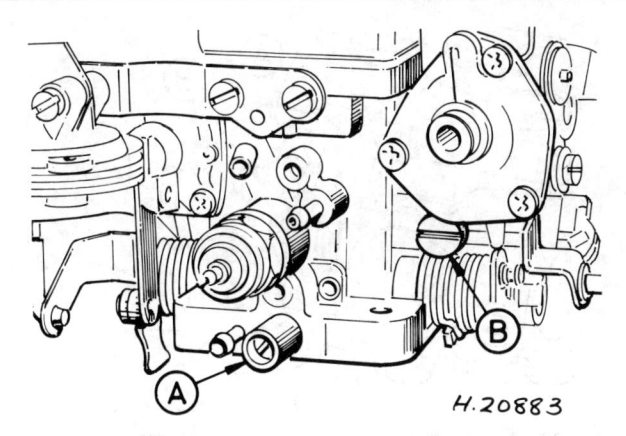

Chapter 3 Fuel, exhaust and emission control systems

Fig. 3.27 Fuel mixture adjusting screw (A) and idle speed adjusting screw (B) (Weber DFTM carburettor) (Sec 29)

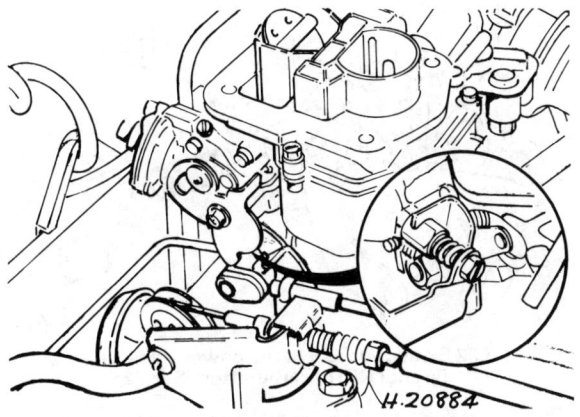

Fig. 3.28 Manual choke fast-idle adjustment. Inset shows fast-idle adjusting screw (Weber DFTM carburettor) (Sec 29)

5   Refer to Fig. 3.28, which shows the method of holding the choke plate open during checking.
6   Ignore the instruction in paragraph 16, starting the engine immediately after fully actuating the choke and holding the choke plate open. Record the fast-idle speed achieved after the tachometer has stabilised, and adjust until the specified speed is obtained.

30 Throttle kicker (Weber DFTM carburettor) – checking and adjustment

1   Refer to Section 24, but connect the throttle kicker directly to the inlet manifold after removing the air filter housing, using a suitable length of vacuum pipe; ignore the instruction given in paragraph 5.
2   Conduct the testing and adjustment as described in paragraphs 6 and 7 of Section 24, then complete the operation by carrying out the instructions given in paragraph 8 (where applicable).

31 Carburettor (Weber DFTM) – removal and refitting

Note: *Refer to Section 2 before proceeding. Note that a new gasket will be required upon refitting, and that a tachometer and an exhaust gas analyser will be required to check the idle speed and fuel mixture upon completion*

1   Disconnect the battery earth terminal.
2   Remove the air filter housing.
3   Disconnect the manual choke inner cable at the carburettor.
4   Carefully prise out the accelerator link retaining clip, remove both securing bolts, then detach the cable and bracket assembly. Position the cable and bracket assembly clear of the carburettor.
5   Disconnect the fuel feed hose at the carburettor, and plug its end to avoid spillage and prevent dirt ingress. If a crimped type hose clip is fitted, cut this free taking care not to damage the hose.
6   Disconnect all relevant vacuum pipes from the carburettor, having labelled them for subsequent refitting.
7   Disconnect the electrical lead from the anti-dieselling (fuel cut-off) solenoid, and the bleed back solenoid (if fitted).
8   Remove the four nuts securing the carburettor to the inlet manifold, then withdraw it from the vehicle.
9   Clean the inlet manifold and carburettor gasket mating faces.
10   Refitting is a reversal of the removal procedure, bearing in mind the following points. A new gasket **must** be used, and the carburettor securing nuts should be tightened in accordance with the Specifications. When reconnecting the fuel feed hose, use a nut and screw type clip to secure. After refitting the cable and bracket assembly and reconnecting the choke inner cable, check the adjustment of the choke cable, and re-adjust as necessary.
11   Upon completion, check the idle speed and fuel mixture settings, and adjust as necessary.

32 Carburettor (Weber DFTM) dismantling, cleaning, inspection and reassembly

Note: *Check parts availability before proceeding. Where possible, obtain an overhaul kit containing all the relevant gaskets and seals required for reassembly*

1   With the carburettor removed from the vehicle, prepare a clean flat work surface prior to commencing dismantling. The procedure given here may be used for partial dismantling, or complete dismantling within practical limits, as required.
2   Clean the exterior of the carburettor and detach all vacuum pipes (where applicable), having noted their fitted location for correct subsequent reassembly.
3   Remove the six screws securing the carburettor upper body, and detach it. Discard the gasket.
4   Dismantle the carburettor upper body as described in the following paragraphs.
5   Remove the three screws securing the choke pull-down diaphragm and detach it.
6   Unscrew the brass nut adjacent to the fuel feed connection, and remove the fuel feed filter.
7   Tap out the float retaining pin, then detach the float and needle valve. Unscrew and remove the needle valve housing.
8   Dismantle the carburettor main body as described in the following paragraphs.
9   Remove the main and idle jets (refer to Fig. 3.30) having noted the size and fitted location of each one to ensure correct subsequent reassembly. Note that the main jets are a combined assembly incorporating air correction jets and emulsion tubes. Carefully prise out the accelerator pump discharge tube.
10   Detach the throttle kicker and its mounting bracket assembly. The throttle kicker may be dismantled as required (refer to Fig. 3.22).
11   Remove the anti-dieselling (fuel cut-off) solenoid.
12   Remove the four screws securing the accelerator pump assembly, and dismantle the assembly noting the fitment of its components.
13   Disconnect the secondary barrel diaphragm operating rod by pulling the lower section of rod downwards and twisting to release it from its retaining socket. Remove its four cover retaining screws and dismantle the assembly.
14   Undo the three screws securing the power valve assembly and remove its diaphragm.
15   Prise out the tamperproof seal covering the fuel mixture screw, then undo and remove the mixture screw.
16   Clean and examine all the carburettor components, including the fuel feed filter, making reference to Section 27 paragraphs 17 and 18 (where applicable).
17   Refit the fuel mixture screw by fully winding the screw in, then unwinding three turns to give an approximate setting.

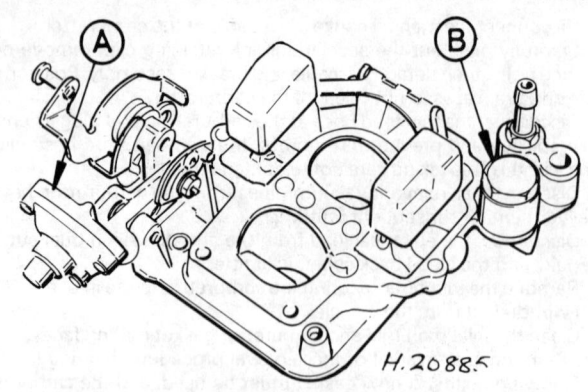

Chapter 3 Fuel, exhaust and emission control systems

Fig. 3.29 Choke pull-down assembly (A) and fuel feed filter housing (B) (Weber DFTM carburettor) (Sec 32)

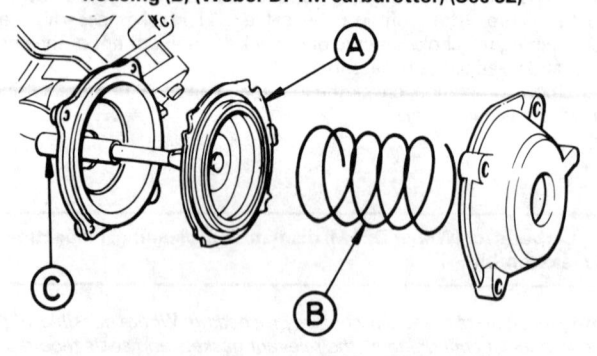

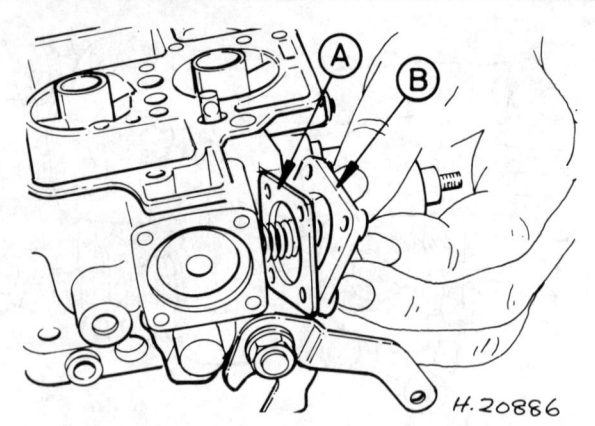

Fig. 3.30 Accelerator pump renewal (Weber DFTM carburettor) (Sec 32)

    A   Diaphragm   B   Housing (cover)

Fig. 3.31 Exploded view of the secondary barrel diaphragm assembly (Sec 32)

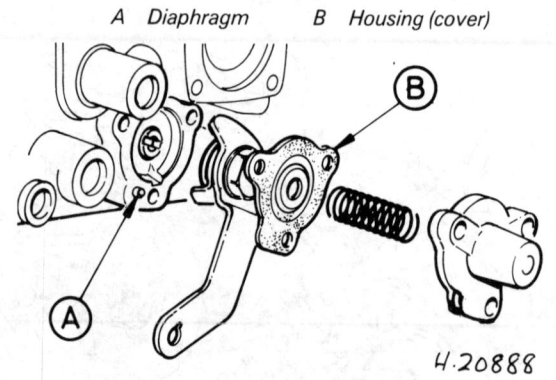

    A   Diaphragm   C   Operating rod
    B   Return spring

Fig. 3.32 Exploded view of the power valve assembly (Weber DFTM carburettor) (Sec 32)

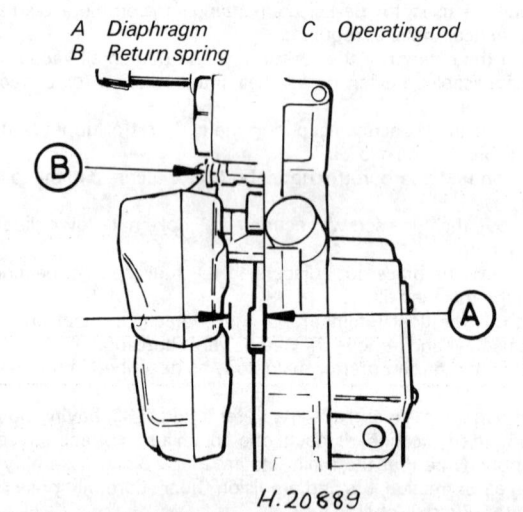

    A   Vacuum gallery   B   Diaphragm

Fig. 3.33 Float level adjustment (Weber DFTM carburettor) (Sec 32)

    A   Float level setting dimension   B   Adjusting tag

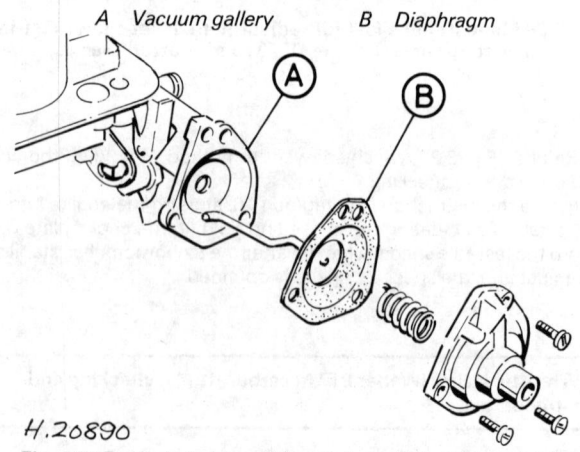

Fig. 3.34 Exploded view of the choke pull-down assembly (Weber DFTM carburettor) (Sec 32)

    A   Housing   B   Diaphragm

18   Refit the power valve assembly ensuring that the diaphragm lies flat, and that the vacuum gallery lines up correctly with the diaphragm and housing (Fig. 3.32).
19   Reassemble the secondary barrel diaphragm into its housing, ensuring that the diaphragm lies flat and that the vacuum gallery lines up with the diaphragm and housing. Also, to assist with installation, do not reconnect its operating rod until the cover has been secured.
20   Refit the throttle kicker mounting bracket.
21   Refit the anti-dieselling (fuel cut-off) solenoid, using a new sealing washer.
22   Refit the accelerator pump assembly, ensuring that the diaphragm lies flat and is not kinked.
23   Reassemble the throttle kicker, ensuring that the diaphragm lies flat and that the relative position of the operating link to the throttle kicker cover is as shown in Fig. 3.22. Fully attach the throttle kicker assembly.
24   Refit the main and idle jets, making reference to the notes taken during dismantling to ensure correct fitted locations (also refer to Fig. 3.25).
25   Reassemble the carburettor upper body as described in the following paragraphs.
26   Refit the needle valve housing (using a new washer if applicable).
27   Refit the needle valve and float assembly, having ensured that the float tag locates below the spring clip on the needle valve. Insert the float retaining pin to secure.
28   Adjust the float level as follows. Ensuring that a new gasket is

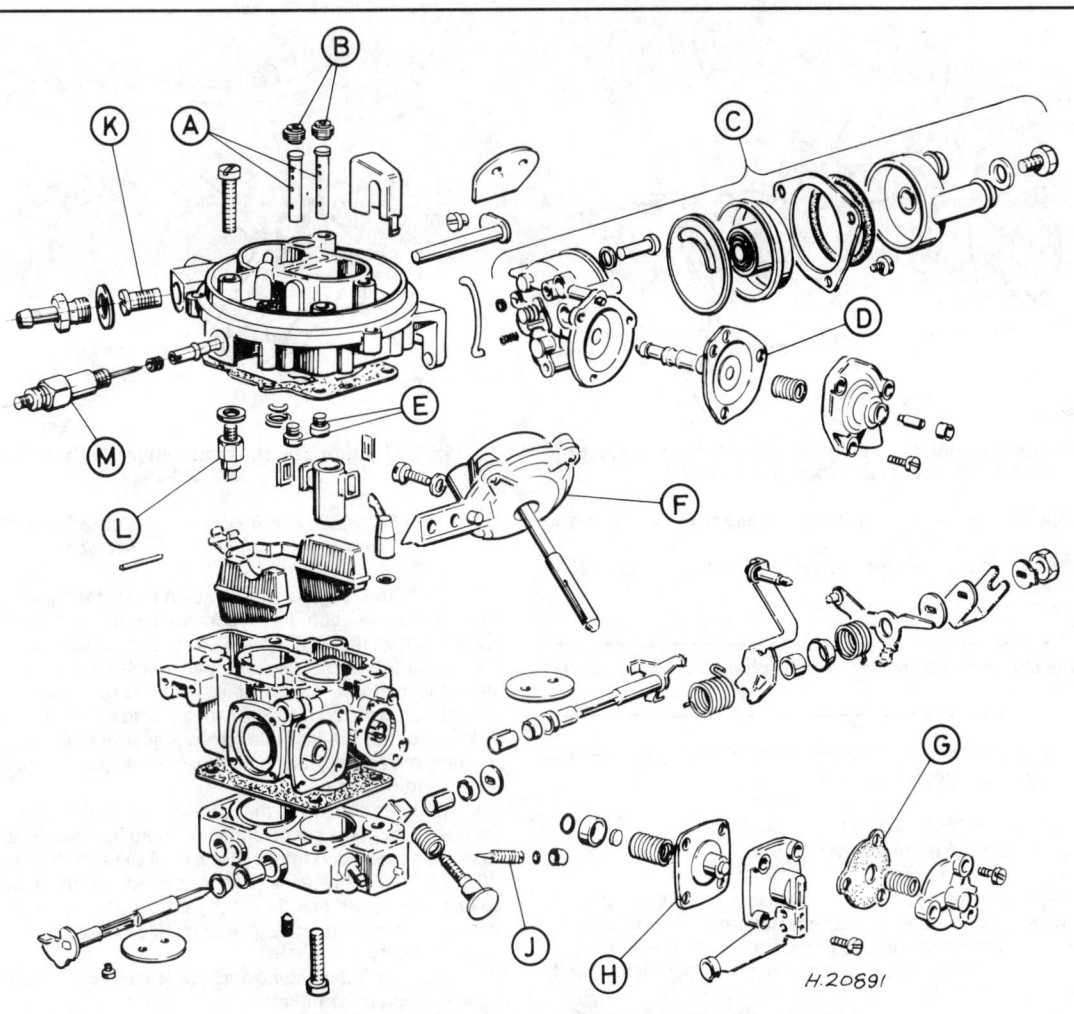

Chapter 3 Fuel, exhaust and emission control systems

Fig. 3.35 Exploded view of the carburettor (Weber TLD) (Sec 33)

- A Emulsion tubes
- B Air correction jets
- C Automatic choke assembly
- D Choke pull-down diaphragm
- E Main jets
- F Secondary barrel diaphragm assembly
- G Power valve diaphragm
- H Accelerator pump diaphragm
- J Fuel mixture adjusting screw
- K Fuel feed filter
- L Needle valve assembly
- M Anti-dieselling (fuel cut-off) solenoid

fitted to the carburettor upper body, hold the upper body in the vertical position; the needle valve must be shut off. Measure the distance shown in Fig. 3.33, and adjust by bending the float tag if the measurement is outside the specification. Recheck the float level adjustment after bending the float tag, as necessary.

29 Refit the choke pull-down diaphragm, ensuring that the diaphragm lies flat and that the vacuum gallery lines up correctly with the diaphragm and housing (Fig. 3.34).

30 Adjust the choke pull-down as follows. Fully close the choke, then manually push the diaphragm operating rod up to its stop; measure the distance between the downdraught side of the choke plate and the venturi, using a gauge rod or the shank of a twist drill bit (of known size). If the measurement is outside specification, remove the tamperproof seal from the housing, and adjust the now revealed adjustment screw accordingly. Fit a new tamperproof seal after successful adjustment.

31 Refit the fuel feed filter, and secure with its brass nut and sealing washer.

32 Refit the carburettor upper body to the carburettor main body, ensuring that its new gasket seats correctly and that the float does not foul during assembly. Insert and tighten the securing screws.

33 Reconnect any vacuum pipes removed during dismantling (where applicable).

34 Once the carburettor has been refitted to the vehicle, check and adjust the idle speed, fuel mixture, fast-idle speed and throttle kicker settings, making reference to the appropriate text. Renew any tamperproof seals after the successful completion of adjustments.

33 Carburettor (Weber TLD) – description

This carburettor incorporates many of the features of the 1.1 litre version (refer to Section 22), except that a throttle kicker is not used, the secondary venturi (barrel) is vacuum-operated and a coolant-heated automatic choke system is fitted.

The choke system is fully automatic. When the engine is cold, the bi-metal spring which controls the position of the choke plate is fully wound up thus holding the choke plate closed. As the engine warms up, the bi-metal spring is heated by the coolant and it begins to unwind, thereby progressively opening the choke plate. A vacuum-operated pulldown mechanism controls the choke plate under certain vacuum conditions, and an internal fast-idle system is incorporated.

Adjustment procedures are given in later text, and it should be understood that *accurate* adjustments can *only* be made using the necessary equipment.

34 Carburettor (Weber TLD) – idle speed and fuel mixture adjustment

1 Refer to Section 17, but note that the fast-idle speed adjustment instructions given later in that Section are **not** applicable to this type of carburettor.

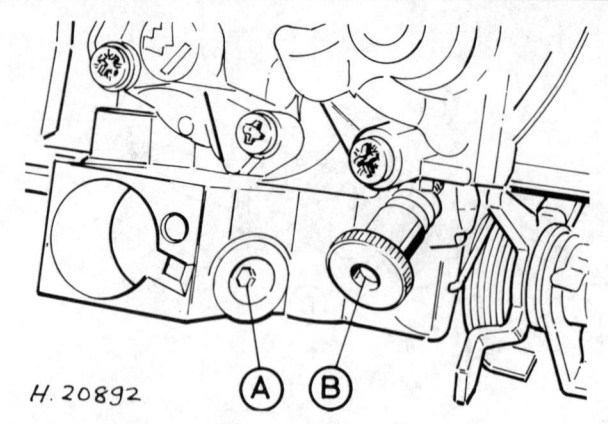

Fig. 3.36 Fuel mixture adjusting screw (A), and idle speed adjusting screw (B) (Weber TLD carburettor) (Sec 34)

2 The idle speed and fuel mixture screw locations are as shown in Fig. 3.36.
3 Fast-idle speed adjustment for this carburettor type is covered in the following Section.

35 Automatic choke and fast-idle speed (Weber TLD carburettor) – adjustment

Note: Refer to Section 2 before proceeding. Note that a tachometer will be required for fast-idle speed adjustment

Vacuum choke plate pull-down
1 Disconnect the battery earth terminal.
2 Remove the air filter housing.
3 Ensure that the cooling system is **not** pressurised (refer to Chapter 2), then disconnect both automatic choke hoses from the choke housing, having identified them for subsequent refitting. Plug their ends, or position as high as possible, to prevent excessive coolant loss. Catch spillage in a suitable container.
4 Note the position of the choke housing alignment marks, then detach the choke housing and bi-metal spring assembly. Remove the internal heat shield.
5 Check and adjust the maximum vacuum choke plate pulldown, as described in the following paragraphs.
6 Fit a rubber band to the choke plate lever, open the throttle to allow the choke plate to close, and then secure the band to keep the plate closed.
7 Using a screwdriver, push the diaphragm open to its stop and measure the clearance between the lower edge of the choke plate and the venturi using a twist drill or other gauge rod. Where the clearance is outside that specified, remove the plug from the diaphragm housing and turn the screw, now exposed, in or out as necessary.
8 Fit a new diaphragm housing plug and remove the rubber band.
9 Refit the heat shield, making sure that the locating peg is correctly engaged in its notch.
10 Refit the choke housing and bi-metal spring assembly, with the spring engaged with the slot in the choke lever which projects through the cut-out in the heatshield. Screw in the retaining screws finger tight, then rotate the housing to set the alignment marks as shown in Fig. 3.40. Fully tighten the retaining screws.
11 Remove the plugs from the end of the automatic choke hoses (if fitted), and reconnect the hoses to their respective locations on the choke housing.
12 Refit the air filter housing.
13 Check and, if necessary, top up the cooling system as described in Chapter 2.
14 Reconnect the battery earth terminal.
15 Check and adjust the fast-idle speed, as described in the following paragraph onwards.

Fast-idle speed
16 Check the idle speed and fuel mixture settings, and adjust as necessary. These **must** be correct before attempting to check or adjust the fast-idle speed.

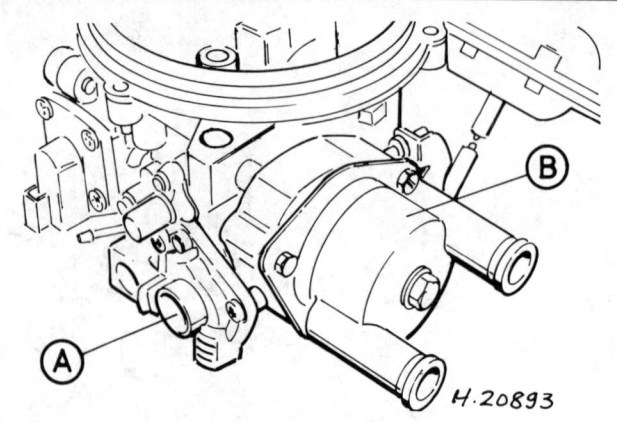

Fig. 3.37 Automatic choke assembly (Weber TLD carburettor) (Sec 35) and 36)

A Pull-down diaphragm housing
B Choke bi-metal spring housing

17 With the engine at its normal operating temperature, and a tachometer connected in accordance with the manufacturer's instructions, remove the air filter housing (if not already done).
18 Hold the throttle linkage partly open, then close the choke plate until the fast-idle speed adjustment screw lines up with the third (middle) step on the fast-idle cam (refer to Fig. 3.41). Release the throttle linkage so that the fast-idle speed adjustment screw rests on the cam. Release the choke plate. The throttle linkage will hold it in the fast-idle speed setting position.
19 Without touching the accelerator pedal, start the engine and record the fast-idle speed achieved. If adjustment is necessary, turn the fast-idle speed adjustment screw until the specified speed is obtained.
20 When the throttle linkage is opened, the choke plate should return to its fully opened position. If this does not happen, either the engine is not at its normal operating temperature or the automatic choke mechanism is faulty.
21 Switch off the engine and disconnect the tachometer. Refit the air filter housing to complete.

36 Automatic choke (Weber TLD carburettor) – removal, overhaul and refitting

Note: Refer to Section 2 before proceeding. Note that a new carburettor upper body gasket will be required during refitting, and that a tachometer will be required for fast-idle speed adjustment

1 Disconnect the battery earth terminal.
2 Remove the air filter housing.
3 Ensure that the cooling system is **not** pressurised (refer to Chapter 2), then disconnect both automatic choke hoses from the choke housing, having identified them for subsequent refitting. Plug their ends, or position as high as possible, to prevent excessive coolant loss. Catch spillage in a suitable container.
4 Loosen the housing bolts or screws, as applicable.
5 Disconnect the anti-dieselling (fuel cut-off) solenoid.
6 Disconnect the fuel feed hose at the carburettor, and plug its end to avoid spillage and prevent dirt ingress. If a crimped type hose clip is fitted, cut this free taking care not to damage the hose.
7 Remove six screws (four of which are Torx head screws) and detach the carburettor upper body assembly.
8 Note the position of the choke housing alignment marks, then detach the choke housing and bi-metal spring assembly. Remove the internal heat shield.
9 Remove the three automatic choke assembly securing screws (see Fig. 3.42), disconnect the choke link at the operating lever and detach the automatic choke assembly.
10 Remove the three vacuum diaphragm housing securing screws and detach the vacuum diaphragm assembly, noting component fitment for subsequent reassembly.
11 Dismantle the remaining choke mechanism, noting component fitment as an aid to reassembly.

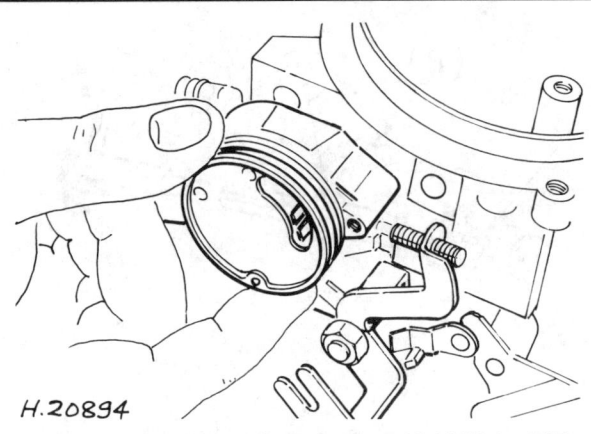

Chapter 3 Fuel, exhaust and emission control systems

Fig. 3.38 Automatic choke internal heatshield (Weber TLD carburettor) (Sec 35 and 36)

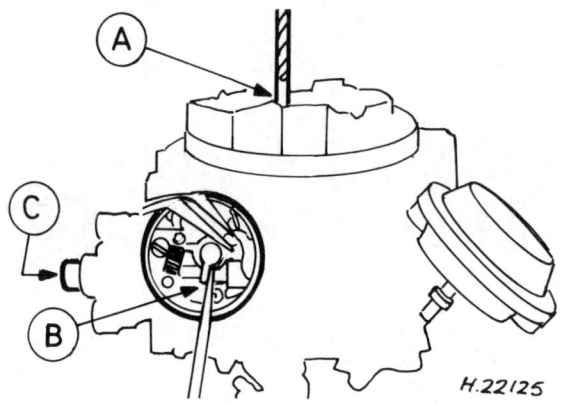

Fig. 3.39 Choke plate pull-down adjustment (Weber TLD carburettor) (Sec 35)

A   Twist drill   C   Adjusting screw
B   Diaphragm held fully open

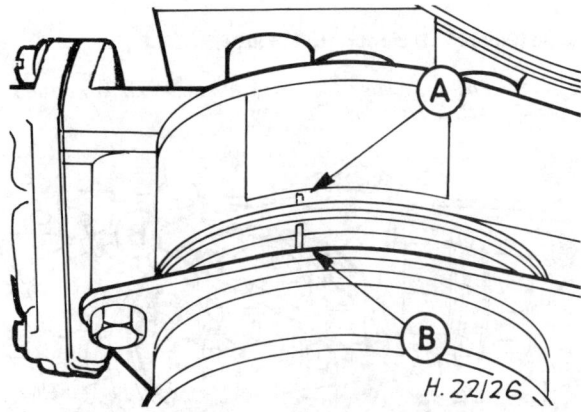

Fig. 3.40 Automatic choke housing alignment (Weber TLD carburettor) (Sec 35) and 36)

A   Dot punch mark   B   Choke housing alignment mark

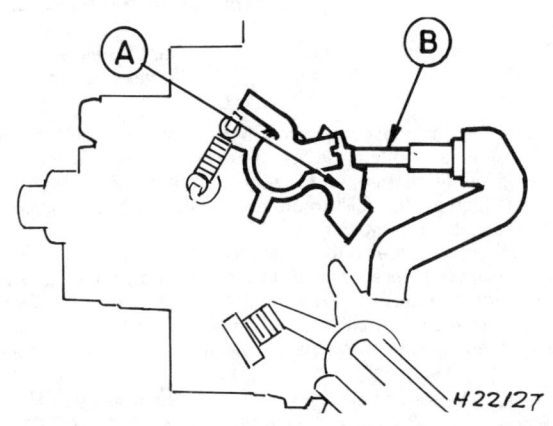

Fig. 3.41 Fast-idle speed adjustment (Weber TLD carburettor) (Housing cut away for illustration clarity) (Sec 35)

A   Fast-idle cam   B   Fast-idle speed adjusting screw

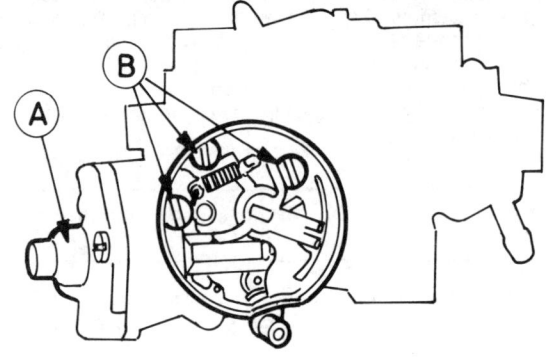

12 Clean and inspect all components for wear, damage, or distortion. Pay particular attention to the condition of the vacuum (pull-down) diaphragm and the choke housing O-ring. Renew any components that are unserviceable or suspect.

13 Reassemble the automatic choke mechanism, making reference to the notes taken during dismantling and to Fig. 3.43. Note that **no** lubricants should be used.

14 Refit the vacuum diaphragm assembly, making reference to the notes taken during dismantling, and ensure that the diaphragm lies flat before the housing is secured.

15 Locate the O-ring, then reconnect the choke link, position the automatic choke assembly and secure it with its screws.

16 Check and adjust the maximum vacuum choke plate pull-down, as described in Section 35 paragraphs 6 to 8 inclusive.

17 Refit the heat-shield making sure that the locating peg is correctly engaged in its notch.

18 Refit the choke housing and bi-metal spring assembly, as described in Section 35 paragraph 10.

19 Refit the carburettor upper body assembly (using a new gasket) and secure with its six screws.

20 Reconnect the fuel feed hose to the carburettor, using a nut and screw type clip to secure.

21 Reconnect the anti-dieselling (fuel cut-off) solenoid.

22 Remove the plugs from the end of the automatic choke hoses (if fitted), and reconnect them to their respective locations on the choke housing.

23 Check and, if necessary, top up the cooling system as described in Chapter 2.

24 Refit the air filter housing, then reconnect the battery earth terminal.

25 Check and adjust the fast-idle speed, as described in the previous Section, paragraph 16 onwards.

Fig. 3.42 Automatic choke assembly (Weber TLD carburettor) (Sec 36)

A   Pull-down diaphragm housing   B   Securing screws

37 Carburettor (Weber TLD) – removal and refitting

Note: *Refer to Section 2 before proceeding. Note that a new gasket will be required upon refitting, and that a tachometer and an exhaust gas analyser will be required to check the idle speed and fuel mixture upon completion.*

1 Disconnect the battery earth terminal.
2 Remove the air filter housing.

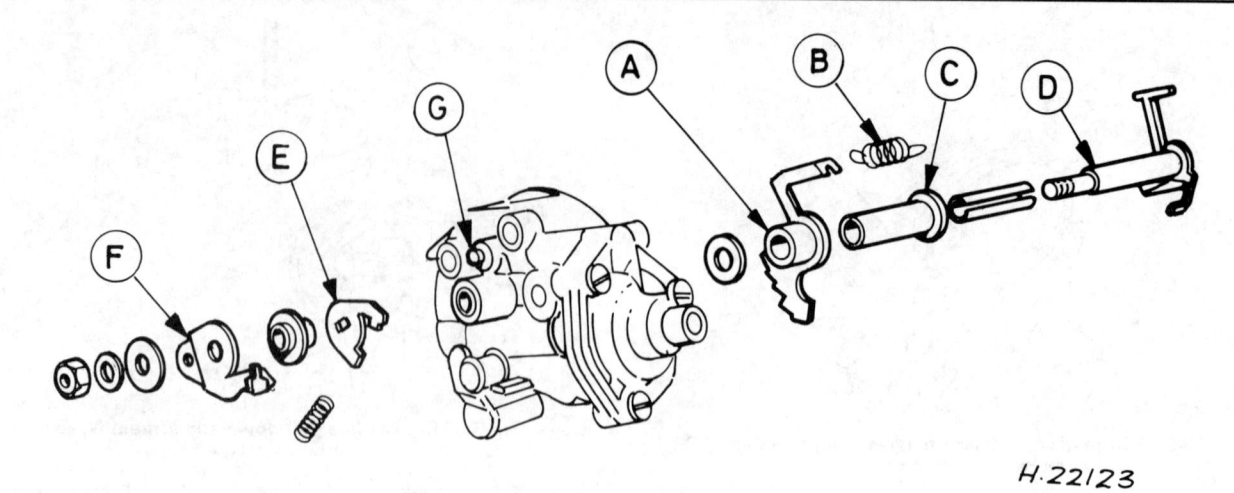

110   Chapter 3 Fuel, exhaust and emission control systems

Fig. 3.43 Exploded view of the automatic choke linkage (Weber TLD carburettor) (Sec 36)

A Operating link/fast-idle cam
B Fast-idle cam return spring
C Spindle sleeve
D Connecting rod and lever assembly
E Pull-down link
F Actuating lever
G Automatic choke housing

3 Ensure that the cooling system is **not** pressurised (refer to Chapter 2), then disconnect both automatic choke hoses from the choke housing, having identified them for subsequent refitting. Plug their ends, or position as high as possible, to prevent excessive coolant loss. Catch spillage in a suitable container.
4 Disconnect the accelerator link at the carburettor.
5 Disconnect the fuel feed hose at the carburettor, and plug its end to avoid spillage and prevent dirt ingress. If a crimped type hose clip is fitted, cut this free taking care not to damage the hose.
6 Disconnect all relevant vacuum pipes from the carburettor, having labelled them for subsequent refitting.
7 Disconnect the electrical lead from the anti-dieselling (fuel cut-off) solenoid.
8 Remove the four Torx head screws securing the carburettor to the inlet manifold, then withdraw it from the vehicle.
9 Clean the inlet manifold and carburettor gasket mating faces.
10 Refitting is a reversal of the removal procedure, bearing in mind the following points. A new gasket **must** be used, and the carburettor securing screws should be tightened to the specified torque. When reconnecting the fuel feed hose, use a nut and screw type clip to secure. After refitting the automatic choke hoses, check and, if necessary, top up the cooling system as described in Chapter 2.
11 Upon completion, check the idle speed and fuel mixture settings, and adjust as necessary.

38 Carburettor (Weber TLD) – dismantling, cleaning, inspection and reassembly

Note: *Check parts availability before proceeding. Where possible, obtain an overhaul kit containing all the relevant gaskets, seals, etc., required for reassembly*

1 With the carburettor removed from the vehicle, prepare a clean flat work surface prior to dismantling. The procedure given here may be used for partial dismantling, or complete dismantling within practical limits, as required.
2 Clean the exterior of the carburettor and detach all vacuum pipes (where applicable), having noted their fitted location for correct subsequent reassembly.
3 Remove the two carburettor upper body securing screws, hold the throttle open and detach the carburettor upper body assembly.
4 Dismantle the upper body assembly, as described in the following paragraphs. First, slide out the float retaining pin, then detach the float and needle valve. Remove the needle valve housing and its washer.
5 Unscrew the fuel feed hose connector, and withdraw its seal (as applicable) and the fuel feed filter.
6 Remove the anti-dieselling (fuel cut-off) solenoid and ensure that its sealing washer comes away with it.

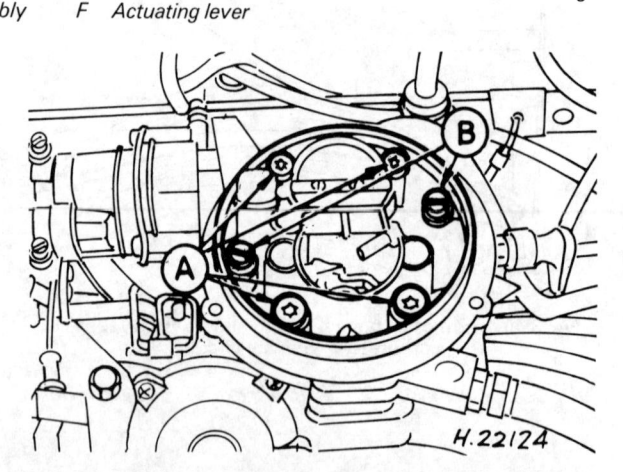

Fig. 3.44 General view of Weber TLD carburettor (Sec 37 and 38)

A Torx head screws securing carburettor to inlet manifold
B Standard screws securing upper body to main body

7 Proceed as described in Section 27, paragraphs 7 to 13 inclusive.
8 Disconnect the secondary barrel diaphragm operating rod by pulling the lower section of rod downwards and twisting to release it from its retaining socket. Remove its four cover retaining screws and dismantle the assembly, noting the components as they are withdrawn.
9 Proceed as described in Section 27, paragraphs 15 to 20 inclusive, but note that the fuel feed filter will also be need to be cleaned.
10 Reassemble the secondary barrel diaphragm into its housing, ensuring that the diaphragm lies flat and that the vacuum gallery lines up with the diaphragm and housing. Also, to assist with installation, do not reconnect its operating rod until the cover has been secured.
11 Proceed as described in Section 27, paragraphs 22 to 28 inclusive.
12 Reposition the fuel feed filter, then refit the fuel feed connector and its seal (as applicable).
13 Refit the needle valve housing, using a new washer.
14 Refit the needle valve and float assembly, having ensured that the float tag hooks under the spring clip on the needle valve. Insert the float retaining pin to secure.
15 Measure and, if necessary, adjust the float level using the method described in Section 27, paragraph 32, making reference to Fig. 3.48 for the required dimension.
16 Hold the throttle partially open and refit the carburettor upper body to the carburettor main body, ensuring that the new gasket seats correctly and that the float does not foul during assembly. Insert and tighten the securing screws.

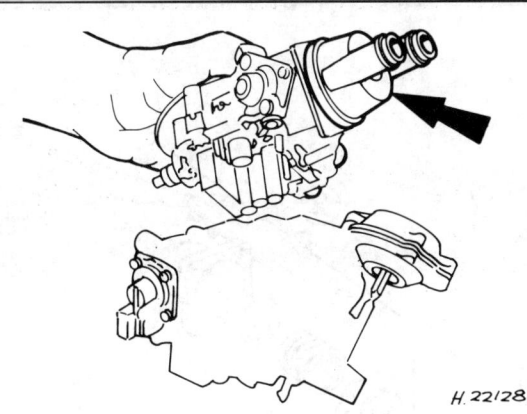

Chapter 3 Fuel, exhaust and emission control systems 111

Fig. 3.45 Removing the carburettor upper body (arrowed) (Weber TLD carburettor) (Sec 38)

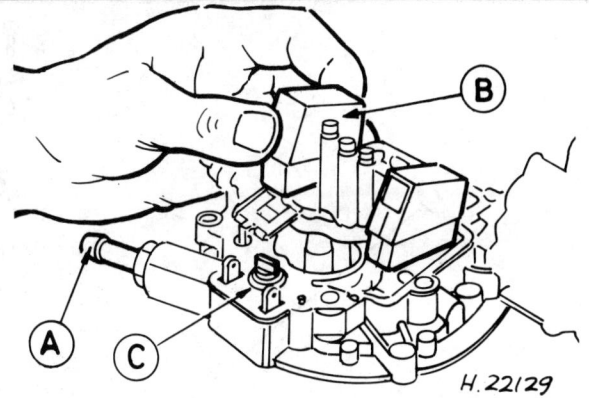

Fig. 3.46 Float and needle valve removal (Weber TLD carburettor) (Sec 38)

A Fuel feed connection   C Needle valve
B Float

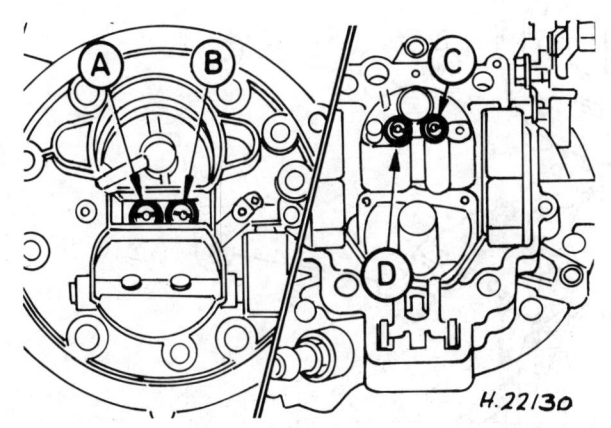

Fig. 3.47 Jet arrangement in carburettor upper body (Weber TLD carburettor) (Sec 38)

A Primary air correction jet   C Secondary main jet
B Secondary air correction jet   D Primary main jet

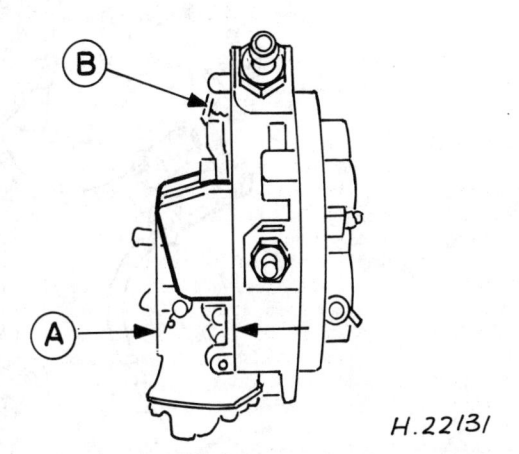

Fig. 3.48 Float level adjustment (Weber TLD carburettor) (Sec 38)

A Float level setting dimension   B Adjusting tag

17 Reconnect any vacuum pipes removed during dismantling (where applicable).
18 Once the carburettor has been refitted to the vehicle, check and adjust the idle speed, fuel mixture and fast-idle speed settings, making reference to the appropriate text. Renew any tamperproof seals after the successful completion of adjustments.

PART B: FUEL INJECTION AND ASSOCIATED FUEL SYSTEM COMPONENTS

39 General description

Two completely different fuel injection systems are available within the Fiesta range, both coming under the overall control of the EEC-IV engine management system. The ignition sub-system operation is covered in Chapter 4.
In both systems, fuel is supplied from the rear mounted fuel tank by an electric fuel pump mounted inside the tank (as a combined assembly with the fuel level sender unit). The fuel passes through a filter mounted in the engine compartment, and up to its individual fuel injection system. The fuel is maintained at the required system operating pressure by a fuel pressure regulator.
The operation of each system is described individually as follows.

1.4 litre CFi models

The CFi unit is a very simple device when compared with a conventional carburettor. Fuel is injected by a single solenoid valve (fuel injector) centrally mounted on the top of the unit. It is from this system feature that the abbreviation CFi is derived (Central Fuel Injection).
The injector, which atomises the fuel entering the engine, is positioned to direct the fuel downwards into the inlet manifold. Electrical signals generated by the EEC-IV engine management module, energise the injector solenoid, thus lifting the pintle valve off its seat and allowing pressurised fuel to flow. The electrical signals take two forms: a high current signal opens the injector, and a low current signal (via a ballast resistor) holds it open for the duration required. At idle speed the injector is pulsed by the EEC-IV engine management module on every other intake stroke, whereas during normal driving it is pulsed on every intake stroke.
The amount of fuel delivered to the engine in relation to the amount of air entering (the fuel mixture) is calculated from values pre-programmed into the EEC-IV engine management module. Other values are obtained from the distributor (engine speed), engine coolant temperature (ECT) sensor, manifold absolute pressure (MAP) sensor, air charge temperature (ACT) sensor, idle-tracking switch, heated exhaust gas oxygen (HEGO) sensor and the throttle position sensor (TPS). No adjustments to the fuel mixture are possible.
The throttle position sensor enables the EEC-IV engine management module to compute not only the throttle position, but also its rate of change. Extra fuel can then be provided for acceleration when the throttle is opened suddenly. Information from the throttle position sensor, in conjunction with the idle-tracking switch, also indicates to the

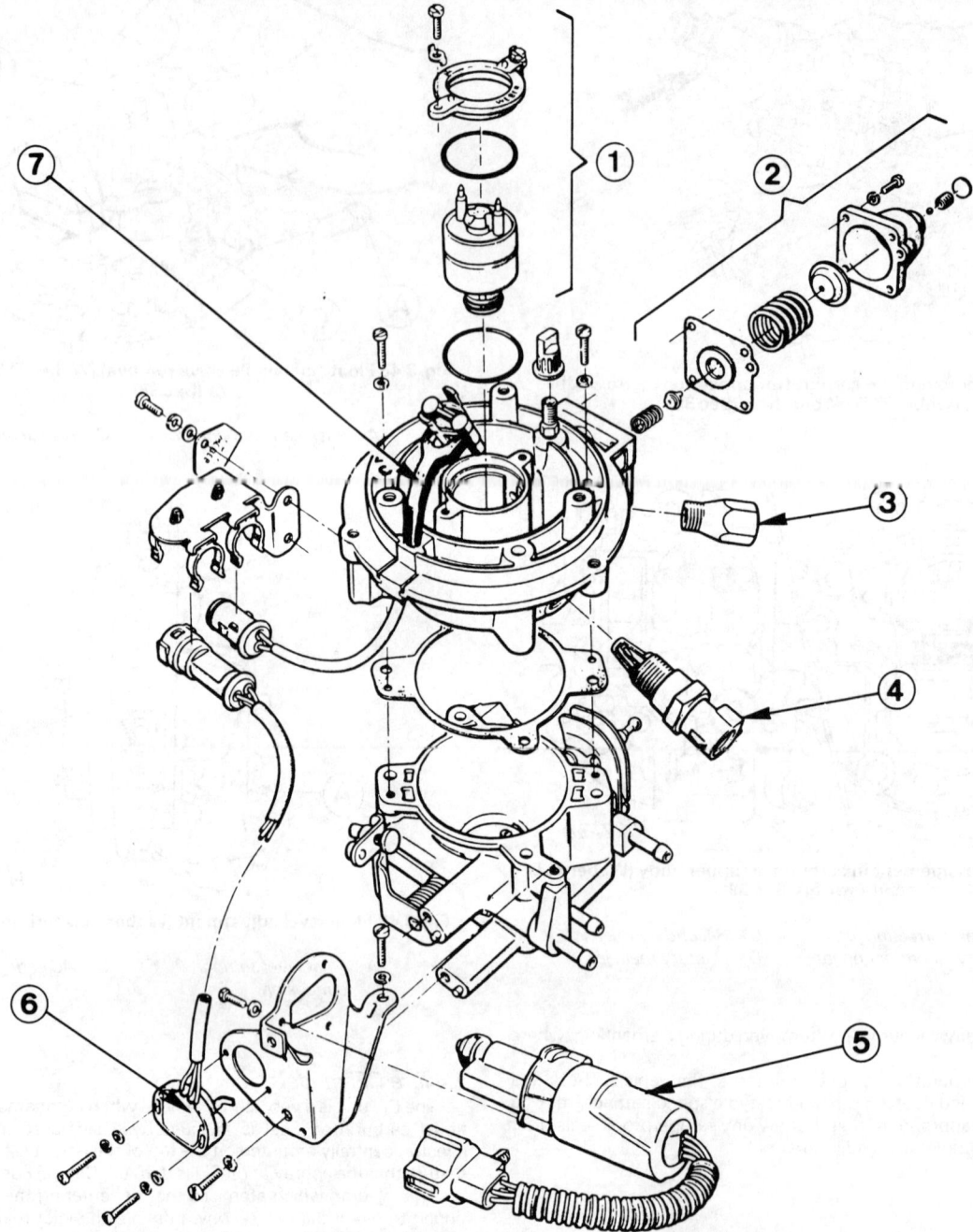

112   Chapter 3   Fuel, exhaust and emission control systems

Fig. 3.49 Exploded view of the CFi unit (1.4 litre models) (Sec 39)

1 Fuel injector assembly
2 Fuel pressure regulator assembly
3 Fuel feed connector
4 Air-charge temperature (ACT) sensor
5 Throttle-plate control motor
6 Throttle position sensor
7 Fuel injector wiring

EEC-IV engine management module when the engine is operating under closed throttle conditions.

Idle speed is controlled by the throttle plate control motor which is mounted on the side of the CFi unit. The motor responds to signals from the EEC-IV engine management module, the signals being calculated from pre-programmed values and information obtained from the distributor (engine speed), road speed sensor, engine coolant temperature (ECT) sensor, idle-tracking switch and the throttle position sensor. When closed throttle conditions are sensed (via the idle-tracking switch on the throttle plate control motor) the EEC-IV engine management module enters the idle speed control mode or the dash pot mode, according to engine speed. In the idle speed control mode, the motor will control the engine speed fluctuations and maintain a constant idle speed, with signals being sent from the EEC-IV engine management module to compensate for any sudden engine loadings (sensed by a drop in engine speed). In the dash pot mode, the EEC-IV engine management module senses the change in engine load and throttle position, then supplies the motor with a series of signals to prevent the throttle plate from instantaneously snapping shut during deceleration. The resultant gradual closing of the throttle plate allows optimum fuelling to be supplied during engine braking, and assists in the reduction of hydrocarbon exhaust emissions by maintaining effective combustion of the fuel mixture.

When the engine is switched off, a signal is sent to the throttle plate control motor, from the EEC-IV engine management module, to fully close the throttle plate and thus prevent 'run-on' or dieselling. The throttle plate is then returned to a basic preset position ready for engine restarting (made possible due to power maintaining circuitry). When the

Chapter 3 Fuel, exhaust and emission control systems

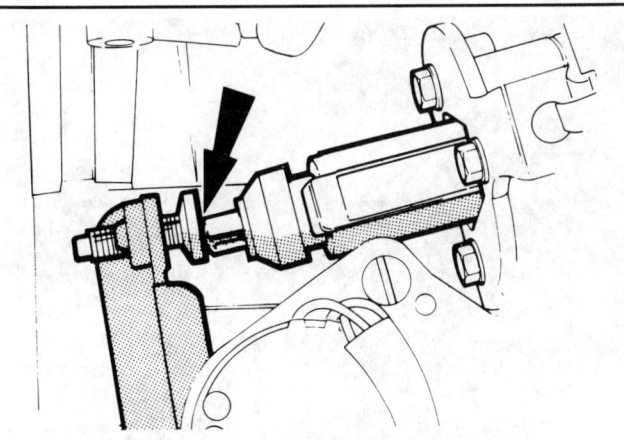

Fig. 3.50 Idle tracking switch (arrowed) in contact with the throttle-plate lever (1.4 litre CFi models) (Sec 39)

ignition is turned on to restart the engine, the motor will reposition the throttle plate, if necessary, depending on the engine temperature.

In the event of a sensor failure, or indeed the failure of the EEC-IV engine management module itself, a 'limited operation strategy' (LOS) allows the vehicle to be driven albeit at reduced power and efficiency.

An inertia switch is fitted to ensure that the fuel supply to the engine is switched off in the event of an accident or similar impact. For further details, refer to Chapter 12.

*As the 1.4 litre CFi model is fitted with a catalytic converter, leaded petrol must **not** be used (refer to the Specifications).* Further details on exhaust and emission control systems may be found later in this Chapter.

1.6 litre EFi models

Fuel enters the distributor rail via a connection on its right-hand side. The distributor rail acts as a pressurised fuel reservoir for the four fuel injectors, and also holds them in position in their inlet manifold tracts. The electro-magnetically actuated injectors have only 'on' or 'off' positions, the volume of fuel injected to meet the engine requirements being determined by the period of time that the injectors remain 'on' (or open).

The volume of fuel required for one power stroke (as calculated by the EEC-IV engine management module) is divided into two equal amounts. The first half of the required volume is injected into the static air ahead of the inlet valve one complete engine revolution before the inlet valve is due to open. After one further engine revolution, the inlet valve opens and the second half of the required volume of fuel is injected into the flow of air as it is drawn into the cylinder. Therefore, fuel will always be injected at two inlet valves simultaneously, at a particular crankshaft position.

The volume of air drawn into the engine is governed by the air filter housing design and variable operational factors. The variables involved are air pressure and density, throttle position, engine speed and the degree of blockage in the air filter element. The EEC-IV engine management module evaluates these variables and produces the signals required to actuate the injectors.

Provision for adjusting the fuel mixture is made via a potentiometer mounted on the bulkhead panel. Adjusting the fuel mixture screw on the potentiometer will alter the injector signals sent by the EEC-IV engine management module accordingly.

An idle speed control valve, itself controlled by the EEC-IV engine management module, stabilises the engine idle speed under all conditions by the opening of an auxiliary air passage which bypasses the throttle. Apart from the base-idle speed adjustment given in a later Section, no adjustments to the operational idle speed can be made.

If the EEC-IV engine management module develops a fault, or one of the sensors fail, a 'limited operation strategy' (LOS) comes into effect.

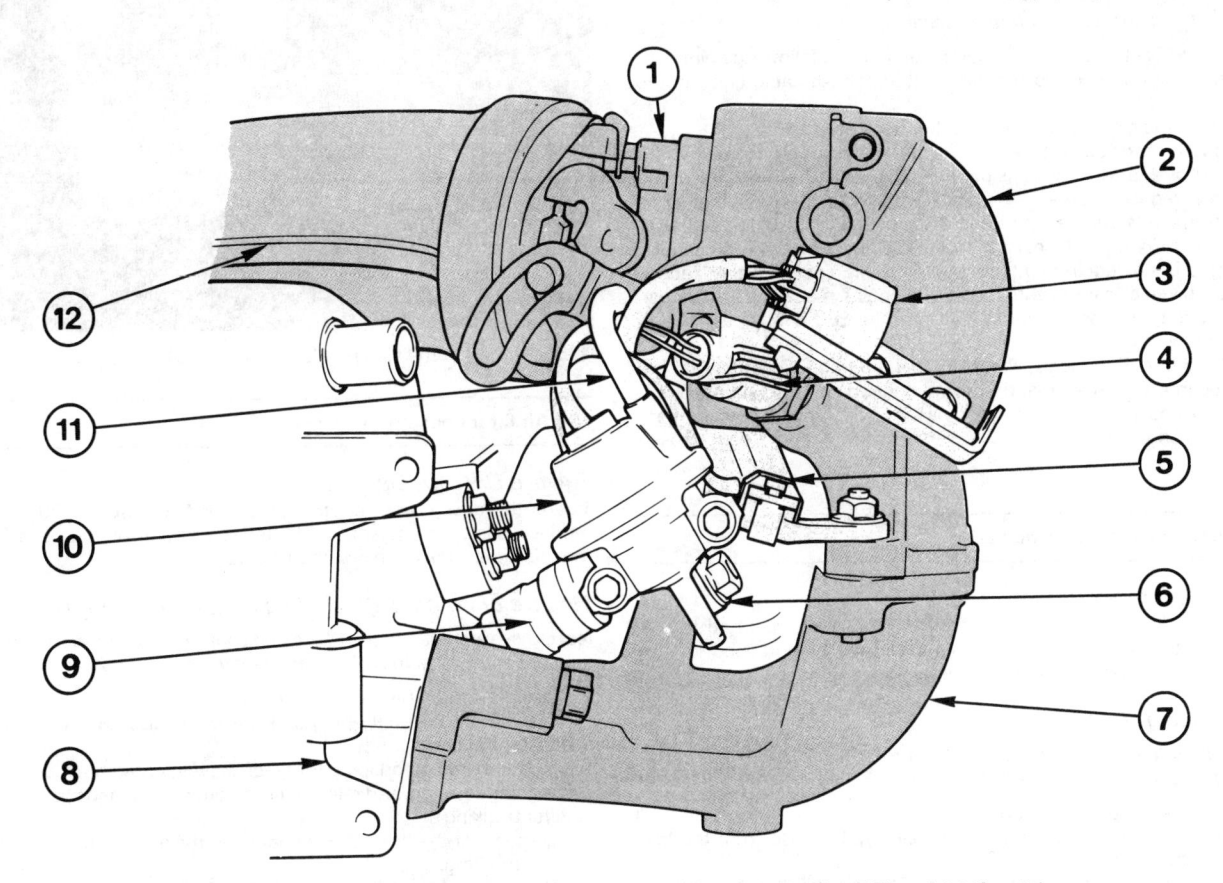

Fig. 3.51 General view of the fuel injection system arrangement (1.6 litre EFi models) (Sec 39)

1 Throttle housing	4 Air charge temperature (ACT) sensor	7 Lower section of inlet manifold	10 Fuel pressure regulator
2 Upper section of inlet manifold	5 Wiring harness ducting	8 Cylinder head	11 Vacuum hose
3 Wiring loom connector	6 Fuel distributor rail	9 Fuel injector	12 Air filter trunking

This allows the vehicle to be driven, albeit at reduced power and efficiency.

As on 1.4 litre CFi models, an inertia switch cuts off the supply of fuel to the engine in the event of an accident, by switching off the fuel pump. For further details refer to Chapter 12.

Details on exhaust and emission control systems may be found later in this Chapter.

40 Fuel system – precautions

Note: *When undertaking any work or repairs to the fuel system it is important that the battery is disconnected. On all models fitted with an EEC-IV engine management system, reference must be made to Chapter 12, Section 5 after battery disconnection and subsequent reconnection*

Refer to Section 2 of this Chapter but note that the fuel injection systems are pressurised, therefore extra care must be taken when disconnecting fuel lines. When disconnecting a fuel line union, loosen the union **slowly** to avoid a sudden release of pressure which may cause fuel spray. Keep a large clean rag to hand, to wrap around the union and soak up any escaping pressurised fuel. After use the rag must be disposed of safely.

On 1.4 litre CFi models, it must be noted that the catalytic converter operates at an *extremely high* temperature. Ensure that the underbody heatshields are always correctly refitted, if removed during an operation.

41 Maintenance and inspection

Note: *Refer to Section 40 before proceeding*

1 Refer to Section 3 of this Chapter, but note the following points.
2 Ignore all reference to the carburettors, choke cable and throttle kicker.
3 Renew the separate fuel filter at the specified intervals.
4 On XR2i models with a Weber-built idle speed control valve mounted on the air filter housing, the valve must be dismantled and cleaned at regular intervals – see Section 62. Note, however, that some of these models will be fitted with Hitachi-built valves (located elsewhere – see Chapter 13, Section 7) which do not require this procedure.
5 There is **no** requirement to check the idle speed at any service interval on vehicles fitted with the EEC-IV engine management system, as idle speed is under the overall control of the EEC-IV engine management module.
6 On 1.6 litre EFi models, check the fuel mixture (CO content) at the specified intervals. Note that this is **not** required on 1.4 litre CFi models, as 'closed loop' fuel mixture control (via the EEC-IV engine management module) is employed.

42 Air filter element – renewal

1.4 litre CFi models
1 As the design and function of the air filter housing fitted to this model is basically similar to that fitted to carburettor equipped 1.4 litre models, refer to Section 4 of this Chapter.

1.6 litre EFi models
Note: *The reference to the idle speed control valve and air bypass hose may not apply to all models – see Chapter 13, Section 7.*

2 Disconnect the battery earth terminal.
3 Disconnect the multi-plug and the air bypass hose from the idle speed control valve.
4 Disconnect the flexible hose between the air filter housing and the air filter trunking. This may be removed with the air filter trunking, as required (refer to the following Section).
5 Unclip and remove the air filter lid then withdraw and discard the old element (photos). A further small pad-filter element is fitted by the air inlet.

42.5A Releasing an air filter lid retaining clip (1.6 litre EFi model)

42.5B Withdraw the air filter element (1.6 litre EFi model)

43 Air filter housing – removal and refitting

1.4 litre CFi models
1 As the design and function of the air filter housing fitted to this model is basically similar to that fitted to carburettor equipped 1.4 litre models, refer to Section 5 of this Chapter.

1.6 litre EFi models (including air filter trunking)
Note: *The reference to the idle speed control valve and air bypass hose may not apply to all models – see Chapter 13, Section 7.*

2 Disconnect the battery earth terminal.
3 Disconnect the multi-plug and the air bypass hose from the idle speed control valve.
4 Withdraw the spark plug HT leads, having noted their locations for subsequent refitting, and release them from the locating channels in the air filter trunking.
5 Disconnect the flexible hose between the air filter housing and the air filter trunking.
6 Release the two bolts securing the air filter trunking to the top of the engine rocker cover, then carefully withdraw from the throttle housing before removing.
7 Disconnect the crankcase ventilation hose from the front of the air filter housing.
8 Unclip and remove the air filter lid, then withdraw the element.

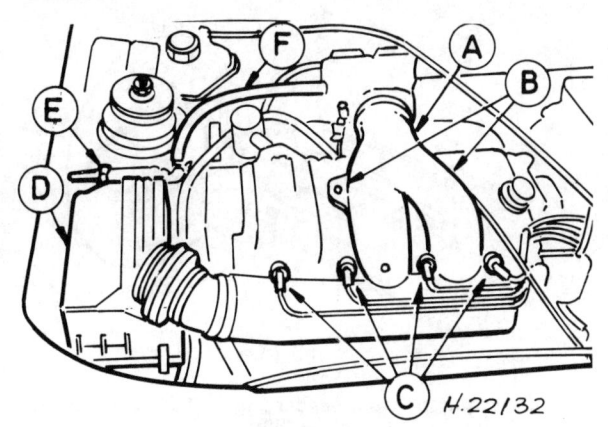

Chapter 3 Fuel, exhaust and emission control systems

Fig. 3.52 General view of the air filter housing and trunking (1.6 litre EFi models) (Sec 43)

- A Air filter trunking
- B Air filter trunking securing bolts
- C Spark plug HT lead connectors
- D Air filter housing cover
- E Idle speed control valve multi-plug
- F Air bypass hose

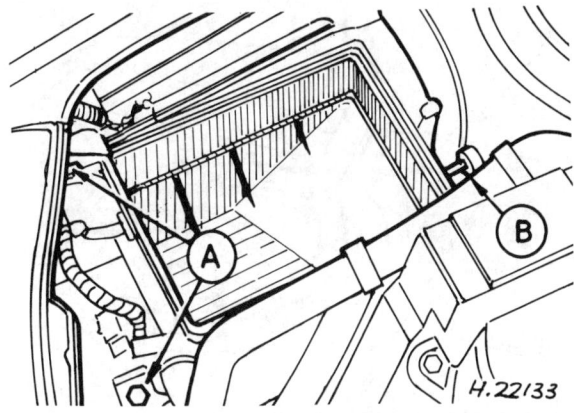

Fig. 3.53 Air filter housing securing arrangement (1.6 litre EFi models) (Sec 43)

- A Bolts
- B Grommet

9 Remove the two bolts securing the forward end of the air filter housing, free the rearward end of the housing from its location and carefully withdraw from the vehicle.
10 Refitting is a reversal of the removal procedure.

44 Accelerator cable – removal, refitting and adjustment

1 Refer to Section 7 of this Chapter, but note that on 1.6 litre EFi models there is no need to remove the air filter housing – the trunking will suffice. Where mention is made of the carburettor throttle-plate linkage, ignore all references to 'carburettor' (photo).

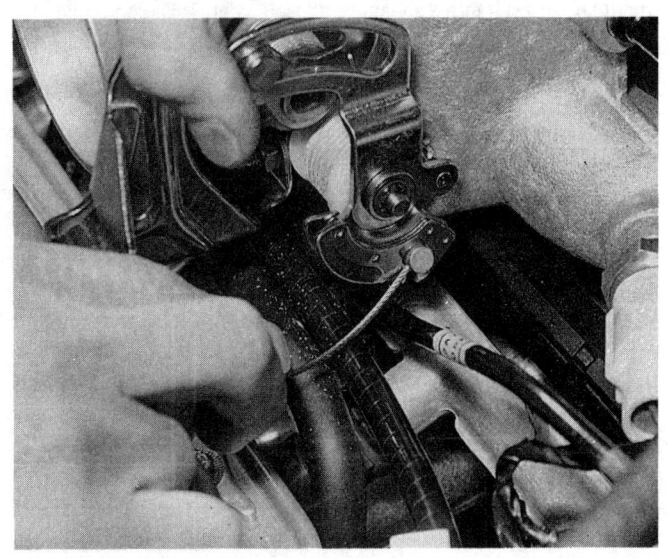

45 Accelerator pedal – removal and refitting

1 Refer to Section 9 of this Chapter, but note that if cable adjustment is required after completing the operation, additional reference must be made to Section 44.

46 Fuel tank ventilation tube – general

Note: *Refer to Section 40 before proceeding*

1 The fuel tank ventilation tube fitted to 1.6 litre EFi models (without evaporative emission control) is basically identical to that fitted to the carburettor equipped models in the range. Reference should therefore be made to Section 11 of this Chapter.
2 The fuel tank ventilation tube on 1.4 litre CFi models (with evaporative emission control) connects to the combined roll-over/anti-trickle-fill valve assembly in much the same manner as on the rest of the range but, instead of venting to atmosphere, a further tube runs the length of the vehicle to a carbon canister in the front right-hand corner of the engine compartment. Further information on this system may be found later in this Chapter.

47 Fuel tank – removal and refitting

Note: *Refer to Section 40 before proceeding*

1 Refer to Section 12 of this Chapter but note that, on 1.4 litre CFi

44.1 Accelerator cable retention arrangement at the throttle linkage (1.6 litre EFi model)

models, the underbody heatshields will need to be repositioned or removed, and the small bore ventilation tube needs to be released from its retaining clips on the fuel tank before the fuel tank securing bolts are removed.

48 Fuel tank filler pipe – removal and refitting

Note: *Refer to Section 40 before proceeding*

1 Refer to Section 13 of this Chapter, but make additional note of the information given in Section 47.

49 Fuel pump – testing

Note: *Refer to Section 40 before proceeding*

1 The combined fuel pump/sender unit assembly is mounted inside the fuel tank.
2 If the pump is functioning, it may be possible to hear it 'buzzing' by listening carefully under the rear of the vehicle when the ignition is switched on.

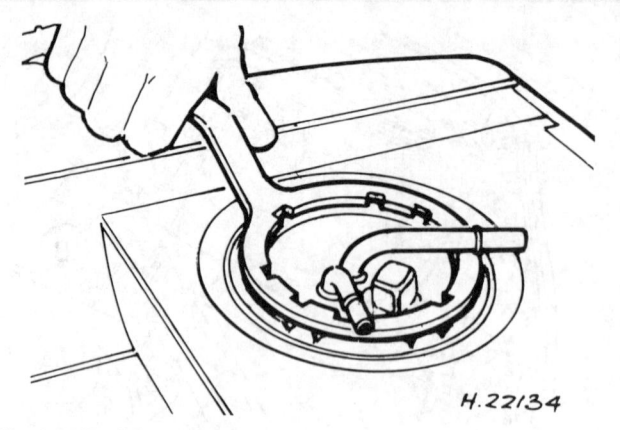

116 Chapter 3 Fuel, exhaust and emission control systems

Fig. 3.54 Ford Special tool engaged on the fuel pump/sender unit (fuel injection models) (Sec 50)

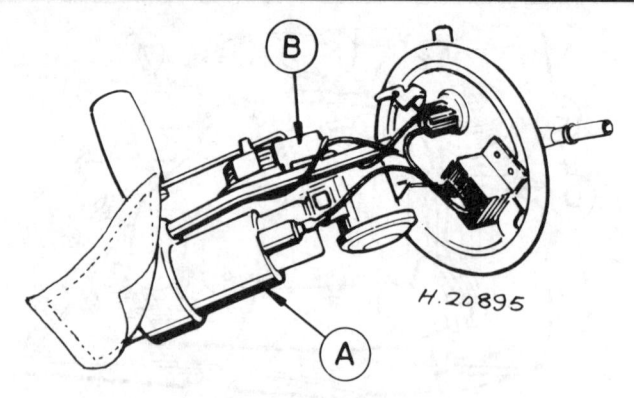

Fig. 3.55 Fuel pump (A) and sender unit (B) (fuel injection models) (Sec 50)

3 If the pump appears to have failed completely, check the appropriate fuse and relay.
4 To test the pump, disconnect a fuel feed pipe in the engine compartment and insert it into a suitable clear container. Take appropriate fire precautions, then have an assistant switch the ignition on for a few seconds (the engine should not be started). Observe the fuel delivery during this period; if in any doubt as to the effectiveness of the delivery, consult your Ford dealer for further testing or renew, as necessary.
5 On completion of the test, securely reconnect the fuel feed pipe. Refer to the Specifications for tightening torque details.

50 Fuel pump/sender unit – removal and refitting

Note: *Ford Special tool 23-026 is required to remove and refit the fuel pump/sender unit. If this is not available, or cannot be suitably fabricated, entrust the tank to your Ford dealer*

1 The procedure is identical to that given in Section 14 of this Chapter, but note that tool application is different.

51 Inertia switch (fuel cut-off) – removal and refitting

1 Refer to Chapter 12, Section 51. Note that additional reference must be made to Chapter 12, Section 5 upon subsequent battery reconnection, to allow the EEC-IV engine management module to re-learn its values.

52 Fuel filter – removal and refitting

Note: *Refer to Section 40 before proceeding*

1 Disconnect the battery earth terminal.
2 Position a suitable container beneath the fuel filter to catch escaping fuel, then slowly slacken the fuel feed pipe union allowing the pressure in the fuel pipe to reduce. When the pressure is fully released, disconnect the fuel feed and outlet pipe unions.
3 Having noted the orientation of the filter in its fitted position (flow direction arrows are marked on the casing), remove the retaining bolt and withdraw the filter assembly from the vehicle (photo). Note that the filter will still contain a small amount of fuel, and care should be taken to avoid spillage.
4 Refitting is a reversal of the removal procedure, but ensure that the flow direction arrows on the filter casing point the right way, and tighten the unions to the specified torque.
5 Switch the ignition on and off at least five times after the battery earth terminal has been reconnected, and check for fuel leakage.

53 Fuel pressure regulator (1.4 litre CFi) – removal and refitting

Note: *Refer to Section 40 before proceeding. Note that the system pressure must be checked by a Ford dealer, or other suitable specialist, on completion*

1 Remove the CFi unit from the vehicle, as described in Section 57.
2 Remove the four screws securing the regulator housing (photo) then dismantle the regulator assembly, noting the fitment of compo-

52.3 Fuel filter viewed from the engine compartment. Note flow direction arrows on the filter casing, and the filter retaining bolt (arrowed)

53.2 Fuel pressure regulator housing securing screws (arrowed) (1.4 litre CFi model, shown *in situ*)

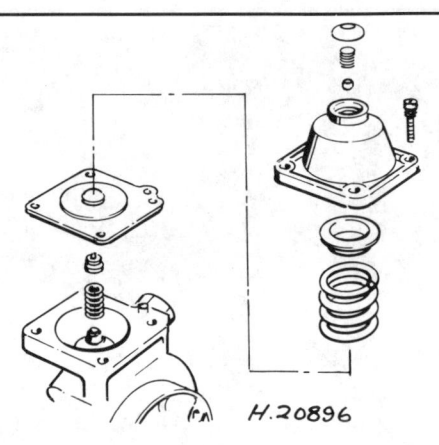

Chapter 3 Fuel, exhaust and emission control systems

Fig. 3.56 Exploded view of the fuel pressure regulator assembly (1.4 litre CFi models) (Sec 53)

nents as they are removed. Do **not** attempt to prise out the plug or adjust the screw (if no plug is fitted) in the regulator housing, as this will alter the system pressure.
3   Examine all the components, and renew any defective items.
4   To refit/reassemble, support the CFi unit on its side so that the regulator components may be fitted from above.
5   Fit the small spring, valve, diaphragm (ensuring that it locates correctly), large spring and the spring cup. Carefully place the ball into position on the spring cup and ensure that it seats correctly. The regulator housing must be fitted with the *utmost* care to avoid disturbing the ball, and once correctly located in position, its screws must be tightened evenly to avoid distortion.
6   Refit the CFi unit in accordance with Section 57, but note that further checks for fuel leaks should be made with the engine running. The fuel system pressure must be checked by a Ford dealer, or other suitable specialist, at the earliest possible opportunity.

54 Fuel injector (1.4 litre CFi) – removal and refitting

Note: *Refer to Section 40 before proceeding*

1   Disconnect the battery earth terminal.
2   Remove the air filter housing.
3   Place a suitable container beneath the fuel feed pipe union on the CFi unit, then slowly slacken the union and allow the pressure in the pipe to reduce.
4   Disconnect the multi-plug from the injector (photo).
5   Bend the injector retaining collar securing bolt locktabs to enable the bolts to be undone. Remove the injector retaining collar (photos).
6   Withdraw the injector from the CFi unit, having noted its orientation, followed by its seals (photos).
7   Withdraw the seal from the injector retaining collar (photo).
8   Refitting is a reversal of the removal procedure, but always use **new** seals (in the CFi unit and the retaining collar) whenever the injector is removed. The seals must be lightly lubricated before fitting (refer to the Specifications).

55 Throttle-plate control motor (1.4 litre CFi) – removal and refitting

Note: *Refer to Section 40 before proceeding. After the completion of this operation, the idle speed must be checked by a Ford dealer or other suitable specialist who has the equipment needed to 'lock' the EEC-IV engine management module during idle speed checking and adjustment operations*

1   Disconnect the battery earth terminal.
2   Remove the air filter housing.
3   Disconnect the throttle position sensor and the throttle-plate control motor multi-plugs (photo).
4   Undo and remove the screws securing the throttle plate control motor and throttle position sensor assembly mounting bracket to the CFi unit, and detach the assembly.

54.4 Disconnect the multi-plug from the injector

54.5A Removing an injector retaining collar securing bolt and its locktab

54.5B Removing the injector retaining collar

54.6A Withdrawing the injector from the CFi unit

54.6B Injector seals in the CFi unit

54.7 Withdraw the seal from the injector retaining collar

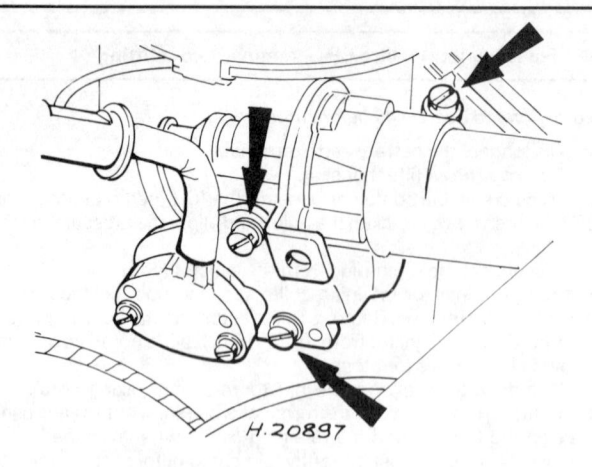

Fig. 3.57 Throttle-plate control motor and throttle position sensor assembly mounting bracket securing screws (arrowed) (1.4 litre CFi models) (Sec 55)

55.3 Disconnecting the throttle plate control motor multi-plug

5   Remove the motor retaining screws and withdraw the motor from its mounting bracket.
6   Refitting is a reversal of the removal procedure, bearing in mind the following points. Ensure that the throttle position sensor is located correctly on its actuating linkage, and that the motor and sensor assembly mounting bracket aligns with its locating pegs. Idle speed must be checked and, if necessary, adjusted (refer to the Note at the beginning of this Section).

57 Fuel injection unit (1.4 litre CFi) – removal and refitting

Note: Refer to Section 40 before proceeding. A new gasket will be required for refitting

1   Disconnect the battery earth terminal.
2   Remove the air filter housing.
3   Slacken the fuel feed union on the CFi unit **slowly**, to relieve the system pressure, then disconnect the fuel feed pipe use a spanner to prevent the main fuel feed connector body from turning (photo). Catch any spillage in a suitable container and plug the pipe.
4   Disconnect and plug the fuel return pipe.
5   Disconnect the accelerator cable.
6   Having ensured that the cooling system is **not** pressurized (refer to Chapter 2), disconnect both coolant pipes and plug them to prevent excessive coolant loss. Catch any spillage in a suitable container.
7   Unclip (where applicable) and disconnect the multi-plugs from the air-charge temperature (ACT) sensor, throttle position sensor, throttle plate control motor and the injector lead, having noted their fitment.
8   Disconnect all applicable vacuum pipes, having noted their locations for subsequent refitting.
9   Undo and remove the four screws securing the CFi unit to the inlet manifold (photo), and withdraw the unit from the vehicle. Recover its gasket.

56 Throttle position sensor (1.4 litre CFi) – removal and refitting

1   Disconnect the battery earth terminal.
2   Remove the air filter housing.
3   Release the throttle position sensor multi-plug from its retaining clip and disconnect it. To disconnect, pull on the multi-plug and not the wiring (photo).
4   Remove both throttle position sensor retaining screws (photo), then withdraw the sensor from the throttle plate shaft.
5   Refitting is a reversal of the removal procedure, ensuring that the sensor actuating arm is correctly located.

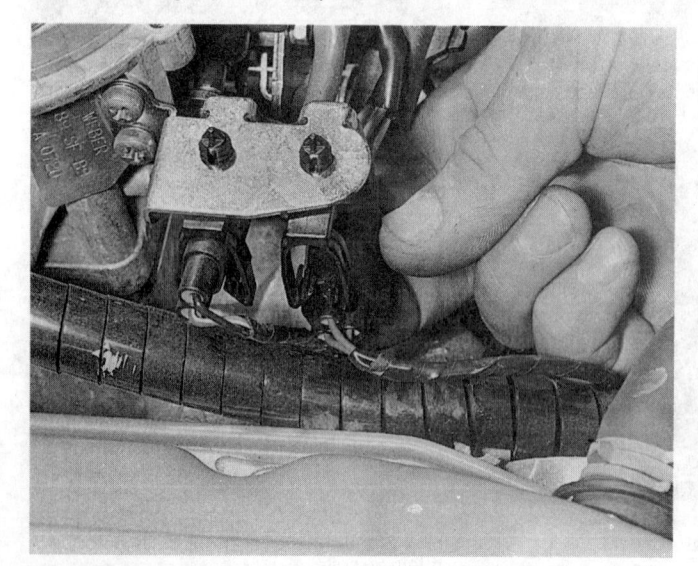

56.3 Release the throttle position sensor multi-plug from its retaining clip and disconnect it

56.4 Throttle position sensor retaining screws (arrowed) (shown with throttle plate control motor removed)

Chapter 3 Fuel, exhaust and emission control systems 119

57.3 General view of CFi unit attachments

A Fuel feed
B Fuel return
C Air-charge temperature sensor multi-plug

10 Clean the CFi unit and inlet manifold gasket mating faces.
11 Refitting is a reversal of the removal procedure, bearing in mind the following points. A new gasket **must** be used, and the CFi securing screws should be tightened to the specified torque. Upon completion, turn the ignition on and off at least five times and check for fuel leaks. Check the coolant level and top up as necessary.

58 Fuel pressure regulator (1.6 litre EFi) – removal and refitting

Note: *Refer to Section 40 before proceeding. A new sealing ring and fuel return hose clip must be used each time the regulator is removed*

1 Disconnect the battery earth terminal.
2 Position a suitable container to catch any fuel spillage, and have a large clean rag to hand to soak up escaping pressurised fuel. **Slowly** undo the fuel feed pipe connection (on the opposite end of the fuel distributor rail to the regulator) to release the system pressure. With the pressure released, retighten the connection.
3 Again position a suitable container to catch any fuel spillage, and disconnect the fuel return hose from the regulator. Insert plugs to prevent excessive fuel loss and avoid dirt ingress.
4 Disconnect the vacuum hose from the regulator.
5 Undo and remove the regulator retaining bolts, then detach the regulator.
6 Refitting is a reversal of the removal procedure, bearing in mind the following points. Lubricate the **new** sealing ring with oil of the specified type before repositioning the regulator and tightening its retaining bolts to the specified torque. Refit the fuel return hose using a new clip. After reconnecting the battery earth terminal, switch the ignition on and off at least five times (without cranking the engine) and check for fuel leaks.

59 Fuel injectors (1.6 litre EFi) – removal and refitting

Note: *Refer to Section 40 before proceeding. New upper and lower seals must be fitted to all injectors even if only one injector is to be removed, and a new throttle housing gasket must be used*

1 Disconnect the battery earth terminal.
2 Remove the air filter trunking.
3 Disconnect the throttle position sensor multi-plug.
4 Disconnect the accelerator cable at the throttle housing, then undo the bolt and nuts securing the cable abutment bracket to the throttle

57.9 CFi unit securing screws (arrowed)

housing and remove it. Undo the remaining throttle housing fixings, and detach the throttle housing. Recover its gasket.
5 Disconnect the multi-plugs from the injectors (photo), and the air-charge temperature and engine coolant temperature sensors.
6 Remove both injector wiring harness securing nuts, and position the harness clear of the fuel distributor rail (photo).
7 Position a suitable container to catch any fuel spillage, and have a large clean rag to hand to soak up escaping pressurised fuel. **Slowly** undo the fuel feed pipe connection on the distributor rail to release the system pressure, then once the pressure is fully released, disconnect the fuel feed pipe. Plug the rail and the pipe to prevent excessive fuel loss and prevent dirt ingress.
8 Disconnect the fuel return hose from the fuel pressure regulator, catching any spillage in a suitable container, and insert plugs to the regulator and the hose.
9 Disconnect the vacuum hose from the fuel pressure regulator, having labelled it to aid subsequent refitting.
10 Remove the fuel distributor rail securing bolts, and withdraw the rail from the vehicle. The injectors should remain with the rail as it is withdrawn.
11 Remove **all** the injectors from the fuel distributor rail, and discard **all** their seals.
12 Refitting is a reversal of the removal procedure, bearing in mind the following points. Make reference to the Specifications for tightening torque details. New upper and lower seals **must** be fitted to **all** injectors, and the seals lubricated with oil of the specified type before fitting the injectors to the fuel distributor rail. When refitting the fuel distributor rail, ensure that the injectors locate correctly. The fuel return hose connec-

59.5 Multi-plug disconnected from a fuel injector. (Fuel feed connection on the distributor rail arrowed)

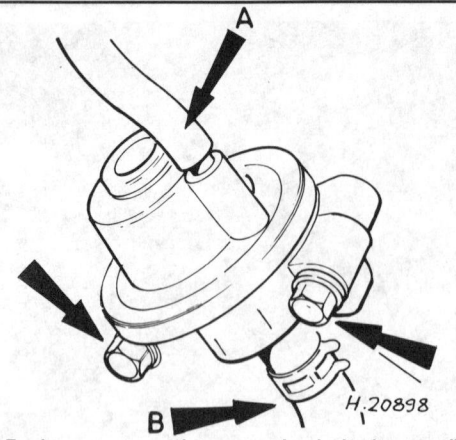

Fig. 3.58 Fuel pressure regulator securing bolts (arrowed) (1.6 litre EFi models) (Sec 58)

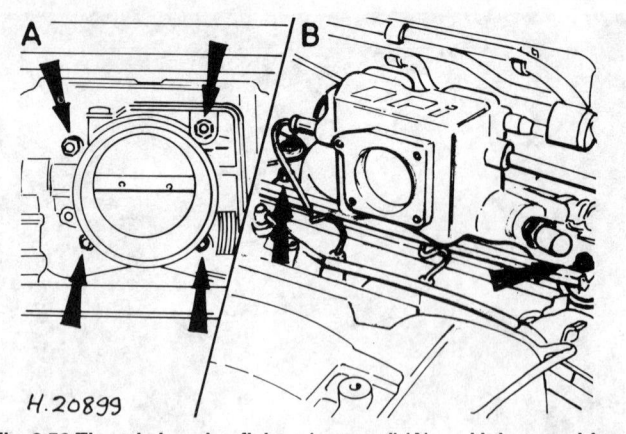

A Vacuum hose   B Fuel return hose

Fig. 3.59 Throttle housing fixings (arrowed) (A), and injector wiring harness ducting retaining nuts (arrowed) (B) (1.6 litre EFi models) (Sec 59)

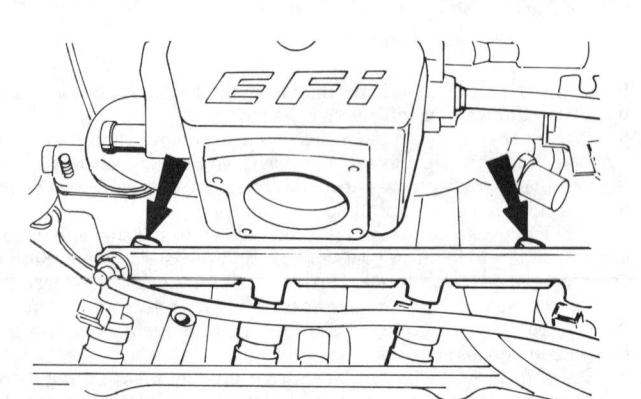

Fig. 3.60 Fuel distributor rail securing bolts (arrowed) (1.6 litre EFi models) (Sec 59)

Fig. 3.61 Fuel injector seal arrangements (1.6 litre EFi models) (Sec 59)

tion to the fuel pressure regulator should be secured with a new clip. Clean the gasket mating faces before refitting the throttle housing using a **new** gasket; tighten all throttle housing related fixings evenly. After reconnecting the battery earth terminal, switch the ignition on and off at least five times (without cranking the engine) and check for fuel leaks. Further checks with the engine running are also advised.

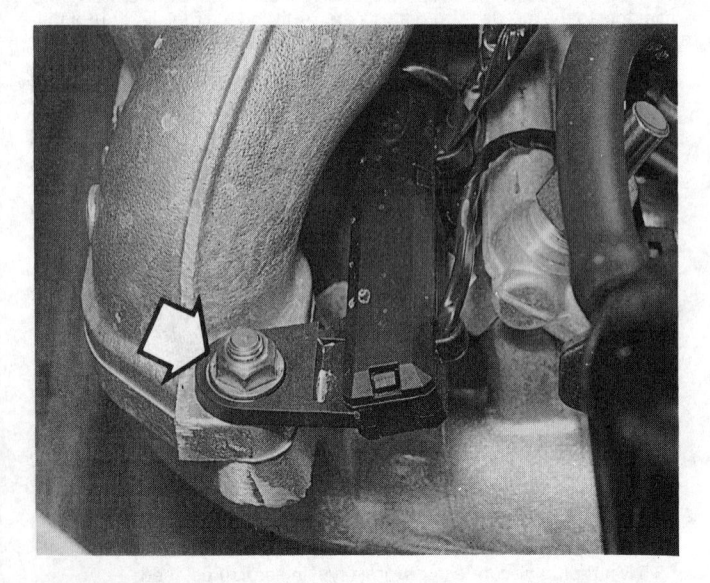

59.6 One of the two injector wiring harness securing nuts (arrowed)

60 Fuel mixture/CO content (1.6 litre EFi) – checking and adjustment

Note: *Refer to Section 40 before proceeding. A tachometer and an exhaust gas analyser (CO meter) will be required. A new tamperproof cap should be fitted after the completion of adjustment*

1 Warm the engine up to normal operating temperature, then switch it off and connect a tachometer and a CO meter in accordance with the manufacturer's instructions.
2 Start the engine and run it at 3000 rpm for approximately fifteen seconds, then allow it to idle, wait for the meters to stabilise, then record the idle speed and CO content readings.
3 If the CO content reading is outside specification, remove the tamperproof cap covering the fuel mixture screw and adjust the screw to give the correct CO content at the specified idle speed (photo); (the operational idle speed is controlled by the EEC-IV engine management module and requires no adjustment). Adjustment must be completed within a time band between 10 and 30 seconds after the meters stabilise. If the time taken is longer than 30 seconds, repeat paragraphs 2 and 3 ignoring the instruction to start the engine.
4 If adjustment has been necessary, repeat the procedure given in paragraph 2. Readjust as necessary.
5 With adjustment satisfactorily completed, switch off the engine and disconnect the tachometer and CO meter. Fit a new tamperproof cap to the fuel mixture screw, as required.

61 Base idle speed (1.6 litre EFi) – checking and adjustment

Note: *Refer to Section 40 before proceeding. A tachometer and an*

Chapter 3 Fuel, exhaust and emission control systems　　　　　121

60.3 Adjusting the fuel mixture screw to obtain the correct exhaust gas CO content

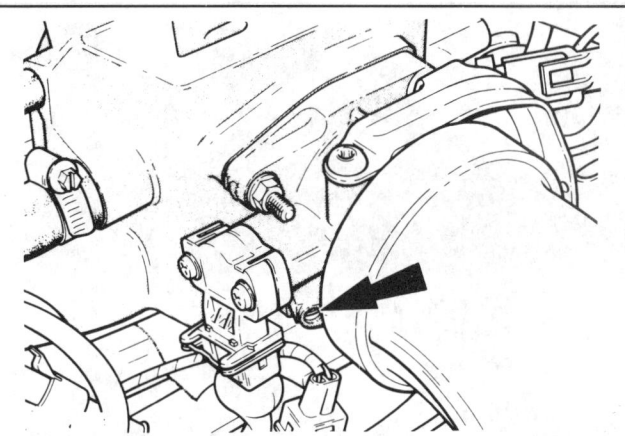

Fig. 3.62 Base idle speed adjustment screw (arrowed) (1.6 litre EFi models) (Sec 61)

exhaust gas analyser (CO meter) will be required. A new tamperproof seal should be fitted after the completion of adjustment

1 Check the fuel mixture/CO content, as described in the previous Section, and adjust as necessary. Do not disconnect the test meters upon completion, as the tachometer is required to determine the base idle-speed setting.
2 Disconnect the multi-plug from the idle-speed control valve.
3 Accelerate the engine to 3000 rpm, and hold it at this speed for approximately thirty seconds before allowing the engine to idle, then check that the base idle-speed is as specified.
4 If the base idle-speed is outside specification, remove the tamperproof seal from the throttle housing (refer to Fig. 3.62) and rotate the now exposed adjustment screw accordingly (anti-clockwise rotation will increase the base idle-speed).
5 If adjustment has been necessary, repeat the procedure given in paragraph 3. Readjust as necessary.
6 With the checking/adjustment satisfactorily completed, refit the idle-speed control valve multi-plug. The engine speed should rise above the operational idle-speed for a short time before returning to normal.
7 Allow the engine to idle for at least five minutes, to enable the EEC-IV engine management module to re-learn its idle values, then switch the engine off and disconnect the test meters. Fit a new tamperproof seal and/or cap, as necessary.

62 Idle speed control valve (1.6 litre EFi)　removal, cleaning and refitting

Note 1: *This procedure is only necessary on models fitted with the Weber-built valve, mounted on the air filter housing. Those fitted with the Hitachi-built valve do not require this procedure*

Note 2: *Refer to Section 40, and to 'Safety first!' at the beginning of this manual, before proceeding. An accurate tachometer will be required, to check the operational idle speed on completion*

1 Disconnect the battery earth terminal.
2 Disconnect the multi-plug from the idle-speed control valve, pulling on the multi-plug itself and not the wiring (photo).
3 Remove the two valve retaining bolts and detach the valve.
4 Immerse the valve head in a suitable container filled with clean petrol, and allow it to soak for approximately three minutes.
5 Clean the valve bore, slots and piston with petrol, using a suitable lint-free cloth, then gently move the piston up and down in its bore using a small screwdriver. Ensure that no cloth particles enter the bore, and do not use the slots to move the piston.
6 Rinse the valve again with clean petrol, then dry it using an air line (or other source of compressed air).

7 Prior to refitting, clean the mating faces of the valve and the air filter housing.
8 Refit the valve to the air filter housing, ensuring that the locating lug is positioned correctly, and tighten its retaining bolts evenly to the specified torque.
9 Reconnect the multi-plug ensuring that its lugs positively locate.
10 Reconnect the battery earth terminal and start the engine. Check that the idle-speed is stable and that no induction leaks are evident.
11 Connect a tachometer in accordance with the manufacturer's instructions, and allow the engine to reach normal operating temperature. Ensure that the EEC-IV engine management module has re-learnt its values (refer to Chapter 12, Section 5).
12 Check that the operational idle speed is within specification, then turn on all available electrical loads (lights, heater blower motor, heated rear window, etc) and check that the system compensates for the increased load (the operational idle speed should still be within specification).
13 With the checks completed, switch off all the electrical loads followed by the engine. Disconnect the tachometer.

63 Throttle position sensor (1.6 litre EFi)　removal and refitting

1 Disconnect the battery earth terminal.
2 Disconnect the multi-plug from the throttle position sensor (photo).

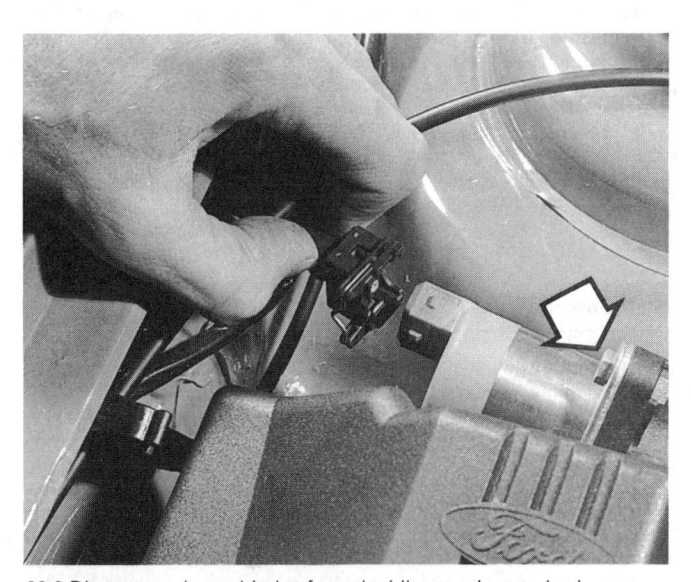

62.2 Disconnect the multi-plug from the idle speed control valve. Upper valve retaining bolt (arrowed)

Chapter 3 Fuel, exhaust and emission control systems

63.2 Disconnect the multi-plug from the throttle position sensor. Sensor retaining screws (arrowed)

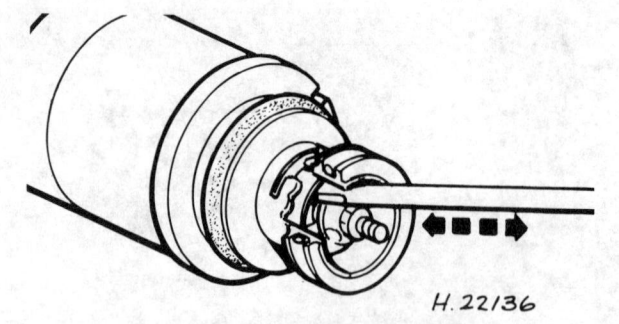

Fig. 3.63 Moving the idle speed control valve piston (1.6 litre EFi models) (Sec 62)

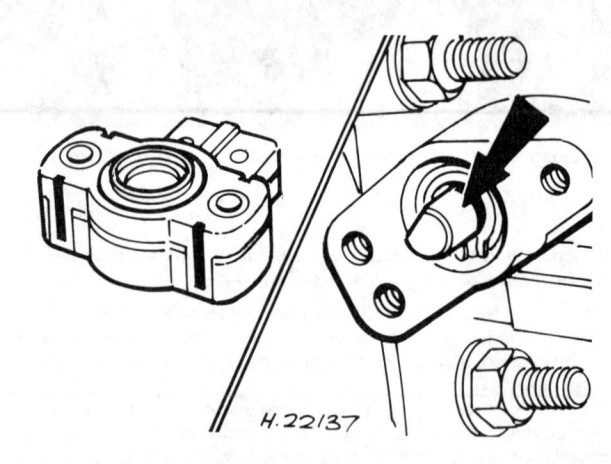

3 Remove the two sensor retaining screws, then withdraw the sensor from its throttle spindle location. Note that the sensor hub should not be rotated past its normal operating sweep.
4 Refitting is a reversal of the removal procedure, but note that reference must be made to Chapter 12, Section 5 upon subsequent battery reconnection, to allow the EEC-IV engine management module to re-learn its values.

PART C: MANIFOLDS, EXHAUST AND EMISSION CONTROL SYSTEMS

Fig. 3.64 General view of the throttle position sensor and its 'D' shaped throttle spindle location (arrowed) (1.6 litre EFi models) (Sec 63)

64 General description

Information on the inlet and exhaust manifolds may be found in later Sections of this Chapter.
The exhaust systems fitted during production are of two-piece construction, whilst those available in service are of three-piece construction. Further, a catalytic converter is fitted to 1.4 litre CFi models, located between the downpipe assembly and the remainder of the exhaust system. Additional features of the exhaust system on 1.4 litre CFi models are the HEGO sensor located in the downpipe assembly, and the underbody heatshields fitted to protect the vehicle underbody and petrol tank from the extreme heat at which the catalytic converter operates.
All exhaust systems are secured at the exhaust manifold, and feature rubber insulator mountings along the system length.
Information on emission control systems may be found in later Sections of this Chapter.

65 Maintenance and inspection

1 At the intervals specified in the *'Routine maintenance'* Section at the beginning of this manual, carry out the following operations.
2 With the vehicle raised on a hoist or supported securely on axle stands, check the exhaust system for signs of leaks, corrosion and damage and check the rubber insulator mountings for condition and security. Where damage or corrosion is evident, renew the section(s) in accordance with the relevant text.
3 If the corrosion is limited, it is often possible to repair the exhaust system. Holts Flexiwrap and Holts Gun Gum can be used for effective repairs to exhaust pipes and silencer boxes, including ends and bends. Holts Flexiwrap is an MOT approved permanent exhaust repair.

4 Check the exhaust manifold nuts for correct tightness.
5 Examine all relevant vacuum hoses for signs of damage and deterioration, renewing where necessary. Also check the security of their connections.
6 On HCS engines, visually check the condition of the oil filler cap mesh filter and renew the cap if necessary.
7 On all models, renew the crankcase ventilation filter/adaptor or the separate filter, as applicable. Visually check the condition and security of all crankcase emission control hoses.

66 Inlet manifold – general

Note: Refer to Section 2 or Section 40 (as applicable) before carrying out any work. If removal is intended, an accurate tachometer will be needed to check the idle speed upon the completion of refitting. Additionally, for all models except the 1.4 litre CFi, an exhaust gas analyser (CO meter) will be required to check the fuel mixture

1 A cast aluminium inlet manifold of one-piece design is used on all models, except the 1.6 litre EFi variant. The inlet manifold fitted to the 1.6 litre EFi model is a two-piece design, the upper section being bolted to the lower section to form an attachment for the throttle housing.
2 The inlet manifold is secured to the cylinder head by bolts and/or nuts (as applicable), and can be removed after disconnection/removal of the components mounted on it (with reference to the relevant text).
3 When refitting the inlet manifold, or bolting the sections together on 1.6 litre EFi models, ensure that all traces of old gasket are removed beforehand, and use **new** gaskets throughout. Refer to the Specifications for tightening torque details.

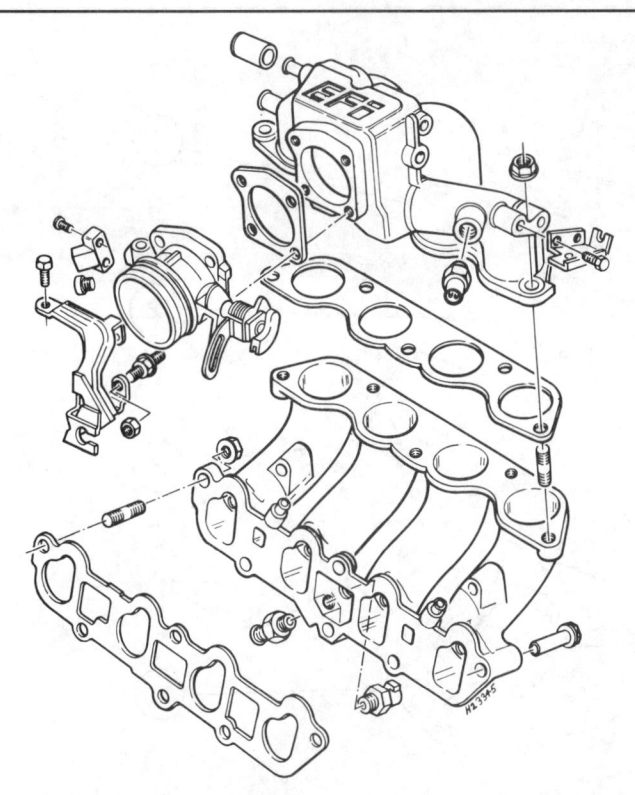

Chapter 3 Fuel, exhaust and emission control systems

Fig. 3.65 Exploded view of the inlet manifold arrangement (1.6 litre EFi models) (Sec 66)

67 Exhaust manifold – general

Note: *Never work on or near a hot exhaust system. Also note that the battery earth terminal should be disconnected prior to commencing removal of the exhaust manifold*

1 The exhaust manifold is secured to the forward side of the cylinder head by studs and nuts, and is similarly attached to the exhaust downpipe assembly. An airbox/heatshield is fitted to provide exhaust heated air during engine warm-up; this is bolted directly onto the manifold.
2 During removal, access from underneath will be required. On certain models, a one-piece undertray is fitted which must be removed. On 1.4 litre CFi models, the HEGO sensor located in the downpipe assembly should be disconnected to prevent its wiring from straining. Always suspend the forward end of the exhaust system to avoid excessive strain on the exhaust mountings.
3 When refitting, note that new exhaust manifold gaskets must be used, and refer to the Specifications for tightening torque details (photos). On vehicles fitted with the EEC-IV engine management system, reference must be made to Chapter 12, Section 5 upon subsequent battery reconnection.

68 Exhaust system – renewal

Note: *Never work on or near a hot exhaust system. Also note that the battery earth terminal should be disconnected prior to commencing any work, and a gasket kit should be obtained where possible*

1 Due to the sectional nature of the exhaust system, complete renewal is not always necessary if damage and/or wear is confined to one part of the system. Two renewal categories have to be considered: production-fitted systems and service-replacement systems (refer to Section 64).
2 On a production-fitted system, the downpipe assembly and the rear

67.3A New exhaust manifold gaskets located to the cylinder head (1.1 litre HCS engine, shown during overhaul)

67.3B Fit the exhaust manifold ...

67.3C ... and tighten its nuts to the specified torque ...

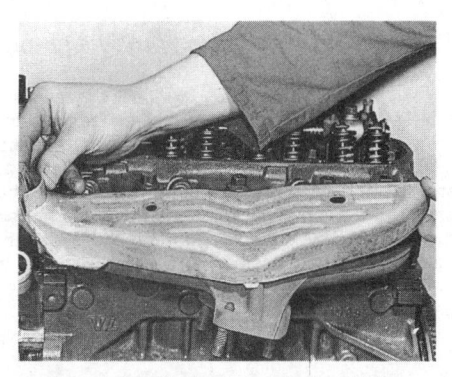

67.3D ... then refit the airbox-heatshield ...

67.3E ... and secure with its bolts (1.1 litre HCS engine, shown during overhaul)

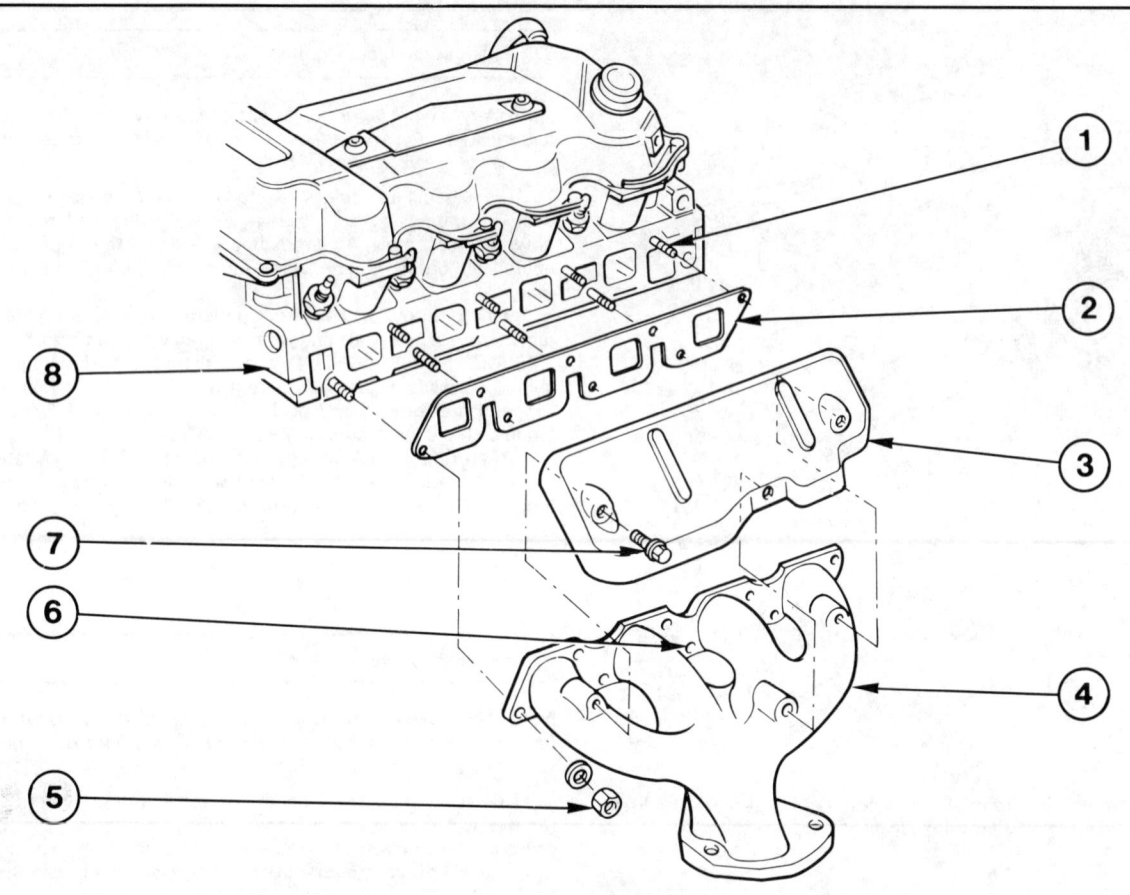

Fig. 3.66 Exploded view of the exhaust manifold arrangement (1.6 litre EFi models) (Sec 67)

1 Retaining stud	3 Heatshield	5 Manifold retaining nut	7 Heatshield retaining bolt
2 Gasket	4 Exhaust manifold	6 Centering hole	8 Cylinder head

silencer can be renewed independently, as required. Additionally, the catalytic converter on 1.4 litre CFi models may also be renewed if necessary. The remaining centre section, comprising the main pipe run, front silencer and resonator box (where applicable), cannot be renewed independently; in this case, service-replacement sections, including the rear silencer, must be obtained.

3 On a service-replacement system, all sections can be renewed independently, as required.

4 Although certain exhaust renewal and sectional removal/refitting operations may be carried out without removing the complete system from the vehicle, occasions will arise when this will be necessary. Indeed, it is often far easier to work on an exhaust system with it removed from the vehicle.

5 Good working access made by raising the vehicle and supporting securely on axle stands, makes the task of exhaust system removal easier. On some models, a one-piece undertray is fitted which must be removed for access to the forward end of the system, and on 1.4 litre CFi models the HEGO sensor wiring must be disconnected before system removal. Remove the exhaust system by releasing the downpipe from the exhaust manifold and unhooking the rubber insulator mountings, then withdraw. The task may be made easier by partial dismantling, but take care not to damage the catalytic converter (if fitted) by dropping it, or by the use of excessive force.

6 Prior to refitting, clean the exhaust manifold flange with emery cloth to eliminate any carbon build-up, and coat the studs with a suitable anti-seize compound. Ensure that the rubber insulator mountings are suitable for re-use, and renew if necessary. Note that the type fitted to 1.4 litre CFi models are made of high temperature resistant natural rubber to withstand the high operating temperatures of this particular system. If renewal is necessary, the **correct** type **must** be obtained.

7 When refitting the exhaust system, use new gaskets where applicable (photo), and refer to the Specifications for tightening torque details. Suitable exhaust jointing compound such as Holts Firegum should be applied to pipe connections (but **never** forward of the catalytic converter, on 1.4 litre CFi models). Loosely refit the exhaust system sections and ensure that all system components are aligned, and that the rubber insulator mountings are not distorted, before fully tightening the flange bolts and U-bolt clamps. When securing the components, start at the exhaust manifold and work towards the rear of the vehicle. Reconnect the HEGO sensor, as applicable.

8 With the exhaust system fully refitted, re-check that the rubber insulator mountings are not distorted. Ensure also that the exhaust does not knock or rub against adjacent components, and that the spring-loaded flexible joint (where applicable) has some free movement. After battery reconnection (on vehicles with EEC-IV engine management systems, refer to Chapter 12, Section 5) check the exhaust system for leaks. This task must be performed in a well ventilated area and **not** from an inspection pit. Refit the one-piece undertray, as applicable. Remove the axle stands and lower the vehicle to the ground.

Rear silencer renewal (on production-fitted exhaust systems)

9 If it becomes necessary to renew the rear silencer of a production-fitted exhaust system, the defective rear silencer must be severed and removed from the remainder of the exhaust system, as shown in Fig. 3.69, and a service-replacement section fitted. Note that the dimension should be checked carefully before cutting, and that the cut line must be at 90° to the remaining pipe. Remove any burrs raised during cutting then, with some Holts Firegum applied, slide the service-replacement rear silencer into position and loosely secure with a U-bolt clamp. Tighten the U-bolt clamp to the specified torque after the section has been supported on its rubber insulator mounting (photo) and correctly aligned.

Catalytic converter (where fitted)

10 Refer to Section 70 for an independent removal and refitting procedure.

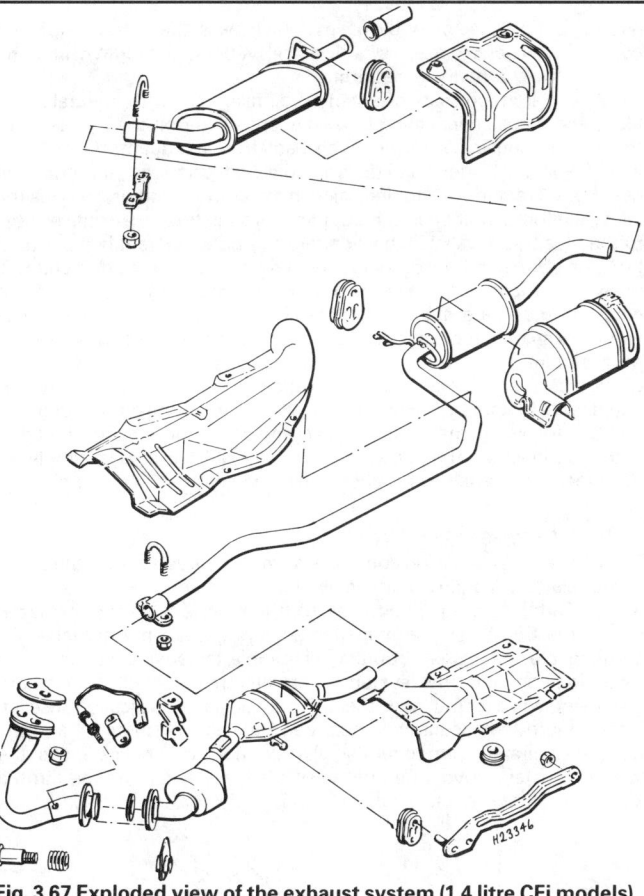

Chapter 3 Fuel, exhaust and emission control systems

Fig. 3.67 Exploded view of the exhaust system (1.4 litre CFi models) (Sec 68)

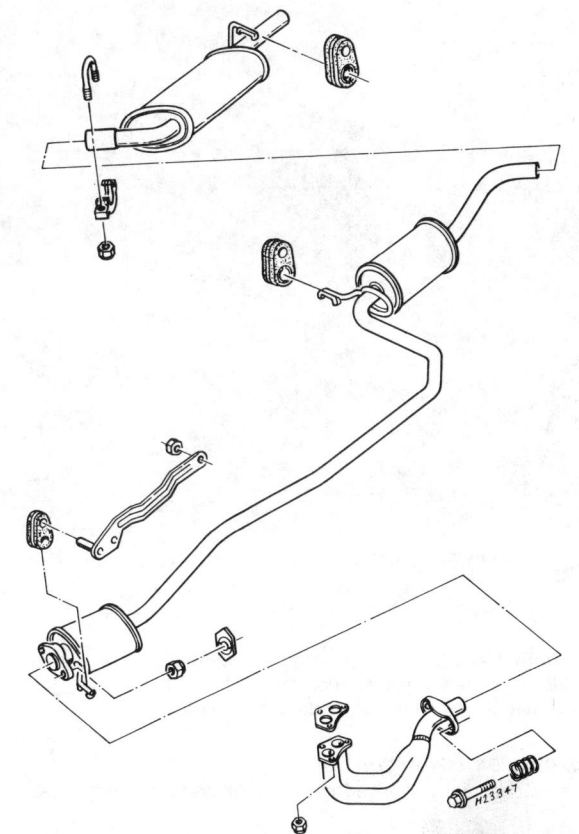

Fig. 3.68 Exploded view of the exhaust system (1.6 litre EFi models) (Sec 68)

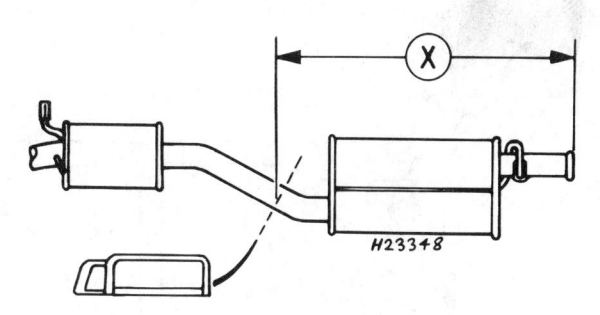

69 Emission control systems – general description

1 Emission control consists of reducing the emission of noxious gases and vapours, which are by-products of combustion, into the atmosphere. The system can be divided into three categories: fuel evaporative emission control, crankcase emission control and exhaust emission control. Dependent on the engine and transmission type, and the emission regulations, various control measures are used. Some of

Fig. 3.69 Cutting dimension for fitting a service-replacement rear silencer to a production-fitted exhaust system (Sec 68)

On 1.1 litre models, X = 651 mm
On 1.4 litre models, X = 664 mm
On 1.6 litre models, X = 711 mm

the control measures are designed into the advanced electronic circuitry employed, whilst others are more physical systems.

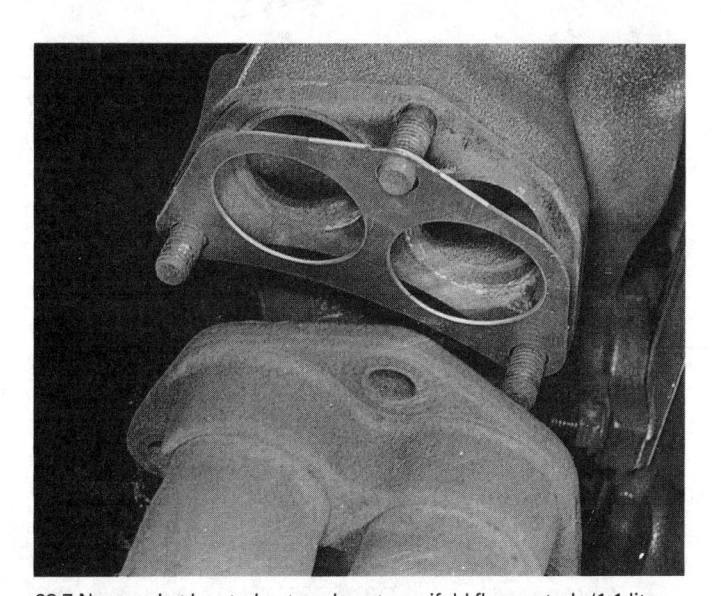

Fuel evaporative emission control

2 Emission control regulations in the UK are far less strict than in certain other countries and fuel evaporative emission control systems are limited on vehicles meeting the 15:04 regulation. Carburettor float chambers are internally vented, whilst fuel tanks vent to atmosphere through a combined rollover/anti-trickle-fill valve.

3 On 1.4 litre CFi models built to meet the 83 US emission regulation, the fuel tank vents through a combined rollover/anti-trickle-fill valve to a carbon canister mounted in the engine compartment. The system functions in the following manner. When the engine is switched off, fuel vapours from the fuel tank are directed to the carbon canister (at this point the canister purge solenoid is closed). After the engine is started, the EEC-IV engine management module controls the canister purge solenoid and signals it to open its valve, thus allowing the vapours to be

68.7 New gasket located onto exhaust manifold flange studs (1.1 litre HCS engine shown)

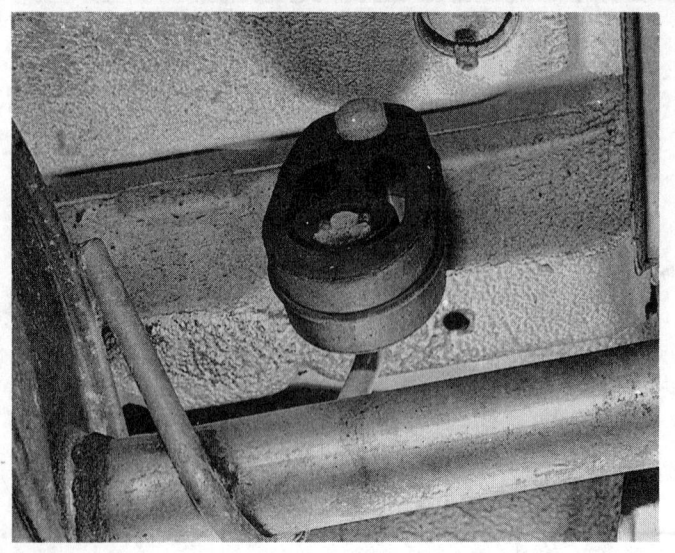

Chapter 3 Fuel, exhaust and emission control systems

68.9 Rearmost exhaust rubber insulator mounting (1.1 litre HCS model shown)

drawn from the carbon canister into the inlet manifold for combustion. A restrictor prevents excess fuel vapour from flowing into the inlet manifold and thus avoids fuel mixture control problems.

Crankcase emission control

4 On HCS engines a closed circuit crankcase ventilation system is used ensuring that blow-by gases which pass the piston rings and collect in the crankcase, as well as oil vapour, are drawn into the combustion chambers to be burnt.

5 The system consists of a vented oil filler cap (with integral mesh filter) and a hose connecting it to an adaptor/filter on the underside of the air filter housing. A further hose leads from the adaptor/filter to the inlet manifold. Under conditions of idle and part load, the emission gases are directed back for inclusion in the combustion process via the inlet manifold. Additional air is supplied through two small orifices next to the mushroom valve in the air filter housing to prevent high vacuum build-up. Under full load conditions, when inlet manifold vacuum is weak, the mushroom valve in the air filter housing opens, and the emission gases are directed via the air filter housing into the engine induction system for inclusion in the combustion process. This arrangement eliminates any fuel mixture control problems.

6 A closed circuit crankshaft ventilation system is also used on CVH engines, its function is broadly similar to that employed on HCS engines, but the following major differences should be noted. The breather hoses connect to the rocker cover, the oil filler cap is simply as described, and certain applications utilise a separate filter (photo).

Exhaust emission control

7 The exhaust emission control systems employed across the range of models vary greatly in complexity and application.

8 On carburettor equipped vehicles, the precise fuel metering facilities of the EEC-IV engine management system are not available. To compensate for this, and obtain acceptable exhaust emission levels (with ignition and fuel systems correctly maintained), a vacuum-modifying valve (or multiples thereof) may be used. Typically, these are inserted into the vacuum line between the vacuum source(s) and the vacuum-operated component(s). An example of application is the vacuum sustain valve, fitted between the carburettor and the throttle kicker on certain manual transmission models (refer to Section 28).

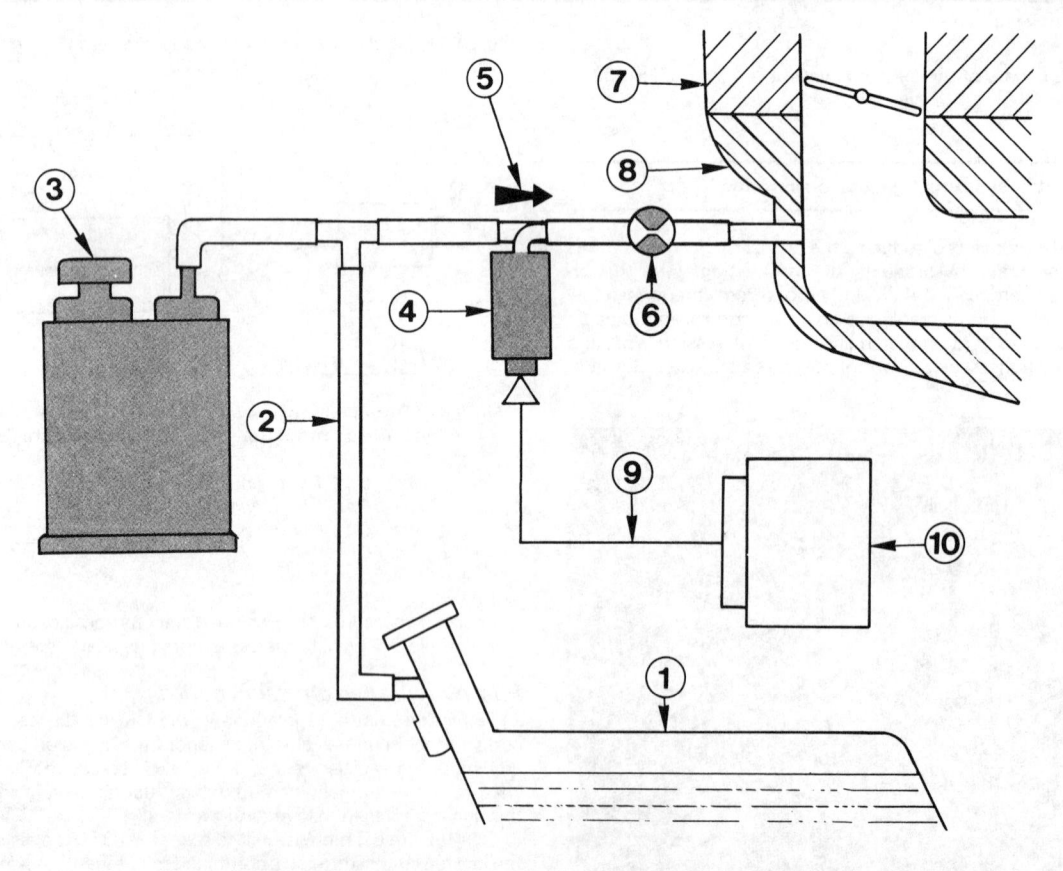

Fig. 3.70 Fuel evaporative emission control system (1.4 litre CFi models) (Sec 69)

1 Fuel tank
2 Fuel vapour pipe
3 Carbon canister
4 Canister purge solenoid
5 Direction of vapour flow into engine
6 Restrictor
7 CFi unit
8 Inlet manifold
9 Signal from the EEC-IV module
10 EEC-IV engine management module

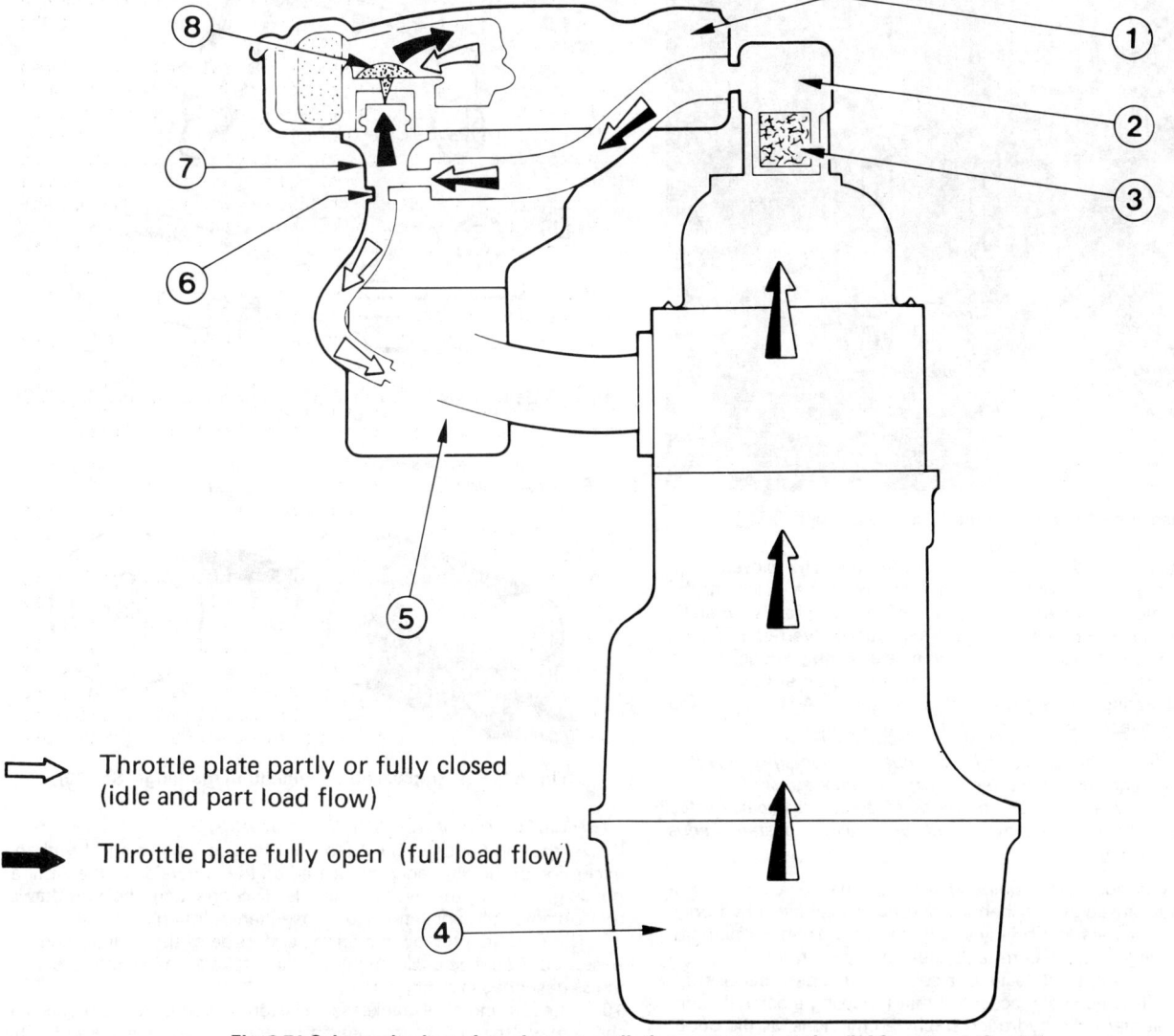

Chapter 3 Fuel, exhaust and emission control systems 127

Fig. 3.71 Schematic view of crankcase ventilation system operation (HCS engines) (Sec 69)

⇨ Throttle plate partly or fully closed (idle and part load flow)

➡ Throttle plate fully open (full load flow)

1 Air filter housing
2 Oil filler cap
3 Mesh filter in oil filler cap
4 Crankcase
5 Inlet manifold
6 Orifice to inlet manifold
7 Adaptor/filter
8 Mushroom valve

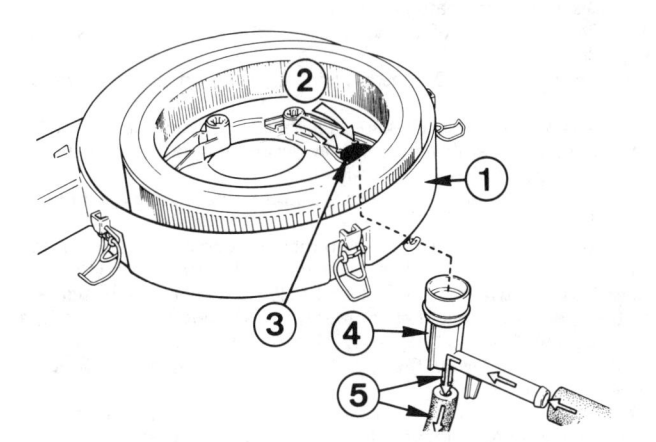

Fig. 3.72 Crankcase ventilation system operation under idle and part load conditions (HCS engines) (Sec 69)

1 Air filter housing
2 Fresh (additional) air supply into adaptor/filter
3 Mushroom valve
4 Adaptor/filter
5 Emission gases from crankcase

These vacuum-modifying valves should **not** be confused with fuel traps, commonly fitted to vacuum lines to prevent the migration of fuel (refer to the following Section).

9 On UK specification 1.6 litre EFi models (built to meet the 15:04 regulation), the precise fuel metering ability of the EEC-IV engine management system, based on the information obtained from the various system sensors, enables acceptable exhaust emission levels to be achieved (again with the ignition and fuel systems correctly maintained).

10 On UK specification 1.4 litre CFi models (built to meet the more stringent US83 regulation), the EEC-IV engine management system controls exhaust emissions by precise fuel metering, control of the throttle plate (under certain engine operating conditions), and exhaust gas monitoring and processing. All the information required for the EEC-IV engine management module calculations is obtained from the various system sensors. As the exhaust gases exit through the downpipe, a heated exhaust gas oxygen (HEGO) sensor detects the amount of oxygen remaining unburnt, and sends an appropriate electrical signal to the EEC-IV engine management module, which then modifies fuel mixture accordingly. This arrangement is termed 'closed loop' control. 'Closed loop' control takes place only when the HEGO sensor and the engine have reached their normal operating temperatures. For this reason HEGO sensor is electrically heated so as to minimise the time that the system is working under 'open loop' conditions (with preset

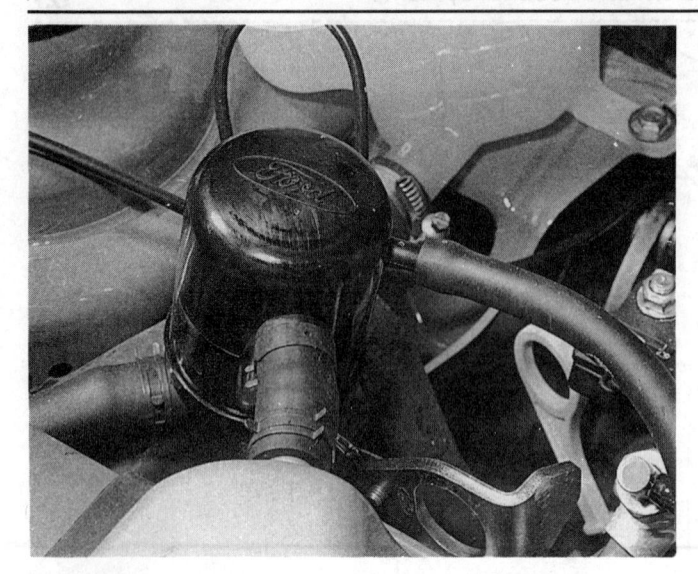

Chapter 3 Fuel, exhaust and emission control systems

69.6 Crankcase ventilation system filter (1.6 litre EFi model)

values being used to determine the fuel mixture). The 'closed loop' control system is necessary to keep the fuel mixture very close to the ideal (stoichiometric) ratio of 14.7 to 1, which it must be for the catalytic converter to operate effectively. The catalytic converter, which is located directly behind the exhaust downpipe assembly, reduces three different exhaust gas pollutant levels at the same time. It works by chemically changing the composition of the gases. Again, operating temperature is a factor, with no initial chemical reaction.

WARNING: *The use of leaded fuel will irreparably damage the HEGO sensor and the catalytic converter, and the internal substrate of the catalytic converter may be rendered useless if subjected to engine misfire (a high speed engine misfire may result in an almost instantaneous substrate melt down)*

Certain catalytic converter equipped vehicles may give off an unpleasant odour caused by the chemical hydrogen sulphide. If hydrogen sulphide emission gets particularly strong, this may be an indication of an engine management system fault causing poor control of the fuel mixture. Refer to your Ford dealer or a suitably equipped specialist for further advice. It is generally accepted that renewing a catalytic converter only heightens the hydrogen sulphide emissions, as the odour should decrease in intensity as the unit 'ages'. A change of fuel brand may help to alleviate the problem.

70 Emission control system components (including vacuum valves and fuel traps) – removal and refitting

Carbon canister (where fitted)
1   Disconnect the battery earth terminal.
2   Detach the vapour pipe from the canister, then undo the canister retaining screw. Withdraw the canister upwards, releasing it from its bracket (photos).
3   Refitting is a reversal of the removal procedure. Refer to Chapter 12, Section 5 upon subsequent battery reconnection, to allow the EEC-IV engine management module to re-learn its values.

Canister purge solenoid (where fitted)
4   Disconnect the battery earth terminal.
5   The canister purge solenoid is located near to the bulkhead panel on the right-hand side of the engine compartment (photo). Disconnect its multi-plug by pulling on the multi-plug itself and not the wiring.
6   Disconnect the hoses, having labelled them for correct subsequent refitting, then withdraw the solenoid from the vehicle.
7   Refitting is a reversal of the removal procedure, but ensure that the multi-plug locking lugs 'snap' into position. Refer to Chapter 12, Section 5 upon subsequent battery reconnection, to allow the EEC-IV engine management module to re-learn its values.

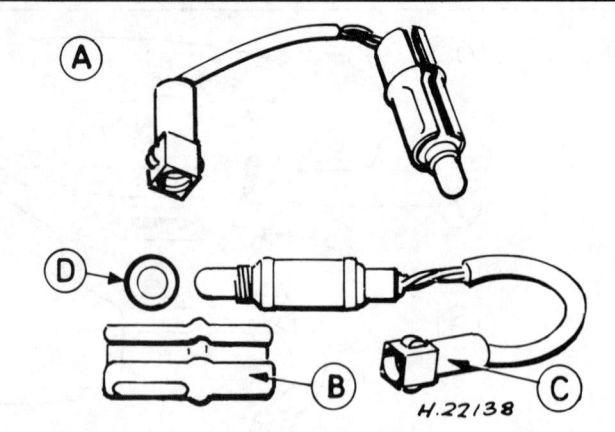

Fig. 3.73 General view of the HEGO sensor components (1.4 litre CFi models) (Sec 70)

A   HEGO sensor assembled
B   Heatshield
C   HEGO sensor multi-plug
D   Sealing ring

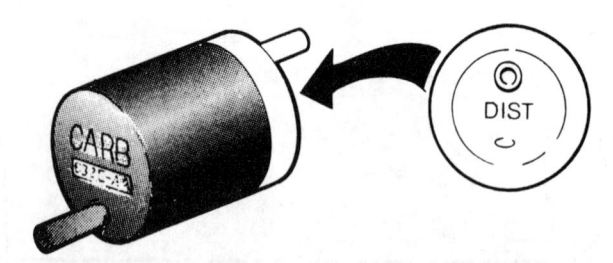

Fig. 3.74 Fuel trap vacuum connection markings (Sec 70)

Crankcase emission control components
8   All the components relating to crankcase emission control, with the exception of the filter/adaptor located on the underside of the air filter housing, may be removed by simple disconnection and withdrawal (having noted all connections for subsequent refitting).
9   The filter/adaptor located on the underside of the air filter housing base, and should be unclipped then pulled to release it before proceeding as described in paragraph 8.
10 The refitting of all crankcase emission components is a reversal of the removal procedure, ensuring that the connections are correctly made. When refitting the filter/adaptor, ensure that its grommet remains correctly seated.

HEGO sensor (where fitted)
11 Disconnect the battery earth terminal.
12 Raise the front of the vehicle and support securely using axle stands.
13 Unclip the HEGO sensor wiring and disconnect its multi-plug (photo).
14 Remove the HEGO sensor heatshield, unscrew the sensor from the downpipe assembly and carefully remove it with its sealing ring. Do **not** touch the tip of the HEGO sensor.
15 Prior to refitting, clean the HEGO sensor threads and the downpipe threads.
16 Refitting is a reversal of the removal procedure, bearing in mind the following points. A suitable sealer must be used on the threads, a **new** sealing ring must be fitted and the HEGO sensor must be tightened to the specified torque. Ensure that the multi-plug locking lugs 'snap' into position. Refer to Chapter 12, Section 5 upon subsequent battery reconnection, to allow the EEC-IV engine management module to re-learn its values.
17 Check to ensure that the joint between the HEGO sensor and the downpipe assembly does not leak, and rectify as necessary.

Catalytic converter (where fitted)
18 Disconnect the battery earth terminal.
19 Raise the vehicle and support securely using axle stands, or position over an inspection pit.
20 Prior to removing the catalytic converter, inspect the rubber

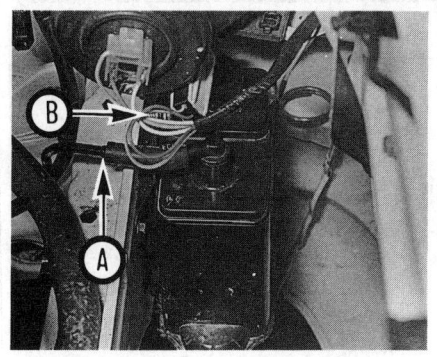

Chapter 3 Fuel, exhaust and emission control systems

70.2A Fuel vapour pipe (A) and canister retaining screw (B) (1.4 litre CFi model)

70.2B Bracket engaging tag on carbon canister (arrowed) (1.4 litre CFi model)

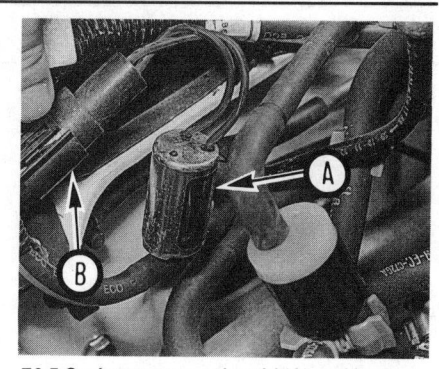

70.5 Canister purge solenoid (A) and its multi-plug (B) (1.4 litre CFi model)

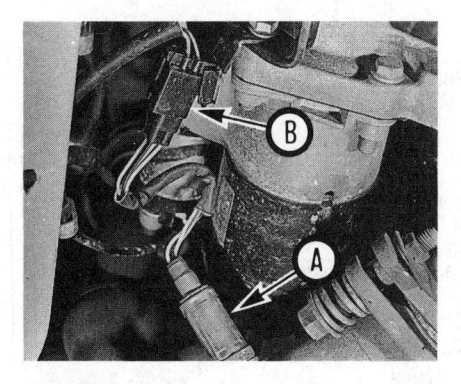

70.13 HEGO sensor (shown with its heatshield removed) (A), and its multi-plug (B) (1.4 litre CFi model)

70.23 Rubber insulator mounting supporting the catalytic converter (1.4 litre CFi model)

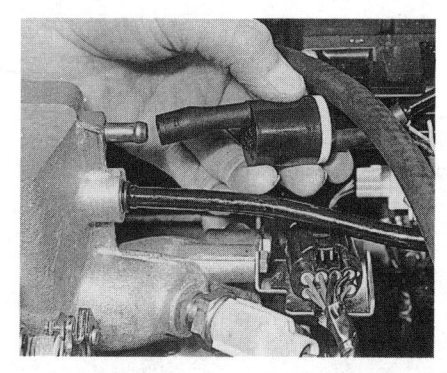

70.28 Fuel trap in vacuum line being correctly connected (1.6 litre EFi model shown)

insulator mountings whilst applying light downward pressure on the exhaust system. If any show signs of cracking or other damage/deterioration, obtain replacements *of the correct type* (refer to Section 68) ready for refitting.
21 Remove the flange bolts securing the catalytic converter to the exhaust downpipe assembly, and separate the joint.
22 Undo and remove the U-bolt clamp securing the rear of the catalytic converter to the exhaust system.
23 Unhook the exhaust system from the rear rubber insulator mountings, then release the catalytic converter from its rubber insulator mounting (photo). The catalytic converter may now be separated from the exhaust system and removed. Take care not to damage the catalytic converter by dropping it, or by the use of excessive force whilst separating it from the exhaust system.
24 Prior to refitting, clean all joint mating faces.
25 Refitting is a reversal of the removal procedure, making reference to the Specifications for tightening torque details and bearing in mind the following points. Suitable exhaust jointing compound should be applied to the joint at the rear of the catalytic converter (but **never**

forward of it). A **new** sealing ring must be used at its downpipe assembly joint. Fully tighten the U-bolt clamp after the system components are correctly aligned, then check that the spring-loaded flexible joint has some free movement. Run the engine, and check the exhaust system for leaks, in a well ventilated area only. Refer to Chapter 12, Section 5.

Vacuum valves and fuel traps
26 Prior to removing any vacuum valve or fuel trap, its orientation should be noted for subsequent refitting, and its hoses labelled.
27 As a means of differentiating between vacuum-modifying valves and fuel traps, various markings and colours are employed. If obtaining a replacement valve or fuel trap, ensure that the replacement component is identical to the original one. If part numbers appear on the components, check that they match.
28 Fuel traps are fitted to prevent the migration of fuel. The fuel trap must be connected with the black end (marked 'CARB') towards the potential source of fuel (inlet manifold or carburettor, as applicable), and the white end (marked 'DIST') towards the protected equipment (module, throttle kicker, etc, as applicable) (photo).

PART D: FAULT DIAGNOSIS

71 Fault diagnosis – carburettor fuel systems

Note: *High fuel consumption and poor performance are not necessarily due to carburettor faults. Make sure that the ignition system is maintained in good order, and that the engine itself is in good mechanical condition, before attempting carburettor adjustments; items such as binding brakes or under-inflated tyres should not be overlooked*

Chapter 3 Fuel, exhaust and emission control systems

Symptom	Reason(s)
Engine will not start	Fuel tank empty Fault in fuel line Fuel pump defective Fuel filter blocked (where applicable) Defective or maladjusted automatic choke (where applicable) Air leak at inlet manifold Carburettor adjustments incorrect Ignition system fault
Poor or erratic idling	Carburettor adjustments incorrect Air leak at inlet manifold Leak in ignition advance vacuum hose Leak in brake servo vacuum hose Ignition system fault
Lack of power and/or stalling	Fault in fuel line Fuel pump defective Air leak at inlet manifold Defective or maladjusted automatic choke (where applicable) Carburettor adjustments incorrect Leak in ignition advance vacuum hose Ignition system fault
Fuel consumption excessive	Leak in fuel system Air filter element choked giving a rich mixture Carburettor float chamber flooding due to worn needle valve or incorrect float level setting Defective or maladjusted automatic choke (where applicable) Carburettor adjustments incorrect Carburettor worn excessively Unsympathetic driving style
Backfiring in exhaust	Ignition timing incorrect (where applicable) Fuel mixture grossly incorrect Exhaust valve(s) burnt, sticking or incorrectly adjusted (where applicable)
Spitting back in inlet manifold or carburettor	Fuel mixture incorrect Ignition timing incorrect (where applicable) Inlet valve(s) burnt, sticking or incorrectly adjusted (where applicable)

Note: *This Section is not intended as an exhaustive guide to fault diagnosis, but summarises the more common faults which may be encountered during a vehicle's life. Consult your Ford dealer or other suitable specialist for more detailed advice, particularly if the fault is intermittent*

72 Fault diagnosis – fuel injection systems

Note: *High fuel consumption and poor performance are not necessarily due to fuel-injection system faults. Make sure that the ignition system is maintained in good order, and that the engine itself is in good mechanical condition, before attempting adjustments (where applicable); items such as binding brakes or under-inflated tyres should not be overlooked*

Symptom	Reason(s)
Engine will not start	Fuel tank empty Defective fuel pump Defective fuel pump relay or blown fuse Fuel pump inertia switch (fuel cut-off) tripped Fault in fuel line Fuel filter blocked Defective fuel pressure regulator Defective idle speed control valve (1.6 litre EFi models) Defective fuel injector(s) Air leak at inlet manifold Ignition system fault Engine management system fault

Chapter 3 Fuel, exhaust and emission control systems

Symptom	Reason(s)
Poor or erratic idling	Defective idle speed control valve (1.6 litre EFi models) Air leak at inlet manifold Defective fuel pressure regulator Leak in MAP sensor vacuum hose Leak in brake servo vacuum hose Base idle speed or fuel mixture incorrectly set (1.6 litre EFi models) Throttle plate control motor defective, or idle tracking switch sticking (1.4 litre CFi models) Ignition system fault Engine management system fault
Lack of power and/or stalling	Fault in fuel line Fuel pump defective Fuel filter partially blocked Defective fuel pressure regulator Defective fuel injector(s) Air filter element choked Air leak at inlet manifold Ignition system fault Engine management system fault
Fuel consumption excessive	Leak in fuel system Air filter element choked giving a rich mixture Defective fuel pressure regulator Fuel mixture adjustment incorrect (1.6 litre EFi models) Defective throttle position sensor Defective idle speed control valve (1.6 litre EFi models) Throttle-plate control motor defective (1.4 litre CFi models) Engine management system fault Unsympathetic driving style
Backfiring in exhaust	Fuel mixture incorrectly set (1.6 litre EFi models) Ignition timing incorrect (1.4 litre CFi models) Exhaust valve(s) burnt or sticking Engine management system fault
Spitting back in inlet manifold or throttle housing	Fuel mixture incorrectly set (1.6 litre EFi models) Ignition timing incorrect (1.4 litre CFi models) Inlet valve(s) burnt or sticking Engine management system fault

Note: *This Section is not intended as an exhaustive guide to fault diagnosis, but is merely a summary of the more common faults which may be encountered during a vehicle's life. Except for the general checks, adjustments and minor overhaul procedures given in this Chapter, the scope for the home mechanic is limited as complete and accurate fault diagnosis is only possible using sophisticated test equipment. Unless a component is obviously defective, or components are freely available for testing by substitution, enlist the aid of your Ford dealer or other suitable specialist particularly where an engine management system fault is suspected, or if a fault occurs intermittently. Do **not** attempt to improvise test meter checks, as the incorrect use of test probes between component connector pins can cause irreparable damage to the internal circuitry of some components*

Chapter 4
Ignition and engine management systems

For modifications, and information applicable to later models, see Supplement at end of manual

Contents

Part A: 1.0 litre and 1.1 litre HCS engines
Air charge temperature (ACT) sensor – removal and refitting	4
Engine coolant temperature (ECT) sensor – removal and refitting	5
Engine speed/crankshaft position sensor (CPS) – removal and refitting	6
Fault diagnosis – ignition system (1.0 litre and 1.1 litre carburettor engines)	10
General description	1
Ignition coil – removal and refitting	7
Ignition system control module (UESC/ESC-H2) – removal and refitting	8
Ignition timing – general	9
Maintenance, inspection and precautions	2
Spark plugs and HT leads – inspection and renewal	3

Part B: 1.4 and 1.6 litre CVH carburettor engines
Distributor – removal and refitting	15
Distributor cap and rotor arm – removal and refitting	14
Fault diagnosis – ignition system (1.4 litre and 1.6 litre CVH carburettor engines)	20
General description	11
Ignition amplifier module – removal and refitting	17
Ignition coil – removal, testing and refitting	19
Ignition timing – adjustment and advance (mechanical and vacuum) checks	16
Maintenance, inspection and precautions	12
Spark plugs and HT leads – inspection and renewal	13
Vacuum diaphragm unit (Lucas) – renewal	18

Part C: 1.4 litre CVH CFi engine
Air charge temperature (ACT) sensor – removal and refitting	24
Distributor – removal and refitting	29
Distributor cap and rotor arm – removal and refitting	28
Engine coolant temperature (ECT) sensor – removal and refitting	25
Engine management module (EEC-IV) – removal and refitting	33
Fault diagnosis – ignition and engine management systems (1.4 litre CVH engine with CFi)	34
General description	21
Ignition coil – removal, testing and refitting	27
Ignition module (TFI-IV) – removal and refitting	32
Ignition timing – general	30
Maintenance, inspection and precautions	22
Manifold absolute pressure (MAP) sensor – removal and refitting	31
Road speed sensor unit – removal and refitting	26
Spark plugs and HT leads – inspection and renewal	23

Part D: 1.6 litre CVH EFi engine
Air charge temperature (ACT) sensor – removal and refitting	38
Engine coolant temperature (ECT) sensor – removal and refitting	39
Engine management module (EEC-IV) – removal and refitting	46
Engine speed/crankshaft position sensor (CPS) – removal and refitting	40
Fault diagnosis – ignition and engine management systems (1.6 litre CVH engine with EFi)	47
Fuel mixture/CO% adjustment potentiometer – removal and refitting	43
General description	35
Ignition coil – removal and refitting	42
Ignition module (EDIS) – removal and refitting	45
Maintenance, inspection and precautions	36
Manifold absolute pressure (MAP) sensor – removal and refitting	44
Road speed sensor unit – removal and refitting	41
Spark plugs and HT leads – inspection and renewal	37

Specifications

Part A: 1.0 litre and 1.1 litre HCS engines
General
System type	Distributorless ignition system (DIS), controlled by UESC or ESC-H2 module
Firing order	1-2-4-3
Location of No 1 cylinder	Crankshaft pulley end

Coil
Type	High output distributorless
Output	37.0 kilovolts (minimum)
Primary resistance (measured at coil tower)	0.5 to 1.0 ohm

Spark plugs (see Note at end of Specifications)
Type	Champion RS9YCC or RS9YC
Electrode gap	1.0 mm (0.040 in)

HT leads
Type	Champion LS-28 (boxed set)
Maximum resistance per lead	30 000 ohms

Chapter 4 Ignition and engine management systems

Torque wrench settings	Nm	lbf ft
Spark plugs	14 to 20	10 to 15
ECT sensor	20 to 25	15 to 18
ACT sensor (if fitted)	20 to 25	15 to 18

Part B: 1.4 litre and 1.6 litre CVH engines (carburettor equipped)

General
System type .... Battery, coil and distributor incorporating electronic amplifier module
Firing order 1-3-4-2
Location of No 1 cylinder .... Crankshaft pulley end

Ignition timing
Initial static advance (all):
    4-star leaded petrol (97 RON) .... 12° BTDC at idle speed (vacuum pipe disconnected and plugged)
    Unleaded petrol (95 RON) .... 8° BTDC at idle speed (vacuum pipe disconnected and plugged)

Distributor
Make .... Lucas
Type .... Breakerless
Automatic advance method .... Mechanical and vacuum control
Direction of rotation .... Anti-clockwise (viewed from cap)
Drive .... Offset dog drive from end of camshaft
Dwell angle .... Governed by module (checking not needed)
Advance characteristics (mechanical) at 2000 rpm engine speed (with vacuum pipe disconnected and plugged):
    1.4 litre .... 3.0° to 9.0° (see text)
    1.6 litre .... 5.5° to 11.5° (see text)
Advance characteristics (total) at 2000 rpm engine speed (vacuum pipe connected):
    1.4 litre .... 18.0° to 32.0° (see text)
    1.6 litre .... 14.5° to 28.5° (see text)
Advance characteristics (vacuum at 2000 rpm engine speed (total advance minus mechanical advance):
    1.4 litre .... 15.0° to 23.0° (see text)
    1.6 litre .... 9.0° to 17.0° (see text)
Heat sink compound for ignition amplifier module .... Ford part number 81 SF-12103-AA

Coil
Type .... High output breakerless
Output .... 30.0 kilovolts (minimum)
Primary resistance .... 0.72 to 0.88 ohms
Secondary resistance .... 4500 to 7000 ohms

Spark plugs (see Note at end of Specifications)
Type .... Champion RC7YCC or RC7YC
Electrode gap:
    RC7YCC .... 0.8 mm (0.032 in)
    RC7YC .... 0.7 mm (0.028 in)

HT leads
Type .... Champion LS-14, boxed set
Maximum resistance per lead .... 30 000 ohms

Torque wrench setting	Nm	lbf ft
Spark plugs	27	20

Part C: 1.4 litre CVH CFi engine

General
System type .... Battery, coil, distributor and TFI-IV ignition module, controlled by EEC-IV engine management module
Firing order 1-3-4-2
Location of No 1 cylinder .... Crankshaft pulley end

Ignition timing
Initial static advance .... 10° (set using STAR test equipment – refer to text)

Distributor
Make .... Bosch
Type .... Breakerless ('Hall effect')
Automatic advance method .... Totally controlled by EEC-IV engine management module
Direction of rotation .... Anti-clockwise (viewed from cap)
Drive .... Offset dog drive from end of camshaft
Dwell angle .... Governed by module; no requirement to check

Coil
Type... High output breakerless
Output.. 30.0 kilovolts (minimum)
Primary resistance... 0.72 to 0.88 ohms
Secondary resistance... 4500 to 7000 ohms

Spark plugs (see Note at end of Specifications)
Type... Champion RC7YCC or RC7YC
Electrode gap.. 1.0 mm (0.040 in)

HT leads
Type... Champion LS-14, boxed set
Maximum resistance per lead... 30 000 ohms

Torque wrench settings

	Nm	lbf ft
Spark plugs	27	20
ECT sensor	20 to 25	15 to 18
ACT sensor	20 to 25	15 to 18

Part D: 1.6 litre CVH EFi engine

General
System type... Distributorless ignition system with EDIS ignition module, controlled by EEC-IV engine management module
Firing order.. 1-3-4-2
Location of No 1 cylinder... Crankshaft pulley end

Coil
Type... High output distributorless
Output.. 37.0 kilovolts (minimum)
Primary resistance... 4.5 to 5.0 ohms

Spark plugs (see Note at end of Specifications)
Type... Champion RC6YC or C6YCC
Electrode gap:
  C6YCC... 0.8 mm (0.032 in)
  RC6YC... 0.7 mm (0.028 in)

HT leads
Type... Champion LS-26, boxed set
Maximum resistance per lead... 30 000 ohms

Torque wrench settings

	Nm	lbf ft
Spark plugs	27	20
ECT sensor	20 to 25	15 to 18
ACT sensor	20 to 25	15 to 18

Note: *Information on spark plug type and electrode gaps is as recommended by Champion Spark Plug. Where alternative types are used, refer to the manufacturers' specifications*

PART A: 1.0 AND 1.1 LITRE HCS ENGINES

1 General description

Principles of operation

The ignition system is responsible for igniting the air/fuel mixture in each cylinder, at the correct moment in relation to engine speed and load, as the electrical spark generated jumps the spark plug gap.

The ignition system is based on feeding low tension (LT) voltage from the battery to the ignition coil where it is converted to high tension (HT) voltage. The high tension voltage is powerful enough to jump the spark plug gap in the cylinders many times a second under high compression pressures, providing that the system is in good condition.

System and operation

The ignition system is divided into two circuits; low tension (primary) and high tension (secondary). The low tension circuit consists of the battery, ignition switch, ignition module and its various outputs, and the coil primary windings. The high tension circuit consists of the coil secondary windings, the HT leads to the spark plugs and the spark plugs themselves.

The ignition module fitted to HCS engine (carburettor) models may be one of two types, dependent on the origin of the vehicle. For UK supplied vehicles, the UESC module (12 pin) is fitted to both models (built to operate on either leaded or unleaded fuel). Imported 1.1 litre carburettor equipped models, built primarily to operate on unleaded fuel and meeting EEC 5th amendment emission levels, may be fitted with an ESC-H2 ignition module (28 pin); if an ESC-H2 module is fitted, note that an air charge temperature (ACT) sensor is located on the underside of the air filter housing. Both types of module incorporate a manifold absolute pressure (MAP) sensor to determine engine load via a vacuum pipe from the inlet manifold. A fuel trap is fitted into the MAP sensor/inlet manifold vacuum line to prevent fuel entering the module (further details in Chapter 3).

The system functions in the following manner. Current flowing through the coil primary windings produces a magnetic field around the coil secondary windings. As the flywheel rotates, the engine speed/crankshaft position sensor receives a signal from the flywheel to indicate that the piston in number one cylinder is at 90° BTDC (this signal is generated from the missing 36th rib on the flywheel). The ignition module receives this signal, as an interruption, in a constant waveform pattern, and is therefore also able to determine the engine speed. With the information that the module now has – engine load, crankshaft position, engine speed, and the additional inputs of engine coolant temperature and air charge temperature (as applicable) – it is able to calculate the precise moment to cut off the supply to the coil primary windings, thus causing the magnetic field around the coil secondary windings to collapse. As the magnetic field collapses, it generates a secondary (high tension) voltage in the ignition coil.

The ignition coil featured in this system may be referred to as a DIS ignition coil (Distributorless Ignition System), and actually comprises two dual ended coil circuits, each one controlled individually by the ignition module. Each time one of the dual ended coils is fired, by the

Chapter 4 Ignition and engine management systems

to the corresponding cylinder on the exhaust stroke. (The spark delivered to the exhaust stroke is ineffective as it produces no power, but does no harm as there is no compression in that cylinder.)

The low tension circuit is switched on again by the ignition module to allow the magnetic field to regenerate before the firing of the next spark plugs, and the timing is automatically advanced or retarded according to engine requirements.

2 Maintenance, inspection and precautions

Warning: *The HT voltage generated by an electronic ignition system is extremely high, and in certain circumstances could prove fatal. Do not handle the HT leads, or touch the DIS ignition coil when the engine is running. If tracing faults in the HT circuit, use well insulated tools to manipulate live leads. On CTX automatic transmission equipped vehicles, refer to the 'Safety notes' in Chapter 6, Section 21, before carrying out any work*

1 At the intervals specified in the 'Routine maintenance' Section at the beginning of this manual, the following tasks should be carried out.
2 Renew the spark plugs, as described in Section 3 of this Chapter. Also clean the HT leads and DIS ignition coil towers, and ensure that the HT leads are securely reconnected to give the correct firing order.
3 Inspect the electrical and vacuum connections of the ignition/engine management system, and make sure that they are clean and secure. Further information regarding vacuum hose valves and fuel traps, as applicable, may be found in Chapter 3.

Module and system precautions

4 Ignition and engine management modules are sensitive components, and certain precautions must be taken to avoid module damage when working on the vehicle.
5 When carrying out welding operations on the vehicle using electric welding equipment, disconnect the battery and the alternator wiring.
6 Although underbonnet mounted modules (except EEC-IV, fitted only to fuel injection equipped vehicles) will tolerate normal underbonnet conditions, they can be adversely affected by excess heat or moisture. If using welding equipment or steam/pressure cleaning equipment in the vicinity of a module, take care not to direct heat or jets of water or steam at the module. If this cannot be avoided, remove the module from the vehicle and protect its wiring multi-plug.

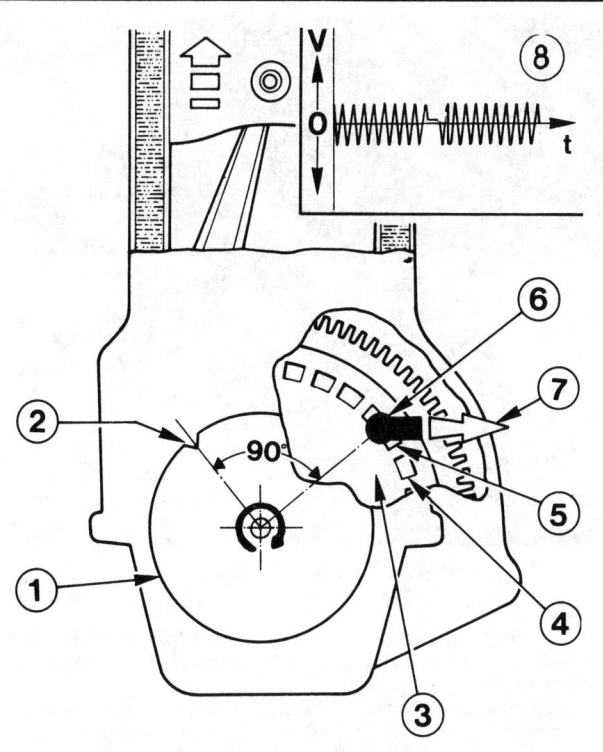

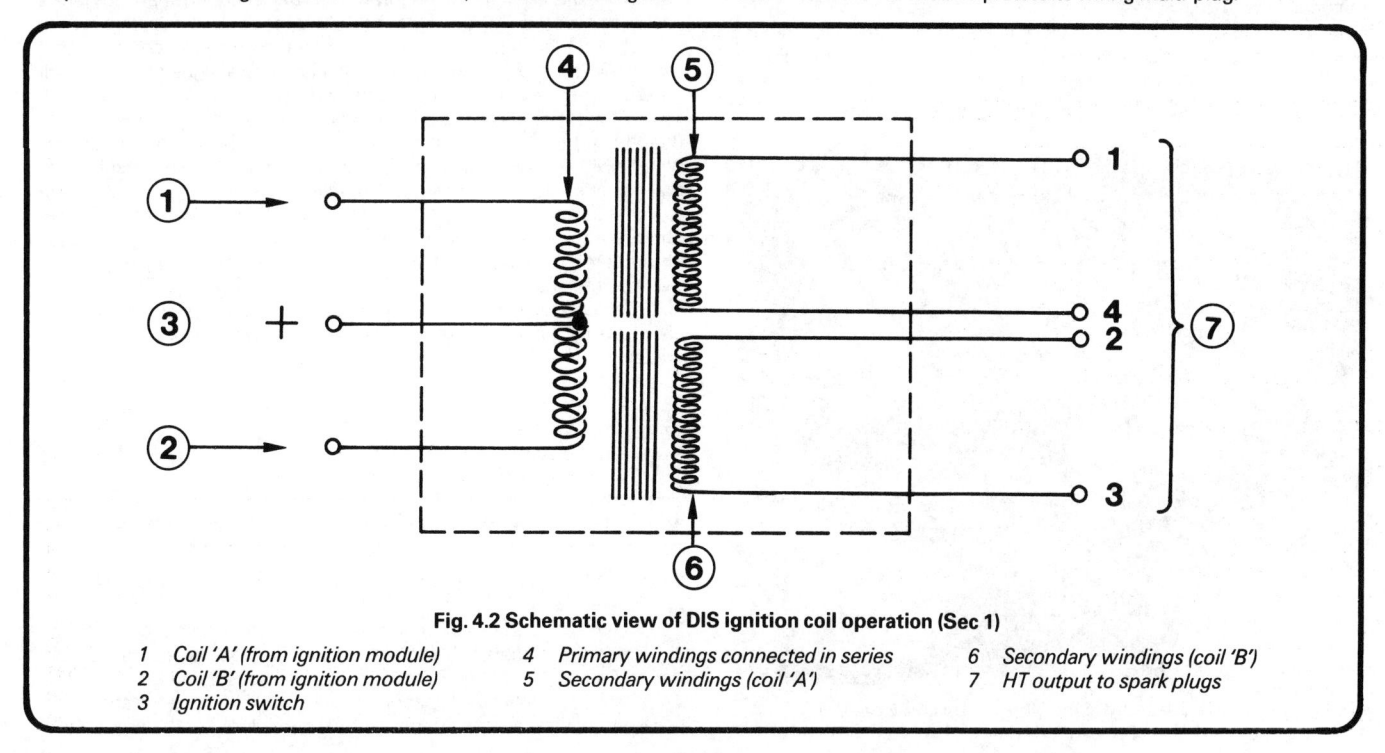

Fig. 4.1 Schematic view of signal generation system (Sec 1)

1 Crankshaft pulley
2 TDC mark on pulley (piston No 1)
3 Flywheel
4 Ribbed pattern around flywheel
5 Missing 36th rib
6 Engine speed/crankshaft position sensor (also taken to represent TDC '0' mark on timing cover)
7 Signal to ignition module
8 Waveform of signal

collapse of the magnetic field around it, the secondary (high tension) voltage is conducted along two spark plug HT leads to produce a spark in two cylinders on opposing strokes; one goes to a cylinder on the compression stroke to ignite the air/fuel mixture, whilst the other goes

Fig. 4.2 Schematic view of DIS ignition coil operation (Sec 1)

1 Coil 'A' (from ignition module)
2 Coil 'B' (from ignition module)
3 Ignition switch
4 Primary windings connected in series
5 Secondary windings (coil 'A')
6 Secondary windings (coil 'B')
7 HT output to spark plugs

Chapter 4 Ignition and engine management systems

7 Before disconnecting any wiring or removing components, always ensure that the ignition is switched off and the battery disconnected (where appropriate).
8 Do not run the engine with any module detached from its fitted position as damage may be caused, either by vibration or by possible internal overheating (loss of heat sink facility).
9 If renewing a module, it must be of the specified type – do not interchange modules between vehicle variants. Always check the part number of a replacement module to ensure that it is similar to the one originally fitted, before installing.
10 Do not attempt to improvise fault diagnosis procedures using a test lamp or multimeter, as irreparable damage could be caused to the module.
11 After working on ignition/engine management system components, ensure that all the wiring is correctly reconnected before reconnecting the battery or switching on the ignition.

3 Spark plugs and HT leads – inspection and renewal

Note: *See Warning in Section 2*

1 The correct functioning of the spark plugs is vital for the correct running and efficiency of the engine. It is essential that the plugs fitted are appropriate for the engine, and the suitable type is specified at the beginning of this Chapter. If this type is used and the engine is in good condition, the spark plugs should not need attention between scheduled replacement intervals. Spark plug cleaning is rarely necessary and should not be attempted unless specialised equipment is available as damage can easily be caused to the firing ends.
2 To remove the plugs, first mark the HT leads to ensure correct refitment, then pull them off the plugs. When removing the leads, pull the terminal connector at the end of the lead – not the lead itself.
3 It is recommended that any dirt present around the spark plug seats be removed, to prevent it dropping into the cylinders as the plugs are removed.
4 Using a suitable spark plug spanner (or a deep socket and extension bar), unscrew the plugs and remove them from the engine one at a time, examining each one as it is withdrawn. The condition of the spark plugs will give a good indication of the overall condition of the engine.
5 If the insulator nose of the spark plug is clean and white, with no deposits, this is indicative of a weak mixture, or too hot a plug. (A hot plug transfers heat away from the electrode slowly – a cold plug transfers it away quickly).
6 If the tip and insulator nose are covered with hard black-looking deposits, then this is indicative that the mixture is too rich. Should the plug be black and oily, then it is likely that the engine is fairly worn, as well as the mixture being too rich.

3.12 Inserting a spark plug to the cylinder head

7 If the insulator nose is covered with light tan to greyish brown deposits, then the mixture is correct and it is likely that the engine is in good condition.
8 The spark plug gap is of considerable importance, as if it is too large or too small, the size of the spark and its efficiency will be seriously impaired. The spark plug gap should be set to the figure given in the Specifications at the beginning of this Chapter.
9 To set it, measure the gap with a feeler gauge, and then bend open, or close, the *outer* plug electrode until the correct gap is achieved. The centre electrode should *never* be bent as this may crack the insulation and cause plug failure, if nothing worse.
10 Special spark plug electrode gap adjusting tools are available from most motor accessory shops.
11 Before fitting the plugs first ensure that the plug threads and the seating area in the cylinder head are clean, dry and free of carbon.
12 Screw in the spark plug by hand initially, then fully tighten to the specified torque. Do **not** overtighten the spark plugs or damage may occur, and they will be extremely difficult to remove in the future (photo).
13 Refit the HT leads in the correct order ensuring that they are a secure fit over the plug ends. Periodically wipe the leads clean to reduce the risk of HT leakage by arcing and remove any traces of corrosion that may occur on the end fittings. If any lead shows signs of cracking or other deterioration, it should be renewed.
14 If the HT leads are to be removed from the DIS ignition coil, note that the retaining lugs are fitted which must be depressed before withdrawing – pull on the terminal connector, not the lead itself. Note that the DIS ignition coil carries a marking adjacent to each connector tower as an aid to correct refitting of the leads (photo). With the HT leads removed from the vehicle, check the resistance of each lead using an ohmmeter connected to each end. If the resistance is any greater than the specified maximum, renewal will be required. When refitting the HT leads, ensure that the terminal connector retaining lugs fully engage at each DIS ignition coil connector tower, and clip the leads into their guides.

3.14 Removing an HT lead from the DIS ignition coil. Note the corresponding markings on the ignition coil and HT lead (arrowed)

4 Air charge temperature (ACT) sensor – removal and refitting

1 The ACT sensor (if fitted) is located on the underside of the air filter housing (Fig. 4.3).
2 To remove, first disconnect the battery earth terminal.
3 Referring to Chapter 3, remove the air filter housing and disconnect the ACT sensor multi-plug (do not pull on the sensor wiring).
4 Using a suitably sized spanner, unscrew and remove the sensor.
5 Refitting is a reversal of the removal procedure.

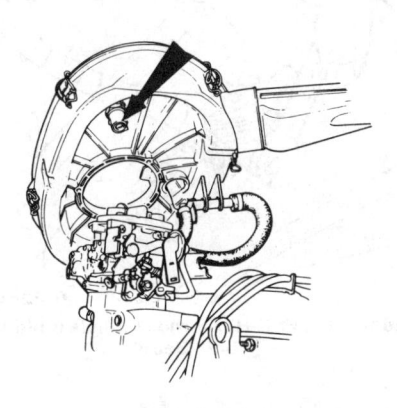

Chapter 4 Ignition and engine management systems

Fig. 4.3 Air charge temperature sensor location on underside of air filter housing (as applicable) (Sec 4)

5 Engine coolant temperature (ECT) sensor – removal and refitting

1 Disconnect the battery earth terminal.
2 Refer to Chapter 2 and drain the cooling system.
3 The ECT sensor is located on the underside of the inlet manifold. Disconnect its multi-plug (do not pull on its wiring), then unscrew the sensor from the inlet manifold.
4 Refitting is a reversal of the removal procedure, tightening the sensor to the specified torque. Do not overtighten and ensure that the multi-plug connection is secure. Refer to Chapter 2 and refill the cooling system.

6 Engine speed/crankshaft position sensor (CPS) – removal and refitting

1 Disconnect the battery earth terminal.
2 Disconnect the multi-plug from the CPS sensor, then undo the single (Torx head) retaining screw and withdraw the sensor (Fig. 4.4).
3 Refitting is a reversal of the removal procedure.

7 Ignition coil – removal and refitting

1 Disconnect the battery earth terminal.
2 Depress the ignition coil multi-plug spring clip, and disconnect the multi-plug. The ignition coil is located on the rearward facing side of the engine.

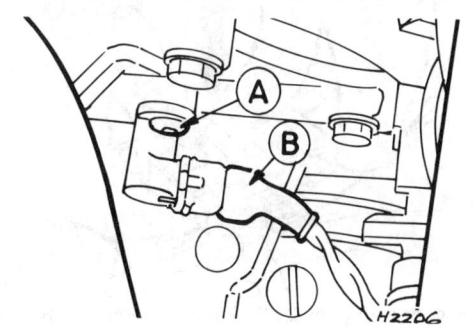

Fig. 4.4 Engine speed/crankshaft position sensor (Sec 6)

A Retaining screw   B Multi-plug

3 Compress the HT lead connector lugs and disconnect the HT leads from the ignition coil. Note HT lead locations on the ignition coil for correct subsequent refitting (the HT connections are marked).
4 Remove the three Torx head screws securing the ignition coil mounting bracket, and withdraw the ignition coil on its mount bracket (photo).
5 Refitting is a reversal of the removal procedure. Ensure that the HT leads are located in their correct ignition coil tower and check the security of the HT lead connectors at the coil.

8 Ignition system control module (UESC/ESC-H2) – removal and refitting

Note: *On earlier vehicles, some difficulties may be experienced with the UESC module multi-plug working loose, leading to water ingress. To eliminate this problem, a self-adhesive foam pad (available from your Ford dealer) can be fitted to the module. The foam pad is fixed as close to the locking tab as possible (see Fig. 4.7). Later vehicles are fitted with a modified multi-plug, identified by a moulded triangle on its surface*

1 Disconnect the battery earth terminal.
2 Disconnect the vacuum pipe from the module (photo).
3 On vehicles fitted with the UESC module, release the multi-plug locking lugs and slide the multi-plug out from the module (see Fig. 4.5).
4 On vehicles fitted with the ESC-H2 module, undo the multi-plug securing bolt and remove the multi-plug (see Fig. 4.6).
5 Remove the screws securing the module and withdraw it.
6 Refitting is a reversal of the removal procedure, ensuring module multi-plug security (on the UESC module, the multi-plug must 'snap' into position).

7.4 DIS ignition coil and its mounting bracket being withdrawn from the rearward facing side of the engine

8.2 Disconnecting the vacuum pipe from the UESC module

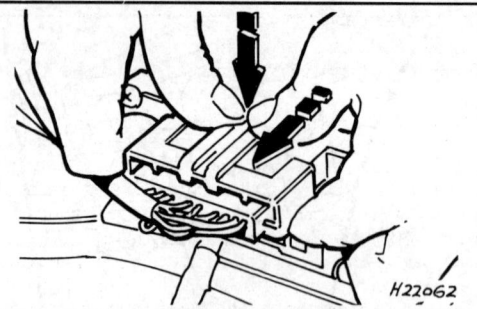

Fig. 4.5 Disconnecting the UESC ignition module multi-plug (Sec 8)

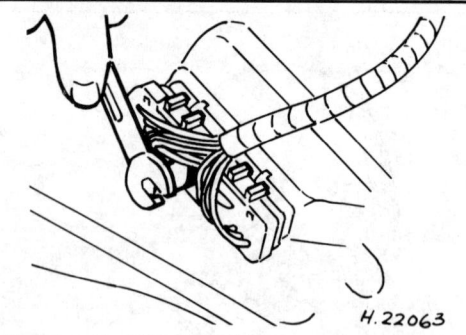

Fig. 4.6 Undoing the ESC-H2 ignition module multi-plug securing bolt (Sec 8)

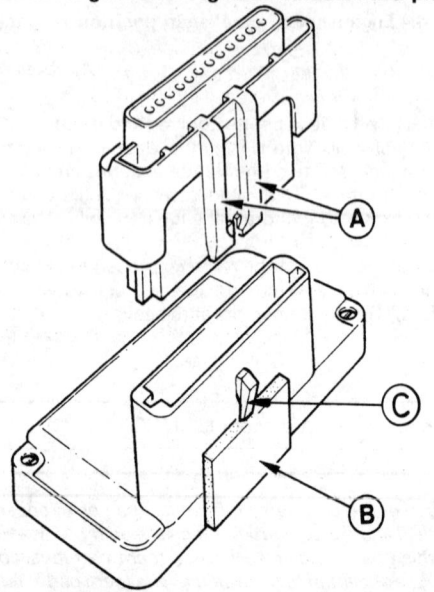

Fig. 4.7 Foam pad fixture location for earlier vehicles (Sec 8)

A Multi-plug locking lugs
B Foam pad
C Locking tab

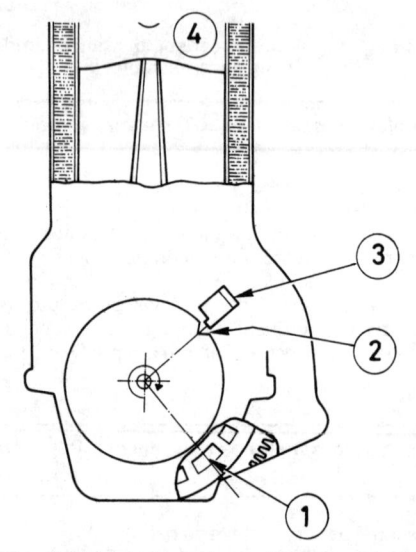

Fig. 4.8 Ignition timing marks – pistons Nos 1 and 4 at TDC (Sec 9)

1 Missing 36th rib
2 Pulley notch
3 Timing chain cover projection and TDC reference mark
4 Piston

9 Ignition timing – general

1 Owners should note that, while it is possible to check the ignition timing roughly, the results cannot be conclusive; there are no proper timing marks or data available, and no adjustment is possible. If the timing is thought to be incorrect, this can only be due to a fault in the system. The vehicle must be taken to a Ford Dealer for full testing on diagnostic equipment until the presence of a fault is confirmed, this equipment can then be used to diagnose the fault so that it can be cured.

2 The 1.0 and 1.1 litre HCS engines are designed to run continuously on unleaded petrol of 95 RON octane rating; this is the 'Premium' grade usually found on sale in the UK. No 'adjustment' (ie, to the ignition timing) is required to allow these engines to use this fuel. Owners who so wish may use the petrol of higher (97/98 RON) octane rating, such as four-star leaded or 'Super/Superplus'-grade unleaded petrol – again, no adjustment is required, and there will be no ill-effects save that on the driver's wallet.

3 If travelling abroad, owners should note that a lower grade (91 RON 'Regular' or 'Normal'-grade) of unleaded petrol is commonly available; depending on the engine and ignition system fitted, a modification may be possible to allow your vehicle to use this petrol. Seek the advice of a Ford dealer, or of one of the motoring organisations, to ensure that you are aware of the grades of petrol that you are likely to encounter, and of the different names used for these grades, compared with the UK terms with which you are familiar.

4 To check the ignition timing, first start the engine and warm it up to normal operating temperature. Stop the engine, and connect a timing light according to the equipment manufacturer's instructions. Use white paint to highlight the notch cut in the crankshaft pulley's inner rim.

5 Restart the engine and allow it to idle, then aim the timing light at the crankshaft pulley; the notch should be reasonably steady, aligned just in front of the fixed reference mark projecting from the timing chain cover (see Fig. 4.8).

6 To check (as far as is possible) that the advance/retard function is operating, increase engine speed, and check that the pulley mark advances smoothly to beyond the beginning of the cover projection, returning to close to the TDC mark when the engine is allowed to idle.

10 Fault diagnosis – ignition system (1.0 litre and 1.1 litre carburettor engines)

Note: *Refer to Section 2 before proceeding*

1 There are two main symptoms indicating ignition faults: either the engine will not start, or the engine is difficult to start and misfires. If a regular misfire is present, the fault is almost certain to be in the HT circuit.

Engine fails to start

2 If the starter motor fails to turn the engine, check the battery and the starter motor with reference to Chapter 12.

3 Disconnect the HT lead from any spark plug and hold the end of the cable approximately 5 mm (0.2 in) away from the cylinder head *using well insulated pliers*. While an assistant spins the engine on the starter motor, check that a regular blue spark occurs. Repeat the test on the remaining HT leads. If the spark test is positive, remove the spark plugs and examine them (refer to Section 3). Re-gap or renew, as necessary.

4 If the engine fails to start due to damp HT leads, a moisture dispersant, such as Holts Wet Start, can be effective. To prevent the problem recurring Holts Damp Start can be used to provide a sealing coat, so excluding any further moisture from the ignition system. In extreme difficulty, Holts Cold Start will help to start a car when only a very poor spark occurs.

Chapter 4 Ignition and engine management systems

5 If no spark occurs, check that the multi-plugs are secure on the ignition module and the DIS ignition coil, and that the HT leads are clean, dry and securely connected to the DIS ignition coil towers. If no spark occurs on individual HT leads yet does occur on others, remove the suspect lead(s) and test as described in Section 3, paragraph 14 – renew as necessary (ensure that a high resistance is not obtained as a result of dirty or corroded terminals).
6 Check that the wiring/multi-plugs are secure on the remaining ignition system sensors.
7 If the above checks reveal no faults but there is still no spark, the DIS ignition coil or the ignition module must be suspect. Consult a Ford dealer for further testing, or test by substitution of a known serviceable unit (with an identical part number).

Engine misfires

8 If the engine misfires regularly, run it at a fast idling speed. Pull off each of the plug HT leads in turn and listen to the note of the engine. Hold the plug leads with a well insulated pair of pliers as protection against a shock from the HT supply.
9 No difference in engine running will be noticed when the lead from the defective circuit is removed. Removing the lead from one of the good cylinders will accentuate the misfire.
10 Remove the HT lead from the end of the plug in the misfiring cylinder and holt it (again using well insulated pliers) about 5 mm (0.2 in) away from the cylinder head. If the spark is fairly strong and regular the fault must lie in the spark plug. If the spark is weak and/or irregular, test the HT lead (as described in Section 3, paragraph 14), wiggling the lead to look for an intermittent fault. Also ensure that the lead is clean, securely connected and dry (see 'Engine fails to start', previous sub heading, paragraph 4).
11 If the fault is traced to the spark plug, the plug may be loose, the insulation may be cracked, or the electrodes may have burnt away creating too wide a gap for the spark to jump. Worse still, one of the electrodes may have broken off. Remove the spark plug and examine it (refer to Section 3). Re-gap or renew, as necessary.
12 A further cause of misfiring may be a split or disconnected vacuum hose.
13 If the above checks do not locate the misfire, the DIS ignition coil or the ignition module may be faulty. Consult a Ford dealer for further testing, or test by substituting a known serviceable unit (with an identical part number).

PART B: 1.4 AND 1.6 LITRE CVH CARBURETTOR ENGINES

11 General description

Principles of operation
Refer to Section 1 of this Chapter.

System and operation
This is the least sophisticated system fitted to the Fiesta range. The ignition system is divided into two circuits; low tension (primary) and high tension (secondary). The low tension circuit consists of the battery, ignition switch, coil primary windings, ignition amplifier module and the signal generating system inside the distributor. The signal generating system comprises the trigger coil, trigger wheel, stator, permanent magnet and trigger coil to ignition amplifier module connector. The high tension circuit consists of the coil secondary windings, the HT lead from the coil to the distributor cap, the distributor cap, the rotor arm, the HT leads from the distributor cap to the spark plugs and the spark plugs themselves.
When the system is in operation, low tension voltage is changed in the coil into high tension voltage by the action of the electronic amplifier module in conjunction with the signal generating system. Any change in the magnetic field force (flux), created by the movement of the trigger wheel relative to the magnet, induces a voltage in the trigger coil. This voltage is passed to the ignition amplifier module which switches off the ignition coil primary circuit. This results in the collapse of the magnetic field in the coil which generates the high tension voltage. The high tension voltage is then fed, via the coil HT lead and the carbon brush in the centre of the distributor cap, to the rotor arm. The voltage passes across to the appropriate metal segment in the cap and via the spark plug HT lead to the spark plug where it finally jumps the spark plug gap to earth.
The distributor is driven by an offset drive dog locating to a correspondingly offset slot in the end of the camshaft.
The ignition advances is a function of the distributor and is controlled both mechanically and by a vacuum operated system. The mechanical governor mechanism consists of two weights which move out from the distributor shaft as the engine speed rises due to centrifugal force. As they move outwards, they rotate the trigger wheel relative to the distributor shaft and so advance the spark. The weights are held in position by two light springs and it is the tension of the springs which is largely responsible for correct spark advancement.
The vacuum control consists of a diaphragm, one side of which is connected via a small bore hose to the carburettor or throttle housing, and the other side to the distributor. Depression in the inlet manifold and/or carburettor, which varies with engine speed and throttle position, causes the diaphragm to move, so moving the stator and advancing or retarding the spark. A fine degree of control is achieved by a spring in the diaphragm assembly. Additionally, one or more vacuum valves may be incorporated in the vacuum line between the inlet manifold or carburettor and the distributor. The function of these is to control the vacuum felt at the distributor and to prevent fuel entering along the vacuum line (as applicable). They are part of the vehicle's emission control systems (further details in Chapter 3).
Note that a separate ballast resistor is not required, as this aspect of the ignition system is incorporated into the ignition amplifier module circuitry.

12 Maintenance, inspection and precautions

Warning: *The HT voltage produced by breakerless ignition systems is approximately 25% greater than conventional ignition systems. Take all possible precautions to avoid receiving electric shocks. Do not touch the HT leads, ignition coil or distributor cap while the engine is running. If the distributor is knocked while the ignition is switched on, a single HT pulse may be generated. If tracing faults in the HT circuit, use well insulated tools to manipulate live leads. On CTX automatic transmission equipped vehicles, refer to the 'Safety notes' in Chapter 6, Section 22, before carrying out any work*

1 At the intervals specified in the *'Routine maintenance'* Section at the beginning of this manual, the following tasks should be carried out.
2 Renew the spark plugs making reference to Section 13.
3 Remove the distributor cap and HT leads and wipe them clean. Also wipe clean the coil tower and make sure that the plastic safety cover is securely fitted, where applicable. Remove the rotor arm, then visually check the distributor cap, rotor arm and HT leads for hairline cracks and signs of arcing (note that the rotor arm has a special lacquer coating to reduce radio interference). Further check the distributor cap to ensure that the spring-tensioned carbon brush in the centre moves freely, and that the HT segments are not worn excessively. Check that the HT leads are securely connected after refitting, to give the correct firing order.
4 Inspect all relevant electrical and vacuum connections, and make sure that they are clean and secure. Further information regarding vacuum hose valves and fuel traps, as applicable, may be found in Chapter 3.

Module and system precautions
5 Refer to Section 2 for applicable information.

13 Spark plugs and HT leads – inspection and renewal

1 Refer to Section 3 of this chapter, but substitute any reference to DIS ignition coil with distributor. The HT lead terminal connectors at the distributor are not fitted with retaining lugs and are simple push fits. An extra HT lead is fitted between the distributor cap central tower and the ignition coil tower, to carry HT voltage from the ignition coil.
2 HT lead connections at the distributor cap, to give the correct firing order, are shown in Fig. 4.9.

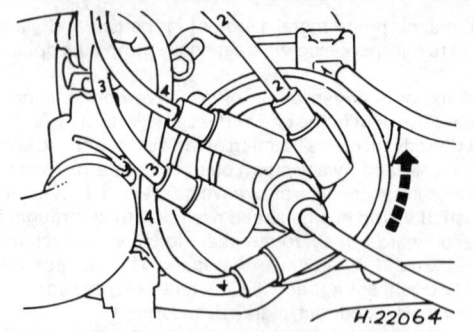

Chapter 4 Ignition and engine management systems

Fig. 4.9 Distributor HT lead connections showing the correct firing order (Sec 13)

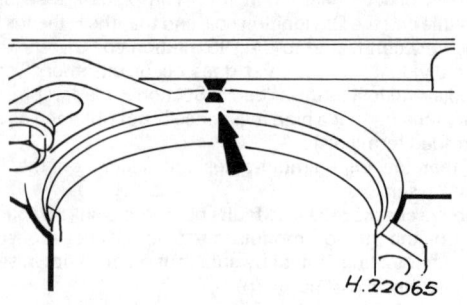

Fig. 4.10 Distributor timing alignment punch mark (arrowed) (Sec 15)

14 Distributor cap and rotor arm – removal and refitting

1 Disconnect the battery earth terminal.
2 Disconnect the HT leads from the distributor cap, having identified them for subsequent refitting. Pull on the connectors, not the leads.
3 Unclip the suppressor shield (where fitted), remove the distributor cap securing screws and detach the cap.
4 Detach the rotor arm.
5 Refer to Section 12 before refitting.
6 Refitting is a reversal of the removal procedure, ensuring that the HT leads are securely connected to give the correct firing order.

15 Distributor – removal and refitting

1 Disconnect the battery earth terminal.
2 If the original distributor is to be refitted to the original cylinder head, check that the punch marks are aligned before removing the distributor (Fig. 4.10). If no punch marks are present, make your own to ensure correct alignment upon subsequent refitting.
3 Refer to the previous Section and remove the distributor cap.
4 Disconnect the vacuum pipe from the distributor vacuum diaphragm unit.
5 Disconnect the loom multi-plug from the distributor.
6 Remove the clamp bolts securing the distributor in position and slide it out.
7 Prior to refitting, check the condition of the distributor oil seal and renew it if necessary.
8 To refit, position the distributor so that its offset drive is engaged with the slot in the end of the camshaft, then loosely insert the two clamp bolts (photos).

9 Where both original punch marks are present, on the cylinder head and distributor body, rotate the distributor body until the punch marks are aligned before tightening the clamp bolts to the specified torque.
10 If one or both of the punch marks are missing (due to component renewal), turn the body of the distributor so that the clamp bolts are centrally located in their slots before tightening.
11 Refit the rotor arm (if removed), distributor cap, suppressor shield (as applicable) and HT leads, making reference to the previous Section, then reconnect the loom multi-plug and vacuum pipe.
12 Reconnect the battery earth terminal.
13 If one or both of the punch marks are missing after component renewal, check the ignition timing, as described in Section 16, and adjust as necessary. This will not be required where the original punch marks have been re-aligned.

16 Ignition timing – adjustment and advance (mechanical and vacuum) checks

Note: When an engine is timed in production, marks are punched into the cylinder head and the distributor body flange to indicate the correct (static) timing position of the distributor (see Fig. 4.10). Therefore, under normal circumstances, ignition timing adjustment will only be necessary if the initial setting has been disturbed. A static advance setting for use with unleaded petrol (95 RON) is given in the Specifications

1 Where the original punch marks are present on the cylinder head and the distributor body flange, correct (static) ignition timing can be set, as necessary, by turning the body of the distributor to align the marks before re-tightening the distributor clamp bolts to their specified torque.
2 If, due to component renewal, one or both of the original punch marks is missing, the following procedure must be carried out.

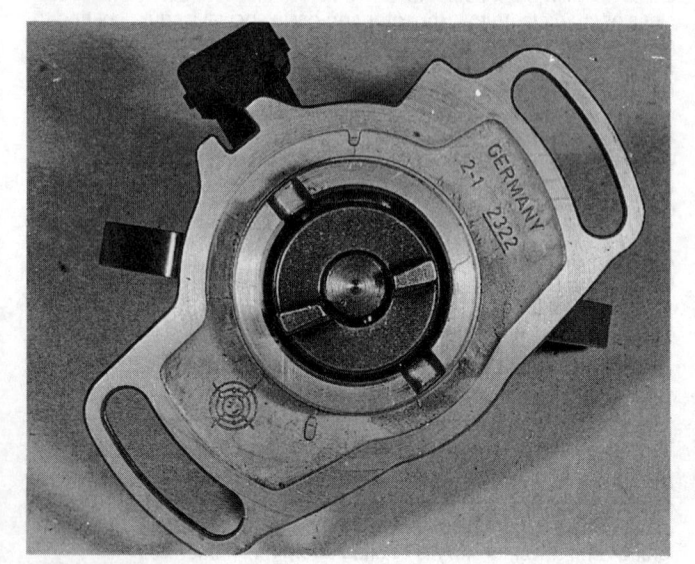

15.8A Distributor drive mechanism. Note the lugs offset from centre (1.4 litre CFi version shown)

15.8B Refitting the distributor (1.4 litre CFi version shown)

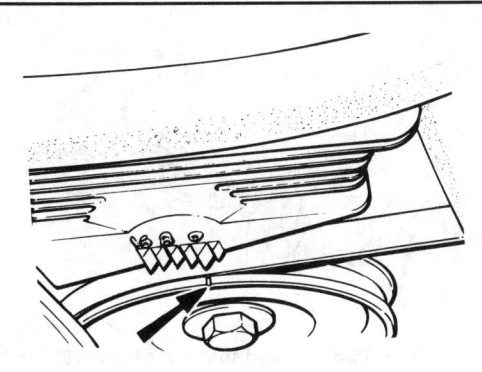

Chapter 4 Ignition and engine management systems

Fig. 4.11 Ignition timing marks on the timing cover scale (Sec 16)

3 Turn the distributor body so that the clamp bolts are located centrally in their slots then tighten the clamp bolts.
4 Increase the contrast of the notch in the crankshaft pulley and the appropriate mark on the timing cover scale (refer to Specifications) by applying a dab of quick-drying white paint (see Fig. 4.11).
5 Connect a timing light (stroboscope) in accordance with the manufacturer's instructions.
6 Start the engine, bring it up to normal operating temperature and allow it to idle (see Specifications in Chapter 3). Ensure that the radiator cooling fan is switched on by disconnecting its thermal switch multi-plug and bridging both contacts inside the multi-plug.
7 Disconnect the vacuum pipe from the distributor and fit blanking plugs.
8 If the timing light is now directed at the engine timing marks, the pulley notch will appear to be stationary and opposite the specified mark on the scale. If the marks are not in alignment, release the distributor clamp bolts sufficient to allow the rotation of the distributor body and turn the distributor body in whichever direction is necessary to align the pulley notch to the appropriate scale mark, then tighten the clamp bolts.
9 Repeat the procedure given in the above paragraph until the pulley notch and appropriate scale mark are aligned, then fully tighten the distributor clamp bolts.
10 Using a suitable punch, re-mark the cylinder head and/or the distributor flange to indicate the **new** distributor timing position for any future repair operations.
11 If it is desired to check the mechanical and vacuum advance refer to paragraph 12. If the operation is limited to obtaining the correct distributor (static) timing position, remove the distributor vacuum pipe blanking plugs and reconnect the vacuum pipe. The bridging wire from the radiator cooling fan thermal switch multi-plug must be removed before reconnecting the multi-plug; also disconnect the timing light.
12 Do not reconnect the vacuum pipe or the radiator cooling fan thermal multi-plug. Ideally, to carry out the advance checks, a timing light with an advance meter incorporated should be used. If one is unavailable the checks will not be precise.
13 With the engine idling, increase the engine speed to 2000 rpm and note the distance which the pulley notch moves out of alignment with its appropriate scale mark.
14 Remove the vacuum pipe blanking plugs, reconnect the vacuum pipe to the distributor and repeat the check when, for the same increase in engine speed, the alignment differential of the timing marks should be greater than previously observed.
15 If the timing marks did not appear to move during the first check a fault in the distributor centrifugal advance mechanism is indicated. No increased movement of the marks during the second test indicate a punctured diaphragm in the vacuum unit, or a leak in the vacuum line.
16 If an advance meter was used to give accurate advance figures, a vacuum advance figure may be obtained by subtracting the mechanical advance figure (obtained in paragraph 13) from the total advance figure (obtained in paragraph 14). Refer to the Specifications, but note that the advance figures quoted are in crankshaft degrees and to obtain distributor degrees they must be divided by two. Also note that the advance figures quoted do not include initial static advance.
17 On completion of the adjustments and checks, switch the engine off, disconnect the timing light, ensure that the distributor vacuum pipe is securely connected and remove the bridging wire from the radiator cooling fan thermal switch multi-plug before reconnecting.

17 Ignition amplifier module – removal and refitting

1 Disconnect the battery earth terminal.
2 Remove the distributor, as described in Section 15.
3 With the distributor on the workbench, remove both screws securing the module to the distributor body, then slide the module from its trigger coil connector and remove it.
4 Check that the rubber grommet is serviceable. If it is not, it must be renewed but ensure that the correct type is obtained.
5 Apply heat sink compound (see Specifications) to the module metal face, ensuring a good earth. This is an essential part of the procedure, protecting the module electronic circuitry from excessive heat build-up and subsequent malfunction.
6 Slide the module into its trigger coil connector and secure with both screws, tightened according to specification.
7 Refit the distributor in accordance with Section 15.
8 Reconnect the battery earth terminal.

18 Vacuum diaphragm unit (Lucas) – renewal

1 Check parts availability before proceeding with this operation.
2 Remove the distributor, as described in Section 15.
3 With the distributor on the workbench, remove the ignition amplifier module (see Section 17) and the distributor cap and rotor arm (if not already done).
4 Remove the three screws securing the distributor body halves and separate the assembly (see Fig. 4.12).
5 Lift out the plastic spacer ring from the upper distributor body half (Fig. 4.13).
6 Remove the trigger coil to ignition amplifier module connector and seal, having noted which way the connector fits for subsequent reassembly (Fig. 4.14).
7 Lift out the trigger coil, before careful not to damage it or its connectors as it is withdrawn.
8 Referring to Fig. 4.15, remove the stator securing circlip and the upper shim.

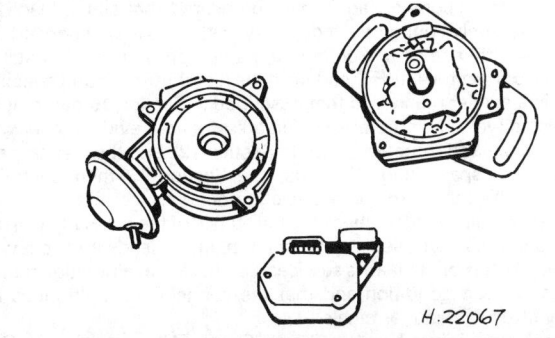

Fig. 4.12 Distributor body halves separated (ignition amplifier module also shown) (Sec 18)

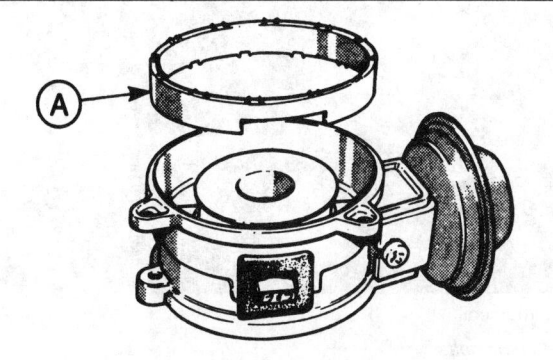

Fig. 4.13 Remove the plastic spacer ring (A) (Sec 18)

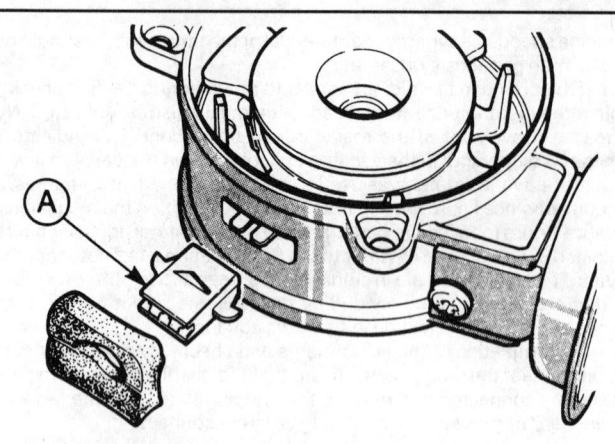

Fig. 4.14 Trigger coil to ignition amplifier module connector (A) (Sec 18)

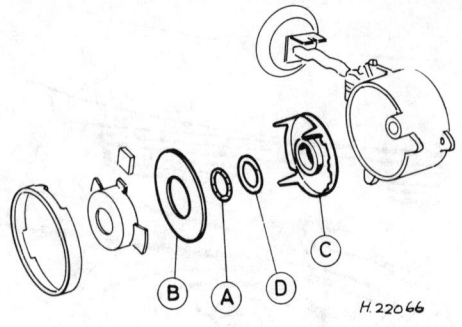

Fig. 4.15 Stator and shim detail (Sec 18)

A   Circlip   C   Stator
B   Upper shim   D   Lower shim

9 Remove the stator and the lower shim.
10 Undo the vacuum diaphragm unit securing screw and detach the unit.
11 This is the limit of dismantling that can be undertaken on these distributors. Should the distributor be worn or unserviceable in any other respect, renewal of the complete unit will be necessary.
12 Reassembling is a reversal of the dismantling procedure. During reassembly, ensure that the pin on the vacuum diaphragm unit arm engaged in the stator as the stator is refitted, and fit the plastic spacer ring so that its cut-out aligns with the trigger coil to ignition amplifier module connector. Additionally, after the distributor body halves' securing screws have been tightened, ensure that the distributor shaft turns easily. Refit the ignition amplifier module in accordance with Section 17.

19 Ignition coil – removal, testing and refitting

1 Disconnect the battery earth terminal.
2 Remove the vehicle jack from its storage position by unscrewing its retainer. The ignition coil is mounted below.
3 Disconnect the HT lead and the low tension (LT) connections from the ignition coil. Note that the LT connections on the ignition coil are of different sizes. As an aid to refitting the positive (+) terminal is larger than the negative (–) terminal (photo).

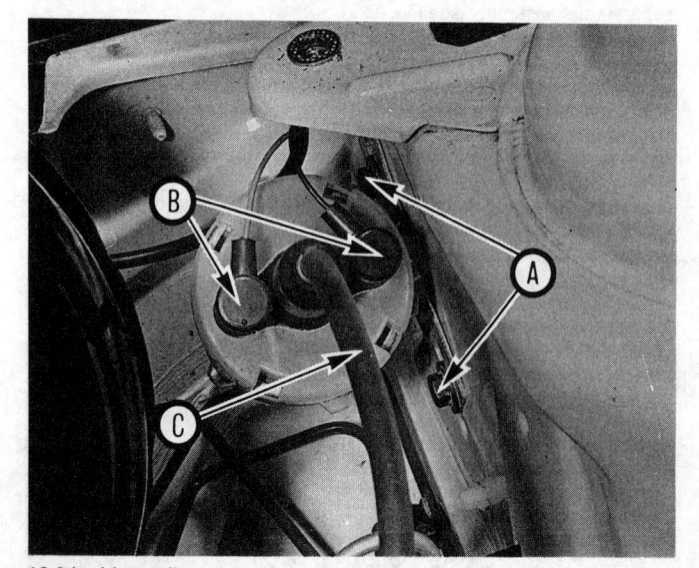

19.2 Ignition coil

A   Retaining bolts
B   LT connections
C   HT lead to distributor cap

4 Remove the two screws or bolts securing the coil mounting bracket to the inner wing panel, and withdraw the coil and mounting bracket assembly.
5 Accurate checking of the coil output requires the use of special test equipment and should be left to a dealer or suitably equipped automotive electrician. It is however possible to check the primary and secondary winding resistance using an ohmmeter as follows.
6 To check the primary resistance (with all leads disconnected if the coil is fitted), connect the ohmmeter across the coil positive and negative terminals. The resistance should be as given in the Specifications at the beginning of this Chapter.
7 To check the secondary resistance (with all leads disconnected if the coil is fitted), connect one lead from the ohmmeter to the coil negative terminal, and the other lead to the centre HT terminal. Again the resistance should be as given in the Specifications.
8 If any of the measured values vary significantly from the figures given in the Specifications, the coil should be renewed.
9 If a new coil is to be fitted, ensure that it is of the correct type. The appropriate Ford supplied ignition coil is identified by a red label, and will be one of three different makes, all of which are fully interchangeable. Bosch and Femsa coils are fitted with protective plastic covers, and Polmot coils are fitted with an internal fusible link. Note that contact breaker ignition coils are not interchangeable with the required breakerless type and could cause electronic module failure if used.
10 Refitting is a reversal of the removal procedure, ensuring correct LT lead polarity.

20 Fault diagnosis – ignition system (1.4 litre and 1.6 litre CVH carburettor engines)

Note: *Refer to Section 12 before proceeding.*

1 Section 10 paragraph 1 categorizes ignition system faults. If the engine fails to start, proceed as described in Section 10 paragraphs 2 and 3 then follow the remainder of the procedure given here. The procedure for tracing a misfire is covered later in this Section.
2 If no spark occurs, try again having ensured that all HT leads are clean and securely connected. If difficulty in starting is experienced due to damp weather refer to Section 10, paragraph 6. If there is still no spark, disconnect the coil HT lead from the distributor cap and check for a spark in a similar manner to that described in Section 10 paragraph 3 (again using well insulated pliers). If sparks are now evident, check the distributor cap and rotor arm (refer to Section 12), and the resistance of the individual spark plug HT leads (see Section 13 then Section 3 paragraph 14) – renew any defective components.
3 If there is still no spark, check the resistance of the coil HT lead in the same manner as that used for the spark plug HT leads and renew as necessary. If the coil HT lead is serviceable, check that all ignition related wiring is in good condition and that the connections/multi-plugs are securely fitted to clean terminals.
4 Test the ignition coil as described in Section 19.
5 If the above checks reveal no faults but there is still no spark, the distributor or its ignition amplifier module must be suspect. Consult a

Chapter 4 Ignition and engine management systems

Ford dealer for further testing, or test by substitution of a known serviceable unit.

Engine misfires

6 Proceed as described in Section 10, paragraphs 7 to 11 inclusive.
7 If the misfire is still present, check the distributor cap and rotor arm (refer to Section 12) and renew as necessary.
8 Refer to Section 16 and check the ignition timing (including advance checks). With the ignition timing set correctly, unless one of the possible faults listed in the following paragraph is applicable, any misfire occurring at this stage is likely to be fuel system-related.
9 Further possible causes of misfiring may be an intermittent fault in the ignition coil HT lead, ignition amplifier module, ignition coil or its LT supply. Additionally, consider the possibility that the distributor triggering mechanism is defective in some manner. A Ford dealer or other competent specialist should be consulted for diagnosis, as necessary, unless components are freely available for testing, using a substitution method.

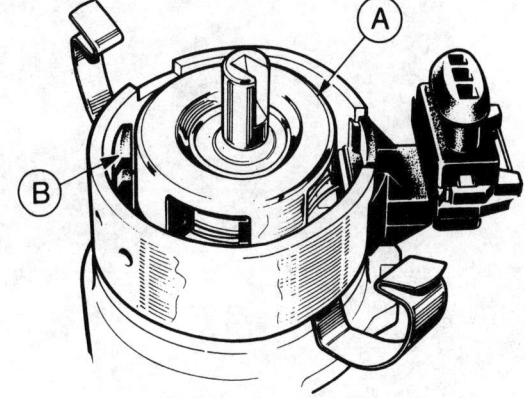

Fig. 4.16 Signal generating system components inside the distributor (1.4 litre CFi models) (Sec 21)

A Trigger vane   B Sensor

PART C: 1.4 LITRE CVH CFi ENGINE

21 General description

The EEC-IV engine management system module contains a powerful microprocessor which is programmed to provide total management of both fuel injection and ignition systems. Further details of the fuel injection system components are container in Chapter 3.
The ignition system consists of a distributor, ignition coil, TFI-IV ignition module, HT leads and spark plugs. A separate ballast resistor is incorporated in the system. The basic principles of operation are as described at the beginning of Section 1.
The ignition system is divided into two circuits; low tension or primary, and high tension or secondary. The low tension circuit consists of the battery, ignition switch, EEC-IV module power relay, TFI-IV ignition module (and EEC- IV module control), ballast resistor, coil primary windings and distributor 'Hall effect' trigger. The high tension circuit consists of the coil secondary windings, coil-to-distributor cap HT lead, distributor cap, rotor arm, spark plug HT leads and spark plugs.
The ignition functions in the following way: The electrical impulses required to switch off the low tension circuit at the coil (thus generating HT voltage) is obtained from the 'Hall effect' trigger system mounted in the distributor. The impulses are sent via the TFI-IV module to the main EEC-IV module where engine speed and crankshaft position are calculated. Additional data is received from the engine coolant temperature (ECT) sensor, manifold absolute pressure (MAP) sensor, air charge temperature (ACT) sensor, and throttle position sensor. Using this information the EEC-IV module then switches off the low tension circuit via the TFI-IV module at the optimum moment to promote the HT spark at the spark plugs. It can be seen that the TFI-IV module functions basically as a high current switch by controlling the low tension supply to the ignition coil primary windings.
In the event of failure of a sensor, the EEC-IV module will substitute a preset value for that input to allow the system to continue to function. In the event of failure of the EEC-IV module, a 'limited operation strategy' (LOS) function allows the vehicle to be driven, albeit at reduced power and efficiency. The EEC-IV module also has a 'keep alive memory' (KAM) function which stores idle and drive values and codes which can be used to indicate any system fault which may occur.

22 Maintenance, inspection and precautions

1 Refer to Section 12, but also inspect the security and condition of the EEC-IV engine management module cover. The module cover is held in position by two nuts, and should be free from any splits, tears or other deterioration.
2 Note that after the battery has been disconnected and subsequently reconnected, the EEC-IV engine management module will need to 're-learn' its values (refer to Chapter 12, Section 5 for details).

23 Spark plugs and HT leads – inspection and renewal

1 Refer to Section 3 of this Chapter, but make additional note of the information given in Section 13.

24 Air charge temperature (ACT) sensor – removal and refitting

1 Disconnect the battery earth terminal.
2 Remove the air filter housing, disconnecting pipes/hoses as necessary.
3 The ACT sensor is located in the rear of the CFi housing. Disconnect its multi-plug (do not pull on its wiring) (photo).
4 Unscrew and remove the sensor.
5 Prior to refitting, apply a smear of suitable sealant to the threads of the sensor.
6 Refitting is a reversal of the removal procedure, tightening the sensor to the specified torque.

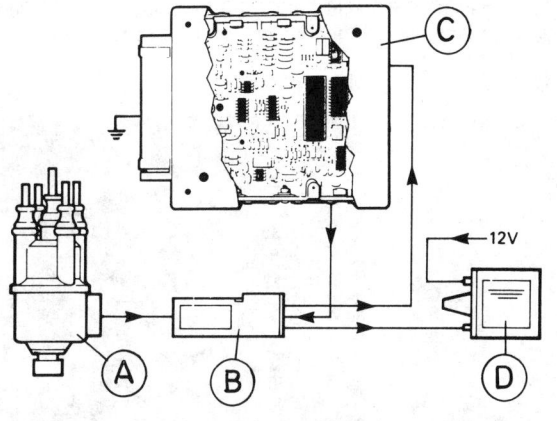

Fig. 4.17 Ignition system control signal routing (1.4 litre CFi models) (Sec 21)

A Distributor
B TFI-IV ignition module
C EEC-IV engine management module
D Ignition coil

Chapter 4 Ignition and engine management systems

24.3 Disconnecting the air charge temperature (ACT) sensor multi-plug (1.4 litre CFi model)

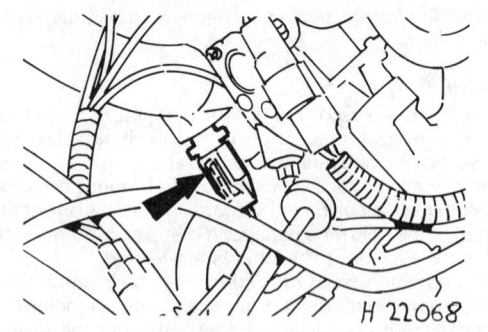

Fig. 4.18 Engine coolant temperature (ECT) sensor location on inlet manifold (1.4 litre CFi models) (Sec 25)

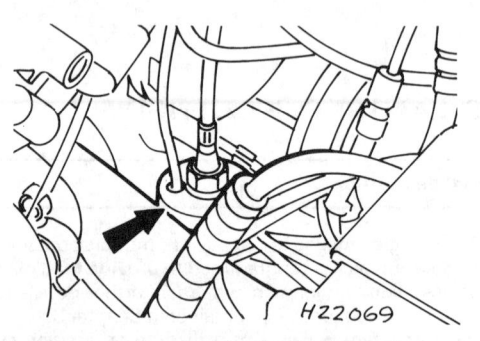

Fig. 4.19 Road speed sensor unit (arrowed) on the transmission housing (Sec 26)

25 Engine coolant temperature (ECT) sensor – removal and refitting

1 The procedure is covered in Section 5 of this Chapter, but the sensor location on the inlet manifold is as shown in Fig. 4.18.
2 As the vehicle is fitted with the EEC-IV engine management control module, reference must be made to Chapter 12, Section 5 upon subsequent battery reconnection.

26 Road speed sensor unit – removal and refitting

1 Disconnect the battery earth terminal.
2 Undo the speedometer drive cable securing nut and withdraw the cable.
3 Unclip and disconnect the sensor lead multi-plug.
4 Remove the securing nut and detach the sensor unit from the transmission housing.
5 Refitting is a reversal of the removal procedure.

27 Ignition coil – removal, testing and refitting

1 The procedure is described in Section 19 of this Chapter.

2 As the vehicle is fitted with the EEC-IV engine management control module, reference must be made to Chapter 12, Section 5 upon subsequent battery reconnection.

28 Distributor cap and rotor arm – removal and refitting

1 Disconnect the battery earth terminal.
2 Disconnect the distributor multi-plug for better access to the rear cap securing clip, then disconnect the central (coil) HT lead – pull on the connector, not the lead (photos).
3 Release the distributor cap securing clips by levering with a screwdriver, withdraw the cap assembly and separate the suppressor shield from the cap (photos).
4 Disconnect the HT leads from the distributor cap, as required, noting cap orientation and lead locations for subsequent refitting.
5 Detach the rotor arm (photo).
6 Refer to Section 12 before refitting.

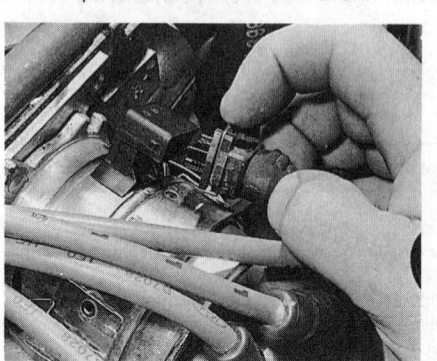

28.2A Disconnecting the distributor multi-plug

28.2B Disconnecting the coil HT lead from the distributor cap

28.3A Releasing one of the distributor cap securing clips by carefully levering with a screwdriver

Chapter 4 Ignition and engine management systems

28.3B Withdrawing the distributor cap and suppressor shield assembly

28.3C Separating the distributor cap and the suppressor shield

28.5 Withdrawing the rotor arm

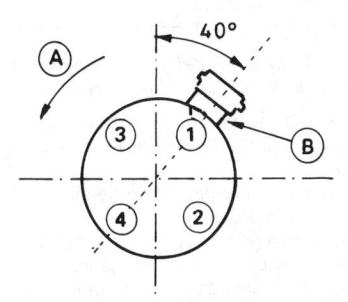

Fig. 4.20 Correct orientation of distributor when refitting (if one or both alignment punch marks missing) (Sec 29)

A   Direction of rotation
B   Centre line through distributor connector (40° to vertical)

7 Refitting is a reversal of the removal procedure, ensuring that the HT leads are securely connected to give the correct firing order (see Fig. 4.9).

29 Distributor – removal and refitting

1 The method of distributor removal and refitting where both original punch marks are present on the cylinder head and distributor body flange, is the same as that described in Section 15 of this Chapter, but ignore any reference to 'vacuum pipe' or 'vacuum diaphragm unit'. Refer to Section 28 when removing or refitting the distributor cap.
2 If, upon refitting, one or both of the punch mark is missing (due to component renewal), rotate the distributor body until the centre line through the distributor multi-plug connector is at 40° to the vertical, as shown in Fig. 4.20, before tightening the distributor clamp bolts to the specified torque. This will give an approximate (static) ignition timing setting to enable starting of the engine after the remaining components have been refitted and the relevant connections made.
3 Refer to Chapter 12, Section 5 when reconnecting the battery earth terminal.
4 Unless specialised STAR test equipment is available, the vehicle will need to be taken to a Ford dealer or other suitably equipped specialist for accurate ignition timing checking and, if necessary, adjustment. An explanation of why this is necessary may be found in the following Section.

30 Ignition timing – general

Note: *When an engine is timed in production it is set, using a microwave timing system, to an accuracy of within half a degree. Unless it is essential, do not remove the distributor or alter the ignition timing. If no distributor (static) timing position punch marks are present on the cylin-*
der head and distributor body flange, make your own before disturbing the setting (see Fig. 4.10)

1 The method of obtaining correct (static) ignition timing, with both original punch marks present, is described in Section 16, paragraph 1.
2 If (due to component renewal) one or both of the punch marks is missing, an approximate (static) ignition timing setting can be obtained to enable starting of the engine by following the instruction given in Section 29, paragraph 2.
3 Accurate ignition timing can only be carried out using specialised equipment – this is a task for your Ford dealer or other suitably equipped specialist. The reason for this is that the EEC-IV engine management module has to 'lock' its internal ignition advance compensations and its idle speed control whilst the timing is set; the 'locking' of the EEC-IV module is performed in one mode of the service STAR test (Self-Test Automatic Readout) equipment, which is also used to access fault codes stored in the module memory and analyse the performance of the system components. New punch marks should be made after accurate timing has been carried out, as necessary.
4 As the EEC-IV engine management module controls all automatic advance characteristics, no detailed advance checks are required. A test that the automatic advance function is operational can be made, if required, by connecting a timing light in accordance with the manufacturer's instructions and accelerating the engine – the crankshaft pulley notch should move correspondingly further away from the appropriate mark on the timing cover scale (highlight the notch for clarity, using a dab of quick-drying white paint).

31 Manifold absolute pressure (MAP) sensor – removal and refitting

1 Disconnect the battery earth terminal.
2 Disconnect the sensor multi-plug (do not pull on the wiring), then withdraw the vacuum hose from the base of the sensor (see Fig. 4.21).

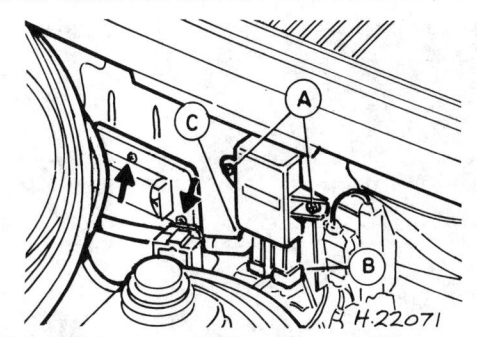

Fig. 4.21 Manifold absolute pressure (MAP) sensor location (Sec 31)

A   Securing screws
B   Multi-plug
C   Vacuum hose
Arrows indicate ignition module retaining screws

146 Chapter 4 Ignition and engine management systems

32.2 Disconnecting the TFI-IV ignition module multi-plug

engine compartment, then carefully draw the cover away from its location (photo).
3 Unscrew the module multi-plug retaining bolt and disconnect the multi-plug from the module (photo).
4 The aid of an assistant will be required at this stage, to support and withdraw the module from inside the passenger compartment as its mounting bracket retaining tags are compressed and released from the engine compartment. Do not allow the module to drop into the passenger compartment as irreparable damage is likely to result (photo).
5 The module may be separated from its mounting bracket by undoing the securing bolts.
6 Refitting is a reversal of the removal procedure, ensuring that the module mounting bracket retaining tags 'snap' into position (see Chapter 12, Section 5).

34 Fault diagnosis – ignition and engine management systems (1.4 litre CVH engine with CFi)

Note: *Refer to Section 12 before proceeding, and make additional reference to the information contained in Section 22*

1 If the engine fails to start, the procedure is the same as that covered in Section 20, but substitute 'ignition amplifier module' with 'TFI-IV ignition module'.
2 If the engine misfires regularly, proceed as described in Section 10, paragraphs 8 to 12 inclusive.
3 If the misfire is still present, check the distributor cap and rotor arm (refer to Section 12) and renew as necessary.
4 Further possible causes of misfiring may be an intermittent fault in the ignition coil HT lead, ignition coil or its LT supply. Additionally, consider the possibility that the ignition timing is incorrect, the distributor trigger mechanism is faulty and/or that the ignition module is defective. Unless components are freely available for testing by substitution (where applicable), consult your Ford dealer or other competent specialist for diagnosis.
5 If no fault has been found in the ignition system, then a fault in the engine management system must be suspected. Refer to Section 47, paragraphs 3, 4 and 5.

3 Undo the two screws securing the MAP sensor to the bulkhead panel and remove the sensor.
4 Refitting is a reversal of the removal procedure, but ensure the security of the hose and multi-plug connections.

32 Ignition module (TFI-IV) – removal and refitting

1 Disconnect the battery earth terminal.
2 Disconnect the module multi-plug. To do this, release the locking lug and withdraw the multi-plug (do not pull on the multi-plug wiring) (photo).
3 Undo the two screws securing the module to the bulkhead panel and remove the module.
4 Refitting is a reversal of the removal procedure, ensuring the security of the multi-plug.

33 Engine management module (EEC-IV) – removal and refitting

Note: *Different versions of the EEC-IV module are not interchangeable; if the wrong module is installed, engine running may be seriously affected and damage may occur. For this reason, always ensure that the part number is correct before installing a replacement*

1 Disconnect the battery earth terminal.
2 Unscrew and remove the two nuts securing the module cover in the

PART D: 1.6 LITRE CVH EFi ENGINE

35 General description

The ignition system fitted to this engine uses the EEC-IV engine management module as previously described for 1.4 litre CVH CFi-equipped models. It consists of the EDIS ignition module (with EEC-IV module control), DIS (distributorless ignition system) ignition coil, HT leads, spark plugs and engine speed/crankshaft position sensor (CPS). The low tension circuit is as described for the aforementioned 1.4 models except that the EDIS ignition module and CPS replace the TFI-IV module and distributor, and no ballast resistor is used. The HT circuit is

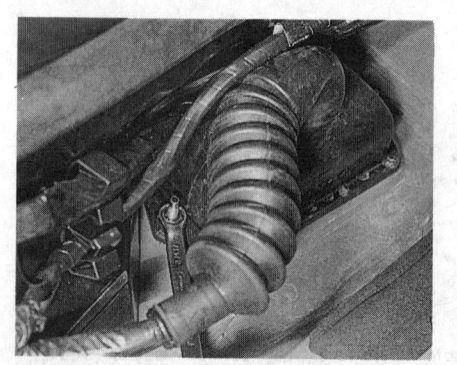

33.2 Undoing the EEC-IV engine management module cover retaining nuts (vehicle jack removed for clarity)

33.3 Undoing the EEC-IV engine management module multi-plug retaining bolt

33.4 Withdrawing the EEC-IV engine management module into the passenger footwell

Chapter 4 Ignition and engine management systems

38.2 Disconnecting the air charge temperature (ACT) sensor multi-plug (1.6 litre EFi model)

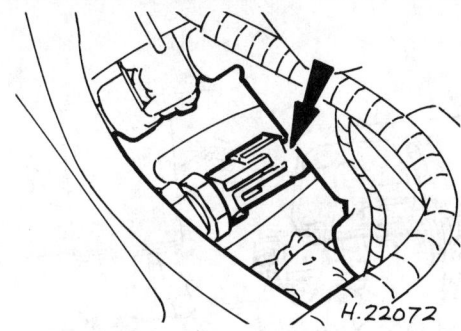

Fig. 4.22 Engine coolant temperature (ECT) sensor location on the inlet manifold (1.6 litre EFi models) (Sec 39)

somewhat simplified, consisting of coil secondary windings, spark plug HT leads and spark plugs.

The system functions in the following way: The engine speed/crankshaft position sensor (CPS) uses the flywheel to send a signal to the EDIS module which indicates the position of the piston in No 1 cylinder. Additional data is received from the various sensors (see Section 21) which, with the CPS data, allows the EEC-IV module to switch off the low tension circuit via the EDIS module at the optimum moment to promote the HT spark at the spark plugs.

The DIS ignition coil functions exactly as described for HCS engined models in Section 1. Sensor failure or failure of the EEC-IV module is accommodated in the manner described for 1.4 CVH models in Section 21.

39 Engine coolant temperature (ECT) sensor – removal and refitting

1 The procedure is covered in Section 5 of this Chapter, but the sensor location is as shown in Fig. 4.22.

40 Engine speed/crankshaft position sensor (CPS) – removal and refitting

1 The removal and refitting procedure is covered in Section 6 of this Chapter, but the CPS sensor on the 1.6 litre EFi model fits to a bushing instead of directly to the cylinder block (see Fig. 1.48 in Chapter 1). The bushing can be removed only with the engine and transmission separated.
2 Reference must be made to Chapter 12, Section 5 upon subsequent battery reconnection, to allow the EEC-IV engine management control module to re-learn its values.

36 Maintenance, inspection and precautions

1 Refer to Section 2, but also inspect the security and condition of the EEC-IV engine management module cover. The module cover is held in position by two nuts, and should be free from any splits, tears or other deterioration.
2 Note that after the battery has been disconnected and subsequently reconnected, the EEC-IV engine management module will need to 're-learn' its values (refer to Chapter 12, Section 5 for details).

41 Road speed sensor unit – removal and refitting

1 The procedure is detailed in Section 26 of this Chapter.

42 Ignition coil – removal and refitting

1 The procedure is basically as detailed in Section 7 of this Chapter, but the ignition coil is located at the rear of the cylinder head (close to the thermostat housing), and its mounting bracket is secured by four Torx head screws (Fig. 4.23).

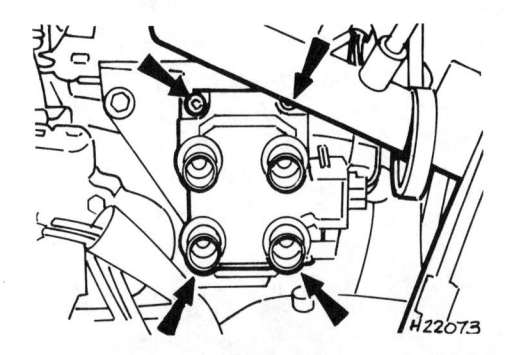

37 Spark plugs and HT leads – inspection and renewal

1 Refer to Section 3 of this Chapter, but note that additional HT lead guides are moulded into the air filter trunking.

38 Air charge temperature (ACT) sensor – removal and refitting

1 Disconnect the battery earth terminal.
2 The ACT sensor is located on the left-hand side of the inlet manifold. Disconnect its multi-plug – do not pull on its wiring (photo).
3 Unscrew and remove the sensor.
4 Prior to refitting, apply a smear of suitable sealant to the threads of the sensor.
5 To refit, screw in the sensor and tighten to its specified torque. Reconnect the sensor multi-plug.

Fig. 4.23 DIS ignition coil retaining screws (arrowed) (1.6 litre EFi models) (Sec 42)

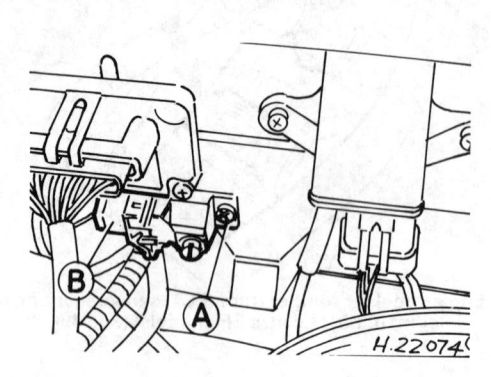

Fig. 4.24 Fuel mixture/CO% adjustment potentiometer (Sec 43)

A   Securing screw   B   Multi-plug

43 Fuel mixture/CO% adjustment potentiometer – removal and refitting

1 Disconnect the battery earth terminal.
2 Disconnect the multi-plug from the potentiometer by first depressing the multi-plug spring clip. Do not pull on the multi-plug wiring (see Fig. 4.24).
3 Remove the single retaining screw and withdraw the potentiometer.
4 Refit by reversing the removal procedure.

44 Manifold absolute pressure (MAP) sensor – removal and refitting

1 The procedure is described in Section 31 of this Chapter.

45 Ignition module (EDIS) – removal and refitting

Note: *On earlier vehicles, some difficulties may be experienced with the EDIS module multi-plug working loose, leading to water ingress (refer to the Note in Section 8)*

1 Disconnect the battery earth terminal.
2 Disconnect the module multi-plug – do not pull on the wiring.
3 Undo both screws securing the module to the bulkhead panel and remove the module (photo).
4 Refitting is a reversal of the removal procedure, ensuring the security of the module multi-plug.

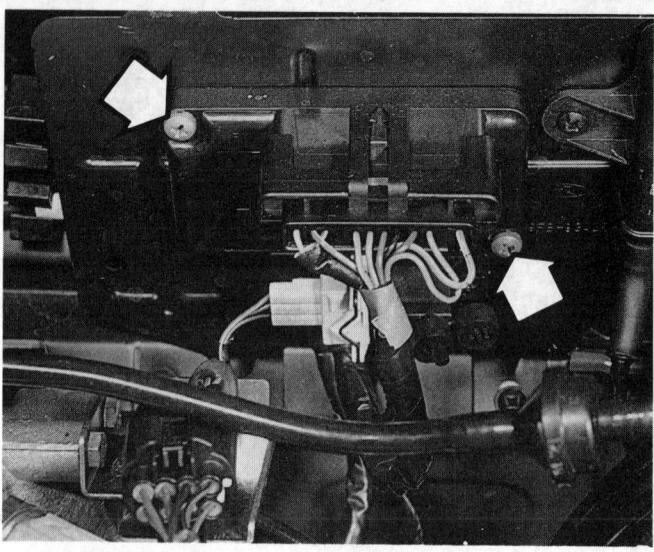

45.2 EDIS ignition module location on bulkhead panel. Note retaining screws (arrowed)

46 Engine management module (EEC-IV) – removal and refitting

1 The procedure is covered in Section 33 of this Chapter.

47 Fault diagnosis – ignition engine management systems (1.6 litre CVH engine with EFi)

Note: *Refer to Section 2 before proceeding, and make additional reference to the information contained in Section 36.*

1 The ignition related fault-finding procedure is described in Section 20.
2 If no fault has been found in the ignition system, then a fault in the engine management system must be suspected.
3 Do not immediately assume that a fault is caused by a defective engine management module. First check that all relevant wiring is in good condition and that all the wiring multi-plugs are securely connected. Similarly check all vacuum hose connections, and thoroughly examine the hose(s) for splits or other damage/deterioration that would allow leakage/blockage. Ensure that the EEC-IV module has fully 're-learnt' its values if the battery has been disconnected (refer to Chapter 12, Section 5).
4 Unless components are freely available for testing by substitution, further investigation should be left to a Ford dealer or other component specialist.
5 Note that engine management system components cannot necessarily be substituted from another vehicle. The modules are dedicated to particular engine, transmission and territory combinations and, if testing by substitution or renewing, **must** bear an identical part number to the original.

Chapter 5 Clutch

For modifications, and information applicable to later models, see Supplement at end of manual

Contents

Clutch – inspection	3
Clutch – refitting	6
Clutch – removal	2
Clutch cable – renewal	7
Clutch pedal – removal, bush renewal and refitting	8
Clutch release bearing – renewal	4
Clutch release shaft – removal, bush renewal and refitting	5
Fault diagnosis	9
General description	1

Specifications

General
Type	Single dry plate with diaphragm spring
Actuation	Self-adjusting cable
Pedal stroke	165 mm ± 5 mm
Grease for use on input shaft splines and release bearing guide sleeve	To Ford specification ESD-MIC220-A (refer to Ford dealer)

Driven plate
Diameter:
1.0 HCS, 1.1 HCS and 1.4 CVH	190 mm
1.6 CVH	220 mm
Lining thickness	3.23 mm

Torque wrench settings
	Nm	lbf ft
Clutch release lever to clutch release shaft	21 to 28	16 to 21
Cover assembly to flywheel	25 to 34	18 to 25

1 General description

The clutch is of single dry plate diaphragm spring type, actuated mechanically, via a self-adjusting cable mechanism, through a pendant-mounted pedal.

The clutch components comprise a steel cover assembly, clutch driven plate, release bearing and release mechanism. The cover assembly which is bolted and dowelled to the rear face of the flywheel contains the pressure plate and diaphragm spring.

The driven plate is free to slide along the transmission input shaft splines and is held in position between the flywheel and pressure plate by the pressure of the diaphragm spring.

Friction material is riveted to the driven plate which has a spring cushioned hub to absorb transmission shocks and to help ensure a smooth take-up of the drive.

Depressing the clutch pedal moves the release lever on the transmission by means of the cable. This movement is transmitted to the release bearing which moves inwards against the fingers of the diaphragm spring. The spring is sandwiched between two annular rings which act as fulcrum points. As the release bearing pushes the spring fingers in, the outer circumference pivots out, so moving the pressure plate away from the flywheel and releasing its grip on the driven plate.

When the pedal is released, the diaphragm spring forces the pressure plate into contact with the friction linings of the driven plate. The plate is now firmly sandwiched between the pressure plate and flywheel, thus transmitting engine power to the transmission.

The self-adjusting mechanism is incorporated in the clutch pedal and consists of a pawl, toothed quadrant and tension spring. When the pedal is released the tension spring pulls the quadrant through the teeth of the pawl until all free play of the clutch cable is taken up. When the pedal is depressed the pawl teeth engage with the quadrant teeth thus locking the quadrant. The particular tooth engagement position will gradually change as the components move to compensate for wear in the clutch driven plate and stretch in the cable.

The only differences between right and left-hand drive variants are the location of the clutch guide in the bulkhead, and the length of the clutch cable.

2 Clutch – removal

1 To obtain access to the clutch for inspection and/or renewal, it is necessary to separate the engine and transmission assemblies. Engine

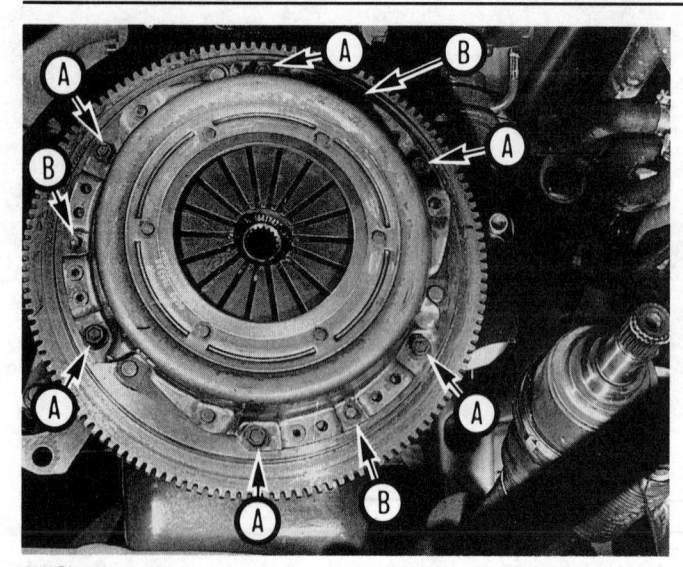

2.2 Clutch cover retaining bolts (A) and cover locating dowels (B)

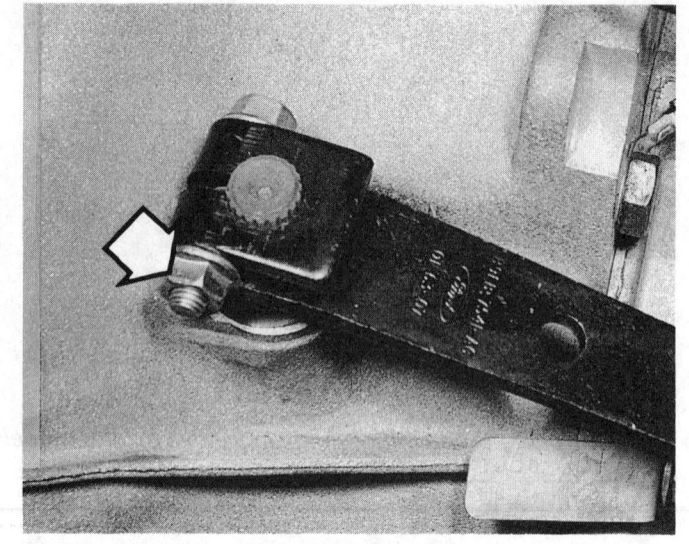

4.2 Clutch release lever-to-clutch release shaft clamp bolt (arrowed)

and transmission removal procedures can be found in Chapters 1 and 6 respectively.
2   Remove the clutch assembly from the flywheel by slackening the bolts securing the cover assembly to the flywheel, half a turn at a time, in a diametrically opposed sequence around its circumference (photo).
3   Once all the bolts are slack, remove them and ease the cover assembly off its locating dowels. The driven plate will drop out as the cover assembly is removed.

3 Clutch – inspection

1   With the cover assembly removed, clean off all traces of asbestos dust using a dry cloth. This should ideally be done outside, or in a well-vented area, as **asbestos dust is harmful and must not be inhaled**.
2   Examine the linings of the driven plate for wear and loose rivets, and the driven plate hub for distortion, cracks, broken torsion springs and worn splines. The surface of the friction linings may be highly glazed, but, as long as the friction material pattern can be clearly seen, this is satisfactory. If there is any sign of oil contamination, indicated by a continuous, or patchy, shiny black discoloration, the plate must be renewed and the source of the contamination traced and rectified. The plate must also be renewed if the lining thickness has worn down to, or just above, the level of the rivet heads. It is often prudent to renew a partly-worn driven plate while the clutch is accessible when other tasks are being performed.
3   The condition of the release bearing is an important factor, and unless it is nearly new, it is a mistake not to renew it during a clutch overhaul. Check the bearing for smoothness of operation. There should be no roughness and no slackness in it. It should spin reasonably easily, bearing in mind it has been pre-packed with grease. The release bearing renewal procedure is described in Section 4.
4   Check that there is no free play in the clutch release shaft bearing bush. If it is worn it will need renewal (see Section 5).
5   Examine the machined faces of the flywheel and pressure plate. If either is grooved or heavily scored, renewal is necessary. If any cracks or splits are apparent on the pressure plate, its springs damaged or its pressure suspect, it must also be renewed.

4 Clutch release bearing – renewal

1   With the engine and transmission separated to provide access to the clutch, attention can be given to the release bearing, located in the bellhousing.

4.3 Removing the release bearing from the release shaft fork

2   To remove the release bearing, first undo the clamp bolt securing the clutch release lever to the clutch release shaft (photo) then, having marked their relative positions, remove the clutch release lever.
3   Draw the release bearing off its guide sleeve, and unhook it from the locating pin on the upper release shaft fork (photo).
4   Refitting is the reverse sequence to removal, ensuring that the clutch release lever-to-clutch release shaft clamp bolt is tightened to the specified torque.

5 Clutch release shaft – removal, bush renewal and refitting

1   With the engine and transmission separated, the clutch release shaft is easily accessible.
2   Remove the clutch release bearing as described in the previous Section.
3   Remove the protective cap from around the top of the release shaft splines to locate the bush.
4   The bush is easily removed by applying gentle leverage, ideally achieved with two screwdrivers, then lift it out and over the splines (photo).
5   With the bush removed the shaft can be withdrawn, if so desired, by

Chapter 5 Clutch

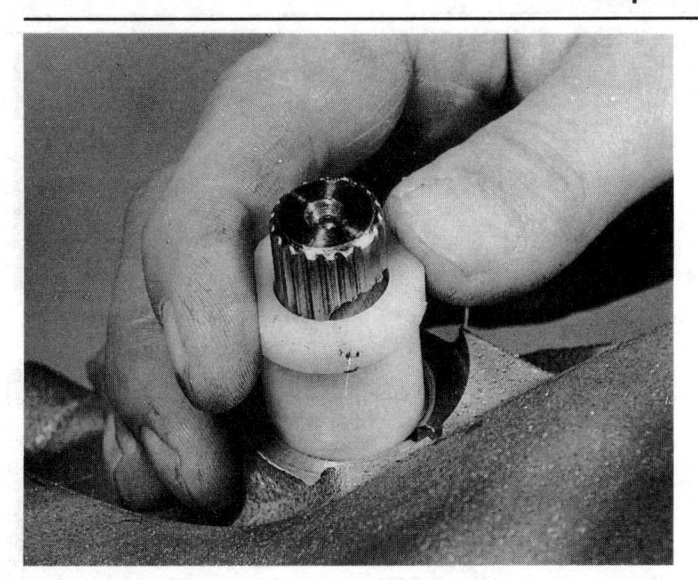

5.4 Removing the nylon bush from the release shaft

5.5 Withdrawing the clutch release shaft

lifting it from its lower bearing bore, tilting it sideways and withdrawing it downwards (photo).
6 Refitting is the reverse sequence to removal, with the plastic bearing bush being slid over the release shaft splines for a flush fit in its upper housing bore.

6 Clutch – refitting

1 Before starting, it is wise to wipe the flywheel and pressure plate faces with a clean dry cloth, to remove all traces of oil or grease, as it is extremely important that the friction linings remain free of contamination. Clean hands help here too.
2 Begin by positioning the driven plate with the flat side, usually marked 'FLYWHEEL SIDE' or 'SCHWUNGRADSEITE' (photo), towards the flywheel. *If the driven plate is fitted the wrong way round, the clutch will not work.*
3 Hold the driven plate against the flywheel, and refit the cover assembly loosely onto the dowels (photo). Refit the bolts securing the cover assembly to the flywheel and tighten them finger-tight so that the driven plate is held, but can still be moved.
4 The clutch must now be centralised so that when the engine and transmission are mated, the transmission input shaft splines will pass through the splines in the centre of the hub.
5 Centralisation can be carried out quite easily by inserting a round bar or long screwdriver through the hole in the centre of the driven plate, so that the end of the bear rests in the hole in the centre of the crankshaft. Moving the bar sideways or up and down will move the plate in whichever direction is necessary to achieve centralisation. With the bar removed, view the driven plate hub in relation to the hole in the end of the crankshaft and the circle created by the ends of the diaphragm spring fingers. When the hub appears exactly in the centre, all is correct. Alternatively, if a clutch aligning tool can be obtained this will eliminate all the guesswork and the need for visual alignment (photo).
6 Tighten the cover retaining bolts gradually, half a turn at a time, in a diametrically opposed sequence around its circumference, to the specified torque wrench setting.
7 Ensure that the input shaft splines, driven plate splines and release bearing guide sleeve are clean. A **thin** film of *special grease* (refer to the Specifications for details) is then applied to the entire surface of the transmission input shaft splines, and to the release bearing **guide sleeve. Only this special grease should be used**. Other lubricants can cause problems after a short time in service. Do not put too much grease onto the transmission shaft splines, as over-application will inevitably cause some grease to find its way onto the friction surfaces when the vehicle is used.
8 The engine and/or transmission can now be refitted, in accordance with the refitting procedures given in Chapters 1 and 6 respectively.

7 Clutch cable – renewal

1 Disconnect the battery earth terminal.
2 Raise the clutch pedal and support it from underneath. A block of wood is ideal.
3 Using pliers, grip the cable on its exposed core near the clutch release lever on the transmission, and unhook it from the claw on the end of the release lever (photo). Take care not to damage the cable if it is to be re-used.
4 Unhook the quadrant tension spring from its pedal location.

6.2 Driven plate marking

6.3 Locating driven plate and cover assembly to flywheel

6.5 Tightening the cover retaining bolts with the alignment tool in position

Chapter 5 Clutch

7.3 Unhooking the clutch cable from its release lever location

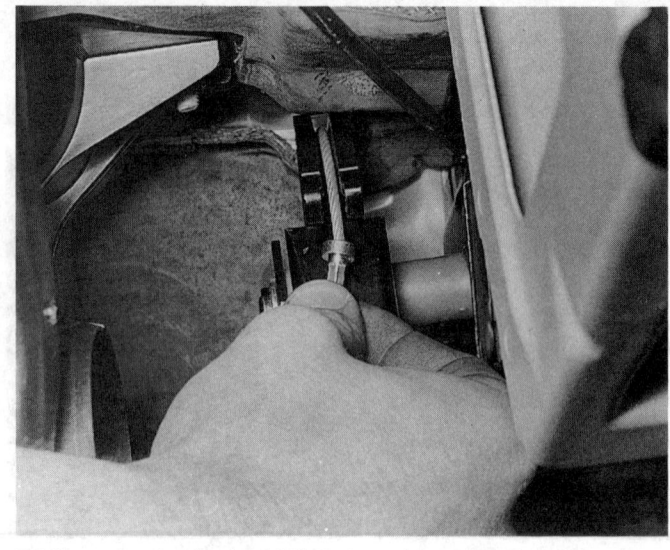

7.5 Removing the clutch cable from its pedal quadrant location

5 Release the cable from the clutch pedal quadrant so that it may be withdrawn through into the engine compartment (photo).
6 In the engine compartment, unhook the cable from its clip location at the base of the brake servo support bracket and withdraw it from the vehicle.
7 Installation is the reverse sequence to removal. With the cable fitted, operate the clutch pedal to ensure correct adjustment.

8 Clutch pedal – removal, bush renewal and refitting

1 Disconnect the battery earth terminal.
2 Remove the clutch cable from its pedal location and pull it through into the engine compartment, as described in the previous Section.
3 Remove the clip securing the brake pedal pushrod to the brake pedal, noting the bush fitted in the pedal, and detach the brake pedal stop-lamp switch loom connection. Unscrew the brake pedal stop-lamp switch from its pedal box location (photo).
4 Remove the accelerator pedal by first disconnecting its cable from the cut-out in the top of the pedal moulding. Detach the clip securing the pedal onto its shaft (photo), then slide the pedal off.
5 Detach the retaining C-clip from the left-hand side of the main pedal shaft (photo).
6 Remove the two nuts securing the clutch end-bracket to the main pedal box.
7 Undo and remove the two nuts in the engine compartment that secure the left-hand side of the pedal box to the bulkhead (photo).
8 Undo and remove the two nuts on the right-hand side of the pedal box that secure the pedal box, and the right-hand servo operating link support bracket, to the bulkhead.
9 The pedal box can be pulled out to free the clutch end-bracket after slackening or removing the two bolts that secure the pedal box vertically to the bulkhead, as necessary.
10 Slide clutch end-bracket from its location, followed by the pedal and pedal spacer.
11 The pedal can now be dismantled as necessary, by prising out the bushes on either side, in order to renew the bushes, tension spring or adjustment mechanism (Fig. 5.1).
12 If required, the main pedal shaft may be withdrawn at this stage by

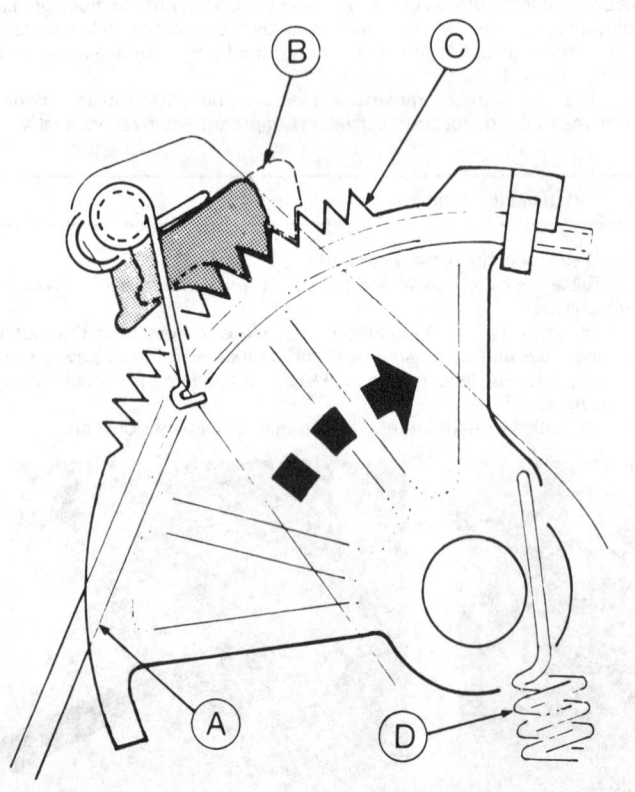

Fig. 5.1 Clutch cable self-adjustment mechanism (Sec 8)

A Clutch cable
B Pawl
C Toothed quadrant
D Tension spring

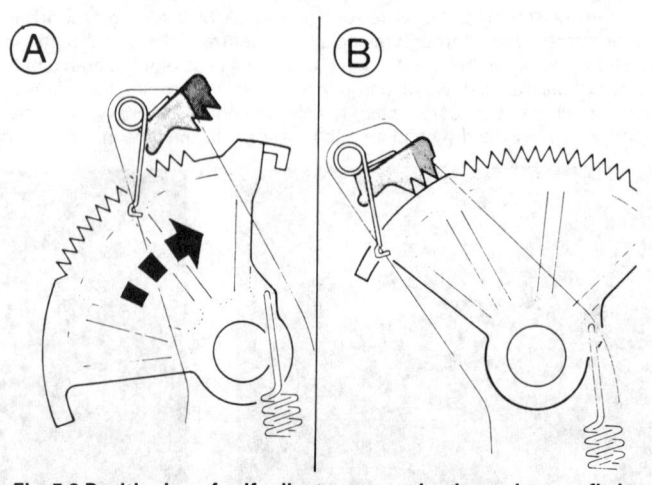

Fig. 5.2 Positioning of self-adjustment mechanism prior to refitting pedal (Sec 8)

A Lift pawl and turn quadrant
B Pawl bearing on smooth section of the quadrant

Chapter 5 Clutch

8.3 Brake pedal pushrod retaining clip (A) and brake pedal stop-lamp switch location (B)

8.4 Accelerator pedal retaining clip (arrowed)

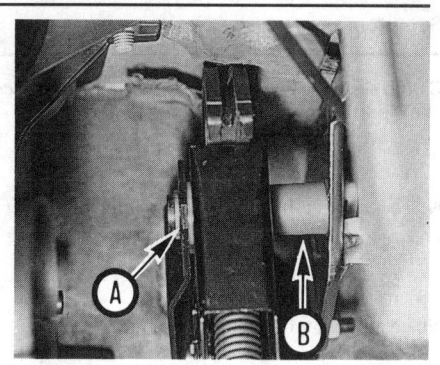

8.5 'C' clip (A) and clutch pedal spacer (B)

8.7 Pedal box left-hand side retaining nuts in engine compartment (arrowed)

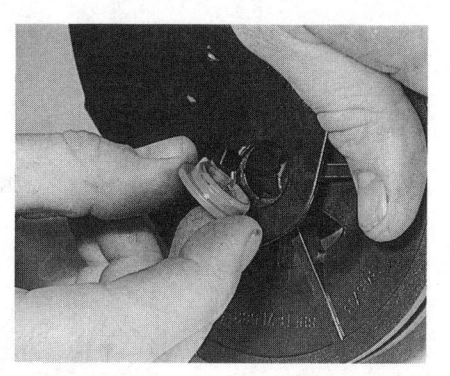

8.15 Clutch pedal bushes going to pedal

removing the C-clip retaining it, and sliding the shaft towards the centre of the vehicle. The C-clip is located next to the brake pedal. As the shaft is slid from its location, the brake pedal and its spacer must be removed. As clearance is tight, the pedal box will have to be manoeuvred to fully withdraw the shaft.

13   To refit the main pedal shaft, reverse the removal procedure, locating the 'D' section of the shaft into the pedal box right-hand support. Ensure that the brake pedal and spacer are correctly located.

14   Lubricate the pedal shaft using molybdenum disulphide grease.

15   Refit the pawl and its associated spring to the clutch pedal, then insert the quadrant, tension spring and bushes (photo), so that the pawl bears on the smooth section of the quadrant (Fig. 5.2). Do not hook the tension spring onto its lower pedal location at this stage.

16   Refit the pedal spacer, the clutch pedal assembly and the clutch end-bracket to the pedal shaft, then manipulate the pedal box assembly back into its correct location before refitting the two nuts to secure the end-bracket to the main pedal box.

17   Refitting is now a reversal of the removal procedure, fully tightening all the nuts once the individual components have been correctly assembled.

18   Upon completion of the operation, operate the clutch pedal to ensure correct cable adjustment.

19   Adjust the brake pedal stop-lamp switch as necessary, but not so that the stop-lamps flicker.

9   Fault diagnosis – clutch

Symptom	Reason(s)
Judder when taking up drive	Loose or worn engine/transmission mountings Weak or broken driven plate torsion spring(s) Worn or oil contaminated friction linings Worn splines on input shaft or driven plate Distorted driven plate or input shaft Clutch cable sticking or otherwise defective
Clutch spin (failure to disengage)	Clutch cable sticking or defective Driven plate seized onto input shaft splines (may happen after long periods standing idle) Distorted or misaligned pressure plate assembly Release bearing fork retaining bolt sheared Excessive free play in cable – check that the self-adjusting mechanism is operational

Symptom	Reason(s)
Clutch slip (engine speed rise does not result in comparable road speed rise – particularly noticeable on gradients)	Driven plate friction linings worn or contaminated with oil Self-adjusting mechanism inoperative Clutch cable sticking or defective Faulty pressure plate, or weak or broken diaphragm spring(s)
Noise evident when depressing clutch pedal	Worn or damaged release bearing Faulty diaphragm spring(s) or pressure plate Worn splines on input shaft or driven plate Worn pedal bushes
Noise evident when releasing clutch pedal	Broken driven plate torsion spring(s) Faulty diaphragm spring(s) or pressure plate Worn pedal bushes Internal wear in gearbox

Chapter 6 Transmission

For modifications, and information applicable to later models, see Supplement at end of manual

Contents

Part A: Four and five-speed manual transmission
Differential unit – overhaul	11
Driveshaft/transmission oil seal – renewal	18
Fault diagnosis	20
Gearchange mechanism – adjustment	3
Gearchange mechanism (five-speed) – removal, overhaul and refitting	14
Gearchange mechanism (four-speed) – removal, overhaul and refitting	4
General description	1
Input shaft (four-speed) – overhaul	10
Mainshaft (four-speed) – overhaul	9
Maintenance and inspection	2
Speedometer driven gear – removal and refitting	13
Transmission (five-speed) – dismantling into major assemblies	15
Transmission (five-speed) – overhaul	16
Transmission (five-speed) – reassembly	17
Transmission (four and five-speed) – removal and refitting	5
Transmission (four-speed) – dismantling into major assemblies	6
Transmission (four-speed) – further dismantling (general)	7
Transmission (four-speed) – reassembly	12
Transmission housing and selector mechanism (four-speed) – overhaul	8
Transmission mountings – removal and refitting	19

Part B: CTX automatic transmission
Automatic transmission – removal and refitting	26
Driveshaft/transmission oil seal – renewal	28
Fault diagnosis	30
Gearchange selector mechanism – cable removal, refitting and adjustment	23
Gearchange selector mechanism – removal, overhaul and refitting	24
General description	21
Maintenance and inspection	22
Starter inhibitor switch – removal and refitting	25
Torsional vibration damper – removal and refitting	27
Transmission mountings – removal and refitting	29

Specifications

Part A: Four and five-speed manual transmission

Transmission type.................... Four or five forward speeds and reverse. Synchromesh on all forward gears

Ratios
Four-speed transmission gear ratios:
   1.0 litre and 1.1 litre HCS engines:
      1st .... 3.58 : 1
      2nd .... 2.04 : 1
      3rd .... 1.32 : 1
      4th .... 0.95 : 1
      Reverse .... 3.77 : 1
Five-speed transmission gear ratios:
   1.1 litre HCS and 1.4 litre CVH CFi engines:
      1st .... 3.58 : 1
      2nd .... 2.04 : 1
      3rd .... 1.32 : 1
      4th .... 0.95 : 1
      5th .... 0.76 : 1
      Reverse .... 3.62 : 1
   1.4 litre and 1.6 litre CVH engines:
      1st .... 3.15 : 1
      2nd .... 1.91 : 1
      3rd .... 1.28 : 1
      4th .... 0.95 : 1
      5th .... 0.76 : 1
      Reverse .... 3.62 : 1
Final drive ratios:
   1.0 litre and 1.1 HCS engines .... 4.06 : 1
   1.4 litre CVH CFi engine .... 3.84 : 1

Ratios (continued)
1.4 litre CVH engine	4.06 : 1
1.6 litre CVH engine (not EFi)	3.82 : 1
1.6 litre CVH engine (EFi)	4.06 : 1

Overhaul data
Snap-ring (circlip) thicknesses available:
- Mainshaft:
  - Gear synchronisers (not 5th) and ball-bearing inner race: 1.5 mm (0.059 in), 1.52 mm (0.060 in), 1.58 mm (0.062 in), 1.61 mm (0.063 in), 1.64 mm (0.065 in)
  - Fifth gear synchroniser: 1.48 mm (0.058 in), 1.53 mm (0.060 in), 1.58 mm (0.062 in)
  - Ball-bearing outer race: 1.89 mm (0.0744 in), 1.97 mm (0.0776 in), 2.04 mm (0.0804 in)
- Input shaft:
  - Large ball-bearing outer race: 1.89 mm (0.0744 in), 1.97 mm (0.0776 in), 2.04 mm (0.0804 in)
  - Fifth gear driving cog: 1.65 mm (0.065 in), 1.70 mm (0.066 in), 1.75 mm (0.068 in)

Lubrication
Lubricant type/specification	High pressure gear oil, viscosity SAE 80 to Ford specification SQM2C-9008-A (Duckhams Hypoid 90)
Oil fill capacity	3.1 litres (5.4 pints)

Assembly greases (refer to Ford dealer):
- Four-speed gearbox:
  - Guide sleeve and input shaft splines: Grease to Ford specification ESD-MIC220-A
  - Selector shaft locking assembly sealer: Anaerobic retaining and sealing compound to Ford specification SM4G-4645-AA or AB
  - All other assembly greasing: Grease to Ford specification SM1C-1020-B
- Five-speed gearbox:
  - Synchroniser cones and mainshaft assemblies: Colloidal molybdenum disulphide in oil to Ford specification SM1C 4504-A
  - Gears, contact and thrust faces: Molybdenum disulphide paste to Ford specification SM1C-4505-A
  - Fifth gear on input shaft: Anti-friction/corrosion paste to Ford specification SAM-1C9107-A
  - Selector shaft locking assembly sealer: Anaerobic retaining and sealing compound to Ford specification SM4G-4645-AA or AB

Torque wrench settings
	Nm	lbf ft
Clutch release lever to release shaft	21 to 28	16 to 21
Flange bolts – transmission to engine	35 to 45	26 to 33
Lower engine adaptor plate (clutch housing cover)	34 to 46	25 to 34
Front suspension crossmember (XR2i only)	80 to 90	59 to 66
Starter motor bolts	35 to 45	26 to 33
Gearchange mechanism stabiliser bar to transmission	50 to 60	37 to 44
Gearchange mechanism shift rod clamp bolt	14 to 17	10 to 12
Final drive gear to differential housing	98 to 128	72 to 94
Small housing section to large housing section	21 to 27	15 to 20
Cover to housing	12 to 14	9 to 10
Cap-nut-selector shaft detent mechanism	20 to 35	15 to 26
Gearchange mechanism to floor	6 to 8	4 to 6
Oil filler/level plug	23 to 30	17 to 22
Reversing light switch	16 to 20	12 to 15
Gearchange stabiliser bar to gearchange housing	5 to 7	4 to 5
Gate to housing	18 to 23	13 to 17
Selector dog to selector shaft	12 to 15	9 to 11
Rear transmission mounting bracket threaded studs to transmission	21 to 27	15 to 20
Rear transmission mounting bracket retaining nuts	41 to 58	30 to 43
Transmission bearer to body	52	38
Transmission bearer mounting nuts	80 to 100	59 to 74

Part B: CTX automatic transmission

Transmission type
Automatic, continuously variable over the entire speed range

Axle ratio
3.842 : 1

Lubrication
Fluid type/specification	ATF to Ford specification ESP-M2C 166-H (Duckhams Uni-Matic)

Fluid fill capacity:
- New transmission:
  - With oil cooler: 3.6 litres
  - Without oil cooler: 3.5 litres
- Overhauled transmission:
  - With oil cooler: 4.2 litres
  - Without oil cooler: 4.1 litres

Transmission fluid (oil) cooling: Twin pipe oil cooler in radiator side tank

Chapter 6 Transmission

Lubrication (continued)
Grease (for assembly only) .... Industrial petroleum jelly to Ford specification SM-1C 115-A (Duckhams LB 10)
Long life grease for selector linkage and CV joints .... To Ford specification SQM-1C 9004 A (Duckhams LBM 10)
Sealer for housing mating face .... To Ford specification SQM 4G 9523-A

Torque wrench settings

	Nm	lbf ft
Transmission fluid cooling pipes to oil cooler	18 to 22	13 to 16
Transmission fluid cooling pipe connectors to the transmission housing	24 to 31	18 to 23
Transmission fluid cooling pipes to their connectors in the transmission	22 to 26	16 to 19
Starter inhibitor switch	10 to 14	7 to 10
Flange bolts – transmission to engine	27 to 50	20 to 37
Torsional vibration damper to flywheel	24 to 33	18 to 24
Lower engine adaptor plate (clutch housing cover)	7.5 to 10	5 to 7
Selector lever housing to floor	8.5 to 11.5	6 to 8
Selector cable bracket to transmission housing	34 to 46	25 to 34
Selector lever rod to selector lever guide	20 to 25	15 to 18
Transmission bearer to body	52 to 64	38 to 47

PART A: FOUR AND FIVE-SPEED MANUAL TRANSMISSION

1 General description

The four and five-speed gearboxes are of the same design as those fitted to the previous Fiesta models, but minor detail changes have been made.

A rubber air deflector is fitted to the transmission bearer on certain models to assist with cooling.

Four-speed gearbox

The gearbox and differential are housed in a two section light alloy casting which is bolted to a transversely mounted engine.

Drive from the engine/transmission is transmitted to the front roadwheels through open driveshafts.

The engine torque is then transmitted to the gearbox input shaft. Once a gear is selected, power is then transmitted to the main (output) shaft. The helically cut forward speed gears on the output shaft are in constant mesh with the corresponding gears on the input shaft.

Synchromesh units are used for 1st/2nd and 3rd/4th gear selection and operate as follows.

When the clutch pedal is depressed and the gearchange lever is moved to select a higher gear, the synchro baulk ring is pressed onto the gear cone. The friction generated causes the faster rotating gear on the input shaft to slow until its speed matches that of the gear on the output shaft. The gears can then be smoothly engaged.

When changing to a lower gear, the principle of operation is similar except that the speed of the slower rotating gear is increased by the action of the baulk ring on the cone.

Reverse gear is of the straight-cut tooth type and is part of the 1st/2nd synchro unit. A sliding type reverse idler gear is used.

The torque from the gearbox output shaft is transmitted to the crownwheel which is bolted to the differential cage and thence through the differential gears to the driveshafts.

Any need for adjustment to the differential and its bearings has been obviated by the inclusion of two diaphragm springs which are located in the smaller half of the transmission housing. Any tolerances which may exist are taken up by the sliding fit of the outer bearing ring in the smaller section of the housing.

Gear selection is obtained by rotary and axial movements of the main selector shaft (transmitted through a selector dog bolted to the

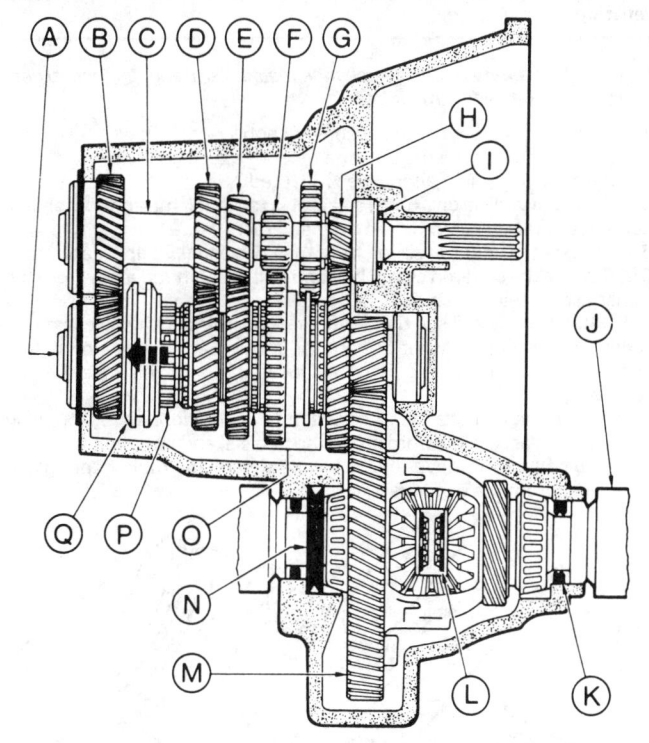

Fig. 6.1 Sectional view of the four-speed transmission (Sec 1)

A Mainshaft
B 4th gear
C Input shaft
D 3rd gear
E 2nd gear
F Reverse gear
G Reverse idler gear
H 1st gear
I Input shaft oil seal
J Driveshaft inner CV joint
K Oil seal
L Driveshaft snap-ring engaged in differential
M Final drive gear (crownwheel)
N Diaphragm springs
O 1st/2nd synchro with reverse gear
P 3rd/4th synchro
Q 3rd/4th synchro ring (4th gear engaged)

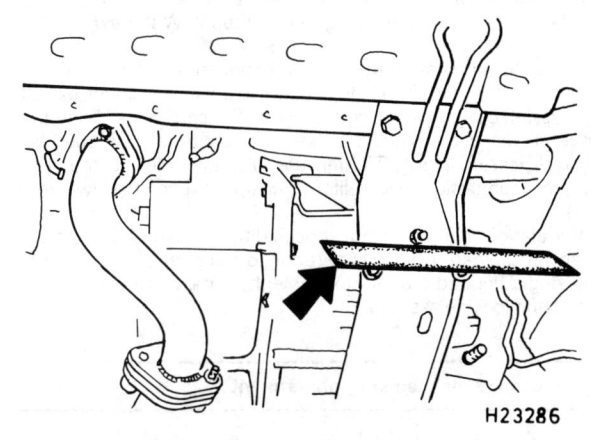

Fig. 6.2 Air deflector on transmission bearer (Sec 1)

Chapter 6 Transmission

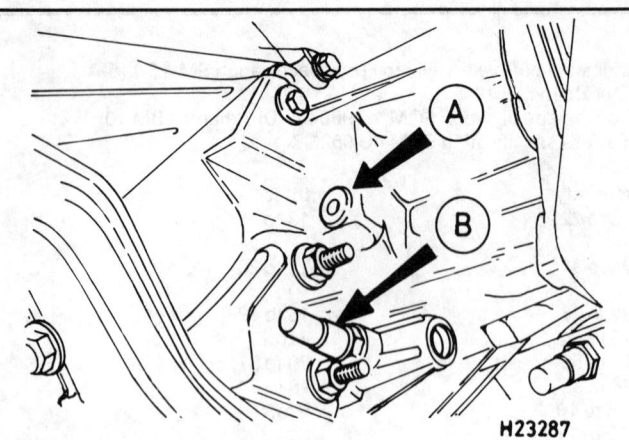

Fig. 6.3 Transmission oil level/filler plug (A), and selector shaft cap nut (B) (Sec 2)

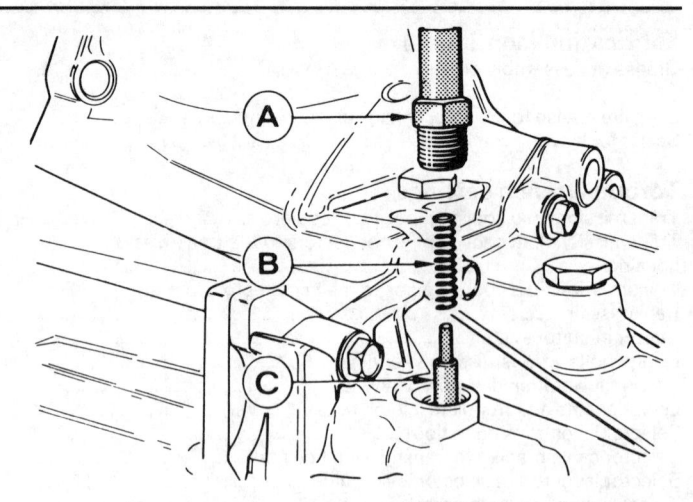

Fig. 6.4 Selector shaft cap nut (A), spring (B) and interlock pin (C) (Sec 2)

selector shaft) and two guide levers to the guide shaft which also carries a selector dog.

Rotary movement of the main selector shaft engages a cam on the guide shaft selector dog either in the cut-out of the 1st/2nd or 3rd/4th gear selector fork or in the aperture in the reverse gear guide lever.

Axial movement of the selector shaft moves the appropriate selector fork on the guide shaft or reverse idler gear through the medium of the guide lever, so engaging the gear.

The selected gear is locked in engagement by a shift locking plate which is carried on the guide shaft selector dog and a spring-loaded interlock pin located in the smaller housing section.

Five-speed gearbox

The transmission is basically the same as the four-speed version with the exception of a modified selector mechanism, and an additional gear and synchro-hub contained in a housing attached to the side of the main transmission casing.

2 Maintenance and inspection

1 The only maintenance required is to check and top-up, if necessary, the oil level in the transmission at the intervals specified in Routine Maintenance at the beginning of this manual.

2 The following procedure should be adopted when checking the oil level. Remove the one-piece undertray, if fitted (XR2i models).

3 Ensure that the car is standing on level ground and has been stationary for some time.

4 Unscrew the combined filler/level plug from the front face of the transmission. The plug is of socket-headed type and a suitable key will be required for removal.

5 With the plug removed, check the coil level. To do this accurately, make up an oil level check dipstick from a short length of welding rod or similar material. Make a 90° bend in the rod, then mark the downward leg in 5 mm increments. The dipstick is then inserted through the filler plug orifice so that the unmarked leg rests flat on the plug orifice threads, with the marked leg dipped in the oil. Withdraw the dipstick and read off the level of oil.

6 The oil level must be maintained between 0 and 5 mm below the lower edge of the filler/level plug hole. Top up (if necessary), using fresh transmission oil of the specified type (see Specifications). Take care not to overfill the unit as this can lead to excessive heat build-up and impaired gear changing. Tighten the filler/level plug to the specified torque on completion, and refit the one-piece undertray (where applicable).

7 Regular oil changing is not specified by the manufacturers, but the oil can be drained if necessary (prior to removal of the unit or after traversing a flooded road for example) by removing the selector shaft cap nut and locking assembly.

3 Gearchange mechanism – adjustment

1 This is not a routine operation and will normally only be required after dismantling, to compensate for wear or to overcome any 'notchiness' evident during gear selection.

2 To adjust the mechanism, refer to Section 4 (four-speed) or Section 14 (five-speed).

4 Gearchange mechanism (four-speed) – removal, overhaul and refitting

Note: *Before removing the gearchange mechanism, engage second gear to ensure correct adjustment later*

1 Unscrew and remove the gearlever knob.
2 Pull up and remove the outer gaiter and retaining frame.
3 Remove the inner gaiter (unless secured).
4 Raise the vehicle on a hoist, or jack it up so that it may be completely supported on axle stands at the front and rear.
5 Unhook the shift rod spring from the floor panel, as applicable.
6 Release the shift rod clamp bolt and pull the shift rod away from the transmission selector shaft.
7 Detach the stabiliser bar from the transmission by releasing its securing bolt (note washer fitment for subsequent reassembly). If an exhaust forward mounting bracket is located over it, this must first be removed.
8 Detach the gearchange mechanism from the floor, by unscrewing the four nuts accessible from inside the vehicle, and remove.
9 To dismantle, remove the circlip and detach the rubber spring and spring cup from the gearlever.

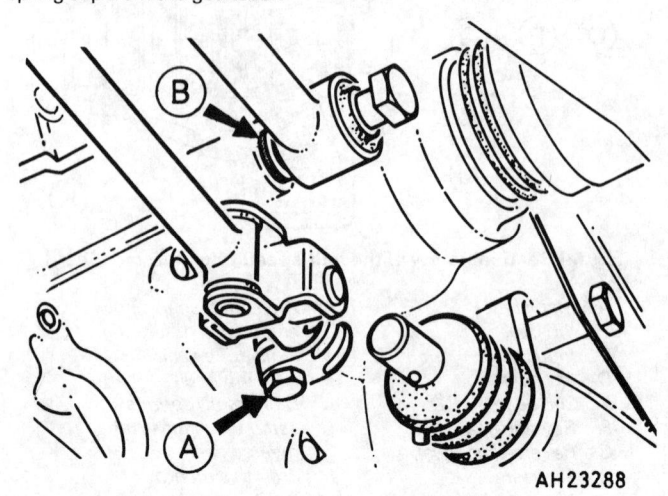

Fig. 6.5 Gearchange mechanism shift rod clamp bolt (A), and stabiliser bar washer fitment (B) (Sec 4)

Chapter 6 Transmission

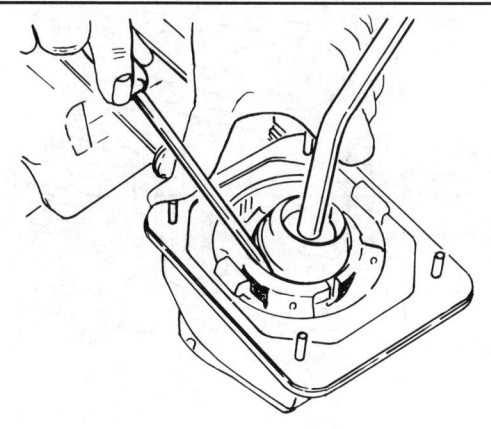

Fig. 6.6 Detach the damping plate from the gearchange mechanism (Sec 4)

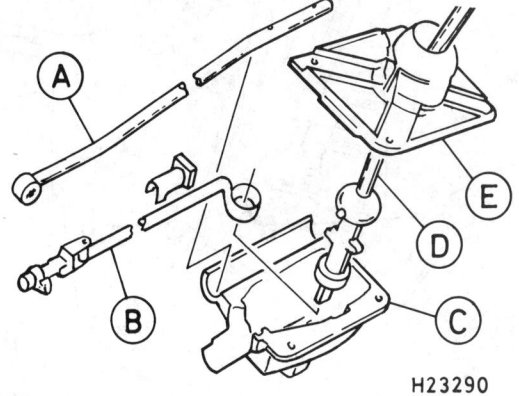

Fig. 6.7 Exploded view of gearchange gate assembly (four-speed) (Sec 4)

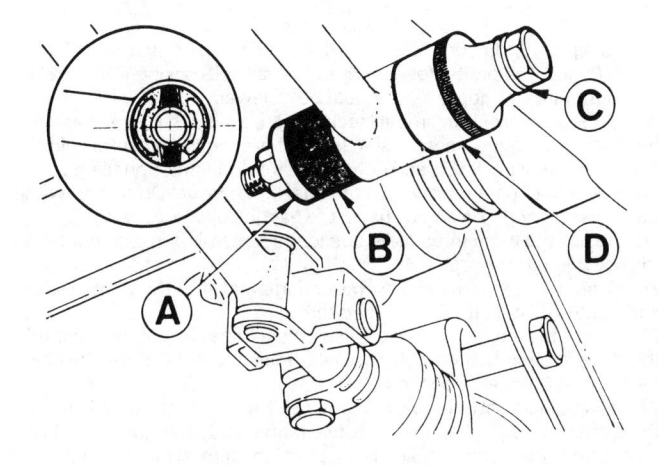

A Stabiliser bar
B Shift rod
C Gearchange gate
D Gear lever
E Gearchange gate cover

Fig. 6.8 Gearchange mechanism stabiliser bar bush fitting method. Note the void positions in the bush (inset) (Sec 4)

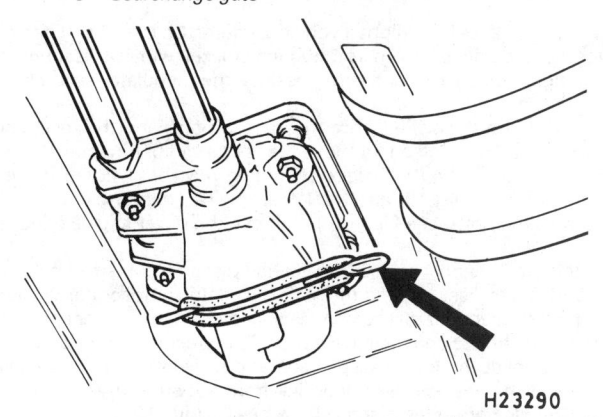

A Washer
B Bush
C Washer
D Socket

Fig. 6.9 Gearlever locked in position using suitable tool (arrowed) inserted through gearchange gate (Sec 4)

10 Turn the damping plate slightly and release the half-shells with a screwdriver. Detach the damping plate from the gearchange mechanism (see Fig. 6.6).
11 Remove the four bolts securing the cover to the gearchange gate, and lift the cover off.
12 Remove the gear lever, shift rod and stabiliser bar from the gearchange gate (see Fig. 6.7).
13 Renew any worn components and reassemble by reversing the dismantling procedure, making reference to the Specifications for tightening torque details. Compress the assembly and fit the circlip. Note that if the stabiliser bar bush is in poor condition, it can cause engine and transmission noises to be transmitted to the vehicle interior; if renewing, note positioning of its voids and take care not to damage or distort it as it is drawn into position (Fig. 6.8).
14 To install, position the gearchange mechanism by reversing the method of removal, but only loosely secure to the floor with the four nuts inside the vehicle.
15 Secure the stabiliser bar to the transmission, having correctly located its washer (see Fig. 6.5), and tighten to the specified torque. Refit the exhaust forward mounting bracket, as applicable.
16 Tighten the nuts securing the gearchange mechanism to the floor to their specified torque.
17 Ensure that the transmission selector shaft is correctly engaged in second gear (rotate to its central position then press in to the stop). Attach the shift rod to the transmission selector shaft, but do not tighten the clamp bolt. Note that the clamping faces must be free of grease.
18 With the gear lever locked in the second gear position, using a suitably-sized rod inserted through the gearchange gate (and held with a sturdy elastic band or similar), tighten the shift rod clamp bolt to the specified torque. Several attempts may have to be made to lock the

gear lever, but this is essential for correct adjustment (refer to Fig. 6.9).
19 Remove the gear lever locking tool.
20 The remainder of the refitting operation is a reversal of the removal procedure.
21 Select each gear in turn to confirm that the mechanism has been correctly set.

5 Transmission (four and five-speed) – removal and refitting

Note: *Before removing the transmission from the vehicle, engage second gear on four-speed transmissions, and fourth gear on five-speed transmissions, to ensure correct adjustment later*

1 Disconnect the battery earth terminal.
2 Unscrew the union nut securing the speedometer cable, and withdraw the cable. On some models it will also be necessary to disconnect the road speed sensor connecting lead.
3 Refer to Chapter 5 and disconnect the clutch cable from the clutch release lever on the transmission.
4 Withdraw the transmission breather hose from the opening in the side-member, as applicable.
5 Unscrew the two upper transmission flange bolts and remove them. On certain models, these bolts also hold wiring loom securing brackets – the loom should be tied up out of the way.
6 On vehicles with CVH engines, tie the thermostat housing-to-heater hose out of the way, otherwise it will get in the way when refitting the transmission.
7 Disconnect the earth leads from the transmission,.
8 Raise the vehicle and support securely on axle stands. On XR2i models, the one-piece undertray (if fitted) should be removed, followed by the front suspension crossmember (Chapter 10).

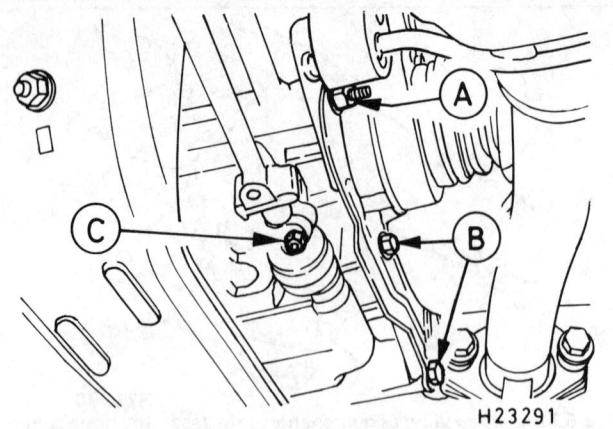

Fig. 6.10 Exhaust forward mounting bracket retaining nut (A), lower engine adaptor plate bolts (B) and gearchange mechanism shift rod clamp bolt (C) (Sec 5)

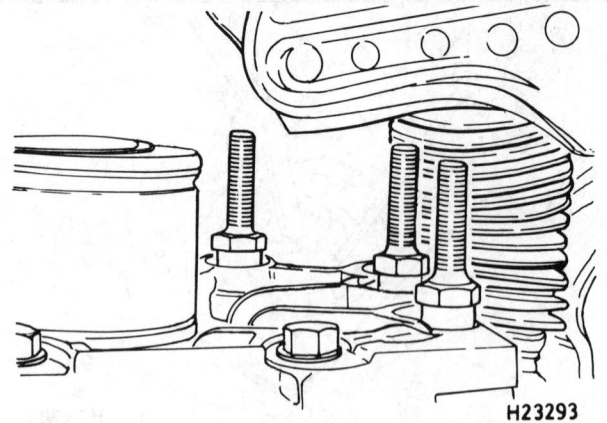

Fig. 6.11 Rear transmission mounting bracket threaded studs in transmission housing (Sec 5)

9   On anti-lock brake equipped vehicles, refer to Chapter 8 and detach the left-hand modulator from its mounting bracket (without disconnecting the rigid brake pipes or return hose); tie the modulator securely to the bulkhead.
10   Disconnect the wiring from the starter motor and the reversing light switch, then unbolt and remove the starter motor (Chapter 12).
11   Detach the exhaust forward mounting bracket (if fitted) by unscrewing its retaining nut and the lower engine adaptor plate (clutch housing cover) bolts (see Fig. 6.10). Remove the lower engine adaptor plate.
12   Refer to Section 4 of this Chapter and disconnect the gearchange mechanism stabiliser bar and the shift rod. Note stabiliser bar washer fitment for subsequent reassembly. Suspend the stabiliser bar and shift rod free ends from a convenient location – the steering rack is ideal.
13   Disconnect the lower suspension arm balljoint from the spindle carrier, and the anti-roll bar upper link from the suspension strut (as applicable), on both sides of the vehicle (see Chapter 10).
14   Refer to Chapter 9 and disconnect the track rod end from the steering arm on the spindle carrier, on both sides of the vehicle.
15   On anti-lock brake equipped vehicles, refer to Chapter 8 and detach the right-hand modulator from its mounting bracket (without disconnecting the rigid brake pipes or return hose); tie the modulator securely to the bulkhead. Additionally, undo the three bolts securing the modulator mounting bracket (Chapter 1).
16   Disconnect the left-hand driveshaft assembly from the transmission using the following method; locate a large lever between the transmission casing and the inner CV joint then, with an assistant applying pressure to the roadwheel (away from the centre of the vehicle), strike the lever sharply with the hand. Insert a suitable plug (an old inner CV joint is ideal) to prevent excessive transmission oil loss, and to prevent the differential from rotating when the right-hand driveshaft is removed. Suspend the driveshaft assembly to avoid straining the CV joints; the inner joint must not be bent more than 20°, and the outer joint by more than 40°.
17   Repeat the above procedure on the right-hand driveshaft assembly and remove the previously-released modulator mounting bracket.
18   Fit an engine support bar (or similar device) and take the weight of the engine/transmission, then detach the transmission bearer from the transmission and the floorpan and remove it (see Chapter 1, Section 13).
19   Lower the engine/transmission as far as the engine support bar allows, then unscrew the three nuts securing the rear transmission mounting bracket to the transmission and remove the bracket.
20   Unscrew the three threaded studs of the rear transmission mounting bracket from the transmission (Fig. 6.11).
21   The transmission is ready to be withdrawn after removing the four remaining engine/transmission flange bolts. The method of supporting the transmission as it is withdrawn is a matter of choice, depending on working clearance and equipment available, but it is advisable to have an assistant at the ready. Do **not** allow the transmission input shaft to bear any weight as the transmission is withdrawn.
22   Prior to refitting, smear the transmission input shaft splines and the clutch release bearing guide sleeve with a thin film of special grease (see Chapter 5). Ensure also that the upper engine adaptor plate is correctly located on its dowels. Note that all self-locking nuts, circlips and snap-rings, as applicable, should be renewed.
23   In a reversal of the removal technique, connect the transmission to the engine, with the transmission input shaft engaging centrally in the clutch driven plate. Insert and tighten the four transmission flange bolts (as removed in paragraph 21) to the specified torque. Do **not** draw the transmission onto the engine using the flange bolts.
24   Insert the three threaded studs to the transmission, and tighten to the specified torque.
25   Refit the rear transmission mounting bracket to the transmission, and tighten its nuts to the specified torque.
26   Raise the engine/transmission on the engine support bar, and refit the transmission bearer, tightening its bolts and nuts to the specified torque (see Chapter 1, Section 13). Remove the engine support bar.
27   On vehicles fitted with the anti-lock braking system, ensure that the belts are located over the driveshafts and that the right hand modulator mounting bracket is in position before reconnecting the driveshaft assemblies.
28   With a **new** snap-ring fitted to both inner CV joint splined shafts, lubricate the splined shafts with transmission oil (see Chapter 7).
29   Remove the plug (or old inner CV joint) from the transmission casing and insert the right-hand driveshaft inner CV joint through the transmission casing into its splined differential location. Apply pressure at the roadwheel to ensure that the snap-ring **fully** locates in the differential.
30   Repeat the above procedure on the left-hand driveshaft assembly.
31   Secure the right-hand modulator mounting bracket, as applicable, then refit the right-hand modulator and associated components (refer to Chapter 8).
32   Refer to Chapter 9 and reconnect the track rod ends. Note that **new** split pins must be used.
33   Refer to Chapter 10 and reconnect the lower suspension arm balljoints to the spindle carriers. Also, as applicable, reconnect the anti-roll bar upper links to the suspension struts.
34   Reconnect the gearchange mechanism stabiliser bar and shift rod and adjust the shift rod, as described in Section 4 or 15 of this Chapter.
35   The remainder of the refitting procedure is a reversal of that used for removal, with reference to the relevant Chapters.
36   Check the transmission fluid level upon completion, and top up as required (see Section 2).
37   Reconnect the battery earth terminal (refer to Chapter 12, Section 5).

6   Transmission (four-speed) – dismantling into major assemblies

1   With the gearbox removed from the vehicle, clean away external dirt and grease using paraffin and a stiff brush or a water-soluble solvent. Take care not to allow water to enter the transmission.
2   Refer to Chapter 5 and remove the clutch release bearing and release shaft.
3   If not removed for draining, unscrew the selector shaft cap nut spring and interlock pin (see Fig. 6.4).

Chapter 6 Transmission

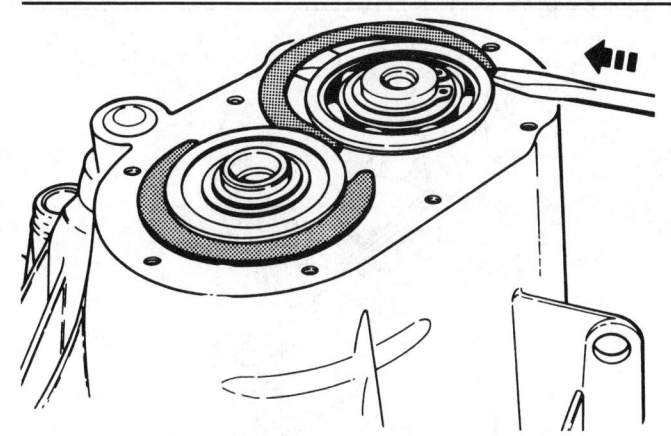

Fig. 6.12 Removing the shaft bearing snap-rings (Sec 6)

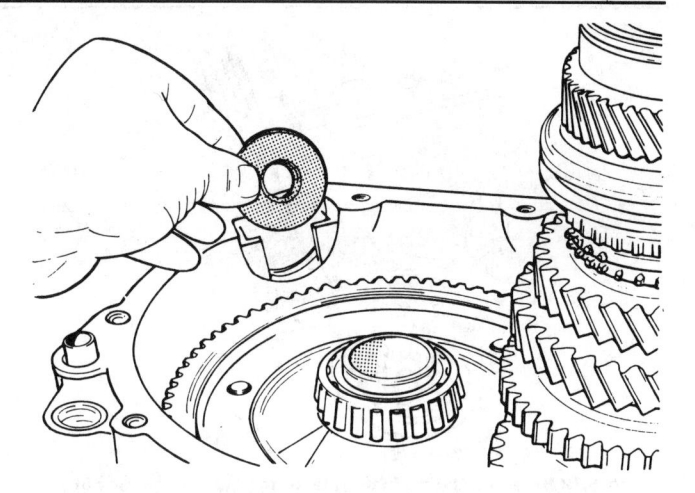

Fig. 6.13 Extract the swarf-collecting magnetic disc (Sec 6)

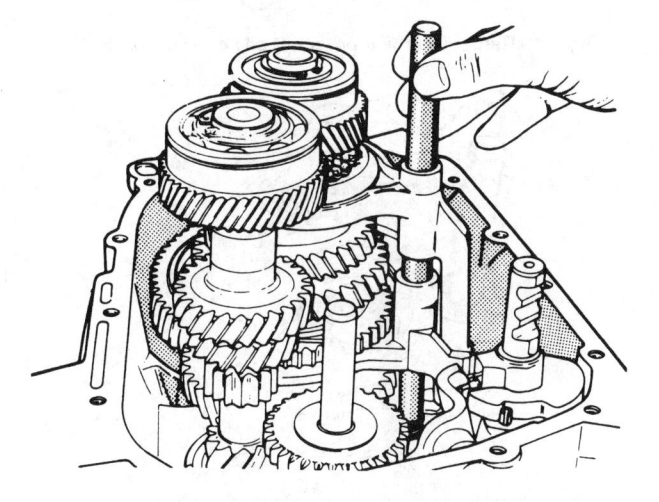

Fig. 6.14 Withdrawing the selector shaft (Sec 6)

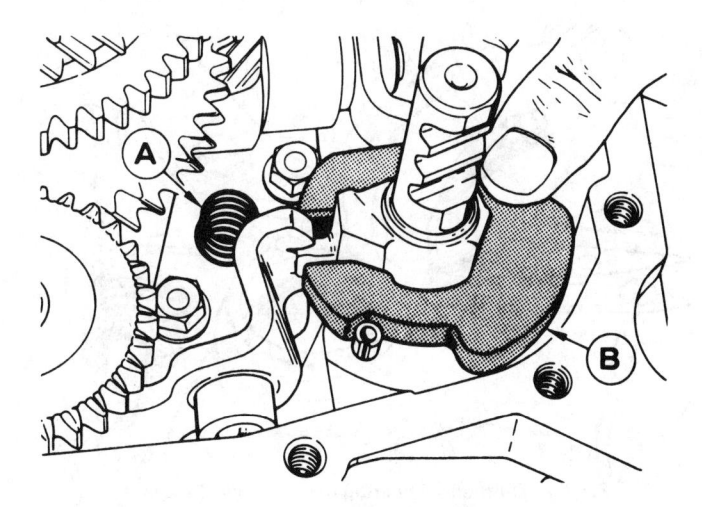

Fig. 6.15 Selector shaft coil spring (A) and shift locking plate (B) (Sec 6)

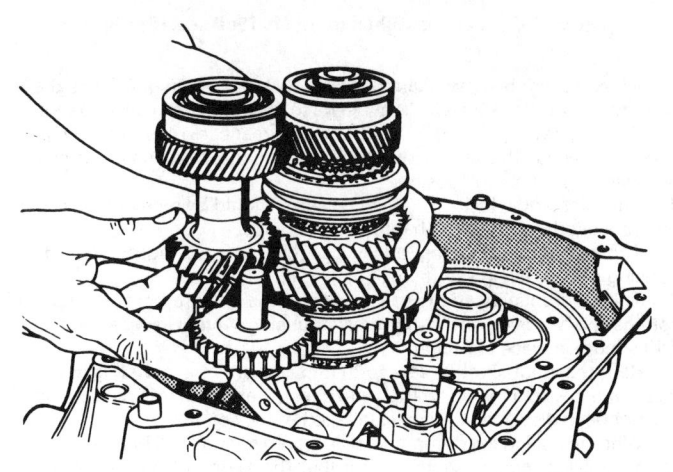

Fig. 6.16 Withdrawing the geartrains (Sec 6)

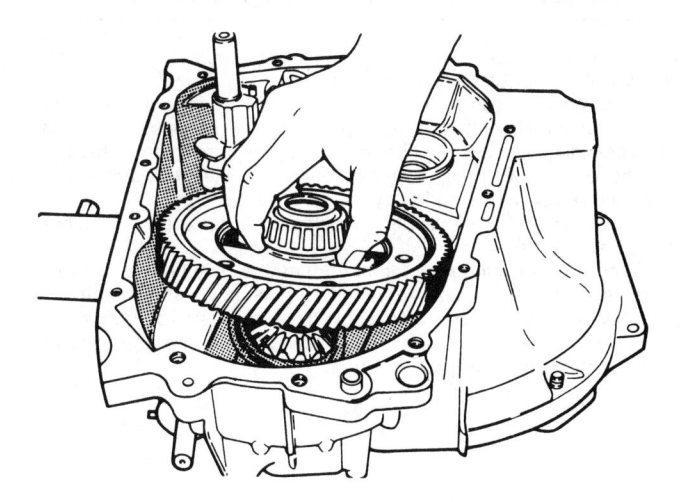

Fig. 6.17 Lifting out the differential assembly (Sec 6)

4 Unbolt and remove the transmission housing cover.
5 Remove the snap-rings from the main and input shaft bearings.
6 Unscrew and remove the connecting bolts and lift the smaller housing from the transmission. If it is stuck, tap it off carefully with a plastic-headed mallet.
7 Extract the swarf-collecting magnet and clean it. Take care not to drop the magnet or it will shatter.
8 Withdraw the selector shaft, noting that the longer portion of smaller diameter is at the bottom as the shaft is withdrawn (Fig. 6.14).
9 Remove the selector shaft coil spring, the selector forks and the shift locking plate. Note the roll pin located in the locking plate cut-out (Fig. 6.15).
10 Withdraw the mainshaft, the input shaft and reverse gear as one assembly from the transmission housing.
11 Lift the differential assembly from the housing.
12 The transmission is now dismantled into its major assemblies.

Fig. 6.18 Breaking the mainshaft bearing plastic cage (Sec 8)

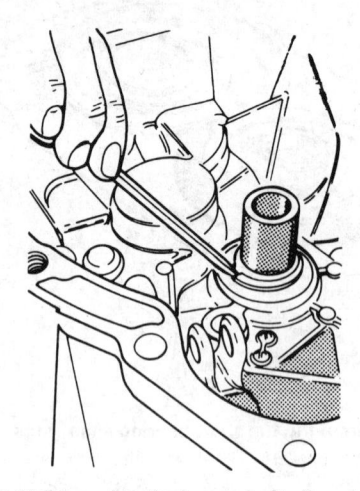

Fig. 6.19 Removing the input shaft oil seal (Sec 8)

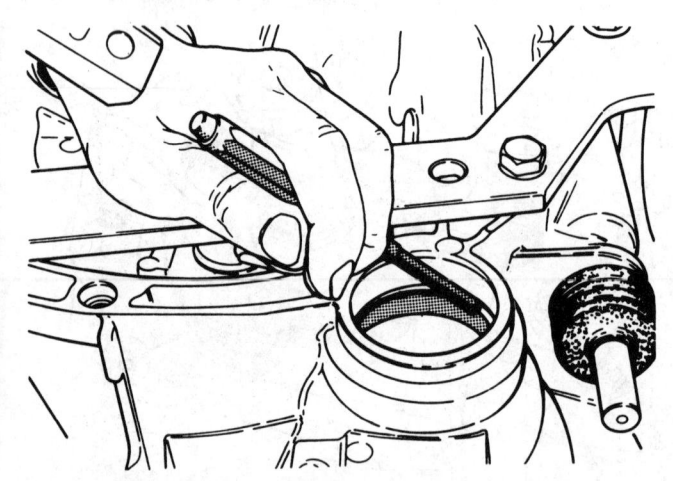

Fig. 6.20 Differential bearing track removal (Sec 8)

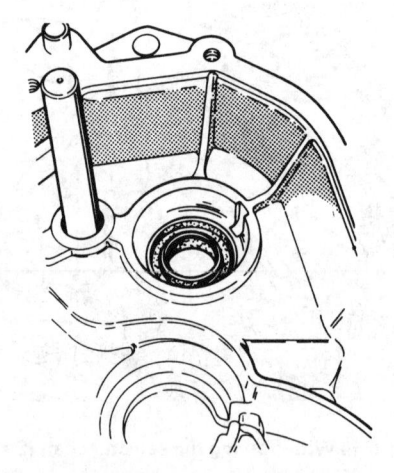

Fig. 6.21 Correctly installed input shaft oil seal (Sec 8)

7 Transmission (four-speed) – further dismantling (general)

1 The need for further dismantling will depend upon the reasons for removal of the transmission in the first place.
2 A common reason for dismantling will be to renew the synchro units. Wear or malfunction in these components will have been obvious when changing over by the noise or by the synchro being easily beaten.
3 The renewal of oil seals may be required, as evident by pools of oil under the vehicle when stationary.
4 Jumping out of gear may mean renewal of the selector mechanism, forks or synchro sleeves.
5 General noise during operation on the road may be due to worn bearings, shafts or gears and when such general wear occurs, it will probably be more economical to renew the transmission complete.
6 When dismantling the geartrains, always keep the components strictly in their originally installed order.

8 Transmission housing and selector mechanism (four-speed) – overhaul

Four-speed transmission

1 To remove the mainshaft bearing, break the plastic roller cage with a screwdriver. Extract the rollers and the cage, the oil slinger and retainers. Remove the bearing outer track.

2 When fitting the new bearing, also renew the oil slinger. Stake the bearing into the housing with light blows to the edge of the bearing seat.
3 When renewing the input shaft oil seal, great care must be taken to avoid damaging the main housing. Drift out the input shaft oil seal rearwards through the housing.
4 The constant velocity (CV) joint oil seals should be renewed at time of major overhaul (refer to Chapter 7, Section 3).
5 The differential bearing tracks can be removed from the large housing section using a drift (Fig. 6.20).
6 The differential bearing outer track and the diaphragm adjustment springs can be driven out of the smaller housing section using a suitable drift such as a piece of tubing.
7 Refit the input shaft oil seal so that its lips are as shown (Fig. 6.21). Apply grease to all the oil seal lips and check that the lip retaining spring has not been displaced during installation of the seal.
8 When installing the differential diaphragm springs and bearing track to the smaller housing section, note that the spring convex faces are towards each other. Stake the track with a light blow from a punch. This is only to hold the track during assembly of the remainder of the transmission.
9 If the selector mechanism is worn, sloppy or damaged, dismantle it by extracting the circlip and taking off the reverse selector lever (photo).
10 Remove the guide lever retaining plate and the guide shaft. Two bolts hold these components in place.
11 Extract the two circlips and detach the guide levers from the retaining plate (Fig. 6.27).
12 To remove the main selector shaft, first detach the gearshift gate from the housing by removing its two securing bolts, then pull the rubber gaiter from the end of the shaft. Extract the single socket screw securing the selector dog to the selector shaft and withdraw the shaft (Fig. 6.28).
13 The selector shaft plastic bushes and oil seal should be renewed.

Chapter 6 Transmission 163

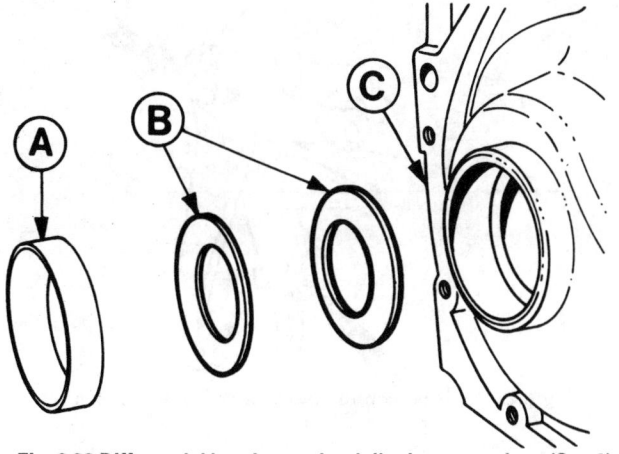

Fig. 6.22 Differential bearing preload diaphragm springs (Sec 8)

A Bearing track
B Diaphragm springs (spring washers)
C Small housing section

8.9 Selector mechanism

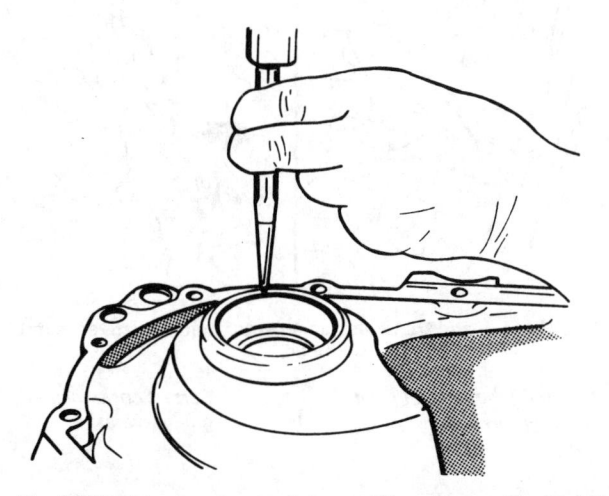

Fig. 6.23 Staking bearing track in small housing section (Sec 8)

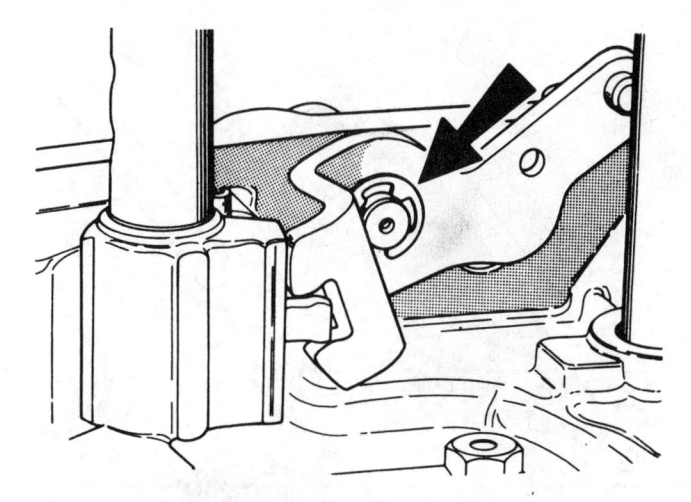

Fig. 6.24 Four-speed transmission reverse selector lever retaining clip (arrowed) (Sec 8)

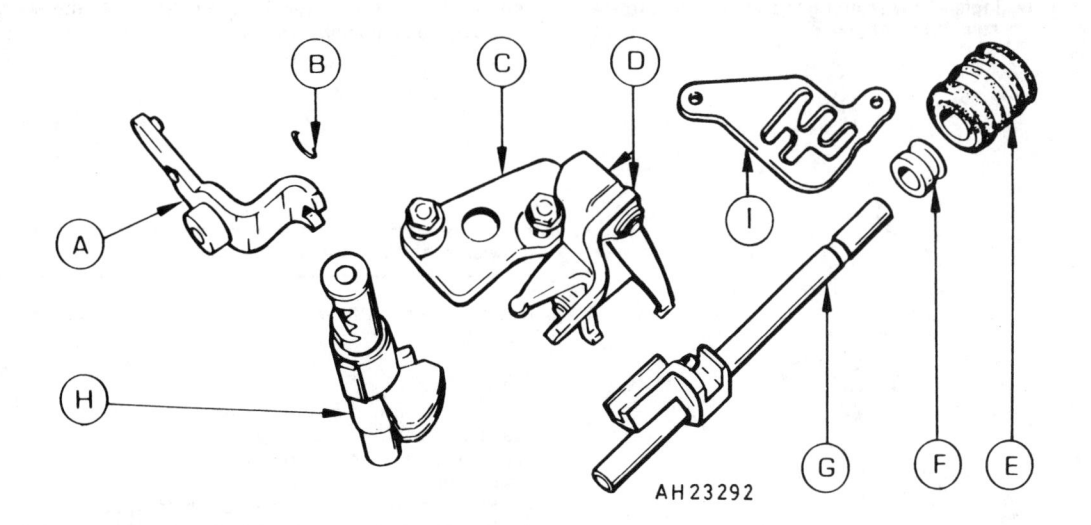

Fig. 6.25 Exploded view of the four-speed transmission selector mechanism (Sec 8)

A Reverse selector lever
B Retaining circlip
C Guide lever retaining plate
D Guide levers
E Selector shaft rubber gaiter
F Oil seal
G Main selector shaft (with dog)
H Auxiliary guide shaft
I Gate

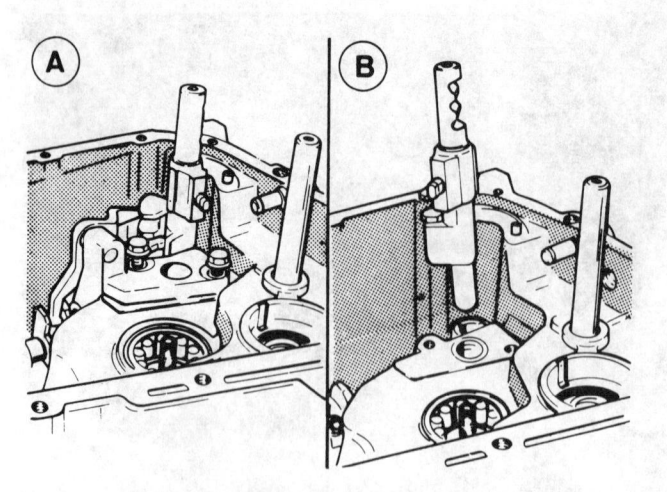

Fig. 6.26 Dismantling selector mechanism (Sec 8)

A Removing retaining plate
B Removing auxiliary guide shaft

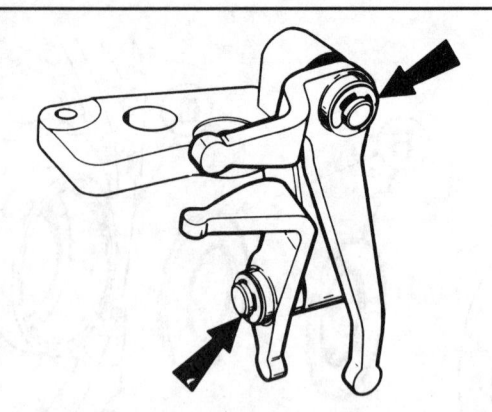

Fig. 6.27 Guide lever retaining circlips (arrowed) (Sec 8)

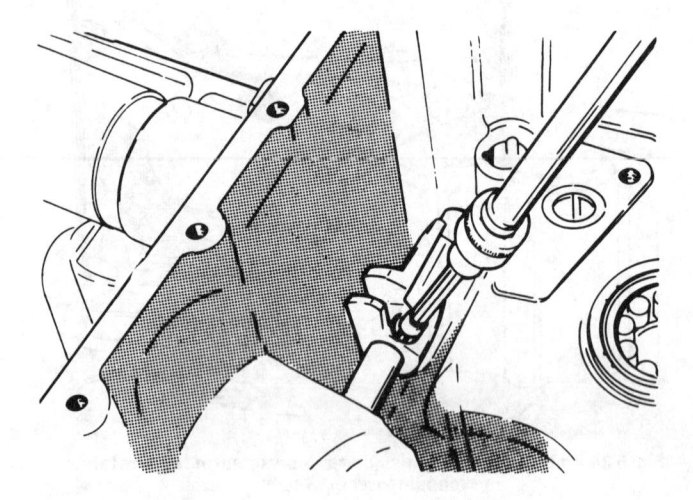

Fig. 6.28 Undoing the single screw securing the selector dog to the selector shaft (Sec 8)

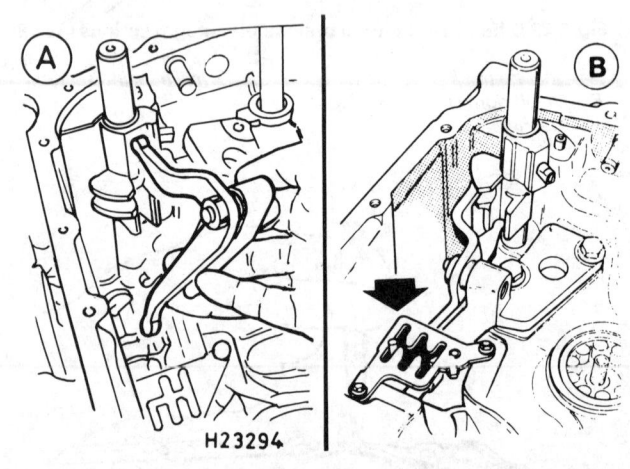

Fig. 6.29 Four-speed transmission selector mechanism reassembly (Sec 8)

A Fitting the retaining plate assembly
B Correct positioning of guide levers

14 Reassembly is a reversal of dismantling, but note that the gaiter vent must face downwards when the transmission is installed in the vehicle. Use new circlips and make reference to the Specifications for details of the tightening torques.

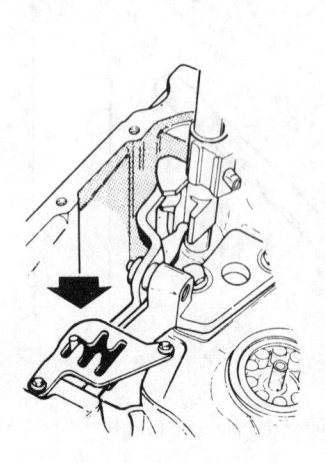

Fig. 6.30 Correct positioning of guide levers (five-speed transmission) (Sec 8)

9 Mainshaft (four-speed) – overhaul

Note: Do not re-use the removed bearing, or any of the circlips or snap-rings

Dismantling

1 Remove the circlip which secures the bearing to the shaft (at the 4th gear end). Insert the outer snap-ring in the peripheral groove in the outer bearing race, and remove the bearing with a standard two-legged puller. Discard the bearing.
2 Slide the 4th gear from the mainshaft, by hand.
3 Extract the snap-ring and remove the 3rd/4th synchro with 3rd gear, using hand pressure only.
4 Remove the thrust ring and the two thrust semi-circular segments, then take 2nd gear from the mainshaft.
5 Extract the snap-ring and take off 1st/2nd gear synchro unit with 1st gear.

Chapter 6 Transmission

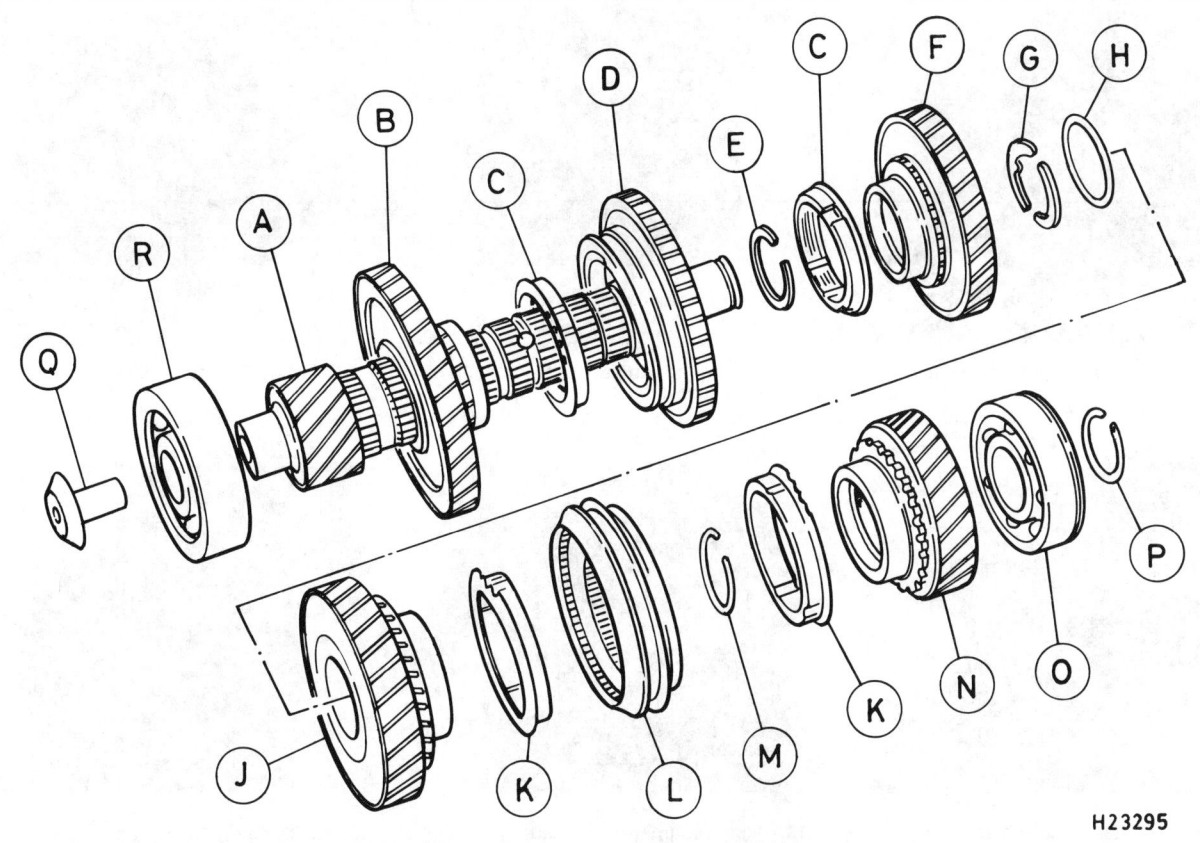

Fig. 6.31 Exploded view of the mainshaft (four-speed transmission) (Sec 9)

- A Mainshaft (with drive pinion gear)
- B 1st gear
- C 1st/2nd gear synchroniser rings
- D 1st/2nd gear synchroniser with reverse gear
- E Snap-ring
- F 2nd gear
- G Thrust segments
- H Thrust ring
- J 3rd gear
- K 3rd/4th gear synchroniser rings
- L 3rd/4th gear synchroniser
- M Snap-ring
- N 4th gear
- O Mainshaft ball bearing
- P Circlip
- Q Oil slinger
- R Roller bearing

6   The mainshaft is now completely dismantled. Do not attempt to remove the drive pinion gear (photo).

Synchronisers

7   The synchro units can be dismantled and new components fitted after extracting the circular retaining springs.
8   When reassembling the hub and sleeve, align them so that the cut-outs in the components are in alignment ready to receive the sliding keys.
9   The two springs should have their hooked ends engaged in the same sliding key, but must run in opposing directions as shown in Fig. 6.33.
10   The baulk rings should be renewed if they do not 'stick' when pressed and turned onto the gear cones, or if a clearance no longer exists between the baulk ring and the gear when pressed onto its cone.

Reassembly

11   With all worn or damaged components renewed, clean the remaining components before reassembly. All components must be lightly lubricated with clean transmission oil before reassembling. When installing new parts (gears, synchroniser hubs or selector rings) the

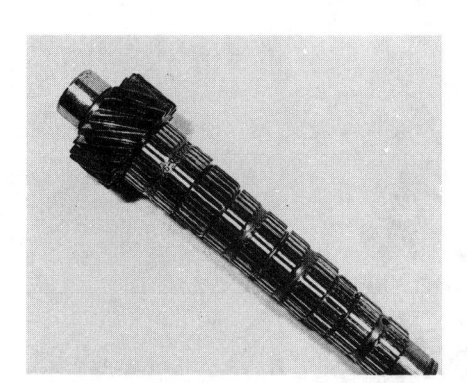

9.6 Mainshaft stripped

9.12 Fitting 1st gear to mainshaft

9.13 Fitting 1st/2nd synchro baulk ring

Chapter 6 Transmission

9.14 Fitting 1st/2nd synchro with reverse gear

9.15 Securing the synchro to the mainshaft with a new snap-ring

9.16 Fitting 1st/2nd synchro baulk ring

9.17 Fitting 2nd gear to mainshaft

9.19A Thrust segments installed

9.19B Fitting the thrust ring

9.20 Fitting 3rd gear to mainshaft

9.21 Fitting 3rd/4th synchro baulk ring

9.22 Fitting 3rd/4th synchro to mainshaft (Note that the serrated edge faces downwards)

9.23 Securing the 3rd/4th synchro to the mainshaft with a new snap-ring

9.24 Fitting the last synchro baulk ring

Chapter 6 Transmission

9.25 Fitting 4th gear to mainshaft

9.26A Fitting the mainshaft bearing

9.26B Using a tubular drift to fit the mainshaft bearing

9.27A Securing the bearing to the mainshaft with a new circlip

9.27B Mainshaft fully assembled (four-speed transmission)

running and side faces of the gears, splined shaft sections, together with the synchroniser cones and mainshaft bearing faces, must be greased (see Specifications). New snap-rings and circlips must always be used, having been selected to give minimum clearance (see Specifications). Do not overstretch the circlips and snap-rings when fitting, and ensure that they locate correctly.

12   Slide the 1st gear onto the shaft so that the gear teeth are next to the pinion drivegear (photo).
13   Fit 1st/2nd synchro baulk ring (photo).
14   Fit 1st/2nd synchro so that reverse gearteeth on the unit are furthest from 1st gear (photo).
15   Fit a new snap-ring to secure the synchro to the mainshaft (photo).
16   Slide on the synchro baulk ring (photo).
17   Slide on 2nd gear (photo).
18   Fit 2nd gear so that the cone is towards the baulk ring.

19   Fit the thrust semi-circular segments and their thrust rung (photos).
20   To the shaft fit 3rd gear so that its teeth are towards 2nd gear (photo).
21   Fit the baulk ring (photo).
22   Slide on 3rd/4th synchro so that its serrated edge is towards the shaft drive pinion gear (photo).
23   Secure the synchro to the mainshaft with a new snap-ring (photo).
24   Fit the baulk ring (photo).
25   Fit 4th gear (photo).
26   Fit the bearing so that its snap-ring groove is nearer the end of the shaft. Apply pressure only to the bearing centre track, using a press or a hammer and a piece of suitable diameter tubing (photos).
27   Fit the circlip to secure the bearing to the shaft. The mainshaft is now fully assembled (photos).

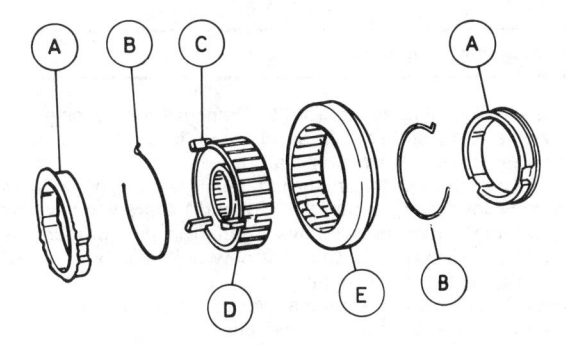

Fig. 6.32 Exploded view of synchroniser (Sec 9)

A   Synchroniser rings
B   Blocker bar retaining springs
C   Blocker bar
D   Synchroniser hub
E   Synchroniser sleeve (selector ring)

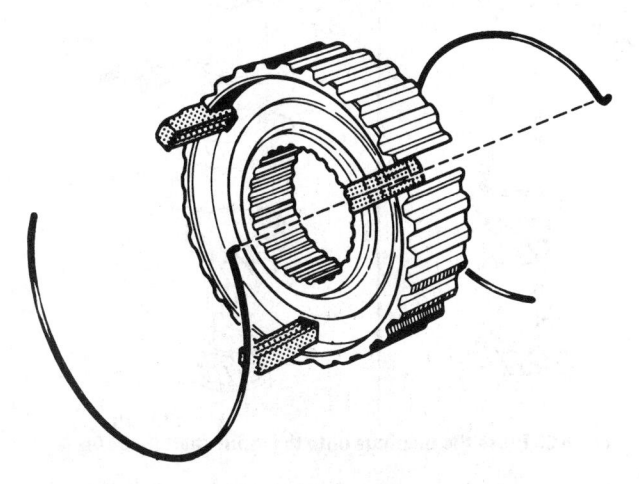

Fig. 6.33 Correct method of fitting blocker bar retaining springs (Sec 9)

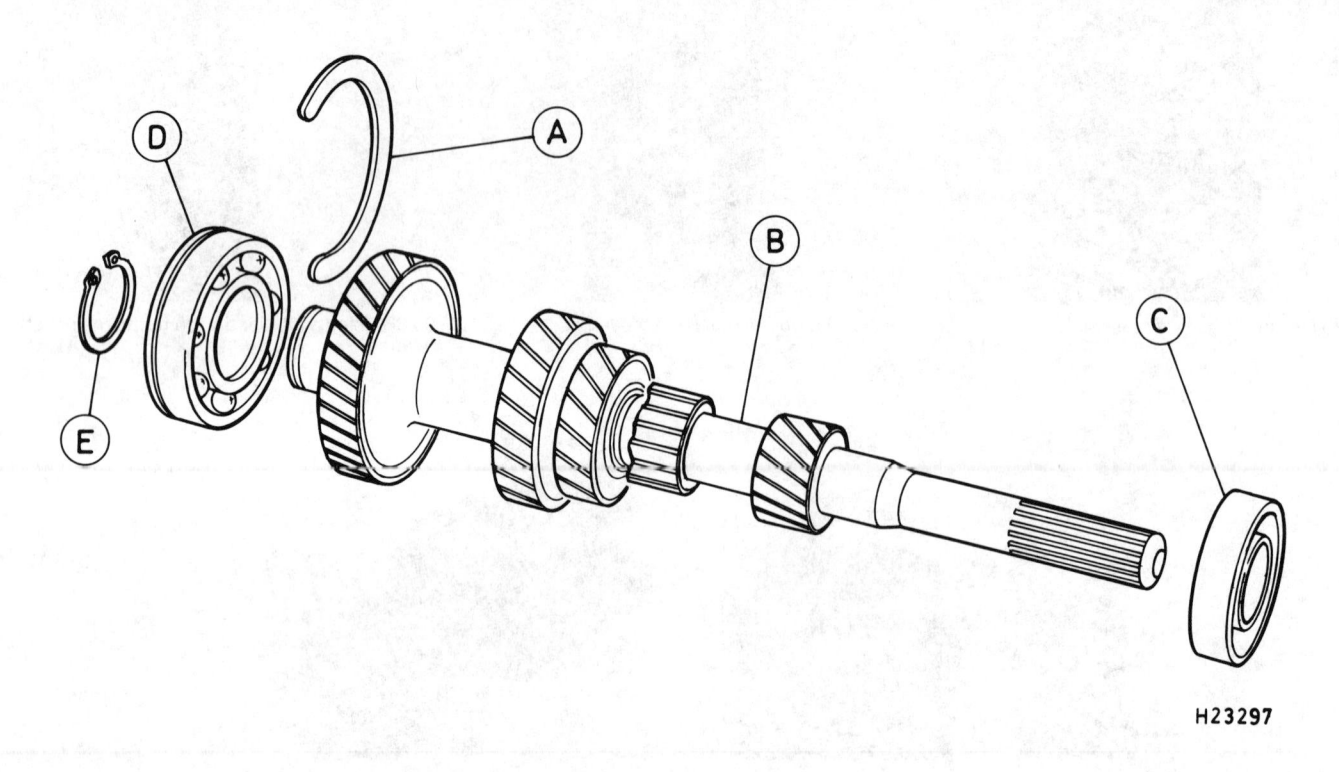

Fig. 6.34 Exploded view of the input shaft (four-speed transmission) (Sec 10)

A Snap-ring
B Input shaft (with gears for 1st to 4th and reverse)
C Small bearing
D Large bearing
E Circlip

10 Input shaft (four-speed) – overhaul

Note: *Do not re-use the ball-bearing races, or any circlips or snap-rings*

1 The only components which can be renewed are the two ball bearing races.
2 Remove the circlip securing the large ball-bearing assembly, and insert the outer snap-ring to the peripheral groove in the outer bearing race. Withdraw the bearing using a standard two-legged puller.
3 Remove the small bearing with the aid of a press or a standard two-legged puller.

4 When fitting the new bearings, apply pressure to the centre track only, using a press or a piece of suitable diameter tubing and a hammer. When installing the larger bearing, make sure that the snap-ring groove is nearer the end of the shaft; fit the circlip. When installing the small bearing the lettering on the bearing should face outwards; press on as far as the stop. Select the snap-ring (for the large ball-bearing assembly outer race) that gives the least play.

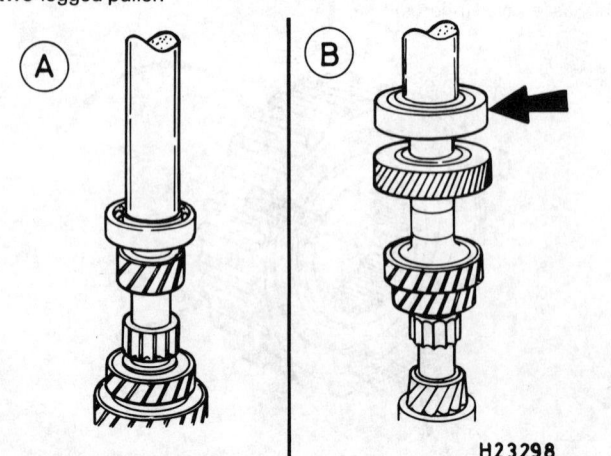

Fig. 6.35 Press the bearings onto the input shaft (Sec 10)

A Small bearing (with lettering facing outwards)
B Large bearing (with selector groove facing outwards)

11 Differential unit – overhaul

1 With the differential removed from the transmission housing, detach the two taper-roller bearings using a two-legged puller, then remove the speedometer worm drivegear.
2 Detach the final drive gear from the differential housing by undoing its securing bolts (all models except the XR2i and 1.4 litre Van have six bolts, whereas these two models have eight). Tap the components apart using a soft-faced mallet. Note that **new** bolts will be required upon reassembly, on five-speed units.
3 Further dismantling is not recommended, and the differential housing will have to be replaced as a unit, if defective.
4 If the final drive gear is to be renewed, then the gearbox mainshaft should be renewed at the same time, as the gearteeth are matched and renewal of only one component will give rise to an increase in noise during operation on the road.
5 Reassembly is a reversal of dismantling, but make sure that the deeply-chamfered edge on the inner circumference of the final drive gear faces the differential housing, and that the securing bolts are tightened to the specified torque (Fig. 6.37).

Chapter 6 Transmission 169

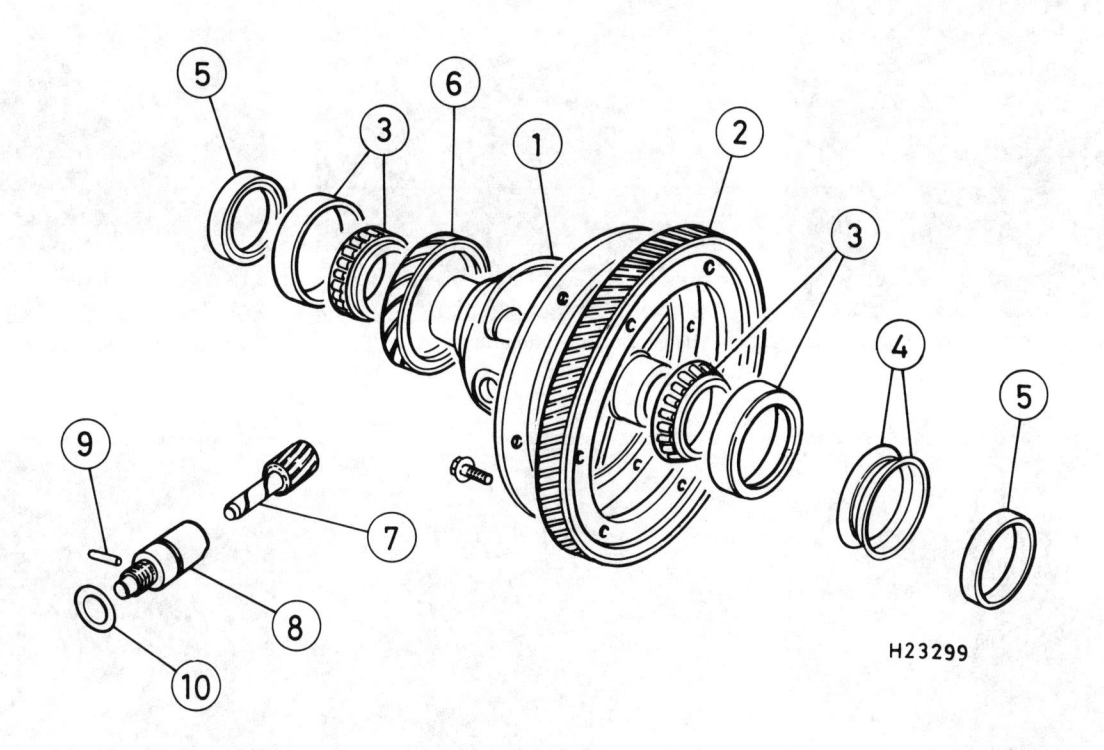

Fig. 6.36 Exploded view of the differential assembly (Sec 11)

1 Differential housing
2 Final drive gear
3 Taper-roller bearing
4 Differential bearing preload diaphragm springs
5 Radial oil seal
6 Speedometer worm drivegear
7 Speedometer pinion
8 Speedometer pinion bearing
9 Roll pin
10 O-ring

6 The sun gears (located within the differential housing) should be held axially to the final drive gear, and plugs or similar inserted so that they remain in correct alignment for the eventual insertion of the driveshaft assembly CV joints.

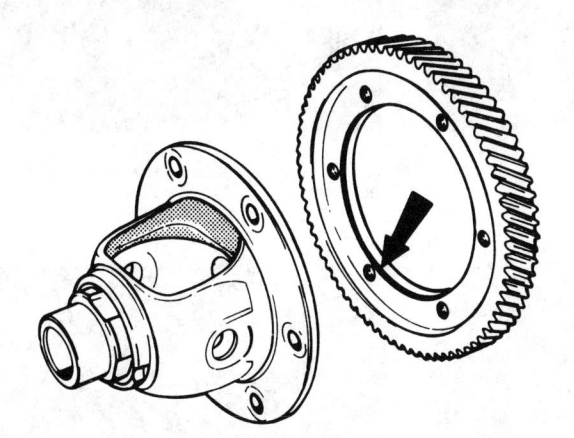

Fig. 6.37 Deeply-chamfered edge of final drive gear (arrowed) facing differential housing (Sec 11)

12 Transmission (four-speed) – reassembly

1 With the larger housing section on the bench, lubricate the differential bearings with transmission oil and insert the differential assembly into the housing (photos).
2 Slide reverse idler gear onto its shaft, at the same time engaging the selector lever in the groove of the gear which should be pointing downwards (photo).
3 In order to make installation of the mainshaft and input shaft easier, lift the reverse idler gear so that its selector lever is held by the reversing lamp switch spring-loaded ball (photo).
4 Mesh the gears of the mainshaft and the input shaft and install both geartrains into the transmission housing simultaneously (photo).
5 Lower the reverse idler gear and its selector lever.
6 Fit the shift locking plate (photo).
7 Engage 1st/2nd selector fork with the groove in the mainshaft synchro sleeve. This fork has the shorter actuating lever (photo).
8 Engage 3rd/4th selector fork with the groove in its synchro sleeve. Make sure that the end of this fork actuating lever is engaged with the shift locking plate (photo).
9 Insert the coil spring in the selector shaft hole and pass the shaft downwards through the holes in the forks. Make sure that the longer section of the reduced diameter of the rod is pointing downwards (photos).
10 Engage 2nd gear for subsequent gearchange mechanism adjustment. To do this, turn the selector shaft clockwise and anti-clockwise as far as the stop, return it to the centre position and press it in as far as the centre position stop. A drilling is provided in the end of the selector shaft which projects from the transmission casing, into which a small screwdriver can be inserted for this purpose (photo).
11 Insert the magnetic swarf collector in its recess, taking care not to drop it (photo).
12 Locate a new gasket on the housing flange, install the smaller housing section and screw in and tighten the bolts to the specified torque (photos).
13 Fit the snap-rings to the ends of the main and input shafts. Cutouts are provided in the casing so that the bearings can be levered upwards to expose the snap-ring grooves, but place a block of wood under the screwdriver to avoid damaging the sealing face (Fig. 6.38). Snap-rings

170　　　　　　　　　　　　　　　　Chapter 6　Transmission

12.1A Interior of transmission larger housing

12.1B Installing the differential

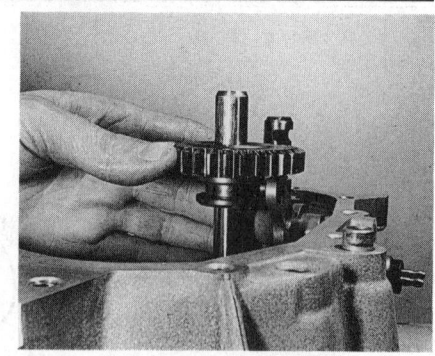

12.2 Fitting the reverse idler gear

12.3 Reverse idler gear supported in raised position

12.4 Installing the geartrains

12.6 Fitting the shift locking plate

12.7 Fitting 1st/2nd selector fork

12.8 Fitting 3rd/4th selector fork

12.9A Inserting the selector shaft coil spring (arrowed)

12.9B Installing the selector shaft

12.10 Turning the selector shaft, having passed a screwdriver through its drilling

12.11 Magnetic swarf collector

Chapter 6 Transmission

12.12A Locate the housing flange gasket ...

12.12B ... then fit the transmission smaller housing section

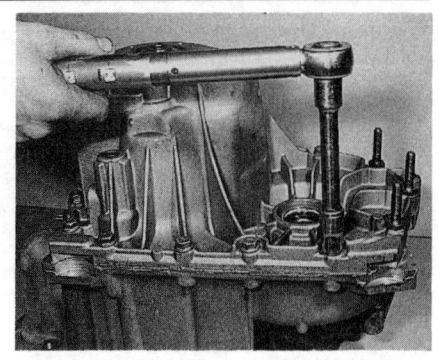

12.12C Tightening the transmission smaller housing securing bolts to the specified torque

12.14 Bearing snap-rings and gasket in position

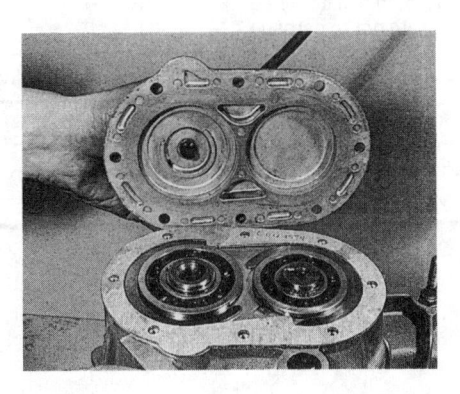

12.15A Fit the transmission cover plate ...

12.15B ... and tighten its bolts to the specified torque – note breather tube

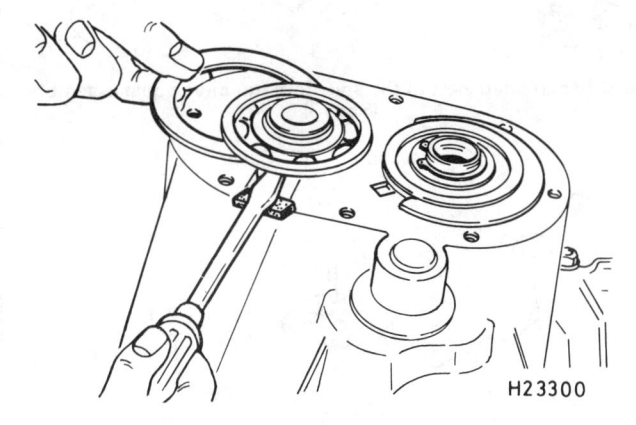

Fig. 6.38 Fitting a snap-ring to its groove, having levered the bearing up to expose the groove (Sec 12)

are available in three thicknesses, and the thickest possible ring should be used which will fit into the groove. If any difficulty is experienced in levering up the bearing on the input shaft, push the end of the shaft from within the bellhousing.
14 Tap out the snap-rings to rotate them so that they will locate correctly in the cut-outs in the cover gasket which should now be positioned on the end of the housing. Fit a new gasket (photo).
15 Fit the cover plate, screw in the bolts and tighten them to the specified torque (photos).
16 Fit the interlock pin, spring and cap nut for the selector shaft locking mechanism. The threads should be coated with jointing compound before installation. Tighten the cap nut to the specified torque (refer to Fig. 6.4).

17 Refer to Chapter 5 and refit the clutch release shaft, bearing and lever.
18 The transmission is now ready for installation in the vehicle. Wait until it is installed before filling with fresh oil of the specified type.

13 Speedometer driven gear – removal and refitting

1 This work may be done without having to remove the transmission from the vehicle.
2 Using a pair of side cutting pliers, lever out the roll pin which secures the speedometer drive pinion bearing in the transmission housing.
3 Withdraw the pinion bearing, together with the speedometer drive cable. Detach the speedometer cable by unscrewing its securing nut.
4 Slide the pinion out of the bearing.
5 Always renew the O-ring on the pinion bearing before refitting.
6 Insert the pinion and bearing into the transmission housing using a back-and-forth twisting motion to mesh the pinion teeth with those of the drivegear. Secure with the roll pin.
7 Reconnect the speedometer cable.

14 Gearchange mechanism (five-speed) – removal, overhaul and refitting

1 The removal procedure is basically the same as that described in Section 4, but the fourth gear position should be selected to ensure correct adjustment later. On XR2i models, remove the one-piece under-

172　　　　　　　　　　　　　　　　　　　　　Chapter 6　Transmission

tray (where fitted) and the front suspension crossmember (see Chapter 10). On 1.4 litre CFi models, remove the underbody heatshields.
2　To dismantle, remove the circlip and detach the rubber spring and spring cup from the gear lever.
3　Turn the damping plate slightly and release the half-shells with a screwdriver. Detach the damping plate from the gearchange mechanism (refer to Fig. 6.6).
4　Remove the side detent screws and springs from the gearchange gate (refer to Fig. 6.41).
5　Remove the five bolts securing the gearchange gate cover, along with their washers. Remove the gearchange gate cover.
6　Remove the gear lever, shift rod, stabiliser bar and latches from the gearchange gate.
7　Renew any worn components and reassemble by reversing the dismantling procedure, making reference to the Specifications for tightening torque details. Compress the assembly and fit the circlip. Note that if the stabiliser bar bush is in poor condition, it can cause engine and transmission noises to be transmitted to the vehicle interior; if it is to be renewed, note the positioning of its voids and take care not to damage or distort it as it is drawn into position (refer to Fig. 6.8).
8　To install, make reference to Section 4, paragraphs 14 to 16 inclusive.
9　Ensure that fourth gear is correctly selected at the transmission by centralising the transmission selector shaft and pressing it right in. The shaft has a drilling which is vertical when the shaft is centralised.

10　Attach the shift rod to the transmission selector shaft but do not tighten the clamp bolt. Note that the clamping faces must be free of grease.
11　With gear lever locked in the fourth gear position, using a 3.5 mm diameter rod inserted through the gearchange gate (and held with a sturdy elastic band or similar – see Fig. 6.9), tighten the shift rod clamp bolt to the specified torque. Ensure that the shift rod does not rotate as the clamp bolt is tightened. Several attempts may have to be made to lock the gear lever, but this is essential for correct adjustment (on 1.4 litre CFi models, a cut-out is provided in the underbody heatshield for access to insert the gear lever locking rod when performing an adjustment-only operation).
12　Remove the gear lever locking tool.
13　The remainder of the refitting operation is a reversal of the removal procedure.
14　Select each gear in turn to confirm that the mechanism has been correctly set.

15　Transmission (five-speed) – dismantling into major assemblies

Note: *Do not re-use circlips and snap-rings*

1　With the gearbox removed from the vehicle, clean away external

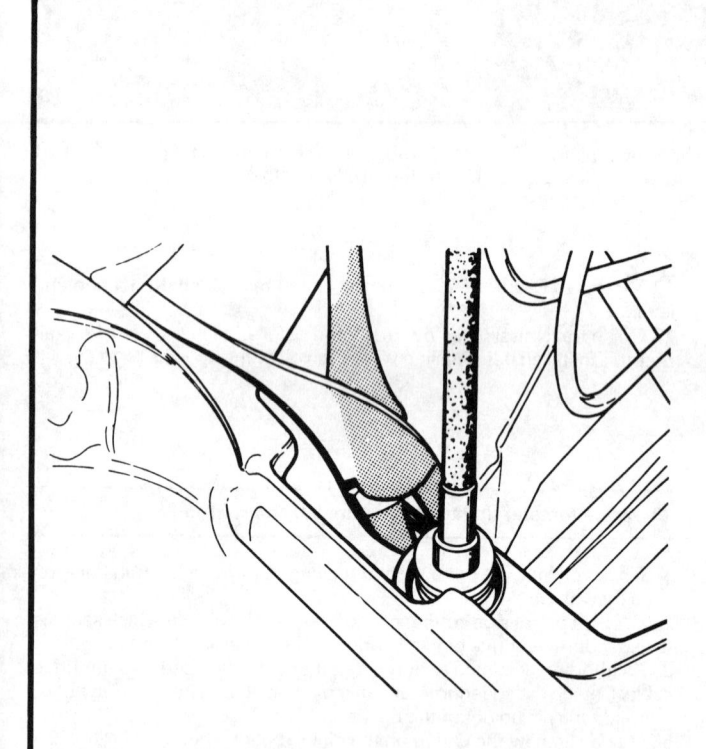

Fig. 6.39 Extracting the roll pin to release the speedometer driven gear assembly (Sec 13)

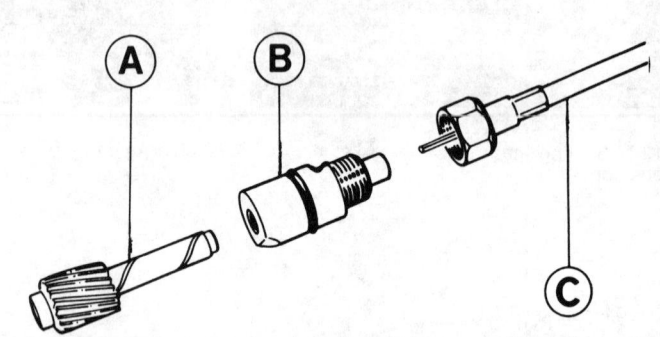

Fig. 6.40 Exploded view of the speedometer driven gear assembly (Sec 13)

A　Gear (pinion)　　C　Cable
B　Bearing

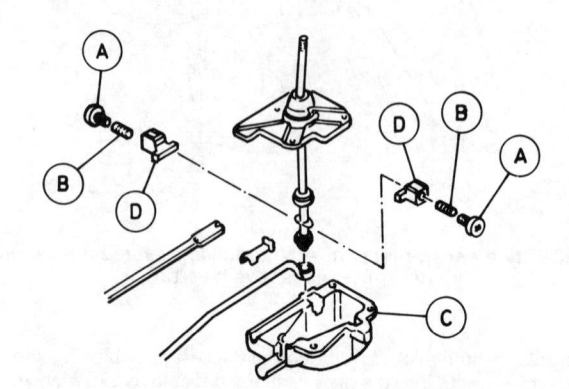

Fig. 6.41 Exploded view of gearchange gate assembly (five-speed) (Sec 14)

A　Detent screws　　C　Gearchange gate
B　Springs　　　　　D　Latches

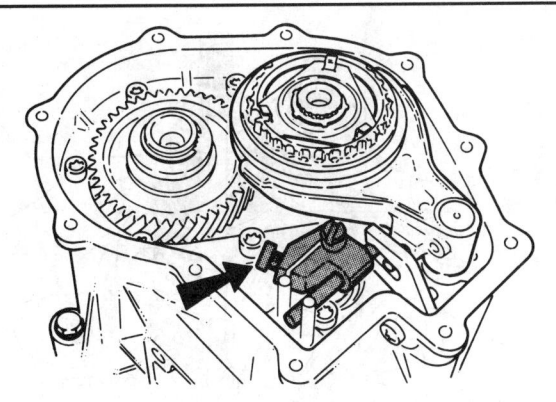

Chapter 6 Transmission

Fig. 6.42 5th gear selector pin assembly clamp bolt (arrowed) (five-speed transmission) (Sec 15)

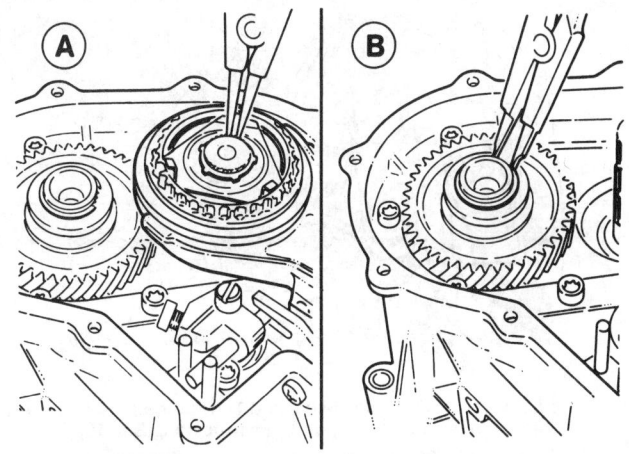

Fig. 6.43 5th gear retaining snap-ring (A) and input shaft circlip (B) locations (five-speed transmission) (Sec 15)

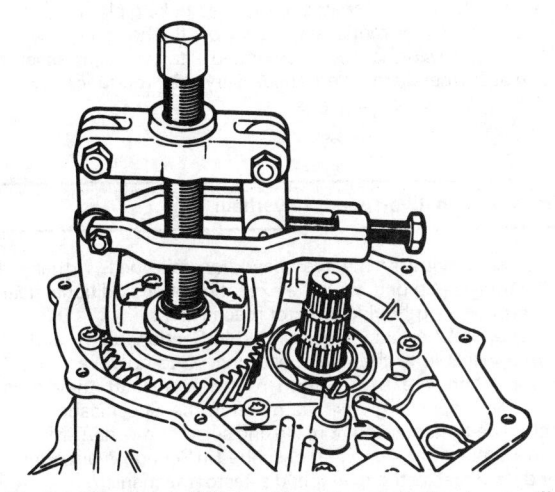

Fig. 6.44 Removal of 5th gear from input shaft using a puller (five-speed transmission) (Sec 15)

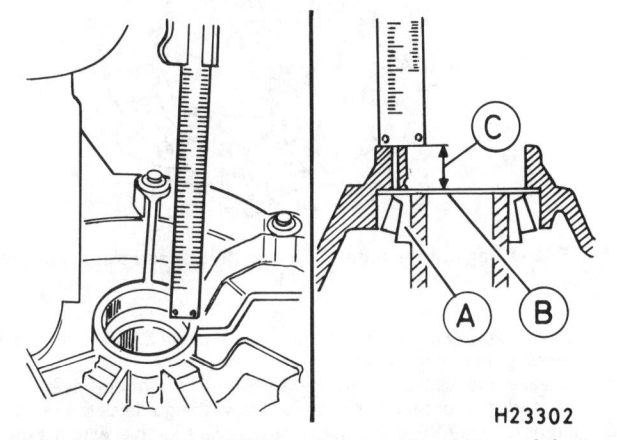

Fig. 6.45 Measuring the distance between the shims and the top edge of the housing (XR2i and 1.4 litre CVH Van models) (Sec 15)

A Differential bearing
B Shim
C Measured distance

dirt and grease using paraffin and a stiff brush, or a water-soluble grease solvent. Take care not to allow water to enter the transmission.
2 Drain off any residual oil in the transmission through a driveshaft opening.
3 Refer to Chapter 5 and remove the clutch release bearing and release shaft.
4 If not removed for draining, unscrew the selector shaft cap nut, spring and interlock pin. Now remove the additional 5th gear selector shaft cap nut, spring and interlock pin.
5 Unbolt and remove the transmission housing cover (photo).
6 Unscrew the clamp bolt and lift the 5th gear selector pin assembly off the shift rod (Fig. 6.42).
7 Using circlip pliers, extract the 5th gear retaining snap-ring, then lift off the 5th gear, complete with synchro assembly and selector fork from the main shaft.
8 Extract the circlip securing the 5th gear driving gear to the input shaft. Using a two-legged puller, draw the gear off the input shaft.
9 Unscrew the nine (Torx-head) bolts securing the 5th gear casing to the main casing and carefully lift it off.
10 Remove the snap-rings from the main and input shaft bearings.
11 Remove the driveshaft inner CV joint oil seal from the transmission small housing section, by prising it out with a screwdriver or other suitable tool. Take care not to raise any burrs which could prevent the new seal from seating correctly during subsequent reassembly.
12 On XR2i and 1.4 litre CVH Van models, establish the distance from the shims to the top edge of the housing. Carry out the measurements using a sliding caliper at three points around the circumference of the driveshaft aperture (see Fig. 6.45). Mark the measurement points. Add the three readings together and calculate the average. Make a note of this figure.
13 Unscrew and remove the connecting bolts then lift the smaller housing from the transmission casing. If it is stuck, tap it off carefully with a plastic-headed mallet.

15.5 Five-speed transmission housing cover

174   Chapter 6   Transmission

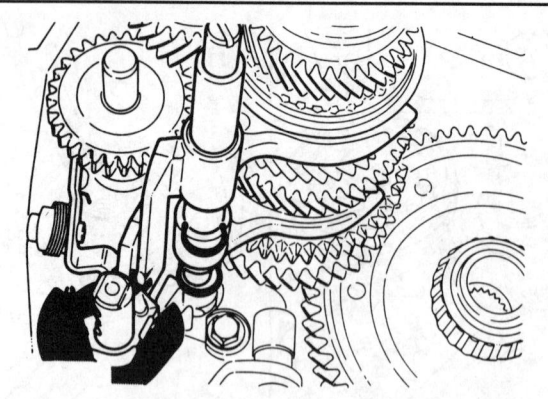

Fig. 6.46 Selector shaft guide sleeve and 1st/2nd gear selector fork circlip locations (five-speed transmission) (Sec 15)

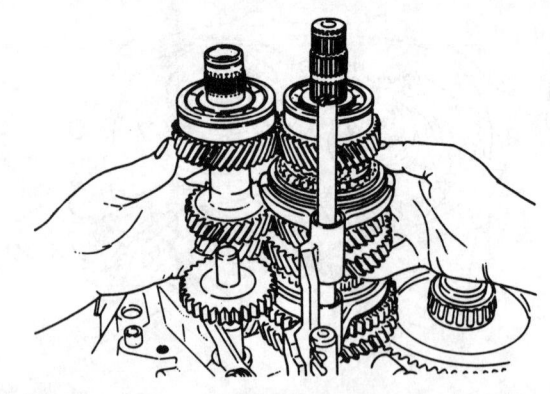

Fig. 6.47 Removal of the mainshaft and input shaft as a complete assembly (five-speed transmission) (Sec 15)

17 Remove the selector shaft and the shift locking plate.
18 Finally lift the differential assembly from the housing.
19 The transmission is now dismantled into its major assemblies, which can be further dismantled if necessary, with reference to Sections 7 and 16.

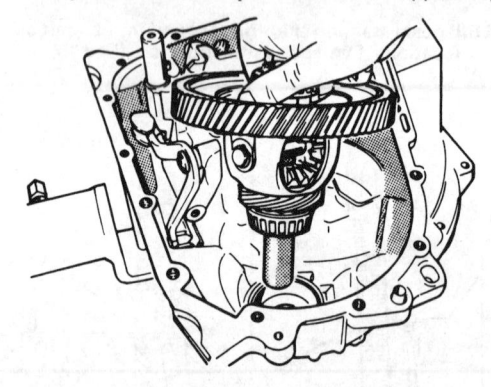

Fig. 6.48 Lifting out the differential unit (five-speed transmission) (Sec 15)

14 Extract the swarf-collecting magnet and clean it. Take care not to drop the magnet or it will shatter.
15 Release the circlips from the selector shaft guide sleeve and 1st/2nd gear selector fork. Carefully withdraw the guide sleeve.
16 Lift out the complete mainshaft assembly, together with the input shaft, selector forks and reverse gear as a complete unit from the transmission housing.

16 Transmission (five-speed) – overhaul

1 As stated previously, the five-speed transmission is virtually identical to the four-speed unit, with the exception of an additional gear and synchro-hub, and a modified selector mechanism.
2 The overhaul procedures described in Sections 9, 10 and 11 are therefore applicable to the five-speed unit. Note however, that if any new components are to be fitted to the mainshaft they must be lubricated with special grease, as given in the Specifications, during assembly. The transmission housing and selector mechanism are overhauled in a similar manner to that described in Section 8, but refer to Fig. 6.49 for differences in the five-speed selector mechanism.
3 When referring to previous Sections for overhaul procedures, it should be noted that all photos, except where indicated, are of the four-speed transmission. Due to the close resemblance of the two transmissions, the photos shown can also be used in most instances for pictorial guidance when working on the five-speed transmission.

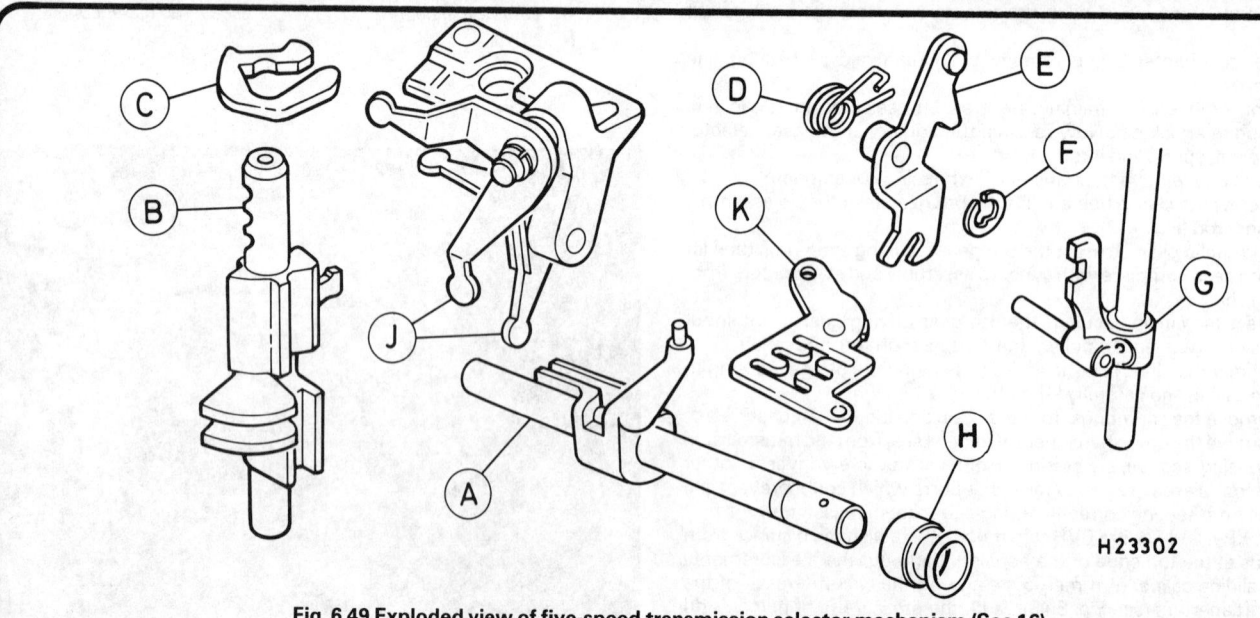

Fig. 6.49 Exploded view of five-speed transmission selector mechanism (Sec 16)

A Main selector shaft (with dog)
B Guide shaft
C Shift locking plate
D Reverse selector lever spring
E Reverse selector lever
F Retaining circlip
G Reverse and 5th gear shift rod
H Oil seal
J Guide levers on retaining plate
K Gate

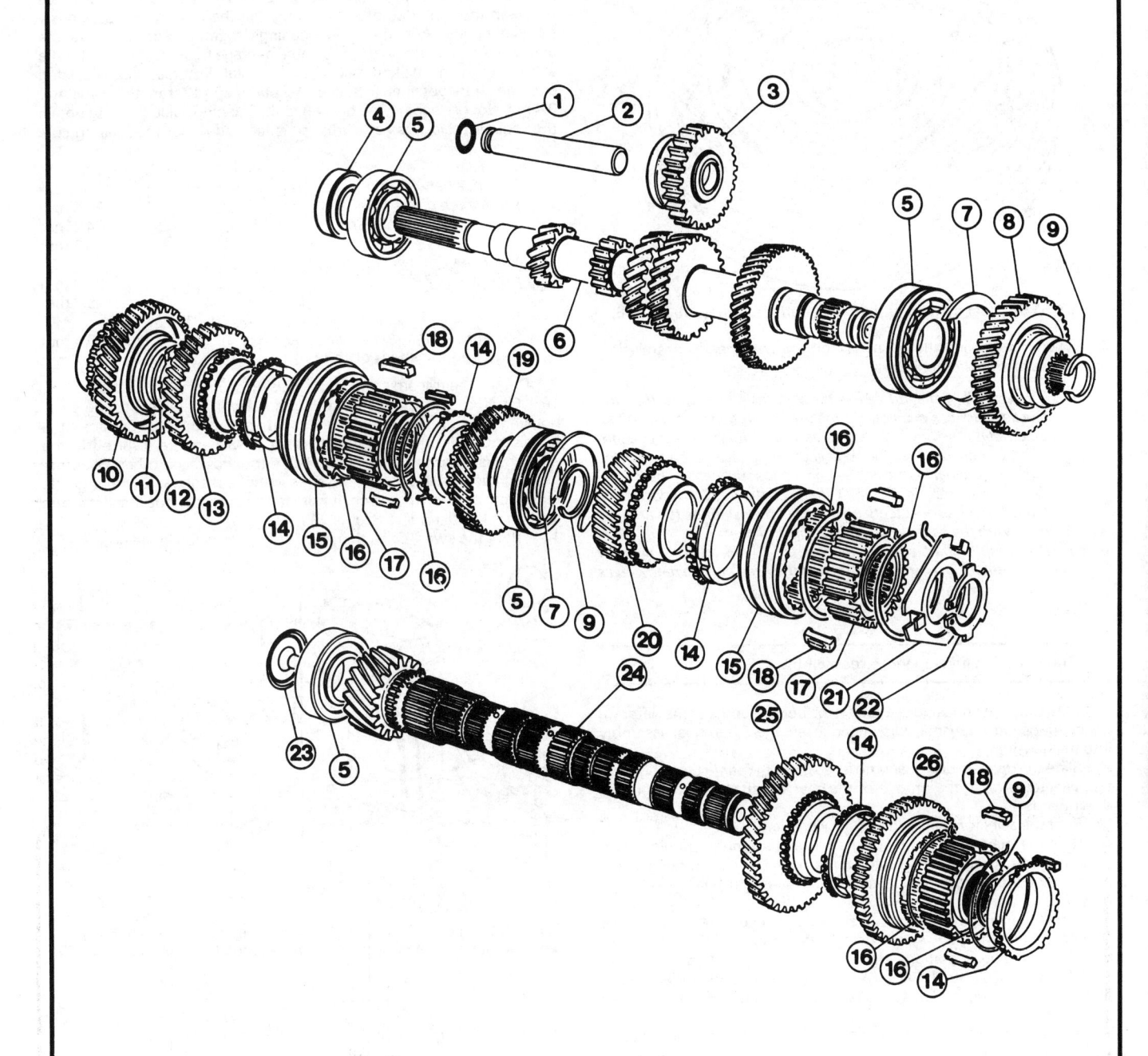

Fig. 6.50 Exploded view of five-speed transmission geartrain arrangement (Sec 16)

1 O-ring	8 5th gear (driving)	14 Synchroniser ring	21 Retaining plate
2 Reverse idler gear shaft	9 Circlip	15 Selector ring	22 Snap-ring
3 Reverse idler gear	10 2nd gear (driven)	16 Retaining spring	23 Oil slinger
4 Radial oil seal	11 Thrust (retaining) ring	17 Synchroniser hub	24 Mainshaft
5 Bearing	12 Semi-circular thrust segments	18 Blocker bar	25 1st gear (driven)
6 Input shaft		19 4th gear (driven)	26 Selector ring with reverse gear
7 Snap-ring	13 3rd gear (driven)	20 5th gear (driven)	

176   Chapter 6   Transmission

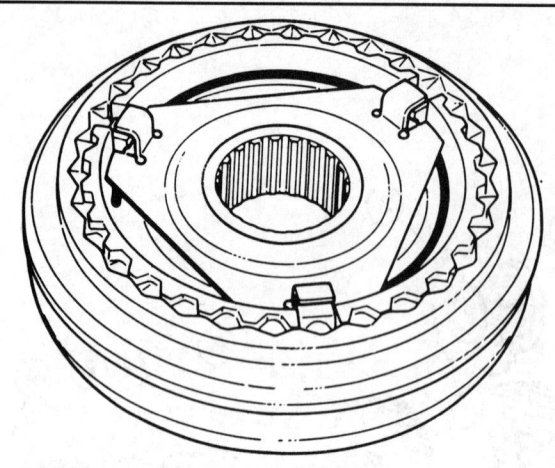

Fig. 6.51 5th gear synchro unit (five-speed transmission) (Sec 16)

4 When overhauling the mainshaft on the five-speed transmission, note the following differences.

(a) If overhauling the fifth gear synchroniser unit, note that the blocker bars are secured by means of a retaining plate. When assembling the unit proceed as described for the other synchro units, but ensure that the retaining spring located between the hub and retaining plate is pressing against the blocker bars
(b) When reassembling the mainshaft, fit the 1st/2nd synchro so that the reverse gear teeth on the unit are positioned towards 1st gear, with the selector groove facing 2nd gear
(c) If the input shaft and/or 5th gear are found to be in need of replacement it should be noted that they can only be renewed as a matching pair

17 Transmission (five-speed) – reassembly

1 With the larger housing section on the bench, lubricate the differential bearings with transmission oil and insert the differential assembly into the housing.
2 Slide the reverse idler gear onto its shaft, at the same time engaging the selector lever in the groove of the gear which should be pointing downwards.
3 Refit the selector shaft and shift locking plate.
4 Refit the mainshaft and input shaft as an assembly complete with selector forks. Guide the selector forks past the shift locking plate, noting that the plate must be turned clockwise to bear against the dowel.
5 Install the selector shaft guide sleeve and secure the 1st/2nd gear selector fork on the guide sleeve using new circlips.
6 The following operations, detailed in paragraphs 7 to 9 inclusive, apply only to XR2i and 1.4 litre CVH Van models, as the differential is fitted with shims instead of spring washers. For other models, proceed to paragraph 10.
7 Insert the bearing cup into the transmission small housing section with a 4.2 mm thick shim. Fit the small housing section (with its gasket) and tighten its retaining bolts to the specified torque.

8 Press the bearing cup and shim onto the bearing, and settle the bearings by rotating the differential through at least five revolutions, from the opposite side. Refer to Fig. 6.45, and measure the distance between the top edge of the housing and the shim at the previously marked points. Add the three readings together and calculate the average. Add 4.2 mm to the calculated average figure to account for the shim already in position, and note the total. Subtract the calculated average figure obtained in Section 15, paragraph 12 from this total. Add 0.4 mm (for preloading and gasket) to this running calculation, to obtain the required thickness of shims to be located under the bearing cup. Example:

Distance between top edge of housing and shim (average)	22.5 mm
Thickness of inserted shim	+ 4.2 mm
	= 26.7 mm
Value determined in Section 15, paragraph 12	−21.0 mm
	= 5.7 mm
Measurement for preloading and gasket	+ 0.4 mm
Required thickness of shims	= 6.1 mm

9 Detach the transmission small housing section again.
10 On XR2i and 1.4 litre CVH Van models, fit the shims to the small housing so that the peripheral groove in the thick shim faces the housing. On all other models, the spring washers are inserted to the small housing with their convex faces towards each other. Slide the bearing cup into the small housing section and secure it with a gentle blow from a punch to prevent it from falling out when the small housing section is fitted (refer to Section 8).
11 Refit the swarf-collecting magnet to its location in the housing.

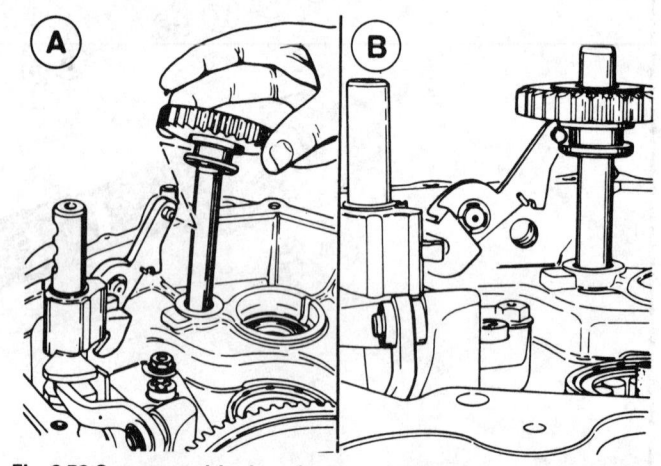

Fig. 6.52 Correct positioning of reverse idler gear on its shaft (A), and engagement of selector lever (B) (five-speed transmission) (Sec 17)

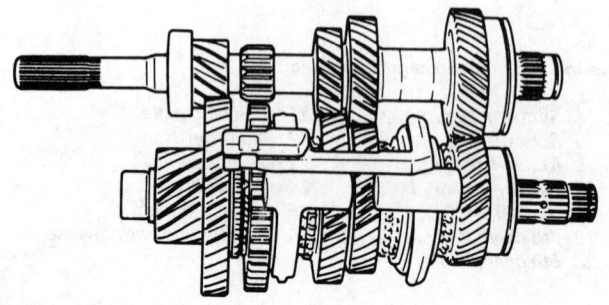

Fig. 6.53 Correct positioning of mainshaft and input shaft assemblies prior to refitting (five-speed transmission) (Sec 17)

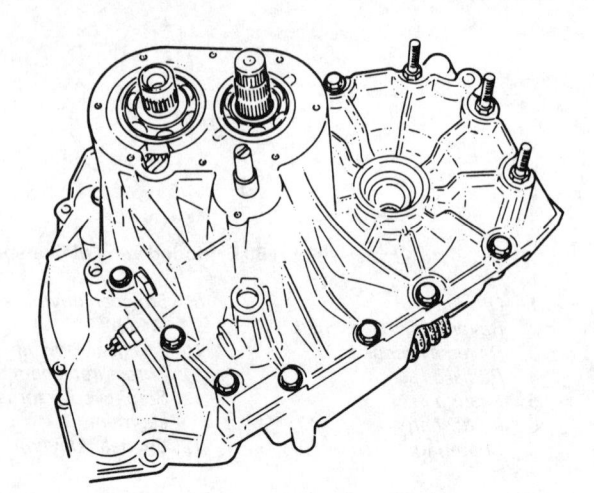

Fig. 6.54 Small housing section in position on main casing (five-speed transmission) (Sec 17)

Chapter 6 Transmission

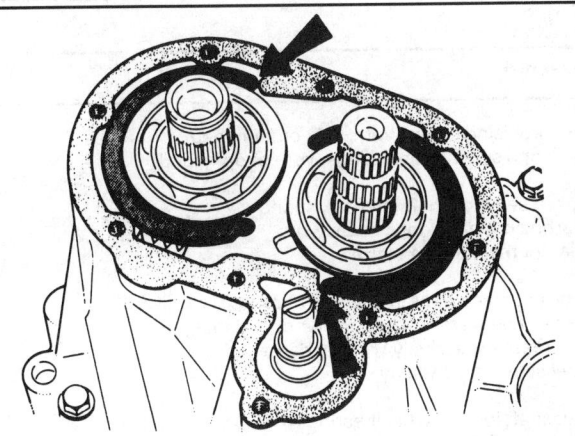

Fig. 6.55 Snap-rings correctly aligned to locate in the cut-outs (arrowed) in the gasket (five-speed transmission) (Sec 17)

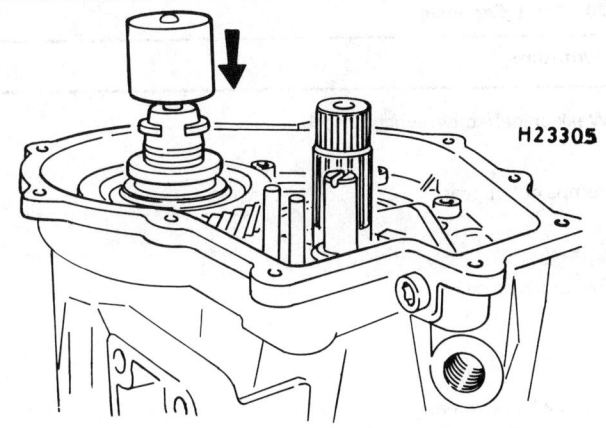

Fig. 6.56 Secure the 5th gear to the input shaft using a suitably-sized section of tubing as a drift (five-speed transmission) (Sec 17)

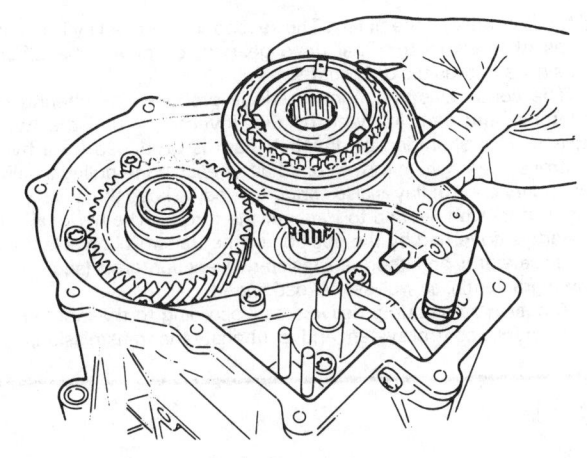

Fig. 6.57 Refitting the 5th gear synchro assembly and selector fork (five-speed transmission) (Sec 17)

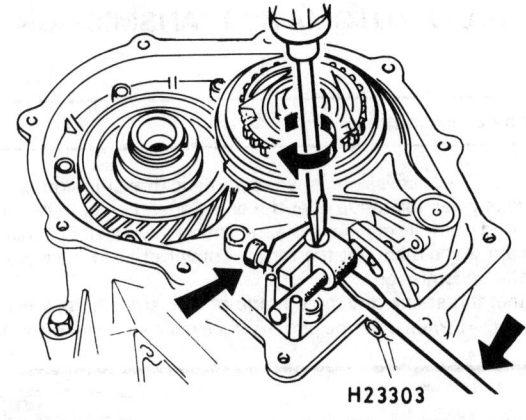

Fig. 6.58 Press the 5th gear selector shaft down and turn clockwise, eliminate any freeplay at the pin (arrowed) then tighten the clamp bolt (five-speed transmission) (Sec 17)

12 Locate a new gasket on the housing flange and place the small housing section in position. Refit and tighten the retaining bolts to the specified torque. The differential should have been immobilised previously (after overhaul) to ensure correct alignment for driveshaft installation; it is important to ensure that this is the case, if not already done (an old driveshaft inner CV joint is ideal).
13 Fit **new** snap-rings to the ends of the main and input shafts. Cutouts are provided in the casing so that the bearings can be levered upwards to expose the snap-ring grooves, but place a block of wood under the screwdriver to avoid damaging the sealing face. Snap-rings are available in three thicknesses, and the thickest possible ring should be used which will fit into the groove. If any difficulty is experienced in levering up the bearing on the input shaft, push the end of the shaft from within the bellhousing.
14 Tap the snap-rings to rotate them so that they will locate correctly in the cut-outs in the 5th gear housing gasket, which should now be placed in position.
15 Fit the 5th gear housing and tighten the retaining bolt to the specified torque.
16 Coat the splines of 5th gear and the input shaft with the special grease (see Specifications). Before fitting the 5th gear, check that the marks on the input shaft and gear web are the same colour.
17 Fit the 5th gear to the input shaft by pressing it into position, whilst supporting the input shaft from underneath. Fit a new circlip (see Specifications) to the input shaft using a tube of suitable diameter as a drift.
18 Fit 5th gear, complete with synchro assembly and selector fork, onto the mainshaft and secure with the snap-ring.
19 Coat the threads of the 5th gear selector shaft locking mechanism cap nut with sealer (see Specifications). Fit the interlock pin, spring and cap nut, then tighten the nut to the specified torque.
20 Fit the 1st/4th and reverse gear selector shaft interlock pin, spring and cap nut after first coating the threads of the cap nut with sealer. Tighten the nut to the specified torque.
21 Refit the 5th gear selector pin assembly to the shift rod, but do not tighten the clamp bolt at this stage.
22 With the main selector shaft in the neutral position, press the 5th gear selector shaft downwards and turn it clockwise to the stop.
23 Engage 5th gear with the main selector shaft, by turning the main selector shaft clockwise as far as the stop and then pulling it fully out.
24 Slide the selector (synchro) ring and selector fork onto 5th gear.
25 Hold the 5th gear selector shaft in the position established in paragraph 22, then raise the 5th gear selector pin assembly so that the play between the pin and the 5th gear locking plate is eliminated. Tighten the clamp bolt to the specified torque in this position, having previously coated its threads with a locking compound (refer to Fig. 6.58).
26 Place a new gasket in position and refit the housing cover, tightening the retaining bolts to the specified torque.
27 At this stage, check the operation of the selector mechanism by engaging all the gears with the selector shaft.
28 Refit the clutch release shaft, bearing and lever into the bellhousing (see Chapter 5).
29 Fit new driveshaft inner CV joint oil seals to the transmission housing, if not already done.
30 The transmission is now ready for installation in the vehicle. Wait until it is installed before filling with oil.

18 Driveshaft/transmission oil seal – renewal

1 Refer to Chapter 7, Section 3.

19 Transmission mountings – removal and refitting

1 Refer to Chapter 1, Section 13.

Chapter 6 Transmission

20 Fault diagnosis

Symptom	Reason(s)
Weak or ineffective synchromesh	Synchronising cones worn, split or damaged Baulk ring synchromesh dogs worn or damaged
Jumps out of gear	Broken selector shaft interlock spring Gearbox coupling dogs badly worn Selector mechanism worn
Excessive noise	Incorrect grade of oil in gearbox or oil level too low Bush or needle roller bearings worn or damaged Gear teeth excessively worn or damaged Excessive shaft endplay
Noise when cornering	Driveshaft joint or wheelbearing worn Differential bearings worn

PART B: CTX AUTOMATIC TRANSMISSION

21 General description

The CTX transmission is an automatic transmission providing continuously-variable drive over the entire speed range. The torque is transmitted from the engine to the transmission via an input shaft and a multi-plate wet clutch (not the torque converter used in conventional automatic transmissions).

A steel thrust-link drivebelt, made of disc-shaped steel elements, transmits the torque from the primary cone pulley (driven by the engine) to the secondary cone pulley. The secondary cone pulley is linked by a series of gears to the final drive gear which drives the differential housing, and thus the driveshafts.

The continuous variation in ratio is produced by altering the diameter of the path followed by the drivebelt around the two cone pulleys. This alteration of drivebelt path is produced by a hydraulic control system which moves one half of each cone pulley axially. The secondary cone pulley is also spring-loaded, to keep the drivebelt at the required tension needed to transmit the torque. The hydraulic control system is governed by the position of the transmission selector lever, the accelerator and the load resistance encountered (such as up or down gradients), as well as the road speed.

A gear-type oil pump delivers oil (according to the input speed) to the hydraulic control system and to lubricate the transmission.

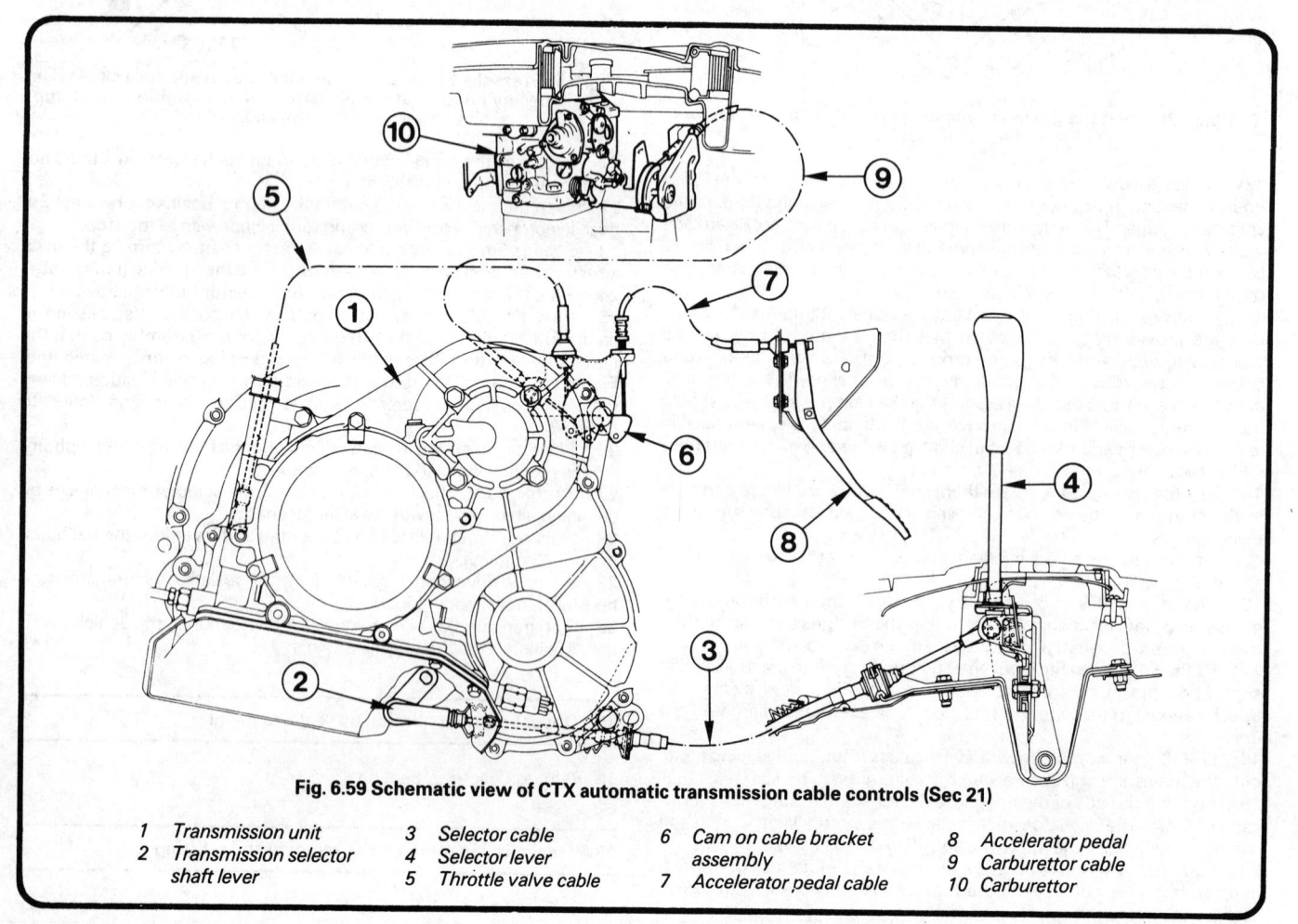

Fig. 6.59 Schematic view of CTX automatic transmission cable controls (Sec 21)

1 Transmission unit
2 Transmission selector shaft lever
3 Selector cable
4 Selector lever
5 Throttle valve cable
6 Cam on cable bracket assembly
7 Accelerator pedal cable
8 Accelerator pedal
9 Carburettor cable
10 Carburettor

Chapter 6 Transmission

Like conventional automatic transmissions, the CTX automatic transmission also has a parking mechanism. The parking pawl engages in the teeth on the outside of the secondary pulley.

When accelerating, the engine speed may sound higher than one would expect for the road speed (similar to a slipping clutch on a vehicle with manual transmission). The reverse is true when decelerating, with the engine speed initially dropping faster than the associated fall in road speed. These are perfectly normal characteristics of the CTX automatic transmission.

A starter inhibitor switch prevents the engine being started in selector positions R, D or L.

A transmission fluid oil cooler is incorporated in the left-hand side tank of the main engine cooling radiator.

The gearchange mechanism selector lever is corrected, by a cable, to a lever on the transmission selector shaft.

Due to the complexity of the CTX automatic transmission, any repair or overhaul work must be left to a Ford dealer or automatic transmission specialist with the necessary equipment for fault diagnosis and repair. The contents of this part of the Chapter are therefore confined to supplying general information and any service information and instructions that can be used by the owner.

Safety notes

When the vehicle is being parked, left with the engine running, or when any checks or adjustments are being carried out, the handbrake must be applied and the selector lever moved to position P.

Do not let the engine speed rise above idle speed when the vehicle is stationary with the selector lever in positions R, D or L.

The engine must not be run at more than 3000 rpm with the selector lever in positions R, D or L, with the driving wheels raised clear of the ground.

If the vehicle is to be towed with the front wheels on the road, move the selector lever to position N. The maximum towing distance is 50 km (30 miles) and the maximum towing speed 50 km h (30 mph), with the front wheels on the road. Ideally, the vehicle should be towed on a trailer or dolly, with the front wheels clear of the road.

22 Maintenance and inspection

1 At the intervals specified in *'Routine maintenance'* at the beginning of this manual, carefully inspect the transmission joint faces, oil seals and fluid pipes for any sign of damage, deterioration or leakage. The selector cable should also be inspected with regard to condition and security.
2 Also at the intervals specified, check the transmission fluid level and top up as necessary, as described in the following procedure.
3 An accurate fluid level check can only be carried out when the transmission fluid has reached normal operating temperature (60 to 70°C).
4 With the vehicle on level ground, apply the handbrake and the footbrake.
5 Move the selector lever through all the positions (with the engine idling), and repeat this three times.
6 Move the selector lever to position P and allow the engine to idle for one minute.
7 With the engine still idling, pull out the transmission fluid dipstick and wipe it with a clean lint-free cloth before re-inserting and withdrawing to read off the level. The top of the fluid film must lie between the MIN and MAX markings on the dipstick. If necessary, add fresh transmission fluid of the specified type through the dipstick tube, with the engine switched off. The use of a funnel or similar device is advisable. If the level is too high, the surplus fluid should be siphoned off.
8 After a new or overhauled transmission has been installed, a fluid level check must be made before and after taking the vehicle on a test drive, as the fluid passages in the transmission have to fill up.
9 The condition of the transmission fluid (colour and odour) is important, and can often indicate if repairs are required (although this is not necessarily conclusive – your Ford dealer will be able to diagnose accurately). Examine the fluid condition when checking the fluid level. Under normal circumstances, the fluid should be clean and red in colour. If the fluid is burnt, discoloured and dark, this is indicative of burnt clutch linings. If the fluid is resinous and sticky, it may be due to an internal leak in the transmission fluid oil cooler, or engine overheating. If solid residues (abraded metal particles) are present, this is indicative of serious internal component wear/failure – all are tasks for your Ford dealer who will additionally flush out the transmission fluid oil cooler and its pipes.
10 The transmission fluid must be changed at the specified interval. A sump drain plug is provided. Use fresh fluid of the specified type when refilling.

23 Gearchange selector mechanism – cable removal, refitting and adjustment

1 Disconnect the battery earth terminal.
2 Refer to Chapter 11 and remove the centre console, if fitted.
3 Move the selector lever to position P (the lever on the transmission selector shaft should also be in position P – the most forward position). This is important for correct subsequent adjustment of the selector cable.
4 Detach the selector gate from the selector lever housing by undoing its two securing screws. Slide it off over the selector lever, having unscrewed the selector lever knob (if not already removed).
5 Release the selector cable from the selector lever by prising off the plastic eye with a screwdriver (photo).
6 Detach the selector cable abutment from the selector lever housing by releasing its retaining clip arrangement. Note the positioning of the retaining washers for subsequent reassembly.
7 Raise the vehicle and support securely on axle stands.
8 Detach the selector cable bracket from the transmission housing by undoing its two securing nuts (photo).
9 Detach the selector cable yoke from the lever on the transmission selector shaft by releasing its retaining clip and withdrawing (photo).
10 Pull the rubber gaiter out from the floor and withdraw the selector cable (photo).
11 To refit, pass the selector cable up through the floor and secure the rubber gaiter.

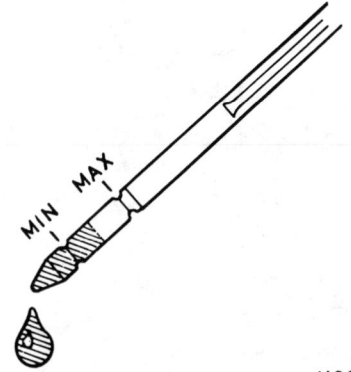

Fig. 6.60 CTX automatic transmission dipstick markings (Sec 22)

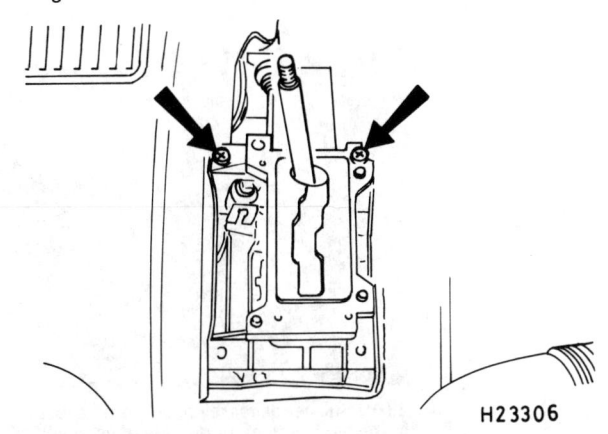

Fig. 6.61 CTX automatic transmission selector gate securing screws (arrowed) (Sec 23)

Chapter 6 Transmission

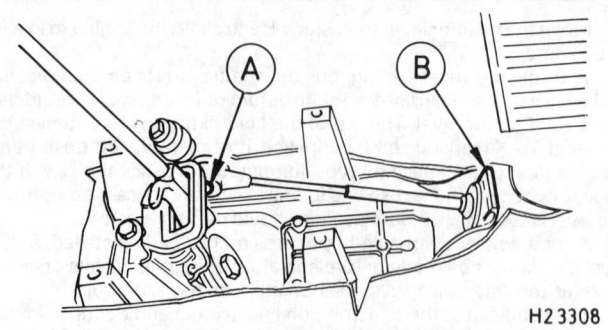

Fig. 6.62 Selector cable fixture to selector lever (A) and selector cable abutment retaining clip arrangement (B) (Sec 23)

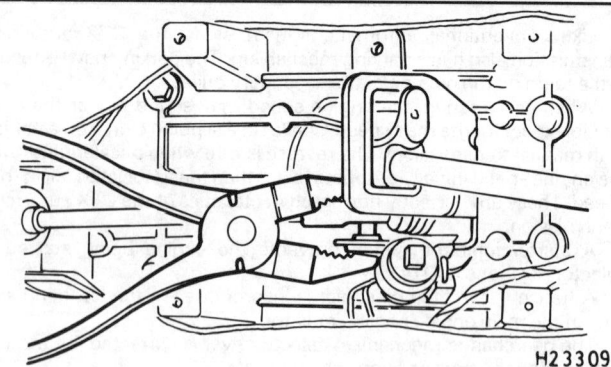

Fig. 6.63 Securing the selector cable to the selector lever (Sec 23)

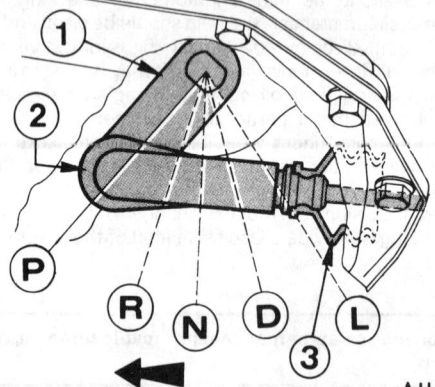

Fig. 6.64 Transmission selector shaft lever positions. Large arrow indicates the front of the vehicle (Sec 23)

1. Transmission selector shaft lever
2. Selector cable yoke
3. Rubber gaiter

12 Remove the axle stands and lower the vehicle to the ground.
13 Reattach the selector cable to the selector lever by using pliers or similar tool to press the eye into engagement. The raised rim of the eye on the selector cable must face the end of the locating pin on the selector lever.
14 Secure the selector cable abutment to the selector lever housing by assembling its retaining clip arrangement.
15 Locate the selector gate to the selector lever housing, align it, then tighten the two front retaining screws to the specified torque.
16 Refit the centre console, selector gate cover and selector lever knob (see Chapter 11).
17 Raise the vehicle again and support it securely on axle stands.
18 Attach the selector cable bracket to the transmission and tighten its securing nuts to the specified torque.
19 Adjust the selector cable, as necessary, in the following manner.
20 With the selector lever held in position P, make sure that the lever on the transmission selector shaft is still in position P (with the front wheels locked). If necessary, rotate the front wheels until the parking pawl engages to lock them. In this position, the holes in the lever on the transmission selector shaft and the yoke on the selector cable must line up, so that the pin on the retaining clip can be inserted easily. If

23.5 Release the selector cable by prising it from the selector lever

23.8 Detaching the selector cable bracket from the transmission housing

23.9 Removing the clip that secures the selector cable yoke to the lever on the transmission selector shaft

23.10 Rubber gaiter released from its location in the vehicle floor (underside of vehicle)

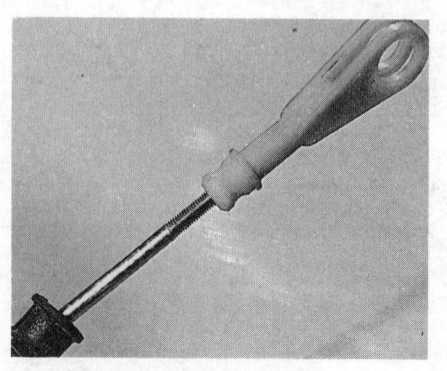

23.20 Selector cable adjustment threads (forward end of cable)

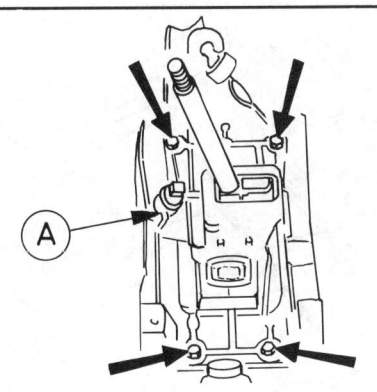

Chapter 6 Transmission

Fig. 6.65 Selector lever housing retaining bolts (arrowed), and selector cover illumination bulbholder (A) (Sec 24)

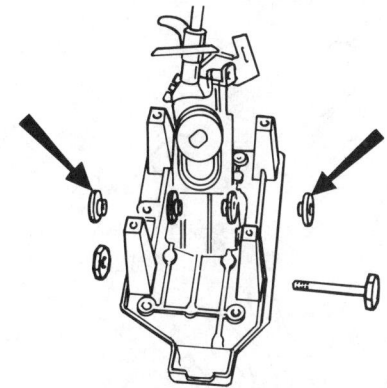

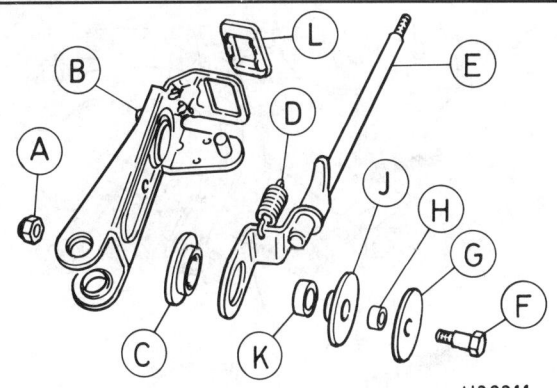

Fig. 6.66 Exploded view of selector lever assembly (Sec 24)

- A Nut
- B Selector lever guide
- C Guide bush
- D Spring
- E Selector lever
- F Retaining pin
- G Steel washer
- H Spacer bush
- J Plastic spacer washer
- K Bush
- L Plastic guide

Fig. 6.67 Inserting the selector lever assembly into its housing. Note pivot bushes (arrowed) (Sec 24)

necessary, slide the rubber gaiter back from the yoke and alter the position of the yoke on the selector cable, by screwing it in or out, until the pin on the retaining clip can be inserted easily. Slide the rubber gaiter back to its original position, if disturbed (photo).
21 With the selector cable correctly adjusted, refit the retaining clip to secure the selector cable yoke to the lever on the transmission selector shaft, then remove the axle stands and lower the vehicle to the ground.
22 Reconnect the battery earth terminal (Chapter 12, Section 5).

24 Gearchange selector mechanism – removal, overhaul and refitting

1 Disconnect the battery earth terminal.
2 Refer to Chapter 11 and remove the centre console, if fitted.
3 Carry out the procedure detailed in paragraphs 3 to 6 inclusive, in the previous Section.
4 Release the selector cover illumination bulbholder by pulling it off, then unscrew the four bolts securing the selector lever housing to the floor. Note earth lead fitment to the left-hand rear bolt. Remove the selector lever housing (photos).
5 To overhaul, begin by removing one of the pivot pin retaining clips, withdraw the pivot pin then remove the selector lever assembly from the housing (photo).
6 Detach the selector lever rod from the selector lever guide. This is achieved by unhooking the spring and unscrewing the nut from the retaining pin before dismantling. Note the order of components as they are removed.
7 Renew any worn or damaged components, then reassemble the selector lever rod and guide assembly by reversing the method of dismantling. Tighten the nut to the specified torque.
8 Fit the selector lever assembly, complete with two new bushes, into the selector lever housing. Ensure that the broad side of the bush guide faces upwards. Insert the pivot pin and secure with its previously-removed clip.
9 Refit the selector lever housing to the floor with its four retaining bolts, reconnecting the earth lead under the left-hand rear bolt. Tighten the bolts to the specified torque.
10 Fit the selector cover illumination bulbholder by reversing the method of removal.
11 Raise the vehicle and support it securely on axle stands, then refer to the previous Section and disconnect the selector cable yoke from the lever on the transmission selector shaft.
12 Remove the axle stands and lower the vehicle to the ground.
13 Carry out the procedure detailed in paragraphs 13 to 17 inclusive in the previous Section.
14 Complete the operation in accordance with paragraphs 19 to 22 inclusive, of the previous Section.

24.4A Unscrewing the selector lever housing bolts – note earth lead (arrowed)

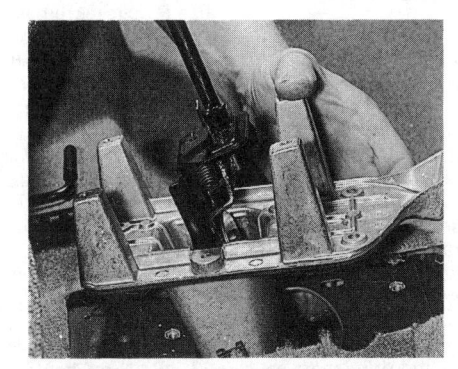

24.4B Withdrawing the selector lever housing

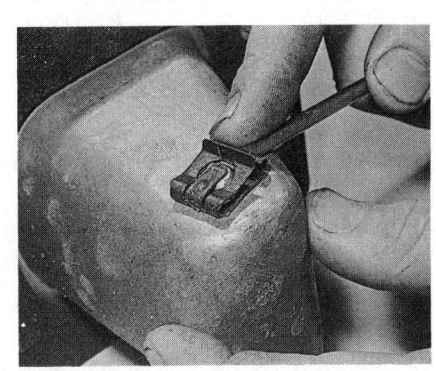

24.5 Removing a pivot pin retaining clip prior to selector lever assembly overhaul

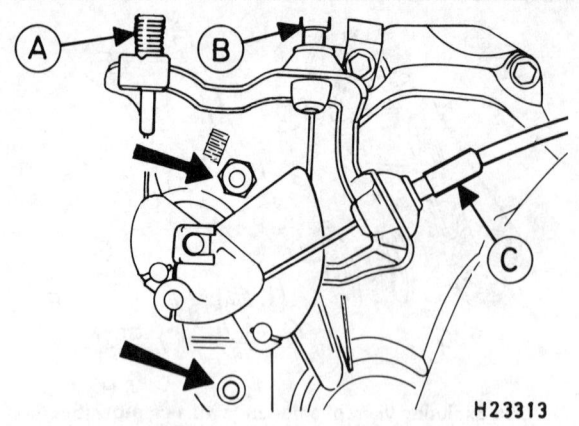

Fig. 6.68 CTX automatic transmission cable bracket assembly. Note retaining nuts (arrowed) (Sec 26)

A Accelerator pedal cable
B Throttle valve cable
C Carburettor cable

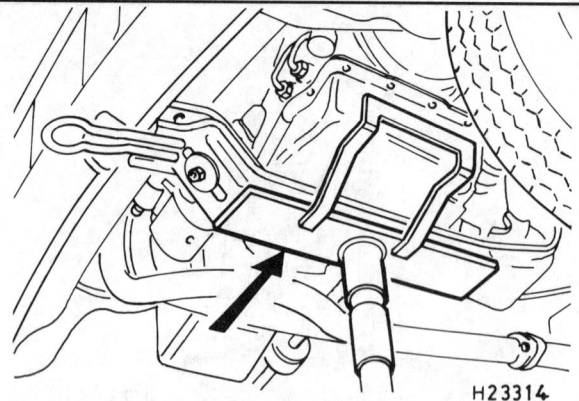

Fig. 6.69 Supporting the transmission. Large support plate (arrowed) to spread the load (Sec 26)

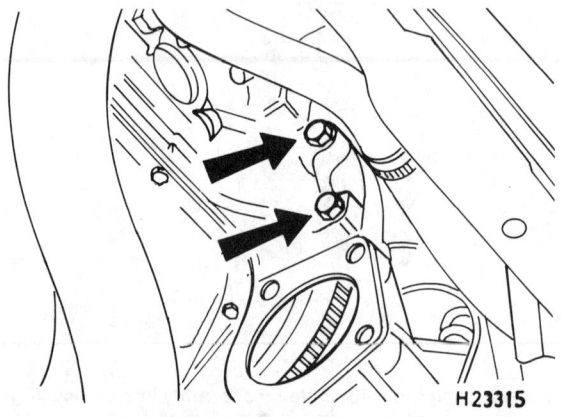

Fig. 6.70 Transmission flange bolts (front shown) (Sec 26)

25 Starter inhibitor switch – removal and refitting

1 Refer to Chapter 12, Section 15.

26 Automatic transmission – removal and refitting

1 Disconnect the battery earth terminal.
2 Refer to Chapter 12, Section 24 and disconnect the speedometer cable. Tie it up out of the way. If fitted, disconnect the lead from the road speed sensor.
3 Detach the cable bracket assembly from the transmission by unscrewing its two securing nuts, then disconnect the throttle valve cable from the assembly. Tie the cable bracket assembly out of the way in a convenient location (see Fig. 6.68).
4 Unscrew the two upper transmission flange bolts and remove them. On certain models, these bolts also hold wiring loom securing brackets – the loom should be tied up out of the way.
5 On vehicles with CVH engines, tie the thermostat housing-to-heater hose back to prevent it getting in the way when refitting the transmission.
6 Raise the vehicle and support securely on axle stands.
7 Refer to Section 23 and disconnect the selector cable yoke from the lever on the transmission selector shaft, and the selector cable bracket from the transmission.
8 Disconnect the wiring from the starter motor and the starter inhibitor switch, then unbolt and remove the starter motor (Chapter 12).
9 Detach the lower engine adaptor plate (clutch housing cover).
10 Detach all earth leads from the transmission, as applicable.
11 Disconnect the automatic transmission fluid oil cooler pipes at the transmission, and move them aside. Fit blanking plugs to prevent excessive fluid loss, and to avoid dirt ingress.
12 Disconnect the lower suspension arm balljoint from the spindle carrier, on both sides of the vehicle (see Chapter 10).
13 Refer to Chapter 9 and disconnect the track rod end from the steering arm on the spindle carrier, on both sides of the vehicle.
14 Disconnect the left-hand driveshaft assembly from the transmission using the following method. Locate a large lever between the transmission casing and the inner CV joint then, with an assistant applying pressure to the roadwheel (away from the centre of the vehicle), strike the lever sharply with the hand. Insert a suitable plug (an old inner CV joint is ideal) to prevent excessive transmission fluid loss, and to prevent the differential rotating when the right-hand driveshaft is removed. Suspend the driveshaft assembly to avoid straining the CV joints; the inner joint must not be bent more than 20°, and the outer joint 40°.
15 Repeat the above procedure on the right-hand driveshaft assembly.
16 Fit an engine support bar (or similar device) and take the weight of the engine/transmission. Detach the transmission bearer from the vehicle body by undoing its four securing bolts. The transmission bearer is removed along with the transmission (refer to Chapter 1, Section 13).
17 Lower the engine/transmission as far as the engine support bar allows.
18 Support the transmission from underneath using a jack, and resting it on a support plate to spread the weight. Support directly on the transmission beater, **not** the sump, and take great care not to dent or damage the sump.
19 At this stage it is advisable to have an assistant at the ready. Undo the four remaining transmission flange bolts (two at the front and two at the rear) and carefully withdraw the transmission from the engine (Fig. 6.70). Lower the jack and remove the transmission.
20 When refitting the transmission do **not** lubricate the transmission input shaft, as this may contaminate and adversely affect the operation of the torsional vibration damper.
21 In a reversal of the removal technique, connect the transmission to the engine, with the transmission input shaft engaging centrally in the torsional vibration damper. Insert and tighten the four transmission flange bolts (as removed in paragraph 19) to the specified torque. Do **not** draw the transmission onto the engine using the flange bolts.
22 Remove the jack and support plate, and raise the engine/transmission on the engine support bar to enable the refitting of the transmission bearer. Tighten the transmission bearer securing bolts to the specified torque.
23 Remove the engine support bar.
24 With a new snap-ring fitted to both inner CV joint splined shafts, lubricate the splined shafts with transmission fluid (see Chapter 7).
25 Remove the plug (or old inner CV joint) from the transmission casing and insert the right-hand driveshaft inner CV joint through the transmission casing into its splined differential location. Apply pressure at the roadwheel to ensure that the snap-ring **fully** locates in the differential.
26 Repeat the above procedure on the left-hand driveshaft assembly.
27 Refer to Chapter 9 and reconnect the track rod ends. Note that **new** split pins must be used.
28 Refer to Chapter 10 and reconnect the lower suspension arm balljoints to the spindle carriers.

Chapter 6 Transmission

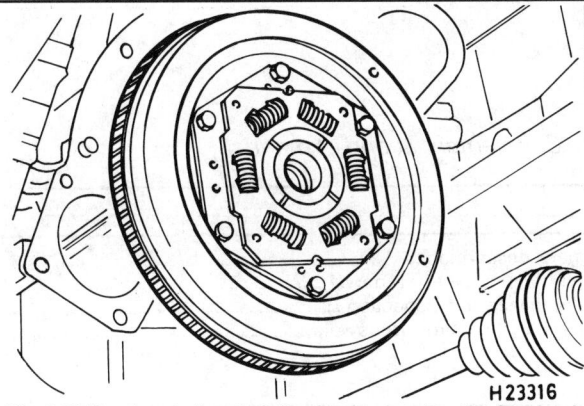

Fig. 6.71 Torsional vibration damper secured to the flywheel (Sec 27)

the other two cables attached to the cams on the cable bracket assembly, will need to be checked by a Ford dealer at the earliest opportunity.

27 Torsional vibration damper - removal and refitting

1 To obtain access to the torsional vibration damper, the engine and transmission assemblies must be separated. Refer to Chapter 1, or Section 26 of this Chapter.
2 Unscrew the six bolts securing the assembly to the flywheel progressively, so as to avoid the risk of distortion, then remove it.
3 Refitting is a straightforward reversal of the removal procedure, tightening the bolts progressively, in a diagonal sequence, to the specified torque.

29 Remove the blanking plugs, and reconnect the automatic transmission fluid oil cooler pipes. Refer to the Specifications for tightening torque details.
30 Refitting is now a reversal of the removal procedure, making reference to the relevant text. Refer to Section 23 of this Chapter when refitting the selector cable, and adjust as necessary. Refer to Chapter 12, Section 5 when reconnecting the battery earth terminal.
31 Upon completion, check the transmission fluid level and top up as necessary (refer to Section 22 of this Chapter).
32 Note that the throttle valve cable adjustment, along with that of

28 Driveshaft/transmission oil seal - renewal

1 Refer to Chapter 7, Section 3.

29 Transmission mounting - removal and refitting

1 Refer to Chapter 1, Section 13.

30 Fault diagnosis

Note: *Due to the complex nature of the CTX automatic transmission and its hydraulic control system, any suspected internal faults should be referred to your Ford dealer for accurate diagnosis and repair. This listing is not exhaustive, and is intended for guidance only.*

Symptom	Reason(s)
Engine can be started in selector positions 'L', 'D' or 'R'	Gearchange selector cable maladjusted Starter inhibitor switch malfunction
Engine cannot be started in selector positions 'N' or 'P' (engine, fuel and electrical systems in order)	Gearchange selector cable maladjusted Starter inhibitor switch malfunction
Very harsh engagement when selecting positions 'L', 'D' or 'R'	Engine idle speed incorrectly adjusted Internal fault
Poor acceleration or driveability	Internal fault Transmission fluid burnt or contaminated (due to internal fault) Fuel, ignition or engine fault
Vibration	Engine/transmission mountings defective or loose Torsional vibration damper defective Driveshaft fault or wheel bearing defective
Engine dies or vehicle creeps excessively when position L, D or R is engaged (engine tuning adjustments correct)	Internal fault
Vehicle does not creep in selector positions L, D or R	Engine idle speed incorrectly adjusted Internal fault
Vehicle creeps in selector position N	Internal fault
Noise evident when pulling away, or sudden power loss	Oil level incorrect Internal fault
Harsh jerk on kickdown or just before vehicle comes to a standstill	Internal fault

Chapter 7 Driveshafts

For modifications, and information applicable to later models, see Supplement at end of manual

Contents

Driveshaft assembly – overhaul	7
Driveshaft assembly – removal and refitting	6
Driveshaft/transmission oil seal – renewal	3
Fault diagnosis – driveshafts	8
General description	1
Inner CV joint gaiter – renewal	4
Maintenance and inspection	2
Outer CV joint gaiter – renewal	5

Specifications

Type .... Unequal length solid shafts, splined to inner and outer constant velocity (CV) joints

Lubrication
Type .... Grease to Ford specification SQM-1C-9004-A (Duckhams LBM 10)
Quantity:
 1.1 litre models .... 30 g (1.05 oz) per joint
 All other models .... 40 g (1.4 oz) per joint

Torque wrench settings

	Nm	lbf ft
Track rod end to spindle carrier steering arm	25 to 30	18 to 22
Hub driveshaft retaining nut	205 to 235	151 to 173
Lower arm balljoint to spindle carrier pinchbolt and nut (Torx)	48 to 60	35 to 44
Brake caliper anchor bracket bolts	50 to 66	37 to 49
Roadwheel nuts	70 to 100	52 to 74

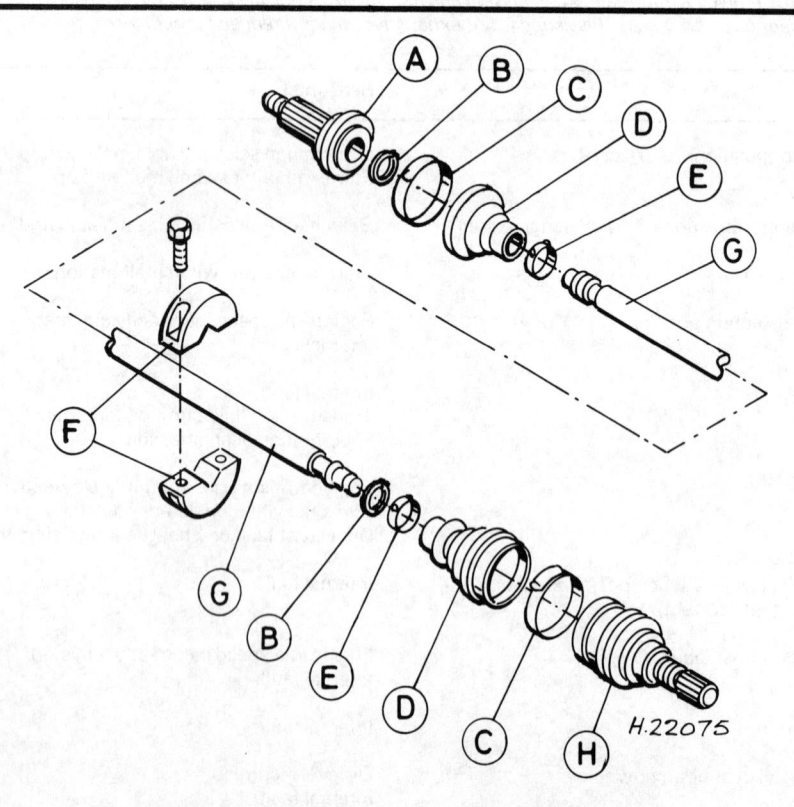

Fig. 7.1 Exploded view of right-hand driveshaft assembly (Sec 1)

A Outer CV joint
B Circlip
C Large gaiter clamp
D Gaiter
E Small gaiter clamp
F Vibration damper
G Driveshaft main member
H Inner CV joint

Chapter 7 Driveshafts

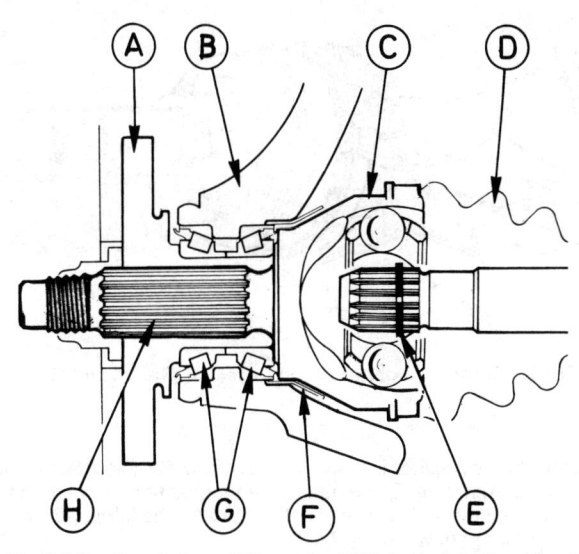

Fig. 7.2 Sectional view of the outer CV joint in location (Sec 1)

- A   Wheel hub
- B   Spindle carrier
- C   Fixed outer CV joint (ball and cage type)
- D   Gaiter
- E   Circlip (driveshaft main member to outer CV joint)
- F   Dust shield
- G   Bearings
- H   Outer CV joint splined shaft section

1 General description

Drive is transmitted from the differential to the front roadwheels by two unequal length driveshaft assemblies. On certain models the longer of the two driveshafts incorporates a vibration damper.

Each driveshaft assembly consists of three main components – a sliding type inner CV joint, a solid driveshaft splined at both ends and a fixed type outer CV joint. The inner CV joints are secured in the differential by the engagement of snap-rings, whilst the outer CV joint are secured in their respective hubs by nuts which are deformed after tightening.

On vehicles fitted with the anti-lock braking system, the inner CV joints drive the modulators through a Kevlar reinforced rubber belt. The main distinguishing feature of these inner CV joints, over those fitted to vehicles with a conventional braking system, is the teeth machined into the outer housings.

On earlier models, all CV joints are of the ball and cage type. Certain later models may be fitted with tripod type inner CV joints (see Chapter 13).

2 Maintenance and inspection

1 At the intervals given in the *Routine maintenance* section at the beginning of this manual, carry out a thorough inspection of the driveshafts and CV joints as follows.

2 Raise the front of the vehicle and support it securely on axle stands. Remove the one-piece undertray, if fitted.

3 Slowly rotate a roadwheel and inspect the condition of the outer CV joint rubber gaiter. Check for splits and tears or other deterioration of the rubber which may allow the grease to escape and lead to water and grit entry into the joint. Also check the security and condition of the retaining clamps. Repeat these checks on the other outer CV joint and both of the inner CV joints. If damage or deterioration is found, the gaiters should be renewed, as described in Section 4 and 5 of this Chapter, as applicable.

4 Continue rotating the roadwheel and check for any distortion or damage to the driveshaft. Ensure that the driveshaft vibration damper is located securely on the driveshaft, as applicable.

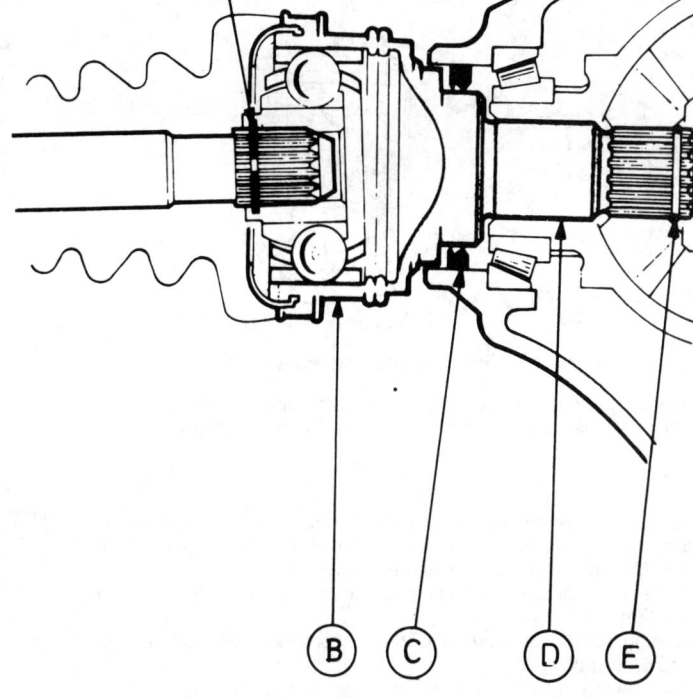

Fig. 7.3 Sectional view of the inner CV joint in location (Sec 1)

- A   Circlip (driveshaft main member to inner CV joint)
- B   Sliding inner CV joint (ball and cage type)
- C   Oil seal
- D   Inner CV joint shaft section
- E   Snap-ring

5 Check for any free play in the outer CV joints by first holding the driveshafts, then attempting to rotate the roadwheels. Repeat this check on the inner CV joints by holding the inner joints and attempting to rotate the driveshafts. Any appreciable movement felt indicates wear in the joints or wear in the driveshaft splines.

6 Road test the vehicle, and listen for a metallic clicking from the front as the car is driven slowly with the steering on full lock. If a clicking noise is heard, this indicates that wear has taken place in an outer CV joint, caused by excessive clearance between the balls in the joint and the recesses in which they operate. Remove and inspect the suspect joint, as described in Section 7.

7 If vibration consistent with road speed is felt through the vehicle, particularly during acceleration, then there is a possibility that wear has taken place within the inner CV joints. Remove and inspect the suspect joint, as described in Section 7.

8 Examine the transmission casing, where the inner CV joints enter, for signs of transmission oil leakage. If this is present, the driveshaft transmission oil seals must be replaced in accordance with the following Section.

3 Driveshaft/transmission oil seal – renewal

1 Withdraw the driveshaft inner CV joint from its differential location, as described in Chapter 8, Section 26, then suspend the driveshaft assembly using a length of strong wire, remembering that the CV joints must not be bent at an angle greater than 20°. Allow escaping transmission oil to drain off into a suitable container. Note that paragraphs 4 and 5 in the referred text apply only to vehicles fitted with the anti-lock braking system. Tackle each oil seal renewal as a separate operation, withdrawing one driveshaft at a time from the transmission casing.

Chapter 7 Driveshafts

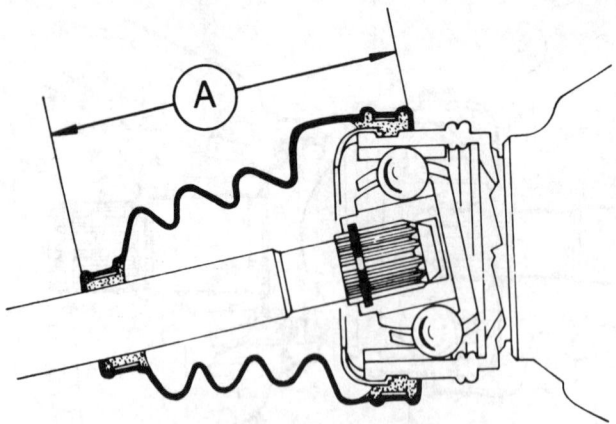

Fig. 7.4 CV joint gaiter fitting (Secs 4 and 5)

A on inner joint (shaft angle 10° to 20°) = 80 to 90 mm (3.1 to 3.5 in)
A on outer joint (shaft horizontal) = 98 to 102 mm (3.8 to 4.0 in)

2 Prise out the old oil seal using a screwdriver or other suitable tool, taking care not to raise any burrs on the transmission casing which would prevent the new seal from seating correctly.
3 The new oil seal can now be tapped into position by using a suitably sized section of tubing as a drift.
4 Refitting should be carried out in accordance with Chapter 8, Section 26, as applicable.

4 Inner CV joint gaiter – renewal

Note: *Check parts availability before commencing as a new circlip, snap-ring and gaiter clamps are required for reassembly.*

1 Proceed as described in the first part of paragraph 1, Section 3, noting that paragraphs 4 and 5 in Chapter 8, Section 26 apply only to vehicles with anti-lock brakes.
2 Unclip and remove the gaiter retaining clamps by prising open the looped section of the clamp with a screwdriver, then slide the gaiter along the shaft to expose the CV (constant velocity) joint.
3 Wipe the surplus grease from the CV joint, then prise open the securing circlip and pull the shaft from the joint.
4 The gaiter can now be withdrawn from the shaft.
5 Fit the new gaiter into position on the shaft and repack the CV joint with grease. Insert the driveshaft back through the joint unit, pushing through until the circlip is felt to engage and secure it.

4.7 Inner CV joint gaiter fitted correctly (crimped section of clamps arrowed)

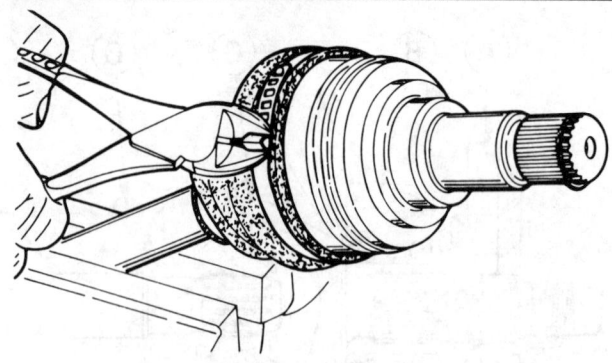

Fig. 7.5 Crimping a gaiter clamp to secure (Secs 4 and 5)

6 Locate the new gaiter into position over the CV joint. The gaiter positioning is important. Check that, when in position with the inner joint fully contracted and at an angle of 10° to 20°, the full length of the gaiter clamps is as shown in Fig. 7.5.
7 Fit and fasten the gaiter clamps by holding them round the gaiter finger tight, then clamp the pin into the next engagement hole. Crimp the clamp to secure it (photo).
8 The remainder of the refitting operation should be carried out in accordance with Chapter 8, Section 26 (as applicable).

5 Outer CV joint gaiter – renewal

Note: *Check parts before commencing as a new circlip, snap-ring and gaiter clamps are required for reassembly.*

1 Unless the driveshaft is to be removed completely for other repair work to be carried out (refer to Section 6), the following method of outer CV joint gaiter renewal is recommended to avoid having to disconnect the driveshaft from the spindle carrier.
2 Remove the inner CV joint gaiter, as described in the preceding Section.
3 On the right-hand driveshaft, mark the relative position of the vibration damper on the shaft then unbolt and remove the damper unit.
4 Release the clamps on the outer CV joint and slide the gaiter along the driveshaft until it can be removed from the inner end of the shaft.
5 Thoroughly clean the driveshaft before sliding on the new gaiter. Replenish the outer CV joint with specified lubricant and slide the gaiter over the joint, setting its overall length to the appropriate dimension (Fig. 7.4).
6 Fit and tighten the gaiter clamps, but make sure that the crimped part of the clamp nearest the hub does not interfere with the spindle carrier as the driveshaft is rotated.
7 Refit the inner CV joint and its gaiter, and connect the driveshaft to the transmission, making reference to the preceding Section and its further referred text.
8 Refit the vibration damper to the right-hand driveshaft and set it in the original position marked during removal or refer to Fig. 7.6 for its setting before fully tightening the retaining bolts.

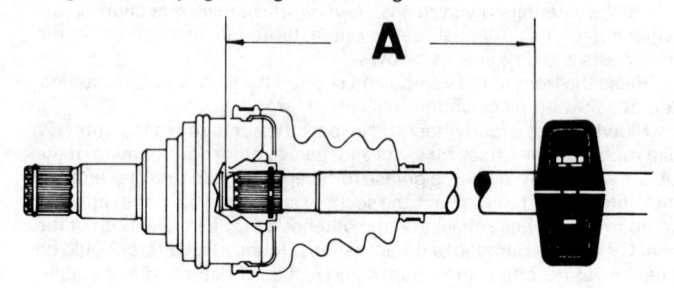

Fig. 7.6 Vibration damper location on right-hand driveshaft (Sec 5)

A = 308 to 312 mm (12.1 to 12.3 in)

Chapter 7 Driveshafts

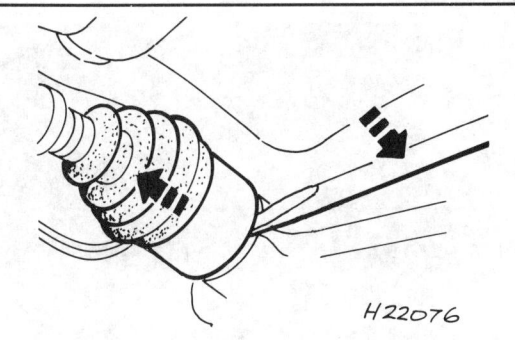

Fig. 7.7 Releasing a driveshaft inner CV joint from its differential location (Sec 6)

6 Driveshaft assembly – removal and refitting

1   Using a small punch, remove the deformity securing the hub driveshaft retaining nut to its shaft, and slacken the nut by half a turn.
2   Disconnect the battery earth terminal.
3   Slacken the relevant roadwheel nuts, then raise the front of the vehicle and support it securely using axle stands. Remove the roadwheel.
4   Remove the one piece undertray where fitted, by turning its bayonet type fasteners and, on XR2i models, remove the front suspension crossmember (see Chapter 10).
5   Remove the bolts securing the brake caliper anchor bracket to the spindle carrier and suspend the caliper assembly out of the working area, using a length of strong wire to prevent the brake hose from straining. Also release the brake hose from its suspension strut location (photo).
6   Remove the track rod balljoint from the steering arm on the spindle carrier, as described in Chapter 9, Section 3, paragraphs 6 and 7.
7   Remove the (Torx head) pinch-bolt and locknut securing the lower suspension arm balljoint to the spindle carrier, then separate the balljoint from the spindle carrier.
8   Disconnect the anti-roll bar link from the anti-roll bar, where fitted, by undoing the nut retaining it.
9   Using a homemade bracket, secure the wheel hub and brake disc to the spindle carrier, retaining the bracket at one end with a wheelnut and at the other end with a brake caliper retaining bolt.
10   Remove the hub driveshaft retaining nut.
11   On vehicles equipped with the anti-lock braking system, remove the belt break switch and drivebelt cover, then slacken the modulator pivot and adjuster bolts to release drivebelt tension before slipping the drivebelt from its modulator pulley location (see Chapter 8, Section 26).
12   Release the driveshaft from its hub location by pulling the spindle carrier outwards, away from the centre of the vehicle. Do not remove it from its location at this stage, but ensure that it is free to be withdrawn when required. If it is held tightly in the hub, gentle tapping on the end of the shaft with a soft-faced hammer may suffice – if this does not work, a proprietary puller should be obtained.
13   Insert a suitable lever between the inner CV joint and the transmission casing, then strike the lever firmly with the flat of the hand to

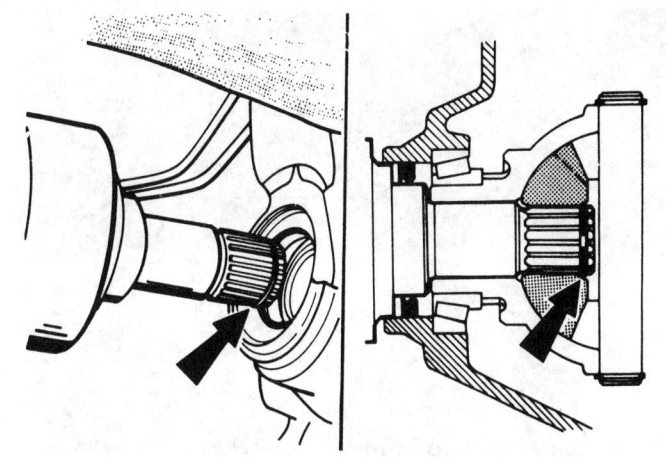

Fig. 7.8 Snap-ring (arrowed) must locate fully in the differential (Sec 6)

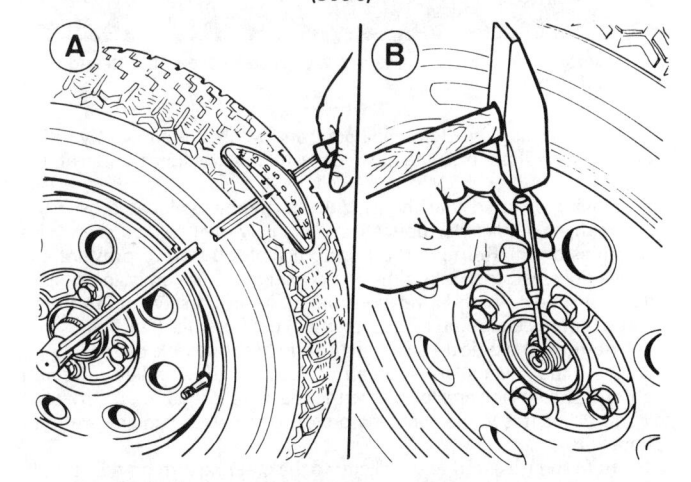

Fig. 7.9 Tighten the hub driveshaft retaining nut to the specified torque (A), then deform the nut into the driveshaft slot (B) (Sec 6)

release the joint from its differential location (Fig. 7.17). Be careful not to damage adjacent components and make provision for escaping transmission oil, if possible plugging the hole to prevent excessive loss. Do not allow the CV joints to bend more than 20° from the horizontal.
14   Withdraw the driveshaft assembly from its hub location and remove the assembly from the vehicle. Note that if both driveshaft assemblies are to be removed at the same time, the differential must be immobilised by inserting an old inner CV joint or suitable shaft, before the other driveshaft is removed.
15   Remove the old snap-ring from the inner CV joint splined shaft.
16   Prior to installation, fit a **new** snap-ring to the inner CV joint splined shaft (photo).
17   Insert the driveshaft assembly to the hub (photo), drawing it fully

6.5 Brake hose location to suspension strut (arrowed) (models without anti-roll bar)

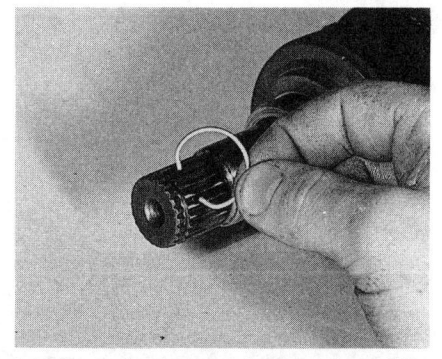

6.16 Renewing the snap-ring on an inner CV joint splined shaft

6.17 Inserting the outer CV joint on the driveshaft assembly to the hub

6.18 Inserting the inner CV joint on the driveshaft assembly to engage in the differential

7.4 Vibration damper installed to right-hand driveshaft assembly

into position using the old retaining nut and washer, then loosely refit a **new** hub driveshaft retaining nut and washer. Do not fully tighten at this stage.

18 Having ensured that the modulator drivebelt is located over the driveshaft, as applicable, lubricate the inner CV joint splined shaft with transmission oil. Remove the temporary plug fitted to prevent excessive fluid loss, then insert the joint, through the transmission casing, into its splined differential location (photo). Apply pressure at the hub to ensure that the snap-ring fully locates in the differential (Fig. 7.8).
19 Remove the bracket securing the hub to the spindle carrier.
20 Refit the lower suspension arm balljoint to the spindle carrier and tighten its pinch-bolt and locknut to the specified torque. Ensure that the pinch-bolt locates to the annular groove on the balljoint spindle (refer to Chapter 10).
21 Refit the track rod end balljoint to the steering arm on the spindle carrier as described in Chapter 9, Section 3, paragraph 16.
22 Reconnect the anti-roll bar link to the anti-roll bar and tighten to the specified torque, as applicable (refer to Chapter 10).
23 Refit the brake caliper to the spindle carrier, fully tightening its anchor bracket bolts to the specified torque.
24 Secure the brake hose to its suspension strut location.
25 On vehicles equipped with the anti-lock braking system, refit the modulator drivebelt, cover and belt break switch, tensioning the drivebelt as described in Chapter 8, Section 25.
26 Refit the front suspension crossmember and the one-piece undertray, as applicable.
27 Refit the roadwheel and lower the vehicle to the ground. Fully tighten the roadwheel nuts to the specified torque with the vehicle resting on its wheels.
28 Fully tighten the new hub driveshaft retaining nut to the specified torque then, using a suitable punch, deform the nut into the driveshaft slot so that it is held securely.
29 Check the level of the transmission oil, and top up as required.
30 Reconnect the battery earth terminal. On vehicles fitted with the EEC-IV engine management module, reference must be made to Chapter 12, Section 5 to allow the module to re-learn its values.

7 Driveshaft assembly – overhaul

Note: *Check parts availability before commencing as new circlips and gaiter clamps are required for reassembly.*

1 Remove the driveshaft, as described in the previous Section.
2 Clean away external dirt and grease, release the gaiter clamps and slide the gaiters from the CV joints.
3 Wipe away enough lubricant to be able to extract the circlip and then separate each CV joint with splined shaft section from the main member of the driveshaft.
4 If removing the vibration damper from the right-hand driveshaft, mark its relative position on the shaft before unbolting it (photo).
5 Thoroughly clean the joint components and examine for wear or damage to the balls, cage, socket or splines. A repair kit may provide a

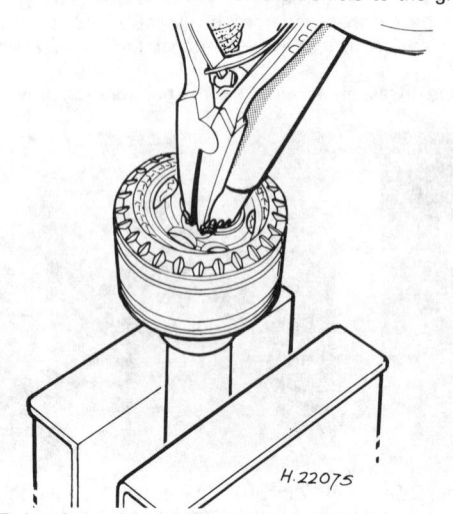

Fig. 7.10 Releasing a driveshaft retaining circlip from a ball and cage type CV joint. Note soft jaw protectors fitted to vice (Sec 7)

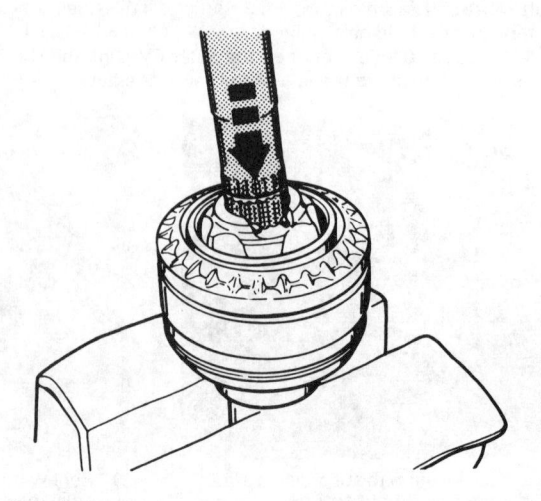

Fig. 7.11 Refitting a driveshaft to a ball and cage type CV joint, with circlip in position. Note soft jaw protectors fitted to vice (Sec 7)

Chapter 7 Driveshafts

solution to the problem, but, if the socket requires renewal, this will of course include the splined section of shaft and will prove expensive. If both joints require renewal of major components, then a new driveshaft or one which has been professionally reconditioned may prove to be more economical.

6 If the vibration damper was removed from the right-hand driveshaft, refit it in the position marked during its removal.

7 Reassemble the joints by reversing the dismantling operations. Use new circlips and pack each joint with the specified quantity of lubricant. When fitting the gaiters set their length in accordance with the information given in Section 4 or 5 according to which joint (inner or outer) is being worked upon. Use new gaiter clamps.

8 Refit the driveshaft, as described in the previous Section.

8 Fault diagnosis – driveshafts

Symptom	Reason(s)
Noise when taking up drive	Worn driveshaft or CV joint splines Worn CV joints Loose driveshaft retaining nut Loose roadwheel nuts
Vibration and/or noise, especially noticeable on slow turns	Lack of lubrication in CV joints, possibly due to split or torn gaiter Worn outer CV joint
Vibration during acceleration or on overrun	Worn inner CV joint Incorrectly positioned or loose driveshaft vibration damper Bent or distorted driveshaft

Chapter 8 Braking system

For modifications, and information applicable to later models, see Supplement at end of manual

Contents

Anti-lock braking system - description	24
Brake warning lights - description and renewal	22
Brake pedal - removal, bush renewal and refitting	23
Brake servo - removal and refitting	17
Brake servo operating link - removal and refitting	19
Brake servo operating link bushes - renewal	18
Brake servo vacuum hose - renewal	16
Fault diagnosis	30
Front brake caliper - removal, overhaul and refitting	5
Front brake disc - inspection, removal and refitting	6
Front brake pads - inspection and renewal	4
General description	1
Handbrake - adjustment	11
Handbrake lever - removal and refitting	12
Handbrake primary cable - renewal	13
Handbrake rear cable - renewal	14
Hydraulic pipes and hoses - removal and refitting	3
Hydraulic system - bleeding (anti-lock braking system)	29
Hydraulic system - bleeding (conventional braking system)	21
Load-apportioning valve - adjustment (anti-lock braking system)	27
Load-apportioning valve - removal and refitting (anti-lock braking system)	28
Maintenance and inspection	2
Master cylinder - removal, overhaul and refitting	15
Modulator - removal and refitting (anti-lock braking system)	25
Modulator drivebelt - removal and refitting (anti-lock braking system)	26
Pressure control valve - removal and refitting	20
Rear brake carrier plate - removal and refitting	10
Rear brake drum/hub assembly - removal, inspection and refitting	7
Rear brake shoes - renewal	8
Rear wheel cylinder - removal, overhaul and refitting	9

Specifications

System type	Servo-assisted, diagonally-split dual-circuit hydraulic, with pressure regulation to the rear brakes. Cable-operated handbrake acting on rear brakes. Anti-lock brakes optional according to model
Fluid type/specification	Brake fluid to Ford specification SAM-6C-9103A Amber (Duckhams Universal Brake and Clutch Fluid)
Hub/brake drum-to-axle flange bolt sealant	To Ford specification SDM-4G 9107A

Front brakes

Type	Solid or ventilated disc with single-piston sliding calipers
Disc diameter	240 mm
Disc thickness:	
Solid disc	10 mm
Ventilated disc	20 mm
Minimum disc thickness:	
Solid disc	8 mm
Ventilated disc	18 mm
Maximum disc run-out (disc installed)	0.1 mm
Minimum brake pad thickness	1.5 mm
Caliper piston diameter:	
Solid disc	48 mm
Ventilated disc	54 mm

Rear brakes

Type	Drums with leading and trailing shoes and automatic adjusters
Drum diameter:	
All models except XR2i and anti-lock variants	180 mm
XR2i and anti-lock variants	203 mm
Wheel cylinder diameter:	
All models except XR2i and anti-lock variants	17.5 mm
XR2i with conventional braking system	19 mm
All models equipped with anti-lock brakes	22 mm
Minimum brake shoe lining thickness	1 mm

Chapter 8 Braking system

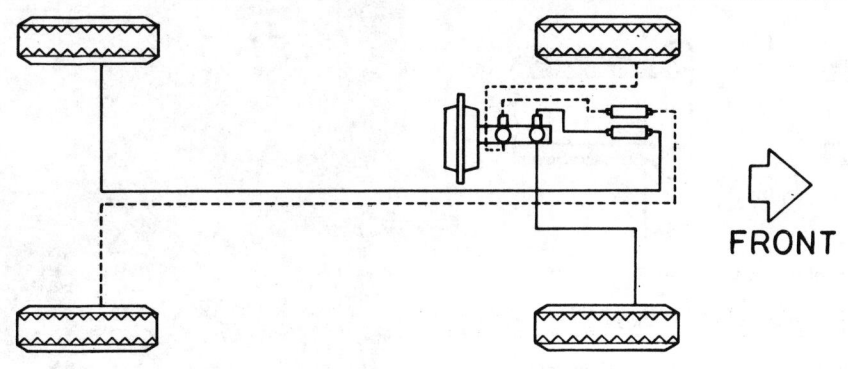

Fig. 8.1 Dual circuit hydraulic system (Sec 1)

Torque wrench settings

	Nm	lbf ft
Master cylinder to brake servo	20 to 25	15 to 18
Servo to mounting bracket	35 to 45	26 to 33
Brake servo operating link brackets to bulkhead	20 to 25	15 to 18
Rear hub to axle	56 to 76	41 to 56
Caliper-to-spindle carrier (anchor bracket) bolts	50 to 66	37 to 49
Caliper piston housing retaining bolts	20 to 25	15 to 18
Load-apportioning valves to bracket	20 to 25	15 to 18
Load-apportioning valve bracket to vehicle	21 to 28	15 to 21
Load-apportioning valve adjustment screw	12 to 16	9 to 12
Load-apportioning valve-to-axle twist beam link rod nut	21 to 28	15 to 21
Modulator pivot and adjusting clamp bolts	22 to 28	16 to 21
Modulator drivebelt cover	8 to 12	6 to 9
Trackrod end balljoint to steering arm	25 to 30	18 to 22
Lower arm balljoint-to-spindle carrier pinch-bolt (Torx)	48 to 60	35 to 44

1 General description

The braking system is of diagonally-split dual-circuit hydraulic type, with servo assistance, employing discs at the front and drums at the rear. The dual-circuit hydraulic system is a useful safety feature, whereby each circuit operates one front and one diagonally opposite rear brake from a tandem master cylinder. Under normal circumstances both circuits operate in unison but, in the event of hydraulic failure in one circuit, the other circuit will still allow full braking force to retard two wheels, and enable the vehicle to be brought to rest. Brake pressure control valves are fitted in-line to each rear brake circuit, to regulate the braking force available at each rear wheel, and reduce the possibility of the rear wheels locking up under heavy braking.

The front brake discs are ventilated on the XR2i, SX and anti-lock equipped models, and solid on the rest, and are acted on by single-piston floating brake calipers. This ensures that equal effort is applied through each pair of brake pads to the discs. The brake pads can be checked for wear without the need to remove the roadwheels.

The rear brake drums can be simply unbolted from their rear axle flanges, along with their integral hub assemblies, and removed to provide unhindered access to the shoe assemblies. The shoe assemblies, which incorporate automatic adjuster mechanisms, are operated by a single wheel cylinder on each rear brake. The leading shoe has a thicker lining than the trailing shoe, as they are designed to wear proportionally.

A cable-operated handbrake acts on the rear brakes to provide an independent means of rear brake operation.

An anti-lock braking system is available as optional equipment, and features many components in common with the conventional braking system. Further information on the anti-lock braking system can be found in later Sections of this Chapter.

2 Maintenance and inspection

1 At weekly intervals, check the fluid level in the translucent reservoir on the master cylinder. The fluid will drop very slowly indeed over a period of time to compensate for lining wear, but any sudden drop in level, or the need for frequent topping-up should be investigated immediately.

2 Always top up with hydraulic fluid which meets the specified standard and has been left in an airtight container. Hydraulic fluid is hygroscopic (absorbs moisture from the atmosphere) and must not be stored in an open container. **Do not** shake the tin prior to topping-up. Fluids of different makes can be intermixed provided they all meet the specification.

3 Inspect the thickness of the friction linings on the disc pads and brake shoes as described in the following Sections, at the intervals specified in 'Routine maintenance'.

4 The rigid and flexible hydraulic pipes and hoses should be inspected for leaks or damage regularly. Although the rigid lines are plastic-coated in order to preserve them against corrosion, check for damage which may have occurred through flying stones, careless jacking or the traversing of rough ground.

5 Bend the hydraulic flexible hoses sharply with the fingers and examine the surface of the hose for signs of cracking or perishing of the rubber. Renew if evident.

6 Renew the brake fluid at the specified intervals and examine all rubber components (including master cylinder and piston seals) with a critical eye, renewing where necessary.

7 Check the adjustment of the handbrake cable periodically, and

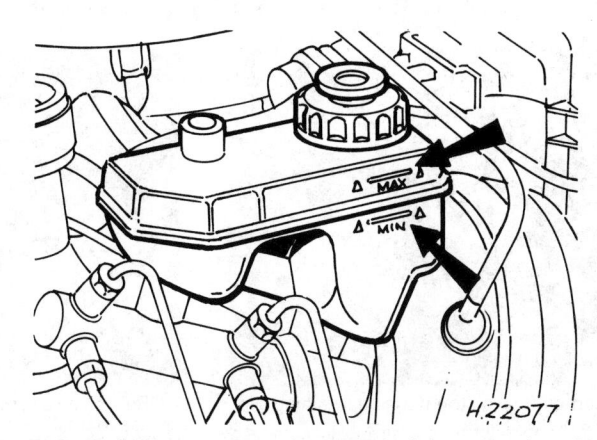

Fig. 8.2 Brake fluid reservoir maximum and minimum level markings (Sec 2)

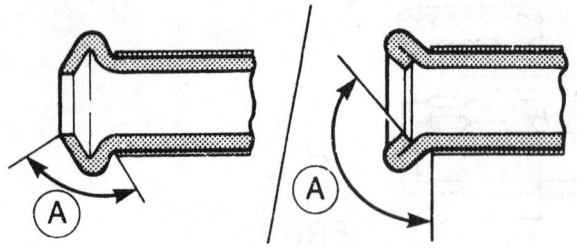

192 Chapter 8 Braking system

Fig. 8.3 Brake pipe flare (Sec 3)

A *Protective coating removed before flaring*

ensure that its associated mechanisms operate freely. Inspect the cable for signs of wear or damage, and ensure that it is located correctly in its clips and guides.

3 Hydraulic pipes and hoses – removal and refitting

1 Inspection has already been covered in the previous Section.
2 Always disconnect a flexible hose by prising out the spring anchor clip from the support bracket (photo) and then, using two close-fitting spanners, disconnect the rigid line from the flexible hose.
3 Once disconnected from the rigid pipe, the flexible hose may be unscrewed from the caliper or wheel cylinder.
4 When reconnecting pipelines, or hose fittings, remember that all union threads are to metric sizes. No copper washers are used at unions and the seal is made at the swaged end of the pipe, so do not try to wind a union in if it is tight yet still stands proud of the surface into which it is screwed.
5 A flexible hose must never be installed twisted, but a slight 'set' is permissible to give it clearance from an adjacent component. Do this by turning the hose slightly before inserting the bracket spring clip.
6 Rigid pipelines can be made to pattern by factors supplying brake components.
7 If you are making up a brake pipe yourself, observe the following essential requirements.
8 Before flaring the ends of the pipe, trim back the protective plastic coating by a distance of 5.0 mm (0.2 in).
9 Flare the end of the pipe as shown (Fig. 8.3).
10 The minimum pipe bend radius is 12.0 mm (0.5 in), but bends of less than 20.0 mm (0.8 in) should be avoided if possible.

4 Front brake pads – inspection and renewal

1 At the intervals specified in *'Routine maintenance'* at the beginning of this manual, place a mirror between the roadwheel and the brake caliper and check the thickness of the brake pad friction material. If the friction material has worn down to the specified minimum or less, the pads must be renewed as an axle set (four pads).
2 To renew the pads, first slacken the roadwheel nuts, then raise the front of the vehicle and support it securely on axle stands.
3 Remove the roadwheels.
4 Remove the piston housing retaining clip from the brake caliper (photo).
5 Remove the rubber blanking plugs protecting the two caliper piston housing (Allen-head) retaining bolts (photo), and undo the bolts. Lift off the caliper piston housing. Do not allow the flexible brake hose to become strained.
6 Carefully separate the inner brake pad from the caliper piston, and then press the piston firmly into its bore using a flat piece of wood or similar. This is necessary as fitting new brake pads requires more space, due to the new thicker friction material. As the piston is pushed into its bore it will displace brake fluid, causing the level in the master cylinder reservoir to rise. It is therefore worthwhile syphoning out some of the fluid, using an old hydrometer or similar device to prevent the fluid reservoir from overflowing. Take care not to drip brake fluid onto the paintwork as it acts as an effective paint stripper.
7 Remove the outer brake pad which is secured into position with an adhesive backing.
8 Before installing new brake pads, ensure that the brake disc faces are clean and free from dirt, grease or rust, **taking care not to inhale any dust** that is present. Also clean the caliper piston housing, taking the same precaution.
9 Fit the inner brake pad to the caliper piston, ensuring that its spring clip is properly located.
10 Peel back the protective paper covering from the rear of the new outer pad and locate it in the jaws of the caliper anchor bracket (photo).

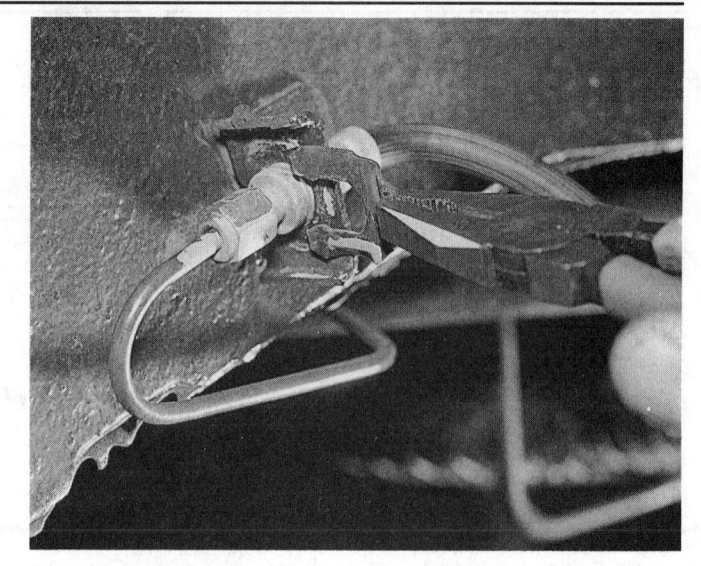

3.2 Prising out a spring anchor clip from a rigid pipe/flexible hose support bracket

4.4 Removing a piston-housing retaining clip from a brake caliper

4.5 Removing a rubber blanking plug for access to a caliper piston-housing retaining bolt

4.10 Refitting caliper piston housing with inner brake pad fitted to caliper piston, and outer pad located in jaws of caliper anchor bracket

Chapter 8 Braking system

11 Refit the caliper piston housing and tighten the Allen bolts according to specification. Refit the protective rubber blanking plugs.
12 Refit the piston housing retaining clip.
13 Repeat the procedure on the other front brake.
14 Apply the footbrake hard several times to position the pads against the disc, then check the fluid in the master cylinder reservoir. Top up the fluid as necessary.
15 Refit the roadwheels and lower the vehicle to the ground, then fully tighten the roadwheel nuts according to specification.
16 To allow the pads to reach full efficiency, a bedding-in period of a hundred miles or so should be observed, before hard use and heavy braking.

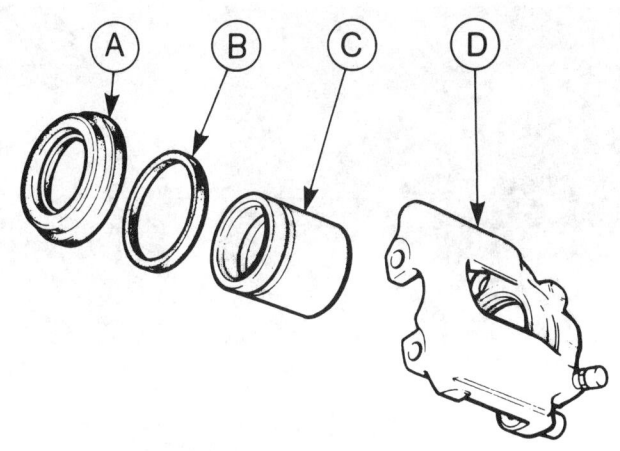

5 Front brake caliper – removal, overhaul and refitting

1 To remove a brake caliper, first slacken the relevant roadwheel nuts, then raise the front of the vehicle and support it securely on axle stands.
2 Remove the roadwheel.
3 Remove the caliper piston housing retaining clip from the brake caliper.
4 Clamp the flexible brake hose using a suitable hose clamp, then slacken the hose connection at the caliper piston housing.
5 Remove the rubber blanking plugs covering the two caliper piston housing (Allen-head) retaining bolts, and undo them.
6 Lift off the caliper piston housing, remove the inner brake pad from its piston location and release the outer pad from its adhesive backing.
7 With the brake pads removed hold the caliper piston housing in one hand and, holding the end fitting of the hose in an open-ended spanner, rotate the caliper piston housing until it is free (photo). Do not allow the hose to twist, and plug its end after removal of the caliper piston housing to prevent dirt ingress.
8 The brake caliper anchor bracket can be removed if required, by undoing the two bolts securing it to the spindle carrier (photo).
9 Remove any external dirt, **taking care not to inhale any dust**, and pull off the piston dust excluder.
10 Remove the piston from its bore by using air pressure from a foot-operated tyre pump applied to the brake hose connection.
11 Using a sharp-pointed hooked instrument, pick out the piston seal from the groove in the cylinder bore. Care **must** be taken not to scratch the surface of the bore.
12 Examine the surfaces of the piston and the cylinder bore. If either is scored or shows evidence of metal-to-metal rubbing, renewal will be necessary. Where the components are in good condition, discard the seal and obtain a repair kit.

Fig. 8.4 Caliper and piston components (Sec 5)

A Dust excluder   C Piston
B Piston seal   D Caliper piston housing

13 Wash the internal components in clean brake fluid or methylated spirit only.
14 Using fingers only, carefully manipulate the new seal into its groove in the cylinder bore.
15 Dip the piston in clean hydraulic fluid and insert it squarely into its bore.
16 Connect the rubber dust excluder between the piston and the piston housing, and then depress the piston fully.
17 Refit the brake caliper anchor bracket to the spindle carrier, as applicable, tightening the two bolts according to specification.
18 Remove the plug from the end of the flexible brake hose, then reunite the hose and the caliper piston housing by reversing the method of removal. Do not fully tighten the hose connection.
19 Refit the brake pads as described in paragraphs 8 to 10 in the previous Section.
20 Refit the caliper piston housing to its anchor bracket and tighten the Allen bolts according to specification. Refit the rubber blanking plugs.
21 Fully tighten the flexible brake hose connection, and remove the brake hose clamp. Ensure that the brake hose is not distorted, and that it will not interfere with any adjacent steering or suspension components.
22 Refit the caliper piston housing retaining clip.
23 Bleed the brake hydraulic circuit in accordance with Section 21 or 29, as applicable.
24 Refit the roadwheel and lower the vehicle to the ground, then fully tighten the roadwheel nuts according to specification.

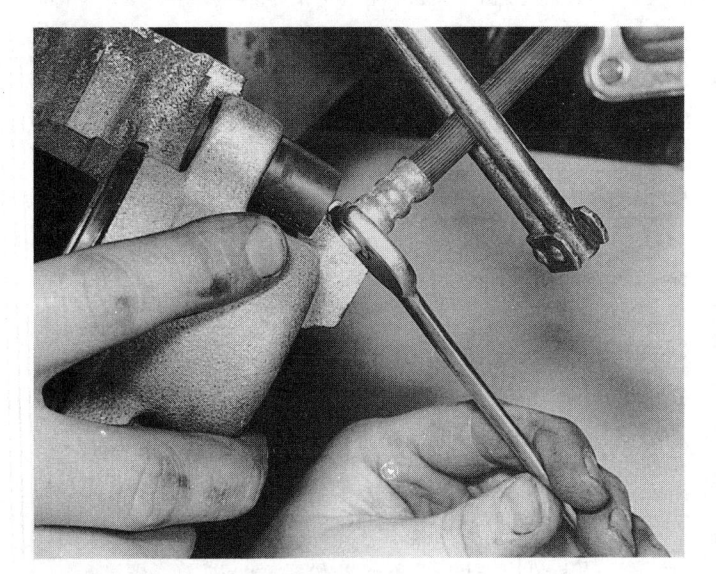

5.7 Holding flexible hose end fitting in spanner jaws, and rotating caliper piston-housing to free it

5.8 Undoing a brake caliper anchor bracket bolt

Chapter 8 Braking system

6.3 Checking a brake disc for run-out using a dial gauge

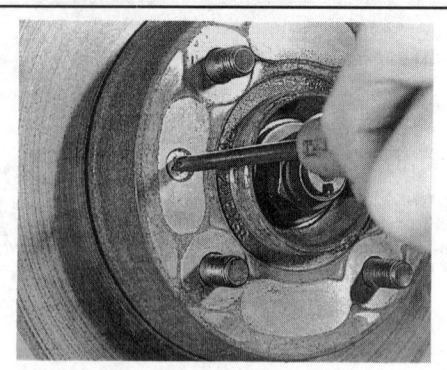

6.5A Extracting the brake disc retaining screw

6.5B Removing the brake disc from the hub

6 Front brake disc – inspection, removal and refitting

1 Slacken the relevant roadwheel retaining nuts, raise the front of the vehicle and support it securely on axle stands. Remove the roadwheel.
2 Examine the surface of the disc. If it is deeply grooved or scored it must be either refinished or renewed. If any cracks are evident it must be renewed. Any refinishing must not reduce the thickness of the disc to below a certain minimum (see Specifications). Light scoring on a brake disc is normal and should be ignored.
3 If disc distortion is suspected, the disc can be checked for run-out (see Specifications) using a dial gauge (photo) or feeler blades located between its face and a fixed point as the disc is rotated. Where run-out exceeds the specified figure, renew the disc.
4 To remove a disc, undo and remove the two bolts securing the caliper anchor bracket to the spindle carrier, and lift the assembly from the disc. Suspend the caliper assembly from the suspension strut using a length of strong wire, to avoid straining the flexible brake hose.
5 Extract the single disc retaining screw and pull the disc from the hub (photos).
6 Refitting is the reverse sequence to removal. If a new disc is being installed, clean its surfaces of any preservative before refitting the caliper. Press the caliper piston slightly into its bore for the extra clearance required when refitting. Tighten the caliper anchor bracket bolts according to specification.
7 Refit the roadwheel and lower the vehicle to the ground, then fully tighten the roadwheel nuts to specification.

7 Rear brake drum/hub assembly – removal, inspection and refitting

1 Slacken the relevant roadwheel nuts then raise the rear of the vehicle and support it securely on axle stands, having chocked the front wheels.
2 Remove the roadwheel.
3 With the handbrake lever released, remove the rubber blanking plug from the rear of the brake carrier plate (photo) and release the automatic brake adjuster by levering the release catch on the adjuster pawl, through the carrier plate.
4 Remove the four bolts securing the hub/brake drum assembly to the axle flange (see Chapter 10, Fig. 10.2) and carefully slide the assembly off over the brake shoes.
5 Remove any dust from inside the drum, **taking care not to inhale it**, and remove any heavy dirt or grease deposits from its external surfaces.
6 Clean away any traces of dirt or grease from the internal friction surface of the drum. If deeply scored or worn, the drums must be renewed. Refinishing of the drums is not recommended as the internal diameter will no longer be compatible with the shoe lining contact diameter, and will cause excessive brake pedal travel and uneven shoe lining wear leading to premature replacement. If renewing the brake drum, new bearings and grease retainer should be fitted rather than attempting to transfer the old components (see Chapter 10).
7 Refitting is the reverse sequence to removal, applying a suitable

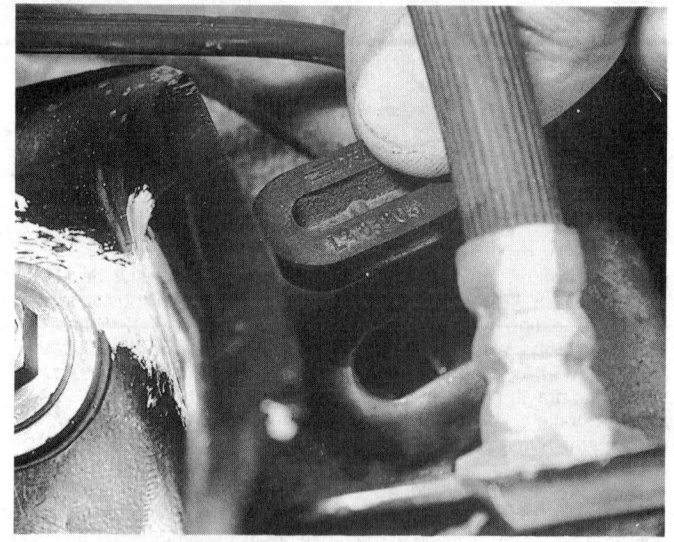

7.3 Removing the rubber blanking plug from the rear of a brake carrier plate

sealant to the threads of the four bolts securing the hub/brake drum assembly to the axle flange before refitting (see Specifications).
8 Tighten the four bolts securing the hub/brake drum assembly to the specified torque then, with the vehicle on the ground, fully tighten the roadwheel nuts to their specified torque. Remove the front wheel chocks.
9 Depress the brake pedal hard several times to actuate the automatic brake adjuster.

8 Rear brake shoes – renewal

1 With the hub/brake drum assembly removed as described in the previous Section, the rear brake shoes are easily accessible.
2 Depress the cups holding the brake shoes in position and rotate them through 90° to release them from the locking pins (photo). Carefully remove the cups and springs, then withdraw the locking pins from the rear of the brake carrier plate.
3 Lift the shoes from their lower pivot and remove the lower pull-off spring (photo).
4 With the shoe assembly pulled away from the wheel cylinder, disengage the handbrake cable from its operating lever on the trailing shoe (photo).
5 Remove the upper pull-off spring, noting method of location.
6 Release the automatic brake adjuster cam and pawl, then remove the brake strut which is held in position by spring tension (photos).
7 Using a screwdriver, lever off the spring clip securing the handbrake operating lever to the trailing shoe (photo) and separate the assembly.

Chapter 8 Braking system

8.2 Depress and turn the cups securing the brake shoes

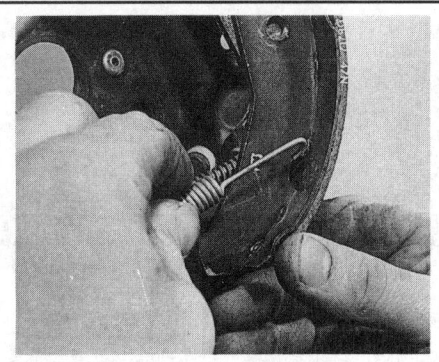

8.3 Detach the lower pull-off spring

8.4 Disengage the handbrake cable from its operating lever on the trailing shoe

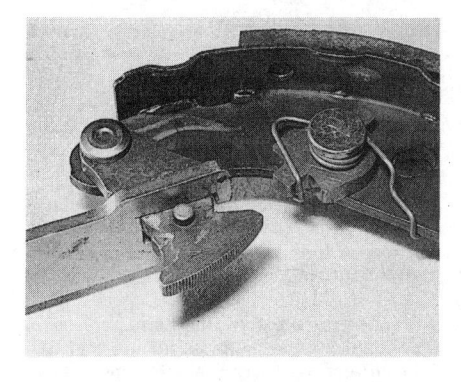

8.6A Release the automatic brake adjuster cam and pawl

8.6B Detach the brake strut

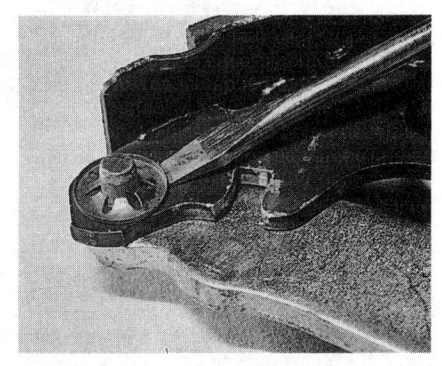

8.7 Lever off the spring clip securing the handbrake operating lever to the trailing shoe

8.9 Brake shoe contact points on brake carrier plate (arrowed)

8.15 Refitting partially-assembled rear brake components to brake carrier plate

8   Detach the automatic brake adjuster cam in a similar manner to that described in the previous paragraph, noting orientation.
9   Before reassembling the brake components, *lightly* coat the brake shoe contact points on the brake carrier plate (photo) with anti-seize compound, taking care not to over-apply.
10 Fit the handbrake operating lever to the trailing shoe, using a **new** spring clip.
11 Fit the automatic brake adjuster cam to the leading shoe, using a **new** spring clip.
12 Apply a small amount of anti-seize compound to the automatic brake adjuster cam and pawl contact faces, and where the cam and handbrake operating lever sweep across their respective brake shoes. Do **not** over-apply, as this may result in lining contamination in use – a **thin smear** will suffice. Take care **not** to get any on the brake linings.
13 Fit the brake strut to the trailing shoe, securing with its spring, then connect the free end of the strut to the automatic brake adjuster cam. Fit the upper pull-off spring between the tops of the two brake shoes.
14 Reconnect the handbrake cable to its operating lever.
15 Position the brake shoes onto the brake carrier plate so that their upper leading edges rest against the wheel cylinder pistons, and their lower leading edges engage either side of the lower pivot (photo). Fit the lower pull-off spring into its locating slots at the bottom end of each brake shoe.
16 Insert the brake shoe locking pins through the rear of the brake carrier plate, then relocate the springs and cups. Depress and turn the cups through 90° to secure.
17 Mark the other hub/brake drum assembly so that it can be refitted to its correct side, then renew the rear brake shoes on that side.
18 Ensure that the brake shoes are centralised and that the automatic brake adjusters are released, to fully retract the shoes, before refitting the hub/brake drum assemblies as described in the previous Section, paragraph 7 onwards.

196 Chapter 8 Braking system

9.6 Rear view of brake carrier plate
A Wheel cylinder-to-brake carrier plate retaining bolt
B Wheel cylinder brake pipe connection
C Bleed screw

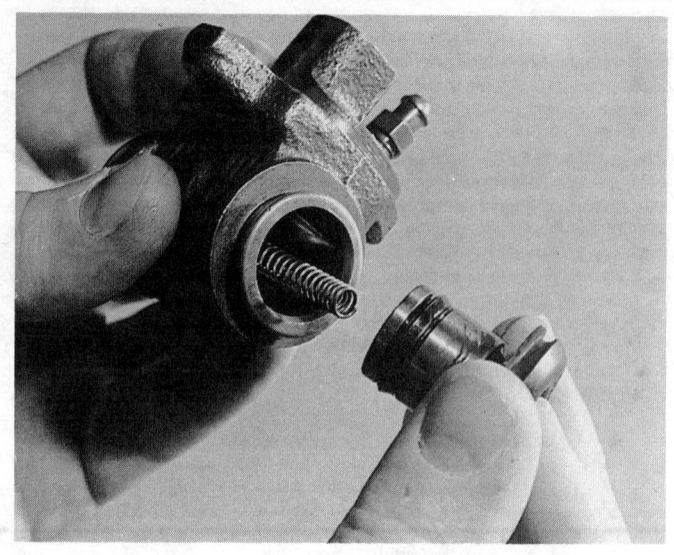

9.15 Inserting piston to wheel cylinder (dust excluder already fitted to piston)

9 Rear wheel cylinder – removal, overhaul and refitting

1   With the hub/brake drum assembly removed as described in Section 7, the following wheel cylinder removal procedure can be followed.
2   Using a suitable hose clamp, isolate the relevant rear brake unit by clamping its flexible brake hose.
3   Disconnect the brake pipe at the wheel cylinder union, and fit a blanking plug to the brake pipe to prevent dirt ingress.
4   Drill out the pop-rivets securing the brake carrier plate to the axle flange, and remove the brake carrier plate with the shoe assembly *in situ* (see Section 10).
5   Expand the brake shoes by pulling their tops away from the wheel cylinder. The automatic brake adjuster will hold the shoes clear of the wheel cylinder for ease of removal.
6   Remove the single bolt securing the wheel cylinder to the brake carrier plate (photo), and withdraw the wheel cylinder.
7   Clean any heavy dirt or grease deposits from the external surfaces of the wheel cylinder, then pull off the dust-excluding covers.
8   The pistons and seals will probably shake out. If they do not, apply air pressure from a foot-operated tyre pump to the brake pipe connection to eject them.
9   Examine the surfaces of the pistons and the cylinder bores for scoring or signs of metal-to-metal rubbing. If evident, renew the complete cylinder assembly.
10   If the cylinder is to be renewed, note that three sizes are used across the Fiesta range, dependent on specification. Ensure that the new cylinder obtained is of the correct size to maintain the rear braking balance.
11   Where the pistons and cylinder bores are in good condition, discard the rubber seals and dust excluders and obtain a repair kit.
12   Any cleaning of the components should be done using clean hydraulic fluid or methylated spirit – nothing else.
13   Reassemble by dipping the first piston in clean hydraulic fluid, then manipulate its seal into position using **fingers only**. Ensure that the seal is fitted correctly with its raised lip facing away from the brake shoe bearing face of the piston.
14   Insert the first piston into the wheel cylinder from the opposite end of the cylinder body. With it located in position, fit a dust-excluding cover to it.
15   Fit the seal to the second piston, as described in paragraph 13, then insert the spring to the wheel cylinder, followed by the second piston (photo). Take care not to damage the lip of the seal when fitting to the wheel cylinder – additional lubrication with clean hydraulic fluid and a slight twisting action may help. Once again, only fingers should be used.
16   Fit the dust excluding cover to the second piston.
17   Refitting is the reverse sequence to removal. Release the automatic brake adjuster so that the brake shoes are brought into contact with the wheel cylinder, and ensure that the newly-inserted rivet heads can locate into the hub flange recesses and allow the hub to seat correctly.
18   Upon completion, bleed the brake hydraulic system in accordance with Section 21 or 29 (as applicable).

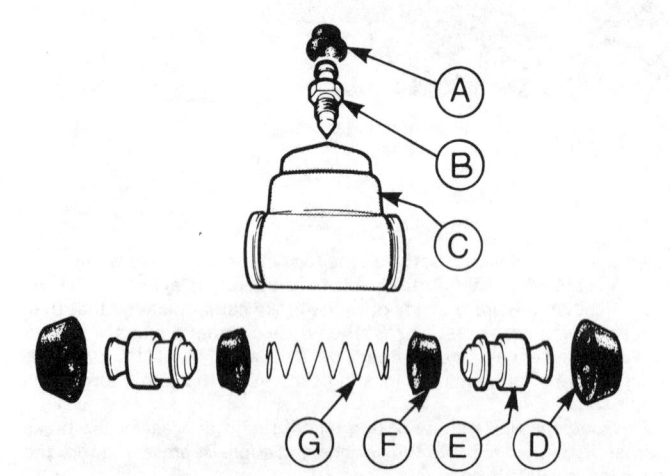

Fig. 8.5 Exploded view of rear wheel cylinder components (Sec 9)

A Dust cap
B Bleed screw
C Wheel cylinder
D Dust-excluding cover
E Piston
F Seal
G Spring

10 Rear brake carrier plate – removal and refitting

1   With the hub/brake drum assembly removed, as described in Section 7, and the rear brake shoes removed, as described in Section 8, lever the handbrake cable retaining clip (photo) over the cable abutment and slide it off. Withdraw the handbrake cable from the rear of the brake carrier plate.
2   Carefully remove the handbrake adjustment plunger from the brake carrier plate by gently prising the spring off, over the plunger abutment, then withdraw the plunger from the brake shoe side. Remove the plunger collar from the rear of the brake carrier plate (photo).
3   Using a suitable hose clamp, isolate the relevant rear brake unit by clamping its flexible brake hose.

Chapter 8 Braking system

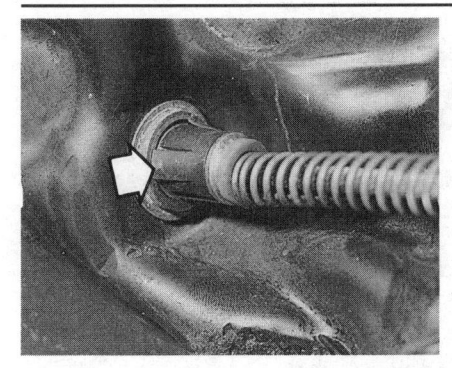

10.1 Handbrake cable retaining clip (arrowed) on brake carrier plate

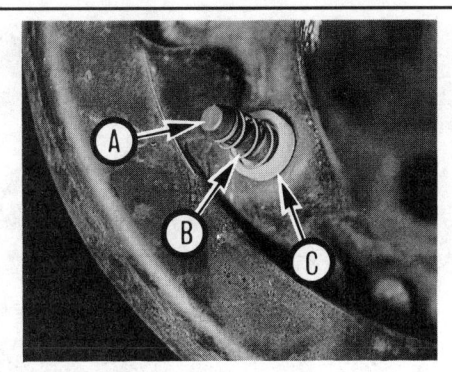

10.2 Handbrake adjustment plunger

A Plunger
B Spring
C Collar

10.5 Drilling out a rivet securing the brake carrier plate to the axle flange

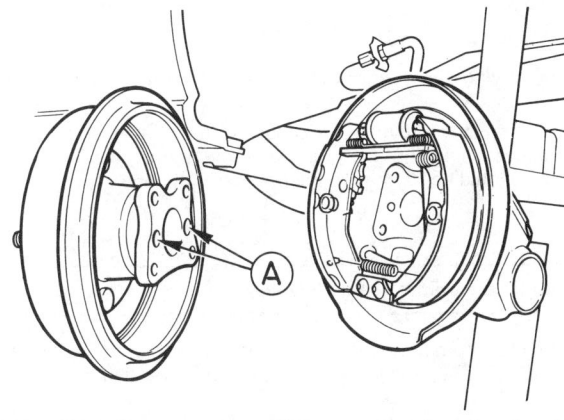

Fig. 8.6 Hub flange recesses (A) to accommodate rivet heads (Sec 10)

4 Disconnect the brake pipe at the wheel cylinder union, and fit blanking plugs to prevent dirt ingress.
5 Drill out the pop-rivets securing the brake carrier plate to the axle flange (photo), and remove the brake carrier plate.
6 Remove the single bolt securing the wheel cylinder to the brake carrier plate, and withdraw the wheel cylinder.
7 Refitting is the reverse sequence to removal, ensuring that the newly-inserted rivet heads can locate into the hub flange recesses and do not prevent the hub from seating correctly (Fig. 8.6).
8 Refit the rear brake shoes and hub/brake drum assembly in accordance with Sections 7 and 8.
9 On completion, bleed the brake hydraulic circuit in accordance with Section 21 or 29 (as applicable).

11 Handbrake – adjustment

1 With the vehicle on a flat surface, lower the handbrake lever ensuring that it is *fully* released.
2 Chock the front wheels then raise the rear of the vehicle and support it securely on axle stands. Note that if adjustment is required on 1.4 litre CFi models, it will be necessary to remove the underbody heatshields.
3 Ensure that the handbrake cable is correctly located in its clips and guides.
4 Examine the handbrake adjustment plunger movement, on the rear of the brake carrier plates (see photo 10.2). If the movement of the two plungers added together is not between 0.5 mm and 2.0 mm, the handbrake is incorrectly adjusted and the procedure below should be followed. Note that if adjustment of the handbrake cable does not affect the movement of the plungers, this indicates that there is a malfunction of the brake mechanism, or that the plungers have seized.
5 To adjust the handbrake, first remove the rubber blanking plugs from the brake carrier plates, then release the automatic brake adjusters by levering the release catches on the adjuster pawls, through the brake carrier plates.

6 With the handbrake still fully released, apply the footbrake to ensure correct automatic rear brake adjustment at each brake unit. Refit the rubber blanking plugs.
7 Remove the lock pin from the handbrake cable adjusting sleeve and slacken the cable locknut (photo). Note that a new lock pin will be needed after adjustment.
8 Adjust the handbrake cable by rotating its adjusting sleeve until the total movement of both handbrake adjustment plungers added together is between 0.5 mm and 2.0 mm.
9 Tighten the handbrake cable locknut as tight as possible by hand (two clicks) then further tighten, using a suitable wrench, by another two clicks (maximum).
10 Fit a new lock pin by tapping it into position.
11 Refit the underbody heatshields (1.4 litre CFi models).
12 Lower the vehicle to the ground and remove the wheel chocks.

12 Handbrake lever – removal and refitting

1 Disconnect the battery earth terminal, and chock the wheels to secure the vehicle.
2 Undo the bolts securing the front seats to the floorpan, and remove both seats from the vehicle (see Chapter 11). Move the seats on their slide mechanisms to expose the mounting bolts, as necessary.
3 Remove the screws securing the rear seat cushion, then raise the

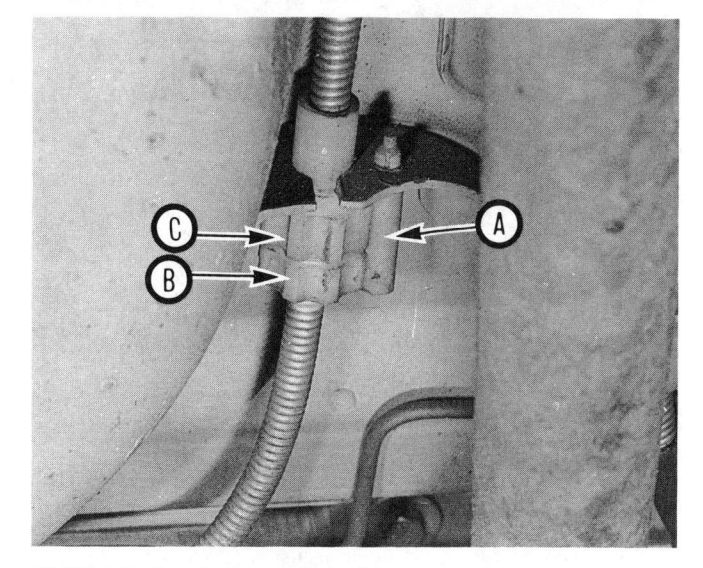

11.7 Handbrake adjustment mechanism

A Lock pin
B Cable locknut
C Cable adjusting sleeve

Chapter 8 Braking system

12.9 Removing the handbrake primary cable clevis pin securing clip

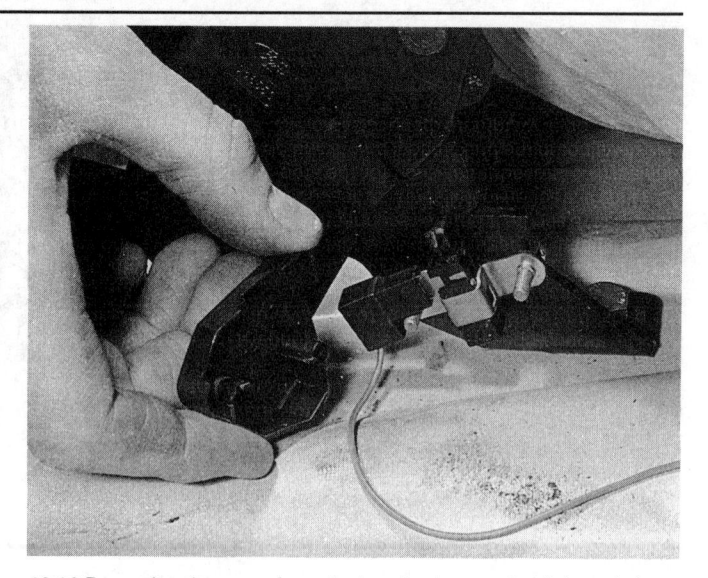

12.10 Removing the cover from the handbrake warning light switch

cushion to obtain access to the carpet retaining screws. Remove the carpet retaining screws.
4 Undo the bolt securing the seat belt clips to the centre of the floorpan, then remove the clip assembly.
5 Remove the seat belt lower anchor bracket bolt from its location at the base of the B-pillar behind the driver's seat.
6 Remove the screws securing the sill scuff plate to the driver's side of the vehicle, then carefully pull the sill scuff plate away from its location so that the carpet is released.
7 Fold the carpet forwards, at the same time carefully easing it out from under the sill scuff plate. Lift the carpet over the handbrake lever.
8 Lift out the noise insulation for access to the lever mounting bolts and the primary cable fixing.
9 Fully release the handbrake lever, then remove the handbrake primary cable clevis pin securing clip (photo). Remove the clevis pin and withdraw the primary cable from the handbrake lever assembly.
10 Remove the cover (photo), then disconnect the handbrake warning light switch loom connection, and undo the two screws securing the switch to the handbrake lever assembly.
11 Undo the handbrake lever mounting bolts, then withdraw the handbrake lever assembly from the vehicle.

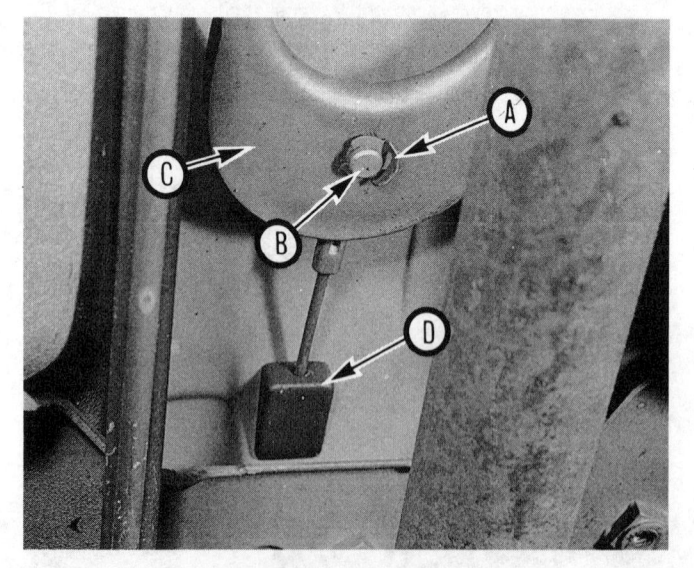

13.3 Handbrake equaliser yoke arrangement

A Clevis pin securing clip
B Clevis pin
C Equaliser yoke
D Primary cable guide

12 Refitting is the reverse procedure to removal, ensuring that the handbrake warning light loom is routed away from the lever ratchet. The loom should be secured to the floorpan with tape.

13 Handbrake primary cable – renewal

1 With the carpet removed to expose the handbrake lever assembly, as described in the previous Section, the forward end of the handbrake primary cable is accessible.
2 Raise the vehicle completely off the ground, supporting it securely on axle stands. Note that 1.4 litre CFi models are fitted with underbody heatshields which must be removed for access.
3 With the handbrake lever in the fully released position, disconnect the primary cable from the handbrake equaliser yoke by removing the spring clip and withdrawing the clevis pin, from underneath the vehicle (photo).
4 Disconnect the forward end of the primary cable from the handbrake lever assembly, as described in the previous Section.
5 Remove the primary cable guide by drifting it out rearwards, through the floorpan, from the inside of the vehicle.
6 Installation is the reverse procedure to removal. Check and adjust the handbrake as necessary, as described in Section 11.

14 Handbrake rear cable – renewal

1 Slacken the rear roadwheel nuts then raise the rear of the vehicle and support it securely on axle stands, having chocked the front wheels. Remove the rear roadwheels.
2 On 1.4 litre CFi models, remove the underbody heatshields.
3 Lower the handbrake lever, ensuring that it is *fully* released.
4 Disconnect the handbrake primary cable from the equaliser yoke by removing the spring clip and withdrawing the clevis pin, as described in the previous Section.
5 Disconnect the handbrake rear cable from its adjuster body location (see Section 11) and its fixed body location (photo), then remove it from its retaining clips.
6 Remove the rubber blanking plugs from the rear of the brake carrier plates, and release the automatic rear brake adjusters by levering on the adjuster pawl catches through the brake carrier plates.
7 Mark the hub/brake drum assemblies so that they can be refitted to their correct sides, then remove them.
8 Remove the brake shoes from the brake carrier plates and disconnect the handbrake cable from the trailing shoe operating levers, as described in Section 8.
9 Detach the cable from the rear brake carrier plates. This is achieved

Chapter 8 Braking system

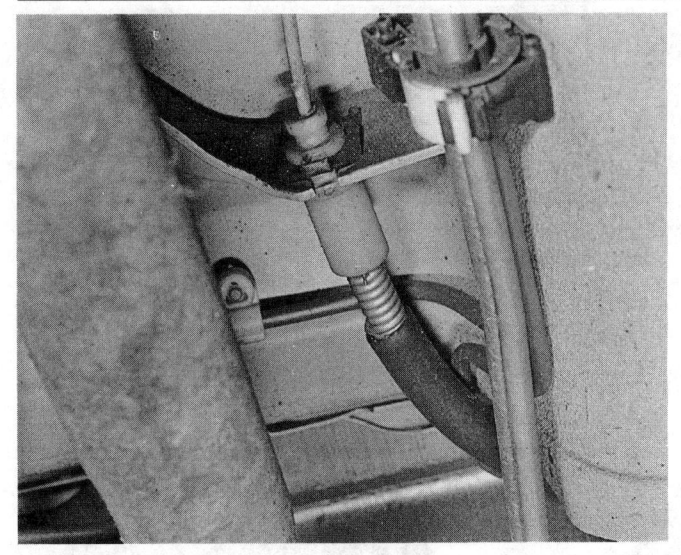

199

14.5 Handbrake rear cable fixed body location

by removing the spring clips around the cable abutments where they pass through the brake carrier plates (see Section 10). The handbrake rear cable can now be withdrawn from the vehicle.
10 Cable refitting is the reverse procedure to removal, ensuring that the hub/brake drum assemblies are fitted to their correct sides, sealant is applied to the four bolts retaining each hub/brake drum assembly, and that the bolts are tightened according to specification.
11 Refit the rear roadwheels.
12 Adjust the handbrake as described in Section 11, then refit the underbody heatshields (if removed).
13 Fully tighten the roadwheel nuts with the vehicle on the ground, and remove the front wheel chocks.

15 Master cylinder – removal, overhaul and refitting

1 Disconnect the battery earth terminal.
2 Disconnect the brake fluid warning indicator loom from the brake fluid reservoir cap (photo), then remove the cap but do not invert it.
3 Syphon out as much hydraulic fluid as possible from the brake fluid

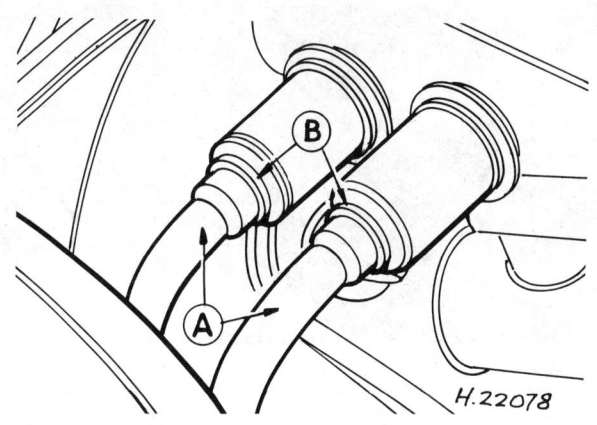

Fig. 8.7 Modulator return hose connections at the brake fluid reservoir (Sec 15)

A Return hoses   B Collars

reservoir, using an old hydrometer or similar device, then disconnect the brake pipes from the master cylinder, fitting blanking plugs to prevent dirt ingress. Additionally, on anti-lock equipped vehicles, disconnect the modulator return hoses from the brake fluid reservoir (Fig. 8.7), collecting fluid spillage from the hoses in a suitable tray. The modulator return hose unions should be disconnected by first pushing the hose into the reservoir, then retaining the collar against the reservoir body whilst withdrawing the hose. Note that the modulator return hoses are colour coded – the left-hand modulator has a black return hose and connector, and should be fitted to the forward section of the reservoir, whilst the right-hand modulator has a grey return hose and connector, and should be fitted to the rear section of the reservoir.
4 Remove the nuts and spring washers securing the master cylinder to the brake servo, then carefully withdraw the master cylinder from the vehicle.
5 Clean any heavy dirt or grease deposits from the exterior surfaces of the master cylinder and the brake fluid reservoir, then carefully separate the brake fluid reservoir from the master cylinder by applying a sideways 'rolling' motion to release the reservoir from its seals (photo).
6 Remove the reservoir seals from the master cylinder (photo).
7 Using a suitable socket, firmly press the primary piston into the master cylinder until the secondary piston retaining pin becomes visible through the secondary piston reservoir opening. Remove the retaining pin using needle-nosed pliers or similar tool (photo).

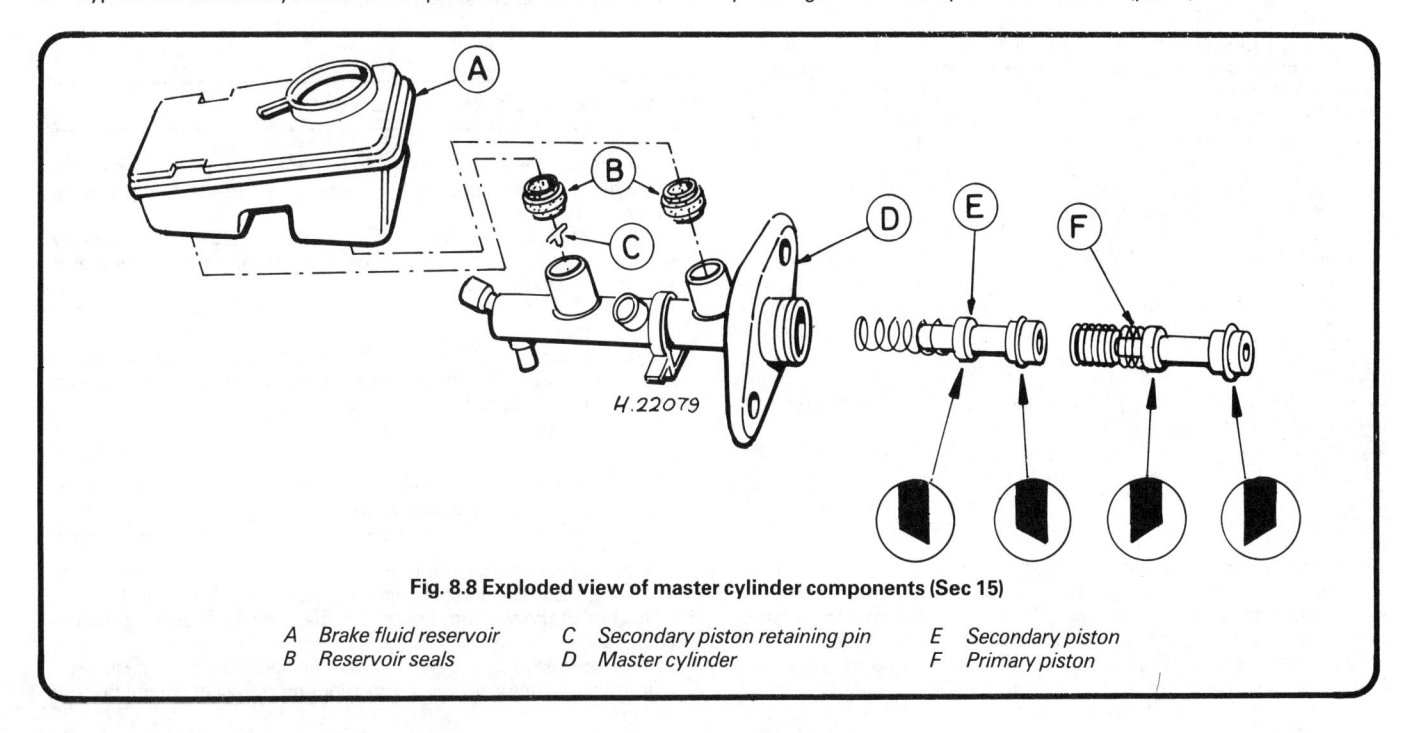

Fig. 8.8 Exploded view of master cylinder components (Sec 15)

A Brake fluid reservoir   C Secondary piston retaining pin   E Secondary piston
B Reservoir seals   D Master cylinder   F Primary piston

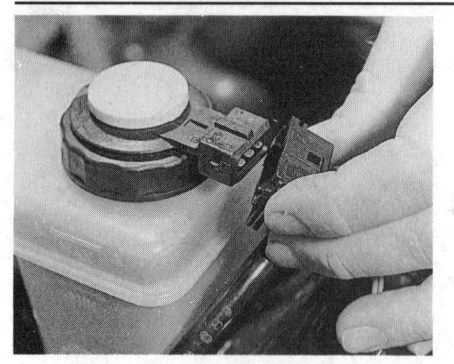

15.2 Disconnect the brake fluid warning indicator loom from the reservoir cap

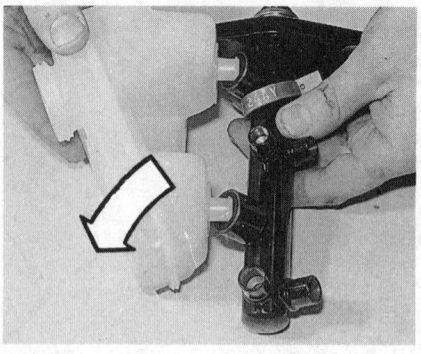

15.5 Applying a rolling motion to release the brake fluid reservoir from its seals in the master cylinder

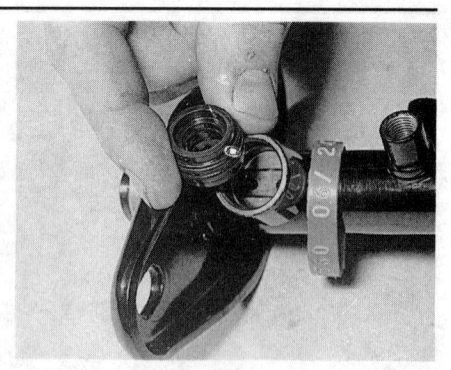

15.6 Brake fluid reservoir seal

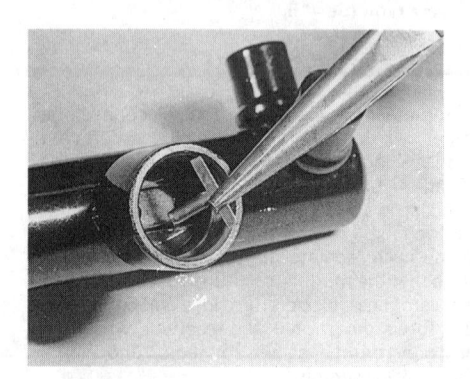

15.7 Removing the secondary piston retaining pin

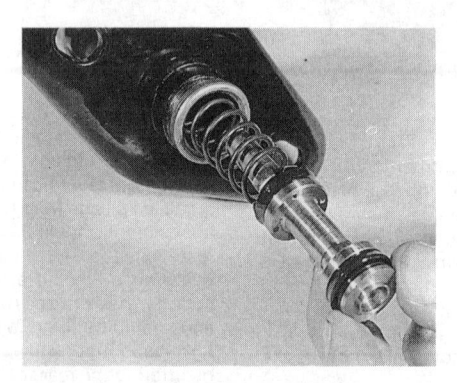

15.12 Inserting the secondary piston assembly to the master cylinder

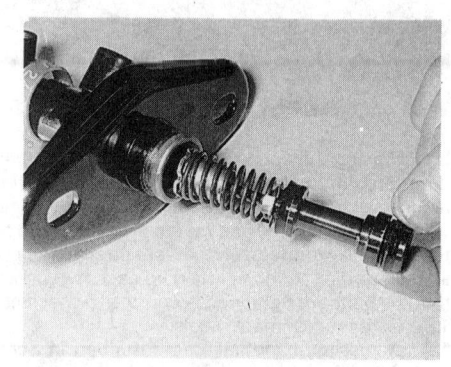

15.14 Inserting the primary piston assembly to the master cylinder

8 Remove the primary and secondary pistons by shaking or gently tapping the master cylinder.
9 Examine the piston and cylinder bore surfaces for scoring or signs of metal-to-metal rubbing. If evident, renew the master cylinder assembly as a complete unit.
10 Cleaning of components should be done using clean brake hydraulic fluid or methylated spirit – nothing else.
11 Obtain a master cylinder repair kit, and brake fluid reservoir seals.
12 Having dipped the secondary piston assembly in clean hydraulic fluid, insert it into the master cylinder (photo). Note that the seals on the secondary piston have raised lips which face away from each other, towards the extremities of the piston. A slight twisting action will assist insertion.
13 Using a suitable socket bar extension, or similar tool, press the secondary piston into the master cylinder to enable fitting of the secondary piston retaining pin.
14 Dip the primary piston assembly in clean hydraulic fluid and, using a similar twisting action to that used for the secondary piston, insert it into the master cylinder (photo). Note that both the seals fitted to the primary piston have raised lips that face the same way – towards its captive spring.
15 Insert new brake fluid reservoir seals to the master cylinder then, using a similar rolling action to that used to remove it, fit the reservoir to the master cylinder.
16 It is recommended that a *small* quantity of fluid is now poured into the reservoir and the pistons depressed several times to prime the unit.
17 Locate the master cylinder to the servo unit, having fitted a new dust cover as applicable. Refit the two spring washers and nuts, then tighten to the specified torque.
18 Remove the blanking plugs fitted on removal of the unit, then reconnect the brake pipes, tightening the unions to seal the system. Additionally, on anti-lock equipped vehicles, reconnect the modulator return hoses to the brake fluid reservoir in accordance with their colour coding (see paragraph 3), pushing the hoses firmly into the reservoir body then levering out the collars to retain.

19 Refill the brake fluid reservoir with fresh fluid of the specified type, then bleed the braking system in accordance with Section 21 or 29, as applicable.
20 Ensure that the brake fluid level is up to the MAX mark on the reservoir before refitting the reservoir cap and warning indicator loom.
21 Reconnect the battery earth terminal.

16 Brake servo vacuum hose – renewal

1 Disconnect the vacuum hose from its servo unit connection by carefully levering between the hose connector and the servo housing collar with a screwdriver.
2 Remove the vacuum hose from its inlet manifold connection by pushing the hose into the connection then, holding the locking collar down, carefully withdraw the hose.
3 To install, first push the vacuum hose connector into its servo housing location.
4 Ensuring correct routeing of the hose, clean its end and insert to the manifold connection. Push the hose fully into the connection, then carefully lever out the locking collar to retain it.

17 Brake servo – removal and refitting

1 Disconnect the battery earth terminal.
2 Remove the brake master cylinder as described in Section 15.
3 Disconnect the vacuum hose from the servo unit, as described in Section 16.
4 Lift up the flap of sound insulation on the bulkhead, in the passenger side footwell, to expose the servo mounting bracket retaining nuts

Chapter 8 Braking system

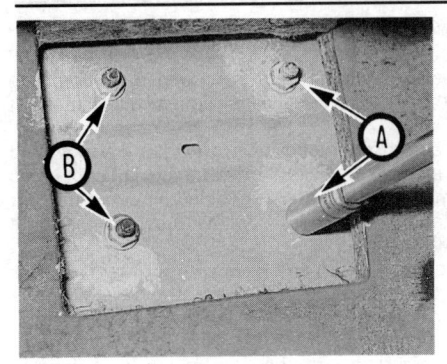

17.4 Servo mounting bracket retaining nuts
A Inner section retaining nuts
B Outer section retaining nuts

17.5 Nuts securing servo unit to its mounting bracket assembly (arrowed)

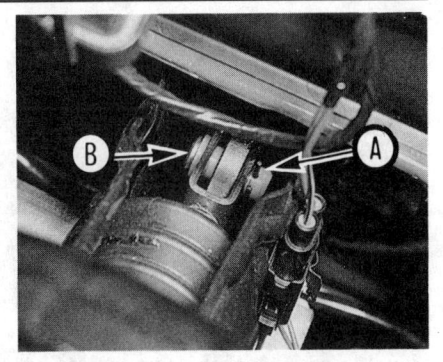

17.6 Spring clip (A) and clevis pin (B) securing servo actuating rod to the operating link

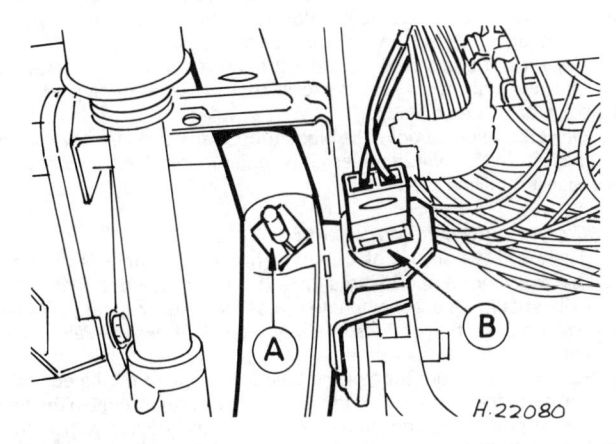

Fig. 8.9 Retaining clip on brake pedal (A) and stop-lamp switch (B) (Sec 17)

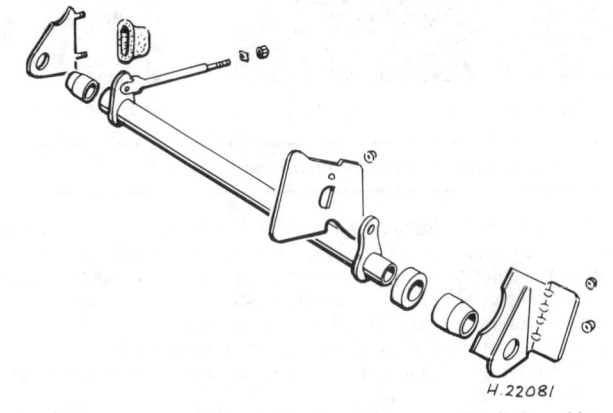

Fig. 8.10 Exploded view of brake servo operating link and its retaining brackets (Sec 18)

(photo). Remove the two innermost nuts to free the inner section of the servo mounting bracket from its bulkhead location. Slacken the other two nuts or remove them, as necessary.
5 Remove the four nuts securing the servo unit to its mounting bracket assembly (photo), then pull the servo forward to remove the inner servo support bracket.
6 Remove the spring clip and clevis pin securing the servo actuating rod to the operating link (photo), then lift out the servo unit.
7 Refitting is the reverse sequence to removal. On completion, bleed the complete brake hydraulic system in accordance with Section 21 or 29 (as applicable).

18 Brake servo operating link bushes – renewal

1 Disconnect the battery earth terminal.
2 Lift up the flap of sound insulation on the bulkhead, in the passenger side footwell, to expose the servo mounting bracket retaining nuts, and remove them (see photo 17.4).
3 Remove the four nuts securing the servo unit to its mounting bracket assembly (see photo 17.5).
4 Disconnect the servo operating link brake rod from its pedal location by removing the retaining clip on the brake pedal, noting the bush fitted in the pedal.
5 Remove the two nuts on the right-hand side of the pedal box assembly that secure the pedal box assembly and the servo operating link right-hand support bracket to the bulkhead.
6 Carefully ease the servo unit and its operating link forward to enable the operating link brackets and their bushes to be withdrawn (see Fig. 8.10).
7 Refitting is the reverse procedure to removal, ensuring that the return spring is located correctly on the servo side of the operating link.

19 Brake servo operating link – removal and refitting

1 Disconnect the battery earth terminal.
2 Disconnect the brake fluid warning indicator loom from the brake fluid reservoir, and remove the cap.
3 Disconnect the servo operating link brake rod from its pedal location by removing the retaining clip on the brake pedal, noting the bush fitted in the pedal.
4 Syphon out as much hydraulic fluid as possible from the brake fluid reservoir, using an old hydrometer or similar device, then disconnect the brake pipes from the master cylinder, fitting blanking plugs to prevent dirt ingress. Additionally, on anti-lock equipped vehicles, disconnect the modulator return hoses from the brake fluid reservoir (see Section 15).
5 Disconnect the vacuum hose from the servo unit, as described in Section 16.
6 Lift up the flap of sound insulation on the bulkhead, in the passenger side footwell, to expose the servo mounting bracket retaining nuts (see photo 17.4), and remove them.
7 Remove the four nuts securing the servo unit to its mounting bracket assembly (see photo 17.5).
8 Pull the servo/master cylinder assembly forward and remove the inner servo support bracket (see Fig. 8.10).
9 Remove the spring clip and clevis pin securing the servo actuating rod to the operating link (see photo 17.6), then lift out the servo/master cylinder assembly.
10 Remove the two nuts on the right-hand side of the pedal box assembly to free the servo operating link right-hand support bracket, then withdraw the servo operating link from the vehicle.
11 Refitting is the reverse procedure to removal, ensuring that the operating link brake rod grommet is seated correctly in the bulkhead and that the brake rod itself locates through the brake pedal before securing the servo operating link support brackets. Ensure correct location of the brake rod bush in the brake pedal.
12 Bleed the complete brake hydraulic system in accordance with Section 21 or 29 (as applicable).

202   Chapter 8 Braking system

20.1 Pressure control valves regulating line pressure to rear brake units

20 Pressure control valve – removal and refitting

1   The pressure control valves are located within the engine compartment and are fixed to the left inner wing panel (photo). The angular offset of the valve assembly from the horizontal determines the 'cut-in' point of the valves, thus reducing line pressure to the rear brake units to a lower pressure than the front brake units when under braking load. The 'cut-in' point is set in production, and is not adjustable.
2   To remove the valve assembly, first disconnect the battery earth terminal.
3   Disconnect the rigid brake pipes from each pressure control valve, and fit blanking plugs to prevent dirt ingress.
4   Remove the two screws securing the valve assembly mounting bracket to the inner wing panel, and withdraw the valve assembly from the vehicle.
5   Unclip the valves from their mounting bracket, as required.
6   Refitting is the reverse procedure to removal.
7   Bleed the complete brake hydraulic system on completion of work, as described in the following Section.

21 Hydraulic system – bleeding (conventional braking system)

Note: *For cars equipped with the Anti-lock Braking System, refer to Section 29*

1   This is not a routine operation but will be required after any component in the system has been removed and refitted or any part of the hydraulic system has been 'opened'. Where an operation has only affected one circuit of the hydraulic system, then bleeding will normally only be required to that circuit (front and rear diagonally opposite). If the master cylinder or the pressure regulating valves have been disconnected and reconnected, then the complete system must be bled.
2   When bleeding the brake hydraulic system on a vehicle fitted with a conventional braking system, the vehicle must maintain a level attitude, ie not tilted in any manner, to ensure that air is not trapped within the hydraulic circuits. During certain operations in this manual, instructions are given to bleed the brake hydraulic system with the front of the rear of the vehicle raised. In such cases raise the rest of the vehicle so that it maintains a level attitude, **but only if it is safe to do so**. If it is not possible to achieve this safely, complete the remainder of the operation and bleed the brake hydraulic system with the vehicle on its wheels.
3   One of three methods can be used to bleed the system.

Bleeding – two-man method
4   Gather together a clean jar and a length of rubber or plastic bleed tubing which will fit the bleed screw tightly. The help of an assistant will be required.
5   Take care not to spill fluid onto the paintwork as it will act as a paint stripper. If any is spilled, wash it off at once with cold water.

6   Clean around the bleed screw on the front right-hand caliper and attach the bleed tube to the screw.
7   Check that the master cylinder reservoir is topped up and then destroy the vacuum in the brake servo by giving several applications of the brake foot pedal.
8   Immerse the open end of the bleed tube in the jar, which should contain two or three inches of hydraulic fluid. The jar should be positioned about 300 mm (12.0 in) above the bleed nipple to prevent any possibility of air entering the system down the threads of the bleed screw when it is slackened.
9   Open the bleed screw half a turn and have your assistant depress the brake pedal slowly to the floor and then, after the bleed screw is retightened, quickly remove his foot to allow the pedal to return unimpeded. Repeat the procedure.
10 Observe the submerged end of the tube in the jar. When air bubbles cease to appear, tighten the bleed screw when the pedal is being held fully down by your assistant.
11 Top up the fluid reservoir. It must be kept topped up throughout the bleeding operations. If the connecting holes to the master cylinder are exposed at any time due to low fluid level, then air will be drawn into the system and work will have to start all over again.
12 Repeat the operations on the left-hand rear brake, the left-hand front and the right-hand rear brake in that order (assuming that the whole system is being bled).
13 On completion, remove the bleed tube. Discard the fluid which has been bled from the system unless it is required for bleed jar purposes, never use it for filling the system.

Bleeding – with one-way valve
14 There are a number of one-man brake bleeding kits currently available from motor accessory shops. It is recommended that one of these kits should be used whenever possible as they greatly simplify the bleeding operation and also reduce the risk of expelled air or fluid being drawn back into the system.
15 Connect the outlet tube of the bleeder device to the bleed screw and then open the screw half a turn. Depress the brake pedal to the floor and slowly release it. The one-way valve in the device will prevent expelled air from returning to the system at the completion of each stroke. Repeat this operation until clean hydraulic fluid, free from air bubbles, can be seen coming through the tube. Tighten the bleed screw and remove the tube.
16 Repeat the procedure on the remaining bleed nipples in the order described in paragraph 12. Remember to keep the master cylinder reservoir full.

Bleeding – with pressure bleeding kit
17 These too are available from motor accessory shops and are usually operated by air pressure from the spare tyre.
18 By connecting a pressurised container to the master cylinder fluid reservoir, bleeding is then carried out by simply opening each bleed

21.18 Bleeding a caliper using a pressure bleeding kit

Chapter 8 Braking system

screw in turn and allowing the fluid to run out, rather like turning on a tap, until no air bubbles are visible in the fluid being expelled (photo).

19 Using this system, the large reserve of fluid provides a safeguard against air being drawn into the master cylinder during the bleeding operations.

20 This method is particularly effective when bleeding 'difficult' systems or when bleeding the entire system at time of routine fluid renewal.

All systems

21 On completion of bleeding, top up the fluid level to the mark. Check the feel of the brake pedal, which should be firm and free from any 'sponginess' which would indicate air still being present in the system.

22 Brake warning lights – description and renewal

1 All models are fitted with a combined low fluid level/handbrake 'ON' warning light, controlled by a fluid level switch in the brake master cylinder fluid reservoir cap, and by a switch at the base of the handbrake lever. On vehicles equipped with the anti-lock braking system, an additional warning light is provided to indicate modulator drivebelt faults; see Section 24.

2 The warning lights are located in the instrument cluster; refer to Chapter 12, Sections 25 and 26 for details of bulb renewal. On switching on the ignition, all braking system warning lights should illuminate as a check that their bulbs are functioning. They should go out as the engine is started and the vehicle is driven away, and should not illuminate again (except whenever the handbrake is applied). If the low fluid level/handbrake 'ON' warning light does not go out, or if it comes on while the vehicle is being driven, then either the handbrake is not fully released or the fluid level has suddenly dropped. In the latter case, immediate attention is required to top-up the fluid level and to establish the cause of the drop; see Section 2.

3 To renew the switches, refer to Section 15, paragraphs 1 and 2 for the fluid level switch, to Section 12 for the handbrake 'ON' switch, and to Section 25, paragraphs 5 and 27, for the modulator belt-break switches on anti-lock equipped vehicles. In all cases, see also Chapter 12, Section 15.

4 The stop-lamps are controlled by a plunger-type switch located on the pedal box assembly (photo); to remove, disconnect its wiring and twist the switch anti-clockwise to release it.

5 On refitting, insert the switch into its retainer, press it lightly against the pedal until all pedal free play is just taken up, then twist the switch clockwise to secure it before reconnecting the wiring.

6 Once it is set on installation, as described in paragraph 5, there is no 'adjustment' procedure as such for the stop-lamp switch. However, as play develops in the brake pedal pivots and pushrod, the pedal can drop under its own weight to hang clear of its stop – this can provide sufficient clearance for the switch plunger to extend, causing the stop-lamps to flicker, especially over bumpy roads, or even to stay on. If this is thought to be happening, the switch position must be reset as described in paragraph 5.

22.4 Brake pedal stop lamp switch *in situ*, on pedal box assembly

23 Brake pedal – removal, bush renewal and refitting

1 Disconnect the battery earth terminal.
2 Remove the stop-lamp switch; refer to Section 22, paragraph 4.
3 Disconnect the servo operating link brake rod from its pedal location, by removing the retaining clip on the brake pedal, noting brake rod bush fitted in the pedal.
4 Remove the C-clip from its pedal shaft location, at the right-hand side of the brake pedal (photo), to allow the shaft to be withdrawn towards the centre of the vehicle.
5 As the pedal shaft is withdrawn, remove the brake pedal from the servo operating link brake rod. The brake pedal spacer can now be slid off the shaft if required, and the brake rod bush removed.
6 Prise the bushes out from both sides of the brake pedal, and renew as necessary.
7 Prior to refitting, apply a small amount of molybdenum disulphide grease to the pedal shaft.
8 Refitting is the reverse sequence to removal, ensuring that the brake rod bush is located correctly, that the pedal shaft 'D' section locates into the pedal box right-hand support, and that the stop-lamp switch is correctly set.

24 Anti-lock braking system – description

A mechanically-driven, two-channel anti-lock braking system is available as a factory-fitted option on certain model variants within the Fiesta range.

The system comprises four main components; two modulators, one for each brake circuit, and two rear axle load-apportioning valves, again, one for each brake circuit. Apart from the additional hydraulic piping, the remainder of the braking system is the same as for conventional models.

The modulators are located in the engine compartment with one mounted on each side of the transmission, directly above the driveshaft inner constant velocity joints. Each modulator contains a shaft which actuates a flywheel by means of a ball and ramp clutch. A rubber toothed belt is used to drive the modulator shaft from the driveshaft inner constant velocity joint.

During driving and under normal braking the modulator shaft and the flywheel rotate together and at the same speed through the engagement of a ball and ramp clutch. In this condition hydraulic pressure from

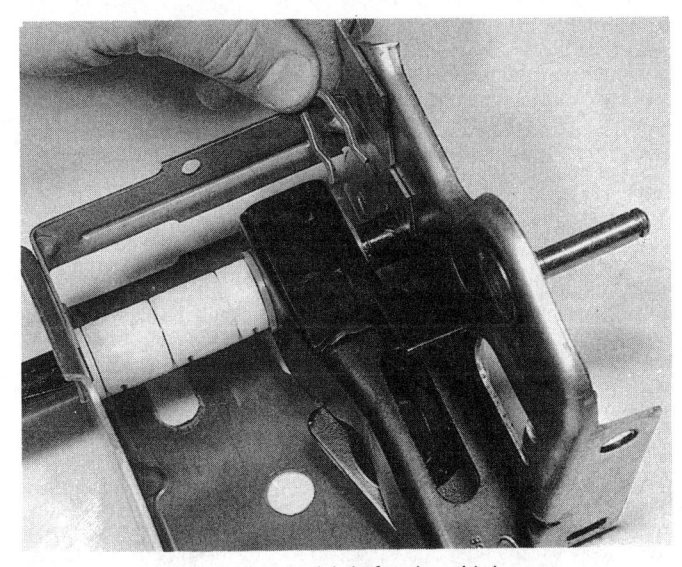

23.4 C-clip removed to allow pedal shaft to be withdrawn

Chapter 8 Braking system

via a de-boost piston allowing the wheel to once again revolve. Fluid passed through the dump valve is returned to the master cylinder reservoir via the modulator return hoses. At the same time hydraulic pressure from the master cylinder causes a pump piston to contact an eccentric cam on the modulator shaft. The flywheel is then decelerated at a controlled rate by the flywheel friction clutch. When the speed of the modulator shaft and flywheel are once again equal the dump valve closes and the cycle repeats. This complete operation takes place many times a second until the vehicle stops or the brakes are released.

The load-apportioning valves are mounted on a common bracket attached to the rear body, just above the rear axle twist beam location, and are actuated by linkages attached to the rear axle twist beam. The valves regulate hydraulic pressure to the rear brakes, in accordance with vehicle load and altitude, so that the braking force available at the rear brakes will always be lower than that available at the front.

A belt-break warning switch is fitted to the cover which surrounds each modulator drivebelt. The switch contains an arm which is in contact with the drivebelt at all times. If the belt should break, or if the adjustment of the belt is too slack, the arm will move out closing the switch contacts and informing the driver via an instrument panel warning light.

25 Modulator – removal and refitting (anti-lock braking system)

1 Disconnect the battery earth terminal.
2 Place a suitable drain tray under the brake fluid reservoir, then disconnect the relevant modulator return hose from it by pushing the hose into the reservoir, retaining the collar against the reservoir body and carefully withdrawing the hose. See Section 15, paragraph 3 for information on the modulator colour-coded connections.

Right-hand side

3 Raise the front of the vehicle, and support it securely using axle stands.
4 Remove the one-piece undertray where fitted, by turning the bayonet type fasteners, and on XR2i models, remove the front suspension crossmember (see Chapter 10).
5 From underneath, remove the belt-break switch from the right-hand drivebelt cover by squeezing its release lever towards the main body of the switch (photo), then carefully withdraw, ensuring that the belt contact arm does not catch on the drivebelt cover.
6 Remove the two bolts securing the modulator drivebelt cover to the modulator mounting bracket (photo), and withdraw the cover.
7 Disconnect the rigid brake pipes from the modulator, fitting blanking plugs to prevent excessive fluid loss and dirt ingress.
8 Remove the modulator pivot bolt and adjuster bolt (photo), then slip the drivebelt from its pulley, and withdraw the modulator unit from the vehicle. Ensure that the modulator return hose does not become kinked as the modulator unit is withdrawn.
9 Disconnect the modulator return hose from the modulator unit, and fit a blanking plug to prevent dirt ingress. Allow for residual fluid spillage as the hose is disconnected.
10 If a new modulator is to be fitted, note that these units are not interchangeable from side to side, and the correct replacement **must** be obtained. The modulator units are colour-coded, and must be fitted with the arrows on top of the casings pointing towards the front of the vehicle.
11 To refit, first connect the modulator return hose to the return outlet on the modulator unit.
12 Locate the modulator unit to its bracket and fit the pivot bolt, having applied a thin smear of anti-seize compound to the belt, but do not *fully* tighten at this stage. Take care not to damage the modulator return hose as it is manoeuvred into position.
13 Fit the drivebelt to its modulator pulley location, ensuring that it sits correctly over the driveshaft pulley, then refit the adjuster bolt but do not fully tighten at this stage.
14 Adjust the tension of the drivebelt by moving the modulator unit, until a belt deflection of 5.0 mm (0.2 in) is obtained under firm finger pressure. Check this using a ruler at a point midway between the two pulleys.
15 With the drivebelt tensioned correctly, tighten the pivot and adjuster belts according to specification. Re-check the tension of the drivebelt after tightening the belts.
16 Reconnect the rigid brake pipes to the modulator, tightening the unions to seal the system.

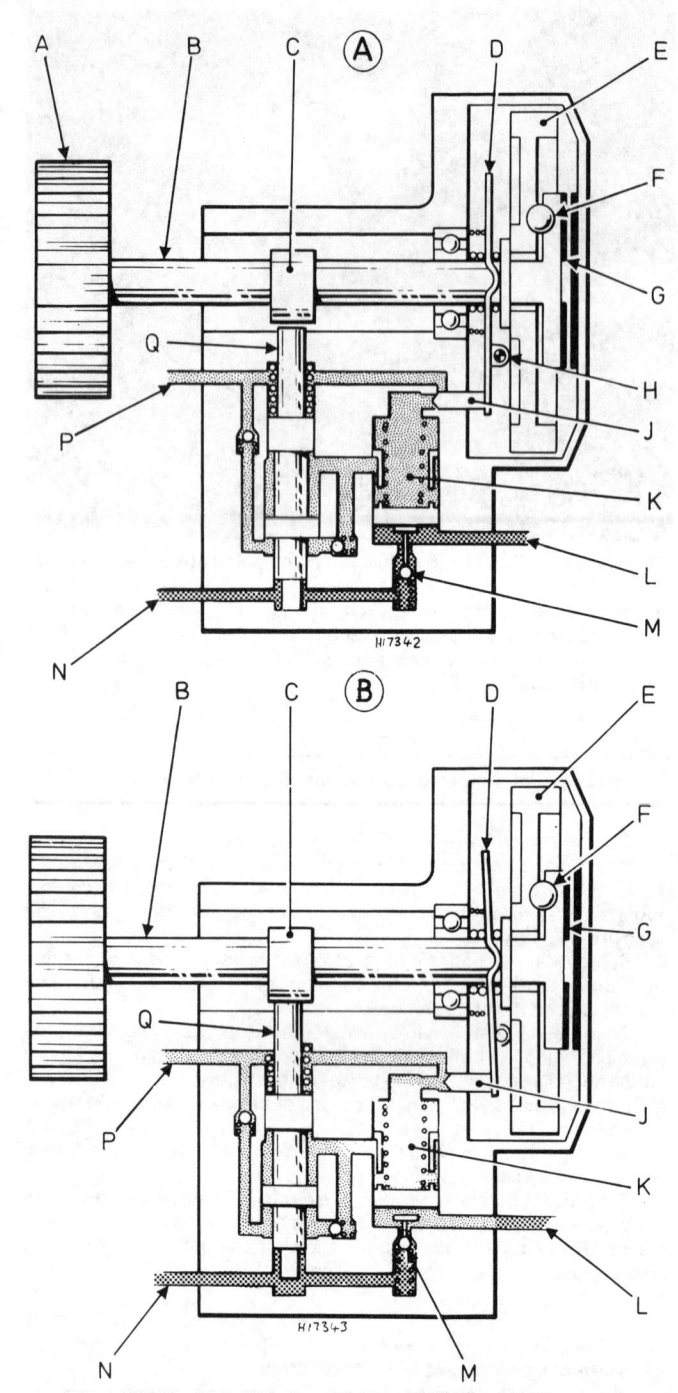

Fig. 8.11 Modulator operational diagram for normal braking (A) and with brakes locked (B) (Sec 24)

A Modulator pulley
B Modulator shaft
C Eccentric cam
D Dump valve lever
E Flywheel
F Ball and ramp drive
G Clutch
H Pivot
J Dump valve
K De-boost piston
L Port to brakes
M Cut-off valve
N From master cylinder
P To master cylinder (brake fluid) reservoir
Q Pump piston

the master cylinder passes to the modulators and then to each brake in the conventional way. In the event of a front wheel locking the modulator shaft rotation will be less than that of the flywheel and the flywheel will overrun the ball and ramp clutch. This causes the flywheel to slide on the modulator shaft, move inward and operate a lever which in turn opens a dump valve. Hydraulic pressure to the locked brake is released

Chapter 8 Braking system

25.5 Belt-break switch in drivebelt cover

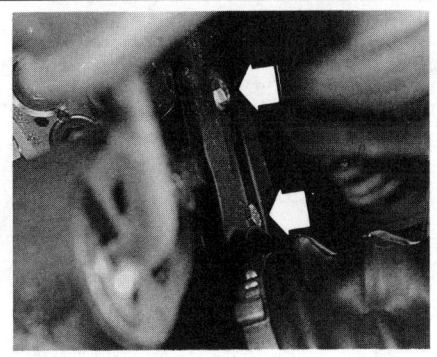

A Main switch body
B Release lever

25.6 Bolts securing modulator drivebelt cover to mounting bracket

25.8 Modulator pivot bolt (A) and adjuster bolt (B)

17 Refit the modulator drivebelt cover to the modulator mounting bracket, and secure with its two retaining bolts.
18 Refit the belt-break switch to the modulator drivebelt cover, taking care not to damage the belt contact arm as it passes through the cover.
19 Reconnect the modulator return hose by pushing the hose firmly into its brake fluid reservoir location, then lever out the collar to retain it. Remove the drain tray from under the brake fluid reservoir.
20 Remove the axle stands and lower the vehicle to the ground.
21 Top up the brake fluid reservoir using fresh fluid of the specified type, then bleed the brake hydraulic system in accordance with Section 29.
22 Refit the front suspension crossmember and the one-piece undertray, as applicable.
23 Reconnect the battery earth terminal.

Left-hand side

24 Repeat the procedures given in paragraphs 1 and 2.
25 Slacken the left-hand front roadwheel nuts, then raise the front of the vehicle and support it securely on axle stands. Remove the roadwheels.
26 Remove the one-piece undertray where fitted, by turning the bayonet type fasteners, and on XR2i models, remove the front suspension crossmember (see Chapter 10).
27 Remove the belt-break switch from the left-hand drivebelt cover in a similar manner to that described in paragraph 5, this time from the engine compartment.
28 Remove the two bolts securing the modulator drivebelt cover to the modulator mounting bracket, then ease the lower portion of the cover over the driveshaft taking care not to damage the driveshaft CV joint gaiter. Withdraw the cover through the engine compartment, manoeuvring it to clear obstructions.
29 Disconnect the rigid brake pipes from the modulator, fitting blanking plugs to prevent excessive fluid loss and dirt ingress.
30 Slacken the modulator pivot and adjuster bolts, then swing the modulator downwards to release the drivebelt tension before slipping the drivebelt from its modulator pulley location.
31 Remove the modulator pivot and adjuster bolts, withdraw the modulator upwards through the engine compartment. Ensure that the modulator return hose does not become kinked as the modulator unit is withdrawn.
32 Disconnect the modulator return hose from the modulator unit, and fit a blanking plug to prevent dirt ingress. Allow for residual fluid spillage as the hose is disconnected.
33 If a new modulator is to be fitted, note that these units are not interchangeable from side to side, and the **correct** replacement must be obtained. The modulator units are colour-coded, and must be fitted with the arrows on top of the casings pointing towards the front of the vehicle.
34 To refit, first connect the modulator return hose to the return outlet on the modulator unit.
35 Locate the modulator unit to its mounting bracket and fit the pivot bolt, having applied a thin smear of anti-seize compound to the bolt, but *do not fully* tighten at this stage. Take care not to damage the modulator return hose as it is manoeuvred into position.
36 Fit the drivebelt to its modulator pulley location, ensuring that it sits correctly over the driveshaft pulley, then refit the adjuster bolt but *do not fully* tighten at this stage.
37 Adjust the tension of the drivebelt by moving the modulator unit, until a belt deflection of 5.0 mm (0.2 in) is obtained under firm finger pressure. Check this using a ruler at a point midway between the two pulleys.
38 With the drivebelt tensioned correctly, tighten the pivot and adjuster bolts according to specification. Re-check the tension of the drivebelt after tightening the bolts.
39 Reconnect the rigid brake pipes to the modulator, tightening the unions to seal the system.
40 Refit the modulator drivebelt cover and secure with its two retaining bolts. Take care not to damage the driveshaft CV joint gaiter as the cover is eased into position.
41 Refit the belt-break switch to the modulator drivebelt cover, taking care not to damage the belt contact arm as it passes through the cover.
42 Reconnect the modulator return hose by pushing the hose firmly into its brake fluid reservoir location, then lever out the collar to retain it. Remove the drain tray from under the brake fluid reservoir.
43 Refit the left-hand front wheel, then remove the axle stands and lower the vehicle to the ground. Fully tighten the wheel nuts according to specification.
44 Top up the brake fluid reservoir using fresh fluid of the specified type, then bleed the brake hydraulic system in accordance with Section 29.
45 Refit the front suspension crossmember and the one-piece undertray, as applicable.
46 Reconnect the battery earth terminal.

26 Modulator drivebelt – removal and refitting (anti-lock braking system)

1 Disconnect the battery earth terminal, and slacken the relevant front roadwheel nuts.
2 Raise the front of the vehicle and support it securely using axle stands. Remove the roadwheel.
3 Remove the one-piece undertray where fitted, by turning its bayonet-type fasteners, and on XR2i models, remove the front suspension crossmember (see Chapter 10).
4 Remove the belt-break switch from the relevant drivebelt cover, then remove the drivebelt cover, as described in the previous Section.
5 Slacken the modulator pivot and adjuster bolts to release drivebelt tension, then slip the drivebelt from the modulator.
6 Remove the track rod end balljoint from the steering arm on the spindle carrier, as described in Chapter 9, Section 3, paragraphs 6 and 7.
7 Disconnect the anti-roll bar link from the anti-roll bar, where fitted, and release the brake hose from its location on the suspension strut.
8 Remove the (Torx head) pinchbolt and nut securing the lower suspension arm balljoint to the spindle carrier, and separate the balljoint from the spindle carrier assembly.

206 Chapter 8 Braking system

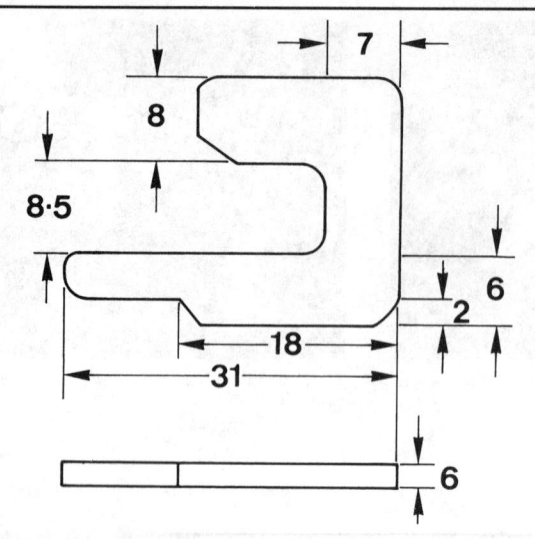

Fig. 8.12 Load-apportioning valve adjustment tool (dimensions given in mm) (Sec 27)

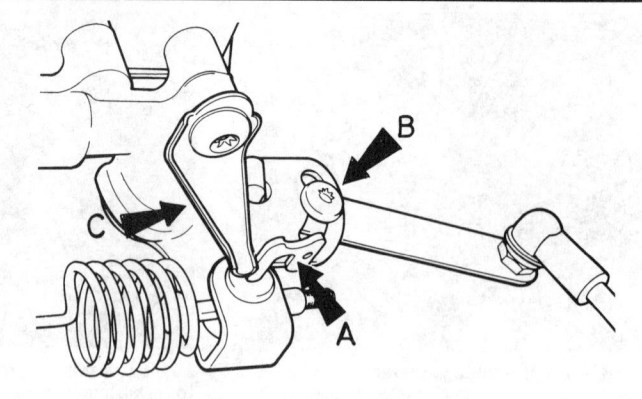

Fig. 8.13 Load-apportioning valve adjustment (Sec 27)

A Setting tool
B Operating link adjustment fixing screw
C Adjustment post

9 To release the driveshaft inner CV joint from the differential, have an assistant pull the spindle carrier away from the centre of the vehicle whilst you insert a lever between the inner CV joint and the transmission casing, then firmly strike the lever with the flat of the hand, but be careful not to damage adjacent components. Make provision for escaping transmission oil, if possible plugging the hole to prevent excessive loss. Do not allow the CV joints to bend more than 20° from the horizontal or internal damage may occur. If both driveshafts are to be removed, immobilise the differential by inserting an old joint or suitable shaft, before the other driveshaft is removed.
10 Slide the drivebelt off the driveshaft.
11 Prior to refitting the drivebelt, thoroughly clean its CV joint pulley location.
12 Fit the drivebelt over the driveshaft then, with a **new** snap-ring fitted to the inner CV joint splines (see Chapter 7), lubricate the splines with transmission oil. Remove the temporary plug and insert the inner CV joint to its transmission casing location. Press against the spindle carrier so that the snap-ring engages **fully** to hold the CV joint splines in the differential.
13 Refitting is now a reversal of the removal procedure, tensioning the drivebelt as described in the previous Section. Ensure that the pinch-bolt securing the lower suspension arm balljoint to the spindle carrier locates in the annular groove on the balljoint spindle. Secure the track rod and balljoint, using a **new** split pin.
14 Check the level of the transmission oil with the vehicle on the ground, and top up as required.

27 Load-apportioning valve – adjustment (anti-lock braking system)

1 Before attempting to adjust the load-apportioning valves, the vehicle must be at its kerb weight, ie with approximately half a tank of fuel and carrying no load. Note that a special setting tool will be required to adjust the valves – this can be fabricated, to the dimensions shown in Fig. 8.12.
2 Raise the vehicle on ramps or drive it over an inspection pit, so that working clearance is obtained with the full weight of the vehicle resting on its roadwheels. Remove the spare wheel and its carrier.
3 To check adjustment, insert the load-apportioning valve setting tool into the nylon sleeve without pre-loading the valve. If unable to insert the tool, carry out the following adjustment procedure.
4 Slacken the operating link adjustment fixing screw then insert the setting tool into the nylon sleeve, applying light pressure to the operating link upper arm, so that the setting tool fully locates. With the setting tool just resting up against the adjustment post (Fig. 8.13), tighten the operating link adjustment fixing screw to the specified torque.
5 Repeat the procedure on the other valve.
6 Refit the spare wheel on completion.

28 Load-apportioning valve – removal and refitting (anti-lock braking system)

1 Raise the vehicle on ramps, or drive it over an inspection pit, so that working clearance may be obtained with the full weight of the vehicle on its roadwheels.
2 Remove the spare wheel and its carrier for access to the load-apportioning valves (photo).
3 Disconnect the load-apportioning valve operating links from the rear axle twist beam, by undoing the nuts securing them.
4 Disconnect the rigid brake pipes from the load-apportioning valves, and fit blanking plugs to prevent dirt ingress. Make provision for escaping fluid as the pipes are disconnected.
5 Remove the bolts securing the valve assembly mounting bracket to the vehicle body, then carefully lower from the vehicle.
6 The valves can now be individually removed from the mounting bracket, as required, by undoing the fixings securing them from the other side of the bracket.
7 Refitting is the reverse sequence to removal, adjusting the load-apportioning valves, as described in the previous Section, before refitting the spare wheel. When fitting a new valve, the plastic tie must be cut off before attempting any adjustment, and the setting tool must be used as in the previous Section.
8 Bleed the brake hydraulic system in accordance with the following Section.

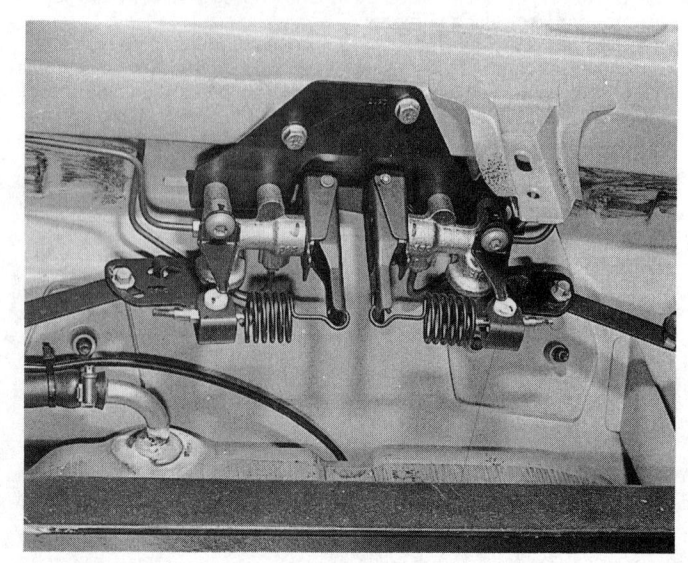

28.2 General view of load-apportioning valve arrangement (spare wheel and carrier removed for access)

Chapter 8 Braking system

29.6 Modulator bypass valve Torx screw (arrowed)

29.7 Auto-bleed plunger (arrowed)

29 Hydraulic system – bleeding (anti-lock braking system)

1 On vehicles equipped with the anti-lock braking system there are two bleed procedures possible, depending on which part of the brake hydraulic system has been disturbed.

2 If any one of the following conditions are present, bleed procedure A should be adopted:

   (a) A modulator has been removed
   (b) A modulator return hose (between modulator and brake fluid reservoir) has been drained
   (c) The rigid brake pipes have been disconnected from a modulator

3 If any one of the following conditions are present, bleed procedure B should be adopted:

   (a) Any condition where the master cylinder has been removed or drained, providing that the modulator return hoses have not lost their head of fluid
   (b) Removal or disconnection of any of the basic braking system components ie, brake caliper, flexible hose or rigid pipe, wheel cylinder, or load-apportioning valve

Bleed procedure A

4 Raise the vehicle on ramps, or drive it over an inspection pit, so that working clearance may be obtained with the full weight of the vehicle on its roadwheels. Remove the one-piece undertray, as applicable, by turning its bayonet-type fasteners and, on XR2i models, remove the front suspension crossmember (see Chapter 10).

5 Top up the brake fluid reservoir to the MAX mark using fresh fluid of the specified type, and keep it topped up throughout the bleeding procedure.

6 Slacken the modulator bypass valve Torx screw, located between the two rigid brake pipe connections on the modulator body (photo), and unscrew it two full turns.

7 Fully depress the auto-bleed plunger on the modulator (photo) and hold it down so that the plunger circlip contacts the modulator body. With the plunger depressed, have an assistant steadily pump the brake pedal at least twenty times whilst you observe the fluid returning to the brake fluid reservoir. Continue this operation until the returning fluid is free from air bubbles.

8 Release the auto-bleed plunger, ensuring that it returns to its normal operational position – pull it out by hand if necessary.

9 Tighten the modulator bypass valve Torx screw.

10 Repeat the operation on the other modulator, if applicable, then refit the one-piece undertray and front suspension crossmember if removed.

11 Now carry out bleed procedure B.

Bleed procedure B

12 This procedure is the same as for conventional braking systems, and reference should be made to Section 21. Note, however, that all the weight of the vehicle must be on the roadwheels, otherwise the load-apportioning valves will not bleed. If problems are encountered whereby the rear brakes will not bleed satisfactorily, ensure that the load-apportioning valves are correctly adjusted (see Section 27). As with the conventional braking system, the brake fluid level must be kept topped up during bleeding.

30 Fault diagnosis

Note: *Apart from checking the condition and adjustment of the modulator drivebelts (where applicable) and all pipes and hose connections, any faults occurring in the anti-lock braking system should be referred to a Ford dealer for diagnosis, due to the nature of the tests involved*

Symptom	Reason(s)
Excessive pedal travel	Low brake fluid level
	Rear automatic brake adjuster(s) inoperative
	Air in the hydraulic system
	Faulty master cylinder
	System component fluid leaks

Chapter 8 Braking system

Symptom	Reason(s)
Excessive pedal pressure required to stop the vehicle	Internal defect in servo unit Disconnected or leaking vacuum hose Wheel cylinder(s) or caliper piston(s) seized Brake pads or shoe linings worn or contaminated New brake pads or shoe linings not yet bedded in (after fitting new ones) Incorrect grade of pads or linings fitted Hydraulic circuit failure
Brake pedal feels spongy	Air in the hydraulic system Faulty master cylinder
Judder felt through brake pedal or steering wheel when under braking load	Excessive run-out or distortion of front discs or rear drums Brake pads or shoe linings worn Brake caliper loose on its mountings Wear in steering or suspension components – see Chapters 9 and 10
Brakes pull to one side	Brake pads or shoe linings worn or contaminated Brake pads or shoe linings renewed on one side only A mixture of friction material grades fitted across the axles Wheel cylinder or caliper piston seized Rear automatic brake adjuster seized
Brakes binding	Wheel cylinder or caliper piston seized Handbrake incorrectly adjusted Broken shoe return spring(s)
Rear wheels lock under normal braking loads	Faulty pressure control valve(s)

Chapter 9 Steering

For modifications, and information applicable to later models, see Supplement at end of manual

Contents

Fault diagnosis	10
General description	1
Maintenance and inspection	2
Steering column – removal overhaul and refitting	6
Steering rack assembly – overhaul	8
Steering rack assembly – removal and refitting	7
Steering rack gaiters – renewal	3
Steering/suspension angles and wheel alignment	9
Steering wheel – removal, refitting and realignment	5
Track rod ends – renewal	4

Specifications

General

Type	Rack-and-pinion
Lubricant type:	
Oil	To Ford specification SLM-1C-9110A (refer to Ford dealer)
Semi-fluid grease	To Ford specification SAM-1C-9106AA (refer to Ford dealer)
Grease	To Ford specification SM-1C-1021-A (Duckhams LB10)
Sealant type	To Ford specification SPM-4G-9112F

Steering/suspension angles

Toe setting:
- Up to 1990 model year:
  - Service check: 3.0 mm toe-out to 3.0 mm toe-in
  - Setting if outside service check: Parallel ± 1.0 mm
- 1990 model year onwards:
  - Service check: 4.5 mm toe-out to 0.5 mm toe-in
  - Setting if outside service check: 2.0 mm toe-out ± 1.0 mm

Suspension angles (non-adjustable for reference only):

	Castor	Camber
Up to 1990 model year:		
1.0 and 1.1 HCS models with manual transmission	0°23′	0°25′
1.1 HCS models with CTX automatic transmission, and 1.4 CVH models	0°18′	0°12′
1.6 S	0°32′	0°08′
XR2i	0°45′	0°13′
Maximum variation (side-to-side)	1°00′	1°15′
1990 model year onwards:		
All models except XR2i	0°53′	0°08′
XR2i	0°56′	0°04′
Maximum variation (side-to-side)	1°00′	1°15′

Torque wrench settings

	Nm	lbf ft
Steering rack to bulkhead	70 to 97	57 to 72
Track rod end to spindle carrier steering arm	25 to 30	18 to 22
Steering wheel bolt to steering shaft	45 to 55	33 to 40
Steering column mounting nuts	10 to 14	7 to 10
Pinion splines-to-lower column universal joint pinch-bolt	45 to 56	33 to 41
Track rod locknut to track rod end	57 to 68	42 to 50
Inner track rod balljoint to rack (staked)	68 to 90	50 to 66
Pinion retaining collar	80 to 90	59 to 66

1 General description

The steering is of conventional rack-and-pinion type, incorporating a safety system of convoluted column tube and double universally-jointed lower steering shaft links.

The steering column tube is supported at its upper end by bracketry, and at its lower end by a nylon support bush. The steering shaft is supported within the column tube by two support bearings, one at either end of the tube.

The steering rack assembly is located on the bulkhead. Steering input is transmitted, via the steering shaft, to the pinion which meshes with the teeth on the rack. The system is of linear gearing type, having equally spaced rack teeth.

Chapter 9 Steering

The pinion transferring the steering input moves the rack within its housing tube, withdrawing and extending the track rods attached to either end of the rack by balljoints. This movement is transferred, by balljoints in the track rod ends, to the steering arms on the spindle carriers which direct the roadwheels.

2 Maintenance and inspection

1 At the intervals specified in *'Routine maintenance'* at the beginning of this manual, carry out the following checks.
2 With the handbrake on, jack the front of the car up and support it securely using axle stands. Remove the one-piece undertray (where fitted) by turning its bayonet-type fasteners.
3 Grasp the roadwheel at the 9 o'clock and 3 o'clock positions, and try to rock it. Movement felt may be attributable to worn hub bearings, or track rod inner or outer balljoints. If the outer balljoint is worn, visual movement will be obvious. If the inner joint is suspect, it can be felt by placing a hand over the steering rack gaiters and gripping the track rod. If the wheel is now rocked, movement will be felt if inner balljoint wear has taken place. Repair procedures are given in Sections 4 and 8 of this Chapter.
4 Ensure regular checks are made regarding the condition of the steering rack gaiters. Splits and tears will allow dirt and water ingress, causing accelerated rack component wear rates. If any damage is present they must be renewed, as described in Section 3.
5 Refit the one-piece undertray as applicable then, with the car back on its wheels, have an assistant turn the steering wheel from side to side, about an eighth of a turn each way, whilst you observe the roadwheels. If any lost movement is apparent, closely observe the inner and outer balljoints, the steering shaft universal joints, steering column mountings, and the rack assembly-to-bulkhead mountings. Any concerns here must be rectified in accordance with the appropriate Sections of this Chapter.
6 Examine the tyres. Uneven wear may be attributable to incorrect tyre pressures or incorrect front wheel alignment. Scrub marks on the tyres can be a sign of misalignment. The wheel alignment procedure is covered in Section 9 of this Chapter.

3 Steering rack gaiters – renewal

1 If the steering rack gaiters are found to be defective, they must be renewed. A suitable balljoint separator will be needed.
2 Slacken the front roadwheel nuts, raise the front of the vehicle and support it securely using axle stands. Remove the roadwheels.
3 Remove the one-piece undertray (where fitted), by turning its bayonet type fasteners, and on XR2i models, remove the front suspension crossmember (see Chapter 10).

3.7 Releasing a track rod end balljoint using a balljoint separator

4 Measure and take note of the amount of thread on the track rods that is exposed near the track rod ends. Mark the relative positions of track rods and track rod ends to ensure correct toe settings upon refitting.
5 Release the track rod end locknut to facilitate track rod end removal.
6 Extract the split pin and remove the nut from the balljoint taper pin.
7 If the taper pin is held fast in the eye of the steering arm use a balljoint separator to free it (photo). Do not attempt to drive it out by hammering on the threaded end.
8 Unscrew the track rod end, and its locknut, from the trackrod.
9 Remove the clamps from both ends of the steering rack gaiter and slide it off the track rod.
10 When ordering new steering rack gaiters and retaining clamps, also specify the diameter of the track rod to ensure that the correct replacements are obtained.
11 If steering lubricant has been lost due to a damaged steering rack gaiter, it will be necessary to drain any remaining lubricant and replace it with fresh. To do this, turn the steering wheel gently to expel as much lubricant as possible from the rack housing. If the opposing gaiter is not being renewed, it is advisable to unclamp and release it from the rack to allow old lubricant to be removed from that end too.
12 Lightly grease the inside surfaces of the gaiter, the steering rack tube and track rod contact faces, and slide the gaiter into position, ensure that the narrow neck locates correctly in the track rod groove.
13 When a new gaiter is being fitted to the pinion end of the rack, leave the gaiter unclamped at this stage. If it is being fitted to the end,

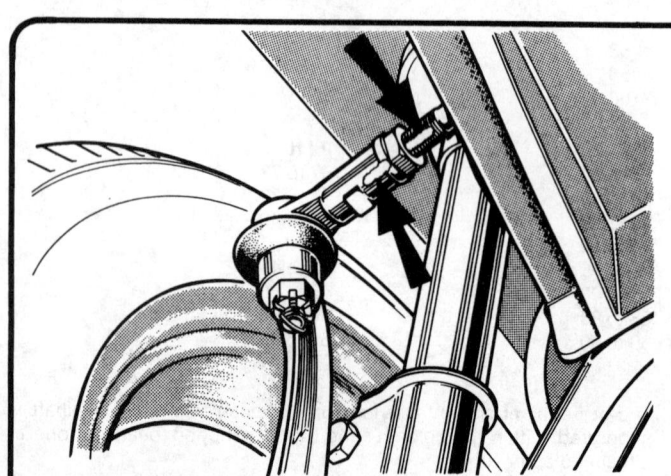

Fig. 9.1 Relative adjustment marks on track rod and track rod end (Sec 3)

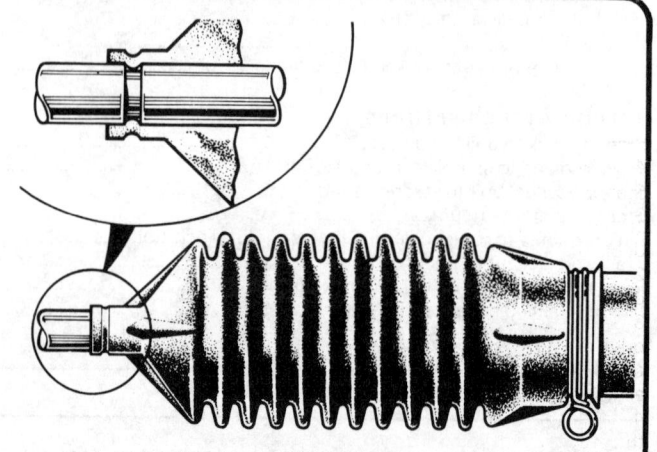

Fig. 9.2 Location of steering rack gaiter narrow neck to track rod groove (Sec 3)

Chapter 9 Steering 211

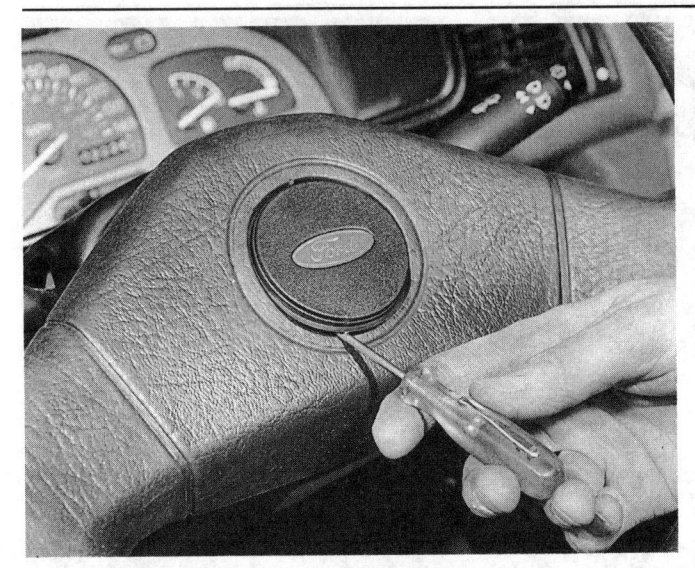

5.2 Removing the motif from the steering wheel

5.4 Tightening the steering wheel retaining bolt

clamp the inner end around the rack tube. New screw-type clamps should be used to secure the gaiters.
14 Clean the threads on the track rod using a wire brush, taking care not to erase the alignment marks.
15 Screw the track rod end locknut into position on the track rod, then follow it with the track rod end, to align with the marks made earlier. It is important that care is taken to refit the track rod end in its original position to maintain original toe setting (assuming that this was correct to start with).
16 Connect the track rod end balljoint taper pin to the steering arm, tighten the nut to the specified torque and insert a **new** split pin to secure.
17 If applicable, renew the steering lubricant as described in Section 8.
18 Fit and tighten inner and outer gaiter clamps as necessary.
19 Tighten the track rod end locknut according to specifications, and check that the alignment marks correspond.
20 Refit the front suspension crossmember and one-piece undertray, as applicable.
21 Refit the roadwheels, then lower the vehicle to the ground. Torque up the roadwheel nuts with the vehicle resting on its wheels. Settle the front suspension by rolling the vehicle backwards and forwards a few times whilst bouncing the front end.
22 Front wheel alignment should be checked at the earliest opportunity (see Specifications, and Section 9).

4 Track rod ends   renewal

1 If the track rod end balljoints are found to be worn, remove them as described in the previous Section.
2 Clean the threads with a wire brush, the apply a thin coating of grease to them.
3 Screw on the new track rod end to take up a similar position to the original. Due to manufacturing tolerances the new component is likely to be slightly different in size, and this means that the front wheel alignment will need to be reset (Section 9).
4 Refit the track rod end balljoint to the steering arm as described in the previous Section, refit the roadwheels and lower the vehicle to the ground.
5 Upon completion of adjustment, ensure that the track rod end-to-track rod locknut is tightened according to specification, and that the steering rack gaiters are properly clamped.

5 Steering wheel – removal, refitting and realignment

1 Drive the vehicle in a straight line onto a level surface, so that the roadwheels are pointing straight ahead.
2 Carefully prise the motif from the centre of the steering wheel (photo).
3 Undo the steering wheel retaining bolt and lift steering wheel from steering shaft, having marked their relative positions.
4 On refitting, prevent the risk of damage to the direction indicator cancelling cam by ensuring that the direction indicator switch is in the neutral position before the steering wheel is refitted. Align the marks on the wheel and shaft that were made on removal, and tighten the retaining bolt to its specified torque wrench setting (photo).
5 If the steering wheel is not centralized when the roadwheels are pointing straight ahead, there is a set procedure to follow, depending on the degree of correction required. This operation must be carried out under the conditions outlined in paragraph 1.
6 Ensure that the toe setting is as specified.
7 If the steering wheel is more than 30° out of alignment, it should be removed and repositioned to reduce the error to less than 30°, and the retaining bolt and motif refitted.
8 Jack the front of the vehicle up, and support it securely on axle stands. Remove the one-piece undertray (where fitted) and on XR2i models, the front suspension crossmember (see Chapter 10).
9 Mark the relative positions of the track rod ends to the trackrods, in the same place for each assembly, then loosen the track rod end-to-track rod locknuts.
10 Release the steering rack gaiter outer clamps.
11 To correct the angular error at the steering wheel, rotate both track rods in the same direction and by the same amount. It is vital that this is done as instructed to avoid disturbing the toe setting. Rotate the track rods by approximately 30° for every 1° of steering wheel angular error. To correct a clockwise angular error at the steering wheel (to move the steering wheel anti-clockwise), rotate the trackrods anti-clockwise when viewed from the left-hand side of the vehicle. An anti-clockwise angular error at the steering wheel (steering wheel needs to move clockwise) can be corrected by rotating the track rods clockwise, viewed from the same side.
12 With the steering wheel centralized, ensure that the roadwheels are still pointing straight ahead, then tighten the track rod end-to-track rod locknuts according to specification. The relative alignment marks on the track rods will probably not line up with the marks made on the track rod ends, but should indicate corresponding rotational movements of the track rods. Exposed thread lengths on the track rods must be equal upon completion of the task.
13 Refit the steering rack gaiter outer clamps.
14 Refit the front suspension crossmember and one-piece undertray, as applicable.

Chapter 9 Steering

6.2 Manual choke control knob and its locating lug (arrowed)

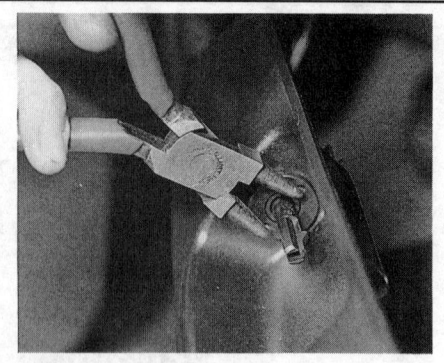

6.3 Removing the manual choke control assembly retaining collar

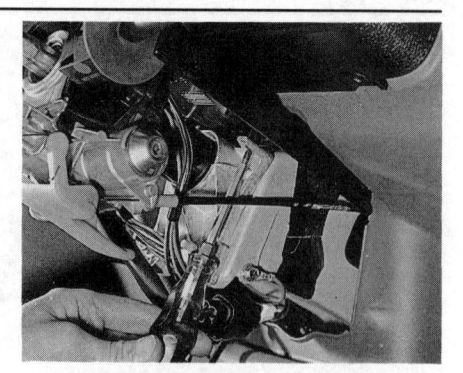

6.4 Removing upper column shroud retaining screws with lower shroud removed (steering wheel removed for clarity)

6.5 Steering column multi-function switch retaining screw (arrowed) (steering wheel removed for clarity)

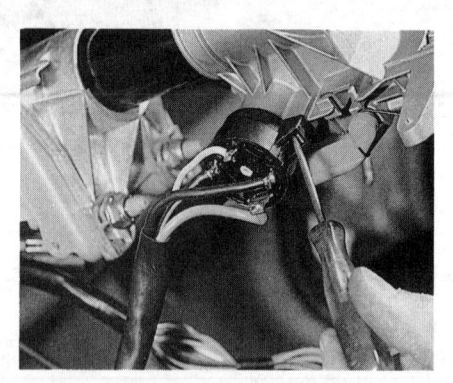

6.6 Removing ignition loom plate from steering column lock housing (steering wheel removed for clarity)

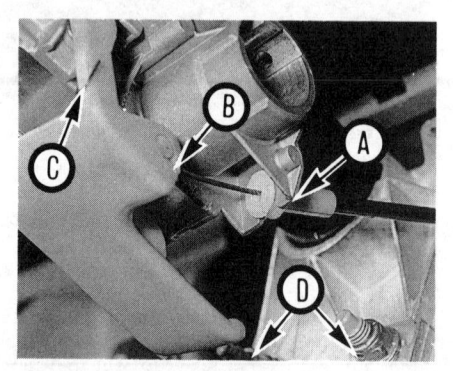

6.7 Bonnet release lever and cable

A Bonnet release cable abutment
B Bonnet release cable slot on release lever
C Bonnet release lever return spring
D Steering column to mounting bracket retaining nuts

15 Lower the vehicle to the ground, and settle the suspension by rolling the vehicle backwards and forwards, bouncing the front end.
16 Although the toe setting should not have altered, check the front wheel alignment at the earliest possible opportunity (Section 9).

6 Steering column - removal, overhaul and refitting

1 Disconnect the battery earth terminal, and refer also to Section 3, paragraph 3.
2 Remove the manual choke control knob, where fitted, by depressing the lug securing it, and pulling it from its shaft (photo). The lug is found on the side of the control knob shank.
3 Remove the lower steering column shroud by undoing its four retaining screws, then detach the choke warning light switch/pull control assembly from the lower shroud by unscrewing its retaining collar (bayonet-type fixing), using a suitable tool to locate in the collar recesses (photo).
4 Remove the two screws securing the upper steering column shroud from above, the two screws securing it from below, the latter accessible only with the lower shroud removed (photo).
5 Detach the electrical connections (multi-plug type) from the steering column multi-function switch assembly, then remove the single screw securing the multi-function switch assembly to the steering column lock housing. This retaining screw is located directly forward of the hazard warning light switch (photo). Remove the multi-function switch assembly.
6 Disconnect the ignition loom multi-plug connector and remove the loom plate from its location on the left-hand side of the steering column (photo).
7 Unclip the bonnet release cable abutment from its location in the steering column lock housing, then detach the cable from the bonnet release lever by aligning the cable core with the slot on the release lever and withdrawing it through that slot. Detach the spring from the release lever arms, then disengage the arms from the steering column lock housing and remove the bonnet release lever (photo).
8 Prise out the steering wheel centre motif using a screwdriver, taking care not to damage the motif or the steering wheel. Ensure that the roadwheels are pointing straight ahead, then loosen the steering wheel retaining bolt *slightly*.
9 Remove the nuts securing the steering column mounting bracket.
10 Remove the pinch-bolt securing the lower steering shaft universal joint to the steering rack pinion splined shaft, located at the rear of the engine compartment (photo), and separate the two as far as possible (see Section 7).
11 Pull the steering column assembly from its bulkhead location and withdraw it from the vehicle, ensuring that the lower steering shaft universal joint and the steering rack pinion splined shaft separate fully. The effort required to remove the column assembly may be quite high, due to the close tolerance of the lower column tube support bush in its location.
12 To overhaul the steering column, first remove the lower column tube support bush (photo).
13 Remove the steering wheel retaining bolt, and make alignment marks for the steering wheel and steering shaft. Insert the ignition key and turn it to position I. Pull the steering wheel off its shaft, using a soft-faced hammer to gently tap the shaft if the steering wheel is held tightly.
14 Slide the upper tolerance ring from the steering shaft, then withdraw the shaft from the column tube assembly. Note that the lower steering shaft universal joint may be separated from the mainshaft if renewal is necessary.
15 With the ignition key in position I, depress the plunger on the side of the lock barrel and withdraw lock mechanism from its housing.
16 Slide the lower tolerance ring and spring off the steering shaft.

Chapter 9 Steering

6.10 Pinch-bolt securing lower steering shaft universal joint to pinion splined shaft (arrowed)

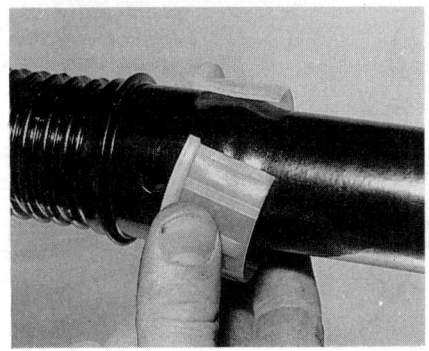

6.12 Removing the lower column tube support bush

6.19 Installing thrust bearing to steering column lock housing, using a socket to bear upon the outer race

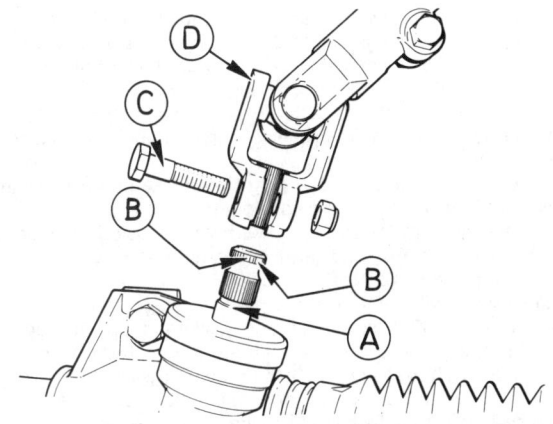

Fig. 9.3 Steering rack-to-column engagement (Sec 6)

A Steering rack pinion splined shaft
B Annular groove to ensure correct location of pinch-bolt
C Pinch-bolt
D Lower steering shaft universal joint

17 Using a suitable implement, prise out the upper and lower thrust bearings from the lock housing and the column tube base.
18 Renew all worn components as necessary.
19 Carefully install new thrust bearings into the lock housing and the column tube base. Use sockets of a suitable size to bear upon the outer bearing races, and tap squarely into position (photo). Do **not** allow the bearings to twist as they are being installed.
20 Fit the spring and lower tolerance ring to the steering shaft, and insert the shaft to the column tube.
21 Refit the ignition lock barrel to the steering lock housing with the ignition key in position I. Slight rotational movements of the barrel may be necessary to align it within its housing.
22 Fit the upper tolerance ring to the steering shaft, with its tapered end locating to the upper thrust bearing.
23 Refit the steering wheel, aligning the marks made previously, but do not fully tighten its retaining bolt at this stage.
24 Fit the lower column tube support bush around the column base.
25 With the help of an assistant, insert the steering column assembly into the vehicle so that the lower steering shaft universal joint and the pinion splined shaft on the steering rack locate correctly, with the steering wheel centralized. Loosely refit the pinch-bolt to secure. Ensure that the steering shaft bulkhead seal seats correctly in its location.
26 With the column assembly located loosely in position, refit the nuts securing it to its mounting bracket, taking care to ensure that the lower column tube support bush seats correctly as the nuts are tightened to specification.
27 Refit the steering column ancillary components, reversing the removal procedure given in paragraphs 2 to 7.
28 Tighten the pinch-bolt securing the lower steering shaft universal joint to the pinion splined shaft according to specification, with the steering wheel centralized and the road wheels in the straight ahead position. Ensure that the pinch-bolt sits in the annular groove on the pinion splined shaft (Fig. 9.3).

29 Tighten the steering wheel retaining bolt according to specification, and refit its centre motif.
30 Reconnect the battery earth terminal, and refit the one-piece undertray and front suspension crossmember (where applicable).

7 Steering rack assembly – removal and refitting

1 Slacken the front roadwheel, then raise the front of the vehicle and support it securely using axle stands. Remove the roadwheels.
2 Remove the one-piece undertray where fitted, by turning its bayonet-type fasteners, and on XR2i models, remove the front suspension crossmember (see Chapter 10).
3 Remove the pinch-bolt securing the pinion splined shaft to the lower steering shaft universal joints, located at the rear of the engine compartment.
4 Separate the track rod end balljoints from the steering arms, as described in Section 3, paragraphs 6 and 7.
5 Remove the lower servo support bracket bolt.
6 Remove the other bolt securing the steering rack assembly to the bulkhead, then withdraw it from the right-hand side of the vehicle, taking care to disengage the pinion splined shaft from the lower steering shaft universal joint as the assembly is moved.
7 On refitting, centralize the rack and steering wheel, then engage the pinion splined shaft to the lower steering shaft universal joint.
8 Refit the steering rack assembly mounting bolts to the bulkhead and tighten to the specified torque, ensuring that the servo support bracket is correctly held.
9 Refit the pinion splined shaft-to-lower steering shaft universal joint pinch-bolt and nut and tighten to specification, ensuring that the pinch-bolt locates to the annular groove as the pinion splined shaft.
10 Refit the track rod end balljoints to the steering arms, as detailed in Section 3, paragraph 16. As long as the track rod end-to-track rod relative positions have not been disturbed, it will not be necessary to reset the front wheel alignment.
11 Refit the front suspension crossmember and one-piece undertray, as applicable.
12 Refit the roadwheels, then lower the vehicle to the ground. Fully tighten the roadwheel nuts with the vehicle on its wheels.
13 Any misalignment of the steering wheel when the roadwheels are in the straight-ahead position can be rectified in accordance with Section 5 of this Chapter.

8 Steering rack assembly – overhaul

1 With the steering rack assembly removed from the vehicle, as described in the previous Section, clean it externally to each overhaul. Before dismantling, it should be noted than when reassembling you will need to use Ford special service tools 13-012 (yoke adjuster), 13-009-A

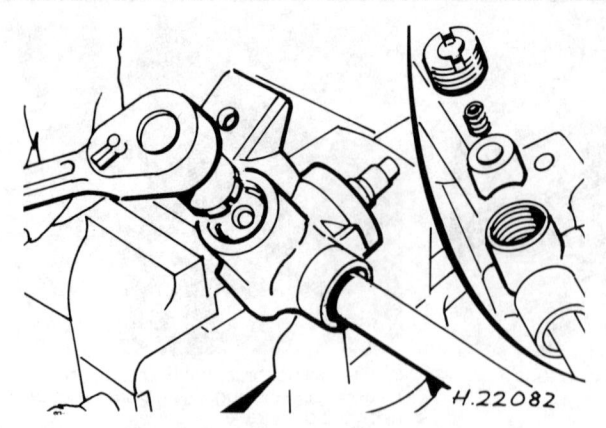

Fig. 9.4 Remove the yoke plug (Sec 8)

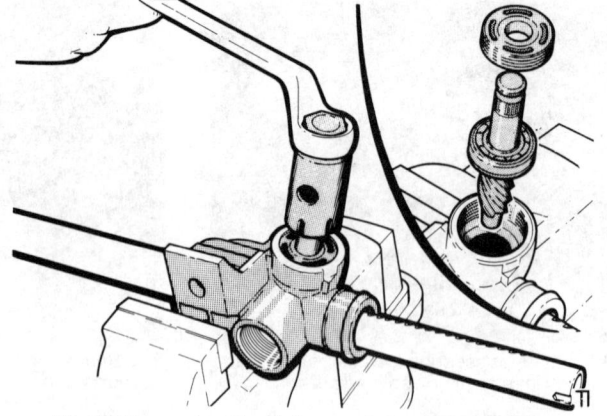

Fig. 9.5 Removing the pinion retaining collar (Sec 8)

(pinion nut adjuster wrench), 13-008-A (steering pinion socket) and 15-041 (torque preload gauge), and also some sealant as quoted in the Specifications. Unless the latter three tools are available, the steering rack assembly overhaul should be entrusted to your Ford dealer, as accurate adjustment of the rack-and-pinion engagement will not be possible. A yoke adjuster can be fabricated. Check parts availability before dismantling.

2 Remove the track rod ends and steering rack gaiters as described in Section 3.

3 Traverse the steering so that the rack slides out of its tube as far as possible. Position the assembly in a vice so that the rack itself is securely held in the vice jaws. Use vice jaw protectors to prevent damage to the rack.

4 If the original track rods are fitted use a pipe wrench to unscrew the balljoints from the rack. If service replacement track rods are fitted, use a spanner on the machined flats. Each track rod must be refitted to the same end of the rack that it was removed from, so keep them in order of fitting.

5 Using service tool 13-012, remove the yoke plug and withdraw the spring and the yoke from the rack tube. As an alternative, you could fabricate a castellated socket to locate in the yoke slots to achieve this (Fig. 9.4).

6 Pull off the pinion dust seal then remove the pinion retaining collar using service tool 13-009-A, and withdraw the pinion and bearing assembly from the rack tube (Fig. 9.5). A slight twisting action will assist the removal of pinion and bearing.

7 Slide the steering rack out of its tube, then carefully prise the rack support bush out from the end of the tube furthest from the pinion housing.

8 Clean and inspect all the components, renewing as necessary any worn or damaged items.

9 Upon reassembly, install the rack support bush into its housing at the end of the rack tube furthest from the pinion housing.

10 Apply a light coating of semi-fluid grease to the steering rack and slide it into the rack tube, taking care not to damage or dislodge the rack support bush.

11 Centralize the steering rack in its housing tube then, after applying a light coating of semi-fluid grease to the pinion and its bearing, insert the assembly into the pinion housing to mesh with the rack teeth.

12 Insert and secure the pinion retaining collar, using service tool 13-009-A, having applied the specified sealant sparingly to the collar threads. Tighten to the specified torque, then stake punch in four places around its housing to secure.

13 Refit the yoke, spring and yoke plug to the rack housing, having applied sealant to the threads of the yoke plug.

14 Ensure that the rack is centralized within its housing then, using service tool 13-012 or home-made equivalent, tighten the yoke plug to the specified torque, then back off 60° to 70°.

15 Using service tool 15-041 (torque preload gauge) in conjunction with service tool 13-008-A (steering pinion socket), the pinion turning torque can be checked. To do this accurately, turn the pinion anti-clockwise half a turn from its central position then measure the torque as the pinion is rotated clockwise one full turn. Rotate the pinion anti-clockwise half a turn to centralize the rack. The pinion turning torque can be adjusted as necessary, until a torque of between 1.05 and 1.7 Nm is reached, by adjusting the yoke plug (Fig. 9.6).

16 When the correct turning torque is achieved, stake punch the yoke plug in three places around its housing to secure.

17 Refit the track rod inner balljoint units to the steering rack, reversing the method used for removal. If re-using the original track rod units, they must be fitted to their original sides and tightened so that the original staking marks align with the steering rack grooves. When tightening, ensure that the rack (**not** the tube) is held in a vice fitted with jaw protectors. If new track rod units are being fitted, they must be tightened to the specified torque wrench setting, using a suitable open-ended torque wrench adaptor, and then staked to the steering rack groove.

18 Refit the steering rack gaiter furthest from the pinion, as described in Section 3.

19 Insert 120 cc of the specified oil into the rack tube, and add 70 cc of the specified semi-fluid grease to the rack teeth and pinion. For lubricant details, see Specifications.

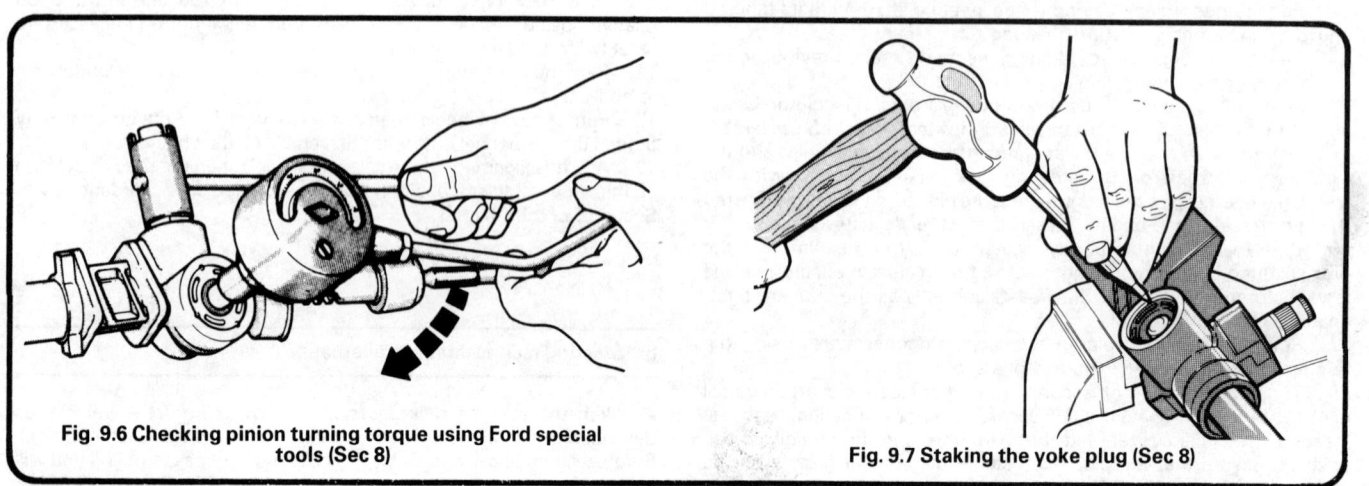

Fig. 9.6 Checking pinion turning torque using Ford special tools (Sec 8)

Fig. 9.7 Staking the yoke plug (Sec 8)

Chapter 9 Steering

20 Refit the steering rack gaiter to the pinion end of the rack, as described in Section 3.
21 If the original track rod ends are being refitted to the original track rods, the method of refitting is as described in Section 3. If either the track rods or track rod ends have been renewed, the procedure is as detailed in Section 4. In either case it is essential that the front wheel alignment is checked, and is necessary adjusted, as detailed in Section 9.
22 Fill the pinion dust seal with grease and fit, over the pinion, to the pinion housing.
23 The steering rack can now be refitted to the vehicle, as described in Section 7.

9 Steering/suspension angles and wheel alignment

1 When reading this Section, reference should also be made to Chapter 10 in respect of front and rear suspension arrangement.
2 Accurate front wheel alignment is essential to good steering and for even tyre wear. Before considering the steering angles, check that the tyres are correctly inflated, that the roadwheels are not buckled, the hub bearings are not worn or incorrectly adjusted and that the steering linkage is in good order.
3 Wheel alignment consists of four factors:
Camber is the angle at which the roadwheels are set from the vertical when viewed from the front of rear of the vehicle. Positive camber is the angle (in degrees) that the wheels are tilted outwards at the top, from the vertical.
Castor is the angle between the steering axis and a vertical line when viewed from each side of the vehicle. Positive castor is indicated when the steering axis is inclined towards the rear of the vehicle at its upper end.
Steering axis inclination is the angle, when viewed from the front or rear of the vehicle, between the vertical and an imaginary line drawn between the lower suspension swivel balljoints and upper strut mountings.
Toe is the amount by which the distance between the front inside edges of the roadwheel differs from that between the rear inside edges. If the distance at the front is less than that at the rear, the wheels are said to toe-in. If the distance at the front inside edges is greater than that at the rear, the wheels toe-out.
4 Due to the need for precision gauges to measure the small angles of the steering and suspension settings, it is preferable to leave this work to your dealer. Camber and castor angles are set in production and are not adjustable. If these angles are ever checked and found to be outside specification then either the suspension components are damaged or distorted, or wear has occurred in the bushes at the attachment point.
5 To check the toe setting yourself, first make sure that the lengths of both track rods are equal when the steering is in the straight-ahead position. This can be measured reasonably accurately by counting the number of exposed threads on the track rods adjacent to the track rod ends.
6 Adjust if necessary by releasing the track rod end locknut and the clamp at the small end of the steering rack gaiters.
7 Obtain a tracking gauge. These are available in various forms from accessory stores, or one can be fabricated from a length of steel tubing, suitably cranked to clear the sump and bellhousing, and having a setscrew and locknut at one end.
8 With the gauge, measure the distance between the two inner rims of the roadwheels (at hub height) at the rear of the wheel. Push the vehicle forward to rotate the wheel through 180° (half a turn) and measure the distance between the wheel inner rims, again at hub height, at the front of the wheel.
9 This last measurement should differ from the first one by the specified toe-in/toe-out (see Specifications).
10 Where the toe setting is found to be incorrect, release the track rod end locknuts and the steering rack gaiter small-end clamps. Turn both track rods by an equal amount, a quarter of a turn at a time, then recheck the alignment. Repeat as necessary. Do not grip the threaded part of the track rod during adjustment. When each track rod is viewed from the rack housing, turning the rods clockwise will increase the toe-out. Always turn the track rods in the same direction when viewed from the centre of the vehicle; failure to comply with this will cause steering wheel misalignment and cause problems with tyre scrubbing, when turning.
11 With adjustment complete, tighten the track rod end locknuts without disturbing the track rod settings.
12 Refit and tighten the steering rack gaiter small-end clamps.
13 Ensure that the lengths of both track rods are equal, as in paragraph 5.
14 Rear wheel alignment is set in production, and is not adjustable.

10 Fault diagnosis

Symptom	Reason(s)
Heavy or stiff steering	Tyres under-inflated Steering column bent or damaged Steering rack bent or damaged Pinion adjusted too tightly Lack of steering rack lubricant Seized steering or suspension balljoint Incorrect front wheel alignment
Vague steering, vehicle wanders	Incorrect tyre pressures Worn track rod end balljoints Steering shaft universal joints worn Worn or loose rack-and-pinion assembly
Wheel wobble and vibration	Roadwheels out of balance Roadwheels buckled Faulty or damaged tyre Worn steering or suspension joints Loose wheel nuts
Vehicle pulls to one side	Incorrect wheel alignment Accident damage to steering or suspension components Wear in steering or suspension components Brakes binding

Chapter 10 Suspension

For modifications, and information applicable to later models, see Supplement at end of manual

Contents

Fault diagnosis	15	General description	1
Front hub bearings - checking, removal and renewal	4	Maintenance and inspection	2
Front suspension anti-roll bar (S, SX and XR2i variants) - removal and refitting	5	Rear axle and suspension strut assembly - removal and refitting	12
Front suspension crossmember (XR2i only) - removal and refitting	6	Rear axle bush - renewal	10
Front suspension lower arm - removal and refitting	7	Rear axle unit - renewal	13
Front suspension spindle carrier - removal and refitting	3	Rear hub bearings - checking, removal and renewal	9
Front suspension strut - removal, overhaul and refitting	8	Rear suspension strut - removal, overhaul and refitting	11
		Roadwheels and tyres - general care and maintenance	14

Specifications

General

Front suspension	Independent MacPherson struts with coil springs and integral double-acting shock absorbers. New design A-shaped lower suspension arms acting through double vertical bushes. Anti-roll bar fitted to S, SX and XR2i variants. Bracing crossmember for XR2i
Rear suspension	Semi-independent with twist beam rear axle. Double-acting shock absorbers with coil springs in strut format similar to front suspension
Track:	
Front	1392 mm to 1406 mm (54.3 to 54.8 in) depending on model
Rear	1376 mm to 1387 mm (53.7 to 54.1 in) depending on model
Wheelbase	2446 mm (95.4 in)
Wheel bearing grease (for rear bearings only)	To Ford specification SAM-1C-9111A (refer to Ford dealer)
Hub/brake drum-to-axle flange bolt sealant	To Ford specification SDM-4G-9107A

Roadwheels and tyres

Roadwheel type:	
Standard models	Pressed-steel
XR2i	Pressed-steel or optional alloy
Roadwheel size:	
Standard models	13 x 4.5 J
S variants	13 x 5 J
XR2i	13 x 5.5 J
Tyre size:	
XR2i	185/60 R 13-H
S variants	165/65 R 13-S
LX/Ghia	155/70 R 13-S
All other models	135 R 13-S or 145 R 13-S

Tyre pressures (cold): bar (lbf/in^2)	Front	Rear
135 R 13-S:		
Normal loading (up to three persons)	2.1 (30)	2.0 (29)
Fully laden (over three persons)	2.3 (33)	2.8 (40)
145 R 13-S:		
Normal loading (up to three persons)	1.8 (26)	1.8 (26)
Fully laden (over three persons)	2.3 (33)	2.8 (40)
155/70 R 13-S:		
Normal loading (up to three persons)	2.0 (29)	1.8 (26)
Fully laden (over three persons)	2.3 (33)	2.8 (40)
165/65 R 13-S:		
Normal loading (up to three persons)	1.8 (26)	1.8 (26)
Fully laden (over three persons)	2.3 (33)	2.8 (40)
185/60 R 13-H:		
Normal loading (up to three persons)	2.0 (29)	1.8 (26)
Fully laden (over three persons)	2.1 (30)	2.3 (33)

Chapter 10 Suspension

Suspension angles
See Specifications in Chapter 13

Torque wrench settings

	Nm	lbf ft
Front suspension		
Hub driveshaft retaining nut	205 to 235	151 to 173
Lower arm balljoint to spindle carrier pinch-bolt and nut (Torx)	48 to 60	35 to 44
Front suspension strut to spindle carrier pinch-bolt	80 to 90	59 to 66
Anti-roll bar link to front suspension strut nut	41 to 58	30 to 43
Anti-roll bar link to anti-roll bar nut	41 to 58	30 to 43
Anti-roll bar retaining clamp bolts to lower arm	20 to 28	15 to 21
Front suspension strut top-mount retaining nut	40 to 52	30 to 38
Front suspension strut spring retaining nut	52 to 65	38 to 48
Track rod end-to-spindle carrier steering arm	25 to 30	18 to 22
Brake caliper anchor bracket bolts	51 to 61	37 to 45
Front suspension crossmember bolts (XR2i only)	80 to 90	59 to 66
Lower arm mounting bracket retaining bolts	80 to 90	59 to 66
Lower arm-to-lower arm mounting bracket bolts (by the torque-to-yield method)	50 ('snug' see text)	37 ('snug' see text)
Rear suspension		
Rear spindle-to-axle beam retaining bolts	56 to 76	41 to 56
Rear hub bearing retaining nut (RH and LH threads respectively)	250 to 290	184 to 214
Body mounting bracket retaining bolts	41 to 58	30 to 43
Rear axle trailing arm void bush bolt (torque taken off bolt head, not nut)	58 to 79	43 to 58
Rear suspension strut top-mount retaining nuts	28 to 40	20 to 30
Rear suspension strut-to-axle mounting bolt	102 to 138	75 to 102
Rear suspension strut spring retaining through-bolt	41 to 58	30 to 42
Load-apportioning valve operating link-to-rear axle twist beam	21 to 28	15 to 21
Wheels		
Roadwheel nuts	70 to 100	52 to 74

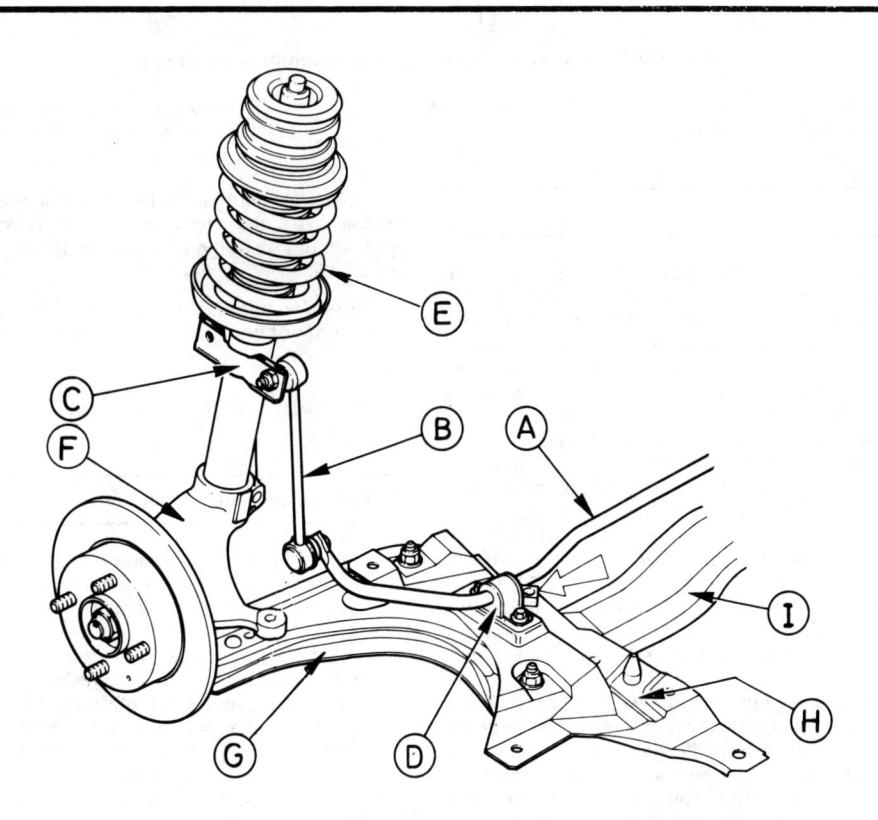

Fig. 10.1 General view of front suspension components (Sec 1)

A Anti-roll bar ('S', 'SX' and XR2i models)
B Anti-roll bar link ('S', 'SX' and XR2i models)
C Anti-roll bar suspension strut bracket ('S', 'SX' and XR2i models)
D Anti-roll bar bush and clamp bracket ('S', 'SX' and XR2i models)
E Coil spring integral with suspension strut
F Spindle carrier
G Lower suspension arm
H Lower suspension arm mounting bracket
I Front suspension (bracing) crossmember (XR2i only)

218 Chapter 10 Suspension

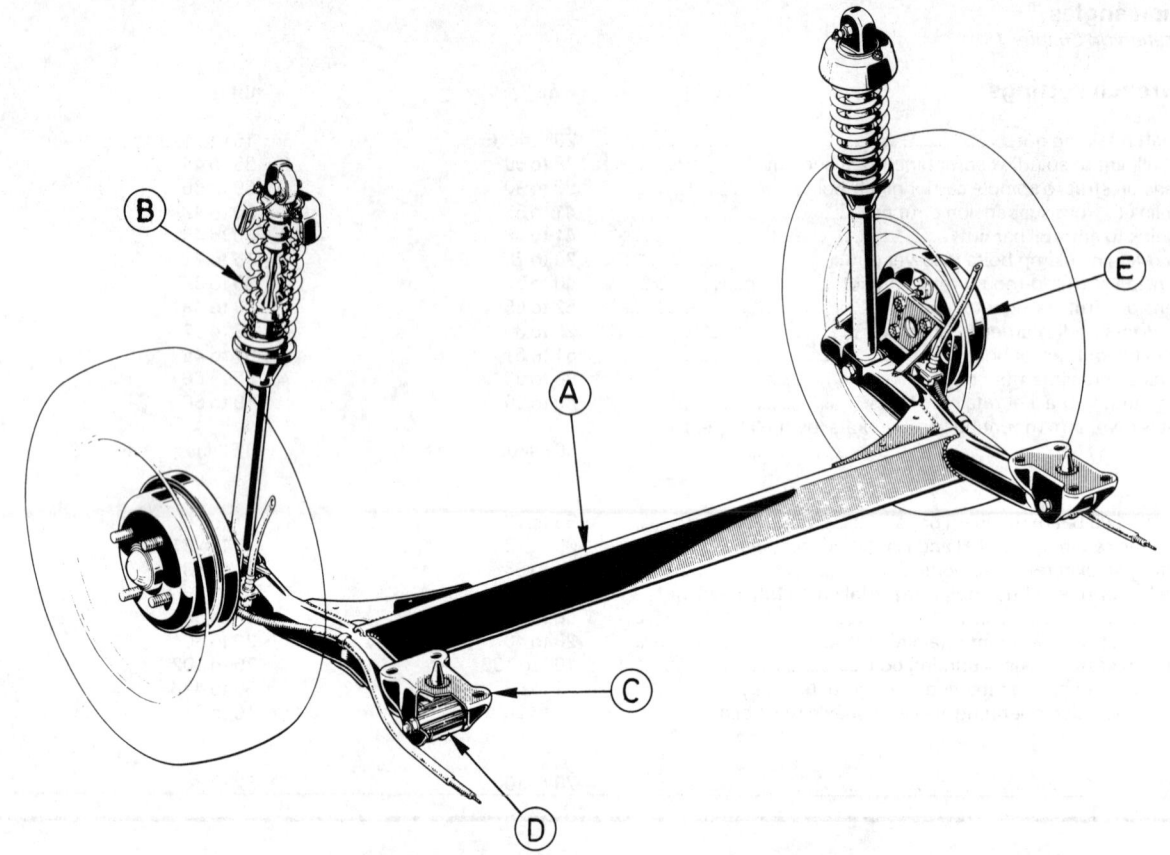

Fig. 10.2 General view of rear suspension components (Sec 1)

A Twist beam rear axle
B Coil spring integral with suspension strut
C Body mounting bracket
D Void bush
E Brake drum, wheel hub and spindle assembly secured to the axle flange by four bolts

1 General description

The front suspension is of independent type, achieved by the use of MacPherson struts. The struts, which incorporate coil springs and integral shock absorbers, are located at their upper mountings by rubber insulators and secured to the inner wing panels by cup seat mountings and locknuts. The lower end of each strut is bolted to the top of a cast spindle carrier. The spindle carriers house non-adjustable hub bearings as a variation of a proven design. The lower mountings of the spindle carriers are attached, via balljoints, to a pressed-steel lower arm assembly. The lower arm assembly consists of two sections. The lower arm mounting bracket is bolted securely to the underside of the vehicle, and has a locating peg and unique outer fixing bolt to ensure correct location. The A-shaped lower arm is attached to its mounting bracket by double vertical bushes and controls both lateral and fore and aft movement of the front wheels. The balljoints connecting the lower mountings of the spindle carriers to the lower arms are riveted to the lower arms, and are not available as separate service items. An anti-roll bar is fitted to S, SX and XR2i models, and the XR2i has an additional bracing crossmember.

The rear suspension is semi-independent, with an inverted V-section beam welded between tubular trailing arms. This inverted V-section beam allows a limited torsional flexibility, giving each rear wheel a certain degree of independent movement, whilst maintaining optimum track and wheel camber control. This type of arrangement is called a 'twist beam' rear axle. The axle is attached to the body by rubber void bushes, through brackets bolted to the underside of the vehicle. Each bracket has a conical locating peg to ensure accurate alignment of the axle assembly. The rear suspension struts, which are similar to the MacPherson struts used at the front, are mounted at their upper ends by nut and captive bolt fixings through the suspension turrets in the luggage compartment. At their lower ends, the struts are attached, close to the wheels, by bolts passed through the trailing arms and lower strut integral bushes. The rear wheel hub/brake drum unit and spindle on each side form an assembly which can be unbolted from the axle without disturbing the hub bearings. The hub bearings are non-adjustable.

Note: *The twist beam rear axle and the bracing crossmember (XR2i only) must **not** be used for jacking or supporting the vehicle.*

2 Maintenance and inspection

1 Regular inspection should be made regarding the condition of the suspension, looking for signs of worn, perished or otherwise unsatisfactory bushes, component damage (especially relevant after kerbing heavily or a minor accident), and worn or damaged balljoints. Renew any unsatisfactory items. Note that balljoints and front suspension lower arm vertical bushes are not available separately, and can only be obtained as an assembly integral with the lower arm.
2 Using a flat bar or a large screwdriver, carefully lever against the anti-roll bar (where fitted) to check for wear in its lower arm mountings. Some movement is to be expected as the mountings are made of rubber, but excessive wear should be obvious. If wear is detected, renew the mountings.
3 During the inspection, examine the struts to ensure that there is no fluid leakage. If fluid leakage is evident, then the affected strut(s) must be renewed.
4 The rear axle trailing arm void bushes should locate the axle assembly firmly. Any excess movement here will be evident in a handling deterioration and noise during operation. They can be checked by carefully levering against each trailing arm using a flat bar. Bearing in

Chapter 10 Suspension

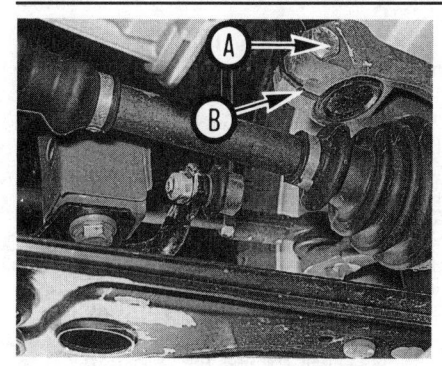

3.12 Pinch bolt securing spindle carrier to suspension strut (A), and spindle carrier slot (B)

3.15 Washer locating over driveshaft stub prior to fitting hub driveshaft retaining nut

3.17 Lower suspension arm balljoint showing annular groove (arrowed)

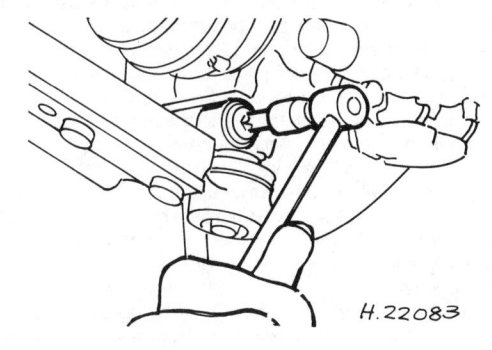

Fig. 10.3 Removing the lower suspension arm balljoint (Torx) pinch-bolt (Sec 3)

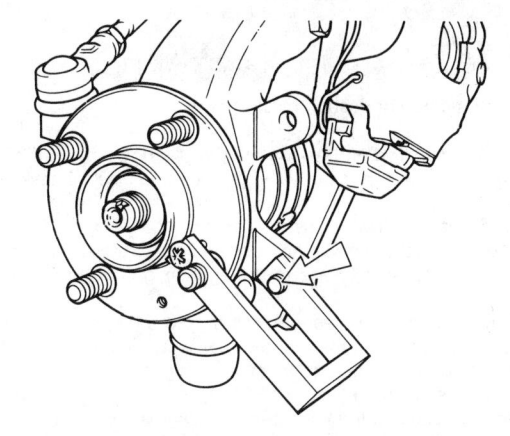

Fig. 10.4 Securing the hub to the spindle carrier using a bracket (Sec 3)

mind the forces dealt with by these bushes, if excess movement is detected, they must be renewed.
5 The security and torque settings of all the bolts and nuts should be checked periodically. Note that when checking the front suspension lower arm vertical bush bolts, the procedure detailed in Section 7 of this Chapter **must** be adhered to. The vehicle **must** be resting on its wheels for tightening the rear suspension strut lower mounting bolts and the rear axle trailing arm void bush bolts to their specified torques, and additionally the void bush bolt torque should be taken off the bolt head and not the nut.
6 Repair procedures can be found in the relevant Sections of this Chapter.
7 For roadwheel and tyre maintenance, see Section 14 of this Chapter.

Note: *Reference should also be made to the 'Routine maintenance' Section at the beginning of this manual, for service intervals and operations.*

3 Front suspension spindle carrier – removal and refitting

1 Before dismantling it should be noted that a balljoint separator, and a home-made bracket to secure the hub to the spindle carrier during removal, will be required. When removing the spindle carrier assembly from the vehicle, in accordance with the procedure detailed, **do not** allow the hub to become separated from the spindle carrier, as this will displace the bearing seals and lead to premature bearing failure. A new hub driveshaft retaining nut will be needed upon reassembly.
2 Using a small punch, relieve the staking securing the hub driveshaft retaining nut to its shaft, and slacken the nut by half a turn.
3 Slacken the relevant roadwheel nuts, then raise the front of the vehicle and support it securely on axle stands.
4 Remove the roadwheel.
5 Remove the bolt securing the brake hose bracket to the front suspension strut.

6 Remove the brake caliper anchor bolts and suspend the caliper assembly from the suspension strut with a length of strong wire, to prevent the flexible brake hose from straining.
7 Remove the track rod end balljoint from the steering arm on the spindle carrier, as described in Chapter 9, Section 3, paragraphs 6 and 7.
8 Remove the (Torx head) pinch-bolt securing the lower suspension arm balljoint to the spindle carrier (Fig. 10.3), and separate the balljoint from the spindle carrier assembly.
9 Remove the single screw securing the brake disc to the hub, and slide the disc off the wheel studs.
10 Install the home-made bracket to retain the hub, with one end secured through the brake caliper anchor bolt holes and the other end located over a wheel stud and held with a wheel nut. Any form of bracket that holds the hub securely to the spindle carrier may be used to achieve this (Fig. 10.4).
11 Remove the hub driveshaft retaining nut.
12 Remove the pinch-bolt securing the spindle carrier to the suspension strut (photo).
13 Using a small crowbar with a thin tip, or a stout screwdriver as an alternative, lever the spindle carrier slot to separate the spindle carrier from the suspension strut. Lower the spindle carrier slightly and carefully pull it off the driveshaft, having supported the driveshaft to prevent damage to the CV joints – the driveshaft must not be bent at an angle greater than 20° from the horizontal. If the assembly is held tightly, use a proprietary puller to free it. Remove the dust sleeve from the inner rim groove of the spindle carrier.
14 To refit, first fit the dust sleeve to its groove, then locate the spindle carrier over the driveshaft, taking care not to dislodge the hub bearings – draw the driveshaft CV joint through the hub using the old retaining nut and washer or, if this is not successful, Ford Special Tool 14-022. Lever the spindle carrier slot open and refit the spindle carrier to the suspension strut. Remove the lever and the driveshaft support, refit the pinch-bolt and tighten according to specification.
15 Fit the new hub driveshaft retaining nut and washer (photo), but do not **fully** tighten.

220   Chapter 10   Suspension

3.22 Staking the hub driveshaft retaining nut

16 Refit the track rod end balljoint to the steering arm on the spindle carrier, tighten the nut to the specified torque and insert a **new** split-pin to secure.
17 Reconnect the lower suspension arm balljoint to the spindle carrier, and tighten the (Torx) pinch-bolt and nut to the specified torque. Note that the bolt must locate to the annular groove on the balljoint spindle (photo).
18 Remove the bracket securing the hub to the spindle carrier, refit the brake disc and tighten the single screw securing the disc to the hub.
19 Refit the brake caliper to the spindle carrier, tightening the bolts according to specification.
20 Refit the bolt securing the brake hose bracket to the front suspension strut.
21 Refit the roadwheel, remove the axle stands and lower the vehicle to the ground.
22 Torque the hub driveshaft retaining nut according to specification then, using a suitable punch, stake the nut into the driveshaft slot so that it is held securely (photo).
23 Tighten the roadwheel nuts according to specification.

4 Front hub bearings – checking, removal and renewal

1 All models are fitted with non-adjustable front hub bearings, which are supplied pre-greased by the manufacturer.
2 To check the bearings for excessive wear, raise the front of the vehicle and support it securely on axle stands.
3 Grip the roadwheel at its top and bottom, and attempt to rock it. If excessive wheel rock is noted, or if there is any roughness, binding or vibration when the wheel is spun, the bearings must be renewed.
4 Before dismantling it should be noted that service tools 14-034 (installation adaptors) 14-035 (inner bearing support), 15-033-01 (adaptor), 15-034 (spindle bearing installer) and 15-068 (pinion bearing cup installer), or suitable alternatives, will be needed. It is possible to fabricate some of the service tools, but try your local tool hire centre first. If you cannot obtain the required equipment, or fabricate it to an acceptable standard, you are advised to entrust the hub bearing renewal to your Ford dealer. Under no circumstances attempt to tap the front hub bearing assemblies into position, as this is certain to cause damage sufficient to render them unserviceable.
5 To dismantle the spindle carrier assembly, with it removed from the vehicle as described in the previous Section, mount it in a vice and remove the home-made bracket holding the hub to the spindle carrier.
6 Slide the hub out of the spindle carrier assembly then, using a suitable punch, remove the inner and outer bearings and cups by tapping at diametrically-opposed points. Do not allow the cups to tilt within the spindle carrier bore, and take care not to raise any burrs on the bore as this may prevent the new cups from seating correctly.
7 Thoroughly clean the spindle carrier bore and the hub before reassembly begins.
8 Insert the outer bearing assembly to the spindle carrier with the sealing lips facing outwards towards its adjacent installation adaptor. Draw it in slowly and carefully (Fig. 10.6), and, once it has been inserted, take care not to dislodge the bearing inner race and seal.
9 Insert the inner bearing assembly to the spindle carrier with the sealing lips facing inwards towards its adjacent installation adaptor. Once again, draw it in slowly and carefully (Fig. 10.7) and take care not to dislodge the inner races and seals.
10 If it is possible to slide the hub through the bearings by hand, the inner bearing must be supported, by using service tool 14-035 or suitable alternative (Fig. 10.8), to prevent separation of the bearing assembly which would render it unserviceable. If the hub cannot be fitted by hand, it should be carefully drawn in as shown in Fig. 10.9.

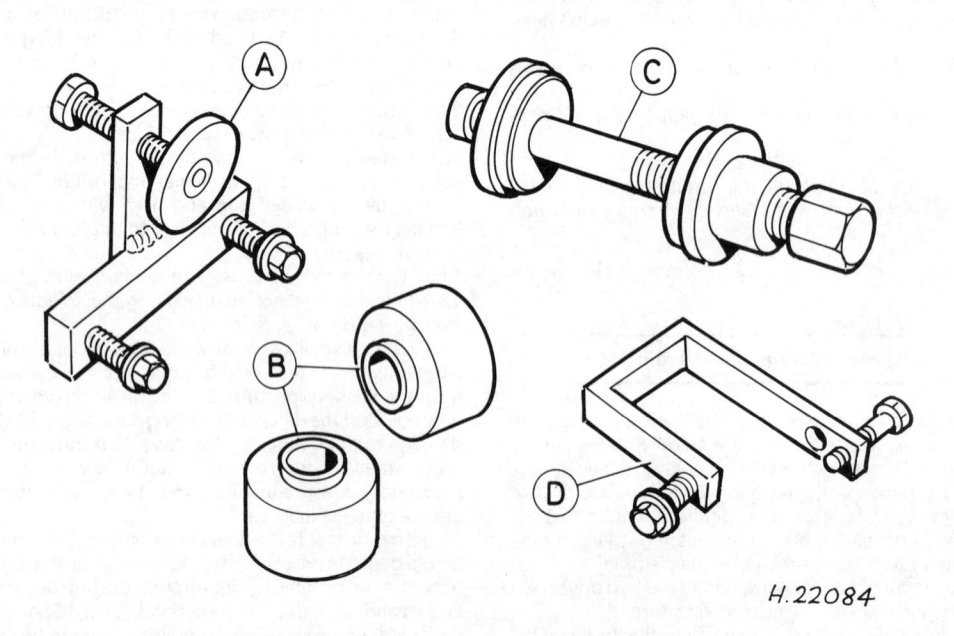

Fig. 10.5 A selection of the service tools used in bearing installation (Sec 4)

A Fixture to support inner bearings to prevent separation of the bearing assembly as hub is refitted
B Hub bearing installation adaptors
C Bearing installation puller
D Bracket to secure hub to spindle carrier

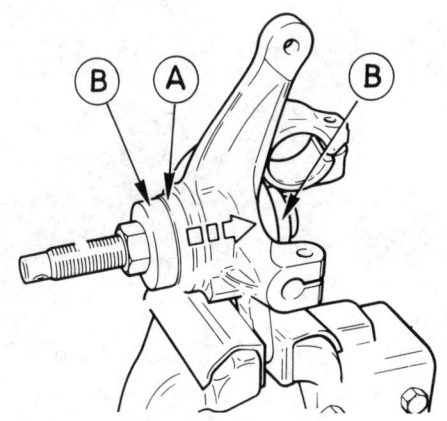

Chapter 10 Suspension 221

Fig. 10.6 Installing the outer bearing assembly to the spindle carrier (Sec 4)

A Bearing assembly   B Installation adaptors

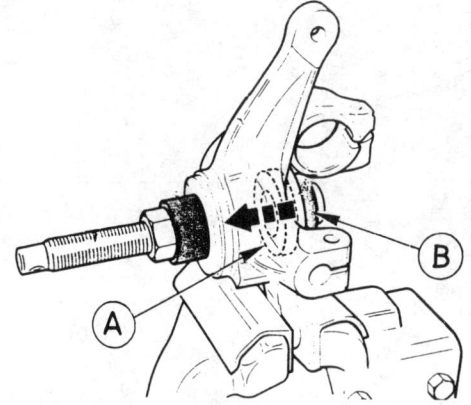

Fig. 10.7 Installing the inner bearing assembly to the spindle carrier (Sec 4)

A Bearing assembly   B Installation adaptor

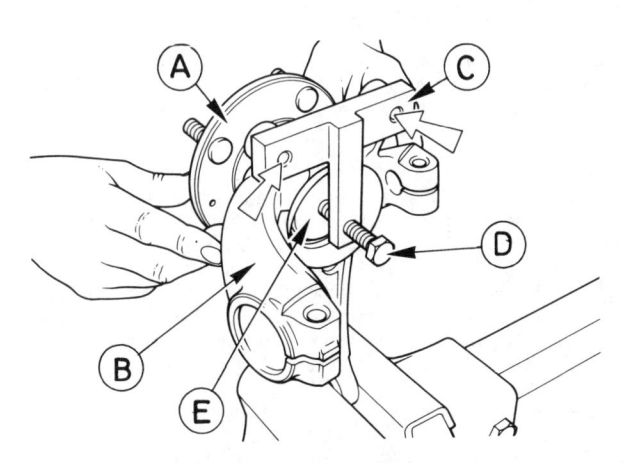

Fig. 10.8 Sliding the hub through the bearings by hand with the inner bearing supported (Sec 4)

A Wheel hub
B Spindle carrier
C Bearing support bracket fastened using brake caliper anchor bolts
D Support plate adjustment
E Support plate against inner bearing

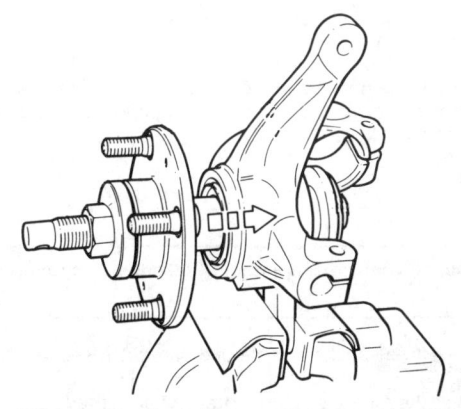

Fig. 10.9 Drawing the hub through the bearings (Sec 4)

11 Fit the home-made bracket to secure the hub into the spindle carrier, dismantling the service tools or equivalent as necessary. Ensure that the hub and spindle carrier do **not** become separated at any time, as this will displace the bearings seals and lead to premature bearing failure.
12 The assembly can now be refitted to the vehicle, as described in the previous Section.

5.3 Anti-roll bar link upper retaining nut (A) and brake hose bracket retaining nut (B)

5 Front suspension anti-roll bar (S, SX and XR2i variants) removal and refitting

1 Loosen the front roadwheel nuts, raise the front of the vehicle and support it securely on axle stands. Remove the front roadwheels.
2 Remove the one-piece undertray where fitted, by turning its bayonet-type fasteners, and on XR2i models, remove the front suspension crossmember as described in the next Section.
3 Undo the nut securing the upper end of the anti-roll bar link to the suspension strut bracket, and disconnect the link (photo).
4 Undo the nut securing the lower end of the anti-roll bar link to the anti-roll bar, and remove the link.
5 Remove the two bolts securing the anti-roll bar brackets to each lower suspension arm (photo), remove the brackets and withdraw the anti-roll bar from the vehicle.
6 The rubber bushes locating the anti-roll bar can be removed by sliding them off over the link connections.

Chapter 10 Suspension

5.5 Anti-roll bar bracket location on lower suspension arm mounting bracket (bolts arrowed)

7 Refitting is the reverse sequence to removal. Tighten the fixings according to specification after all the components have been loosely fitted.

6 Front suspension crossmember (XR2i only) – removal and refitting

1 Raise the front of the vehicle and support it securely using axle stands.
2 Remove the one-piece undertray where fitted, by turning its bayonet-type fasteners.
3 The crossmember, which serves as an additional bracing component, is located between the lower suspension arm mounting brackets, and is secured to these brackets by four bolts. To remove it, simply undo the four bolts and lower it from the vehicle.
4 Refitting is the reverse sequence to removal, ensuring that the bolts are tightened to the specified torque.

7 Front suspension lower arm – removal and refitting

1 Slacken the relevant roadwheel nuts.
2 Raise the front of the vehicle and support it securely on axle stands. Remove the relevant roadwheel.
3 Remove the one-piece undertray where fitted, by turning its bayonet-type fasteners.
4 Undo the lower suspension arm balljoint pinch-bolt (Torx) and separate the balljoint from the spindle carrier assembly.
5 If lower arm mounting brackets are to be removed on XR2i models, remove the front suspension crossmember as described in the previous Section.
6 On XR2i, SX and S model variants, remove the anti-roll bar retaining brackets, as described in Section 5 of this Chapter.
7 To remove a lower arm from its mounting bracket, simply undo the two bolts that pass through the vertical bushes (Fig. 10.10), remove the bolts and pull the arm clear.
8 The lower arm mounting brackets are each retained by five bolts. To remove a mounting bracket, simply undo the five bolts and lower it from the vehicle.
9 To refit, insert the lower arm into its mounting bracket and fit the two bolts that pass through the vertical bushes **to finger tightness only**.
10 Fit the lower arm mounting bracket to the vehicle, ensuring that the locating dowel is correctly located in its recess. One of the five bolts

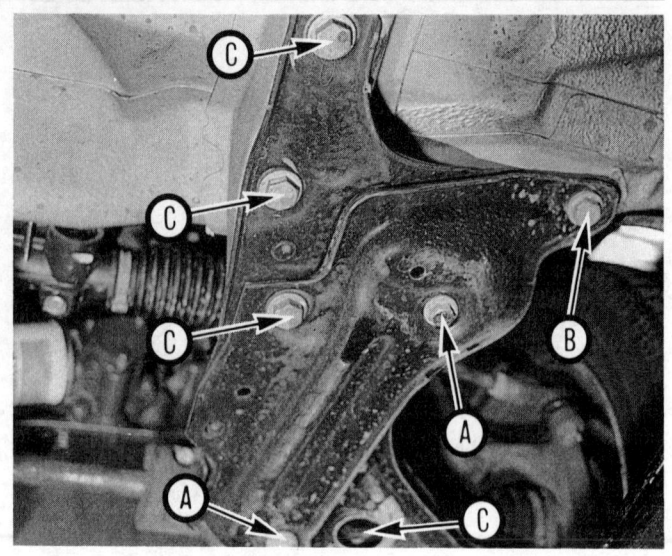

7.10 Lower suspension arm and mounting bracket arrangement

A Arm-to-mounting bracket bolts
B Special 'shouldered' locating bolt
C Mounting bracket retaining bolts

securing the mounting bracket has a locating shoulder and a larger thread diameter, and this should be fitted first to ensure correct alignment of the mounting bracket to the vehicle (photo). Refit the other four bolts and tighten all five according to specification.
11 Refit the anti-roll bar brackets and front suspension crossmember, as applicable, in accordance with the relevant Sections of this Chapter.
12 Locate the lower suspension arm balljoint into the spindle carrier assembly and tighten the (Torx head) pinch-bolt and nut to the specified torque. Note that the bolt must locate to the annular groove on the balljoint spindle.
13 Refit the roadwheel, remove the axle stands and lower the vehicle to the ground.
14 Tighten the wheel nuts according to specification.
15 Tighten the lower arm-to-lower arm mounting bracket bolts, that pass through the vertical bushes, by the torque-to-yield method as

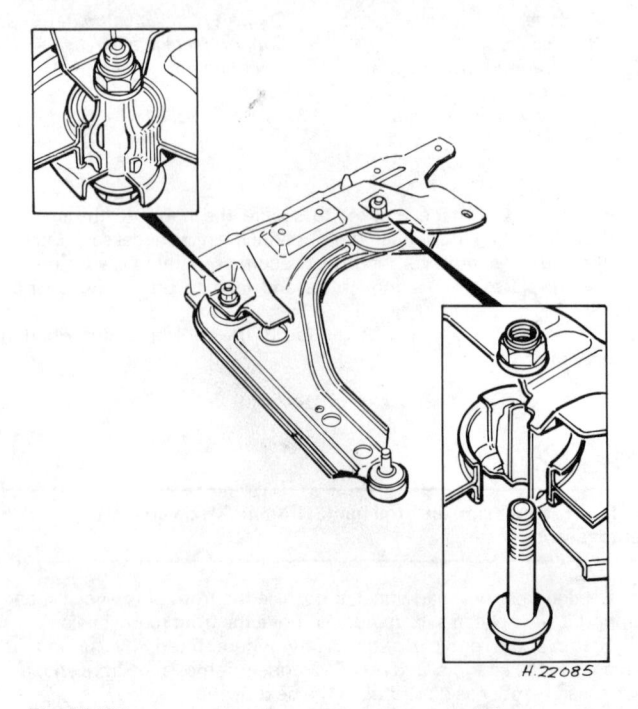

Fig. 10.10 Lower suspension arm arrangement in its mounting bracket (Sec 7)

Chapter 10 Suspension

8.4A Removing cap from front suspension strut top-mount retaining nut

8.4B Slackening the front suspension strut top-mount retaining nut whilst preventing the piston rod from turning

8.15 Correct spring location in its lower seat

follows. Note that the weight of the vehicle **must** be on the roadwheels for these procedures and new bolts must be used. Tighten the bolts to the 'snug' torque, then back off to zero torque. Retighten to the 'snug' torque then apply further tightening force to rotate the bolts through 90°. When checking these bolts for tightness, back the bolts off to zero torque first, then retighten to the 'snug' torque before applying further force to rotate the bolts through 90°. It is **vitally important** that these procedures are followed and that the bolts are not subjected to further rotation which could result in them failing. The torque-to-yield method **must** be followed every time that these bolts are disturbed.

16   Raise the front of the vehicle again and support it securely on axle stands to refit the one-piece undertray (where applicable).

Fig. 10.11 Typical pair of coil spring compressors in use (Sec 8)

5   Remove the bolt securing the brake hose bracket to the front suspension strut.
6   Remove the brake caliper anchor bolts and suspend the brake caliper within the wheelarch with a length of strong wire, to prevent the flexible brake hose from straining.
7   Disconnect the anti-roll bar link from the suspension strut, where applicable.
8   Remove the pinch-bolt securing the spindle carrier to the suspension strut then, using a suitable lever in the spindle carrier slot – a stout screwdriver is ideal, prise the joint apart slightly so that the two can be separated by pulling the spindle carrier gently downwards. Support beneath the lower suspension arm to avoid strain on the CV joints.
9   Remove the suspension strut top-mount retaining nut and its upper cup seat mounting, and carefully withdraw the strut from the vehicle.
10   To overhaul the suspension strut, with it removed from the vehicle, hold the one-piece coil clamping tool in a vice. Mount the suspension strut into this tool, ensuring that the spring is correctly engaged in the tool's spring retaining arms. If using a pair of coil spring compressors as an alternative, engage the spring retaining arms in such a manner that the spring is held securely on both sides (Fig. 10.11), and control the movement of the spring by adjusting each compressor unit

8 Front suspension strut – removal, overhaul and refitting

1   It should be noted before proceeding that, to dismantle a front suspension strut, a pair of suitable coil spring compressor units or a one-piece coil clamping tool, will be required. They will not be required, however, if simply replacing a front suspension strut as an assembled unit.
2   Slacken the relevant roadwheel nuts, raise the front of the vehicle and support it securely on axle stands.
3   Remove the roadwheel.
4   Remove the cap located over the suspension strut top-mount retaining nut (photo). Using an Allen key to prevent the piston rod within the strut assembly from rotating, slacken the suspension strut top-mount retaining nut (photo), but do not remove it at this stage.

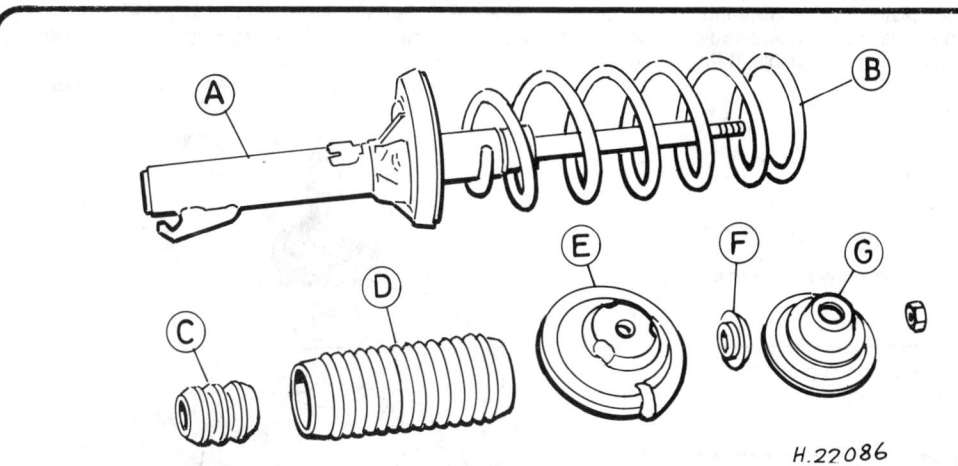

A   Suspension strut
B   Coil spring
C   Bump-stop
D   Gaiter
E   Upper spring seat
F   Thrust bearing
G   Lower cup seat mounting

Fig. 10.12 Exploded view of front suspension strut assembly (Sec 8)

224   Chapter 10   Suspension

9.7 Removing the outer grease cap from the hub centre

9.8 Undoing the hub bearing retaining nut

9.9 Slide off the wheel hub/brake drum unit

half a turn at a time. Whichever type of tool is used, ensure that spring pressure is **fully** relieved from the spring seat before attempting to separate the strut components. While the spring is compressed it is under extreme load, therefore **great care must be taken** during this operation.
11 Compress the spring and remove the spring retaining nut. The lower cup seat mounting, thrust bearing, upper spring seat, gaiter and bump-stop can be removed by sliding them over the piston rod (see Fig. 10.12).
12 The suspension strut and coil spring may now be separated. If a new suspension strut is to be fitted there is no need to release the coil spring from compression, but if a new coil spring is to be fitted, release the compressor/compressors gently until the spring is in its released state, then remove it.
13 If a new spring is to be fitted, engage the compressor/compressors as during removal, and compress the spring sufficiently to enable suspension strut reassembly.
14 Reunite the spring and suspension strut, and refit the bump-stop gaiter, spring seat, thrust bearing and lower cup seat mounting, renewing components as necessary. Refit the spring retaining nut and tighten according to specification.
15 Carefully release the spring tension, ensuring that the spring locates correctly into its upper and lower spring seats (photo).
16 Remove the spring compressor/compressors.
17 The rubber insulator fitted to the top of the inner wing is a simple push fit, and is easily replaceable. Ensure when replacing this, that the lip sits evenly around the locating hole.
18 To refit the suspension strut to the vehicle, locate it through its inner wing position and refit the upper cup seat mounting and top-mount retaining nut. Do not tighten the nut at this stage.
19 Apply leverage to the spindle carrier slot so that the spindle carrier can be refitted to the base of the suspension strut. Refit the suspension strut to spindle carrier pinch-bolt and tighten according to specification.
20 Tighten the suspension strut top-mount retaining nut according to specification, using an Allen key to prevent the piston rod from rotating. The *final* torque will have to be applied without the use of the Allen key unless a suitable open-ended torque wrench adaptor is available. Refit the cap over the nut.
21 Refit the brake caliper to the spindle carrier, and tighten the caliper anchor bolts according to specification.
22 Refit the bolt to secure the brake hose bracket to the front suspension strut, and fully tighten.
23 Remove the support from under the lower suspension arm.
24 Reconnect the anti-roll bar link to the suspension strut, where applicable, and tighten the nut according to specification.
25 Refit the roadwheel, remove the axle stands and lower the vehicle to the ground.
26 Tighten the roadwheel nuts according to specification.
27 Ideally, any replacement of suspension strut or coil spring should be matched by a replacement fitted to the corresponding unit on the other side of the vehicle, to maintain balanced handling characteristics.

9 Rear hub bearings – checking, removal and renewal

1 All models are fitted with non-adjustable rear hub bearings, the bearing play being set when the hub bearing retaining nut is tightened to its specified torque.
2 To check the bearings for excessive wear, chock the front wheels then raise the rear of the vehicle and support it securely on axle stands.
3 If there is any roughness, binding or vibration when the wheel is spun, the bearings must be renewed. If excessive wheel rock is noticed when the wheel is gripped at the top and bottom, this will probably mean that the bearings require renewal, although it is worthwhile loosening the hub bearing nut and re-tightening it to the specified torque and re-checking. If excessive rock is still evident, renewal of the bearings is the only solution.
4 Before dismantling it should be noted that service tools 14-028 (seal installer) and 15-051 (hub bearing cup installer), or suitable alternatives, are required for rear hub bearing renewal. Bearings and bearing cups **must** be supplied from the same source as matched units.
5 To renew, first slacken the relevant roadwheel nuts then raise the rear of the vehicle and support it securely on axle stands, having chocked the front wheels. Remove the roadwheel.
6 With the handbrake lever released, remove the rubber blanking plug from the brake carrier plate, then release the automatic rear brake adjuster by levering the catch on the adjuster pawl, through the carrier plate (see Chapter 8).
7 Remove the outer grease cap from the hub centre – it is necessary to destroy the grease cap to obtain access (photo).
8 Remove the hub bearing retaining nut – **right-hand thread** on the vehicle right-hand side, and **left-hand thread** on the vehicle left-hand side (photo).
9 Slide the wheel hub/brake drum unit off its spindle (photo).
10 Using a stout screwdriver, or other suitable lever, prise the grease retainer out of the hub bore, taking care not to damage the bore (photo).
11 Remove the inner and outer bearing cones from the hub bore then, using a suitable punch (preferably brass), drive the bearing cups out of the bore at their respective ends by tapping alternately at diametrically-opposed points on each cup (photos). Do not allow the cups to tilt within the bore as they are removed, and ensure that no damage is done to the bore itself, as any burrs incurred may prevent the new cups from seating correctly.
12 Clean the hub bore and spindle thoroughly before reassembly.

Fig. 10.13 Inserting bearing cups into hub bore (Sec 9)

Chapter 10 Suspension

9.10 Prise the grease retainer from the hub bore

9.11A Removing a bearing cone from the hub bore

9.11B Removing bearing cups

13 Upon reassembling, tap the new bearing cups squarely into the hub bore, using service tool 15-051, making sure that they are in full contact with their locating shoulders within the bore. A piece of tubing of *slightly* smaller diameter than the bearing cup may suffice in the absence of the correct tool.

14 Pack the inner bearing cone with grease (see Specifications) and insert to its cup then, using service tool 14-028, fit the new grease retainer to the inner end of the hub bore, having greased its inner lip for ease of fitment. A piece of tubing of the same diameter as the grease retainer may suffice in the absence of the correct tool.

15 Pack the outer bearing cone with grease and insert to its cup, then refit the wheel hub/brake drum unit to its spindle.

16 Refit the hub bearing retaining nut and tighten to specification, rotating the hub whilst tightening.

17 Fit a new outer grease cap to the hub centre, gently tapping around its circumference to locate.

18 Refit the rubber blanking plug to the brake carrier plate, refit the roadwheel and lower the vehicle to the ground. Remove the wheel chocks.

19 Apply the footbrake to actuate the automatic rear brake adjuster assembly, then fully tighten the roadwheel nuts to specification.

10 Rear axle bush – renewal

1 It should be noted before proceeding with this operation that service tool 15-084 (pivot bush installer), or a suitable equivalent, will be needed. If a suitable tool is not available the task is best entrusted to your Ford dealer, as attempting the operation without it will almost certainly lead to greatly reduced bush service life. Bushes should be replaced as a set.

2 Chock the front wheels, then raise the rear of the vehicle and support it securely on axle stands.

3 Position a suitable support (preferably adjustable) under the axle twist beam, so that the axle assembly can be lowered from the vehicle for working clearance. **Note:** *This only entails the axle twist beam carrying its own weight,* **not** *the full weight of the vehicle.*

4 On vehicles equipped with the anti-lock braking system, undo the nuts securing the load-apportioning valve operating links to the rear axle twist beam, and disconnect the links from the twist beam (see Chapter 8).

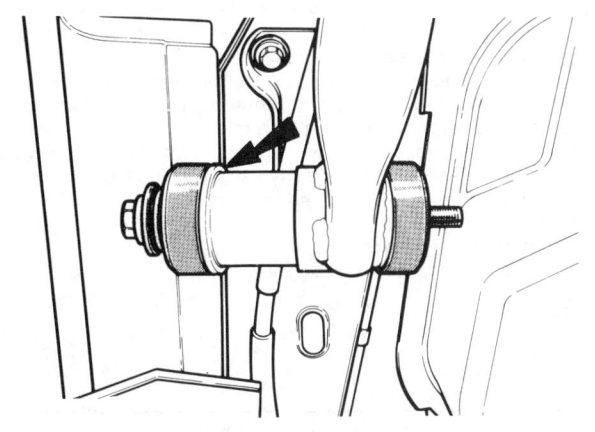

Fig. 10.14 Fitting void bush with its collar (arrowed) towards the outer edge of the vehicle (Sec 10)

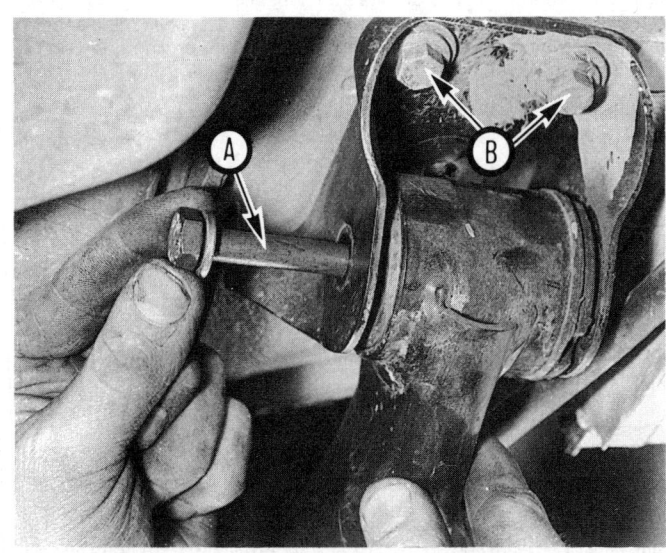

10.5 Void bush bolt (A) and body mounting bracket bolts (B)

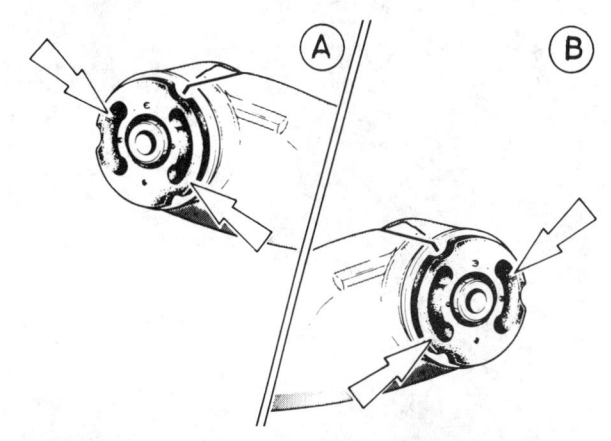

Fig. 10.15 Correct void bush positioning in trailing arm (Sec 10)

    A   Left-hand side   B   Right-hand side

5 Remove the two void bush bolts locating the forward ends of the trailing arms to the body mounting brackets (photo), and lower the axle assembly so that the bushes are clear of any obstructions. Support the axle twist beam securely. If necessary, clamp and disconnect the flexible brake hoses from the axle assembly to prevent them from straining (see Chapter 8).
6 Undo the body mounting bracket bolts and remove the brackets.
7 Using a soft-faced hammer and a suitable punch or drift, drive the bushes from their locations, taking care not to raise any burrs on the trailing arm eyes.
8 On installation, first position the bush in the tool with its collar nearest to the outer edge of the vehicle (Fig. 10.14). The bush **must** be installed with its voids as shown in Fig. 10.15. Draw the bush inwards towards the centre of the vehicle. Care should be taken to avoid damage to the bush and to obtain correct positioning of the voids. Replace both bushes.
9 With the new bushes in position, refit the body mounting brackets loosely to the trailing arms – do **not** fully tighten the void bush bolts.
10 Raise and support the axle assembly so that the conical locating pegs on the body mounting brackets engage in their body locations. Refit the body mounting bracket bolts and tighten to specification.
11 Remove the axle assembly support.
12 On vehicles equipped with the anti-lock braking system, reconnect the load-apportioning valve operating links to the rear axle twist beam, and tighten the nuts according to specification.
13 If the flexible brake hoses were removed during this operation, reconnect them and bleed the brake hydraulic system (see Chapter 8).
14 Lower the vehicle to the ground.
15 Fully tighten the void bush bolts to specification, applying the torque to the bolt head, **not** the nut.
16 Remove the chocks from the front wheels.

11 Rear suspension strut – removal, overhaul and refitting

1 It should be noted before proceeding that, to dismantle a rear suspension strut, a pair of suitable coil spring compressor units or a one-piece coil clamping tool, will be required. They will not be required, however, if simply replacing a rear suspension strut as an assembled unit.
2 Chock the front wheels then raise the rear of the vehicle, supporting it securely on axle stands.
3 Support the relevant trailing arm section of the axle assembly securely, then remove the lower suspension strut to axle mounting bolt. The trailing arm may need to be raised slightly, by light hand pressure, to fully withdraw the bolt.

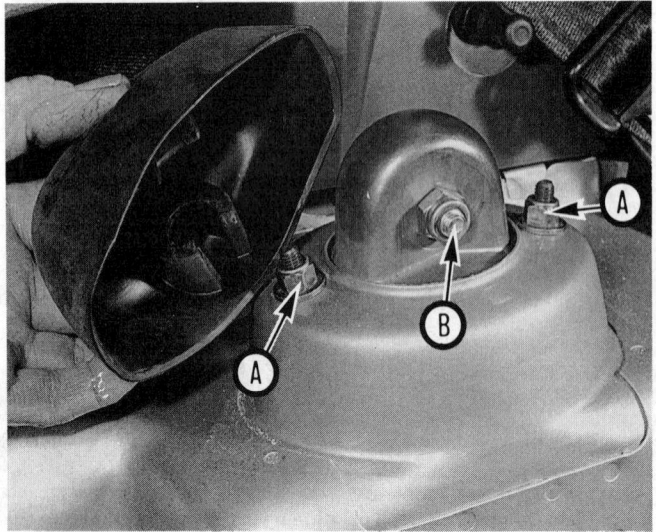

11.4 Suspension strut top-mount nuts (A), and spring retaining through-bolt fixing (B)

4 Lift off the suspension cap then remove the two upper suspension strut top-mount nuts, located on either side of the suspension turret in the luggage compartment (photo). Do **not** attempt to remove the through-bolt fixing on the top of the suspension strut mount-cup.
5 Lower the trailing arm slightly to allow the lower suspension strut mounting to clear its axle location, and withdraw the suspension strut from the vehicle. On vehicles equipped with the anti-lock braking system, do not allow the load-apportioning valve operating links to become strained when lowering the trailing arm – disconnect them as necessary.
6 To overhaul the suspension strut, with it removed from the vehicle, hold the one-piece clamping tool in a vice. Mount the suspension strut into this tool, ensuring that the spring is correctly engaged in the tool's spring retaining arms. If using a pair of coil spring compressors as an alternative, engage the spring retaining arms in such a manner that the spring is held securely on both sides (see Fig. 10.11), and control the movement of the spring by adjusting each compressor unit half a turn at a time. Whichever type of tool is used, ensure that spring pressure is *fully* relieved from the spring seats before attempting to separate the strut components. While the spring is compressed it is under extreme load, therefore **great care must be taken** during this operation.
7 Compress the spring and remove the through-bolt fixing from the strut mount-cup. Remove the strut mount-cup, which incorporates the upper spring seat, and its rubber insulator.
8 The suspension strut and the coil spring may now be separated. If a new suspension strut is to be fitted there is no need to release the coil spring from compression, but if a new coil spring is to be fitted, release the compressor/compressors gently until the spring is in its released state, then remove it.
9 If a new spring is to be fitted, engage the compressor/compressors as during removal, and compress the spring sufficiently to enable suspension strut reassembly.
10 Reunite the spring and suspension strut, and refit the rubber insulator and the strut mount-cup. Refit the through-bolt to the strut mount-cap, ensuring that it locates through the strut piston rod fixing. Tighten the through-bolt according to specification.
11 Carefully release the spring tension, ensuring that both upper and lower spring 'tails' locate correctly in their seats.
12 Remove the spring compressor/compressors.
13 To refit the suspension strut to the vehicle, first locate its upper mountings through the suspension turret, then refit the top-mount nuts. Tighten the top-mount nuts according to specification, then refit the suspension cap.
14 Position the lower suspension strut mounting into its axle location, and refit its mounting bolt but do **not** fully tighten.
15 Reconnect the load-apportioning valve operating links, as appropriate, and tighten the nuts according to specification.
16 Remove the support from underneath the trailing arm and lower the vehicle to the ground.
17 Fully tighten the lower suspension strut to axle mounting bolt, according to specification.
18 Remove the chocks from the front wheels.
19 Ideally, any replacement of suspension strut or coil spring should be matched by a replacement fitted to the corresponding unit on the other side of the vehicle, to maintain balanced handling characteristics.

12 Rear axle and suspension strut assembly – removal and refitting

1 Slacken the rear roadwheel bolts, then raise the vehicle completely off the ground and support it securely on axle stands. Remove the rear roadwheels.
2 On 1.4 litre CFi models, remove the underbody heatshields for access to the handbrake cable.
3 Disconnect the handbrake primary cable from the equaliser yoke, and remove the handbrake rear cable from its adjuster and its fixed body locations (see Chapter 8).
4 Release the rear automatic brake adjuster through the brake carrier plate (see Chapter 8) and undo the rear spindle to axle beam bolts on one side of the vehicle. Remove the hub/brake assembly.

Chapter 10 Suspension

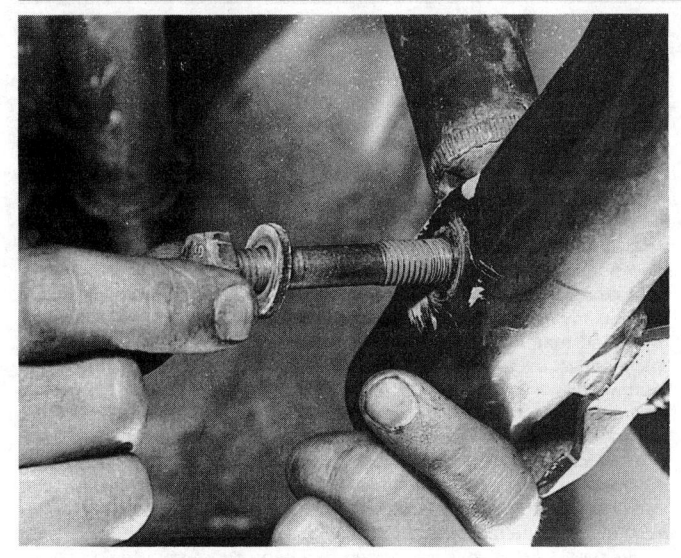

12.15 Lower suspension strut-to-axle mounting bolt

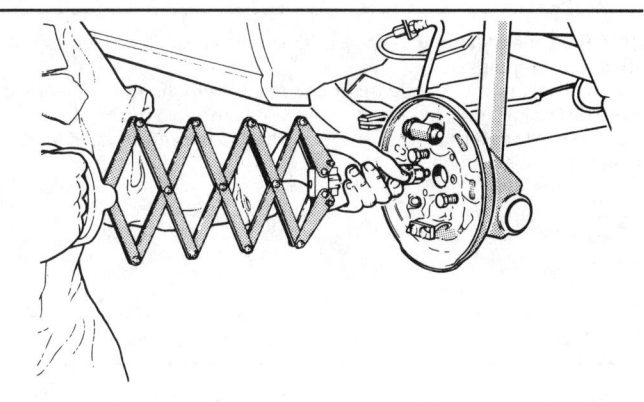

Fig. 10.16 Rivetting brake caliper plate to axle flange (Sec 13)

5 Disconnect the handbrake rear cable from the exposed rear brake assembly (see Chapter 8), and withdraw it as far as possible to the opposite side of the vehicle. Note cable routing and the cable ties fitted.
6 Using suitable hose clamps, isolate the rear brake assemblies by clamping the flexible brake hoses on both sides of the axle, then disconnect the hoses from their steel brake pipe axle connections.
7 On vehicles equipped with the anti-lock braking system, undo the nuts securing the load-apportioning valve operating links to the rear axle twist beam, and disconnect the links from the twist beam.
8 Support the rear axle assembly securely, then remove the rear axle body mounting bracket retaining bolts on both sides of the vehicle. Lower the front end of the axle.
9 Remove the lower suspension strut-to-axle mounting bolt on both sides of the vehicle.
10 Fully lower the axle assembly, and remove it from the vehicle towards the side with the handbrake rear cable still attached.
11 Lift off the suspension strut caps then remove the suspension struts by undoing the two top-mount retaining nuts, located on each suspension turret in the luggage compartment, and lowering the struts from their locations. Do **not** attempt to remove the through-bolt fixing on the top of the suspension strut mount cup.
12 To refit, first locate the suspension strut upper mountings through the suspension turrets, then refit the top-mount nuts. Do not fully tighten the top-mount nuts at this stage.
13 Position the axle assembly under the vehicle, raise the front end and support it.
14 Attach the rear axle body mounting brackets to the body, on both sides of the vehicle, and tighten the body mounting bracket bolts according to specification.
15 Raise the rear end of the axle assembly and support it. Attach the suspension struts to the trailing arms, but do not fully tighten the lower suspension strut-to-axle mounting bolts (photo).
16 Refit the handbrake rear cable to its rear brake unit, ensuring correct routeing of the cable, and attach the cable ties as necessary. Reassemble the rear brake components (see Chapter 8).
17 Apply sealant to the threads of the four bolts that secure the hub/brake drum assembly to the axle (see Specifications) and refit, tightening the bolts according to specification.
18 Refit the handbrake rear cable to its fixed and its adjuster body locations, and reconnect the equaliser yoke to the handbrake primary cable (see Chapter 8).
19 Reconnect the flexible brake hoses on both sides of the vehicle, tightening the unions to seal the system, then remove the hose clamps.
20 Refit the automatic rear brake adjuster rubber blanking plug to the brake carrier plate.
21 On vehicles equipped with the anti-lock braking system, reconnect the load-apportioning valve operating links to the rear axle twist beam, and tighten the nuts according to specification.
22 Bleed the brake hydraulic system, then adjust the handbrake (see Chapter 8).

23 On 1.4 litre CFi models, refit the heatshields to the vehicle underside.
24 Refit the rear roadwheels, remove the axle assembly supports and lower the vehicle to the ground.
25 Tighten the suspension strut top-mount nuts, the lower suspension strut-to-axle mounting bolts and the rear roadwheel nuts according to specification. If the void bush bolts securing the forward ends of the axle trailing arms to the body mounting brackets have been disturbed, for example during axle unit replacement, it will also be necessary to tighten these according to specification at this stage, taking the torque setting from the bolt head, not the nut.
26 Refit the suspension strut caps.

13 Rear axle unit – renewal

1 It should be noted, before dismantling the old axle assembly that new void bushes will have to be fitted to the new axle unit. Access to service tool 15-084 (pivot bush installer), or suitable equivalent, will therefore be required. A pop-rivet gun will also be needed, to secure the brake carrier plates.
2 With the rear axle assembly removed from the vehicle as described in the previous Section, support is securely on axle stands and prepare to dismantle.
3 Remove the void bush bolts securing the body mounting brackets to the forward ends of the axle trailing arms, and remove the brackets.
4 Remove the clips securing the steel brake pipes to their axle mounting brackets, located underneath the axle mounting brackets. Remove the steel brake pipes from their axle brackets, but leave them attached to their respective wheel cylinders through the brake carrier plate.
5 Remove the rubber blanking plug from the brake carrier plate and release the rear automatic brake adjuster on the assembled rear brake unit. Release the four rear spindle-to-axle beam retaining bolts and remove the hub/brake drum assembly. Mark the assembly so that it can be refitted to the same side of the vehicle.
6 Using a suitable drill, remove the two rivets securing each brake carrier plate to the axle flanges. Remove the brake carrier plates, having marked them for correct fitting to their respective sides.
7 Remove the old axle unit from the axle stands and securely support the new axle unit in its place, ready for reassembly.
8 Ensure that the brake carrier plates are refitted to their respective sides. Align the rivet locating holes on the brake carrier plates with their corresponding holes in the rear axle flanges. Insert two pop-rivets through each brake carrier plate and secure the brake carrier plates to their rear axle flanges, ensuring that all mating faces are clean (Fig. 10.16).
9 Apply sealant (see Specifications) to the threads of the four bolts that secure the previously marked hub/brake drum assembly and refit it to the brake carrier plate with the shoe components attached, having ensured that the mating faces are clean. Tighten the four bolts according to specification, then refit the automatic brake adjuster blanking plug to that brake unit.

10 Refit the steel brake pipes to their axle mounting brackets with their retaining clips.
11 Loosely refit the body mounting brackets to the forward ends of the axle trailing arms, but do not fully tighten the void bush bolts.
12 The axle assembly can now be refitted to the vehicle as described in the previous Section.

14 Roadwheels and tyres – general care and maintenance

Wheels and tyres should give no real problems in use provided that a close eye is kept on them with regard to excessive wear or damage. To this end, the following points should be noted.

Ensure that tyre pressures are checked regularly and maintained correctly. Checking should be carried out with the tyres cold and not immediately after the vehicle has been in use. If the pressures are checked with the tyres hot, an apparently high reading will be obtained owing to heat expansion. Under no circumstances should an attempt be made to reduce the pressures to the quoted cold reading in this instance, or effective underinflation will result.

Underinflation will cause overheating of the tyre owing to excessive flexing of the casing, and the tread will not sit correctly on the road surface. This will cause a consequent loss of adhesion and excessive wear, not to mention the danger of sudden tyre failure due to heat build-up.

Overinflation will cause rapid wear of the centre part of the tyre tread coupled with reduced adhesion, harsher ride, and the danger of shock damage occurring in the tyre casing.

Regularly check the tyres for damage in the form of cuts or bulges, especially in the sidewalls. Remove any nails or stones embedded in the tread before they penetrate the tyre to cause deflation. If removal of a nail *does* reveal that the tyre has been punctured, refit the nail so that its point of penetration is marked. Then immediately change the wheel and have the tyre repaired by a tyre dealer. Do *not* drive on a tyre in such a condition. If in any doubt as to the possible consequences of any damage found, consult your local tyre dealer for advice.

Periodically remove the wheels and clean any dirt or mud from the inside and outside surfaces. Examine the wheel rims for signs of rusting, corrosion or other damage. Light alloy wheels are easily damaged by 'kerbing' whilst parking, and similarly steel wheels may become dented or buckled. Renewal of the wheel is very often the only course of remedial action possible.

The balance of each wheel and tyre assembly should be maintained to avoid excessive wear, not only to the tyres but also to the steering and suspension components. Wheel imbalance is normally signified by vibration through the vehicle's bodyshell, although in many cases it is particularly noticeable through the steering wheel. Conversely, it should be noted that wear or damage in suspension or steering components may cause excessive tyre wear. Out-of-round or out-of-true tyres, damaged wheels and wheel bearing wear/maladjustment also fall into this category. Balancing will not usually cure vibration caused by such wear.

Wheel balancing may be carried out with the wheel either on or off the vehicle. If balanced on the vehicle, ensure that the wheel-to-hub relationship is marked in some way prior to subsequent wheel removed so that it may be refitted in its original position.

General tyre wear is influenced to a large degree by driving style harsh braking and acceleration or fast cornering will all produce more rapid tyre wear. Interchanging of tyres may result in more even wear, but this should only be carried out where there is no mix of tyre types on the vehicle. However, it is worth bearing in mind that if this is completely effective, the added expense of replacing a complete set of tyres simultaneously is incurred, which may prove financially restrictive for many owners.

Front tyres may wear unevenly as a result of wheel misalignment. The front wheels should always be correctly aligned according to the settings specified by the vehicle manufacturer.

Legal restrictions apply to many aspects of tyre fitting and usage, and in the UK this information is contained in the Motor Vehicle Construction and Use Regulations. It is suggested that a copy of these regulations is obtained from your local police if in doubt as to the current legal requirements with regard to tyre type and condition, minimum tread depth, etc.

15 Fault diagnosis

Symptom	Reason(s)
Steering feels vague, vehicle wanders at speed	Tyre pressures uneven Faulty or worn front suspension struts Weak coil springs Steering geometry incorrect due to wear or accident damage
Vehicle pulls to one side	Steering geometry incorrect due to wear or accident damage Brakes binding
Heavy or stiff steering	Seized steering or suspension balljoint Tyres under-inflated Steering geometry incorrect due to wear or accident damage
Wheel wobble and vibration	Roadwheels out of balance Roadwheels buckled Faulty or damaged tyre Loose wheel nuts Worn hub bearings Weak coil springs Faulty or worn suspension struts
Excessive rolling or pitching motion when cornering or under braking load	Faulty or worn suspension struts Weak coil springs

Chapter 10 Suspension

Symptom	Reason(s)
Tyre wear uneven	Toe settings incorrect (front tyres) Tyre pressures incorrect Steering geometry incorrect due to wear or accident damage Roadwheels out of balance

Chapter 11 Bodywork and fittings

For modificatins, and information applicable to later models, see Supplement at end of manual

Contents

Body adhesive emblems – renewal	9
Body trim mouldings	10
Bonnet assembly – removal and refitting	6
Bonnet latch – adjustment	8
Bonnet release mechanism – removal and refitting	7
Bumper assembly (front) – removal and refitting	11
Bumper assembly (rear) – removal and refitting	12
Bumper mouldings – renewal	13
Bumper overriders – removal and refitting	14
Central locking system components – removal and refitting	21
Centre console – removal and refitting	50
Door aperture weatherseal – removal and refitting	15
Door assembly (front) – removal and refitting	16
Door assembly (rear) – removal and refitting	17
Door handle (exterior) – removal and refitting	22
Door latch and interior release handle – removal and refitting	24
Door lock barrel – removal and refitting	23
Door mirror – removal and refitting	20
Door mirror glass – removal and refitting	19
Door trim panel – removal and refitting	18
Door window glass (fixed rear quarter) – removal and refitting	29
Door window glass (front) – removal and refitting	30
Door window glass (rear) – removal and refitting	28
Door window regulator (electric, front) – removal and refitting	27
Door window regulator (manual, front) – removal and refitting	26
Door window regulator (rear) – removal and refitting	25
Facia – removal and refitting	52
Fixed rear side window glass – removal and refitting	31
Front seat and slide assembly – removal and refitting	42
Front seat belt and clip (five-door models) – removal, refitting and maintenance	46
Front seat belt and clip (three-door models) – removal, refitting and maintenance	47
Front seat belt and clip (Van models) – removal, refitting and maintenance	48
Front seat belt height adjuster – removal and refitting	45
General description	1
Glove compartment lid – removal and refitting	51
Interior trim panels – removal and refitting	44
Load compartment dividers (Van models) – removal and refitting	63
Maintenance – bodywork and underframe	2
Maintenance – upholstery and carpets	3
Major body damage – repair	5
Minor body damage – repair	4
Parcel shelf support/rear loudspeaker housing – removal and refitting	53
Passenger grab handle – removal and refitting	54
Rear seat – removal and refitting	43
Rear seat belt and clip – removal, refitting and maintenance	49
Rear view mirror – removal and refitting	55
Sill extension moulding – removal and refitting	58
Spare wheel carrier – removal and refitting	57
Sun visor – removal and refitting	56
Sunroof panel – removal and refitting	39
Sunroof panel seal – renewal	40
Sunroof weatherseal – removal and refitting	41
Tailgate – removal and refitting	35
Tailgate lock – removal and refitting	38
Tailgate spoiler – removal and refitting	36
Tailgate support strut – removal and refitting	34
Tailgate trim panel – removal and refitting	37
Tailgate window glass – removal and refitting	33
Wheelarch liner (front) – removal and refitting	59
Wheelarch moulding (front) – removal and refitting	60
Wheelarch moulding (rear) – removal and refitting	61
Wind deflector/radiator grille slat – removal and refitting	62
Windscreen – removal and refitting	32

Specifications

Torque wrench settings

	Nm	lbf ft
Front seat frame-to-slide mechanism bolts	21	15
Front seat slide mechanism-to-floor mounting bolts	25 to 32	18 to 24
Rear seat backrest hinge to backrest	3.7 to 4.6	2.7 to 3.4
Rear seat backrest hinge to body	21 to 25	15 to 18
Rear seat belt anchors and clips to body	25 to 45	18 to 34
Rear seat belt retractor assembly to body	25 to 45	18 to 34
Front seat belt clips to floorpan (all models)	25 to 45	18 to 34
Front seat belt lower anchor bolt to body	25 to 45	18 to 34
Front seat belt lower anchor slide bar bolt to body	25 to 45	18 to 34
Front seat belt retractor to B-pillar bolt (all models)	25 to 45	18 to 34
Front seat belt upper anchor bolt to B-pillar (non-adjustable)	25 to 45	18 to 34
Front seat belt upper anchor adjuster plate-to-B-pillar bolts	21 to 28	14 to 22
Front seat belt upper anchor bolt to adjuster slide	25 to 45	18 to 34

Chapter 11 Bodywork and fittings

1 General description

The body is an all-steel monocoque of welded construction, incorporating impact-absorbing front and rear sections. It is available in three-door, five-door and Van configurations.

Rust and corrosion protection is applied to all new vehicles in a multi-stage anti-corrosion process, including zinc phosphating, wax injection of box sections, and PVC and wax underbody coating. Plastic liners are fitted to front wheelarches for extra protection.

All body panels are welded, including the front wings, so in the event of major body damage, refer to Section 5 of this Chapter.

On certain models a one-piece undertray may be fitted beneath the engine compartment. In such cases, access to the engine bay from underneath will require the removal of the undertray, by releasing its bayonet-type retaining clips.

2 Maintenance – bodywork and underframe

The general condition of a vehicle's bodywork is the one thing that significantly affects its value. Maintenance is easy but needs to be regular. Neglect, particularly after minor damage, can lead quickly to further deterioration and costly repair bills. It is important also to keep watch on those parts of the vehicle not immediately visible, for instance the underside, inside all the wheel arches and the lower part of the engine compartment.

The basic maintenance routine for the bodywork is washing – preferably with a lot of water, from a hose. This will remove all the loose solids which may have stuck to the vehicle. It is important to flush these off in such a way as to prevent grit from scratching the finish. The wheel arches and underframe need washing in the same way to remove any accumulated mud which will retain moisture and tend to encourage rust. Paradoxically enough, the best time to clean the underframe and wheel arches is in wet weather when the mud is thoroughly wet and soft. In very wet weather the underframe is usually cleaned of large accumulations automatically and this is a good time for inspection.

Periodically, except on vehicles with a wax-based underbody protective coating, it is a good idea to have the whole of the underframe of the vehicle steam-cleaned, engine compartment included, so that a thorough inspection can be carried out to see what minor repairs and renovations are necessary. Steam cleaning is available at many garages and is necessary for removal of the accumulation of oily grime which sometimes is allowed to become thick in certain areas. If steam cleaning facilities are not available, there are excellent grease solvents available, such as Holts Engine Degreasant, which can be brush applied. The dirt can then be simply hosed off. Note that these methods should not be used on vehicles with wax-based underbody protective coating or the coating will be removed. Such vehicles should be inspected annually, preferably just prior to winter, when the underbody should be washed down and any damage to the wax coating repaired using Holts Undershield. Ideally, a completely fresh coat should be applied. It would also be worth considering the use of such wax-based protection for injection into door panels, sills, box sections, etc, as an additional safeguard against rust damage where such protection is not provided by the vehicle manufacturer.

After washing paintwork, wipe off with a chamois leather to give an unspotted clear finish. A coat of clear protective wax polish, like the many excellent Turtle Wax polishes, will give added protection against chemical pollutants in the air. If the paintwork sheen has dulled or oxidised, use a cleaner/polisher combination such as Turtle Wax Hard Shell to restore the brilliance of the shine. This requires a little effort, but such dulling is usually caused because regular washing has been neglected. Care needs to be taken with metallic paintwork, as special non-abrasive cleaner/polisher is required to avoid damage to the finish. Always check that the door and ventilator opening drain holes and pipes are completely clear so that water can be drained out. Bright work should be treated in the same way as paint work. Windscreens and windows can be kept clear of the smeary film which often appears by the use of a proprietary glass cleaner like Holts Mixra. Never use any form of wax or other body or chromium polish on glass.

3 Maintenance – upholstery and carpets

Mats and carpets should be brushed or vacuum cleaned regularly to keep them free of grit. If they are badly stained remove them from the vehicle for scrubbing or sponging and make quite sure they are dry before refitting. Seats and interior trim panels can be kept clean by wiping with a damp cloth. If they do become stained (which can be more apparent on light coloured upholstery) use a little liquid detergent and a soft nail brush to scour the grime out of the grain of the material. Do not forget to keep the headlining clean in the same way as the upholstery. When using liquid cleaners inside the vehicle do not over-wet the surfaces being cleaned. Excessive damp could get into the seams and padded interior causing stains, offensive odours or even rot. If the inside of the vehicle gets wet accidentally it is worthwhile taking some trouble to dry it out properly, particularly where carpets are involved. *Do not leave oil or electric heaters inside the vehicle for this purpose.*

4 Minor body damage – repair

The colour bodywork repair photographic sequences between pages 32 and 33 illustrate the operations detailed in the following sub-sections.

Note: *For more detailed information about bodywork repair, Haynes Publishing produces a book by Lindsay Porter called the Car Bodywork Repair Manual. This incorporates information on such aspects as rust treatment, painting and glass fibre repairs, as well as details on more ambitious repairs involving welding and panel beating.*

Repair of minor scratches in bodywork

If the scratch is very superficial, and does not penetrate to the metal of the bodywork, repair is very simple. Lightly rub the area of the scratch with a paintwork renovator like Turtle Wax Color Back, or a very fine cutting paste like Holts Body + Rubbing Compound, to remove loose paint from the scratch and to clear the surrounding bodywork of wax polish. Rinse the area with clean water.

Apply touch-up paint to the scratch using a fine paint brush; continue to apply fine layers of paint until the surface of the paint in the scratch is level with the surrounding paintwork. Allow the new paint at least two weeks to harden; then blend it into the surrounding paintwork by rubbing the scratch area with a paintwork renovator or a very fine cutting paste, such as Holts Body + Plus Rubbing Compound or Turtle Wax Color Back. Finally, apply wax polish from one of the Turtle Wax range of wax polishes.

Where the scratch has penetrated right through to the metal of the bodywork, causing the metal to rust, a different repair technique is required. Remove any loose rust from the bottom of the scratch with a penknife, then apply rust inhibiting paint, such as Turtle Wax Rust Master, to prevent the formation of rust in the future. Using a rubber or nylon applicator fill the scratch with bodystopper paste such as Holts Body + Plus Knifing Putty. If required, this paste can be mixed with cellulose thinners, such as Holts Body + Plus Cellulose Thinners, to provide a very thin paste which is ideal for filling narrow scratches. Before the stopper-paste in the scratch hardens, wrap a piece of smooth cotton rag around the top of a finger. Dip the finger in cellulose thinners, such as Holts Body + Plus Cellulose Thinners, and then quickly sweep it across the surface of the stopper-paste in the scratch; this will ensure that the surface of the stopper-paste is slightly hollowed. The scratch can now be painted over as described earlier in this Section.

Repair of dents in bodywork

When deep denting of the vehicle's bodywork has taken place, the first task is to pull the dent out, until the affected bodywork almost attains its original shape. There is little point in trying to restore the original shape completely, as the metal in the damaged area will have stretched on impact and cannot be reshaped fully to its original contour. It is better to bring the level of the dent up to a point which is about 3 mm below the level of the surrounding bodywork. In cases where the dent is very shallow anyway, it is not worth trying to pull it out at all. If the underside of the dent is accessible, it can be hammered out gently from behind, using a mallet with a wooden or plastic head. Whilst doing this, hold a suitable block of wood firmly against the outside of the panel to absorb the impact from the hammer blows and thus prevent a large area of the bodywork from being 'belled-out'.

Should the dent be in a section of the bodywork which has a double skin or some other factor making it inaccessible from behind, a different technique is called for. Drill several small holes through the metal inside the area – particularly in the deeper section. Then screw long self-tapping screws into the holes just sufficiently for them to gain a good purchase in the metal. Now the dent can be pulled out by pulling on the protruding heads of the screws with a pair of pliers.

The next stage of the repair is the removal of the paint from the damaged area, and from an inch or so of the surrounding 'sound' bodywork. This is accomplished most easily by using a wire brush or abrasive pad on a power drill, although it can be done just as effectively by hand using sheets of abrasive paper. To complete the preparation for filling, score the surface of the bare metal with a screwdriver or the tang of a file, or alternatively, drill small holes in the affected area. This will provide a really good 'key' for the filler paste.

To complete the repair see the sub-section on filling and re-spraying.

Repair of rust holes or gashes in bodywork

Remove all paint from the affected area and from an inch or so of the surrounding 'sound' bodywork, using an abrasive pad or a wire brush on a power drill. If these are not available a few sheets of abrasive paper will do the job just as effectively. With the paint removed you will be able to gauge the severity of the corrosion and therefore decide whether to renew the whole panel (if this is possible) or to repair the affected area. New body panels are not as expensive as most people think and it is often quicker and more satisfactory to fit a new panel than to attempt to repair large areas of corrosion.

Remove all fittings from the affected area except those which will act as a guide to the original shape of the damaged bodywork (eg headlamp shells etc). Then, using tin snips or a hacksaw blade, remove all loose metal and any other metal badly affected by corrosion. Hammer the edges of the hole inwards in order to create a slight depression for the filler paste.

Wire brush the affected area to remove the powdery rust from the surface of the remaining metal. Paint the affected area with rust inhibiting paint like Turtle Wax Rust Master; if the back of the rusted area is accessible treat this also.

Before filling can take place it will be necessary to block the hole in some way. This can be achieved by the use of aluminium or plastic mesh, or aluminium tape.

Aluminium or plastic mesh or glass fibre matting is probably the best material to use for a large hole. Cut a piece to the approximate size and shape of the hole to be filled, then position it in the hole so that its edges are below the level of the surrounding bodywork. It can be retained in position by several blobs of filler paste around its periphery.

Aluminium tape should be used for small or very narrow holes. Pull a piece off the roll and trim it to the approximate size and shape required, then pull off the backing paper (if used) and stick the tape over the hole; it can be overlapped if the thickness of one piece is insufficient. Burnish down the edges of the tape with the handle of a screwdriver or similar, to ensure that the tape is securely attached to the metal underneath.

Bodywork repairs – filling and re-spraying

Before using this Section, see the Sections on dent, deep scratch, rust holes and gash repairs.

Many types of bodyfiller are available, but generally speaking those proprietary kits which contain a tin of filler paste and a tube of resin hardener are best for this type of repair, like Holts Body + Plus or Holts No Mix which can be used directly from the tube. A wide, flexible plastic or nylon applicator will be found invaluable for imparting a smooth and well contoured finish to the surface of the filler.

Mix up a little filler on a clean piece of card or board – measure the hardener carefully (follow the maker's instructions on the pack) otherwise the filler will set too rapidly or too slowly. Alternatively, Holts No Mix can be used straight from the tube without mixing, but daylight is required to cure it. Using the applicator apply the filler paste to the prepared area; draw the applicator across the surface of the filler to achieve the correct contour and to level the filler surface. As soon as a contour that approximates to the correct one is achieved, stop working the paste – if you carry on too long the paste will become sticky and begin to 'pick up' on the applicator. Continue to add thin layers of filler paste at twenty-minute intervals until the level of the filler is just proud of the surrounding bodywork.

Once the filler has hardened, excess can be removed using a metal plane or file. From then on, progressively finer grades of abrasive paper should be used, starting with a 40 grade production paper and finishing with 400 grade wet-and-dry paper. Always wrap the abrasive paper around a flat rubber, cork, or wooden block – otherwise the surface of the filler will not be completely flat. During the smoothing of the filler surface the wet-and-dry paper should be periodically rinsed in water. This will ensure that a very smooth finish is imparted to the filler at the final stage.

At this stage the 'dent' should be surrounded by a ring of bare metal, which in turn should be encircled by the finely 'feathered' edge of the good paintwork. Rinse the repair area with clean water, until all of the dust produced by the rubbing-down operation has gone.

Spray the whole repair area with a light coat of primer, either Holts Body + Plus Grey or Red Oxide Primer – this will show up any imperfections in the surface of the filler. Repair these imperfections with fresh filler paste or bodystopper, and once more smooth the surface with abrasive paper. If bodystopper is used, it can be mixed with cellulose thinners to form a really thin paste which is ideal for filling small holes. Repeat this spray and repair procedure until you are satisfied that the surface of the filler, and the feathered edge of the paintwork are perfect. Clean the repair area with clean water and allow to dry fully.

The repair area is now ready for final spraying. Paint spraying must be carried out in a warm, dry, windless and dust-free atmosphere. This condition can be created artificially if you have access to a large indoor working area, but if you are forced to work in the open, you will have to pick your day very carefully. If you are working indoors, dousing the floor in the work area with water will help to settle the dust which would otherwise be in the atmosphere. If the repair area is confined to one body panel, mask off the surrounding panels; this will help to minimise the effects of a slight mis-match in paint colours. Bodywork fittings (eg chrome strips, door handles etc) will also need to be masked off. Use genuine masking tape and several thicknesses of newspaper for the masking operations.

Before commencing to spray, agitate the aerosol can thoroughly, then spray a test area (an old tin, or similar) until the technique is mastered. Cover the repair area with a thick coat of primer; the thickness should be built up using several thin layers of paint rather than one thick one. Using 400 grade wet-and-dry paper, rub down the surface of the primer until it is really smooth. While doing this, the work area should be thoroughly doused with water, and the wet-and-dry paper periodically rinsed in water. Allow to dry before spraying on more paint.

Spray on the top coat using Holts Dupli-Color Autospray, again building up the thickness by using several thin layers of paint. Start spraying in the centre of the repair area and then, with a side-to-side motion, work outwards until the whole repair area and about 50 mm of the surrounding original paintwork is covered. Remove all masking material 10 to 15 minutes after spraying on the final coat of paint.

Allow the new paint at least two weeks to harden, then, using a paintwork renovator or a very fine cutting paste such as Turtle Wax Color Back or Holts Body + Plus Rubbing Compound, blend the edges of the paint into the existing paintwork. Finally, apply wax polish.

Plastic components

With the use of more and more plastic body components by the vehicle manufacturers (eg bumpers, spoilers, and in some cases major body panels), rectification of more serious damage to such items has become a matter of either entrusting repair work to a specialist in this field, or renewing complete components. Repair of such damage by the DIY owner is not really feasible owing to the cost of the equipment and materials required for effecting such repairs. The basic technique involves making a groove along the line of the crack in the plastic using a rotary burr in a power drill. The damaged part is then welded back together by using a hot air gun to heat up and fuse a plastic filler rod into the groove. Any excess plastic is then removed and the area rubbed down to a smooth finish. It is important that a filler rod of the correct plastic is used, as body components can be made of a variety of different types (eg polycarbonate, ABS, polypropylene).

Damage of a less serious nature (abrasions, minor cracks etc) can be repaired by the DIY owner using a two-part epoxy filler repair material like Holts Body + Plus or Holts No Mix which can be used directly from the tube. Once mixed in equal proportions (or applied direct from the tube in the case of Holts No Mix), this is used in similar fashion to the bodywork filler used on metal panels. The filler is usually cured in 20 to 30 minutes, ready for sanding and painting.

Chapter 11 Bodywork and fittings

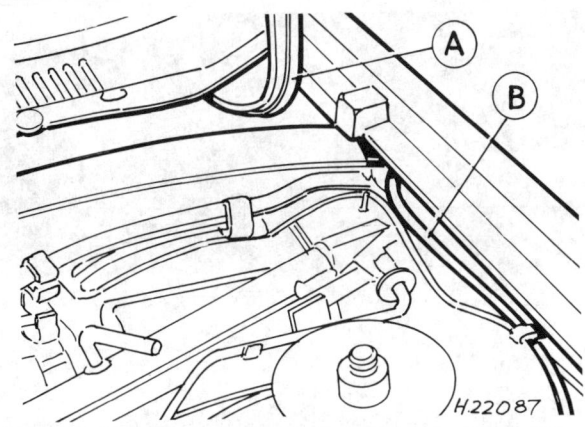

Fig. 11.1 Windscreen washer jet hose in engine compartment (Sec 6)

A Hose located to bonnet hinge clip
B Position of cut

If the owner is renewing a complete component himself, or if he has repaired it with epoxy filler, he will be left with the problem of finding a suitable paint for finishing which is compatible with the type of plastic used. At one time the use of a universal paint was not possible owing to the complex range of plastics encountered in body component applications. Standard paints, generally speaking, will not bond to plastic or rubber satisfactorily, but Holts Professional Spraymatch paints to match any plastic or rubber finish can be obtained from dealers. However, it is now possible to obtain a plastic body parts finishing kit which consists of a pre-primer treatment, a primer and coloured top coat. Full instructions are normally supplied with a kit, but basically the method of use is to first apply the pre-primer to the component concerned and allow it to dry for up to 30 minutes. Then the primer is applied and left to dry for about an hour before finally applying the special coloured top coat. The result is a correctly coloured component where the paint will flex with the plastic or rubber, a property that standard paint does not normally possess.

5 Major body damage – repair

Where serious damage has occurred or large areas need renewal due to neglect, it means that completely new sections or panels will need welding in, and this is best left to professionals. If the damage is due to impact, it will also be necessary to completely check the alignment of the bodyshell structure. Due to the principle of construction, the strength and shape of the whole car can be affected by damage to one part. In such instances the service of a dealer with specialist checking jigs are essential. If a body is left misaligned it is first of all dangerous as the car will not handle properly, and secondly uneven stress will be imposed on the steering, engine and transmission, causing abnormal wear or complete failure. Tyre wear may also be excessive.

6 Bonnet assembly – removal and refitting

1 Raise the bonnet and support it on its stay.
2 Using a felt tip marker pen or similar, mark around the hinge positions on the bonnet.
3 Cut the windscreen washer jet hose in the engine compartment, or release it from its one-way valve (if already fitted), then release the hose from the bonnet hinge clip (Fig. 11.1).
4 With the aid of an assistant, support the bonnet assembly and remove the four bolts securing it to its hinges (photo). Remove the bonnet assembly, taking care to disengage the stay before the bonnet is moved.
5 To refit, first align the marks made on the bonnet with the hinges, then refit and fully tighten the four securing bolts. Support the bonnet on its stay.
6 Refit the windscreen washer jet hose into the bonnet hinge clip, and join it up using a one-way (non-return) valve, having ensured correct routeing. Ensure that the valve is installed the correct way round, allowing flow to the jets but resisting return flow back to the reservoir.
7 Close the bonnet and ensure that there is an equal gap at each side, between the bonnet and the wings, and that it sits flush in relation to its surrounding panels.
8 The bonnet should close smoothly and positively with no excessive pressure being applied. If this is not the case, adjustment will be necessary.
9 To adjust the bonnet closure, adjustable bump stops are fitted to the closure panel (photo). These may be raised or lowered by screwing in or out, as necessary. The bonnet latch may also be adjusted, as required, and this is covered in Section 8 of this Chapter.

7 Bonnet release mechanism – removal and refitting

1 Remove the screws securing the lower steering column shroud to its location and detach the choke warning light switch/pull control assembly, from it, as described in Chapter 9, Section 6, as applicable.
2 Operate the bonnet release lever then raise and support the bonnet. If the release cable is broken, it will be necessary to detach the latch

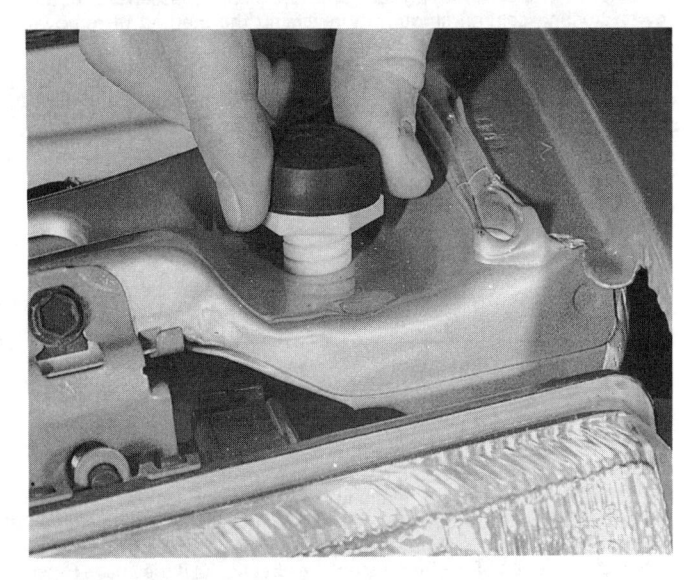

6.4 Removing the bonnet assembly

6.9 Altering the setting of a bonnet closure bump stop

Chapter 11 Bodywork and fittings

7.3A Removing the bonnet release latch

7.3B Bonnet release latch

A *Outer cable attachment*
B *Inner cable attachment*

from its body location by undoing the three latch retaining screws through the gap between the leading edge of the bonnet and the radiator grille slot.

3   With the bonnet open, remove the three screws securing the latch to the body. Disengage the release cable from the latch (photos).
4   Pull the latch end of the cable into the engine compartment, noting cable routeing and clips fitted. Remove the cable from its clips.
5   Detach the cable from its release lever on the steering column (Chapter 9, photo 6.7), by aligning the cable core with the slot on the release lever and withdrawing the end fixing. Detach the cable from its outer core abutment on the steering column lock housing.
6   Unclip the cable from its pedal box location, then detach the bulkhead grommet and pass the cable through into the engine compartment. Withdraw the cable from the vehicle.
7   The release lever on the steering column may be removed, if required, by unhooking the spring from its retaining arms, then disengaging its retaining arms from the steering column lock housing.
8   Refit the release lever, if removed, by reversing the method of removal.
9   To install the release cable, first pass the latch end of the cable down the right-hand side of the steering column, through its bulkhead location, and out into the engine compartment.
10   Fit the cable to its clip on the pedal box assembly, then reconnect the cable to the release lever and the steering column lock housing abutment by reversing the method of removal.
11   Refit the choke warning light switch/pull control assembly, to the lower steering column shroud, by reversing the method of removal (where applicable). Refit the shroud.
12   Seat the release cable grommet into the bulkhead.
13   Route and secure the release cable in the engine compartment.
14   Reconnect the release cable to the latch, then refit the latch to the body, setting the latch at its maximum height position, and tightening only the bottom retaining screw.
15   Adjust the latch for flush bonnet closure in accordance with Section 8.

8 Bonnet latch – adjustment

1   To adjust the bonnet latch, remove the two upper latch retaining screws, then with the latch raised to its maximum height position and secured with the lower retaining screw, close the bonnet.
2   Slacken the lower latch retaining screw, through the gap between the leading edge of the bonnet and the radiator grille slot, then set the bonnet so it sits flush with its surrounding panels – it may be necessary to adjust the height of the bump stops (see Section 6) if they have been moved in any way, or if fitting a new bonnet.

3   With the desired bonnet closure obtained, fully tighten the lower latch retaining screw, then open the bonnet and refit the two upper latch retaining screws, tightening to the specified torque.

9 Body adhesive emblems – renewal

1   Using a length of strong thin diameter cord (fishing line is ideal), break the adhesive bond between the emblem and the panel.
2   Thoroughly clean all traces of the old adhesive from the emblem location, using methylated spirit, taking all normal safety precautions. Allow the emblem location to dry.
3   Gently heat the new emblem until it is warm to the touch.
4   Peel the protective backing paper from the emblem then, taking care not to touch the adhesive, position the emblem on the panel. Maintain hand pressure evenly for at least thirty seconds to ensure a good bond.

10 Body trim mouldings

Roof drip rail moulding – removal and refitting
1   Remove the drip rail moulding by gently raising the forward end from its retaining flange, taking care not to bend or kink it, then carefully pull it off the retaining flange (Fig. 11.2).

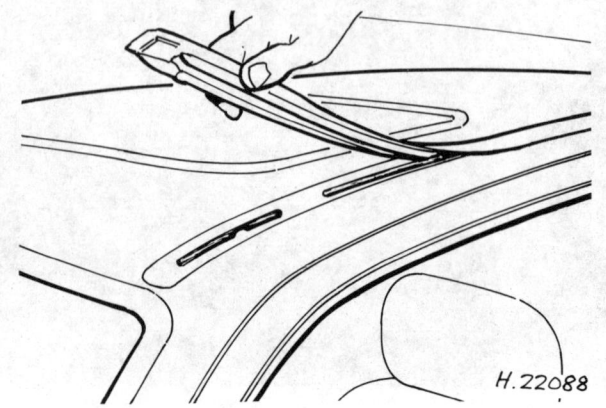

Fig. 11.2 Removing a roof drip rail moulding (Sec 10)

Chapter 11 Bodywork and fittings

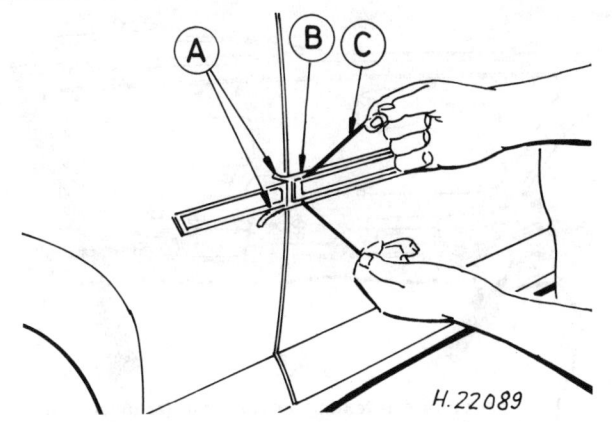

Fig. 11.3 Removing a door side moulding (Sec 10)

A Masking tape
B Moulding
C Nylon cord (fishing line)

2   To refit the drip rail moulding, first align the rear of the moulding to the roof panel edge by the tailgate, then, using the flat palm of the hand, gently tap the moulding down.
3   If fitting the Ford roof rack, the drip rail mouldings on both sides must be removed and replaced by a ten-piece moulding kit, available from Ford dealerships.

Door side moulding – renewal

4   Apply masking tape, as an alignment guide and to protect the paintwork, just above and just below the moulding to be renewed (Fig. 11.3).
5   Using a length of strong thin diameter cord (fishing line is ideal), break the bond between the moulding and the panel, and remove the moulding.
6   Thoroughly clean the moulding location of any trace of old adhesive, using methylated spirit, taking all normal safety precautions. Allow the moulding location to dry.
7   Continue to proceed using a similar technique to that described in Section 9, paragraphs 3 and 4 taking care to align the moulding correctly.
8   To improve the adhesive bond, apply pressure over the whole length of the moulding using a roller.
9   Remove the masking tape carefully.

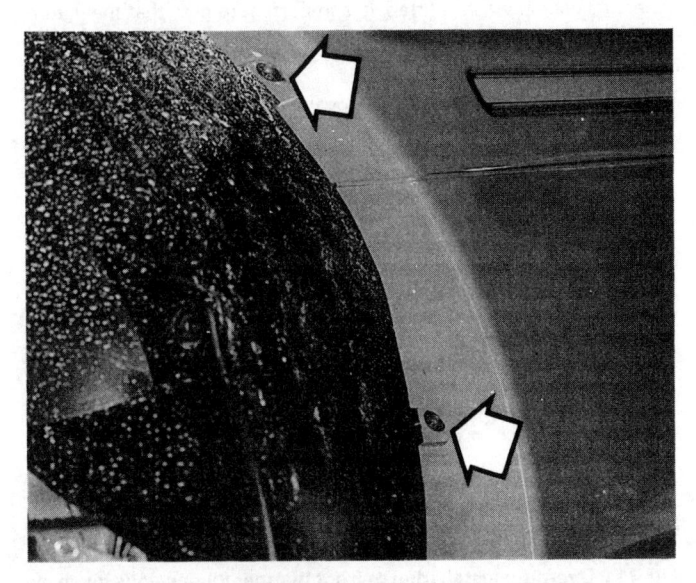

11.2 Front bumper-to-wheelarch retaining screws (arrowed)

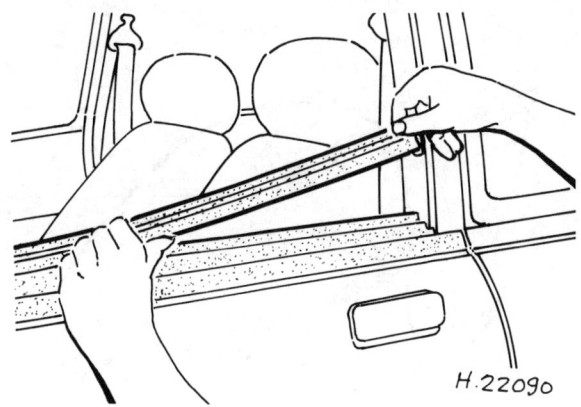

Fig. 11.4 Removing a door belt weatherseal moulding (Sec 10)

Door belt weatherseal moulding – removal and refitting

10   Remove the door mirror, as described in Section 20 of this Chapter, then, using a screwdriver, carefully prise up the moulding and remove it (Fig. 11.4). Do **not** bend or kink the moulding, as this will permanently deform it.
11   To refit, align the moulding to its rearward location (latch end of the door), then carefully tap it into position by hand.
12   Refit the mirror in accordance with Section 20 of this Chapter.

11 Bumper assembly (front) – removal and refitting

1   Open the bonnet and disconnect the auxiliary lamp multi-plugs, as applicable.
2   Remove the bumper retaining screws from the leading edge of the wheelarch flanges (photo), then ease the bumper away from its wheelarch location.
3   Remove the bumper retaining nuts from the reverse side of the body panel beneath the headlights (Fig. 11.5) then, with the help of an assistant, remove the bumper assembly from the vehicle.
4   To refit, (again with assistance) position the bumper onto its panel, ensuring that the retaining studs pass through their body locations, and that its ends align to the wheelarches.
5   Loosely refit the bumper retaining nuts and the bumper-to-wheelarch retaining screws.
6   Ensuring that the bumper is level, and that an even gap is maintained between it and surrounding body panels, tighten the retaining nuts to the specified torque.
7   Tighten the bumper-to-wheelarch retaining screws.
8   Refit the auxiliary lamp multi-plugs, as applicable.
9   The alignment of lamp units requires the use of optical beam setting equipment so, where applicable, entrust this task to a Ford dealer.

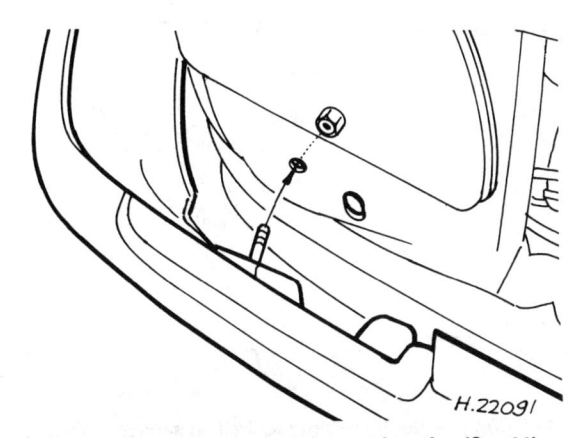

Fig. 11.5 Front bumper retaining nut location (Sec 11)

236　　　　　　　　　　　　　　　　Chapter 11　Bodywork and fittings

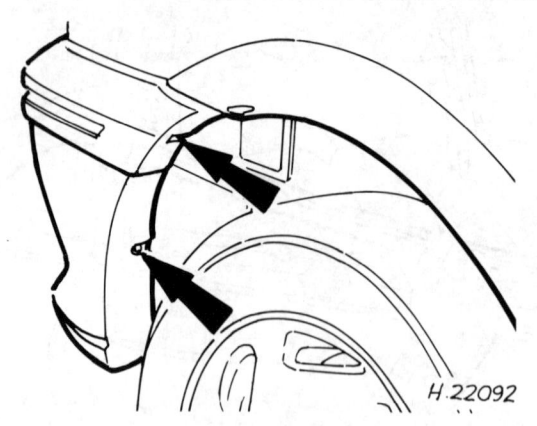

Fig. 11.6 Rear bumper-to-wheelarch retaining screws (arrowed) (Sec 12)

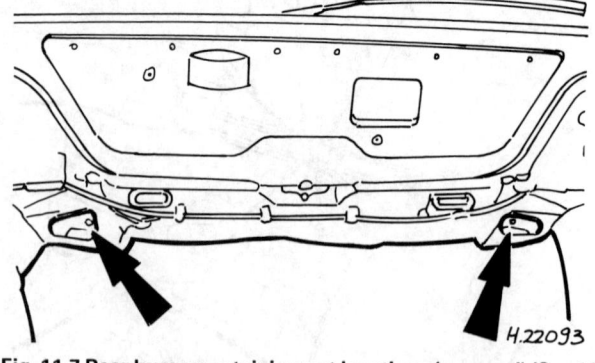

Fig. 11.7 Rear bumper retaining nut locations (arrowed) (Sec 12)

12 Bumper assembly (rear) – removal and refitting

1　Using a screwdriver or similar tool, prise up the number plate lamp unit from the rear bumper, being careful not to damage the bumper. Disconnect the bulbholder and remove from the bumper.
2　Remove the bumper-to-wheelarch inner rim retaining screws (Fig. 11.6), as necessary.
3　Open the tailgate then, using a suitably-sized socket, remove the bumper retaining nuts located inside the luggage compartment (Fig. 11.7).
4　Carefully remove the bumper from its location.
5　To refit, first align the bumper to the vehicle body, ensuring that the ends engage correctly at the wheelarches and the securing studs enter through the body panel.
6　Refit the bumper-to-wheelarch inner rim retaining screws, as applicable.
7　Refit the bumper retaining nuts, tightening to the specified torque.
8　Reconnect the number plate lamp bulbholder, then refit the lamp unit to the bumper.

13 Bumper moulding – renewal

1　With the bumper removed from the vehicle, place it on a clean flat surface, then remove the clips from the moulding ends.
2　Prise up and remove the old moulding from its bumper recess.
3　Remove all trace of old adhesive from the bumper recess, using methylated spirit, taking all normal safety precautions. Allow the recess to dry.
4　If fitting a new moulding to a service replacement bumper, slots must be cut into the bumper to accommodate the moulding end fixings (Fig. 11.8).
5　Apply adhesive primer (from a Ford dealer) to the recess, and allow it to dry.
6　Gently heat the moulding until it is warm to the touch, then insert the first moulding end clip into the bumper and press the moulding to its recess, secured with its moulding end, peeling the protective backing paper off as it is applied.
7　Apply the moulding smoothly and carefully across the bumper, rolling into its recess using a plastic roller or similar tool, then, when the other end is reached, secure with the other moulding end and clip.

14 Bumper overriders – removal and refitting

1　With the bumper removed as described in Section 11, remove the two nuts securing each overrider from their locations on the bumper inner face. Remove the overriders.
2　If overriders are being fitted to a service replacement bumper, it will be necessary to drill holes in the bumper to accommodate the mounting studs (Fig. 11.9). The large 24 mm diameter hole should be drilled only where headlamp wash is fitted. It is important to note that the datum points represented by black dots in Fig. 11.9 coincide with the top of the flat moulding recess, and that the fixing hole positions are symmetrically opposite.
3　Refitting is a reversal of the removal procedure.

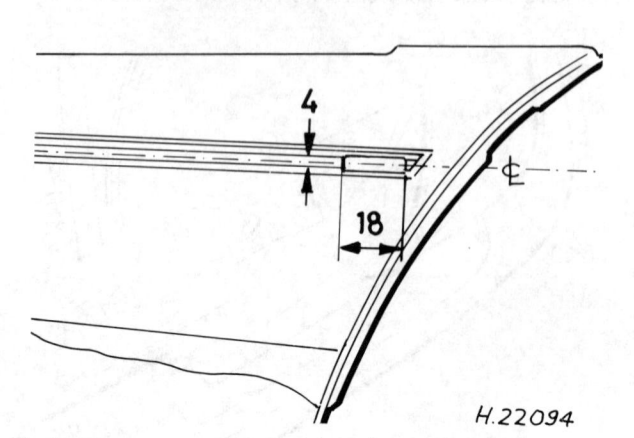

Fig. 11.8 Bumper moulding slot required on service replacement bumpers (dimensions given in mm) (Sec 13)

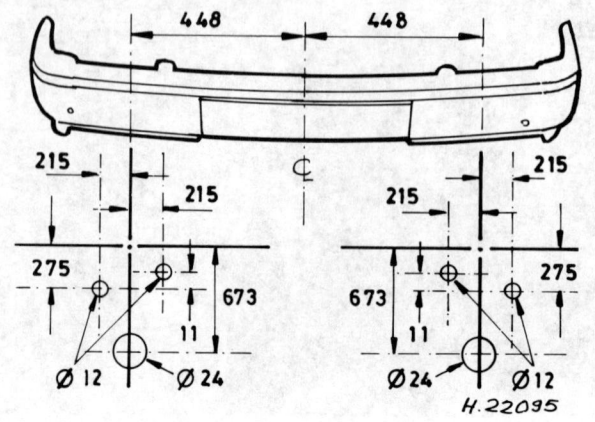

Fig. 11.9 Overrider installation to front bumper (dimensions given in mm) (Sec 14)

Chapter 11 Bodywork and fittings 237

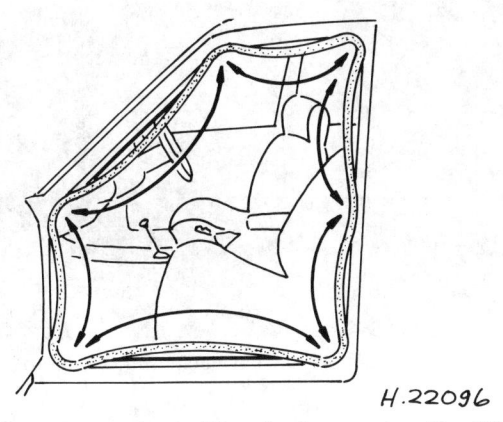

Fig. 11.10 Loop the weatherseal into the door aperture (Sec 15)

15 Door apertures weatherseal – removal and refitting

1 To remove, pull the weatherseal off the door aperture flange, starting with one end of the joint and working around to the other end.
2 To refit, roughly align the weatherseal joint so that it lies in the centre of the bottom (sill panel) flange.
3 Loop the weatherseal into the corners of the door aperture (Fig. 11.10).
4 With all the corners roughly in position, work around the aperture from one end of the weatherseal, pressing the seal fully home. Ensure that it follows the contours of the corners without wrinkling, and that it sits over any interior trim edgings.
5 Seal the weatherseal joint with a little caulking compound applied to the body flange, to prevent water entering by capillary action.
6 Check that the door closes properly, without excessive effort being required. If the door requires excessive effort to close, the door striker plate may be adjusted as necessary.

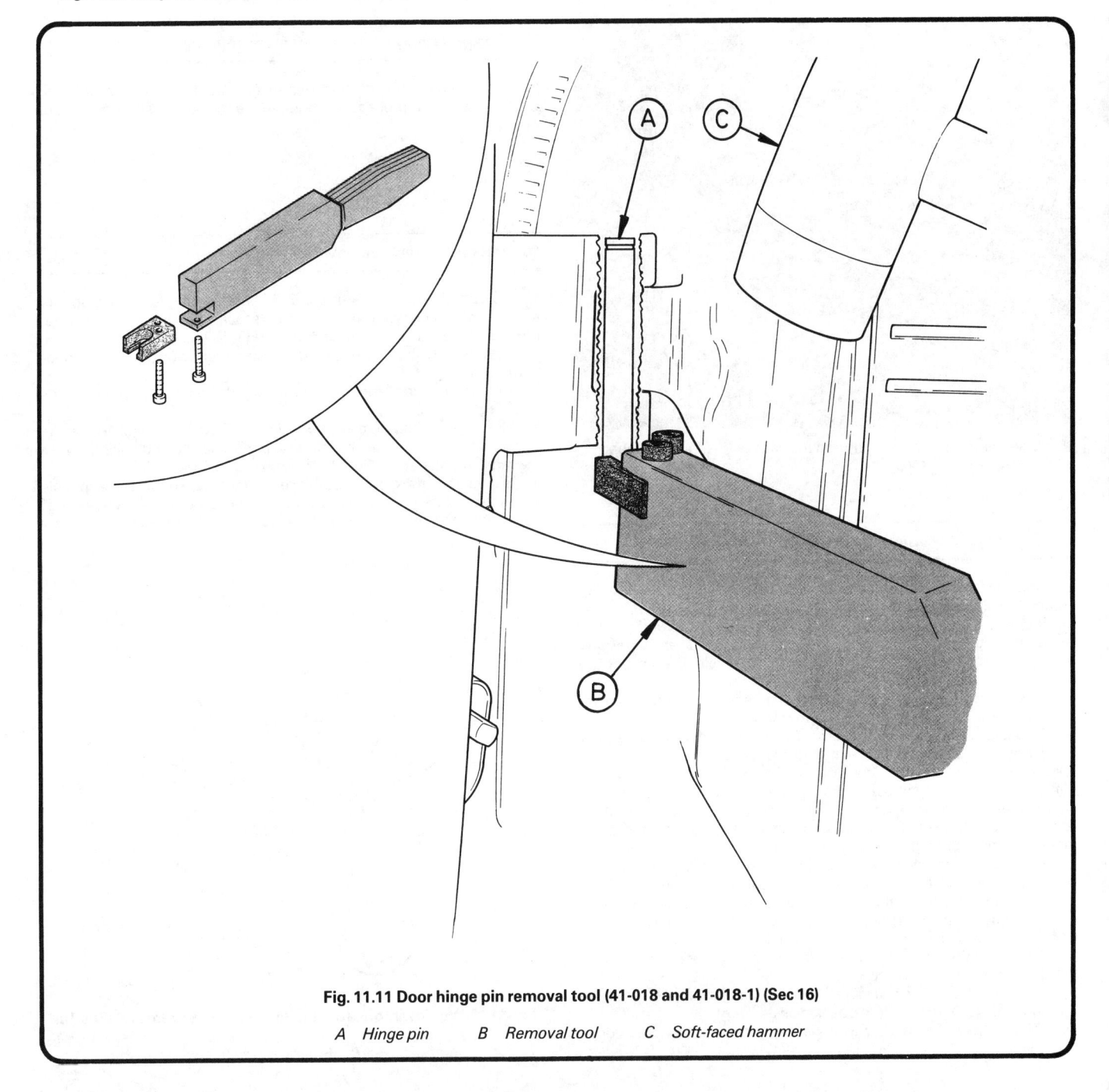

Fig. 11.11 Door hinge pin removal tool (41-018 and 41-018-1) (Sec 16)

A Hinge pin   B Removal tool   C Soft-faced hammer

238 Chapter 11 Bodywork and fittings

18.2 Removing the door interior release handle bezel (armrest/doorpull shown removed)

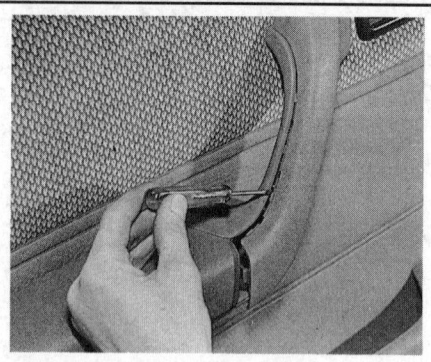

18.4A Detaching armrest/doorpull removable panel

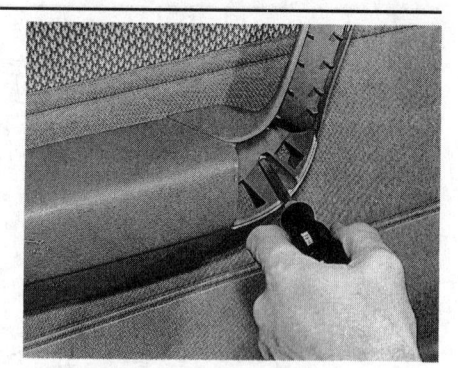

18.4B Undoing armrest/doorpull retaining screws

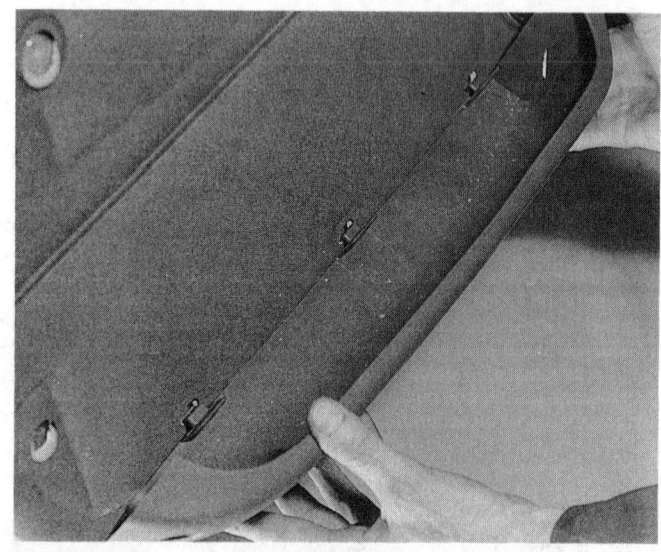

18.5 Disengaging the hook fixings on the base of the door stowage pocket/speaker grille moulding

16 Door assembly (front) – removal and refitting

1 Open the door and detach the door aperture weatherseal from around the hinge area.
2 Remove the screw securing the door check arm to the body.
3 Squeeze the ears together on the electrical multi-plug and withdraw it from its body location on the A-pillar, as applicable. Disconnect the multi-plug.
4 Using a screwdriver, remove the door hinge pin retaining clips from the top of both hinges, by levering them off.
5 Ford dealerships use a special tool for removing door hinge pins. If this is not available, it is possible to fabricate an alternative. Engage the special tool (see Fig. 11.11), then strike the main body of the tool with a soft-faced hammer to extract the lower hinge pin from its location.
6 With the help of an assistant to support the door, remove the upper hinge pin from its location, in a similar manner to that used for the lower hinge pin, and then remove the door assembly.
7 To refit, align the door hinge sections to the hinge sections on the body, and insert both hinge pins as far as possible, by hand. Ensuring that the hinge pins have entered all three sections of hinge, tap them gently through, with a soft-faced hammer, until their retaining clips can be fitted. **New** retaining clips should be used.
8 Reconnect the multi-plug and insert it to its body location on the A-pillar, ensuring that it seats correctly.
9 Refit the door check arm to its body location. Press the door aperture weatherseal back into position on its flange.
10 There should be no need to adjust the door striker, or check door alignment.

17 Door assembly (rear) – removal and refitting

1 For removal and refitting of the rear door assemblies, refer to Section 16, as the procedures required are the same as those for front doors.

18 Door trim panel – removal and refitting

1 Using a screwdriver with a broad flat blade, carefully prise up and remove the interior weatherstrip from its location at the top of the door trim panel. Be careful not to kink or bend it as it is being removed. On front doors, pull the end of the weatherstrip out from under the door mirror trim.
2 Detach the door interior release handle bezel by removing its retaining screw (photo).
3 Using a small thin screwdriver, release the window regulator handle (where fitted) by releasing its spring clip from behind and pulling it from the regulator shaft (see Fig. 11.12). Remove its bezel also.
4 Detach the armrest/doorpull assembly from the door. Some models are fitted with an armrest/doorpull secured by three screws, two of which are concealed behind a removable panel (photos). Certain lower

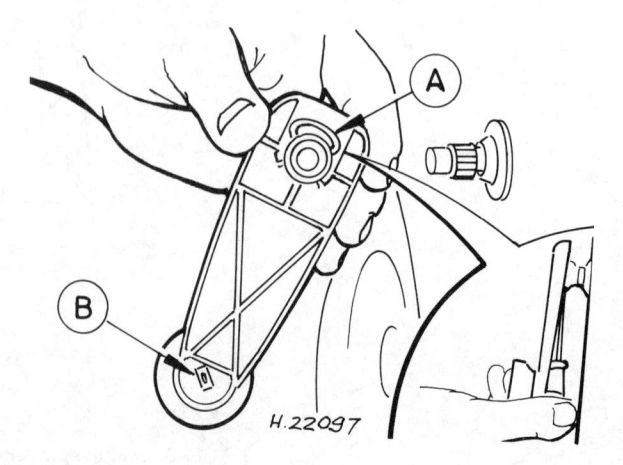

Fig. 11.12 Window regulator handle. Inset shows removal (Sec 18)

A Spring clip   B Rotate the tag to remove the knob

Chapter 11 Bodywork and fittings

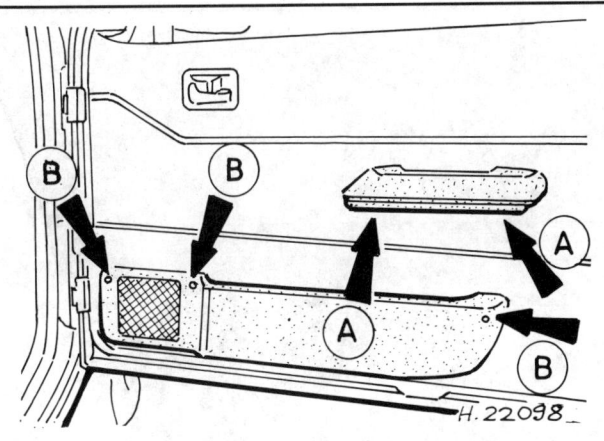

Fig. 11.13 Front door interior trim (Sec 18)

A Armrest/doorpull securing screws
B Door stowage pocket/speaker grille moulding securing screws

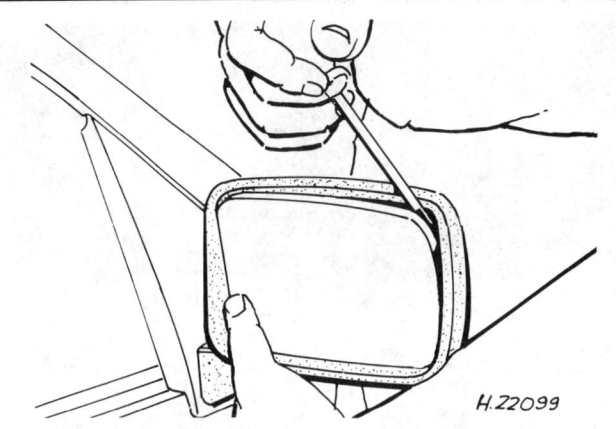

Fig. 11.14 Removing the door mirror glass from the mirror housing (Sec 19)

7 To refit, first align the trim panel to the door, secure its upper edge into its hook fastenings, then press into position to engage the trim clips.
8 If necessary, refit the clip to the window regulator handle, then refit the handle to the regulator shaft by pushing it on. The clip should engage positively, in the annular groove on the shaft.
9 Refit the door stowage pocket/speaker grille moulding, armrest/doorpull and door interior release handle bezel, as applicable, by reversing the method of removal.
10 Refit the interior weatherstrip to its location, tapping it down into position, being careful not to bend or kink it. On front doors, it will be necessary to slide its forward edge under the door mirror trim before tapping into position.

19 Door mirror glass – removal and refitting

1 Insert a thin flat-headed screwdriver behind the mirror glass, at the upper outermost corner, then carefully lever the glass assembly forward to disconnect the outer pivot (Fig. 11.14).
2 Disengage the inner pivot then, using a screwdriver, disconnect the operating link to allow the glass assembly to be withdrawn (see Fig. 11.15).
3 To refit, first reconnect the operating link and engage the inner pivot.
4 Align the outer pivot in the mirror housing, then push the glass assembly carefully into position so that the outer pivot engages fully.

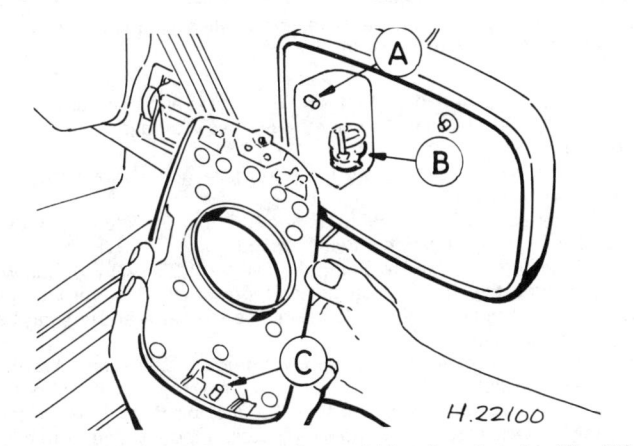

Fig. 11.15 Mirror glass connections within mirror housing (Sec 19)

A Inner pivot
B Operating link
C Outer pivot

specification vehicles have the door pull secured with only two screws (Fig. 11.13).
5 On front doors, remove the three screws securing the door stowage pocket/speaker grille moulding, then pull the moulding upper edge away from the door, and disengage the hook fixings on its base from their locations (photo).
6 Using a trim clip tool, release the retaining clips around the base and the sides of the panel, then raise it to free it from the door. If a trim clip tool is unavailable, two thin flat-bladed screwdrivers carefully inserted between the trim panel and the door may suffice. Place the screwdrivers on each side of the clips and then apply gentle leverage.

20 Door mirror – removal and refitting

1 Remove the door trim panel as described in Section 18.
2 Remove the door mirror operating knob by lifting its retaining clip, using a small screwdriver (photo), then pulling it from its shaft.
3 Remove the mirror trim retaining screw (photo), then ease the

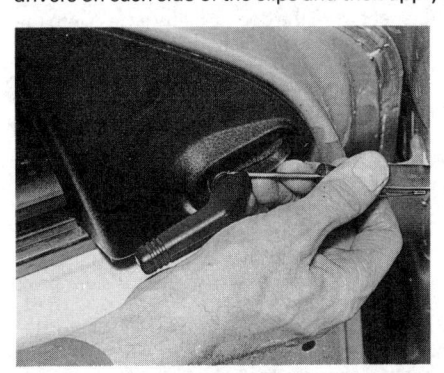

20.2 Lift the retaining clip, then pull the door mirror operating knob from its shaft

20.3 Removing the mirror trim retaining screw

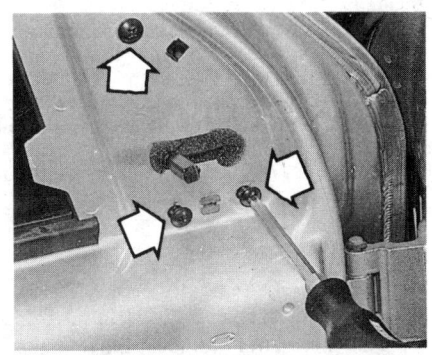

20.4 Removing the three mirror retaining screws (arrowed)

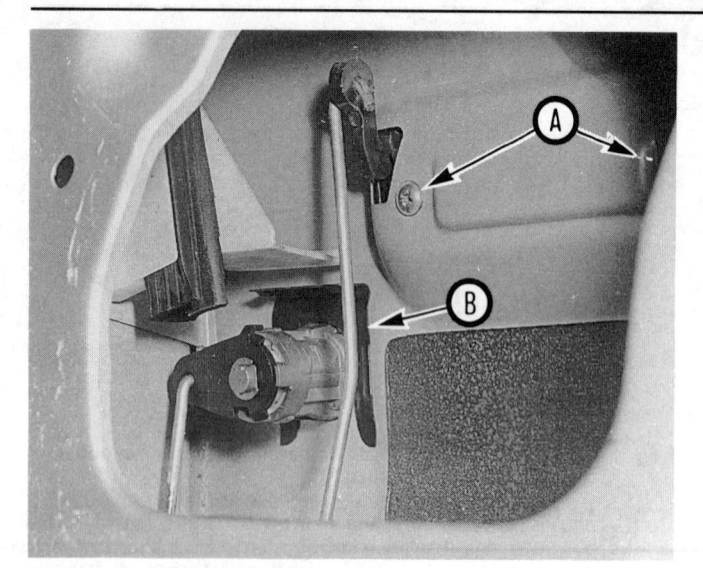

240 Chapter 11 Bodywork and fittings

22.4A Door exterior handle and lock barrel securing arrangements (front door)

A Handle screw locations
B Lock barrel retaining clip

22.4B Removing the door exterior handle, guiding its operating rod out through the doorskin (front door)

mirror trim out from its forward lug location and slide it out towards the rear of the vehicle, to clear its rearward fixings.
4 Remove the three mirror retaining screws (photo), and manoeuvre the mirror from the door.
5 Refitting is the reverse sequence to removal.

21 Central locking system components – removal and refitting

1 Refer to Chapter 12 for a description of the system, and component removal and refitting procedures.

22 Door handle (exterior) – removal and refitting

1 Open the door and remove the door trim panel, as described in Section 18.
2 Locally detach the PVC sheet from the door to allow access into the door cavity. Do not tear the sheet; cut closely around the clips, as required.
3 Disconnect the handle operating rod from the latch assembly, by twisting the clip and withdrawing the rod, as applicable.
4 Remove the two screws securing the handle then, from the outside, withdraw the handle and rod, guiding the rod out through the doorskin as necessary (photos). The handle and rod may be disconnected as required.
5 Refitting is a reversal of the removal procedure.

23 Door lock barrel – removal and refitting

1 Open the door and remove the door trim panel, as described in Section 18.

2 Locally detach the PVC sheet from the door to allow access into the door cavity. Do not tear the sheet; cut closely around the clips, as required.
3 Drill out the rivets securing the door glass guide above the lock barrel, and remove it.
4 Disconnect the lock barrel operating rod from the latch assembly, by twisting the clip and withdrawing the rod.
5 Release the lock barrel retaining clip (see photo 22.4A) by withdrawing it upwards, then from the outside, withdraw the lock barrel and rod, guiding the rod out through the doorskin. The lock barrel and rod may be disconnected as required.
6 Refitting is a reversal of the removal procedure, ensuring that the lock barrel gasket is located correctly between the lock barrel outer surround and the doorskin, and that the door glass guide is riveted securely. Check that the window operates smoothly and correctly, before refitting the PVC sheet and door trim panel, by temporarily refitting the regulator handle or operating the electric window motor.

24 Door latch and interior release handle – removal and refitting

1 Open the door and remove the door trim panel, as described in Section 18. On vehicles fitted with central locking, disconnect the battery earth terminal.
2 Locally detach the PVC sheet from the door to allow access into the door cavity. Do not tear the sheet, cut loosely around the clips, as required.
3 On rear doors fitted with central locking, undo the two screws securing the rod guard (Fig. 11.6) and remove it.
4 Disconnect the operating rods from the latch assembly (photo), as necessary. On rear doors with central locking, the motor should be released to disconnect its operating rod (see Chapter 12, Section 35).
5 On front doors, vehicles fitted with central locking have a latch with an integral motor. If central locking is fitted, prise up and twist the rearward clips securing the motor loom to the door, to free the loom, disconnect the multi-plugs and withdraw the loom (motor side) back into the door cavity.
6 Remove the interior release handle surround, then prise up the forward end of the handle unit. Release the handle from its rearward location by sliding it forward to disengage the hooks (photo).
7 Detach the door interior release cable from its retaining clips on the door panel.
8 Remove the latch securing screws from the upper rear edge of the

Chapter 11 Bodywork and fittings

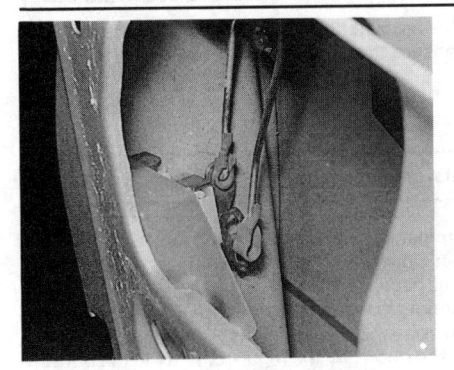

24.4 Operating rod connections at latch assembly (front door)

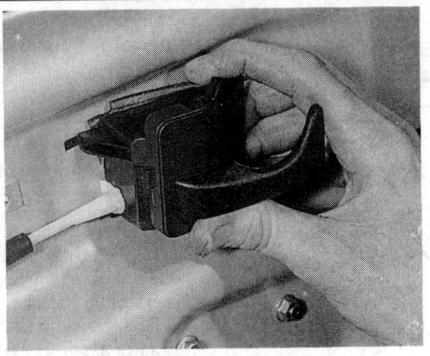

24.6 Detaching the interior door release handle

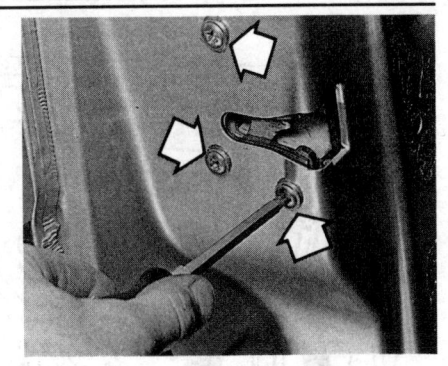

24.8 Removing the latch securing screws (arrowed)

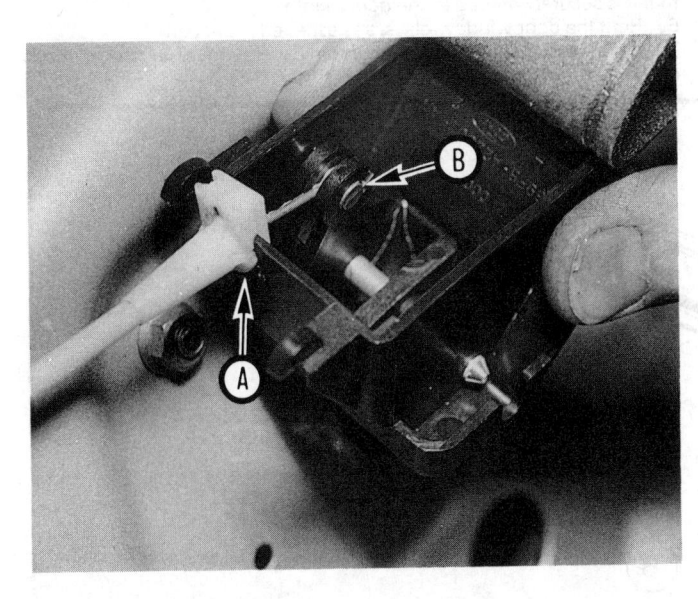

24.10 Interior door release handle-to-cable connections

A Outer cable attachment
B Inner cable attachment

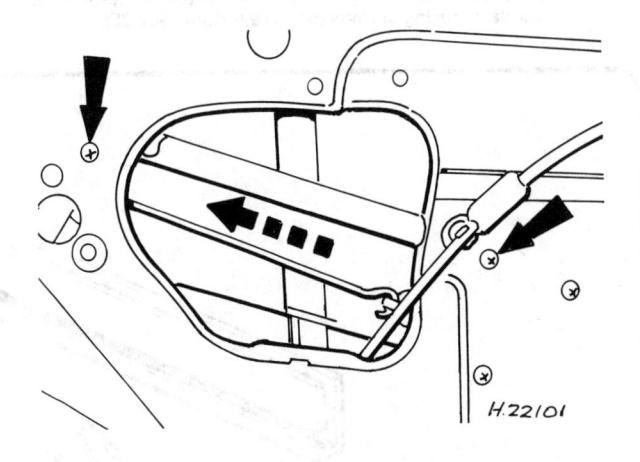

Fig. 11.16 Rod guard securing screws (arrowed) on rear doors with central locking only (Sec 24)

door (photo), and withdraw the latch and interior release mechanism from the vehicle.

9 Remove the cable cover from the latch assembly, then detach the cable from the latch (Fig. 11.17).

10 Peel the sponge pad from the reverse side of the interior release handle, and remove the outer cable from its abutment in the body of the handle. Remove the inner cable core by aligning the core with the slot by its end fixing, the slide the end fixing out (photo).

11 Refitting is a reversal of the removal procedure.

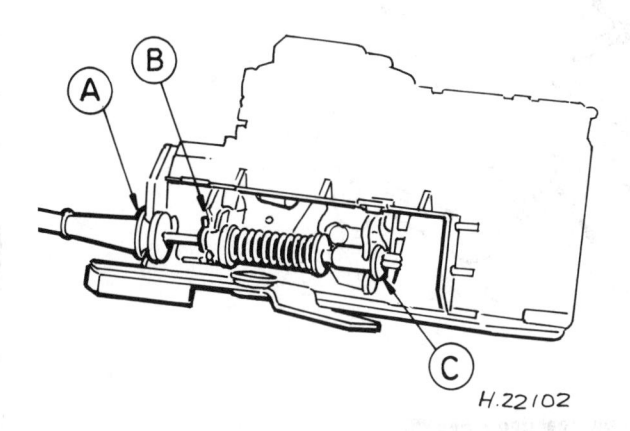

Fig. 11.17 Cable connection at door latch (Sec 24)

A Outer cable abutment   C Latch lock
B Latch release

25 Door window regulator (rear) – removal and refitting

1 Remove the door trim panel, as described in Section 18.
2 Detach the PVC sheet from the door, taking care not to tear it. Cut closely around the clips, as necessary.
3 Drill the heads off the four regulator securing rivets, and remove the two lower window glass guide securing screws.
4 Support the window glass then, having pushed the regulator into the door cavity, disconnect the regulator arm from it (see Fig. 11.19). Remove the regulator assembly.
5 Refitting is a reversal of the removal procedure, ensuring that the window regulator is securely riveted to the door. Check the operation of the window regulator, before refitting the PVC sheet and the door trim,

242 Chapter 11 Bodywork and fittings

by temporarily refitting the regulator handle and, if necessary, adjust the window glass guide to ensure that the window glass does not stick or judder.

26 Door window regulator (manual, front) – removal and refitting

1 Remove the door trim panel, as described in Section 18.
2 Detach the PVC sheet from the door, taking care not to tear it. Cut closely around the clips, as necessary.
3 Remove the door window glass, as described in Section 30.
4 Drill the heads off the regulator securing rivets and manoeuvre the regulator assembly from the door. Lay the regulator body in the bottom of the door, then withdraw the gear mechanism and cable followed by the rigid 'pillar'.
5 Refit the regulator by reversing the method of removal, ensuring that it is securely riveted to the door (photo).
6 Refit the door window glass, as described in Section 30.
7 Refit the PVC sheet and the door trim panel by reversing the method of removal.

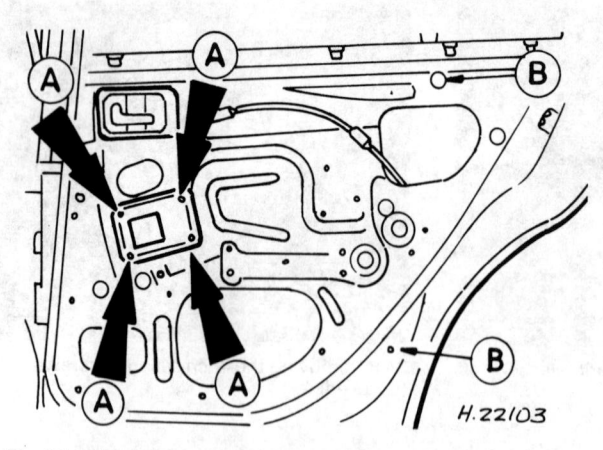

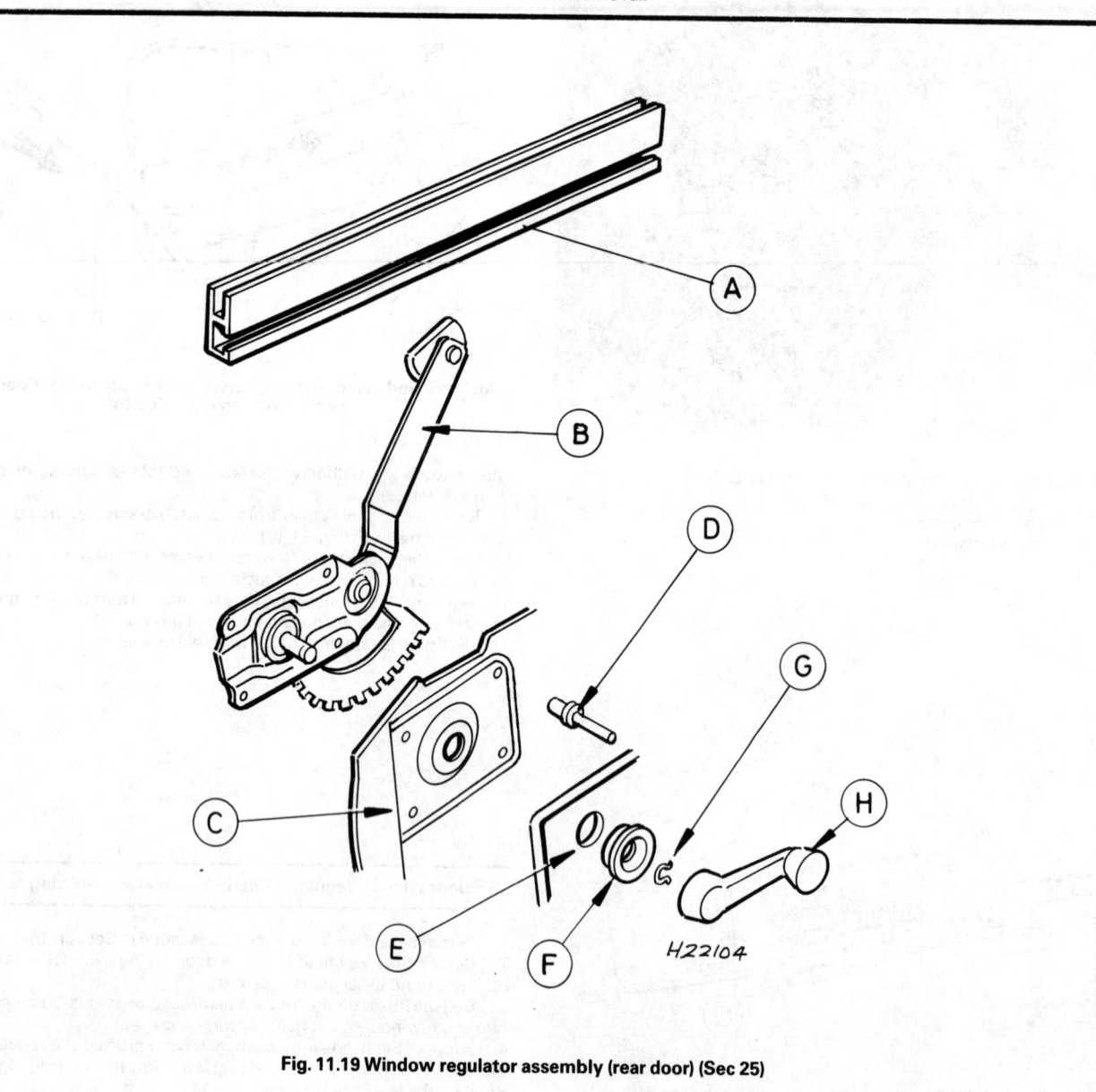

Fig. 11.18 Regulator retaining rivets (A) and lower window glass guide securing screws (B) on rear door (Sec 25)

Fig. 11.19 Window regulator assembly (rear door) (Sec 25)

A Window glass lifting channel
B Regulator assembly
C Interior door panel
D Rivet
E Door trim panel
F Bezel
G Spring clip
H Window regulator handle

Chapter 11 Bodywork and fittings 243

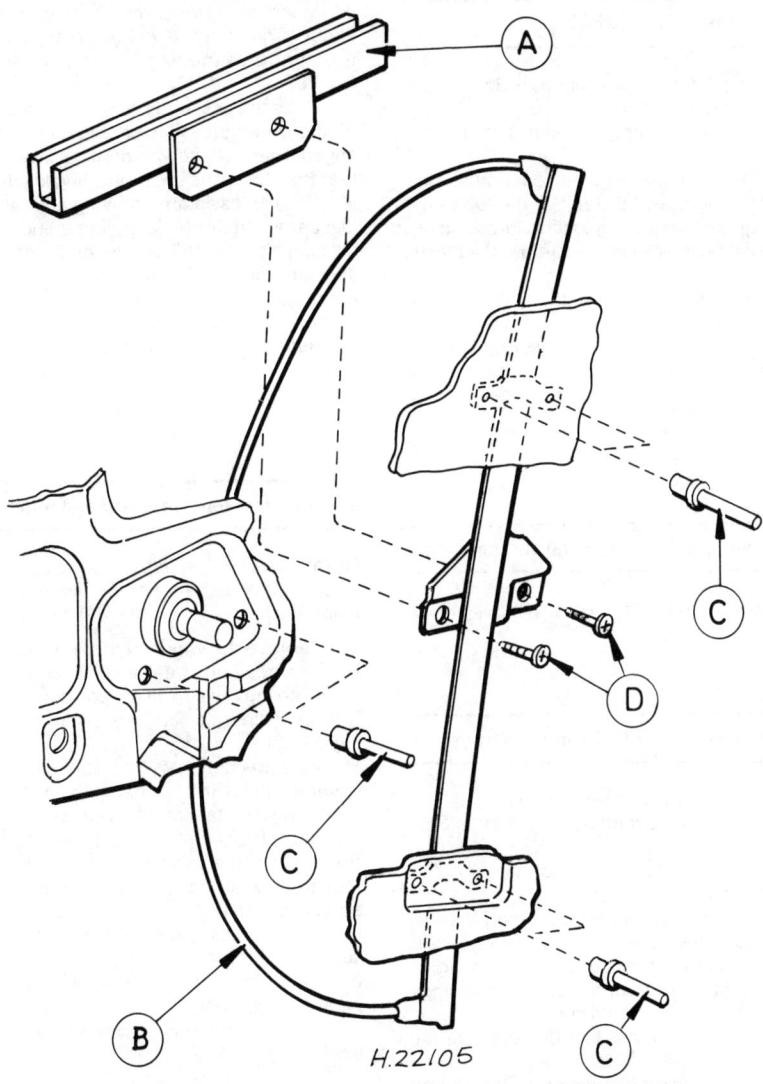

Fig. 11.20 Window regulator assembly (manual, front) (Sec 26)

A Window glass lifting channel   B Regulator assembly   C Rivets   D Screws

26.5 Riveting upper pillar section of regulator assembly to a front door. Note window glass lifting channel retaining screw upper access holes (arrowed)

27 Door window regulator (electric, front) – removal and refitting

Note: *For motor removal and refitting procedure, refer to Chapter 12.*

1   Remove the door trim panel, as described in Section 18.
2   Detach the PVC sheet from the door, taking care not to tear it. Cut closely around the clips, as necessary.
3   Remove the door window glass, as described in Section 30.
4   Disconnect the multi-plug on the regulator motor body, and remove the three nuts securing the motor/winding mechanism section of the assembly to the door.
5   Drill the heads off the rivets securing the pillar section of the regulator assembly to the door, then remove the regulator assembly from the door, as described in Section 26.
6   To refit, first position the regulator assembly in the door, and refit the three nuts securing the motor/winding mechanism section.
7   Re-rivet the pillar section of the regulator assembly securely to door (see photo 26.5).
8   Reconnect the multi-plug to the regulator motor body.
9   Refit the door window glass, as described in Section 30.
10 Refit the PVC sheet and the door trim panel by reversing the method of removal.

28 Door window glass (rear) – removal and refitting

1 Remove the door window regulator, as described in Section 25, then fully lower the window glass into the door.
2 Remove the door belt weatherseal mouldings in a similar manner to that described in Section 10.
3 Remove the remaining two window glass guide securing screws from their location at the top of the door, and disengage the top of the guide. Manoeuvre the fixed rear quarter window glass forwards and out from its location, along with the window glass guide. Lift out the plastic strip.
4 Manoeuvre the door window glass up and out of the door, from the exterior side of the door.
5 Refitting is a reversal of the removal procedure, ensuring that the fixed rear quarter window seats correctly in its location, and that the window glass guide holds the window glass correctly (refer to Section 25).

29 Door window glass (fixed rear quarter) – removal and refitting

1 The removal and refitting procedure is contained within Section 28.

30 Door window glass (front) – removal and refitting

1 Remove the door trim panel, as described in Section 18.
2 Remove the door belt weatherseal moulding as described in Section 10 of this Chapter.
3 Detach the PVC sheet from the door, taking care not to tear it. Cut closely around the clips, as necessary.
4 Remove the two screws securing the window glass lifting channel to the regulator assembly, through either their upper or lower access holes (see photo 26.5). Support the glass as the screws are removed.
5 Remove the window glass by tilting it forwards towards the hinge end of the door, then withdraw it towards the latch end.
6 To refit, position the window glass into the door by reversing the method of removal.
7 Loosely refit the two screws to hold the window glass lifting channel to the regulator assembly, but do not tighten them at this stage.
8 Raise the window glass fully, then tighten the retaining screws through their upper access holes. This ensures correct positioning of the window glass in its aperture.
9 Refit the door belt weatherseal moulding in accordance with Section 10.
10 Refit the PVC sheet and the door trim panel by reversing the method of removal.

31 Fixed rear side window glass – removal and refitting

1 The lip of the weatherseal surrounding the window glass must be released from the top and sides of the body aperture flange, using a suitable tool, before exerting pressure from the inside of the vehicle to remove the glass. Any surrounding interior trim should be removed. It is important that an assistant helps and holds the glass from the outside as it is pressed out from the inside.
2 With the assembly removed from the vehicle, detach the weatherseal from the window glass.
3 If the weatherseal is to be re-used it should be cleaned. Do **not** use petrol, white spirit or other similar substances, as they may cause rapid rubber deterioration. Ensure that the glass groove in the weatherseal is free from any sealant or glass fragments.
4 Ensure that the body aperture flange is free from any sealant also.
5 Commence refitting by attaching the weatherseal to the window glass, ensuring that it is seated correctly.
6 Insert a length of nylon or terylene cord into the body aperture flange groove of the weatherseal, so that the cord ends emerge at the top centre of the window, and overlap by approximately 150 mm (6.0 in).
7 Offer the assembly to the body aperture, and engage the lower lips of the weatherseal over the body aperture flange. Ensure that the engagement of the weatherseal lips is not hampered in any way, and that the cord ends are protruding inside the vehicle.
8 With an assistant applying a gentle even pressure on the window glass from the outside, pull one end of the cord at right-angles to the glass. This will pull the inner lip of the weatherseal over the body aperture flange. When the cord has been pulled halfway around the aperture, repeat the procedure with the other end of the cord. The cord should release when the window glass is fully fitted.
9 Refit interior trim as necessary.

32 Windscreen – removal and refitting

Note: *The DIY mechanic is advised to leave windscreen removal and refitting to an expert. For the owner who insists on doing it himself the following paragraphs will prove helpful.*

1 Remove the windscreen wiper arms (see Chapter 12).
2 Remove both A-pillar trims, and release the front of the headlining by removing the sun visors and courtesy light. If a heated windscreen is fitted, disconnect its electrical contacts at the windscreen (see Fig. 11.23).
3 As a precautionary measure to prevent damage to the paintwork, cover the cowl and bonnet with an old blanket or similar.
4 Using a suitable tool, release the windscreen weatherseal from the top and sides of the body aperture flange before exerting pressure from the inside of the vehicle to remove the assembly. It is strongly advised that an assistant outside the vehicle is ready to receive the windscreen as it is pushed out.
5 With the assembly removed from the vehicle, detach the weatherseal from the windscreen glass.
6 Clean the weatherseal, making particular reference to Section 31, paragraph 3, then refit in accordance with the rest of that Section.
7 Any trim and equipment removed during preparation should now be refitted.

33 Tailgate window glass – removal and refitting

1 Tailgate window glass removal and refitting procedures are very similar in execution to those for the windscreen, therefore reference should be made to Section 32.
2 Disconnect the heated rear window, and remove the tailgate wiper arm (as applicable).

34 Tailgate support strut – removal and refitting

1 Open the tailgate and support it securely, using a length of timber, or have an assistant hold it open.
2 Detach the support strut by raising the retaining clip (photo) and pulling the strut away from its ball stud fixing on the tailgate. Do **not** raise the clip more than 4 mm, or the support strut will be damaged.
3 Repeat the operation on the other end of the support strut.
4 To refit, align the thicker (cylinder) end of the support strut to the ball stud fixing on the tailgate, then push until it snaps into engagement.
5 Repeat the operation with the thinner (piston) end of the support strut to the ball stud fixing on the body.

Chapter 11 Bodywork and fittings

34.2 Raising a tailgate support strut retaining clip (4 mm maximum)

35 Tailgate – removal and refitting

1 Open the tailgate and mark around the hinge positions on the tailgate panel itself.
2 Remove the central blanking plug from the upper portion of the tailgate to expose the washer jet, and disconnect the washer hose from the jet base, as applicable. Free the washer hose grommet and withdraw the hose from the tailgate.
3 Disconnect the support strut from the tailgate, as described in Section 34.
4 With the aid of an assistant, undo the four bolts securing the tailgate to the vehicle body and remove the tailgate.
5 To refit, align the hinges on the body with the marks made on the tailgate panel, then refit and tighten the four bolts to secure. Refit the support strut as described in Section 34.
6 The remainder of the refitting operation is a reversal of the removal procedure.
7 Close the tailgate and check the alignment. Adjust as necessary to obtain an even gap between the tailgate and all adjacent panels.

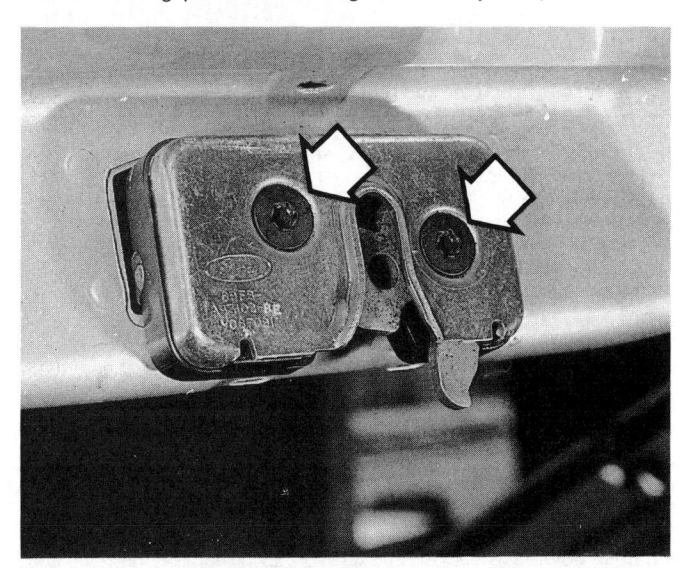

38.2 Tailgate latch and its Torx securing screws (arrowed)

36 Tailgate spoiler – removal and refitting

1 Drill out the rivets securing the outer edges of the spoiler to the tailgate.
2 Open the tailgate and remove the four blanking plugs covering the spoiler retaining nuts. Undo the nuts and remove the spoiler.
3 To refit, first align the spoiler mounting studs to their tailgate fixing holes, then rivet the outer edges of the spoiler to the tailgate.
4 Refit the four retaining nuts and tighten to the specified torque.
5 Refit the blanking plugs.

37 Tailgate trim panel – removal and refitting

1 Open the tailgate then, using a screwdriver, remove the seven plastic retaining screws from the trim panel.
2 Remove the square plastic clip from the trim panel.
3 Disengage the clips from the panel edge nearest the window, and manoeuvre the panel to release the hooks on the other panel edge. Withdraw the trim panel from the vehicle.
4 Refitting is a reversal of removal.

38 Tailgate lock – removal and refitting

Note: *For vehicles equipped with remote tailgate release, see Chapter 12, Section 37.*

1 Open the tailgate and remove its trim panel, as described in Section 37.
2 Remove the two Torx screws securing the tailgate latch (photo), then pull it down, disconnect its operating rod and remove the latch from the vehicle.
3 Remove the bolt securing the lock barrel retainer to the tailgate (photo), then withdraw the retainer.
4 Remove the lock assembly from the tailgate by partially withdrawing the lock, twisting it so that the lock rod can be removed, then fully withdrawing.
5 To refit, first ensure that the lock gasket is in position, then partially insert the lock into the tailgate and reconnect the lock rod. Fully insert the lock and refit its retainer and securing bolt.
6 Using an electrician's thin screwdriver, prise up the lever on the latch and fit the track rod to it (Fig. 11.21), ensuring that it fully engages.
7 Align the latch with its location and refit its two securing screws.
8 Refit the tailgate interior trim panel by reversing the method of removal.
9 Check for correct lock operation.

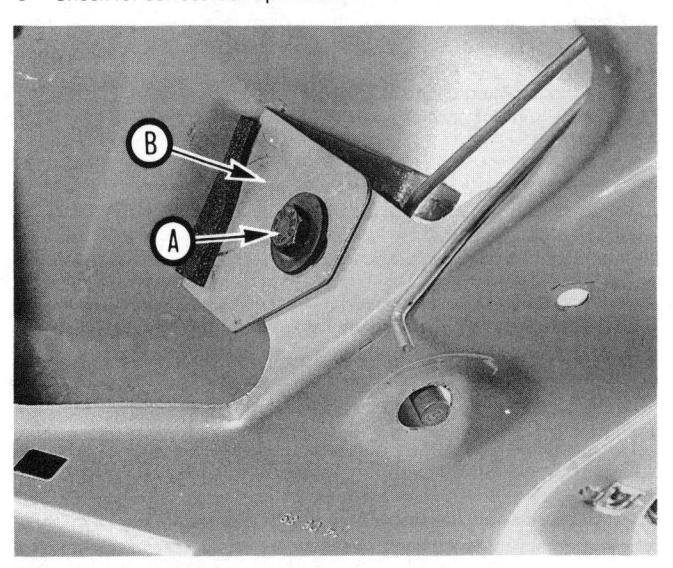

38.3 Tailgate lock retention arrangement

A *Lock barrel retainer securing bolt*
B *Lock barrel retainer*

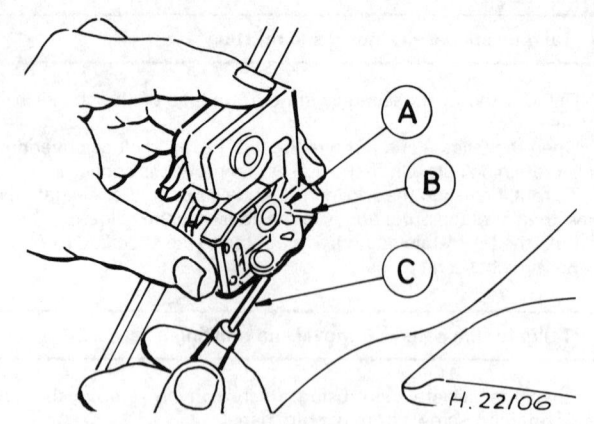

Fig. 11.21 Refitting the lock rod to the tailgate latch (Sec 38)

A Rod clip/latch lever
B Lock rod
C Screwdriver

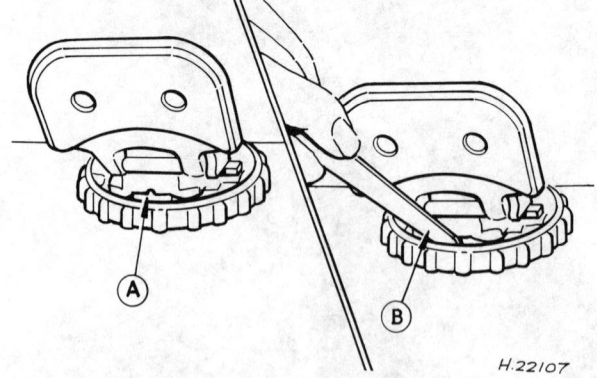

Fig. 11.22 Sunroof handwheel locking mechanism (Sec 39)

A Locking bar engaged   B Releasing locking bar

39 Sunroof panel — removal and refitting

1 Open the sunroof fully using the handwheel.
2 Detach the sunroof from its rearward mounting point by pressing in the red locking bar on the handwheel (Fig. 11.22), then raise it and remove.
3 To refit, engage the hinges in their locations, in the forward edge of the sunroof opening, then lower the sunroof. Allow the handwheel to click into its safety lock.
4 Check that the safety lock mechanism retains the panel securely.
5 Close the sunroof fully using the handwheel.

40 Sunroof panel seal — renewal

1 Remove the sunroof panel, as described in Section 39, then lay the panel on a soft cloth.
2 Remove the two hinge plate assemblies from the panel by undoing the hinge plate retaining screws.
3 Pull the seal from the panel edge, noting the position of its join.
4 Starting from the position of the original join, press the seal into position, ensuring that it sits evenly right around. Adjust as necessary so that the ends butt tightly together.
5 Refit the two hinge plate assemblies to the panel.
6 Refit the sunroof panel, as described in Section 39.

41 Sunroof weatherseal — removal and refitting

1 Remove the sunroof panel, as described in Section 39.
2 Starting from the joint on the rearward opening edge, pull the seal up and remove.
3 Refitting is a reversal of the removal procedure, ensuring that the weatherseal does not deform at the corners, or split at the joint.

42 Front seat and slide assembly — removal and refitting

1 Move the seat to its furthest rearward adjustment to expose the two front slide assembly to floorpan bolts. Remove all four bolts securing the slides to the floorpan, and remove the seat from the vehicle.

2 Detach the seat cushion by removing the two plastic cap nuts, releasing the wire tabs, lifting the front of the cushion and pulling it forwards.
3 Remove the four cushion springs from the rear of the sprung platform, then remove the sprung platform.
4 Detach the seat slides as required, by removing the two nuts securing each one to the seat base.
5 If desired the head restraint may be removed by depressing the single catch on each head restraint adjustment socket and then pulling the head restraint up and out of the seat backrest.
6 To refit the seat and its components, first align the components then insert the slide studs through the surround, into the tubular seat frame.
7 Refit the nuts securing the slides to the seat base, and tighten to the specified torque.
8 Refit the sprung platform and the four cushion springs.
9 Engage the two hooks on the back of the seat cushion under the bar which forms the rear of the sprung platform.
10 Lower the seat cushion, and engage the wire tabs attached to the front of it into the plastic seat surround.
11 Holding the seat cushion down, refit the two plastic cap nuts to the seat cushion studs.
12 Position the seat in the vehicle, and align the slide assemblies with their mounting bolt locations in the floorpan. Refit the two forward securing bolts first, followed by the two rear securing bolts, to ensure that the seat runs smoothly in its slide mechanism. Tighten the bolts to the specified torque.

43 Rear seat — removal and refitting

1 To remove the backrest, release the backrest catch/catches, then fold the backrest forwards.
2 Remove the hinge-to-backrest securing screws (photo), then remove the backrest.
3 The hinges may be removed at this stage if required, by undoing the hinge-to-body securing screws.
4 To install the backrest, align the hinges to the body and backrest, then loosely refit the securing screws.
5 Fold the backrest into its upright position and engage the backrest catch/catches. Adjust the alignment of the components then, when correct, tighten the securing screws to their specified torques.
6 Check for correct catch engagement.
7 If desired, the rear seat backrest catch may be removed by removing the catch retaining bolts (photo). Refitting the catch is the reverse of removal.
8 To remove the cushion, remove the three screws securing the forward edge of the seat cushion to the raised floorpan section.
9 Push down and back on the cushion, to disengage the hook and catch on the rear underside of the seat.
10 Remove the seat from the vehicle after guiding the seat belt clips through the slits in the seat cushion.
11 Refitting the cushion is a reversal of the removal procedure.

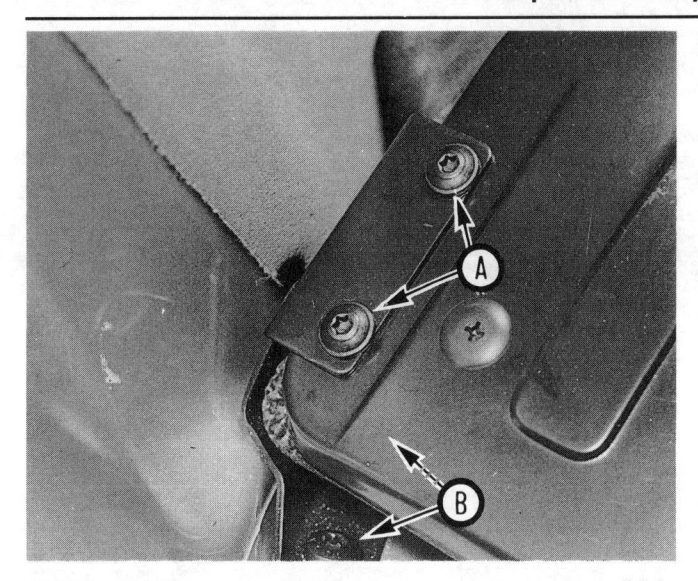

Chapter 11 Bodywork and fittings 247

43.2 Rear seat hinge arrangement

A Hinge-to-backrest screw locations
B Hinge-to-body screw locations (removed after backrest detached)

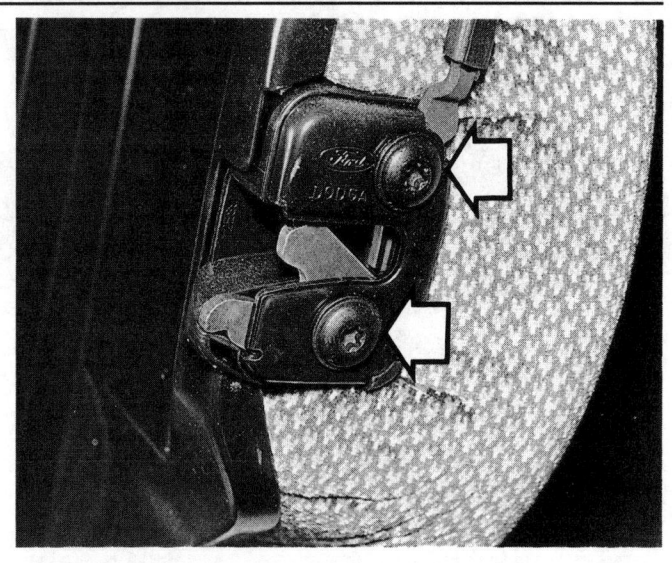

43.7 Rear seat backrest catch (retaining bolts arrowed)

anchor bolt to the specified torque. Ensure that the anchor is free to rotate.

C-pillar trim

9 Open the tailgate, then detach the loudspeaker from its location by disconnecting its multi-plug connection, removing its two retaining screws and unhinging it.
10 Detach the tailgate weatherseal from its flange, around the trim panel.
11 Remove the parcel shelf, then slacken the parcel shelf support/rear loudspeaker housing retaining screws, and remove the single screw from the base of the C-pillar trim (Fig. 11.24).
12 Remove the upper seat belt anchor bolt cover, and remove the bolt and anchor, noting any spacer fitment.
13 Remove the screw(s) from the top of the C-pillar trim (see Fig. 11.24), then remove the trim from the vehicle.
14 To refit, position the base of the trim behind the parcel shelf support/rear loudspeaker housing, then loosely refit the upper and lower securing screws.
15 Refit the seat belt anchor, bolt, and spacer, as applicable, tightening the bolt to the specified torque. Refit the anchor bolt cover, and ensure that the anchor is free to rotate.
16 Tighten the trim and parcel shelf support/rear loudspeaker housing screws, then refit the parcel shelf.
17 Refit the loudspeaker by reversing the method of removal, then press the tailgate weatherseal back into place.

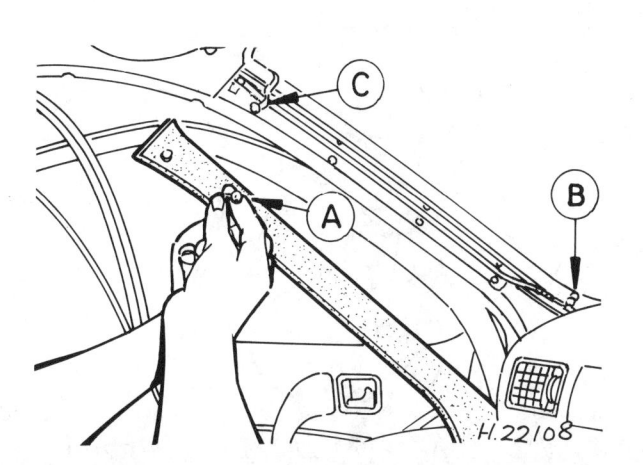

44 Interior trim panels – removal and refitting

A-pillar trim

1 Remove the securing screw from the trim panel, located near to the windscreen top corner (Fig. 11.23).
2 Detach the door aperture weatherseal from around the trim panel, then pull the trim panel from the pillar.
3 To refit, align the trim fixing lugs to the pillar and press into position. Refit the securing screw then push the door aperture weatherseal back onto its flange.

B-pillar trim

4 Remove the front seat belt height adjuster knob, as applicable, by inserting a small screwdriver into the aperture on the underside of the knob and levering it off.
5 Remove the upper seat belt anchor bolt cover, and remove the bolt, as required.
6 Open the doors and detach the door aperture weatherseals from the area around the B-pillar trim.
7 Remove the securing screws and detach the trim.
8 Refitting is the reverse sequence to removal, tightening the seat belt

Fig. 11.23 A-pillar trim retaining screw (A), heated windscreen live wire (B) and heated windscreen earth wire (C) (Sec 44)

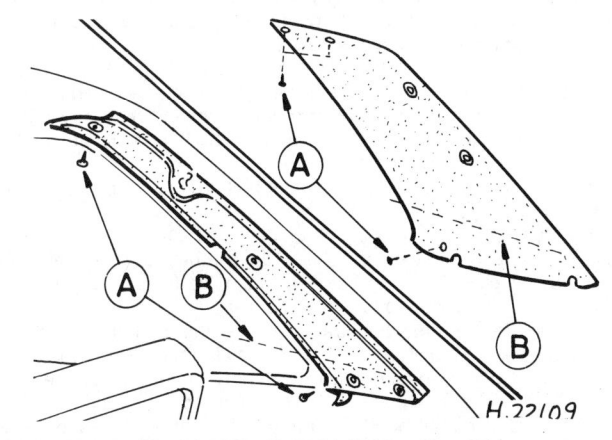

Fig. 11.24 C-pillar trim fixings (Sec 44)

A Screw locations
B Parcel shelf support/rear loudspeaker housing location

248　Chapter 11　Bodywork and fittings

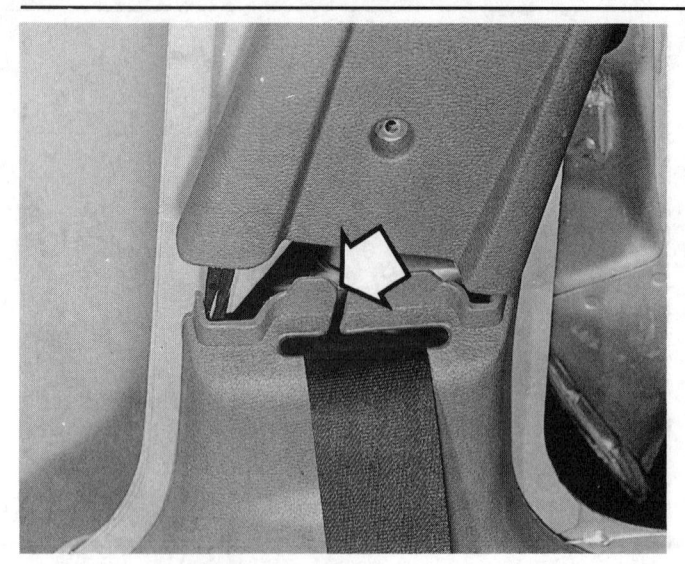

44.23 B-pillar trim being removed to expose seat belt removal/refitting slot in sill scuff plate (arrowed)

45.2 Removing a seat belt upper anchor cover

Sill scuff plate

18 Remove the front seat adjacent to the scuff plate to be removed.
19 Detach the door aperture weatherseal(s) from around the scuff plate location.
20 Remove the B-pillar trim, as applicable.
21 Remove the lower seat belt anchor bolt on five-door models, and detach the anchor from its location.
22 On three-door models, remove the bolt securing the forward end of the seat belt slide bar, then manoeuvre the slide bar to free it from its rear location.
23 Remove the plastic screws securing the sill scuff plate, then slip the seat belt through its slot (photo) on five-door models, and remove the sill scuff plate from the vehicle. Should the soft plastic screws round off during attempts to remove them, force an electrician's thin screwdriver into the body of the screw and try again, fitting a new screw upon reassembly.
24 Refitting is a reversal of the removal procedure, tightening the seat belt anchor bolts to the specified torque.

Rear quarter trim panel

25 Detach the door aperture weatherseal from around the trim location.
26 Remove the rear seat cushion, as described in Section 43.
27 Remove the screw from the upper rear corner of the trim panel.
28 Remove the plastic stud from the lower rear corner of the trim panel.
29 Detach the seat belt slide bar by removing the bolt from its forward end then manoeuvring it to free it from its rear location (see photo 53.4). Remove the seat belt from the slide bar.
30 Release the trim clips from the front and top edges of the panel using a trim clip tool. This may be achieved by carefully inserting two thin flat-bladed screwdrivers between the panel and the body - one each side of the clip being released, and applying gentle leverage, if a trim clip tool is not available.
31 Remove the seat belt upper anchor bolt cover and undo the bolt (see photo 51.2).
32 Carefully prise up the seat belt guide from its trim panel location, and remove it from the seat belt. Allow the seatbelt to retract through the trim panel, clamping a clothes peg, or similar item, onto its end to prevent it being fully wound into the retractor.
33 Manoeuvre the trim panel out from under the scuff plate, slackening or removing the rearward scuff plate retaining screws if necessary.
34 Upon installation, engage the trim panel under the scuff plate, and refit and tighten the scuff plate retaining screws as necessary.
35 Pull the seat belt through the trim panel, refit the guide and press the panel onto the body to re-engage the trim clips.
36 Refit the upper seat belt anchor, tightening the bolt to the specified torque, and refit its cover. Ensure that the anchor is free to rotate.
37 Refit the plastic stud and the screw to the lower and upper rear corners of the trim, respectively.

38 Refit the seat belt to the slide bar, then refit the slide bar by reversing the method of removal. Ensure that the seat belt is not twisted as it is located on the slide bar, and that the bolt is tightened to the specified torque.
39 Refit the rear seat cushion by reversing the method of removal.
40 Press the door aperture weatherseal back into position.

45 Front seat belt height adjuster - removal and refitting

Note: *If a height adjuster mechanism is not fitted, it is possible to lower the upper anchor position from its production setting, by removing the plug from the lower adjuster plate hole and bolting the anchor into this (see Fig. 11.26). The plug can then be fitted to the production setting hole.*

1 Remove the height adjuster knob by carefully inserting a small screwdriver into the aperture on the underside of the knob, and gently levering it off (Fig. 11.25).
2 Remove the anchor cover (photo), again using a screwdriver, then remove the anchor bolt and anchor.
3 Remove the door aperture weatherseal(s) around the B-pillar location, then remove the B-pillar trim, as described in Section 44.
4 Remove the bolts securing the adjuster plate to the B-pillar (Fig. 11.26), noting washer fitment, then remove the adjuster plate from the vehicle.
5 Refitting is the reverse sequence to removal, ensuring that the adjuster plate bolts are tightened to the specified torque, and that the anchor is free to rotate.
6 Check operation of adjuster mechanism.

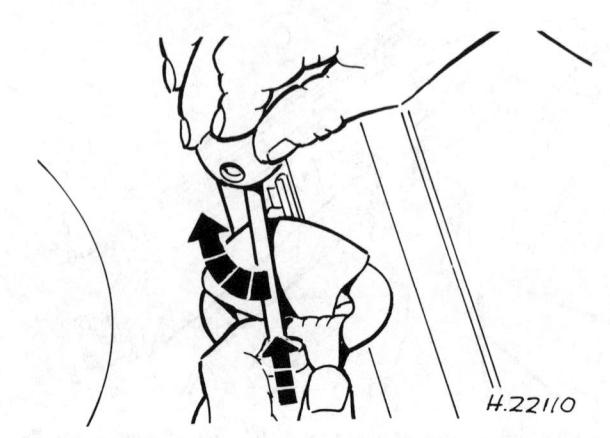

Fig. 11.25 Front seat belt height adjuster knob removal (Sec 45)

Chapter 11 Bodywork and fittings 249

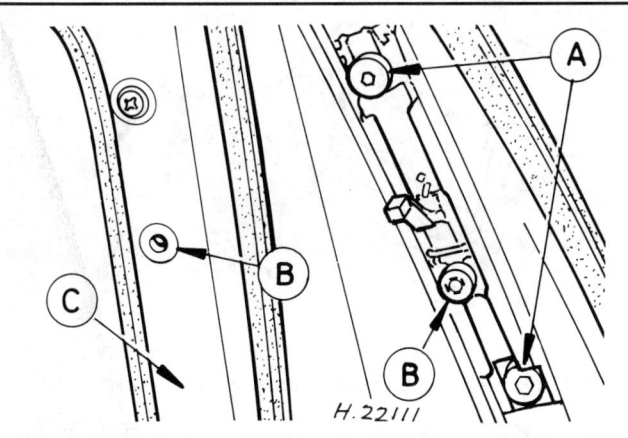

Fig. 11.26 Seat belt upper anchor position (fixed and adjustable) (Sec 45)

A Adjuster plate securing bolt
B Anchor bolt position (production setting)
C Lower setting for fixed type (if required)

46 Front seat belt and clip (five-door models) – removal, refitting and maintenance

1 Remove the bolt securing the seat belt clip assembly to the floor, between the two front seats (Fig. 11.27), then remove the clip assembly from the vehicle.
2 Remove the cover from the upper anchor position and remove the anchor bolt (see photo 51.2).
3 Remove the lower anchor bolt from its location on the floor by the base of the B-pillar. Prevent the retractor unit from reeling in too great a quantity of seat belt by attaching a clothes peg, or similar item, to the seat belt.
4 Remove the B-pillar trim and sill scuff plate as described in Section 44.
5 Remove the bolt securing the seat belt retractor unit to its location in the base of the B-pillar (photo), then remove the retractor unit and the seat belt from the vehicle.
6 Refitting is a reversal of the removal procedure, ensuring that the tag on the retractor unit engages in its cut-out at the base of the B-pillar, that the anti-rotation plate engages to its floorpin to locate the clip assembly correctly, and that the seat belt upper anchor is free to rotate. Tighten all bolts to their specified torque.

Maintenance

7 Periodically check the belts for fraying or other damage. If evident, renew the belt.
8 If the belts become dirty, wipe them with a damp cloth using a little liquid detergent only.

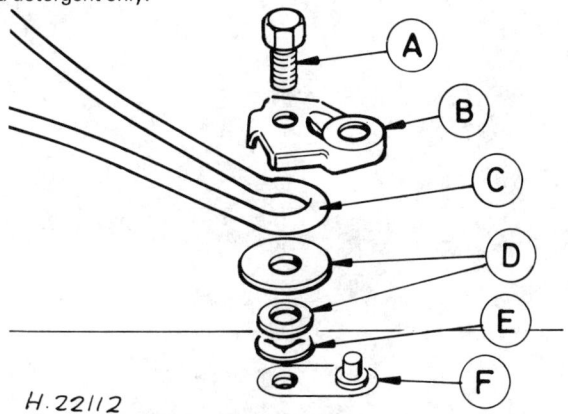

Fig. 11.27 Front seat belt clip assembly (Sec 46)

A Bolt
B Anti-rotation plate
C Clip
D Metal washers
E Paper washer
F Floorpin

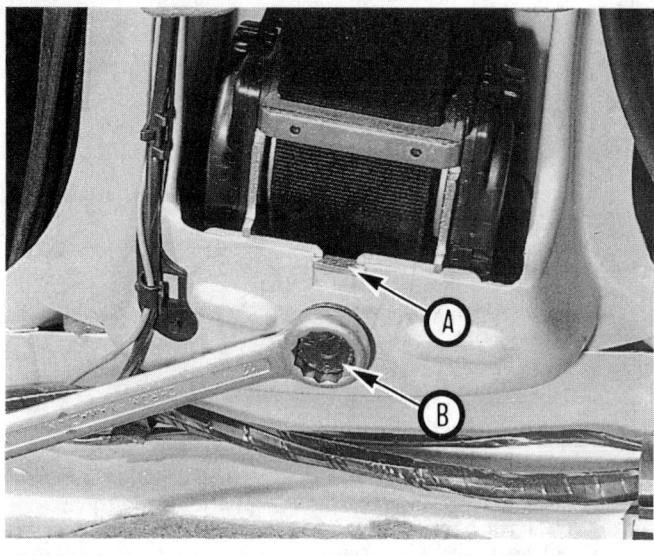

46.5 Seat belt retractor unit location in base of B-pillar (five-door models)

A Retractor unit locating tag
B Retractor unit securing bolt

9 Check the tightness of the anchor bolts and if they are ever disconnected, make quite sure that the original sequence of fitting of washers, bushes and anchor plate is retained.
10 Never modify the belt or alter its attachment point to the body.

47 Front seat belt and clip (three-door models) – removal, refitting and maintenance

1 Remove the bolt securing the seat belt clip assembly to the floor, between the two front seats (see Fig. 11.27), then remove the clip assembly from the vehicle.
2 Remove the rear quarter trim panel, as described in Section 44.
3 Undo the bolt securing the seat belt retractor unit to its location, then remove the retractor unit and seat belt from the vehicle.
4 Refitting is a reversal of the removal procedure, ensuring that the tag on the retractor unit engages in its body panel cut-out, that the anti-

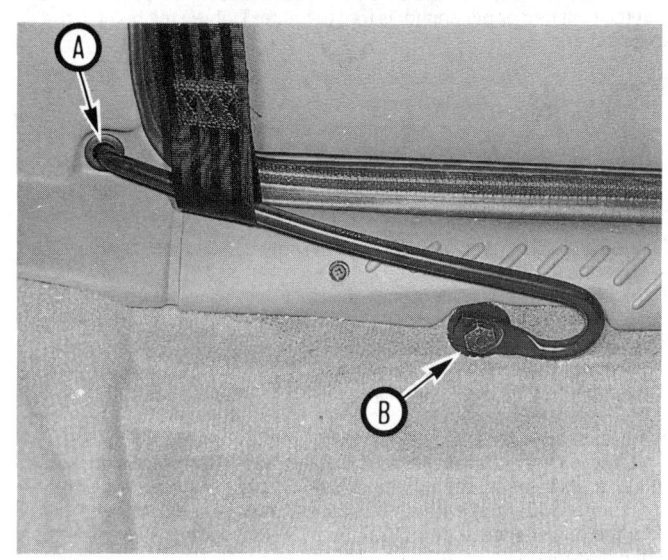

47.4 Seat belt lower slide bar anchor arrangement (three-door models)

A Bush location
B Anchor bolt

Chapter 11 Bodywork and fittings

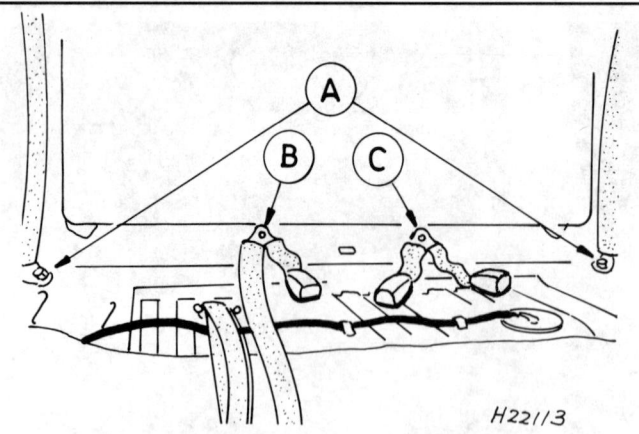

Fig. 11.28 Rear seat belt clip securing arrangements (rear seat cushion removed) (Sec 49)

A Seat belt lower anchor retaining bolts
B Centre lap belt/single clip assembly securing bolt
C Dual clip assembly securing bolt

rotation plate engages to its floorpin to locate the clip assembly correctly, and that the seat belt upper anchor is free to rotate. Before engaging the seat belt slide bar (photo), ensure that its rearward bush fixing is correctly located in the panel. Tighten all bolts to their specified torque.

Maintenance
5 Refer to Section 46.

48 Front seat belt and clip (Van models) – removal, refitting and maintenance

1 Remove the bolt securing the seat belt clip assembly to the floor, between the two front seats (see Fig. 11.27), then remove the clip assembly from the vehicle.
2 Remove the cover from the upper anchor position and remove the anchor bolt (see photo 51.2).
3 Undo the bolt securing the retractor unit and integral lower anchor position, then remove the retractor unit and seat belt from the vehicle.
4 Refitting is a reversal of the removal procedure, ensuring that the tag on the retractor unit engages into its location, that the anti-rotation plate engages to its floorpin to locate the clip assembly correctly, and that the seat belt upper anchor is free to rotate. Tighten the bolts to their specified torque.

Maintenance
5 Refer to Section 46.

49 Rear seat belt and clip – removal, refitting and maintenance

1 Remove the rear seat cushion, as described in Section 43, then also remove the parcel shelf.
2 Undo the bolt securing each seat belt clip assembly to its body location (Fig. 11.28), and remove the clip assemblies from the vehicle, as required.
3 Prevent the retractor units from reeling in too much seat belt by attaching a clothes peg or similar item to the belt, close to the retractor unit.
4 Detach the seat belt lower anchor by removing its retaining bolt.
5 Remove the upper anchor bolt cover and undo the bolt securing the anchor to its location. Detach the anchor, noting any spacer fitment.
6 On vehicles fitted with a parcel shelf, remove the seat belt guide from the parcel shelf support/rear loudspeaker housing, then pass the seat belt through it.
7 On vehicles without a parcel shelf as standard, remove the trim panel covering the seat belt retractor unit, where fitted, by undoing its retaining screws. Remove the seat belt guide from the trim panel, then pass the seat belt through it.

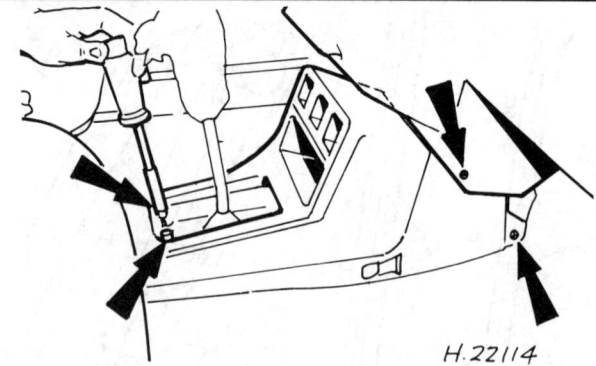

Fig. 11.29 Centre console securing screws (arrowed) (Sec 50)

8 Remove the bolt securing the seat belt retractor unit to its body location, then withdraw the retractor unit and the seat belt from the vehicle.
9 Refitting is a reversal of the removal procedure, noting the following points. Ensure that the tag on the retractor unit engages in its body location, and that it does not trap the wiring loom between itself and the body. Having fitted the upper anchor to its location, ensure that it can rotate freely. The clip assembly anchor(s) must engage to the anti-rotation floorpin(s) and, when refitting a centre lap belt/clip assembly, the lap belt must be fitted to lie on the right-hand side of the vehicle. Tighten all bolts to their specified torque.

Maintenance
10 Refer to Section 46.

50 Centre console – removal and refitting

1 Disconnect the battery earth terminal.
2 Carefully prise up the console switches, as necessary, using a flat-bladed screwdriver, then disconnect their multi-plugs.
3 Unscrew the gear lever knob, then raise the gaiter from its location (photo) and lift it off over the gear lever. A similar method is also used to remove the selector cover on CTX automatic transmission equipped vehicles.
4 Undo the four screws securing the centre console to the floor pan (Fig. 11.29). On automatic transmission equipped models, ensure that the bulb assembly does not restrict centre console removal. Remove the centre console.
5 Refitting is a reversal of the removal procedure, ensuring that the gaiter (or selector cover) locates correctly to the centre console (as applicable).

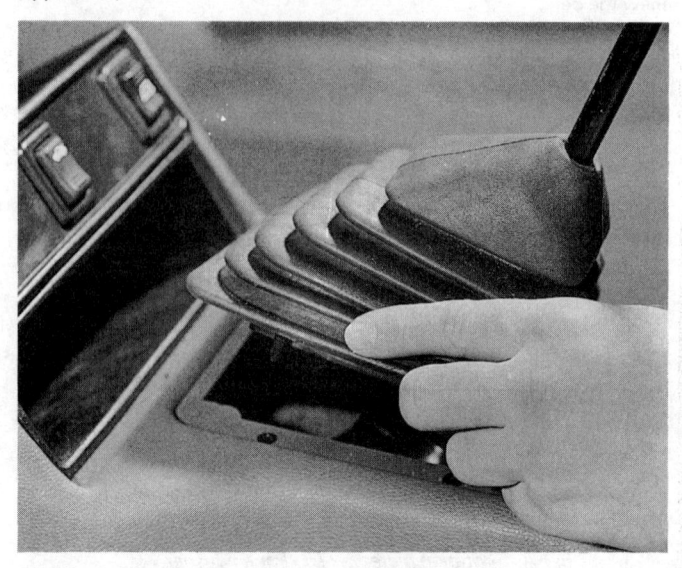

50.3 Removing the gear lever gaiter from the centre console

Chapter 11 Bodywork and fittings

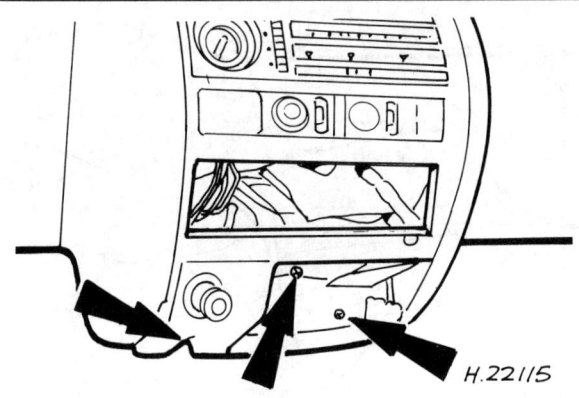

Fig. 11.30 Centre panel retaining screws (arrowed) (Sec 52)

51 Glove compartment lid – removal and refitting

1 Open the glovebox and remove the hinge retaining screws.
2 Disengage the arms on either end of the lid and remove the lid from the vehicle.
3 Refitting is a reversal of the removal procedure. Check the alignment of the lid on completion, and adjust as necessary.

52 Facia – removal and refitting

1 Disconnect the battery earth terminal.
2 Remove the upper and lower steering column shrouds, and the steering wheel (as Chapter 9).
3 Remove the two screws securing the instrument cluster bezel from its underside, and carefully detach the bezel (see Chapter 12).
4 Disconnect the steering column multi-function switch assembly and remove its single retaining screw, as described in Chapter 9, Section 6. Remove the assembly.
5 Disconnect the ignition loom multi-plug on the steering column.
6 Disconnect the brake pedal stop-lamp switch loom connection.
7 Disconnect the speedometer cable at the transmission casing, to allow easier removal of the instrument cluster (see Chapter 12).
8 Remove the four screws securing the instrument cluster to its location, then carefully pull it out to allow access to the speedometer cable and multi-plug connections. Disconnect the speedometer cable and multi-plug, then remove the instrument cluster from the vehicle (see Chapter 12).
9 Remove the radio assembly and loudspeaker balance control, where fitted (see Chapter 12).
10 Remove the centre console, where fitted, as described in Section 50.
11 Pull the heater fan motor control knob off, then move the air distribution and temperature controls fully to the right. Unclip and remove the heater slide facia towards the left-hand side of the vehicle, removing the slide control knobs only as necessary, and disconnecting its bulbholder (bayonet type) as it is withdrawn.
12 Remove the ashtray, then undo the three screws from the base of the centre panel (Fig. 11.30). Detach the centre panel, disconnecting the cigarette lighter connections as it is withdrawn.
13 Squeeze the two release tabs together on the heater fan motor control switch, and remove it, disconnecting its multi-plug as it is withdrawn (see Chapter 12). Remove the three heater control panel securing screws (see Chapter 2).
14 Remove the switches from the centre panel and disconnect their multi-plugs.
15 Using a thin flat-bladed screwdriver, prise the clock from its location and disconnect its multi-plug, as applicable.
16 Remove the fusebox lid, then remove the two retaining screws and detach the fusebox from the facia.
17 Disconnect the earth strap on the right-hand side of the steering column mounting bracket, by removing its securing bolt, and remove any cable ties fitted.
18 Open both front doors and disconnect the multi-plugs in the A-pillars, where fitted, by squeezing their ears and withdrawing.
19 Detach the door aperture weatherseals from the A-pillar and along the base of the door aperture, on both front doors.
20 Remove both front door courtesy light switches, disconnecting their loom connections as they are withdrawn.
21 Remove the sill scuff plate retaining screws on both sides of the vehicle.
22 Release the wiring loom from its securing clips, under the sill scuff plates.
23 Remove the right-hand adjustable side vent from the facia by carefully prising it out, using a thin flat-bladed screwdriver, then release the wiring loom loop from its facia retaining clip through the resultant opening (Fig. 11.31).
24 Prise up the cover obscuring the central facia retaining screw (photo), using a thin flat-bladed screwdriver then remove the seven facia retaining screws (Fig. 11.32).
25 Gently ease the facia from its location, having ensured that all wires are clear to move, then remove the cable ties securing the loom to the facia. Ensure that the loom is free, then remove the facia from the vehicle. The aid of an assistant, at this stage, is recommended.
26 Refitting is a reversal of the removal procedure, noting the following points. Secure the wiring loom loop to its clip and cable tie before refitting the facia retaining screws, tightening its cable tie, along with the rest, when the viewing loom connections have been pulled out through their relevant facia openings. New cable ties should be used. Ensure that the multi-plugs seat correctly in their A-pillar locations. When refitting the instrument cluster, ensure that the tape mark on the speedometer cable is positioned at the bulkhead grommet (as applicable).

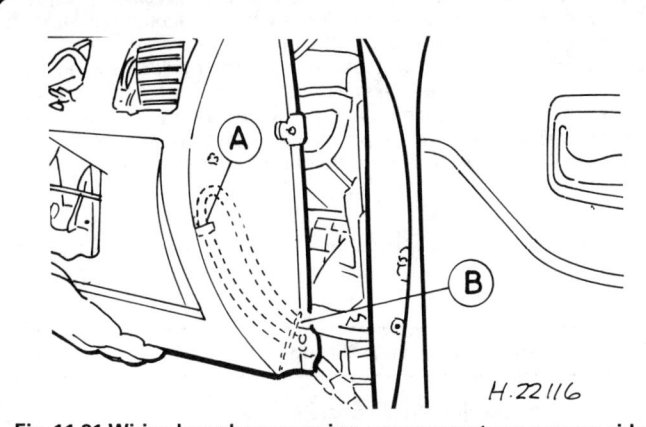

Fig. 11.31 Wiring loom loop securing arrangements on reverse side of facia (Sec 52)

A Retaining clip   B Cable tie

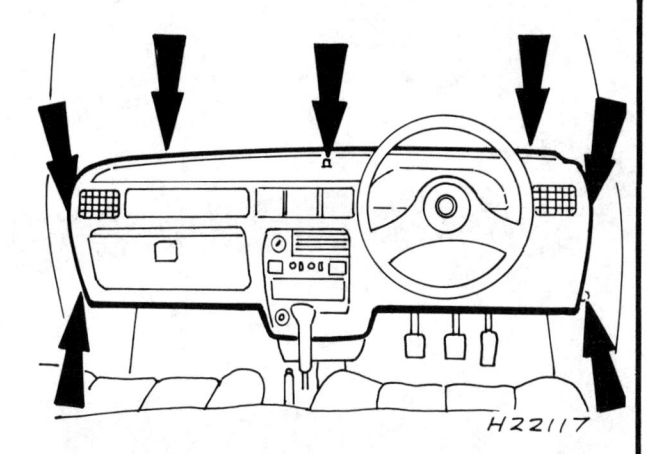

Fig. 11.32 Facia retaining screw locations (arrowed) (Sec 52)

252 Chapter 11 Bodywork and fittings

52.24 Central facia retaining screw cover

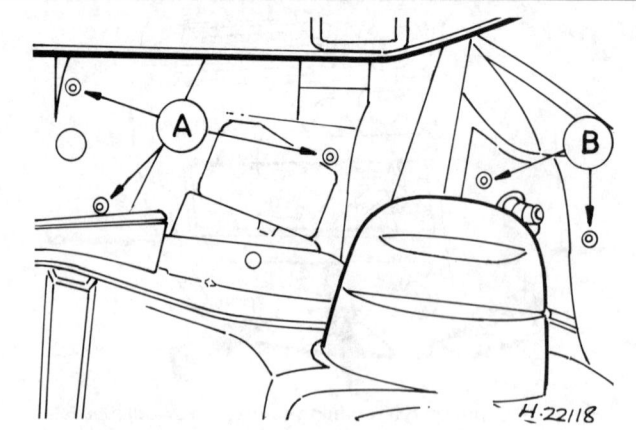

Fig. 11.33 Parcel shelf support/rear loudspeaker housing retaining screw locations (Sec 53)

A Hexagonal head fixings   B Phillips head screw fixings

53 Parcel shelf support/rear loudspeaker housing – removal and refitting

1 Remove the rear seat bet and retractor unit from the appropriate side of the vehicle, as described in Section 49.
2 Disconnect the luggage compartment (courtesy) lamp, where fitted, by prising the lamp assembly from its location using a thin flat-bladed screwdriver, then twist the bulbholder anti-clockwise to remove.
3 Detach the loudspeaker, where fitted, by removing its retaining screws, disengaging its locating tags and disconnecting its multi-plug.
4 Fold the seat backrest forward. Remove the parcel shelf support/rear loudspeaker housing retaining screws (Fig. 11.33), then manoeuvre it out from under the quarter panel trim as necessary, to clear the seat backrest catch striker pin.
5 Refitting is a reversal of the removal procedure (refer also to Section 49).

54 Passenger grab handle – removal and refitting

1 Carefully prise up the trim flaps on either end of the handle to expose the two mounting screws (photo).

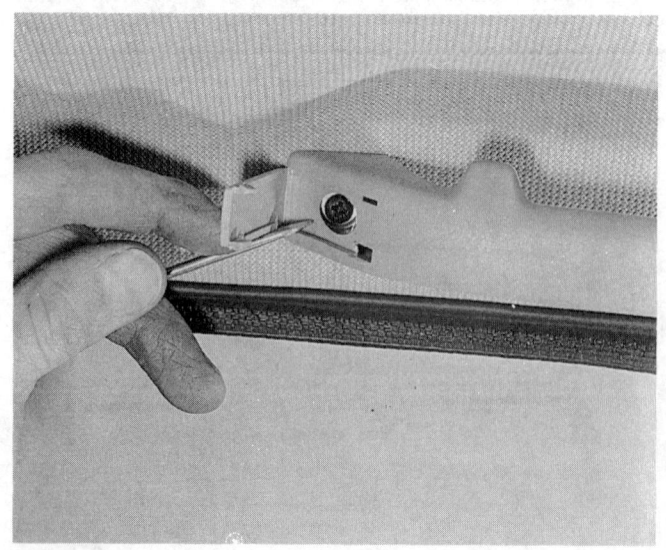

54.1 Prising up a trim flap to expose a passenger grab handle mounting screw

2 Undo the mounting screws and remove the grab handle.
3 Refitting is the reverse procedure to removal.

55 Rear view mirror – removal and refitting

1 Using a length of strong thin cord or fishing line, break the adhesive bond between the mirror base and the windscreen, then remove the mirror from the vehicle.
2 During installation, it is important to note that the mirror base, windscreen black patch and the adhesive patch should not be touched, other than for cleaning, or the adhesive bond may be adversely affected.
3 Remove all traces of old adhesive from the mirror base, using a lint-free cloth and methylated spirit, taking all normal safety precautions. Allow the mirror base to dry.
4 Clean the windscreen black patch in a similar manner.
5 Note that, before installing the mirror, the vehicle should have been in an ambient temperature of approximately 20°C (68°F) for at least an hour.
6 With the contact surfaces scrupulously clean, remove the protective tape from one side of the adhesive patch and press firmly into contact with the mirror base.
7 Note that when fitting a mirror to a new windscreen, the protective tape must be removed from the windscreen black patch.
8 Warm the mirror base and adhesive patch for about thirty seconds, to a temperature of 50 to 70°C (122 to 158°F).
9 Remove the protective tape from the other side of the adhesive patch on the mirror base, then align the mirror base and windscreen black patch accurately, and firmly press the mirror base into position. Hold the mirror base firmly in position for at least two minutes.
10 Wait at least thirty minutes before adjusting the mirror.

56 Sun visor – removal and refitting

1 Unclip the sun visor from its catch.
2 Remove the two screws securing the sun visor hinge plate to the body (photo), and remove the sun visor assembly from the vehicle.
3 The sun visor catch is removed by carefully prising open the trim flap covering its retaining screw, and removing the screw (photo).
4 Refitting is a reversal of the removal procedure.

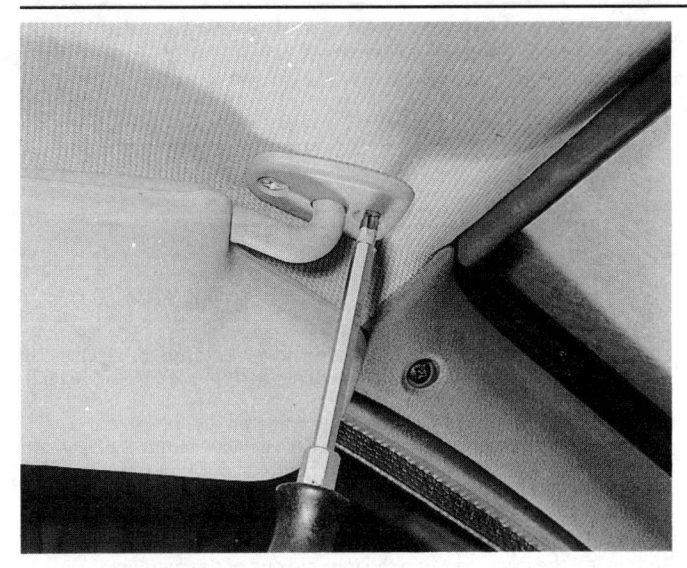

Chapter 11 Bodywork and fittings

56.2 Removing the sun visor hinge plate securing screws

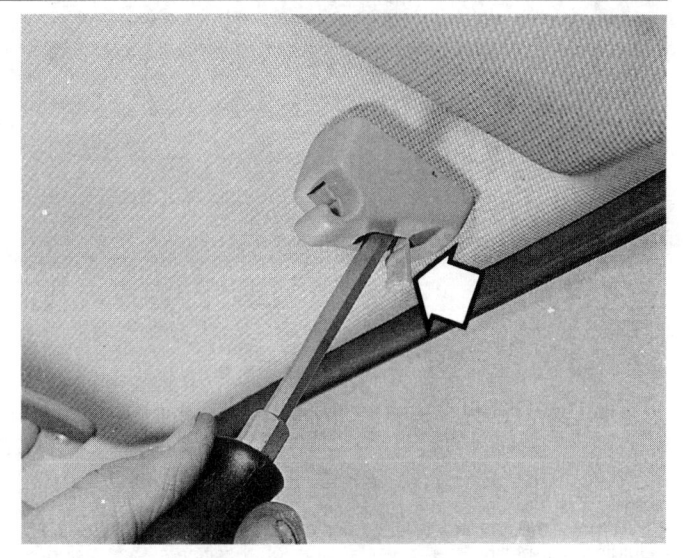

56.3 Removing the sun visor catch retaining screw (trim flap arrowed)

57 Spare wheel carrier – removal and refitting

1 Open the tailgate and fold back the luggage compartment floor covering to expose the carrier retaining hook bolt (photo). Slacken the bolt six to eight turns.
2 Disengage the carrier from its retaining hook, then lower the carrier and remove the spare wheel.
3 The forward ends of the carrier framework locate to brackets on the vehicle underside. Disengage them and remove the carrier from the vehicle.
4 Refitting is a reversal of the removal procedure.

58 Sill extension moulding – removal and refitting

1 The sill extension moulding fitted to XR2i models is secured with retaining studs and rivets, therefore a rivet gun will be required upon reassembly.
2 To remove, open the door and prise out the four retaining studs from the upper surface of the moulding.
3 From underneath, drill out the five securing rivets then remove the moulding from the vehicle.

4 To refit, first align the moulding to its location, centring it between the two wheelarch mouldings, then refit the four retaining studs to secure.
5 Insert the rivets to secure the moulding from underneath.

59 Wheelarch liner (front) – removal and refitting

1 Slacken the relevant roadwheel nuts.
2 With the handbrake on, raise the front of the vehicle and support it securely on axle stands.
3 Remove the roadwheel.
4 Release the fasteners securing the wheelarch liner in position (Fig. 11.34), then remove the wheelarch liner from the vehicle, manoeuvring it to clear obstructions as necessary.
5 Refitting is a reversal of the removal procedure, tightening the roadwheel nuts to the specified torque (see Chapter 10 Specifications).

60 Wheelarch moulding (front) – removal and refitting

1 Remove the wheelarch liner, as described in Section 59.
2 From underneath the wheelarch, remove the four fixing nuts securing the upper part of the moulding.

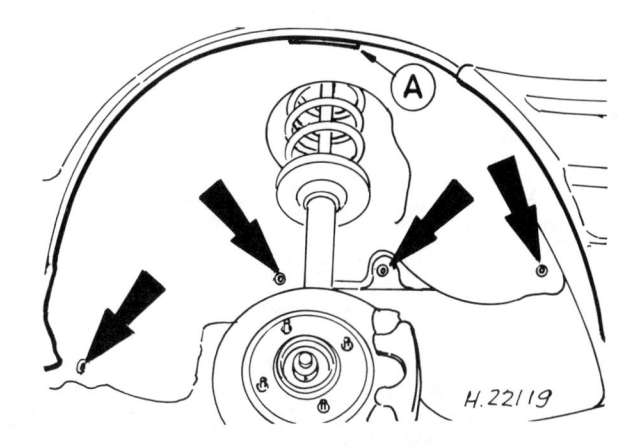

57.1 Spare wheel carrier retaining hook bolt (at rear of luggage compartment floor)

Fig. 11.34 Front wheelarch liner fixings (Sec 59)

A Locating lug at top of wheelarch

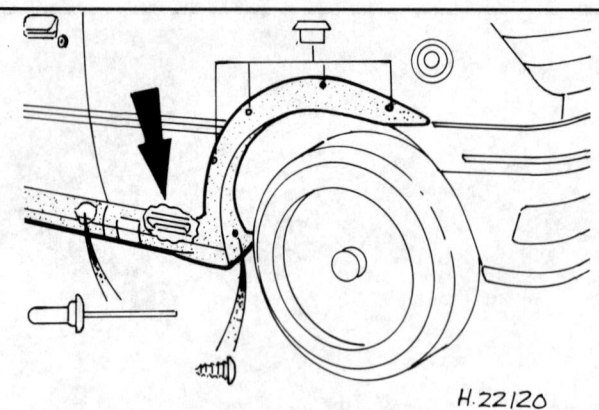

Fig. 11.35 Rear wheelarch moulding fixings (clamp cutaway arrowed) (Sec 61)

3 Remove the plastic stud from the lower edge of the wheelarch flange.
4 Remove the forward jacking position cover from the sill extension moulding, by pulling the lower section of the cover, then using a suitably-sized drill, remove the rivet securing the rear edge of the wheelarch moulding.
5 Carefully detach the wheelarch moulding from the vehicle, sliding its rear out from under the sill extension moulding.
6 Refitting is a reversal of the removal procedure, adjusting alignment as necessary before riveting the rear of the moulding. Tighten its fasteners to the specified torque.

61 Wheelarch moulding (rear) – removal and refitting

1 Remove the sill extension moulding, as described in Section 58.
2 Drill out the rivet securing the forward end of the wheelarch moulding (Fig. 11.35).
3 Remove the plastic stud from the lower edge of the wheelarch flange.
4 From underneath the wheelarch, remove the four fixing nuts securing the upper part of the moulding.
5 Carefully pull the wheelarch moulding away from the body, disengage it from the clamp, and remove.
6 To refit, engage the wheelarch moulding into the clamp, align the moulding studs with their wheelarch locations, then position the moulding onto the wheelarch. Refit the four fixing nuts to secure the upper part of the moulding, but do not fully tighten.

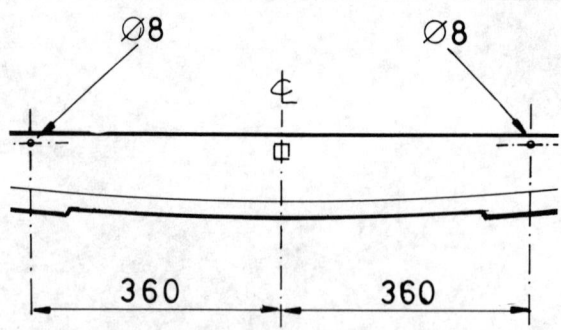

Fig. 11.36 Radiator grille slot fixing holes required on service replacement bumpers (dimensions given in mm – see text) (Sec 62)

7 Refit the plastic stud, but do not fully tighten.
8 Offer the sill extension moulding to its location, centring it between the front and rear wheelarches to check the rear wheelarch moulding alignment. Adjust the alignment as necessary.
9 With the rear wheelarch moulding alignment correct, fully tighten the fixing nuts and the plastic stud.
10 Insert the rivet to secure the forward end of the moulding.
11 Refit the sill extension moulding, as described in Section 58.

62 Wind deflector/radiator grille slat – removal and refitting

1 The radiator grille slat is secured to the front bumper by three clips. To remove it, simply slide the clips rearwards to release them, then withdraw the grille slat from its bumper locating holes.
2 If fitting a radiator grille slat to a service replacement bumper, two 8 mm (0.32 in) diameter holes will need to be drilled to accommodate the grille slat end fixings and, in addition, a 12 mm (0.5 in) square central hole must also be made (Fig. 11.36).
3 Refitting is a reversal of the removal procedure.

63 Load compartment dividers (Van models) – removal and refitting

1 The dividers are bolted into position. Undo the bolts securing them, then remove them separately from the vehicle.
2 Refitting is a reversal of the removal procedure.

Chapter 12 Electrical system

For modifications, and information applicable to later models, see Supplement at end of manual

Contents

Aerial removal and refitting	21
Alternator description	6
Alternator removal and refitting	9
Alternator testing in the car	8
Alternator brushes and regulator renewal	10
Alternator drivebelt removal, refitting and adjustment	7
Auxiliary lamps removal and refitting	28
Battery maintenance, testing and charging	4
Battery removal and refitting	5
Bulbs (exterior lamps) renewal	30
Bulbs (interior lamps) renewal	31
Central locking system description	34
Central locking system components removal and refitting	35
Choke warning light switch removal and refitting	50
Cigarette lighter removal and refitting	22
Clock removal and refitting	23
Electrical system precautions	2
Electrically operated windows description	32
Electrically operated windows removal and refitting	33
Exterior lamp units removal and refitting	27
Fault diagnosis	56
Fuel tank sender unit removal and refitting	52
Fuses and relays general	14
General description	1
Headlamps and auxiliary lamps beam alignment	29
Heated rear window element general	47
Heated windscreen general	48
Heater fan motor and resistor assembly removal and refitting	49
Horn removal and refitting	53
In-car entertainment equipment general	16
Inertia switch (fuel cut-off) removal and refitting	51
Instrument cluster removal and refitting	25
Instrument cluster components removal and refitting	26
Loudspeaker balance control joystick removal and refitting	20
Loudspeaker (front) removal and refitting	18
Loudspeaker (rear) removal and refitting	19
Maintenance and inspection	3
Radiator cooling fan motor removal and refitting	54
Radio or radio/cassette player removal and refitting	17
Radio equipment (non-standard) general	55
Speedometer cable removal and refitting	24
Starter motor overhaul	13
Starter motor removal and refitting	12
Starter motor testing in the car	11
Switches removal and refitting	15
Tailgate electrical contact switch components removal and refitting	36
Tailgate remote release motor removal and refitting	37
Tailgate washer jet and hose removal and refitting	44
Tailgate wiper motor removal and refitting	39
Windscreen/tailgate washer pump removal and refitting	45
Windscreen/tailgate washer reservoir removal and refitting	46
Windscreen/tailgate wiper blades and arms removal and refitting	38
Windscreen washer jets and hose removal and refitting	43
Windscreen wiper linkage removal and refitting	41
Windscreen wiper motor removal and refitting	40
Windscreen wiper pivot shaft renewal	42

Specifications

System type .... 12 volts, negative earth

Battery
Type .... 12 volt lead-acid (capacity dependent on model
Charge condition:
  Poor .... 12.4 volts or less
  Normal .... 12.6 volts
  Good .... 12.7 volts and above

Alternator

	Bosch	**Magneti Marelli**	**Mitsubishi**
Nominal rated output (13.5v at an engine speed of 6000 rpm)	55 A (K1-55A) 70 A (K1-70A)	55 A (A127/55) 70 A (A127/70)	55 A (A5T)

Minimum brush length (all types) .... 5.0 mm
Regulated voltage at 4000 rpm with 3 to 7 amp load (all types) .... 13.7 to 14.6 volts

Starter motor
Type .... Pre-engaged
Make and model .... Bosch DM (0.8, 0.9 or 1.0 kW)
Bosch DW (1.0 or 1.4 kW)
Bosch EV (2.2 kW)
Magneti Marelli M79 (0.8 or 0.9 kW)
Nippondenso (0.6 or 0.8 kW)

Starter motor (continued)

Number of brushes (all types)	4
Brush material (all types)	Carbon
Minimum brush length:	
Bosch and Magneti Marelli	8.0 mm
Nippondenso	10.0 mm
Minimum diameter of commutator:	
Nippondenso and all Bosch types	32.8 mm
Magneti Marelli	Not available at time of writing
Armature endfloat:	
Bosch	0.3 mm
Magneti Marelli	0.25 mm
Nippondenso	0.6 mm

Fuses

Circuits

Circuits	Fuse number	Fuse rating (amps)
Electronic engine control system	1	3
Interior light, cigarette lighter, clock and radio memory	2	15
Central locking system	3	20
Heated rear window element	4	30
Dim-dip lighting	5	10
Left-hand side lamps and rear foglamp (RHD)	6	10
Right-hand side lamps	7	10
Left-hand dipped beam	8	10
Right-hand dipped beam	9	10
Left-hand main beam and right-hand auxiliary driving lamp	10	15
Right-hand main beam and left-hand auxiliary driving lamp	11	15
Heater fan motor and reversing light	12	20
Radiator cooling fan motor	13	30
Front foglamps (XR2i only)	14	15
Horn	15	15
Wiper motor and windscreen/tailgate washer pump	16	20
Brake lights, instrument illumination and instrument warning	17	10
Electrically operated windows	18	30
Electric fuel pump	19	20
HEGO sensor (vehicle with catalytic converter)	20	10
Left-hand direction indicators	21	10
Right-hand direction indicators	22	10
Unused	23	
Unused	24/25	–
Tailgate remote release	26	15
Heated windscreen	27/28	30

Relays (on fusebox)

Relay notation	Circuit
I	Heated rear window
II	Windscreen wiper delay
III	Fuel injection/power delay
IV	Headlamp wash (not UK spec)
V	Ignition switch
VI	Automatic transmission/electric fuel pump (bridge fitted to carburettor engines without CTX automatic transmission)
VII	Main beam
VIII	Dim-dip lighting
IX	Heated windscreen (LHD versions)
X	Daytime running lights (not UK spec)
XI	Heated windscreen (RHD versions)
XII	Anti-lock braking system
A	Idle speed (CTX automatic transmission) two-tone horn (XR2i)
B	Heated seats (not UK spec)
C	Front foglamps (XR2i)
D	Dipped beam
E	Dim-dip sidelamps

Wiper blades

Front	Champion X-4803, with SP01 spoiler, where required
Rear	Champion X-4103

Bulbs

Headlamp (halogen)	H4, 60/55W
Sidelamp (front)	5W
Direction indicators (main)	21W
Side direction indicator repeaters	5W
Auxiliary driving and foglamps (S, SX and XR2i)	H3, 55W
Stop/tail lamp	21/5W
Rear foglamp	21W
Reversing lamp	21W

Chapter 12 Electrical system

Bulbs (continued)
Number plate lamp	10W
Interior lamp	10W
Luggage compartment lamp	5W
Instrument warning lamps	1.3 or 2.6W
Panel illumination	1.3 or 2.6W
Cigarette lighter illumination	1.4W
CTX selector cover illumination	2W

Lubricants
Grease for windscreen wiper linkage and pivots	To Ford specification SAM-1C-9111-A (refer to Ford dealer)

Torque wrench settings
	Nm	lbf ft
Starter motor	35 to 45	26 to 33
Wiper motor to bracket	8 to 10	6 to 7
Wiper motor bracket to bulkhead/tailgate	6 to 8	4 to 6
Windscreen wiper crank to driving shaft nut	22 to 24	16 to 18
Windscreen/tailgate wiper arm retaining nut (see text)	17 to 18	12 to 13
Windscreen wiper pivot shaft nut	10 to 12	7 to 9
Windscreen/tailgate washer reservoir securing bolts	2.5 to 3.5	2 to 3
Headlamp retaining bolt	5.4 to 7.0	4 to 5
Tail lamp securing nuts	1.5 to 2.5	1 to 2
Auxiliary lamp retaining nut (S models)	6.8 to 9.2	5 to 7
Horn bracket retaining bolt	24 to 33	18 to 24
Starter inhibitor switch	10 to 14	7 to 10
Temperature gauge sender:		
HCS engine	4 to 8	3 to 6
CVH engine	5 to 7	4 to 5
Reversing light switch	16 to 20	12 to 15

1 General description

The electrical system is of the 12 volt negative earth type, and consists of a 12 volt battery, alternator, starter motor and related electrical accessories, lighting equipment and a host of minor components. The battery is charged by an alternator, which is belt-driven from the crankshaft pulley. The starter motor is of the pre-engaged type, incorporating an integral solenoid. On starting the solenoid moves the drive pinion into engagement with the flywheel ring gear before the starter motor is energised. Once the engine has started, a one-way clutch prevents the motor armature being driven by the engine until the pinion disengages from the flywheel.

When servicing starter motors and/or alternators, note that Magneti Marelli units are simply renamed Lucas components; either name may be found on units fitted as standard or supplied as a replacement. All procedures, replacement parts and exchange or replacement units are completely interchangeable.

Further details of the major electrical systems are given in the relevant Sections of this Chapter. For the reader whose interests extend beyond component renewal, a copy of *'Automobile Electrical and Electronic Systems'* is available from the publishers of this book.

Caution: *Before carrying out any work on the vehicle electrical system, read through the precautions given in 'Safety First!' at the beginning of this manual and in Section 2 of this Chapter.*

2 Electrical system – precautions

1 It is necessary to take extra care when working on the electrical system to avoid damage to semi-conductor devices (diodes and transistors), and to avoid the risk of personal injury. In addition to the precautions given in *'Safety first!'* at the beginning of this manual, observe the following items when working on the system.

2 *Always remove rings, watches, etc before working on the electrical system.* Even with the battery disconnected, capacitive discharge could occur if a component live terminal is earthed through a metal object. This could cause a shock or nasty burn.

3 *Do not reverse the battery connections.* Components such as the alternator or any other having semi-conductor circuitry could be irreparably damaged.

4 If the engine is being started using jump leads and a slave battery, connect the positive terminals of both batteries with one of the jump leads. Connect one end of the remaining jump lead to the negative (earth) terminal of the slave battery, and the other end to a good earth on the vehicle to be started – not the negative (earth) terminal on the discharged battery.

5 Never disconnect the battery terminals, or alternator multi-plug connector, when the engine is running.

6 The battery leads and alternator multi-plug must be disconnected before carrying out any electric welding on the car.

7 Never use an ohmmeter of the type incorporating a hand cranked generator for circuit or continuity testing.

8 When using a battery charger, always ensure that the battery is connected in accordance with the battery charger manufacturer's instructions.

3 Maintenance and inspection

1 At regular intervals (see *Routine maintenance*) carry out the following maintenance and inspection operations on the electrical system components.

2 Check the operation of all the electrical equipment, ie wipers, washers, lights, direction indicators, horn, etc. Refer to the appropriate Sections of this Chapter if any components are found to be inoperative.

3 Visually check all accessible wiring connectors, harnesses and retaining clips for security, or any signs of chafing or damage. Rectify any problems encountered.

4 Check the alternator drivebelt for cracks, fraying or other damage. Renew the belt if worn. If the bolt is in satisfactory condition check its tension, adjusting as necessary in accordance with the relevant Section of this Chapter.

5 Check the condition of the wiper blades. If they are cracked or show signs of deterioration, renew them in accordance with the relevant Section of this Chapter.

6 It is recommended that headlamp and auxiliary lamp aim be adjusting using optical beam setting equipment. This work should be entrusted to a Ford dealer.

7 Top up the windscreen/tailgate washer reservoir, and check the security of the pump multi-plug and washer hoses.

8 While carrying out a road test, check the operation of all the instruments and warning lights, including the operation of the direction indicator self-cancelling mechanism.

4 Battery – maintenance, testing and charging

Maintenance
1 Check the battery terminals, and if there is any sign of corrosion disconnect and clean them thoroughly. Smear the terminals and battery posts with petroleum jelly before refitting the plastic covers. If there is any corrosion on the battery tray, remove the battery, clean the deposits away and treat the affected metal with an anti-rust preparation. Repaint the tray in the original colour after treatment.
2 The maintenance-free type battery does not require routine topping-up with distilled water. The only maintenance requirement is to ensure that the cable terminals are secure and corrosion-free.
3 Where the original battery has been replaced by a low-maintenance type, the electrolyte level should be maintained just above the tops of the cells, or up to the mark on the battery case where applicable. If topping-up is necessary, distilled water must be used.

Testing
4 The condition of a maintenance-free type battery is tested by connecting a voltmeter between the terminals, and comparing the reading obtained with the charge condition figures in the Specifications. This test is only accurate if the battery is in a stable condition, with no charge or discharge having taken place for at least 6 hours. If the battery has been charged, it can be stabilised before the test by switching the headlights on for 30 seconds, then waiting 4 to 5 minutes after switching the headlights off (with all other electrical components switched off). Charging information, if required, is given later in this Section.
5 If the vehicle covers a small annual mileage and is fitted with a battery type other than of maintenance-free design, it is worthwhile checking the specific gravity of the electrolyte every three months to determine the state of charge of the battery. Use a hydrometer to make the check and compare the results with the following table.

	Ambient temperature above 25°C (77°F)	Ambient temperature below 25°C (77°F)
Fully charged	1.210 to 1.230	1.270 1.290
70% charged	1.170 to 1.190	1.230 1.250
Fully discharged...	1.050 to 1.070	1.110 1.130

Note that the specific gravity readings assume an electrolyte temperature of 15°C (60°F; for every 10°C (18°F) below 15°C (60°F) subtract 0.007. For every 10°C (18°F) above 15°C (60°F) add 0.007. If the battery condition is suspect first check the specific gravity or electrolyte in each cell. A variation of 0.040 or more between any cells indicates loss of electrolyte or deterioration of the internal plates. If the specific gravity variation is 0.040 or more, the battery should be renewed. If the cell variation is satisfactory but the battery is discharged, it should be charged as described later in this Section.
6 Whichever type of battery is fitted, remove it from the vehicle (Section 5) if charging is required.

Charging
Note: *Battery charging must be carried out in a well-ventilated area.*
7 The maintenance-free type battery takes considerably longer to fully recharge than the standard type, the time taken being dependent on the extent of discharge, but it can take anything up to three days. A constant voltage type charger is required, to be set, when connected, to 13.9 to 14.9 volts with a charger current below 25 amps. Using this method the battery should be usable within three hours, giving a voltage reading of 12.5 volts, but this is for a partially discharged battery and, as mentioned, full charging can take considerably longer. If the battery is to be charged from a fully discharged state (condition reading less than 12.2 volts) have it recharged by your Ford dealer or local automotive electrician as the charge rate is higher and constant supervision during charging is necessary.
8 Batteries other than of maintenance-free design should be charged at a rate of 3.5 to 4.0 amps, over a four hour period, until no further rise in specific gravity is noted. Alternatively, a trickle charger charging at the rate of 1.5 amps can be safely used overnight. Specially rapid 'boost' charges which are claimed to restore the power of the battery in 1 to 2 hours are not recommended as they can cause serious damage to the battery plates through overheating. While charging the battery note that the temperature of the electrolyte should never exceed 37.8°C (100°F).

5 Battery – removal and refitting

1 The battery is located forward on the left-hand side of the engine compartment, on a platform welded to the vehicle structure.
2 Disconnect the leads from the battery terminals, removing the earth (negative) first.
3 Release the clamp securing the battery to its platform and remove it. Lift the battery from its location, keeping it in an upright position to avoid the possibility of corrosive electrolyte spilling onto the paintwork.
4 Refitting is a reversal of the removal procedure. Smear the battery terminals with petroleum jelly when refitting, and always connect the positive lead first and the earth (negative) lead last. Ensure that the battery is located securely.

Note: *Vehicles fitted with the EEC-IV (electronic engine control) module will lose the 'keep alive memory' (KAM) information stored in the module when the battery is disconnected. This includes idling and operating valves, and any fault codes detected. This may cause surge, hesitation, erratic idle or a generally deteriorated standard of performance. To allow the module to 're-learn' its values, on subsequent reconnection of the battery, the vehicle should be allowed to idle for three minutes at normal operating temperature. The engine speed should be increased to 1200 rpm and maintained at this level for approximately two minutes. It may be necessary for the vehicle to be driven to allow the module to complete its 're-learning' phase – the distance required is dependent upon the type of driving encountered, but is normally about five miles. In addition, after reconnecting the battery, the 'Keycode' audio unit (if fitted, see Section 17) and clock will need to be reset.*

6 Alternator – description

1 One of a number of different makes of alternator may be fitted, dependent upon model and engine capacity. The maximum output of the alternator varies similarly.
2 The alternator is belt-driven from the crankshaft pulley, it is fan cooled and incorporates a voltage regulator.
3 The alternator provides a charge to the battery at very low engine revolutions and basically consists of a stator in which a rotor rotates. The rotor shaft is supported in bearings, and slip rings are used to conduct current to and from the field coils through carbon brushes.
4 The alternator generates ac (alternating current) which is rectified by an internal diode system to dc (direct current) which is rectified by an internal diode system to dc (direct current) which is the type of current needed for battery storage.

7 Alternator drivebelt – removal, refitting and adjustment

1 Dependent on model, remove the air filter housing trunking (see Chapter 3) then raise the front of the vehicle and support it securely using axle stands.
2 The alternator, dependent on type/model, has either a single pivot bolt or a dual pivot bolt securing arrangement. This is accessible from underneath the vehicle, and should be slackened off; a drivebelt shield may be fitted – undo its three bayonet type fasteners and remove it (photos).
3 Slacken off the alternator adjuster bolt (photo) and tilt the alternator to release the drivebelt tension. If necessary, remove the adjuster bolt. Remove the drivebelt from its pulley locations.
4 If the conventional vee drivebelt is to be re-used, examine it carefully for signs of wear, damage or deterioration. If a new drivebelt is readily available, it is advisable to fit this – the old drivebelt may be kept for emergency use.

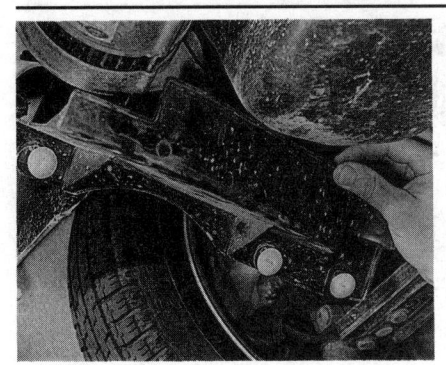

Chapter 12 Electrical system

7.2A Removing the alternator driveshaft shield (where fitted)

7.2B Alternator pivot bolt arrangement (alternator released from operating position to expose mounting bracket bolts)

A Pivot bolts
B Mounting bracket to engine bolts

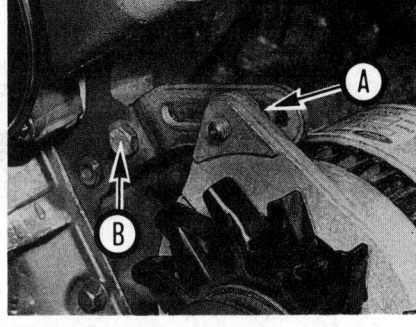

7.3 Alternator adjuster position

A Adjuster bolt
B Adjuster slide bracket to engine bolt

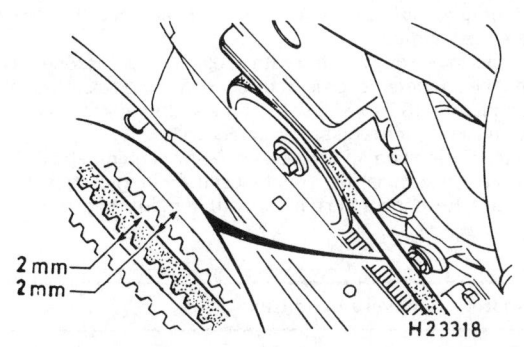

Fig. 12.1 Adjusting alternator drivebelt tension (Sec 7)

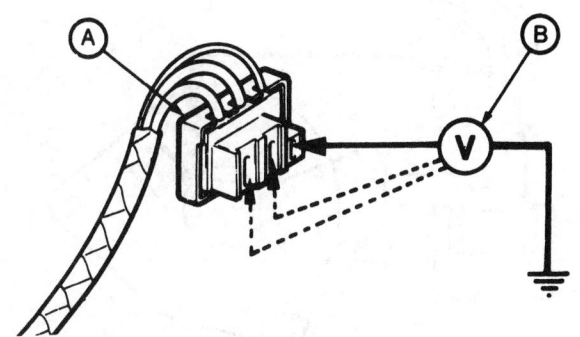

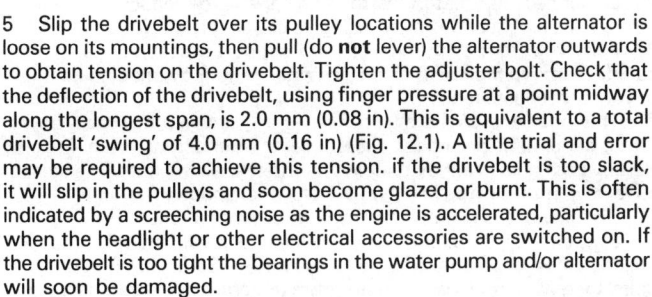

Fig. 12.2 Testing charging circuit continuity (Sec 8)

A Wiring connections   B 0 to 20 volt voltmeter

5 Slip the drivebelt over its pulley locations while the alternator is loose on its mountings, then pull (do **not** lever) the alternator outwards to obtain tension on the drivebelt. Tighten the adjuster bolt. Check that the deflection of the drivebelt, using finger pressure at a point midway along the longest span, is 2.0 mm (0.08 in). This is equivalent to a total drivebelt 'swing' of 4.0 mm (0.16 in) (Fig. 12.1). A little trial and error may be required to achieve this tension. if the drivebelt is too slack, it will slip in the pulleys and soon become glazed or burnt. This is often indicated by a screeching noise as the engine is accelerated, particularly when the headlight or other electrical accessories are switched on. If the drivebelt is too tight the bearings in the water pump and/or alternator will soon be damaged.

6 When the desired tension is achieved, tighten the pivot bolt(s) and refit the drivebelt shield (where fitted), then lower the vehicle to the ground.

7 If a new drivebelt has been fitted, the tension should be rechecked and, if necessary, adjusted again after the engine has run for at least ten minutes.

8 Refit the air filter housing trunking if removed (see Chapter 3).

8 Alternator – testing in the car

Note: *To carry out the complete test procedure, a 0 to 20 volt voltmeter/multimeter, a 0 to 1 volt voltmeter/multimeter, a 100 amp (min) ammeter/multimeter and a 30 amp (min) load rheostat will be required.*

1 Check that the drivebelt tension is correct (Section 7), and that the battery is well charged (Section 4), before conducting any of the following tests.

Wiring continuity test
2 Disconnect the battery earth terminal, then disconnect the wiring loom multi-plug from the alternator.
3 Reconnect the battery earth terminal, switch the ignition on and connect the 0 to 20 volt voltmeter to a good earth point.
4 Check the voltage reading on each of the multi-plug terminals in turn. The voltmeter should indicate battery voltage in all cases – a zero reading indicates an open-circuit in the wiring.

Alternator output test
5 Connect the ammeter, rheostat and 0 to 20 volt voltmeter, as shown in Fig. 12.3.
6 Switch on the headlamps, heater fan motor and heated rear window, then start the engine and run it at 3000 rpm. Vary the resistance to increase the current loading, and check that the alternator rated output is reached without the voltage dropping below 13.5 volts.
7 Switch off the ignition, headlamps, heater fan motor and heated rear window, then disconnect the test equipment.

Positive side voltage drop test
8 Connect the test equipment, as shown in Fig. 12.4
9 Switch on the headlamps, heater fan motor and heated rear win-

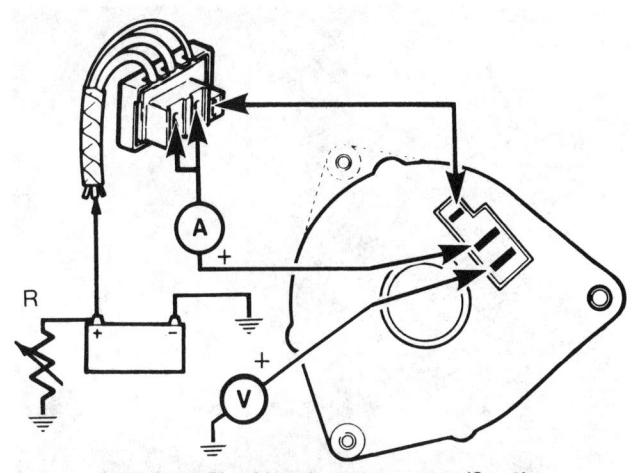

Fig. 12.3 Checking alternator output (Sec 8)

A Ammeter   R Rheostat
V Voltmeter

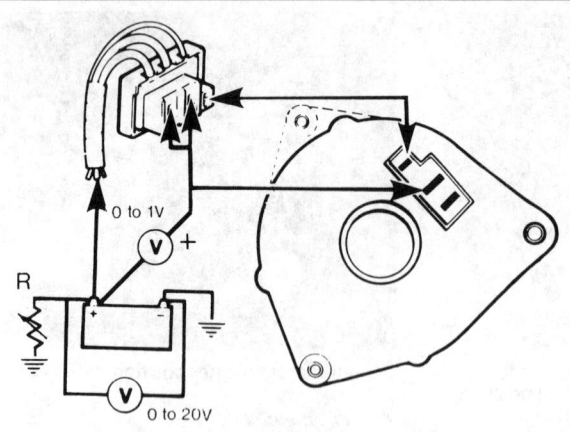

Fig. 12.4 Checking charging circuit voltage drop – positive side (Sec 8)

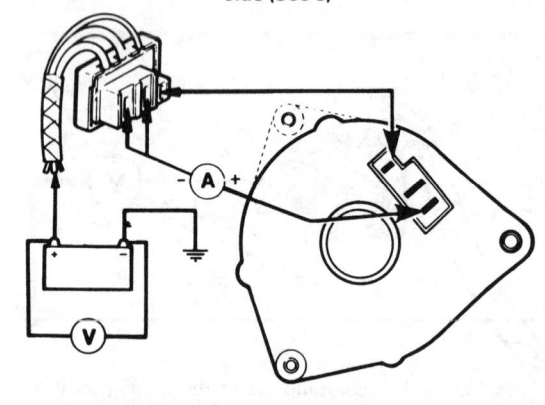

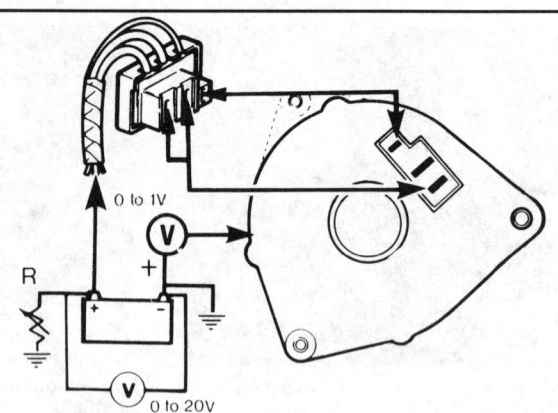

Fig. 12.5 Checking charging circuit voltage drop – negative side (Sec 8)

Fig. 12.6 Checking regulator control voltage (Sec 8)

dow, then start the engine and run it at 3000 rpm. Adjust the resistance to give 13.0 volts across the battery, then measure the voltage drop on the 0 to 1 volt voltmeter. If the voltage drop indicated is in excess of 0.5 volts, this means that there is a high resistance in the positive side of the charging circuit – this must be located and rectified.
10   Switch off the ignition, headlamps, heater fan motor and heated rear window.

Negative side voltage drop test

11   Proceed as described in paragraphs 9 and 10, having connected the test equipment as shown in Fig. 12.5. A voltage drop in excess of 0.25 volts means that there is a high resistance on the negative side of the charging circuit – this must be located and rectified.

Alternator voltage regulator test

12   Referring to Fig. 12.6, connect the ammeter and the 0 to 20 volt voltmeter as shown.
13   Run the engine at 4000 rpm with all electrical accessories switched off. When the ammeter records a current of 3 to 7 amps, check that the voltmeter records 13.7 to 14.6 volts. If the voltmeter reading is outside the specified limits, the regulator is defective.
14   Switch the ignition off and detach the test equipment. Disconnect the battery earth terminal, then reconnect the wiring loom multi-plug to the alternator. Reconnect the battery earth terminal to complete.

9   Alternator – removal and refitting

1   The operations are similar for each type of alternator.
2   Disconnect the battery earth terminal and, dependent on model, remove the air filter housing trunking (see Chapter 3).
3   Raise the front of the vehicle and support it securely using axle stands. Remove the drivebelt shield (where fitted) by undoing its three bayonet type fasteners.
4   The inner end of the alternator has a heatshield or splash cover fitted, dependent on model, accessible from underneath the vehicle. Remove the heatshield by undoing its three retaining nuts. The splash cover is removed by pulling it from its location (photo).
5   Disconnect the wiring loom from the alternator (photo).
6   Slacken the alternator adjuster bolt and its pivot bolt/bolts, then tilt the alternator to facilitate drivebelt removal (See Section 7). Remove the bolts, then withdraw the alternator from underneath the vehicle.
7   Refitting is a reversal of the removal procedure, adjusting the drivebelt tension as described in Section 7.

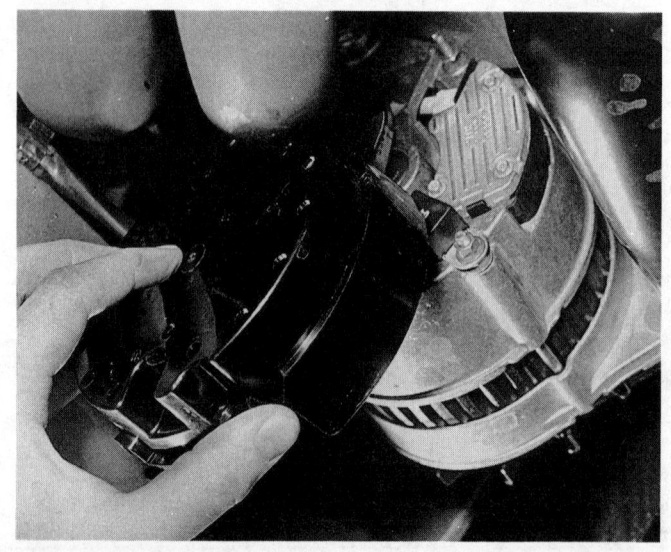

9.4 Removing the splash cover from the inner end of the alternator

9.5 Wiring loom routeing on rear of alternator (CVH engine)

261

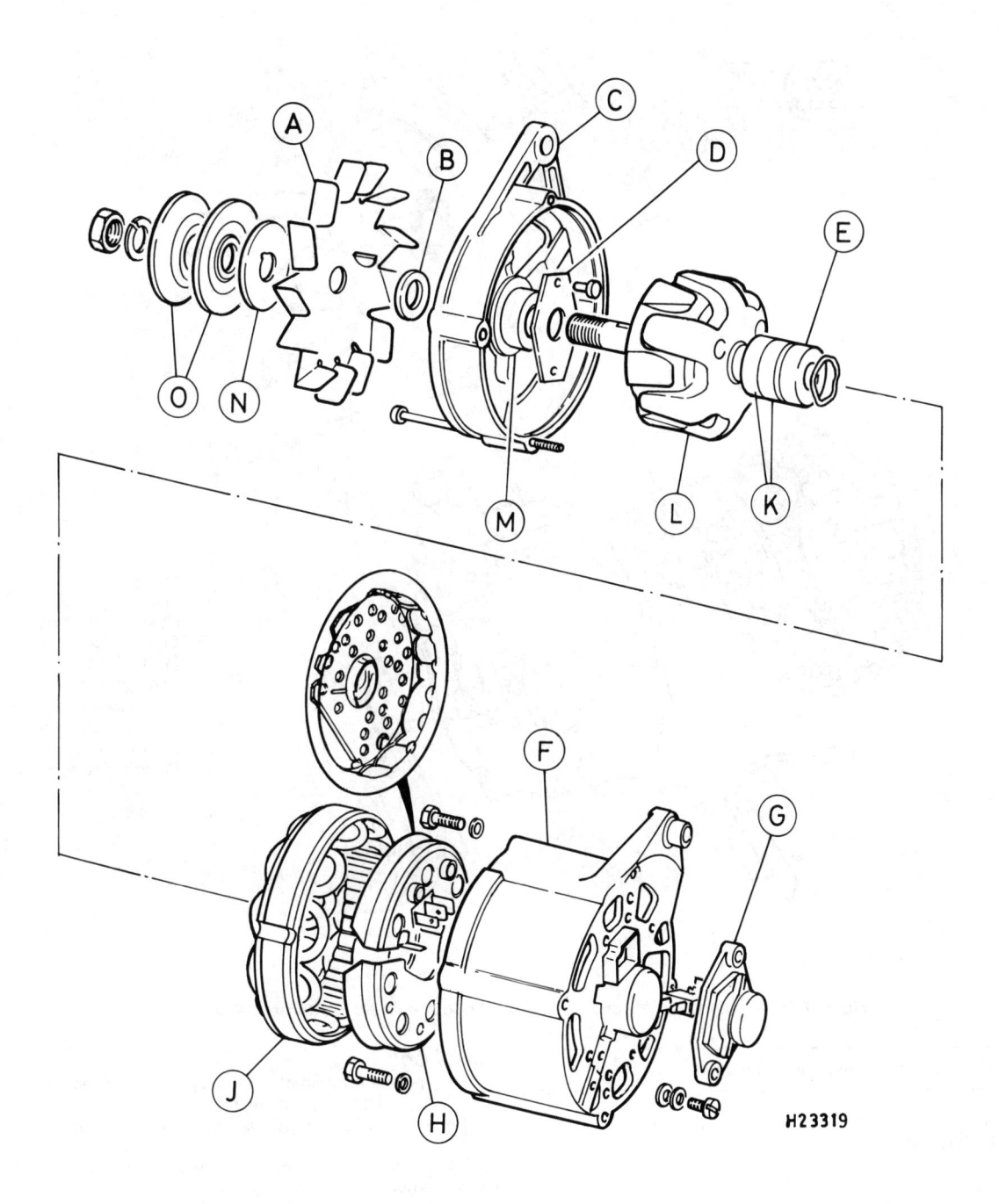

Fig. 12.7 Exploded view of Bosch K1-55A and K1-70A alternators (Sec 10)

- A Fan
- B Spacer
- C Drive end housing
- D Drive end bearing retaining plate
- E Slip ring end bearing
- F Slip ring end housing
- G Brushbox and regulator
- H Rectifier (diode) pack
- J Stator
- K Slip rings
- L Rotor
- M Drive end bearing
- N Spacer
- O Pulley

262　Chapter 12　Electrical system

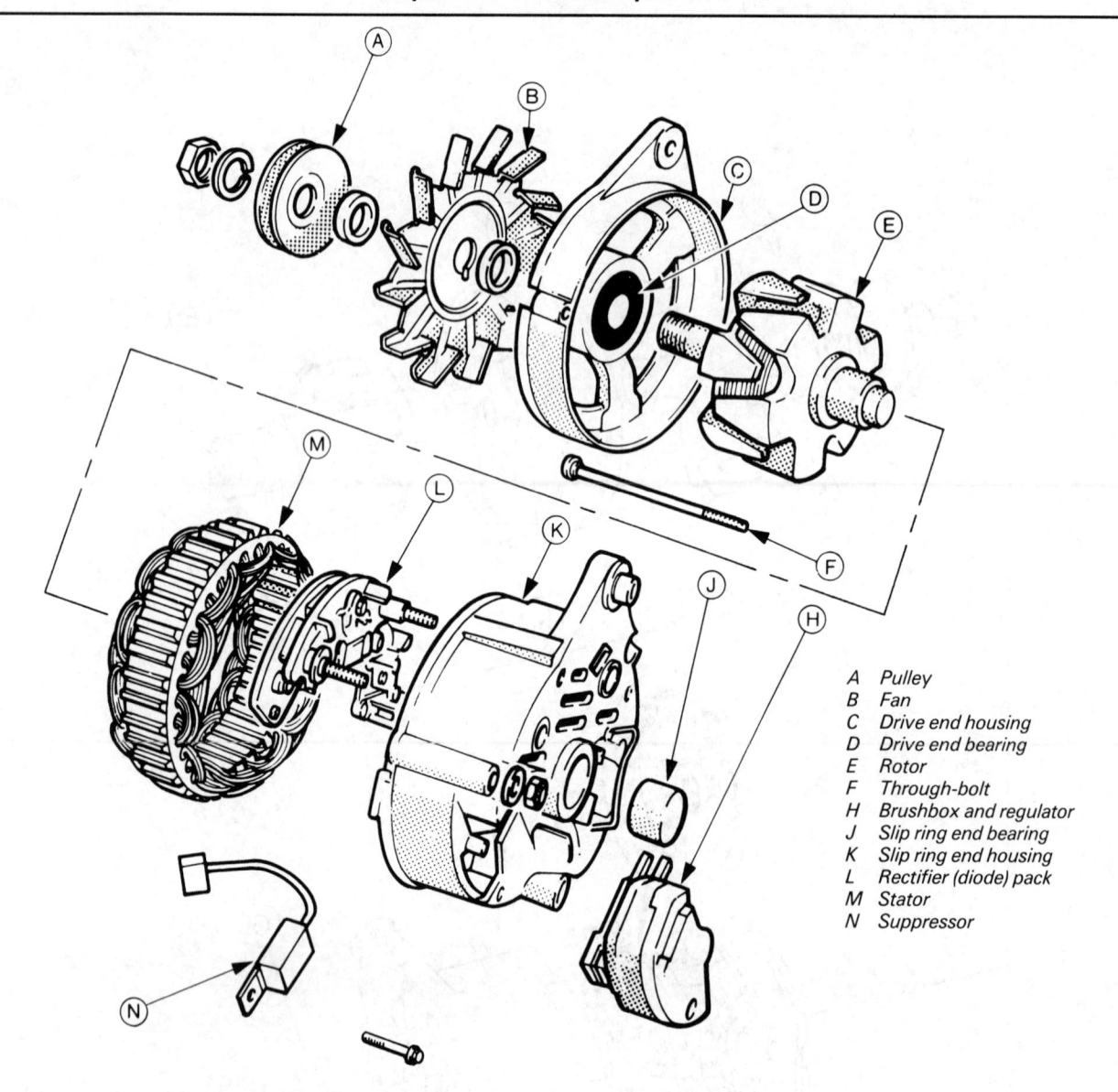

Fig. 12.8 Exploded view of Magneti Marelli A127/55 and A127/70 alternators (Sec 10)

A　Pulley
B　Fan
C　Drive end housing
D　Drive end bearing
E　Rotor
F　Through-bolt
H　Brushbox and regulator
J　Slip ring end bearing
K　Slip ring end housing
L　Rectifier (diode) pack
M　Stator
N　Suppressor

10　Alternator brushes and regulator – renewal

1　With the alternator removed from the vehicle, as described in the previous Section, clean its external surfaces.

Bosch

2　Remove the screws securing the regulator and brush box assembly, and withdraw the assembly from the rear of the alternator.
3　Check the brush length. If either is less than, or close to, the specified minimum, renew them by unsoldering the brush wiring connectors and removing the brushes and springs.
4　Refit by reversing the removal operations.

Magneti Marelli

5　Remove the screws securing the regulator and brushbox assembly, then partially withdraw the assembly, disconnect its field connector and remove it from the rear of the alternator (photos).
6　If the brushes are worn beyond the minimum length specified, a new regulator and brushbox assembly must be obtained, as the brushes are not serviced separately.
7　Refit by reversing the method of removal.

Mitsubishi

8　Holding the rotor shaft stationary with an 8 mm hexagonal Allen key, unscrew the pulley nut and remove the washer.
9　Remove the pulley, cooling fan, large spacer and dust shield from the rotor shaft.
10　Mark the front housing, stator and rear housing so that they may be reassembled in their original relative positions.
11　Unscrew the through-bolts, then remove the front housing from the rotor shaft, followed by the dust seal and thin spacer.
12　Remove the rotor from the rear housing and stator assembly. If difficulty is experienced, heat the rear housing with a 200 watt soldering iron for 3 or 4 minutes (Fig. 12.10).
13　Unbolt the rectifier (diode pack)/brush box and stator assembly from the rear housing.
14　Unsolder the stator and brush box from the rectifier (diode pack), using the minimum heat assembly. Use a pair of pliers as a heat sink to reduce the spread of heat to the diodes – overheating may cause diode failure.
15　Renew the brush box and brushes if they are worn below the specified minimum.
16　Refit by reversing the removal operations. Insert a suitable piece of wire through the access hole in the rear housing, to hold the brushes in the retracted position as the rotor is refitted. **Do not** forget to release the brushes (see Fig. 12.11).

263

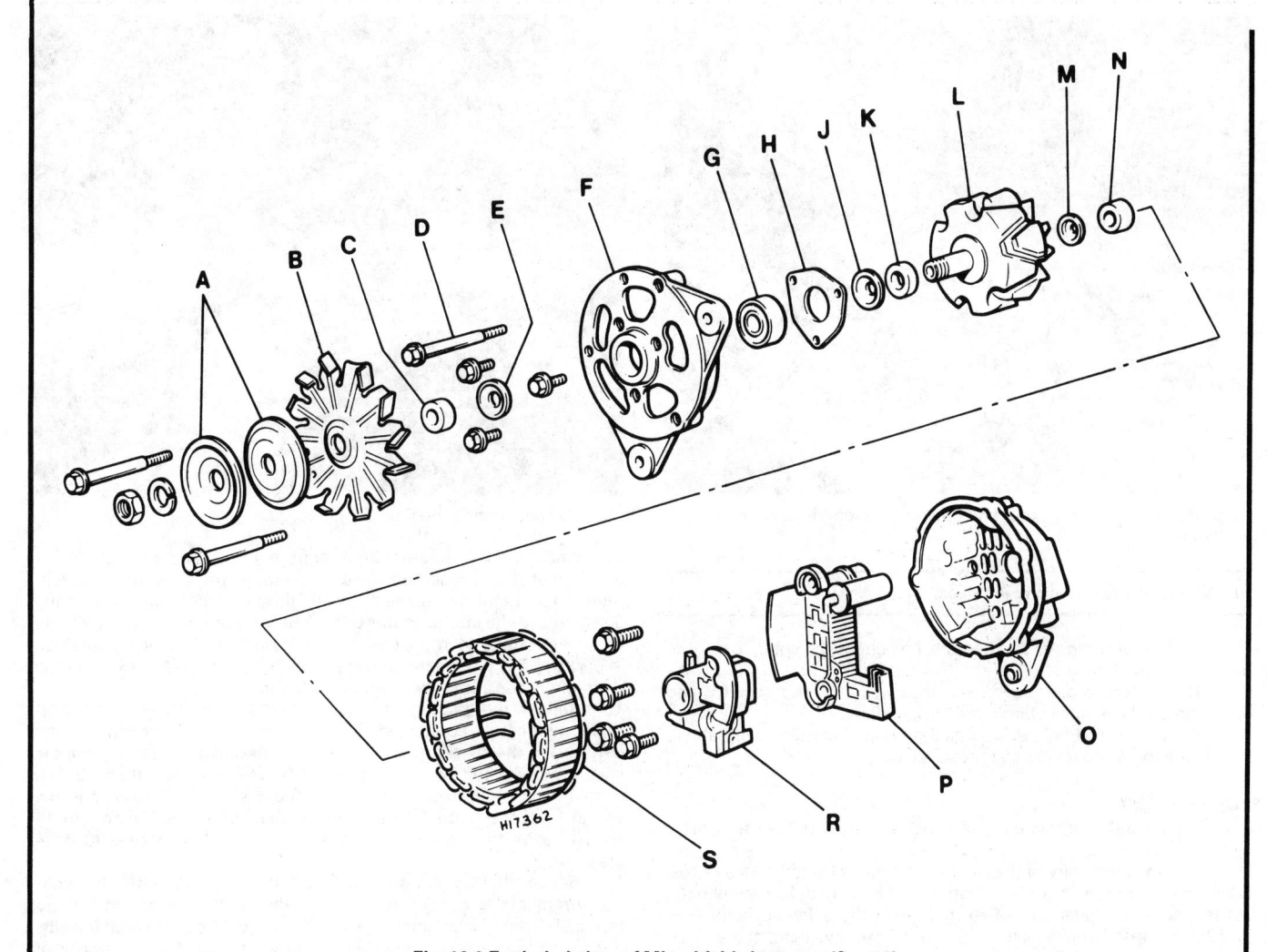

Fig. 12.9 Exploded view of Mitsubishi alternator (Sec 10)

A Pulley	F Drive end housing	K Thin spacer	O Slip ring end housing
B Fan	G Bearing	L Rotor	P Diode pack
C Thick spacer	H Bearing retainer	M Seal	R Brush box
D Through-bolt	J Dust cap	N Bearing	S Stator
E Dust shield			

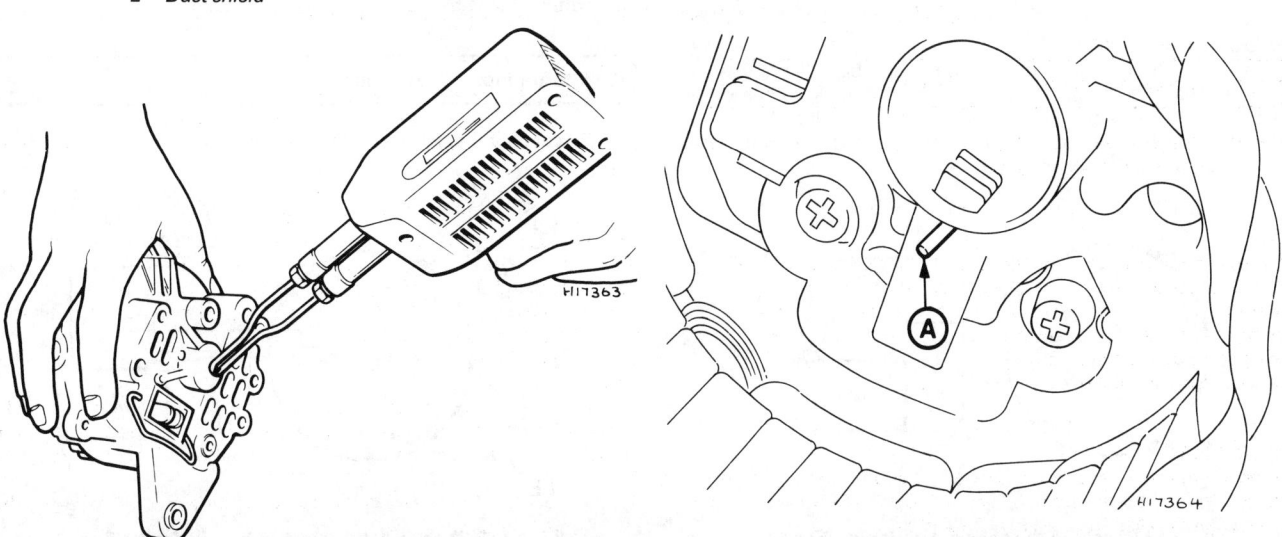

Fig. 12.10 Using a soldering iron to heat the slip ring end housing, as necessary – Mitsubishi alternator (Sec 10)

Fig. 12.11 Using a length of wire (A) to hold the brushes in the retracted position – Mitsubishi alternator (Sec 10)

Chapter 12 Electrical system

10.5A Removing the regulator and brushbox assembly securing screws

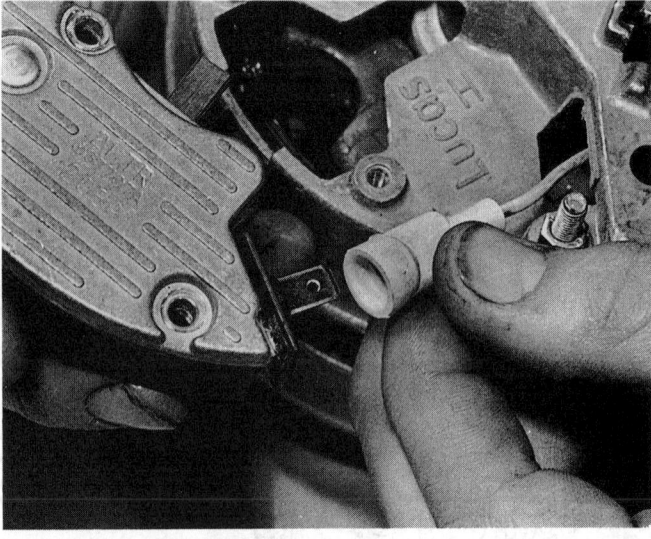

10.5B Disconnecting the field connector

11 Starter motor – testing in the car

1 If the starter motor fails to operate, first check the condition of the battery (see Section 4).
2 If the battery proves to be satisfactory, check the security and condition of all relevant cables (earth included).
3 If the starter still fails to operated, having ascertained the above points, the following tests must be carried out.

Solenoid test

4 Disconnect the battery earth terminal, and all leads from the solenoid.
5 Check the continuity of the solenoid windings by connecting a test lamp circuit, comprising a 12 volt battery and a 3 watt bulb, between the starter motor terminal on the solenoid, and the solenoid body (Fig. 12.12). The test lamp should light – if it does not light, there is an open-circuit in the solenoid windings.
6 Construct a new test circuit, as shown in Fig. 12.13, with an 18 watt bulb connected between the solenoid starter motor and battery terminals. Energise the solenoid with a further lead to its switch terminal. The solenoid should be heard to operate, and the test lamp should light.
7 Disconnect the test lamp circuit and reconnect the solenoid leads and battery earth terminal.

On-load voltage test

8 Connect a voltmeter/multimeter (0 to 20 volts) directly between the battery terminals. Disconnect the coil positive terminal to immobilise the ignition, then operate the starter motor by turning the ignition switch. The reading on the voltmeter should be not less than 10.5 volts.

Disconnect the voltmeter/multimeter from the battery terminals.
9 Repeat the procedure given in paragraph 8, with the voltmeter/multimeter connected between the starter motor terminal on the solenoid and the starter motor body. The voltmeter/multimeter should register a voltage not more than 1.0 volt lower than the reading obtained in paragraph 8. If a greater voltage drop is present, a fault exists in the wiring from the battery to the starter motor.
10 Repeat the procedure given in paragraph 8, with the voltmeter/multimeter connected between the positive terminal on the battery and the starter motor terminal on the solenoid. Battery voltage should be indicated initially, dropping down to a level less than 1.0 volt. If the reading is in excess of 1.0 volt, there is a high resistance in the wiring from the battery to the starter, and the test in paragraph 11 should be made. If the reading is less than 1.0 volt, proceed to paragraph 12.
11 Repeat the procedure given in paragraph 8, with the voltmeter/multimeter connected between the starter motor and battery terminals on the solenoid. Battery voltage should be indicated initially, dropping down to a level less than 0.5 volts. If the reading is in excess of 0.5 volts, the ignition switch and/or connections may be faulty.
12 Repeat the procedure given in paragraph 8, with the voltmeter/multimeter connected between the earth terminal on the battery and the starter motor body. If the earth return is satisfactory, the reading should not be in excess of 0.5 volts. If the reading is higher, this may be due to a dirty, loose or corroded terminal, and the earth connections to the battery and vehicle body should be checked.

12 Starter motor – removal and refitting

1 Disconnect the battery earth terminal.

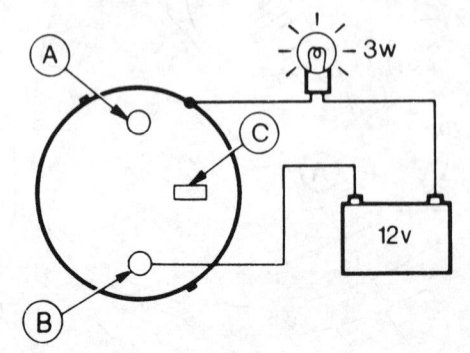

Fig. 12.12 Checking the solenoid windings (Sec 11)

A Battery terminal C Switch terminal
B Starter motor terminal

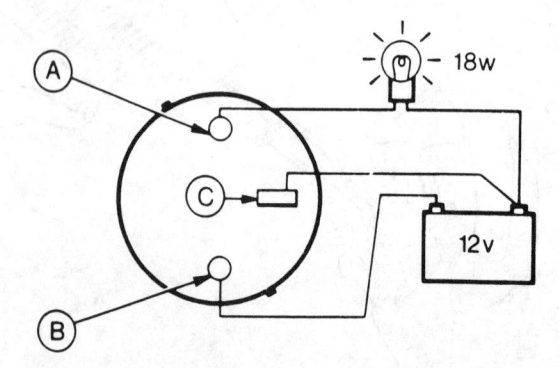

Fig. 12.13 Checking the solenoid for continuity (Sec 11)

A Battery terminal C Switch terminal
B Starter motor terminal

Chapter 12 Electrical system

12.2 Disconnecting the wires from the starter motor solenoid

12.3A Starter motor retaining bolts

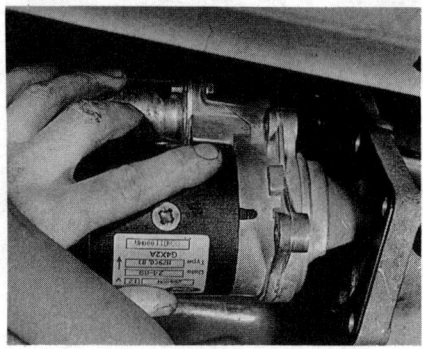

12.3B Withdrawing the starter motor

2 Working underneath the vehicle, having supported it securely on axle stands, disconnect the wires from the solenoid terminals (photo).
3 Remove the three bolts securing the starter motor, noting the earth cable fitment to the upper bolt, and the bracket to the centre bolt, and withdraw it from its location (photos).
4 Refitting is a reversal of the removal procedure, tightening the bolts to the specified torque (see Specifications).

13 Starter motor – overhaul

1 Having removed the starter motor from the vehicle, as described in the previous Section, clean its external surfaces of heavy grease and dirt deposits.

2 Clamp the starter motor in a vice fitted with protective jaw covers, and prepare to dismantle. Do **not** apply excessive force to hold the starter motor or damage may occur.

Bosch DM

3 Undo the three screws and remove the solenoid yoke, having ensured that its starter motor brushplate connection is disconnected.
4 Remove the two screws securing the commutator end plate cap, then remove the cap and the rubber seal.
5 Wipe the grease from the armature shaft and remove the C-clip and shim(s). Remove the securing screws, then lift off the commutator end plate.
6 Release the brush holders, complete with brushes and springs, by pushing them towards the commutator and unclipping from the brush plate. Withdraw the brush plate.
7 Separate the drive end housing and the armature from the yoke (main casing) by tapping apart with a soft-faced hammer. Remove the rubber block.
8 Disconnect the solenoid armature from the actuating arm.
9 To remove the drive pinion assembly from the armature shaft, use a socket or tube of suitable dimension as a drift to release the thrust collar from over the C-clip location (Fig. 12.15). Remove the C-clip and slide the thrust collar and drive pinion assembly off. Do **not** grip the one-way clutch in the vice whilst carrying out the operation, or damage will result.
10 Examine the components and renew as necessary.
11 If the brushes have worn to less than the specified minimum, renew them as a set. To renew the brushes, the leads must be unsoldered from the terminals on the brush plate, and the leads of the new brushes soldered to these terminals.
12 The commutator face should be clean and free from burnt spots. Where necessary, burnish with fine glass paper (**not** emery) and wipe with a petrol-moistened cloth. If the commutator is in very bad condition it can be skimmed, providing its diameter is not excessively reduced. Any skimming should be followed by surface polishing with fine glass paper and wiping with a petrol-moistened cloth. If the commutator is in very bad condition it can be skimmed, providing its

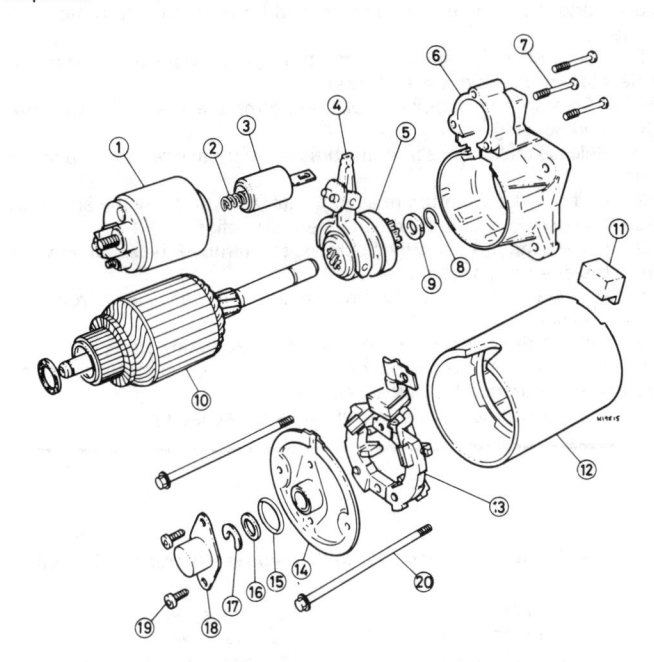

Fig. 12.14 Exploded view of Bosch DM type starter motor (Sec 13)

1	Solenoid yoke	11	Rubber block
2	Solenoid return spring	12	Yoke
3	Solenoid armature	13	Brush plate
4	Actuating arm	14	Commutator end plate
5	Drive pinion and clutch assembly	15	Seal
		16	Shim
6	Drive end housing	17	C-clip
7	Solenoid securing screws	18	Commutator end plate cap
8	C-clip	19	Securing screw
9	Thrust collar	20	Commutator end plate securing screw
10	Armature		

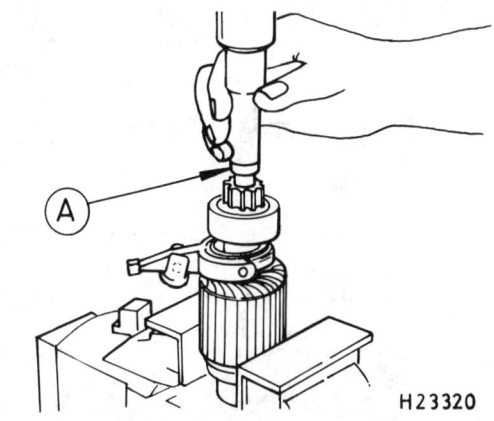

Fig. 12.15 Releasing the thrust collar (A) from over the C-clip location (Sec 13)

266　Chapter 12　Electrical system

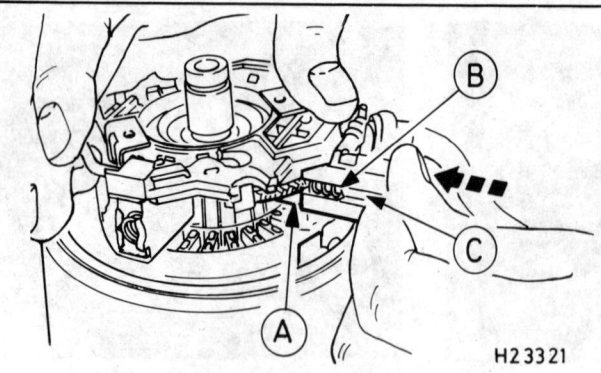

Fig. 12.16 Inserting the brushes – Bosch DM type starter motor (Sec 13)

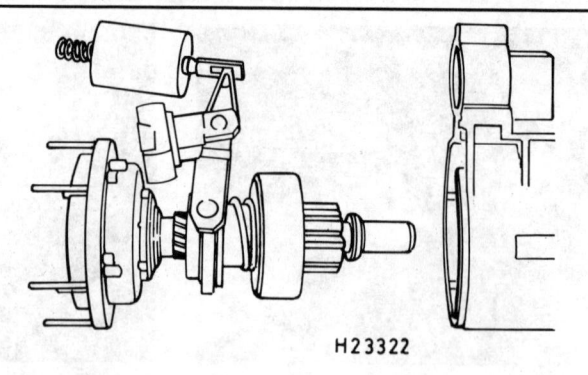

A Brush
B Spring
C Brush holder

Fig. 12.17 Withdrawing the complete pinion/roller clutch assembly from the drive end housing – Bosch DW type starter motor (Sec 13)

diameter is not excessively reduced. Any skimming should be followed by surface polishing with fine glass paper and wiping with a petrol-moistened cloth. If recutting the insulation slots, take care not to cut into the commutator metal.

13 Renew the end housing bushes, which are of self-lubricating type, and should have been soaked in clean engine oil for at least twenty minutes before installation. Drive out the old bushes, whilst supporting the endplate/housing, using a suitable drift.

14 Accurate checking of the armature, commutator, and field coil windings and insulation is time-consuming, and requires the use of some special test equipment. If the starter motor was inoperative when removed from the vehicle and the previous checks have not highlighted the problem, then it can be assumed that there is a continuity or insulation faulty, and the unit should be renewed.

15 Commence reassembly by sliding the drive pinion assembly and the thrust collar onto the armature shaft. Fit the C-clip into its groove then use a two-legged puller (or home-made alternative), as necessary, to draw the thrust collar over the C-clip.

16 Smear lithium-based grease onto the solenoid armature end, and attach it to the actuating arm. Position it in the drive end housing.

17 Ensuring that the rubber block is (and remains) correctly located, guide the yoke (main casing) over the armature and abut to the drive end housing before tapping home with a soft-faced hammer.

18 Position the brush plate over the commutator, then assemble the brush holder, springs and brushes, ensuring that the brush holder clips are securely located (Fig. 12.16). The brush plate will be positively located when the commutator end plate securing screws are fitted, and it should be aligned for this purpose.

19 Ensuring that the brush plate connection insulator is correctly located in the cut-out in the yoke (main casing), fit the commutator end plate and secure with its two screws.

20 With the solenoid armature return spring in position, guide the solenoid yoke over the solenoid armature, align with the drive end housing and secure with its three screws.

21 Slide the armature in its bearings so that the shaft protrudes as far as possible at the commutator bearing end.

22 Fit sufficient shims to the end of the armature shaft to eliminate endfloat when the C-clip is in position, then fit the C-clip.

23 Fit the commutator end plate cap seal, smear a small quantity of lithium-based grease onto the end of the armature shaft, then fit the commutator end plate cap and secure it with two screws.

24 Refit the brush plate connection to the solenoid and secure.

Bosch DW

25 Refer to paragraphs 1 and 2 of this Section.

26 Remove the two securing screws and withdraw the commutator end plate cap.

27 Remove the C-clip and shim from the end of the armature shaft.

28 Undo the two securing screws and lift off the commutator end plate.

29 Disconnect the starter motor brushplate connection from the solenoid terminal (if not already done).

30 Withdraw the complete yoke and armature assembly from the drive end housing.

31 Release the brushes from the brush holders, and remove the brush plate.

32 Remove the armature retaining plate, and withdraw the armature from the yoke, overcoming the magnetic attraction.

33 Extract the three securing screws, and remove the solenoid yoke from the drive end housing.

34 Withdraw the complete pinion/roller clutch assembly from the drive end housing (Fig. 12.17).

35 Unhook the solenoid armature from the actuating arm.

36 To remove the drive pinion from the output shaft, proceed as described in paragraph 9 of this Section.

37 Examine the components and renew as necessary.

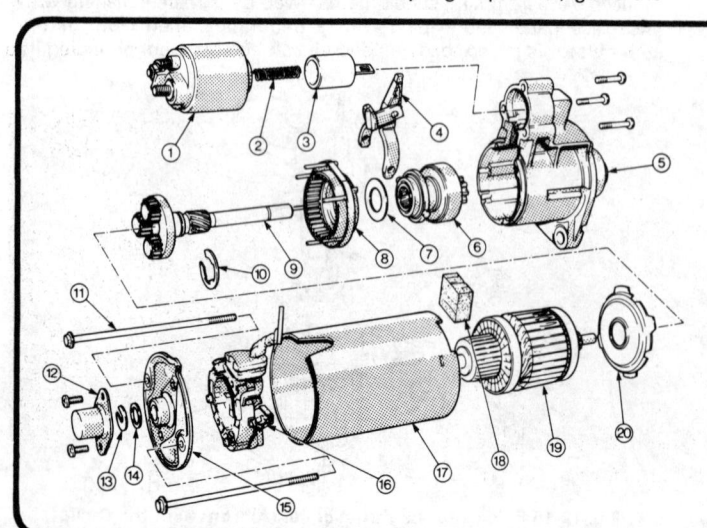

Fig. 12.18 Exploded view of Bosch DW type starter motor (Sec 13)

1 Solenoid yoke
2 Solenoid return spring
3 Solenoid armature
4 Actuating arm
5 Drive end housing
6 Drive pinion and clutch assembly
7 Spacer
8 Ring gear and carrier
9 Output shaft and planet gear assembly
10 Circlip
11 Commutator end plate securing screw
12 Commutator end plate cap
13 C-clip
14 Shim
15 Commutator end plate
16 Brush plate
17 Yoke
18 Rubber block
19 Armature
20 Armature retaining plate

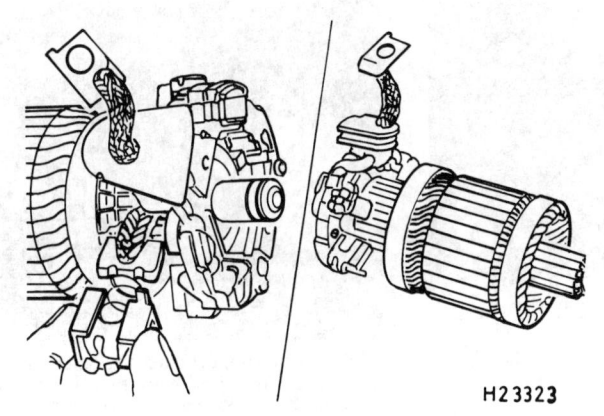

Chapter 12 Electrical system

Fig. 12.19 Assembly of brush plate to commutator – Bosch DW type starter motor (Sec 13)

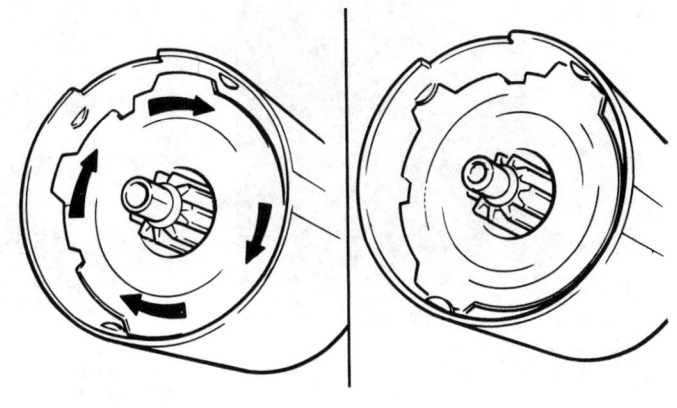

Fig. 12.20 Twist the armature retaining plate to position under the retaining tags in the yoke – Bosch DW type starter motor (Sec 13)

38 If the brushes are to be renewed, it will be necessary to replace the complete brush plate assembly.
39 Proceed as described in paragraphs 12 to 14 inclusive.
40 Commence reassembly by refitting the drive pinion to the output shaft as described in paragraph 15 of this Section.
41 Apply a little lithium-based grease to the end of the solenoid armature, and reconnect it to the actuating arm.
42 Refit the pinion and clutch assembly into the drive end housing, ensuring that the ring gear carrier and rubber block are correctly located.
43 Ensure that the solenoid armature return spring is correctly positioned, then guide the solenoid yoke over the armature and secure with its three screws.
44 Position the brush plate over the commutator, then assemble the brush holders, springs and brushes, ensuring that the brush holder clips are securely located (Fig. 12.19).
45 Insert the armature into the yoke, making sure that the brush plate stays in place, and engage the brush plate connection insulator into the yoke cut-out.
46 Refit the armature retaining plate, positioning it as shown in Fig. 12.20.
47 Refit the yoke and armature assembly to the drive end housing, aligning the sun gear with the planet gears.
48 Refit the commutator end plate and secure with its two screws.
49 Refit the brush plate connection to the solenoid, and secure.
50 Refit the shim and C-clip to the end of the armature shaft, then smear the end of the shaft with a little lithium-based grease.
51 Fit the commutator end plate cap and secure with its two screws.

Bosch EV

52 The procedure is basically as described for the Bosch DW type starter motor except that a commutator end housing is fitted in place of the end plate.

Magneti Marelli M79

53 Refer to paragraphs 1 and 2 of this Section.

54 Disconnect the brush plate connection from the solenoid.
55 Undo the two screws securing the commutator end housing cap, and remove the cap and its seal (photo).
56 Remove the C-clip and shim(s) from the end of the armature shaft (photos).
57 Remove the two commutator end housing securing screws and separate the end housing from the yoke.
58 Unscrew the two securing screws and withdraw the solenoid yoke, then unhook the solenoid armature from the actuating arm and remove the armature.
59 Remove the two drive end housing securing screws (photo), and withdraw the housing from the yoke and armature assembly.
60 Separate the armature from the yoke, taking care not to damage the brushes. The actuating arm will be withdrawn with the armature along with the plastic support block and rubber pad.
61 Separate the brush components from the yoke – see Fig. 12.23.
62 Remove the drive pinion from the armature shaft, as described in paragraph 9 of this Section.
63 To remove the actuating arm from the drive pinion, remove the retaining circlip and slide the lever and pivot assembly from the pinion.
64 Examine the components and renew as necessary.
65 If the brushes have worn to less than the specified minimum, renew them as a set. To renew the brushes, cut the leads at their mid-point and make a good soldered joint when connecting the new brushes.
66 Proceed as described in paragraphs 12 to 14 inclusive.
67 Commence reassembly by refitting the actuating arm to the drive pinion and securing with the circlip.
68 Refit the drive pinion to the armature shaft as described in paragraph 15 of this Section (photo).
69 Fit the armature assembly to the yoke, and locate the actuating arm in the drive end housing, with the plastic support block and rubber pad. Note that the support block locates in the cut-out in the face of the yoke (photo).
70 Connect the solenoid armature to the actuating arm, then refit the solenoid yoke and secure with its two screws.

13.55 Removing the commutator end housing cap

13.56A Removing the C-clip from the end of the armature shaft ...

13.56B ... followed by the shim

268 Chapter 12 Electrical system

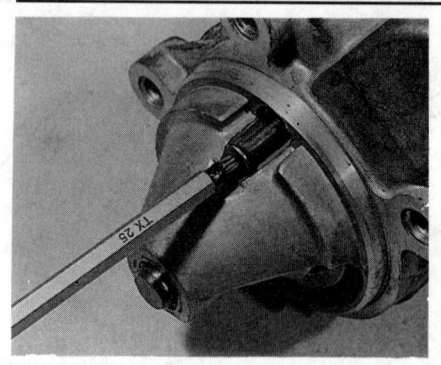

13.59 Removing one of the two drive end housing securing screws

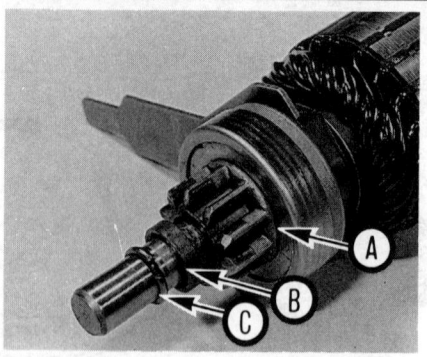

13.68 Drive pinion unit (A), thrust collar (B) and C-clip (C) in position on the armature shaft prior to drawing the thrust collar over the C-clip

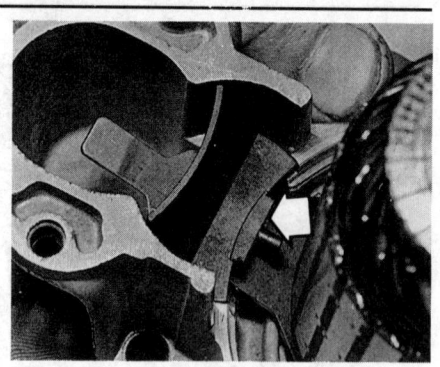

13.69 Plastic support block in position in drive end housing, with yoke cut-out tag on block arrowed (yoke not shown for clarity)

71 Align the drive end housing with the yoke and armature assembly, and fit its two securing screws.
72 Locate the brush plate over the commutator, then position the brushes and fit the nylon insulator cover. Route the brush wiring into the locating channel, then secure the brushes with the springs and locking clips (photos).
73 Refit the commutator end housing, locating the brush plate connection insulator to the cut-out in the housing (photo). and secure with its two screws.
74 Refit the shim(s) and the C-clip to the end of the armature shaft, then fit the commutator end housing cap and seal and secure with its two screws.
75 Refit the brush plate connection to the solenoid and secure.

Nippondenso

76 Refer to paragraphs 1 and 2 of this Section.
77 Disconnect the brush plate connection from the solenoid terminal.
78 Unscrew the two securing nuts and withdraw the solenoid yoke, then unhook the solenoid armature from the actuating arm and remove the armature.
79 Remove the two screws securing the commutator end housing cap and remove the cap.
80 Remove the C-clip from the groove in the armature shaft, and remove the spring.
81 Unscrew and remove the two commutator end housing securing bolts and their washers, then withdraw the commutator end housing.
82 Withdraw the two field brushes from the brush plate, then remove the brush plate.
83 Withdraw the armature and drive end housing from the yoke.
84 Separate the armature and actuating arm from the drive end housing, then unhook the actuating arm from the drive pinion flange.
85 Remove the drive pinion from the armature shaft, as described in paragraph 9 of this Section.
86 Examine the components and renew as necessary.
87 For brush renewal, refer to paragraph 65 of this Section.
88 Proceed as described in paragraphs 12 to 14 inclusive.
89 Commence reassembly by refitting the drive pinion to the armature shaft, as described in paragraph 15 of this Section.
90 Align the actuating arm in the drive end housing. Guide the armature into position in the drive end housing, simultaneously locating the actuating arm over the drive pinion flange.
91 Guide the yoke over the armature, and abut to the drive end housing. Tap into position with a soft-faced hammer.
92 Position the brush plate over the commutator, aligning the cut-outs in the brush plate with the loops in the field windings. The brush plate will be positively located when the commutator end housing securing bolts are fitted.
93 Fit the brushes to their locations in the brush plate, and retain with the springs.
94 Fit the commutator end housing, making sure that the brush plate connection insulator locates to the cut-out in the housing, and secure with its two bolts and washers.
95 Fit the spring and the C-clip to the end of the armature shaft, then smear the end of the shaft with a little lithium-based grease and refit the commutator end housing cap, securing with its two screws.

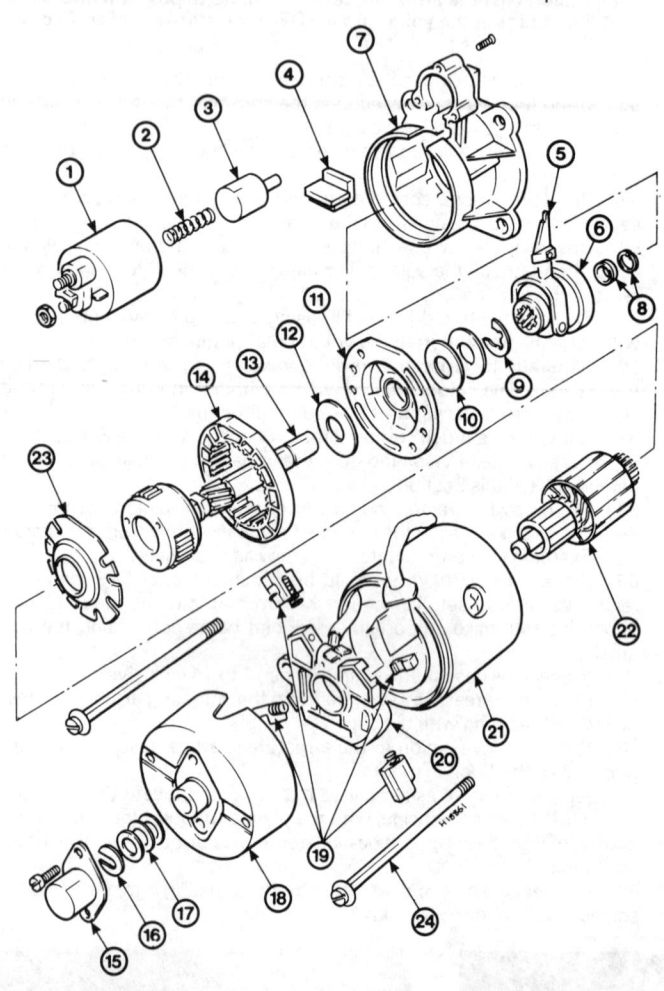

Fig. 12.21 Exploded view of Bosch EV type starter motor (Sec 13)

1 Solenoid yoke
2 Solenoid return spring
3 Solenoid armature
4 Rubber block
5 Actuating arm
6 Drive pinion and clutch assembly
7 Drive end housing
8 C-clip and thrust collar
9 Circlip
10 Spacers
11 Cover plate
12 Spacer
13 Output shaft and planet gear assembly
14 Ring gear
15 Commutator end housing cap
16 C-clip
17 Shims
18 Commutator end housing
19 Brushes
20 Brush plate
21 Yoke
22 Armature
23 Armature retaining plate
24 Commutator end housing securing screw

Chapter 12 Electrical system

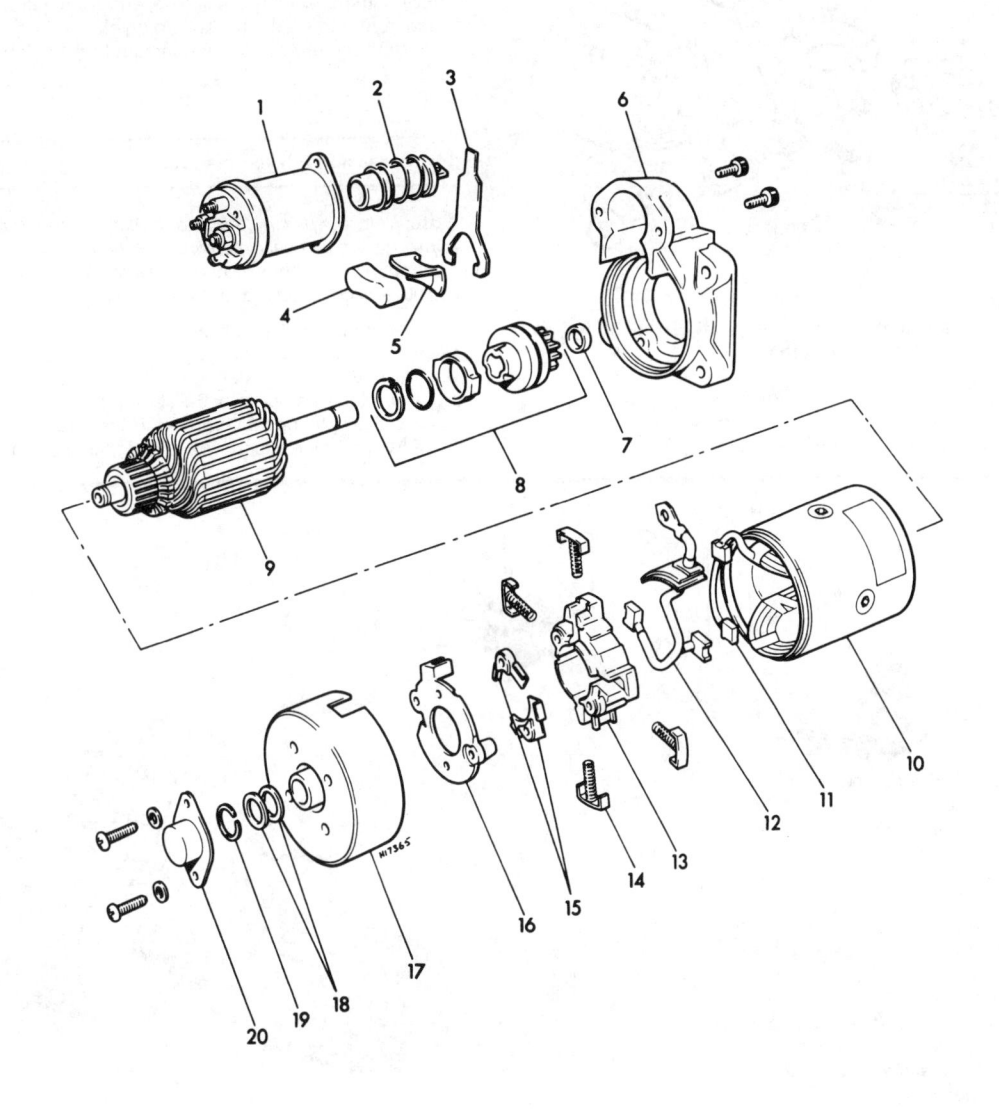

Fig. 12.22 Exploded view of Magneti Marelli M79 type starter motor (Sec 13)

1 Solenoid yoke
2 Solenoid armature
3 Actuating arm
4 Rubber pad
5 Plastic support block
6 Drive end housing
7 Thrust collar
8 Drive pinion and clutch assembly
9 Armature
10 Yoke
11 Brush
12 Brush link
13 Brush plate
14 Brush holder and spring
15 Insulators
16 Brush plate insulator
17 Commutator end housing
18 Shims
19 C-clip
20 Commutator end housing cap

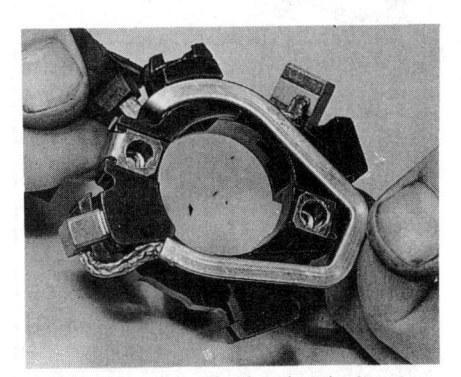

13.72A Brush link going into brush plate

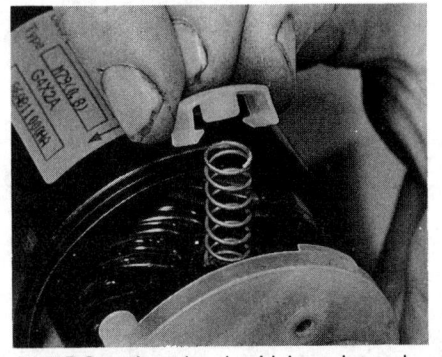

13.72B Securing a brush with its spring and locking clip

13.73 Brush plate connection insulator locating to cut-out in the commutator end housing

Chapter 12 Electrical system

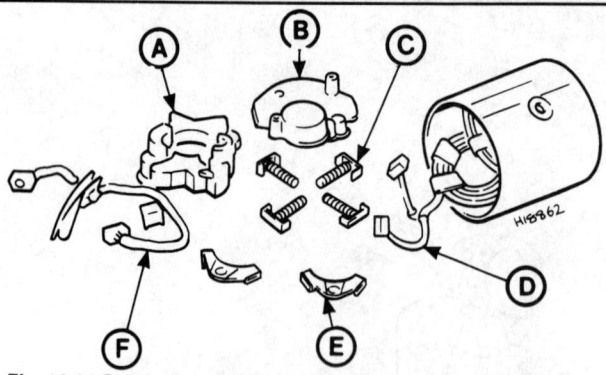

Fig. 12.23 Brush plate components – Magneti Marelli M79 type starter motor (Sec 13)

- A Brush plate
- B Brush plate insulator
- C Brush holders and springs
- D Brushes
- E Insulators
- F Brush link

96 Hook the solenoid armature over the actuating arm in the drive end housing then, with the solenoid armature return spring correctly located, refit and secure the solenoid yoke.
97 Refit the brush plate connection to the solenoid and secure.

14 Fuses and relays – general

Note: *Before replacing a fuse or a relay, ensure that the ignition and the appropriate electrical circuit are switched off. In addition, if it is necessary to remove the fusebox from its location, the battery earth (negative) cable must be disconnected prior to removal*

1 The fuses and relays, with the exception of the indicator flasher relay, are mounted in one unit behind the facia. The indicator flasher relay is located on the steering column multi-function switch assembly, and is hidden from view by the steering column shrouds (see Chapter 9, Section 6 for steering column shroud removal).
2 Always renew a fuse with one of identical rating, and if the replaced

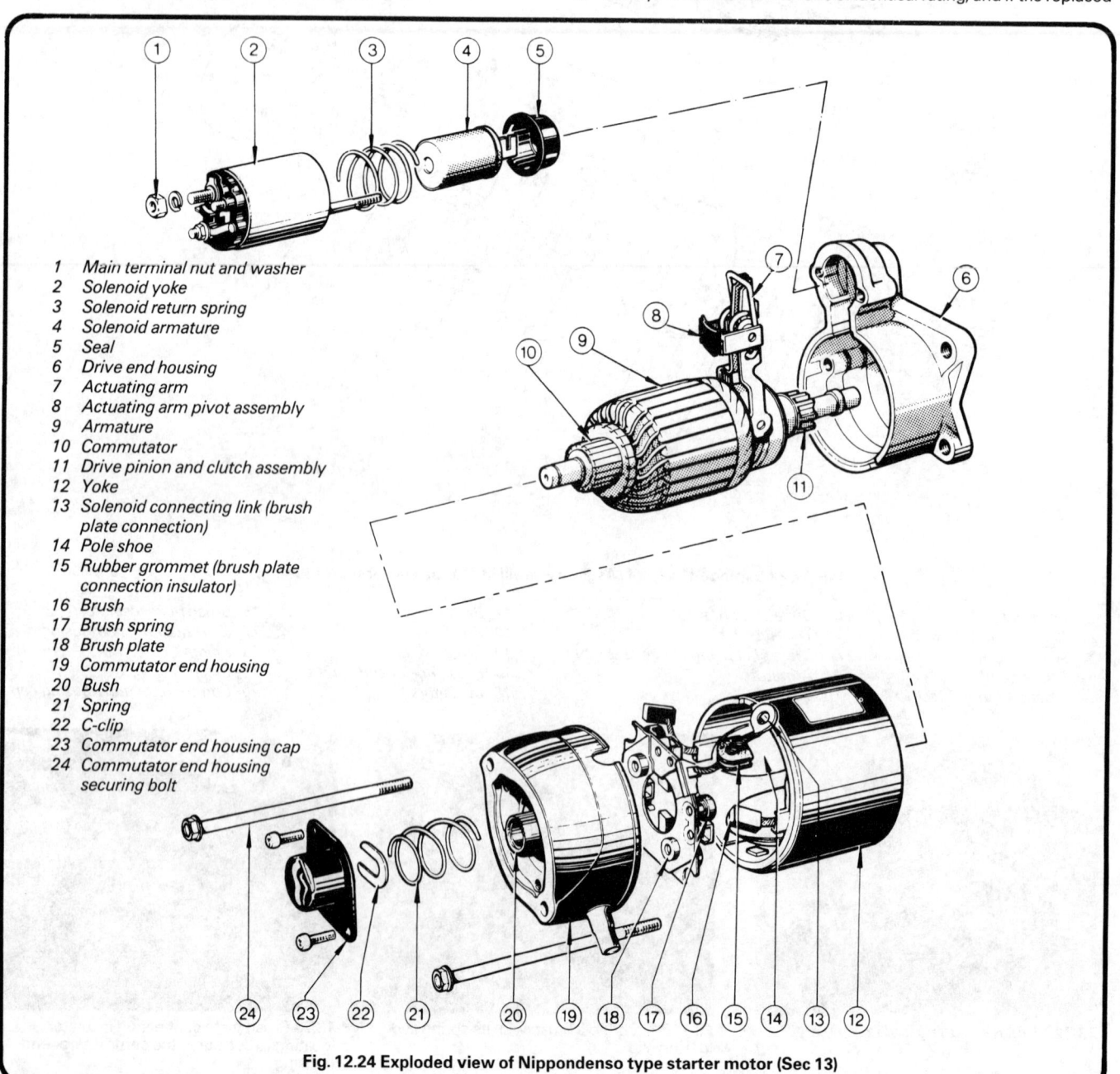

1 Main terminal nut and washer
2 Solenoid yoke
3 Solenoid return spring
4 Solenoid armature
5 Seal
6 Drive end housing
7 Actuating arm
8 Actuating arm pivot assembly
9 Armature
10 Commutator
11 Drive pinion and clutch assembly
12 Yoke
13 Solenoid connecting link (brush plate connection)
14 Pole shoe
15 Rubber grommet (brush plate connection insulator)
16 Brush
17 Brush spring
18 Brush plate
19 Commutator end housing
20 Bush
21 Spring
22 C-clip
23 Commutator end housing cap
24 Commutator end housing securing bolt

Fig. 12.24 Exploded view of Nippondenso type starter motor (Sec 13)

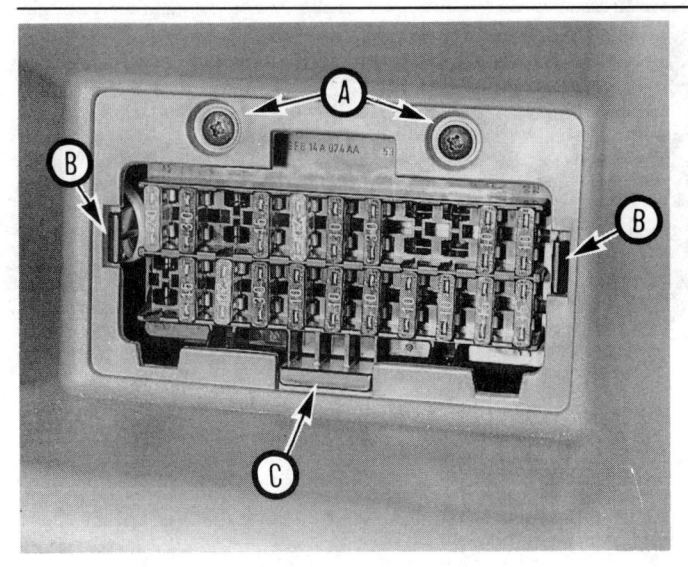

Chapter 12 Electrical system

14.3A Method of fusebox retention

A Retaining screws
B Retaining lugs
C Support

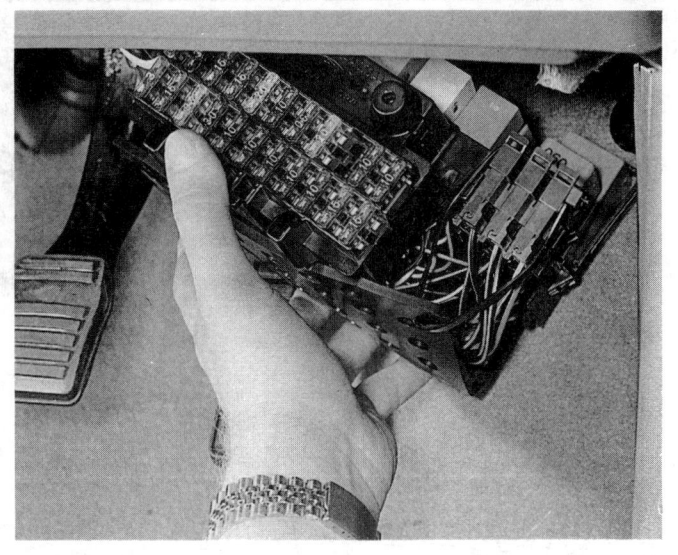

14.3B Withdrawing the fusebox downwards into the driver's footwell

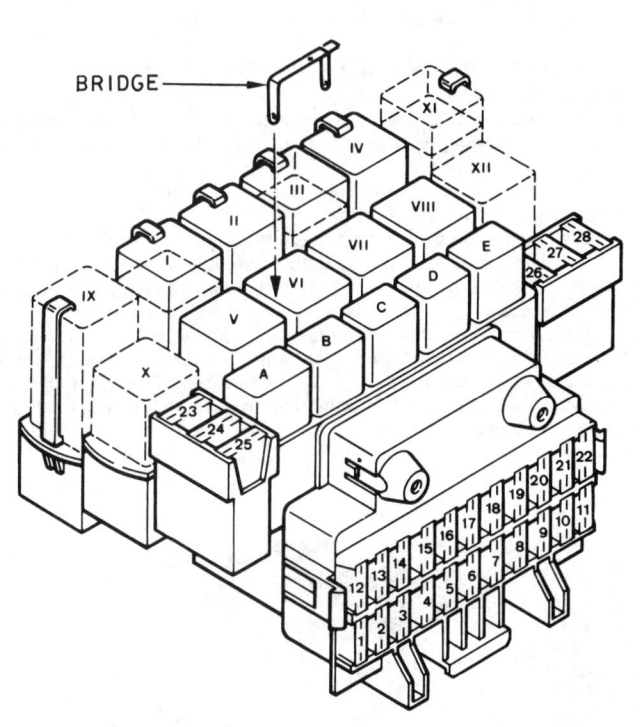

Fig. 12.25 Fusebox layout (see Specifications) (Sec 14)

fuse blows, find and rectify the source of the problem before renewing again. **Never** bypass a fuse using tinfoil or any other conductive material – fuses are there for a purpose.
3 Access to the main fuse plate is gained by removing the hinged fusebox lid, located in the lower recess to the right of the steering wheel. The remaining fuses and relays may be accessed by removing the two fusebox retaining screws, releasing the retaining lugs on either side of the main fuse plate and withdrawing the fusebox downwards into the driver's footwell (photos).
4 Fuses and relays are a push fit, and are laid out as shown in Fig. 12.25. Listings of both appear in the Specifications, for identification purposes. A bridge is fitted to carburettor equipped vehicles without CTX automatic transmission.

15 Switches – removal and refitting

1 Disconnect the battery earth terminal prior to removing any switches.

Steering column multi-function switch assembly
2 The steering column multi-function switch assembly controls the main lighting, indicators, hazard flashers, horn and wash/wipe functions. Removal and refitting procedures are covered in Chapter 9, Section 6.

Facia centre panel switches (below heater controls)
3 These switches individually control the front and rear foglamps, heated windscreen and heated rear window element. Where these features are not fitted to the vehicle, blanking plates are installed instead of switches.
4 Remove the radio assembly (see Sections 16 and 17).
5 Remove the facia centre panel (see Chapter 11, Section 52, paragraphs 11 and 12).
6 Push the required switch/switches out from behind, disconnecting the multi-plug before removing from the vehicle.
7 Refitting is a reversal of the removal procedure.

Centre console switches
8 The switches mounted on the centre console control the electrically operated windows and the tailgate remote release mechanism, where fitted.
9 To remove a switch, carefully prise it from its location using a thin flat-bladed screwdriver, then disconnect the multi-plug (photos).
10 To refit, connect the multi-plug then push home to secure.

Heater fan motor control switch
11 Pull the heater fan motor control knob off, then move the air distribution and temperature controls fully to the right. Unclip and remove the heater slide facia towards the left-hand side of the vehicle, removing the slide control knobs only as necessary, and disconnecting its bulbholder (bayonet type) as it is withdrawn.
12 Squeeze the two release tabs together on the heater fan motor control switch, and remove it, disconnecting its multi-plug as it is withdrawn.
13 Refit by reversing the removal procedure.

Brake pedal stop-lamp switch
14 To remove, see Chapter 8, Section 22, paragraph 4.
15 Refitting is a reversal of the removal operations, setting the switch position as described in Chapter 8, Section 22, paragraph 5.

Handbrake-on warning light switch
16 See Chapter 8, Section 12 for removal procedure.

Chapter 12 Electrical system

15.9A Prise the switch up from its location ...

15.9B ... then disconnect its multiplug and remove

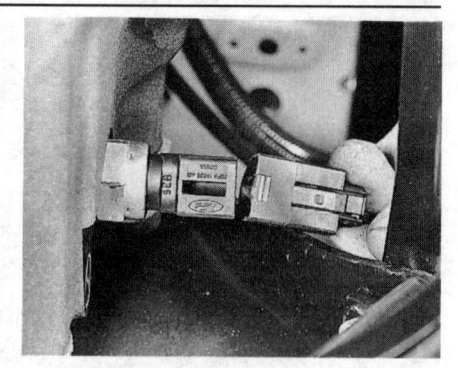

15.28 Disconnecting the reversing light switch multi-plug (shown from below)

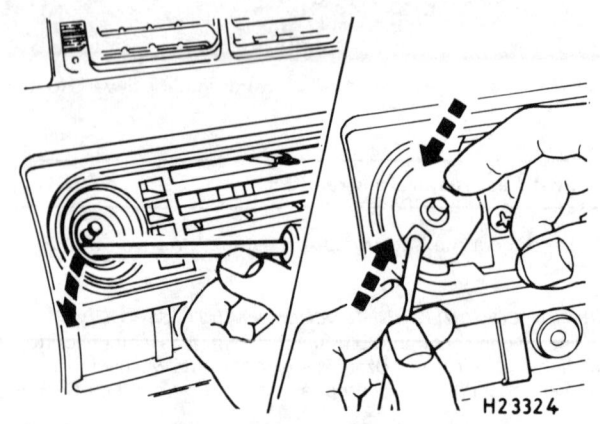

Fig. 12.26 Heater fan motor control switch removal (Sec 15)

17 Refitting is a reversal of the removal procedure. Ensure that the loom is routed away from the lever ratchet and is secured to the floorpan with tape.

Modulator belt-break switches (anti-lock braking system, equipped vehicles only)

18 Belt-break switches are fitted to each of the two modulator drive-belt covers, and monitor the tension of the belts, providing an early warning against belt breakage and/or belt slackness.
19 To renew the switches, proceed as described in Chapter 8, Section 25, paragraphs 5 and 27, then trace their wiring back to their common connector and disconnect it to release the switches.
20 Refitting is the reverse of the removal procedure.

Brake fluid warning light switch

21 This is incorporated into the brake fluid reservoir cap, and senses fluid level in the reservoir. It cannot be renewed separately from the cap.
22 To remove, disconnect the warning indicator loom multi-plug and unscrew the reservoir cap.
23 Refit by reversing the removal procedure.

Courtesy light switch

24 The courtesy light switches fitted to the door pillars control the interior light.
25 To remove a switch, first undo and remove its retaining screw. Carefully withdraw the switch from its location, disconnecting it from its loom connection as it is withdrawn.
26 Refitting is a reversal of the removal procedure.

Reversing light switch

27 The reversing light switch is a screw-in fitting, located on the forward side of the transmission casing.
28 To remove, disconnect its multi-plug and unscrew from its location (photo).

29 Refitting is a reversal of the removal procedure. See Specifications for tightening torque.

Ignition switch (loom plate and lock barrel)

30 To remove, first carry out the procedure detailed in Chapter 9, Section 6, paragraphs 1 to 6. Turn the ignition key to position I, then depress the lock barrel plunger through the steering column lock housing. As the lock barrel plunger is depressed, pull on the ignition key to remove the lock barrel (photo).
31 Refitting is a reversal of the removal procedure.

Starter inhibitor switch (CTX automatic transmission equipped models)

32 A starter inhibitor switch, located on the transmission casing, is fitted to prevent the possibility of starting the engine with the selector lever in a drive position.
33 To remove, first raise the front of the vehicle and support securely using axle stands.
34 Disconnect the switch multi-plug, then unscrew the switch from its location and remove its O-ring. Catch escaping transmission fluid in a suitable container as the switch is removed, plugging the hole to prevent excessive loss.
35 Refitting is a reversal of the removal procedure, using a **new** O-ring, and tightening the switch to the specified torque. Check that the engine only starts with the selector lever in positions P or N, and that the reversing light comes on (with ignition on) in position R. These functions should not occur in any other position. Top up the transmission fluid on completion, using fresh fluid of the specified type (see Chapter 6).

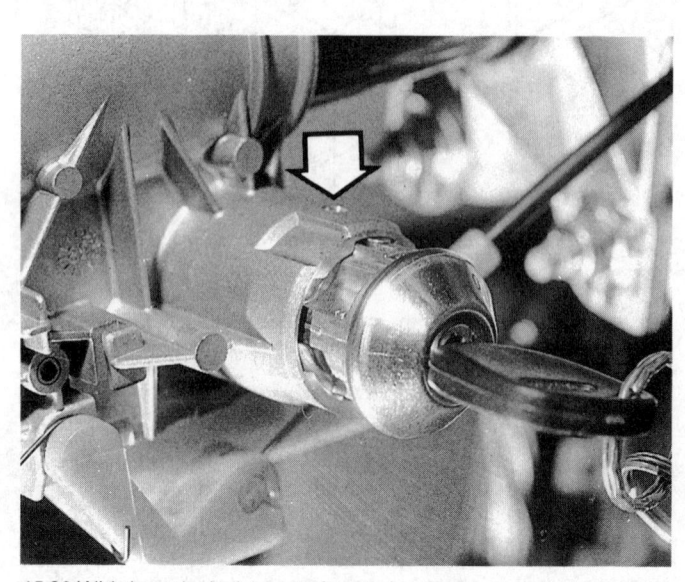

15.30 Withdraw the lock barrel after depressing its plunger through the aperture in the steering column lock housing (arrowed)

Chapter 12 Electrical system 273

17.2 Unscrewing the securing pins from the audio unit

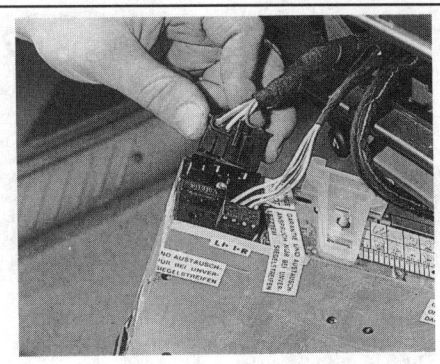

17.4A Disconnect the multi-plugs from the rear of the audio unit ...

17.4B ... followed by the aerial lead

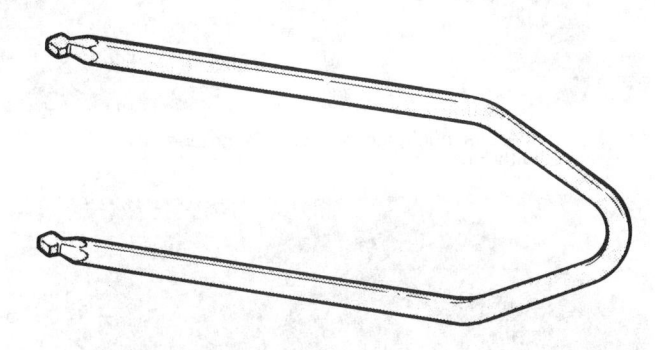

Fig. 12.27 Radio/cassette player extractor tool (Sec 17)

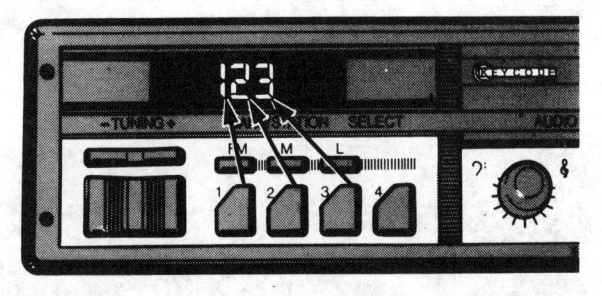

Fig. 12.28 Using station preset buttons to enter security code on reconnecting battery or radio – Ford 'Keycode' audio units (Sec 17)

Radiator cooling fan thermal switch
36 See Chapter 2 for removal and refitting procedures.

Engine oil pressure switch
37 The oil pressure switch is located on the rear of the cylinder block, close to the oil filter assembly, on both HCS and CVH engine types. It serves to warn the driver, via a warning light on the instrument cluster, if the engine oil pressure drops below a predetermined level (see Specifications in Chapter 1).
38 To remove, first raise the front of the vehicle and support securely using axle stands.
39 Remove the switch connection and unscrew the switch from the cylinder block.
40 Refitting is a reversal of the removal procedure, tightening the switch to its specified torque (see Specifications in Chapter 1).

16 In-car entertainment equipment   general

1 The following Sections (17 to 21) cover Ford audio equipment, as fitted during production, either as standard or optional equipment.
2 Where equipment is to be installed at a later date and is not necessarily of Ford make, reference should be made to Section 55 of this Chapter.

17 Radio or radio/cassette player   removal and refitting

1 Disconnect the battery earth terminal (also see paragraph 9).
2 Unscrew the four hexagonal head securing pins from the corners of the unit with an Allen key (photo).
3 Insert the U-shaped removal tools into each side of the unit until they engage positively (Fig. 12.27).
4 Applying a little opposing sideways pressure to the removal tools, pull the tools outwards to withdraw the unit. As the unit is withdrawn, disconnect the multi-plugs and aerial lead from its rear (photos).
5 Remove the tools from the unit by gently wiggling and pulling.
6 To refit, first connect the multi-plugs and aerial lead then, having located its support slide to the support rail, push the unit home to secure.
7 Refit the securing pins to the corners of the unit.
8 Reconnect the battery earth terminal.
9 If a Ford 'Keycode' radio (or radio/cassette) is fitted, and the unit and/or the battery is disconnected, the unit will not function again on reconnection until the correct security code is entered. Details of this procedure, which varies according to the unit and model year, are given in the 'Ford Audio Systems Operating Guide' supplied with the vehicle when new; the code itself being given in a 'Radio Passport' and/or a 'Keycode Label' at the same time.
10 For obvious security reasons, the procedure is not given in this manual – if you do not have the code or details of the correct procedure, but can supply proof of ownership and a legitimate reason for wanting this information, the vehicle's selling dealer may be able to help.
11 Note that these units will allow only ten attempts at entering the code – any further attempts will render the unit permanently inoperative until it has been reprogrammed by Ford themselves. At first, three consecutive attempts are allowed; if all three are faulty, a 30-minute delay is required before another attempt can be made, and each of any subsequent attempts (up to the maximum of ten) can be made only after a similar delay.

18 Loudspeaker (front)   removal and refitting

1 Disconnect the battery earth terminal.
2 Remove the door trim panel (Chapter 11, Section 18).
3 Remove the four screws securing the loudspeaker assembly to the door and withdraw from the door. Disconnect its multi-plug as it is withdrawn (photos).
4 Refitting is a reversal of the removal procedure.

274   Chapter 12 Electrical system

18.3A Remove the screws securing the loudspeaker assembly to the door

18.3B Disconnect its multi-plug (arrowed) as the loudspeaker assembly is withdrawn

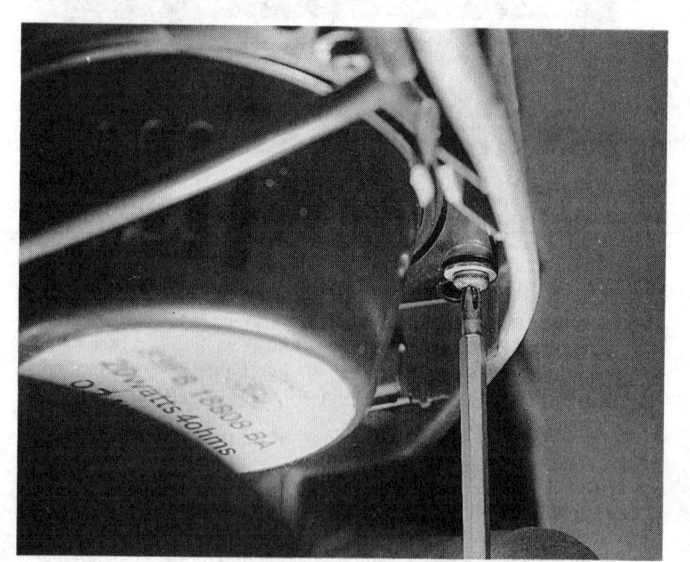

19.3A Remove the loudspeaker securing screws ...

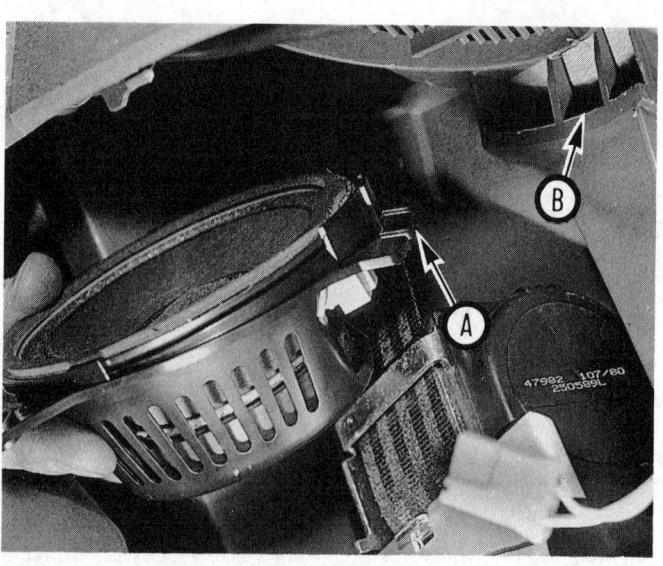

19.3B ... then disengage its locating tags (A) from their location on the parcel shelf support (B)

19 Loudspeaker (rear) – removal and refitting

1 Disconnect the battery earth terminal.
2 The rear loudspeakers are suspended beneath the parcel shelf supports. They are secured in position by locating tags and screws.
3 Remove the securing screws then lower the loudspeaker, disengaging its locating tags from the parcel shelf support (photos). Disconnect its multi-plug as it is withdrawn.
4 Refitting is a reversal of the removal procedure.

20 Loudspeaker balance control joystick – removal and refitting

1 Disconnect the battery earth terminal.
2 Using a thin flat-bladed screwdriver, carefully prise the joystick assembly out of the facia. Use a piece of card or similar to prevent damage to the facia. Withdraw the assembly so that its multi-plug may be disconnected, then remove it from the vehicle (photos).
3 To refit, first connect its multi-plug then push home to secure.
4 Reconnect the battery earth terminal.

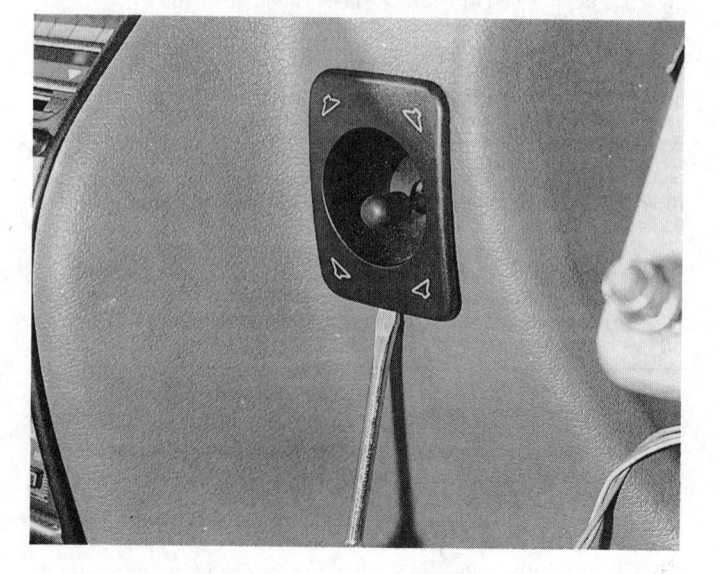

20.2A Prise up the loudspeaker balance control joystick ...

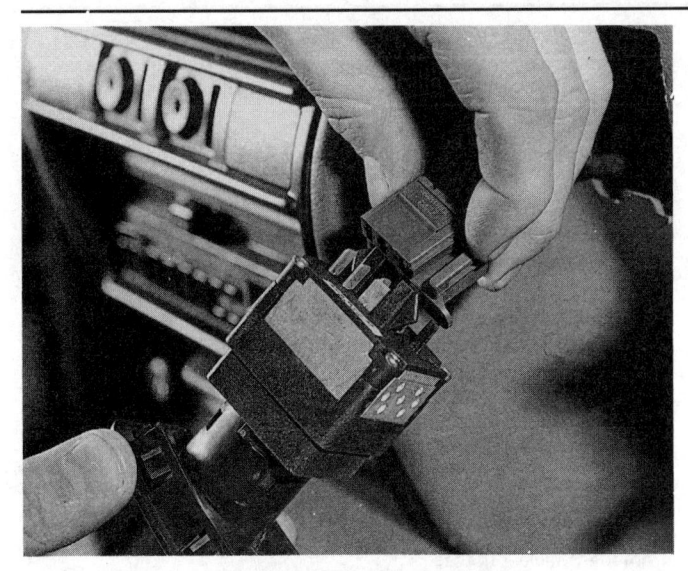

Chapter 12 Electrical system 275

20.2B ... then disconnect its multi-plug as it is withdrawn

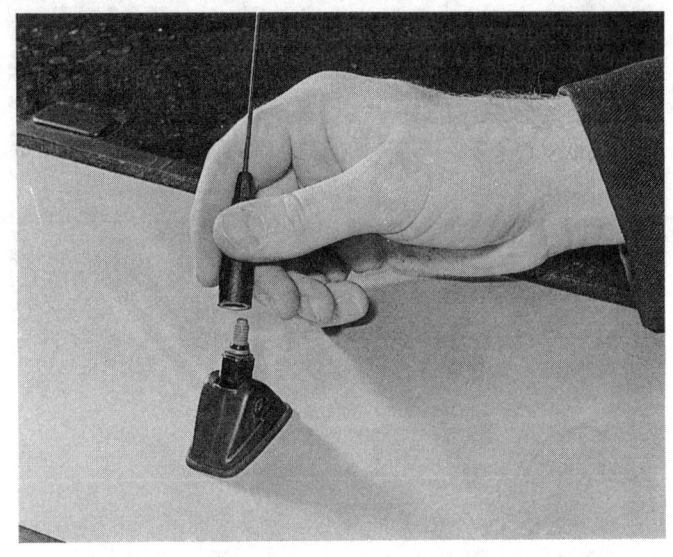

21.1 Unscrew the aerial mast section from its base, and remove

23.2A Carefully prise the clock out of the facia ...

23.2B ... then disconnect its multi-plug and remove

21 Aerial – removal and refitting

Note: *The roof-mounted aerial mast section should be removed prior to using an automatic carwash. This is achieved by unscrewing it from the aerial base.*

1 Unscrew the aerial mast section (photo), then remove the base section as follows. Insert a thin flat-bladed screwdriver into the slot in the interior (courtesy) lamp assembly and carefully lever the assembly out (see photos 31.16A and 31.16B).
2 Through the resultant opening, the aerial base securing screw is accessible. Remove the screw and detach the aerial base from the roof of the vehicle, having noted the aerial lead fitment.
3 Refitting is a reversal of the removal procedure, ensuring that the aerial base sits squarely on the roof. Insert the switch end of the lamp assembly to its aperture first, then pivot the lamp upwards and push home to secure.

22 Cigarette lighter – removal and refitting

Note: *If the cigarette lighter socket may only be used to power 12 volt appliances not exceeding a current rating of 10 amps*

1 Disconnect the battery earth terminal.
2 Remove the cigarette lighter element (heated section).
3 Carefully prise the element barrel out from the illuminated surround, disconnecting its multi-plug as it is withdrawn.
4 Hinge the illuminated surround out carefully, removing its bulb feed connector as it is withdrawn.
5 Refitting is a reversal of the removal procedure.

23 Clock – removal and refitting

1 Disconnect the battery earth terminal.
2 Using a thin flat-bladed screwdriver, carefully prise the clock out of the facia. Use a piece of card or similar to prevent damage to the facia. Withdraw the clock so that its multi-plug may be disconnected, then remove it from the vehicle (photo).
3 To refit, first connect its multi-plug then push home to secure.
4 Reconnect the battery earth terminal, then reset the clock to the correct time.

Chapter 12 Electrical system

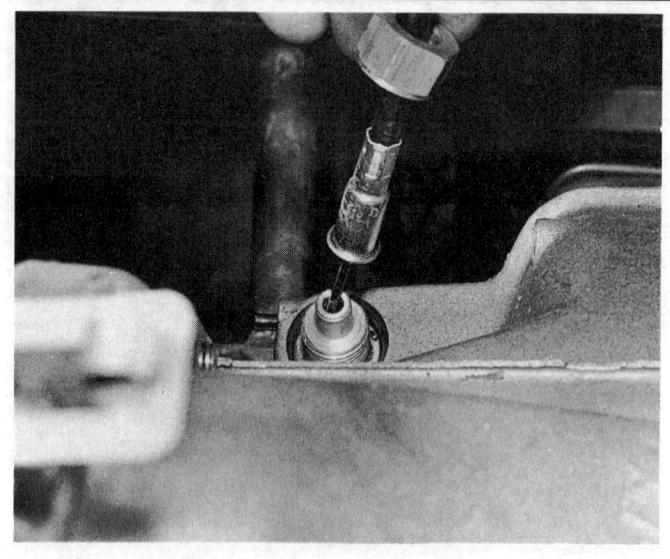

24.2 Detaching the speedometer cable at the transmission casing

24 Speedometer cable – removal and refitting

1 Disconnect the battery earth terminal.
2 Detach the speedometer cable at the transmission casing, having undone its securing nut (photo).
3 Remove the instrument cluster, as described in the following Section, then, having ensured that the speedometer cable is released from its clips and grommet, withdraw it through the bulkhead.
4 Refitting is a reversal of the removal procedure, ensuring that the bulkhead grommet seats correctly, and that the tape mark on the speedometer cable is positioned at the grommet (as applicable).
5 The speedometer driven gear in the transmission casing is covered in Chapter 6.

25 Instrument cluster – removal and refitting

1 Disconnect the battery earth terminal.
2 Disconnect the speedometer cable at the transmission casing (see Section 24).
3 Remove the two screws securing the instrument cluster bezel from its underside. Remove the instrument cluster bezel.

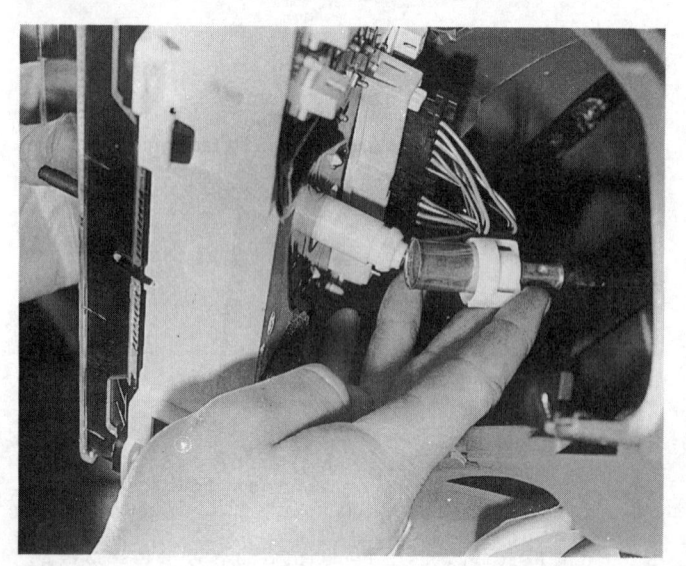

25.4A Disconnect the speedometer cable from the rear of the instrument cluster ...

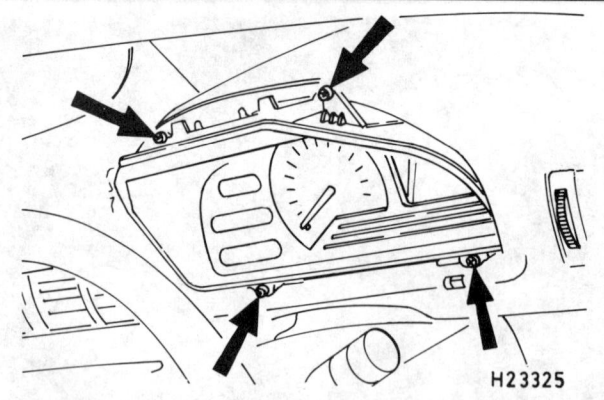

Fig. 12.29 Instrument cluster securing screws (arrowed) (Sec 25)

4 Remove the four screws securing the instrument cluster to its location (Fig. 12.29), then carefully pull it out to allow access to the speedometer cable and multi-plug connections. Disconnect the speedometer cable and multi-plug, then remove the instrument cluster from the vehicle (photos).
5 Refitting is a reversal of the removal procedure, ensuring that the tape mark on the speedometer cable is positioned at the bulkhead grommet (as applicable).

26 Instrument cluster components – removal and refitting

1 Remove the instrument cluster, as described in the previous Section.

Panel illumination and warning light bulbs
2 All bulbholders are a bayonet fit, requiring a 'twist and withdraw' removal technique (photo). Bulbs cannot be removed from their holders – they are renewed complete.
3 Refitting is a reversal of the removal procedure.

Printed circuit
4 Insert a thin flat-bladed screwdriver into the multi-plug retainer, as shown in Fig. 12.30, and carefully unclip it, having noted its orientation.
5 Remove all panel illumination and warning light bulbholders.
6 Using a suitable tool (a trim clip removal tool is ideal), carefully prise the printed circuit off its instrument terminals, and release it from its retainers before removing.
7 Refitting is a reversal of the removal procedure, ensuring that the multi-plug retainer is securely located, and the printed circuit terminal connections are pushed fully home.

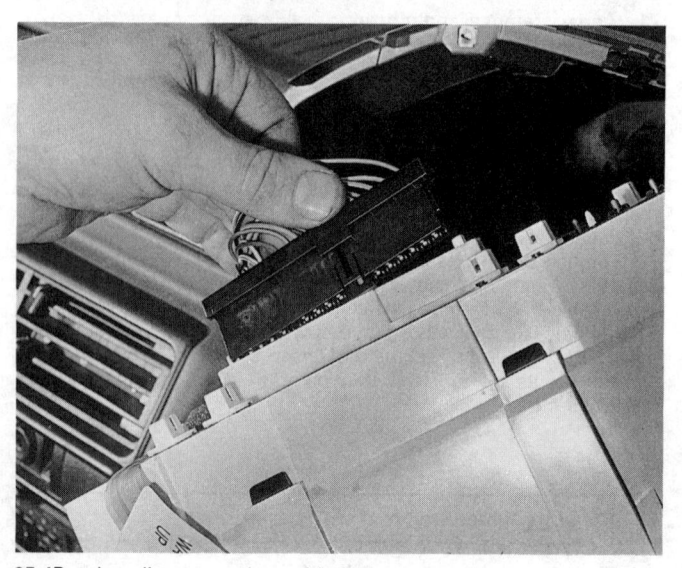

25.4B ... then disconnect the multi-plug

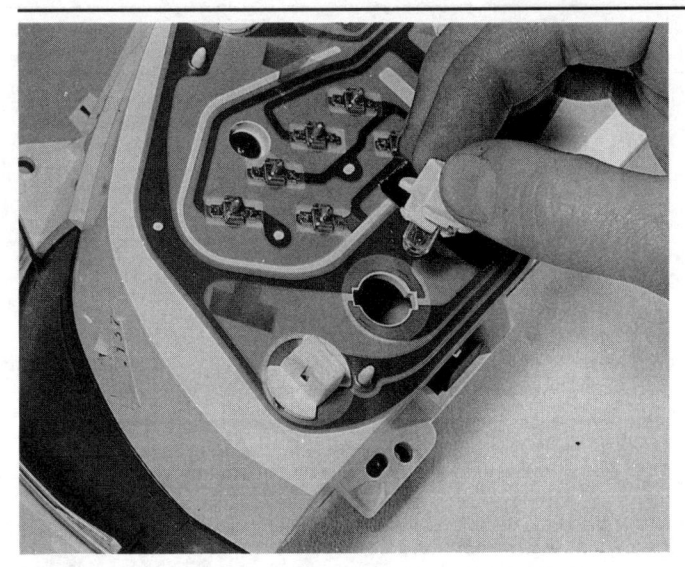

Chapter 12 Electrical system

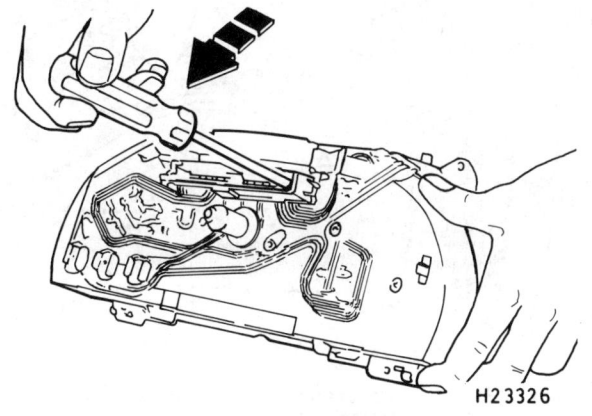

Fig. 12.30 Unclipping the multi-plug retainer from the rear of the instrument cluster (Sec 26)

26.2 Removing a bulbholder from the rear of the instrument cluster (bayonet type fitting)

Fuel and temperature gauge assembly, and temperature gauge sender

16 Removal and refitting procedures are similar in method to those for the tachometer, but only one Torx-head screw retains the assembly.
17 The temperature gauge sender is screwed into the cylinder head near the thermostat housing (photos). With the engine cold, unscrew the pressure cap on the expansion tank, then refit it. This will release any residual pressure in the system and minimise coolant loss when the sender unit is removed. Disconnect the wiring and unscrew the sender unit from its location. To refit, smear the threads of the sender unit with jointing compound and tighten to the specified torque. Reconnect the wiring and top up the cooling system (see Chapter 2, Sections 3 and 4) to complete.

Speedometer

8 Remove the two bulbholders on the top of the cluster assembly, and release their strip of printed circuit from its retainers, so that the instrument cluster halves may be separated.
9 Separate the cluster halves by releasing the retaining tags (photo), taking care to avoid damaging or losing the warning light graphic strips.
10 Remove its two Torx-head retaining screws, then detach and withdraw the speedometer from the front of the assembly.
11 Refitting is a reversal of the removal procedure.

27 Exterior lamp units – removal and refitting

Note: *Disconnect the battery earth terminal before carrying out any of the following procedures*

Headlamp

1 Disconnect the multi-plug from the back of the headlamps (photo).
2 Twist the sidelamp bulb holder to release, then withdraw it.
3 Remove the headlamp securing bolt and depress the headlamp retaining spring, to allow the headlamp assembly to be hinged forwards and out of its location (Fig. 12.31).

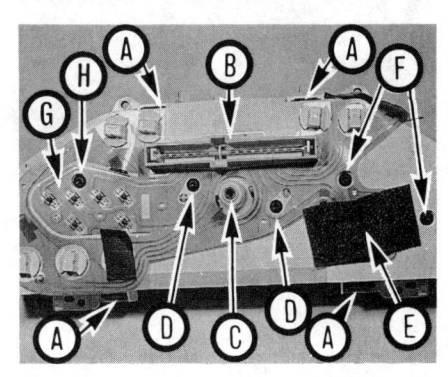

Tachometer

12 Carry out the procedure given in paragraphs 8 and 9.
13 Carefully prise the printed circuit from the tachometer terminals, using a similar method to that described in paragraph 6, releasing it from its retainers as necessary.
14 Remove its two Torx-head retaining screws, then unclip and withdraw the tachometer from the front of the assembly.
15 Refitting is a reversal of the removal procedure, ensuring that the printed circuit terminal connections are pushed fully home.

26.9 Rear view of instrument cluster

A Retaining tags
B Multi-plug retainer
C Speedometer cable connection
D Speedometer gauge retaining screws
E Tachometer terminals (obscured by protective pad)
F Tachometer gauge retaining screws
G Fuel and temperature gauge assembly terminals
H Fuel and temperature gauge assembly retaining screw

26.17A Temperature gauge sender unit location (HCS engine)

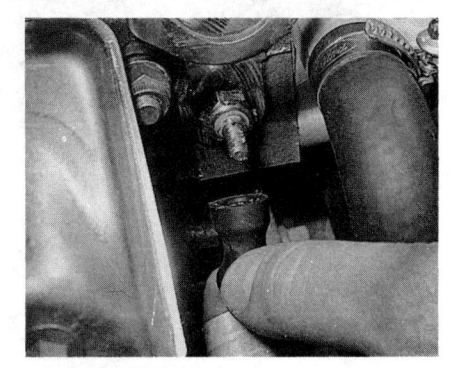

26.127B Temperature gauge sender unit location (CVH engine)

278　Chapter 12　Electrical system

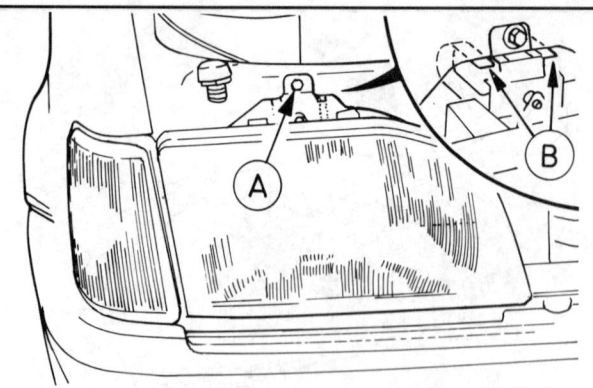

Fig. 12.31 Headlamp retaining bolt (A), and retaining spring arrangement (B) (Sec 27)

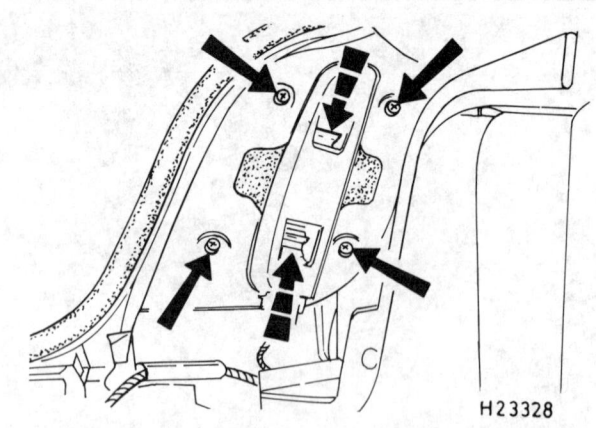

Fig. 12.32 Press the retaining lugs on the bulbholder together (broken arrows) to release the bulbholder (Sec 27)

4　Refitting is a reversal of the removal procedure, ensuring that the lower headlamp mounting guide is inserted correctly (photo), and that the retaining spring engages fully. Tighten the securing bolt to the specified torque.
5　Refer to Section 29 for beam alignment.

Front direction indicator lamp

6　To remove, release the indicator lamp retaining spring from its body location, then pull the lamp assembly forwards and disconnect its multi-plug as it is withdrawn (photos).
7　Refitting is a reversal of the removal procedure.

Side direction indicator repeater lamp

8　Remove the appropriate front wheel arch liner (Chapter 11, Section 65).
9　Remove the appropriate sill scuff plate (Chapter 11, Section 50), and release the clip securing the insulation to the panel forward of the lower A-pillar.

10　Disconnect the supply lead connector and the earth lead, then release their grommet from its panel location.
11　From outside the vehicle, twist the lamp assembly to release it, then withdraw it and its leads (photo).
12　Refitting is a reversal of the removal procedure, ensuring that the grommet is seated correctly in its panel location.

Tail lamp

13　Disconnect the multi-plug from the bulbholder, then press the retaining lugs on the bulbholder together and remove it (Fig. 12.32).
14　Unscrew the four nuts securing the tail lamp unit, then remove the unit and its seal (Fig. 12.33).
15　Refitting is a reversal of the removal procedure. Tighten the tail lamp securing nuts to their specified torque.

Rear number plate lamp

16　Insert a thin flat-bladed screwdriver between the lamp assembly

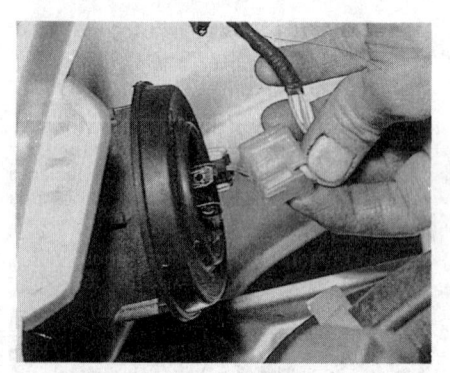

27.1 Disconnecting the multi-plug from the back of the headlamp

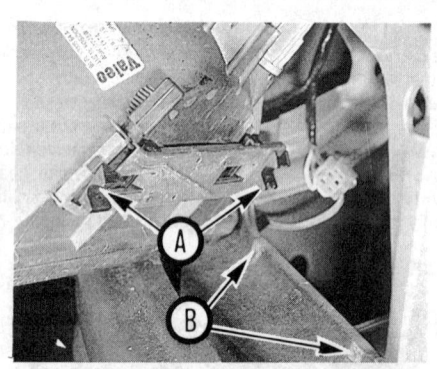

27.4 Refitting the headlamp. Mounting guides (A) to fit to panel (B)

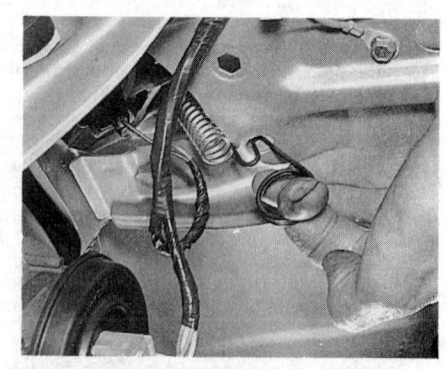

27.6A Release the front direction indicator lamp retaining spring from its body location ...

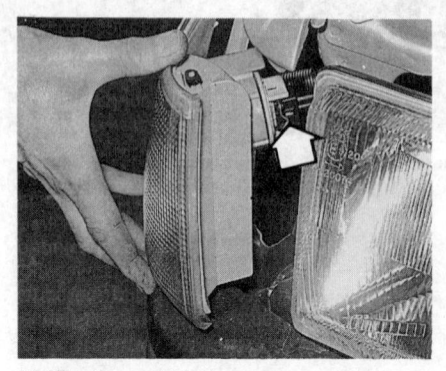

27.6B ... then pull the lamp out to enable its multi-plug (arrowed) to be disconnected

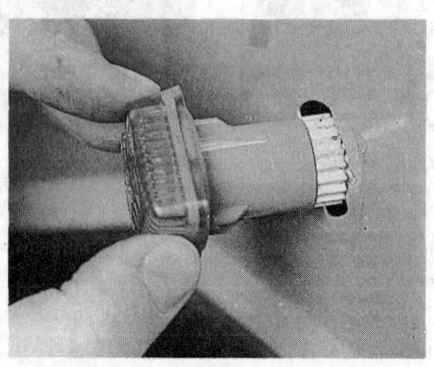

27.11 Withdrawing a side direction indicator repeater lamp assembly, having twisted it to release from its panel location

Chapter 12 Electrical system

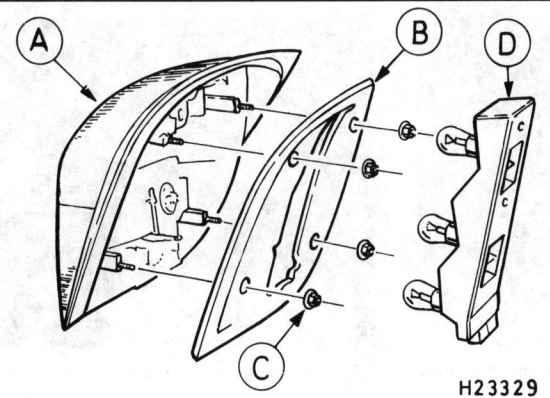

Fig. 12.33 Exploded view of the tail lamp assembly (Sec 27)

A Lamp unit/lens
B Seal
C Retaining nuts
D Bulbholder

and the bumper, and carefully prise the lamp out. Use a rag, or a piece of card, between the screwdriver and the bumper, to prevent damage to the bumper.
17 Detach the connections on the underside of the lamp assembly.
18 Refitting is a reversal of the removal procedure.

28 Auxiliary lamps – removal and refitting

S and SX models
1 Disconnect the battery earth terminal.
2 To remove an auxiliary lamp, first disconnect its multi-plug then unscrew its retaining nut, withdraw the bolt and remove the lamp unit.
3 Refitting is a reversal of the removal procedure, tightening the retaining nut to the specified torque. Refer to Section 29 for beam alignment.

XR2i models
4 Disconnect the battery earth terminal.
5 Undo the four Torx retaining screws securing the relevant dual lamp assembly to its bumper location. Note that the retaining and adjusting screws are captive within the lamp assembly – they cannot be removed from the assembly (Fig. 12.35).
6 Withdraw the lamp assembly from its location, then remove the caps protecting the bulbs and disconnect the wiring.

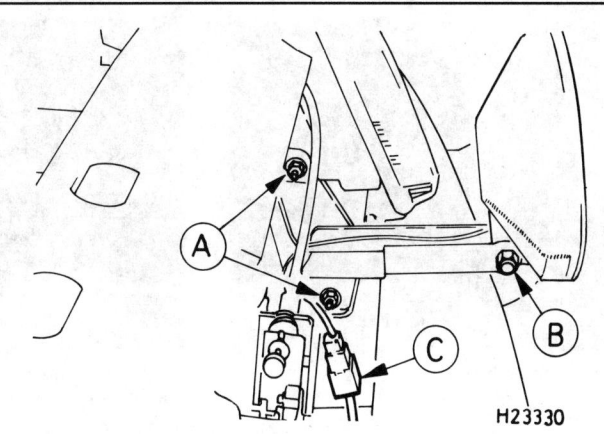

Fig. 12.34 Auxiliary lamp fixture (S and SX models) (Sec 28)

A Bracket retaining nuts
B Auxiliary lamp retaining nut
C Auxiliary lamp multi-plug

7 If required, the lamps may be removed individually from their housing at this stage. Each lamp is secured to its housing unit by a combination of two types of clips – foglamp retention differs from driving lamp retention. The adjusting/retaining clips are removed by undoing the adjustment screws on the front of the housing unit, then turning the clips using pliers or similar tool, before withdrawing. To remove a retaining-only clip, lift the lug on the side of the clip using a screwdriver, then turn the clip using pliers or similar tool, before withdrawing.
8 Refitting is a reversal of the removal procedure. Refer to Section 29 for beam alignment.

29 Headlamps and auxiliary lamps – beam alignment

1 The headlamps have provision for individual horizontal and vertical adjustment, accessible from within the engine compartment (see Fig. 12.36).
2 S and SX model auxiliary lamps are adjusted by slackening the lamp retaining nuts and swivelling the lamp assemblies (see Fig. 12.34). Tighten the nuts upon completion.
3 XR2i model auxiliary lamps are individually adjustable within their housings, provision being made for both vertical and horizontal adjustment of the driving lamps, and vertical adjustment only of the foglamps.

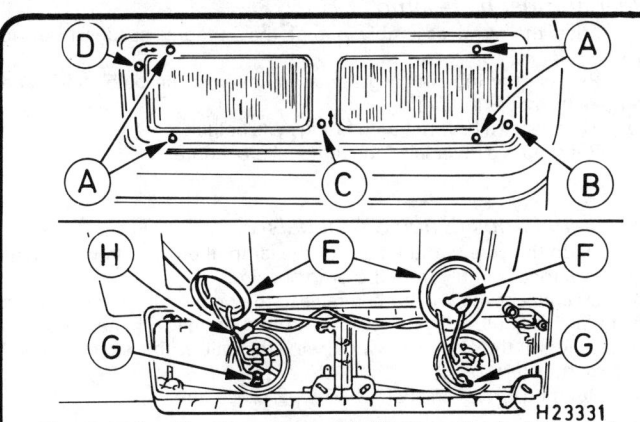

Fig. 12.35 Auxiliary lamp assembly (XR2i models) – left-hand unit shown (Sec 28)

A Lamp assembly retaining screws
B Foglamp vertical adjustment screw
C Driving lamp vertical adjustment screw
D Driving lamp horizontal adjustment screw
E Bulb protective caps
F Fog lamp bulb connector
G Bulb earth leads
H Driving lamp bulb connector

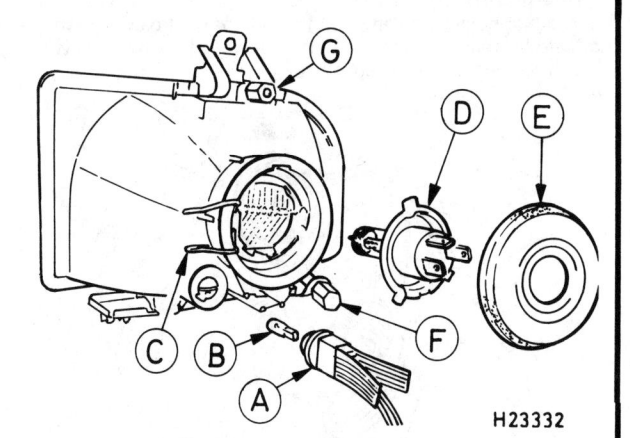

Fig. 12.36 Exploded view of the headlamp unit (Sec 29)

A Sidelamp bulb holder
B Sidelamp bulb
C Headlamp bulb retainer
D Headlamp bulb
E Bulb protective cap
F Horizontal adjustment screw
G Vertical adjustment screw

Chapter 12 Electrical system

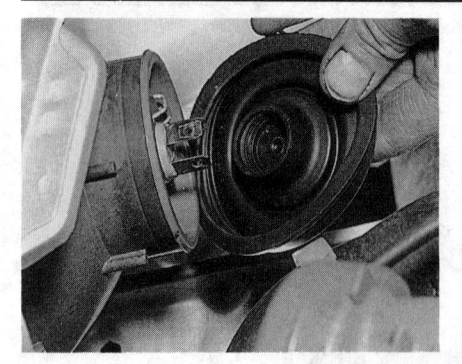

30.2A Remove the protective cap ...

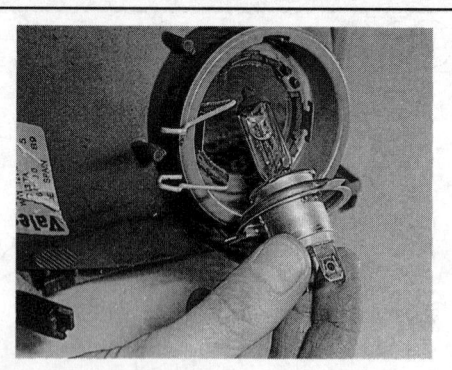

30.2B ... then unlock the bulb retaining spring and withdraw the bulb

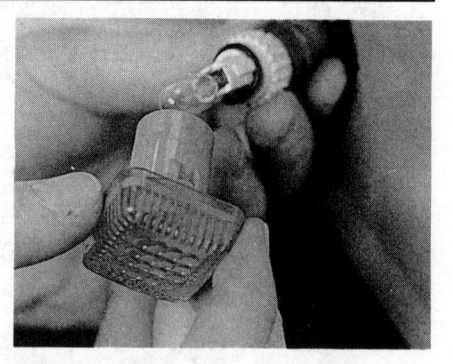

30.11 Separate the lens from the bulbholder assembly ...

Adjustment is made via Torx-head captive screws on the front of the housings (see Fig. 12.35).
4 Holts Amber Lamp is useful for temporarily changing the headlight colour to conform with the normal usage in Continental Europe.

All lamps
5 The above paragraphs enable approximate adjustments only, and accurate alignment can only be carried out using optical beam setting equipment. This work should be entrusted to a Ford dealer.

30 Bulbs (exterior lamps) renewal

Note: *Disconnect the battery earth terminal before carrying out any of the following procedures.*

Headlamp
1 From within the engine compartment, disconnect the multi-plug from the back of the headlamp (see photo 27.1).
2 Remove the rubber bulb protective cap, then unlock the bulb retaining spring clip or retaining ring (according to type) and withdraw the bulb (photos). If the bulb is to be re-used, take care not to touch the glass with your fingers. If the glass is touched, wipe the bulb with a rag moistened with methylated spirit.
3 Refitting is a reversal of the removal procedure.

Front sidelamp
4 The bulbholder is located on the side of the headlamp unit, and is removed by twisting anti-clockwise and withdrawing from within the engine compartment (see Fig. 12.36).

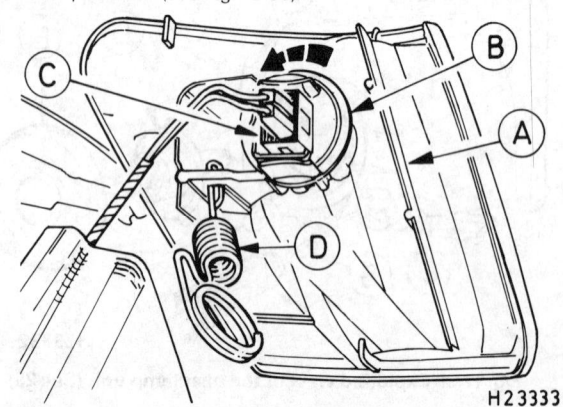

Fig. 12.37 Remove the front direction indicator lamp bulbholder by turning it anti-clockwise (Sec 30)

A Lamp housing   C Multi-plug
B Bulbholder   D Retaining spring

5 Withdraw the push-fit bulb from its holder.
6 Refitting is a reversal of the removal procedure.

Front direction indicator lamp
7 Remove the indicator lamp assembly, as described in Section 27.
8 Turn the bulbholder anti-clockwise and remove it (Fig. 12.37). The bulb is a bayonet type fitting in its holder.
9 Refitting is a reversal of the removal procedure.

Side direction indicator repeater lamp
10 Twist the lamp assembly clockwise to release it from the panel, then withdraw it.
11 Separate the lamp lens from the bulbholder by twisting it free (photo).
12 Withdraw the push-fit bulb from its holder (photo).
13 Refitting is a reversal of the removal procedure.

Tail lamp
14 Access to the tail lamp bulbholder is gained from the luggage compartment. Disconnect the multi-plug from the bulbholder, then press the retaining lugs on the bulbholder together and remove it (photos).
15 All bulbs are of bayonet type fitting.
16 Refitting is a reversal of the removal procedure.

Rear number plate lamp
17 Prise the lamp assembly out of the bumper, as described in Section 27.
18 Release the tab securing the lamp cover to the bulbholder, and remove the cover (photo).
19 The bulb is a bayonet type fitting in its holder.
20 Refitting is a reversal of the removal procedure.

Auxiliary lamps (S and SX models)
21 Undo the screw at the base of the lamp, then withdraw the lens and reflector assembly from the lamp housing.
22 Disconnect the wiring then release the bulb retainer and remove the bulb. If the bulb is to be re-used, take care not to touch the glass with your fingers. If the glass is touched, wipe the bulb with a rag moistened with methylated spirit.
23 Refitting is a reversal of the removal procedure.

Auxiliary lamps (XR2i model)
24 Remove the auxiliary lamp assembly from its bumper location, as described in Section 28, paragraphs 4 to 6.
25 Release the bulb retainer (photo), then remove the bulb. If the bulb is to be re-used, take care not to touch the glass with your fingers. If the glass is touched, wipe the bulbs with a rag moistened with methylated spirit.
26 Refitting is a reversal of the removal procedure.

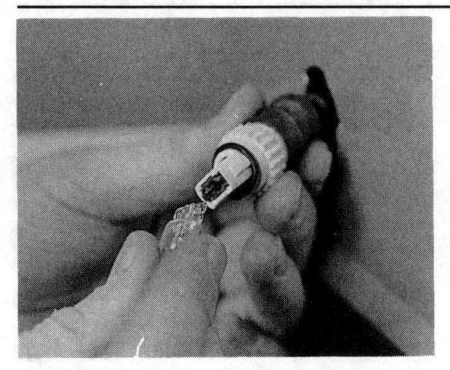

Chapter 12 Electrical system

30.12 ... then withdraw the push-fit bulb

30.14A Disconnect the multi-plug from the bulbholder ...

30.14B ... then having pressed its retaining lugs, withdraw the bulbholder for access to the bulbs

30.18 Release the securing tab (arrowed) and separate the bulbholder and lamp cover

30.25 Releasing the bulb retainer (XR2i auxiliary lamp)

31 Bulbs (interior lamps) renewal

Panel illumination and warning light bulbs
1 Refer to Section 26.

CTX selector cover illumination bulb
2 Using a thin flat-bladed screwdriver, prise up the selector cover and remove it.
3 Pull the bulb assembly to release it from the selector lever, then remove its cover. The bulb is a bayonet fit in its holder (photos).
4 Refitting is a reversal of the removal procedure.

Hazard warning switch bulb
5 To renew the bulb without risk of damage to the steering column multi-function switch assembly, depress the switch knob and allow it to rise to the 'ON' position, then pull off the knob and pull out the bulb.
6 Refitting is a reversal of the removal procedure.

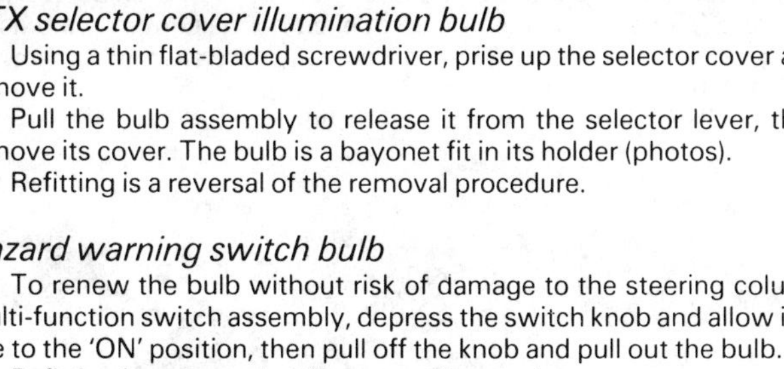

Fig. 12.38 Heater fan illumination arrangement (Sec 31)

Clock backlight bulb
7 Carefully prise the clock out of its facia location, as described in Section 23.
8 Twist the bulbholder and withdraw it from the rear of the clock (photo). Note that the bulb and bulbholder cannot be separated — they are replaced as an assembly.
9 Refitting is a reversal of the removal procedure.

Cigarette lighter illumination surround bulb
10 Remove the cigarette lighter, as described in Section 22.
11 Pull the bulbholder from its location in the illuminated surround, then pull the bulb out of the bulbholder.
12 Refitting is a reversal of the removal procedure.

Heater facia illumination bulb
13 Remove the heater facia, as described in Section 15, paragraph 11.
14 The bulb is removed by pulling it from its bulbholder (Fig. 12.38).
15 Refitting is a reversal of the removal procedure.

Interior (courtesy) lamp bulb
16 Insert a thin flat-bladed screwdriver into the slot in the lamp assembly, and carefully lever it out of its aperture. The bulb is a bayonet fit (photos).
17 Refit the bulb by reversing the method of removal, then insert the switch end of the lamp to the aperture, pivot the lamp upwards and push home to secure.

Luggage compartment (courtesy) lamp bulb
18 Carefully prise the luggage compartment (courtesy) lamp assembly out of its location, using a thin flat-bladed screwdriver. Twist the bulbholder anti-clockwise to remove.
19 The bulb is removed by pulling it from its bulbholder.
20 Refitting is a reversal of the removal procedure.

Chapter 12 Electrical system

31.3A Remove the bulb cover ...

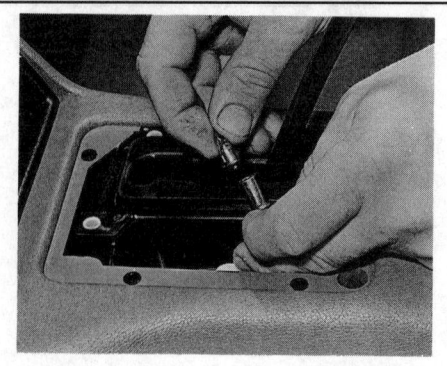

31.3B ... followed by the (bayonet fitting) bulb

31.8 Withdrawing the clock backlight bulb and its integral holder from the rear of the clock

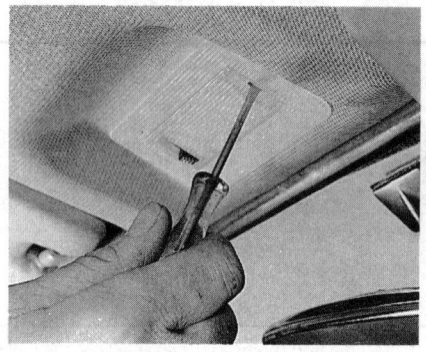

31.16A Releasing the interior (courtesy) lamp from its location

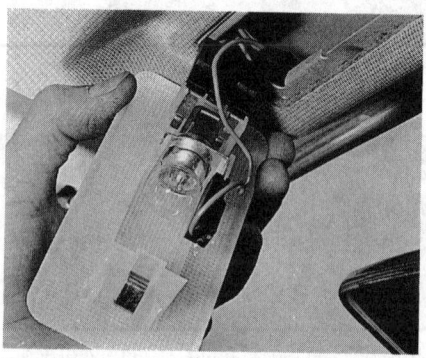

31.16B Interior (courtesy) lamp pivoted for access to bulb

32 Electrically operated windows – description

1 Electrically operated front windows are available as standard or optional equipment, according to model.
2 The electric motor drivegear engages directly with the window regulator mechanism.
3 A circuit-breaker type of overload protection is provided.
4 The motors are individually controlled by switches mounted on the centre console.

33 Electrically operated windows – removal and refitting

1 Remove the window regulator from the vehicle, as described in Chapter 11, Section 27.
2 To remove the motor from the regulator mechanism, undo and remove the two Torx head bolts securing it (photo), then carefully separate by unscrewing.
3 To refit, carefully screw the motor shaft into the regulator mechanism. Temporarily connect the multi-plug, switch on the ignition and activate the motor, to engage and pull the motor *fully* into the regulator mechanism.
4 Ensure that the multi-plug connection is located on top of the motor (as if the window regulator is in position in the door), before securing the motor to the regulator mechanism with its two Torx-lead bolts.
5 Switch off the ignition and disconnect the multi-plug.
6 Refit the window regulator to the vehicle, in accordance with Chapter 11, Section 27.

34 Central locking system – description

1 This system is available as standard or optional equipment, according to model.

2 The system can be activated from the outside using the key, or from the inside by depressing the interior door release handles, but only when all doors are closed.
3 A tailgate remote release mechanism is also fitted. This is activated from the inside by a switch mounted in the centre console (only with engine off), or the tailgate may be opened normally from the outside using the key. See Section 37 for tailgate remote release motor removal and refitting procedures.
4 Circuit-breaker type overload protection is provided.

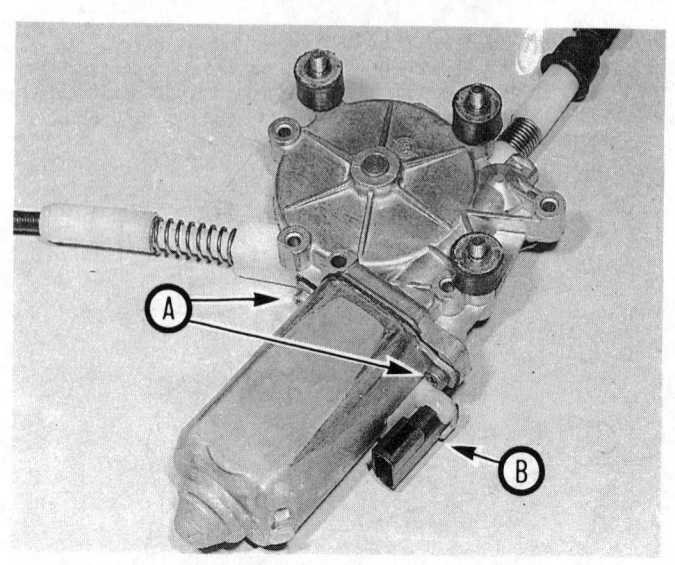

33.2 Electrically operated window, motor-securing bolts (A), and multi-plug connection (B)

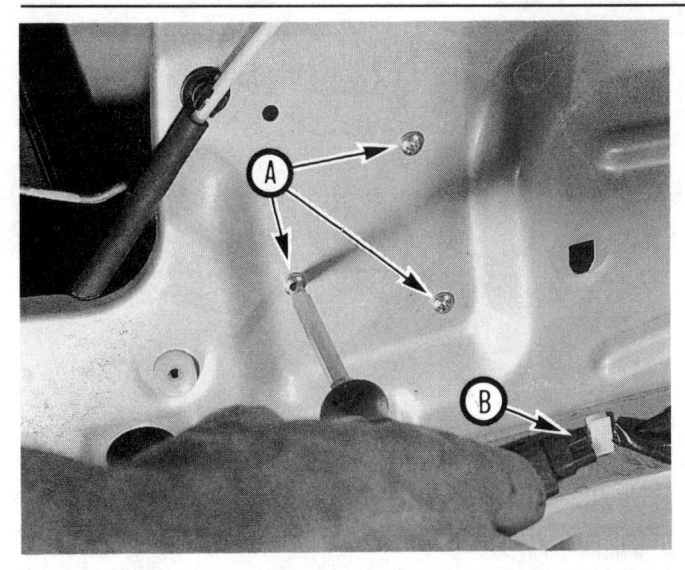

Chapter 12 Electrical system

35.4A Central locking motor retaining screws (A), and loom multi-plug assembly (B) (rear door)

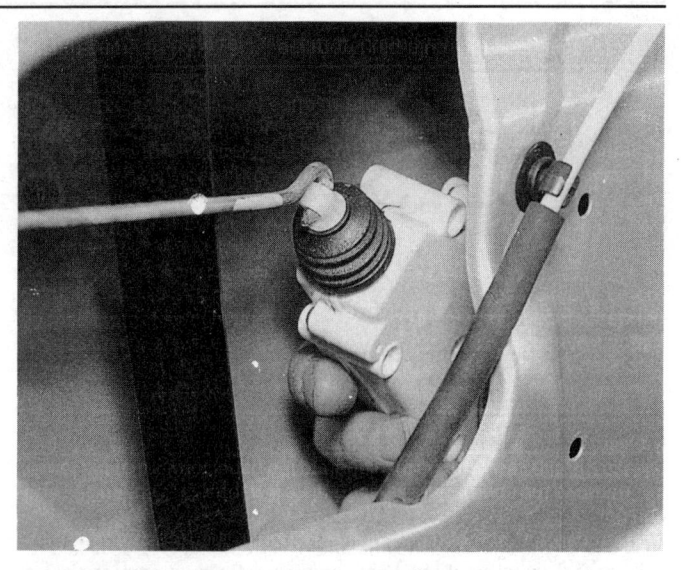

35.4B Central locking motor operating rod disengagement, during motor removal (rear door)

35 Central locking system components - removal and refitting

1 On vehicles fitted with central locking, there is a latch with an integral motor located in the front doors. See Chapter 11, Section 24 for removal and refitting procedure.
2 On rear doors, carry out the procedures detailed in Chapter 11, Section 24, paragraphs 1 to 3. The motor removal procedure is as follows.
3 Disconnect and release the motor loom multi-plug assembly, and withdraw the motor side of the loom back into the door cavity.
4 Undo the three screws securing the motor assembly, disengage its operating rod and remove it from the vehicle (photos).
5 Refitting is a reversal of the removal procedure.

36 Tailgate electrical contact switch components - removal and refitting

1 Disconnect the battery earth terminal.

Tailgate side (contact fingers)
2 Detach the tailgate trim panel (see Chapter 11, Section 37).
3 Undo the two screws securing the contact finger section of the switch and withdraw it from its tailgate location.
4 Unclip and disconnect the switch multi-plug, and disconnect the earth wiring (see photo 39.4). Remove the switch.
5 Refitting is a reversal of the removal procedure.

Body side (contact plate)
6 Insert two thin flat-bladed screwdrivers into the slots on the rearward edge of the contact plate section of the switch, and carefully prise it out of its body location.
7 Disconnect the multi-plugs and remove the switch (photo).
8 To refit, reconnect the multi-plugs and push home to secure.

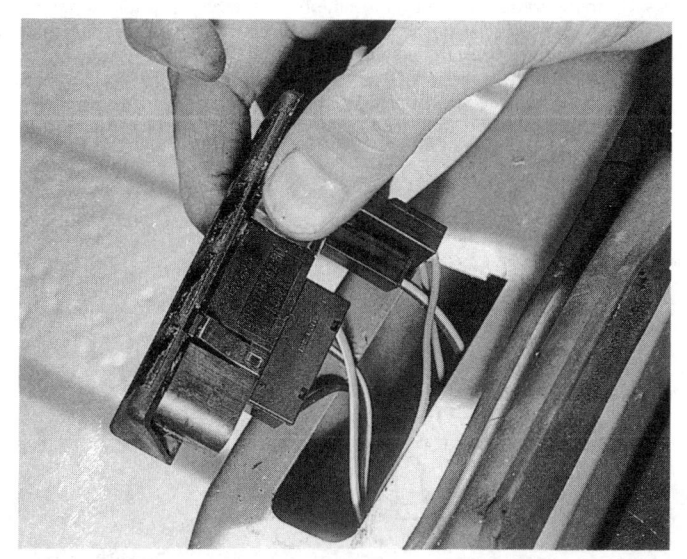

37 Tailgate remote release motor - removal and refitting

1 Disconnect the battery earth terminal.
2 Remove the tailgate trim panel (see Chapter 11, Section 37).
3 Remove the two motor securing screws, then twist the operating rod retaining clip and withdraw the operating rod from it. Disconnect the wiring and remove the motor assembly.
4 The motor may be separated from its bracket by removing two further screws (Fig. 12.39).
5 Refitting is a reversal of the removal procedure.
6 If a fault develops in the tailgate remote release circuit, before suspecting the switch, check that the tailgate electrical contact spring pins move freely, are aligned correctly with their respective contact plates, and that the contact surfaces are absolutely clean. Check also that there is continuity between the contact pins and motor, and that the motor is correctly earthed to the tailgate by the earth cable.

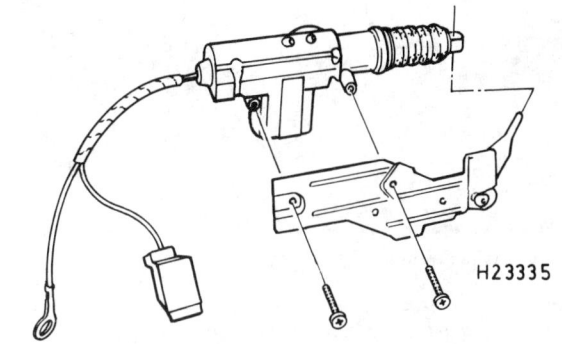

36.7 Withdrawing the switch contact plate for access to disconnect its multi-plugs

Fig. 12.39 Tailgate remote release motor and bracket (Sec 37)

284   Chapter 12 Electrical system

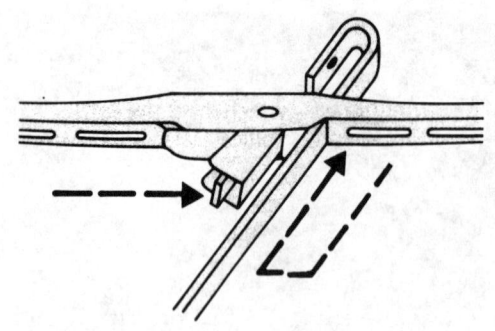

Fig. 12.40 Releasing a wiper blade from its arm (Sec 38)

38 Windscreen/tailgate wiper blades and arms – removal and refitting

1 With the wiper arms in the parked position, mark their blade positions on the windscreen with lengths of masking tape as an aid to refitting. This is relevant only when removing the arms from the pivot shafts.
2 To remove a wiper blade from its arm, simply depress the retaining clip and slide the blade out of the hooked part of the arm (Fig. 12.40).
3 To remove a wiper arm (with blade removed), lift the plastic cap covering its retaining nut and unscrew the nut approximately two turns. Lift the wiper arm against its spring pressure until it is standing upright, then ease it free of its pivot shaft taper. Remove the wiper arm retaining nut, washer and wiper arm.
4 Refitting is a reversal of the removal procedure. Tighten the wiper arm retaining nut to its specified torque, then wet the glass and operate the wipers before retightening to the specified torque.

39 Tailgate wiper motor – removal and refitting

1 Disconnect the battery earth terminal.
2 Remove the wiper arm, as described in the previous Section.
3 Remove the tailgate trim panel (see Chapter 11, Section 37).
4 Disconnect the wiper motor multi-plug and earth lead, then undo the three bolts securing the wiper motor bracket, and remove the assembly from the vehicle (photo).

39.4 Tailgate wiper motor

A Wiper motor bracket retaining bolts
B Wiper motor and tailgate 'contact fingers' multi-plugs
C Earth connection

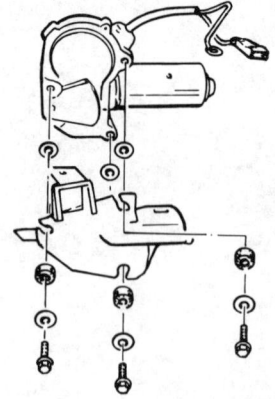

Fig. 12.41 Tailgate wiper motor fixture to bracket (Sec 39)

5 The wiper motor may be separated from its bracket by undoing the three mounting bolts securing it. Note washer/spacer and rubber insulator fitment (Fig. 12.41).
6 Refitting is a reversal of the removal procedure, ensuring that the wiper motor shaft locates through its collar on the exterior panel surface. Tighten all bolts to their specified torque.
7 If a fault develops in the tailgate wiper motor circuit, before suspecting the switch, check that the tailgate electrical contact spring pins move freely, are aligned correctly with their respective contact plates, and that the contact surfaces are absolutely clean. Check also that there is continuity between the contact pins and motor, and that the motor is correctly earthed to the tailgate by the earth cable.

40 Windscreen wiper motor – removal and refitting

1 Disconnect the battery earth terminal.
2 Dependent on model, disconnect and remove the air filter housing to allow access to remove the bulkhead panel.
3 Remove the expansion tank from the right-hand rear corner of the engine compartment (see Chapter 2).
4 Release the wiring loom, any connectors, cable ties and hoses from the right-hand half of the bulkhead panel, then remove its rubber seal.
5 The right-hand half of the bulkhead panel is secured by screws and a single nut. The nut is located behind the panel at the bonnet hinge end. Release the right-hand half of the panel and, having ensured that it is free to move, remove it (photos).
6 Unscrew the nut from the driving shaft (photo), and pull the crank off the driving shaft taper.
7 Undo the three wiper motor retaining bolts and withdraw the motor assembly. Remove the motor cover, then disconnect the multi-plug and remove the motor from the vehicle (photos).
8 Refitting is a reversal of the removal procedure, ensuring that reference is made to the Specifications for tightening torque details. Top up the expansion tank to its 'MAX' mark upon completion (see Specifications in Chapter 2).

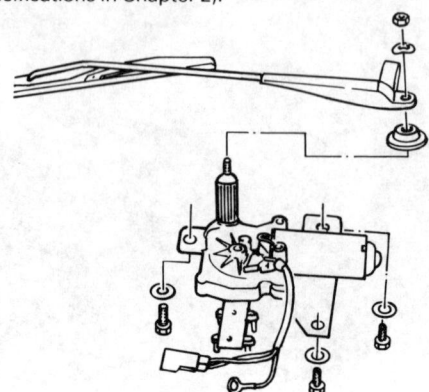

Fig. 12.42 General view of tailgate wiper mechanism (Sec 39)

Chapter 12 Electrical system

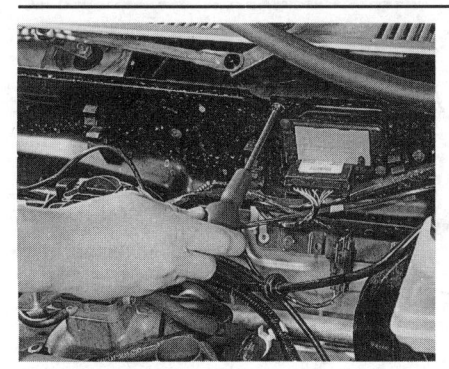

40.5A Remove all retaining screws from the right-hand half of the bulkhead panel (upper centre screw shown) ...

40.5B ... then disengage the right-hand half of the panel from under its single retaining nut (left-hand shown)

40.6 Unscrew the nut from the driving shaft

40.7A With the crank removed from the driving shaft taper, undo the three wiper motor retaining bolts (arrowed)

40.7B Withdraw the wiper motor assembly

40.7C Remove the wiper motor cover

40.7D Disconnect the multi-plug from the wiper motor

41 Windscreen wiper linkage – removal and refitting

1 Pull the rubber seal off the bulkhead panel.
2 Bring the windscreen wiper linkage to an accessible position, using the ignition switch as a means of stopping the wiper motor returning to the parked position.
3 Disconnect the battery earth terminal.
4 Prise the linkages off their ball pivots as required, using a suitably sized open-ended spanner. The crank has a dual (vertically stacked) ball

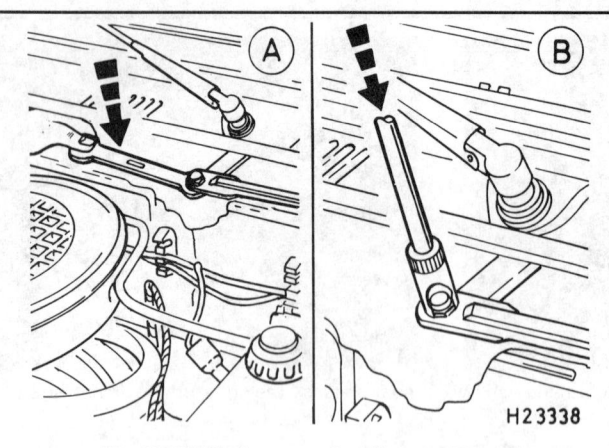

286   Chapter 12   Electrical system

Fig. 12.43 Windscreen wiper linkage (Sec 41)

A   Removal   B   Refitting

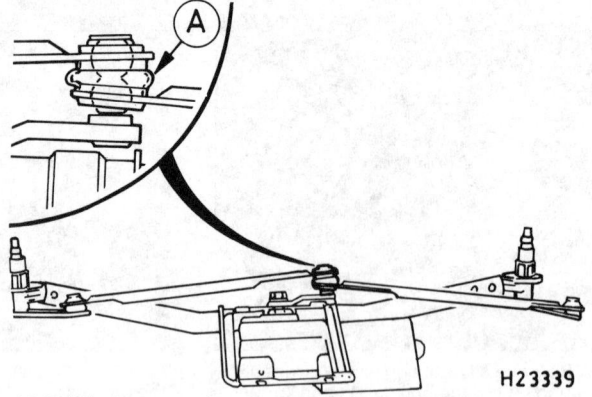

Fig. 12.44 Windscreen wiper motor bracket with linkage and motor. Inset shows bellows arrangement (A) (Sec 41)

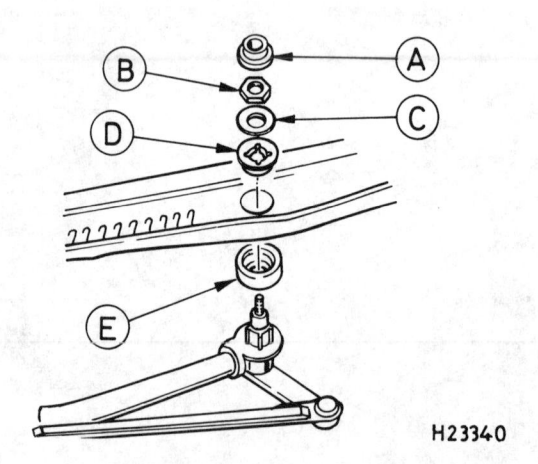

Fig. 12.45 Pivot shaft fixture (Sec 42)

A	Collar	D	Outer rubber bush
B	Pivot shaft securing nut	E	Inner rubber bush
C	Washer		

pivot with a bellows separating the two linkages (see Fig. 12.43 and Fig. 12.44).

5   If the running faces of the ball pivots are damaged, the pivot shaft(s) or crank must be renewed.

6   Prior to refitting, grease the ball pivot sockets on the linkages, then locate them on their appropriate ball pivots and press into position using a suitably-sized socket (refer to Fig. 12.43). The bellows between the two linkages on the crank must also be greased (see Specifications).

7   Refit the bulkhead panel rubber seal.

8   Reconnect the battery earth terminal and check the wiper operation.

42   Windscreen wiper pivot shaft – renewal

1   Remove the windscreen wiper arms (see Section 38).

2   Carry out the procedure detailed in Section 40, paragraphs 1 to 5.

3   Remove the collars from the pivot shafts, unscrew the pivot shaft securing nuts and remove the washers and outer rubber bushes (Fig. 12.45).

4   Undo the two wiper motor bracket securing bolts and remove the bracket assembly from the vehicle, disconnecting the multi-plug from the motor as the assembly is withdrawn (photo).

5   Remove the inner rubber bush from the pivot shaft being renewed, and disconnect its linkage as described in paragraph 4 of the previous Section.

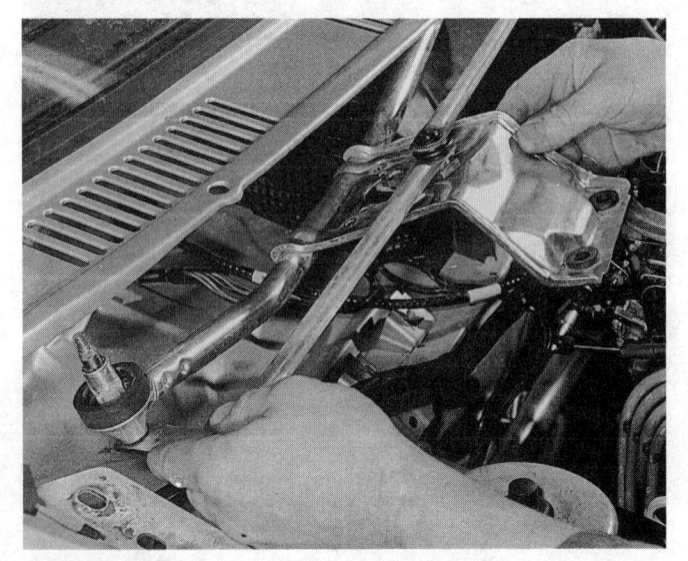

6   Remove the circlip, washers, shims and O-ring, keeping them in order for refitting, and withdraw the pivot shaft from the wiper motor bracket assembly.

7   Refitting is a reversal of the removal procedure, greasing the pivot shaft (see Specifications) before inserting it to its location. Refer to paragraph 6 in the previous Section before refitting the linkage. Reference must also be made to the Specifications for details of tightening torques. Refit the windscreen wiper arms in accordance with Section 38. Top up the expansion tank to its MAX mark upon completion (see Specifications in Chapter 2).

43   Windscreen washer jets and hose – removal and refitting

1   Disconnect the battery earth terminal.

2   With the bonnet raised and supported on its stay, release the fasteners securing its insulation panel (where fitted). Remove the insulation panel.

3   Carefully press in the retaining lugs on the washer jets using a flat-bladed screwdriver, then raise the washer jets from the exterior surface of the bonnet and separate them from their hoses (photo).

4   The windscreen washer jet hose may have been fitted with a one-way (non-return) valve (see Chapter 11, Section 6). If this is the case, the main hose run sections can be removed from either side of the valve as required.

5   Raise the front of the vehicle and support it securely using axle stands.

6   Disconnect the windscreen washer hose (marked with adhesive tape) from the washer pump (see Fig. 12.47). Withdraw the hose from the reservoir guide, and into the engine compartment.

42.4 Removing the wiper motor bracket and linkage assembly (wiper motor removed)

Chapter 12 Electrical system

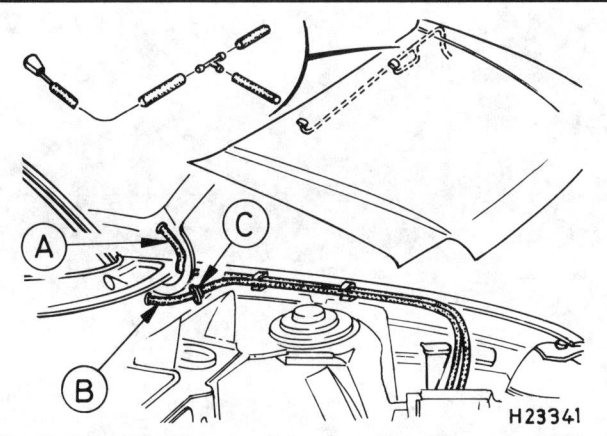

Fig. 12.46 Routing of washer hoses in engine compartment (Sec 43)

A Windscreen washer hose
B Tailgate washer hose
C Tailgate washer hose one-way valve

7 Release the hose from its clips in the engine compartment, including the bonnet hinge clip, release the hose grommet from the bonnet (where fitted) and withdraw the hoses from the bonnet (Fig. 12.46).
8 Refitting is a reversal of the removal procedure. The water jets are a push fit into the bonnet. If the hose has been renewed, mark it (near the pump) using adhesive tape for future identification. Adjust the washer jets, as required, by inserting a pin into the ball adjusting nozzle and swivelling to the required position.

44 Tailgate washer jet and hose – removal and refitting

1 Disconnect the battery earth terminal.
2 On XR2i models, remove the tailgate spoiler (Chapter 11, Section 36).
3 Remove the central blanking plug from the upper interior surface of the tailgate, to expose the washer jet base.
4 Depress the washer jet retaining lug using a flat-bladed screwdriver, then push the washer jet out through the panel. From the outside, fully withdraw the washer jet and disconnect it from its hose. Note washer jet seal fitment.
5 Remove the left-hand sun visor (Chapter 11, Section 56).
6 Remove the left-hand A-pillar trim (Chapter 11, Section 44).
7 Release the left-hand side of the headlining by removing the retaining clips/grab handles/coat hooks, as applicable.

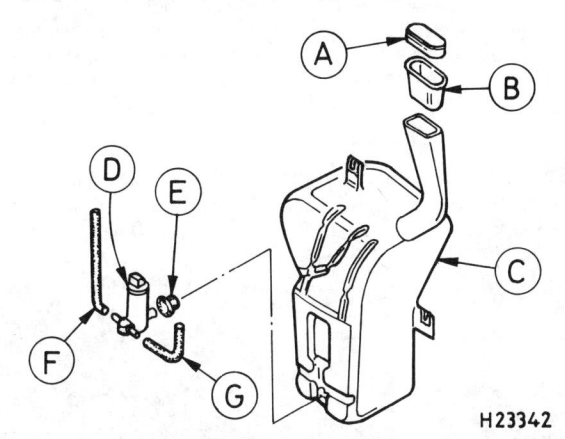

Fig. 12.47 Exploded view of windscreen/tailgate washer reservoir and pump components (Sec 45)

A Cap
B Filter
C Reservoir
D Washer pump
E Pump seal
F Tailgate washer hose
G Windscreen washer hose (marked with adhesive tape)

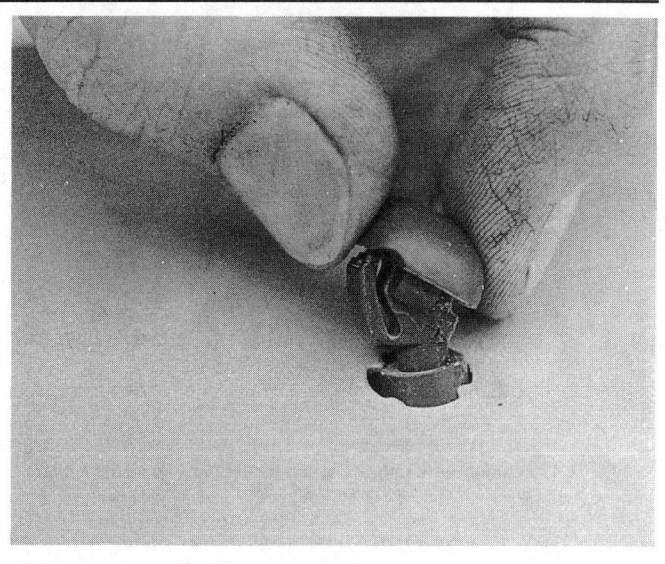

43.3 Removing a windscreen washer jet

8 In the engine compartment, disconnect the tailgate washer hose from its valve (see Fig. 12.46). The forward hose run may be removed, if required, in a manner similar to that described in Section 43, paragraphs 5 and 6, releasing it from its clips in the engine compartment.
9 Remove the tailgate washer hose grommet, then withdraw the hose through the bulkhead into the passenger compartment.
10 Release the hose from its A-pillar and roof frame locations. Release the grommet (hose protector) from its tailgate and body locations, and withdraw the hose from the vehicle.
11 Refitting is a reversal of the removal procedure. The washer jet is a push fit into the tailgate, having ensured that its seal is seated correctly. The nozzle is not adjustable.

45 Windscreen/tailgate washer pump – removal and refitting

1 Disconnect the battery earth terminal.
2 Raise the front of the vehicle and support it securely using axle stands.
3 Withdraw the washer pump from the reservoir, collecting the fluid in a suitable container as the pump is removed. Note seal fitment.
4 Having noted hose fitment to pump connectors, remove the multi-plug and hoses, then remove the pump from the vehicle.
5 Examine the pump seal. If it is perished or otherwise damaged, it should be replaced.
6 Refit the seal then the windscreen washer pump to the reservoir, and ensure that it is securely located. The refitting procedure is a reversal of removal. Refill the washer reservoir upon completion.

46 Windscreen/tailgate washer reservoir – removal and refitting

1 Disconnect the battery earth terminal.
2 Loosen the left-hand front roadwheel nuts, then raise the front of the vehicle and support it securely using axle stands. Remove the roadwheel.
3 Remove the wheelarch liner (Chapter 11, Section 59).
4 Withdraw the windscreen/tailgate washer pump from the reservoir with its hoses and multi-plug attached, collecting the fluid in a suitable container. Note seal fitment.
5 Release the hoses and wiring from the reservoir guide and remove the three reservoir securing bolts (Fig. 12.48). Remove the reservoir.
6 Refitting is a reversal of the removal procedure, inserting the seal to the reservoir prior to fitting the pump. Refer to the Specifications for tightening torque details. Refill the washer reservoir upon completion.

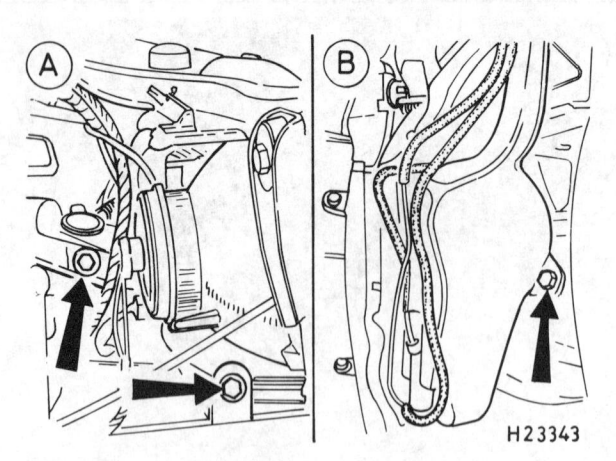

Fig. 12.48 Windscreen/tailgate washer reservoir securing bolts (arrowed) (Sec 46)

A In engine compartment   B In wheelarch (wheelarch liner removed)

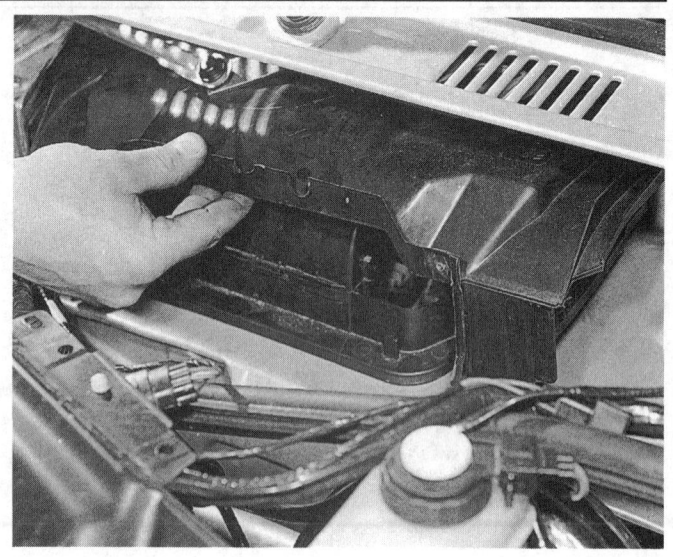

49.6 Removing the cover from the heater fan motor assembly

47 Heated rear window element – general

1 The rear window heater element is fixed to the interior surface of the glass.
2 When cleaning the window use only water and a leather or soft cloth, and avoid scratching with rings on the fingers.
3 Avoid sticking labels over the element and packing luggage so that it can rub against the glass.
4 In the event of the element being damaged, it can be repaired using one of the special conductive paints now available.
5 If a fault develops in the heated rear window circuit, before suspecting the switch, check that the tailgate electrical contact spring pins move freely, are aligned correctly with their respective contact plates, and that the contact surfaces are absolutely clean.

48 Heated windscreen – general

1 The factory-fitted option of an electrically heated front windscreen is available on certain models.
2 For quick de-icing or demisting, a fine wire mesh through which electric current passes, is incorporated into the glass of the windscreen.
3 The heated windscreen operates only when the engine is running and, unless switched off manually, automatically switches itself off after approximately four minutes.
4 To obtain access to its electrical contacts, remove the A-pillar trim (see Chapter 11, Section 44).

49 Heater fan motor and resistor assembly – removal and refitting

1 Disconnect the battery earth terminal.
2 Dependent on model, disconnect and remove the air filter housing to allow access to remove the bulkhead panel.
3 Remove the expansion tank from the right-hand rear corner of the engine compartment (see Chapter 2), and the jack and wheelbrace from the left-hand rear corner.
4 Disconnect the modules on the bulkhead panel. Release the wiring loom and any connectors, cable ties and hoses from the bulkhead panel, and remove its rubber seal.
5 Remove the bulkhead panel. The panel is secured by screws, with a nut at either end (behind the panel), and is removed in two sections (see photos 40.5A and 40.5B).
6 Detach and remove the cover from the heater fan motor assembly (photo).
7 Disconnect the wiring from the heater fan motor and resistor assembly (photo).
8 Undo the two nuts securing the assembly to the bulkhead and remove it from the vehicle (Fig. 12.49).
9 Detach the heater fan cover, having released its two retaining clips.
10 Release the heater fan motor retaining strap from underneath, using a flat-bladed screwdriver, and remove it. Remove the resistor assembly and/or heater fan motor, as required (photos).
11 Refitting is a reversal of the removal procedure, ensuring that the heater fan motor wires are routed under the retaining strap, but are not pinched by it. Ensure that the expansion tank is topped up to the MAX mark upon completion (see Specifications in Chapter 2).

49.7 Disconnecting the wiring from the heater fan motor and resistor assembly (arrowed)

49.10A Releasing the heater fan motor retaining strap (from the underside of the assembly)

49.10B Heater fan motor retaining strap released to enable motor removal

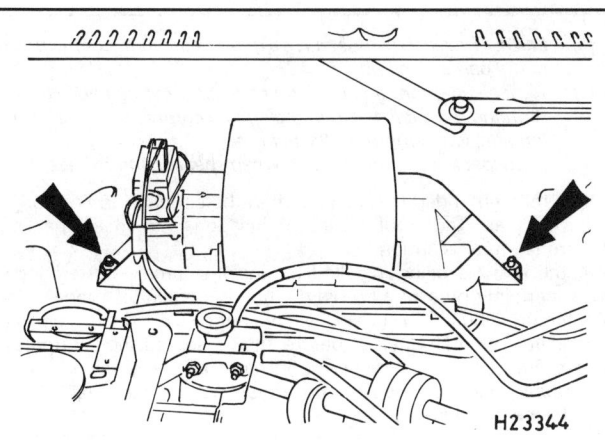

Chapter 12 Electrical system

Fig. 12.49 Heater fan motor assembly securing nuts (arrowed) (Sec 49)

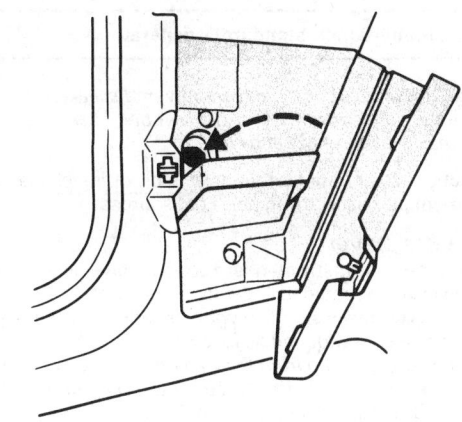

Fig. 12.50 Inertia switch/fuel cut-off reset button – arrowed (fuel injection) (Sec 51)

50 Choke warning light switch – removal and refitting

1 The choke warning light switch is an integral part of the choke cable.
2 See Chapter 3 for choke cable removal and refitting procedures.

51 Inertia switch (fuel cut-off) – removal and refitting

Note: *Fuel injection equipped vehicles are fitted with an inertia switch to automatically cut off the current flow to the electric fuel pump in the event of a collision. The switch may be reset by depressing the button on its upper surface, which is accessible through the passenger footwell map stowage pocket (Fig. 12.50)*

1 Disconnect the battery earth terminal.
2 Remove the left-hand sill scuff plate (see Chapter 11, Section 44).
3 Undo the single screw securing the inertia switch bracket. Remove the bracket assembly, disconnecting the multi-plug from the switches as it is withdrawn (photos).
4 The switch may now be separated from its mounting bracket by removing its two securing screws.
5 Refitting is a reversal of the removal procedure, ensuring that the multi-plug locking tabs snap into position, and that the base of the bracket locates correctly to the panel. Depress the button on top of the switch after refitting the bracket, to ensure that it is reset.

52 Fuel tank sender unit – removal and refitting

For fuel tank sender unit removal and refitting procedure, refer to Chapter 3.

53 Horn – removal and refitting

1 The horn is mounted forward on the left-hand side of the engine compartment, near the battery.
2 Disconnect the battery earth terminal.
3 Remove the wiring loom connection from the horn(s).
4 Both single and dual horns are mounted to a bracket which is secured to the bodywork by a single bolt. Remove the bolt and withdraw the horn(s) and bracket from the vehicle. The horn(s) may be separated from the bracket, as required, by removing the retaining nut(s).
5 Refitting is a reversal of the removal procedure. Tighten the bolt securing the horn bracket to the bodywork to its specified torque.

54 Radiator cooling fan motor – removal and refitting

Refer to Chapter 2 for radiator cooling fan motor removal and refitting procedures.

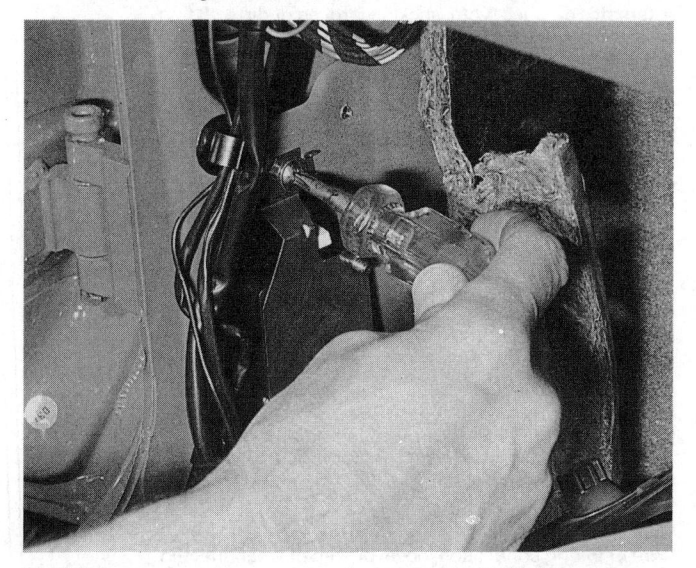

51.3A Remove the screw securing the inertia switch bracket ...

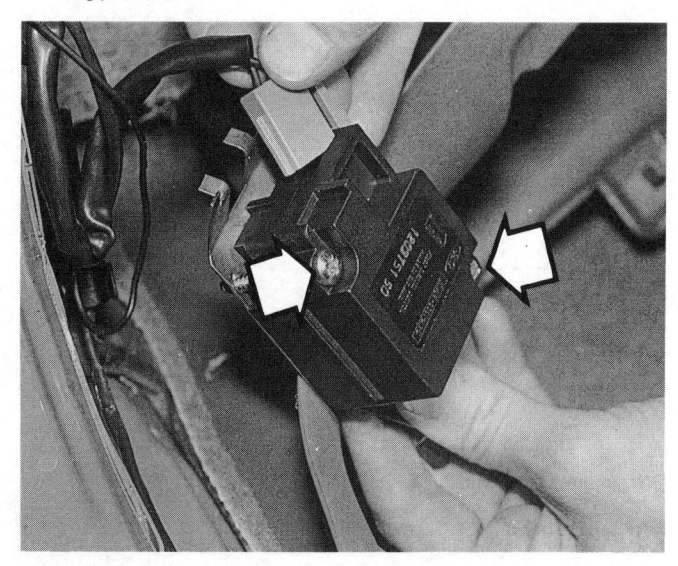

51.3B ... and disconnect its multi-plug as it is withdrawn. Switch to bracket retaining screws (arrowed)

55 Radio equipment (non-standard) – general

Warning: *The siting of uprated high-powered speakers close to seat belt inertia reels can adversely affect the seat belt operation, due to the influence of the speakers' magnetic field.*

1 This Section covers briefly the installation of in-vehicle entertainment equipment purchased from non-Ford sources.

Radio/cassette player

2 It is recommended that a standard sized receiver is purchased and fitted into the location provided in the facia panel.
3 A fitting kit is normally supplied with the radio or cassette player.
4 Connections will be required as follows
   (a) *Power supply, taken from the ignition switch so that the radio is only operational with the ignition key in position I or II. Always insert a 2A in-line fuse in the power lead*
   (b) *Earth. The receiver must have a good clean earth connection to a metal part of the body*
   (c) *Aerial lead. From an aerial which itself must be earthed. Avoid routeing the cable through the engine compartment or near the ignition, wiper motor or flasher relay*
   (d) *Loudspeaker connections, between speaker and receiver*

5 It is recommended that aerial fitment be confined to Ford original equipment as the roof-mounted location provides the most interference-free reception.
6 If the radio is being installed for the first time, interference will almost certainly present a problem when the engine is running. Most manufacturers build in a certain level of suppression, but it is advisable to consult your Ford dealer before fitting any further suppression devices. Further information on suppression may be obtained from the *'Automobile Electrical and Electronic Systems'* manual, available from the publishers of this manual.

56 Fault diagnosis – electrical system

Symptom	Reason(s)
Starter fails to turn engine	Battery discharged or defective Battery terminals loose or corroded Starter motor terminals loose or corroded Engine earth cable loose or broken Starter motor faulty or solenoid not functioning Starter motor brushes worn or sticking Dirty or worn commutator Armature faulty
Starter turns engine very slowly	Battery discharged Starter motor brushes badly worn or sticking Loose wires in starter motor circuit
Starter spins but does not turn engine	Pinion or flywheel gear teeth broken or worn
Starter motor noisy or rough in engagement	Pinion or flywheel gear teeth broken or worn Starter motor retaining bolts loose
Battery will not hold charge for more than a few days	Battery defective internally Electrolyte level too low Battery terminals loose or corroded Alternator drivebelt slipping Alternator not charging Short circuit causing continual battery drain Integral regulator unit faulty
Ignition light fails to go out, battery runs flat in a few days	Alternator drivebelt loose and slipping or broken Alternator brushes worn, sticking, broken or dirty Alternator brush springs weak or broken Internal fault in alternator
Failure of component motor	Fuse blown Wire or wiring connection loose, broken or disconnected Brushes sticking or worn Armature bearings seized Armature faulty Dirty or burnt commutator Field coils faulty
Lights do not come on	If engine not running, battery discharged Wire or wiring connection loose, broken or disconnected Bulb blown Fuse blown Switch faulty
Lights give poor illumination	Lamp glasses dirty Lamps badly out of alignment
Horn operates continually	Horn push earthed or stuck down
Horn fails to operate	Fuse blown Wire or wiring connection loose, broken or disconnected Internal fault in horn

Ford Fiesta RS Turbo

Ford Courier

Chapter 13 Supplement:
Revisions and information on later models

Contents

Introduction 1
   General
   Project vehicles
Specifications 2
Buying spare parts and vehicle identification numbers 3
   DOHC 16-valve engine number location
   VIN plate emission standard codes
Routine maintenance 4
   Vehicles built up to (and including) 1991 model year, and all
   RS Turbo models
   Vehicles built from 1992 model year onwards, except RS
   Turbo
Engine 5
Part A: All engines
   Compression test – description and interpretation
Part B: HCS and CVH engines
   1.0 and 1.1 HCS engines – modifications
   1.1 HCS CFi engine
   1.3 HCS engines
   1.4 and 1.6 (non-turbo) CVH engines – modifications
   1.6 CVH EFi (turbocharged) engine
Part C: DOHC 16-valve engine – general
   General description
   Lubrication system – description
   Maintenance and inspection
Part D: Operations possible with DOHC 16-valve engine in vehicle
   Major operations possible with the engine in the vehicle
   Compression test – description and interpretation
   Cylinder head cover – removal and refitting
   Timing belt covers – removal and refitting
   Timing belt – renewal
   Timing belt tensioner – removal and refitting
   Camshaft oil seals – renewal
   Camshafts and hydraulic tappets – removal, examination and refitting
   Cylinder head – removal and refitting
   Cylinder head and pistons – decarbonising
   Cylinder head – dismantling, overhaul and reassembly
   Sump – removal and refitting
   Crankshaft oil seals – renewal
   Oil pump – removal, inspection and refitting
   Piston/connecting rod assemblies – removal and refitting
   Flywheel – removal and refitting
   Engine/transmission mountings – removal and refitting
Part E: DOHC 16-valve engine and transmission removal
   Major operations requiring engine removal
   Engine – method of removal
   Engine/transmission assembly – removal
   Engine and transmission – separation
   Engine – refitting to transmission
   Engine/transmission assembly – refitting
Part F: DOHC 16-valve engine dismantling, overhaul and reassembly
   Engine dismantling – general

   Engine – complete dismantling
   Examination and renovation – general
   Engine – component examination and renovation
   Engine reassembly – general
   Engine – complete reassembly
   Engine – initial start-up after overhaul
   Fault diagnosis – engine
Cooling, heating and ventilation systems 6
   Coolant renewal – general
   Cooling system modifications – 1.6 litre EFi (turbocharged) engine
   Cooling system – DOHC 16-valve engine
Fuel, exhaust and emission control systems 7
Part A: Carburettor fuel system – modifications
   Tamperproof mixture adjusting screws
Part B: 1.1 and 1.3 litre Central Fuel injection (CFi) system
   General description
Part C: 1.4 litre Central Fuel injection (CFi) system – modifications
   General description – engines from September 1990 onwards
   Quick-release fuel line couplings – general
   Fuel starvation and/or fuel tank collapse problems
   Fuel filter – removal and refitting
   Inlet manifold heater – removal and refitting
   Throttle position sensor wiring
   Injector ballast resistor – removal and refitting
Part D: 1.6 litre (non-turbo) Electronic Fuel injection (EFi) system – modifications
   Quick-release fuel line couplings – general
   Fuel filter – removal and refitting
   CO potentiometer – adjustment, removal and refitting
   Base idle speed – checking and adjustment
   Idle speed control valve – general
Part E: 1.6 litre (turbocharged) Electronic Fuel injection (EFi) system
   General description
   Quick-release fuel line couplings – general
   Fuel filter – removal and refitting
   Fuel injectors – removal and refitting
   CO potentiometer – adjustment, removal and refitting
   Base idle speed – checking and adjustment
   Idle speed control valve – removal and refitting
   Intercooler – removal and refitting
   Turbocharger – removal, examination and refitting
   Turbocharged engines – notes and precautions
   Boost control valve – removal and refitting
   Boost pressure – check and adjustment
Part F: DOHC 16-valve engine EEC-IV engine management system – fuel system components
   General description
   Fuel system – precautions
   Quick-release fuel line couplings – general
   Fuel starvation and/or fuel tank collapse problems
   Maintenance and inspection
   Idle speed and mixture adjustment – general
   Air filter element – renewal

Chapter 13 Supplement: Revisions and information on later models

Air filter housing – removal and refitting
Accelerator cable – removal and refitting
Accelerator pedal – removal and refitting
Fuel tank – removal and refitting
Fuel tank filler pipe – removal and refitting
Fuel pump – testing
Fuel pump/fuel gauge sender unit – removal and refitting
Inertia switch (fuel cut-off) – removal and refitting
Fuel filter – removal and refitting
Fuel pressure regulator – removal and refitting
Fuel distributor rail and injectors – removal and refitting
Throttle position sensor – removal and refitting
Idle speed control valve – removal and refitting
Throttle housing – removal and refitting
Air mass meter – removal and refitting
Fault diagnosis – general
Part G: Manifolds, exhaust and emission control systems (except DOHC 16-valve engine)
1.1, 1.3 and 1.4 litre CFi engines – general
Pulse-air system – 1.3 litre HCS CFi engines
1.6 litre EFi (non-turbo) engines – general
1.6 litre EFi (turbocharged) engine – general
Part H: Manifolds, exhaust and emission control systems (DOHC 16-valve engine)
Inlet manifold – general
Exhaust manifold – general
Exhaust system – renewal
Emissions control systems – general description
Emissions control system components – removal and refitting
Ignition and engine management systems 8
Part A: HCS and CVH carburettor engines
1.0 and 1.1 litre HCS engines – modifications
1.3 litre engine – general
1.4 litre CVH engine – general
Part B: 1.1 and 1.3 litre HCS CFi engines
General description
Component removal and refitting
Part C: 1.4 CFi and 1.6 EFi (turbocharged) engines
1.4 CFi engines – modifications (September 1990-on)
1.4 CFi engines (September 1990-on) – component removal and refitting
1.6 EFi (turbocharged) engine – general
Part D: DOHC 16-valve engine EEC-IV engine management system – ignition system components
General description
Maintenance, inspection and precautions
Spark plugs and HT leads – inspection and renewal
Air Charge Temperature (ACT) sensor – removal and refitting
Engine Coolant Temperature (ECT) sensor – removal and refitting
Engine speed/crankshaft position sensor – removal and refitting
Camshaft position sensor – removal and refitting
Road speed sensor unit – removal and refitting
Ignition coil – removal and refitting
EDIS ignition module – removal and refitting
EEC-IV engine management module – removal and refitting
Fault diagnosis
Clutch 9
'Low-lift' clutch components – all 1.6 and 1.8 litre engines
Clutch release bearing guide sleeve
Driveshafts 10
Tripod-type inner CV joint
Vibration damper Torx-type bolts
Braking system 11
Master cylinder pressure-conscious reducing valves – removal and refitting
Rear brake shoe shims
Anti-lock braking system – modulator drivebelt renewal
Light-laden valve (Courier)
Suspension and steering 12
Front hub bearings – renewal (all models, June 1990-on)
Rear axle bush – renewal (all models except Courier, April 1990-on)
Rear suspension anti-roll bar – removal and refitting

Rear suspension (Courier) – general description
Rear suspension (Courier) – ride height adjustment
Rear suspension (Courier) – component removal and refitting
Steering wheel (September 1993-on models) – removal and refitting
Bodywork and fittings 13
Bonnet vents (RS Turbo) – removal and refitting
Roof mouldings (Courier) – removal and refitting
Rear quarter mouldings (Courier) – removal and refitting
Rear bumper (Courier) – removal and refitting
Front door trim panel (1992-on models) – removal and refitting
Opening rear side windows (three-door Hatchbacks) – removal and refitting
Rear sliding windows (Courier Kombi) – removal and refitting
Rear door window glass (Courier) – removal and refitting
Rear door check assembly (Courier) – removal and refitting
Rear doors (Courier) – removal and refitting
Rear door interior handle (Courier) – removal and refitting
Rear door exterior handle and lock assembly (Courier) – removal and refitting
Rear door upper and lower catches (Courier) – removal and refitting
Full-length sunroof ('Calypso') – general
Rear seats (Courier Kombi) – removal and refitting
Headlining (Courier) – removal and refitting
Load compartment interior trim panels (Courier) – removal and refitting
Electrical system 14
Alternator drivebelt (DOHC 16-valve engine) – check, renewal and adjustment
Starter motor – removal and refitting
Nippondenso reduction-gear starter motor – overhaul
Lucas/Magneti Marelli M80R starter motor – overhaul
Fuses and relays – revisions
'Lights-on' warning module (1993-on models) – removal and refitting
Electric front window switch (1992-on models) – removal and refitting
Full-length sunroof switch ('Calypso') – removal and refitting
Temperature gauge sender unit (DOHC 16-valve engine) – removal and refitting
Tail lamp assembly (Courier) – removal and refitting
Rear number plate lamp (Courier) – removal and refitting
Tail lamp (Courier) – bulb renewal
Rear number plate lamp (Courier) – bulb renewal
Load compartment lamp (Van and Courier) – bulb renewal
Heated rear door window glass (Courier) – general
Anti-theft immobilisation system – general

1 Introduction

General

This Supplement contains information which is additional to, or a revision of, the material contained in the preceding twelve Chapters of this manual. Since the manual was first written, the Fiesta range has undergone few significant changes or revisions, apart from the introduction of fuel-injected, catalytic converter-equipped, 1.1 and 1.3 litre models, the RS Turbo model, the DOHC 16-valve engine and the Courier. Whilst primarily intended to cover models produced from late 1990 onwards, additional or revised information is included on earlier models. To use the Supplement to its best advantage, it is therefore recommended that it is referred to before the main Chapters of this manual.

Project vehicles

The vehicle used in the preparation of this Supplement, and appearing in some of the photographic sequences, was a 1992-model Fiesta XR2i 16V (105 PS 1.8 litre DOHC 16-valve engine). Additional work was carried out and photographed on a 1993-model Courier 1.3 Standard (1.3 HCS carburettor engine); also used was a 1993-model Fiesta 1.1 LX (1.1 HCS CFi engine).

Engine and underbonnet component locations – 1.8 litre DOHC 16-valve engine (XR2i) shown

1. Engine oil filler cap
2. Engine oil level dipstick
3. Cooling system expansion tank
4. Braking system fluid reservoir
5. Windscreen/tailgate washer fluid reservoir cap
6. Battery
7. VIN plate
8. Thermostat housing
9. Timing belt top cover
10. Top of suspension strut mounting assembly
11. Windscreen wiper motor mounting bracket
12. Jack and wheelbrace retaining bolt
13. Ignition coil
14. Fuel filter
15. Air filter housing
16. Air intake trunking
17. Idle speed control valve
18. Fuel pressure regulator
19. Throttle housing
20. Inlet manifold
21. Throttle position sensor
22. Fuel system pressure release/test point
23. EEC-IV engine management module cover
24. Air mass meter
25. EDIS ignition module

Chapter 13 Supplement: Revisions and information on later models

Front underbody view – 1.8 litre DOHC 16-valve engine (XR2i) shown

1. Engine oil drain plug
2. Front suspension lower arm
3. Brake caliper assembly
4. Driveshaft
5. Alternator
6. Alternator drivebelt shield
7. Horn
8. Windscreen/tailgate washer pump
9. Carbon canister
10. HEGO sensor
11. Front suspension crossmember
12. Catalytic converter
13. Underbody heat shields
14. Gearchange mechanism shift rod
15. Gearchange mechanism stabiliser bar

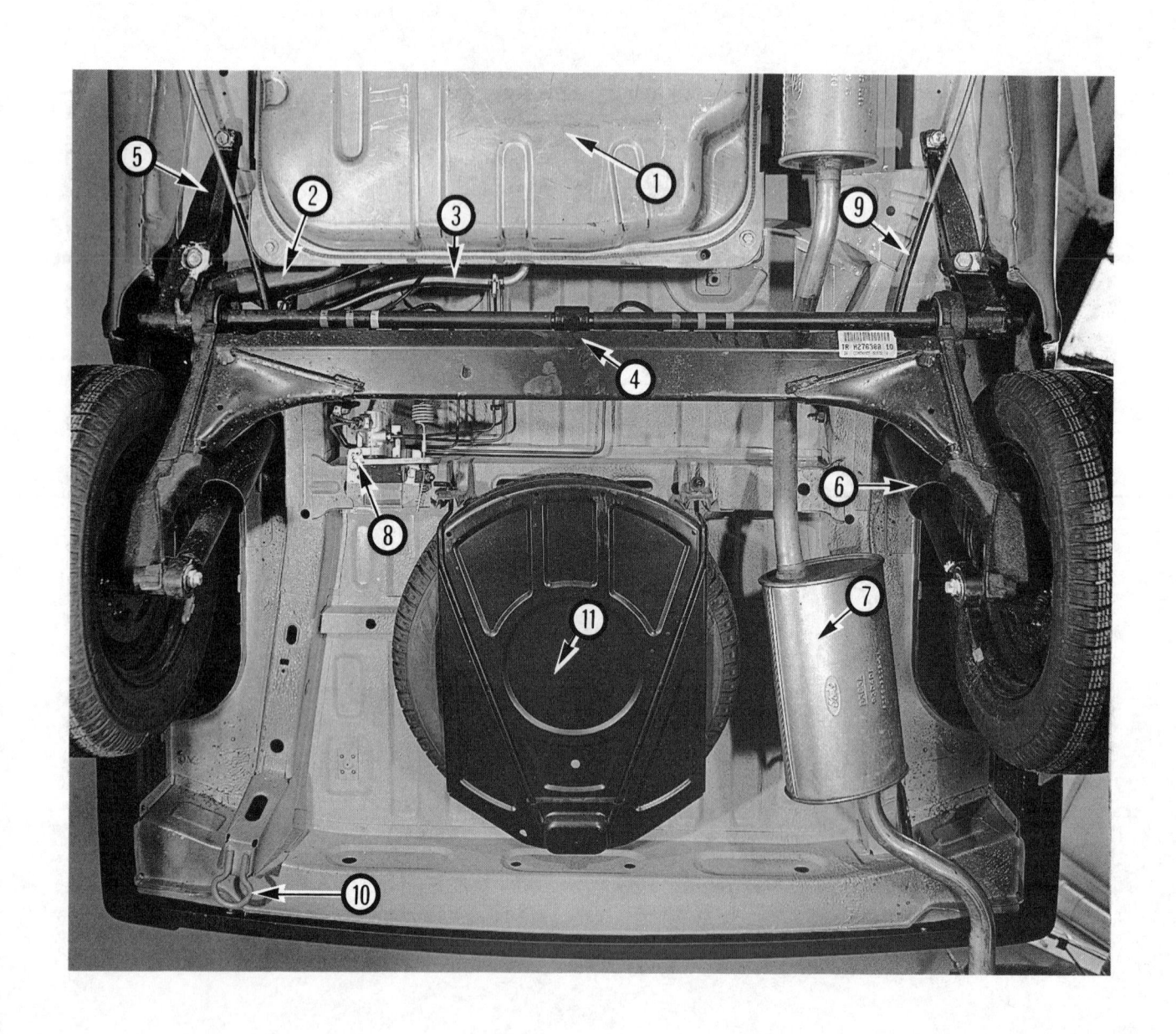

Chapter 13 Supplement: Revisions and information on later models

Rear underbody view – Courier shown

1 Fuel tank
2 Fuel filler pipe
3 Fuel tank ventilation hose
4 Rear axle assembly – spring torsion bars visible
5 Rear axle pivot brackets
6 Rear suspension dampers
7 Exhaust system rear silencer
8 Braking system light-laden valve
9 Handbrake cables
10 Rear towing eye
11 Spare wheel carrier

Chapter 13 Supplement: Revisions and information on later models

2 Specifications

The Specifications given below are revisions of, or supplementary to, those at the beginning of the preceding Chapters. Unless otherwise specified, for information on 1.3 litre HCS-engined models, refer to that given for 1.1 litre HCS models. Similarly, for information on RS Turbo and RS 1800 models, refer to that given for the XR2i (1.6 litre CVH EFi).

General dimensions, weights and capacities

Dimensions

Overall length – Courier	4053 mm (159.6 in)
Width (excluding mirrors) – Courier	1650 mm (64.9 in)
Overall height – Courier	1812 to 1840 mm (71.3 to 72.4 in)
Wheelbase – Courier	2700 mm (106.3 in)
Front track:	
RS Turbo, RS 1800, XR2i – 1.8 litre model	1430 mm (56.3 in)
Courier	Not available
Rear track – Courier	Not available

Weights

Kerb weight:	
3-door Hatchback:	
1.0 litre models	770 to 780 kg
1.1 litre models	785 to 825 kg
1.3 litre models	825 to 850 kg
XR2i – 1.8 litre model	955 to 974 kg
RS Turbo	910 to 930 kg
RS 1800	954 to 965 kg
5-door Hatchback:	
1.1 litre models	810 to 845 kg
1.3 litre models	845 to 870 kg
1.4 litre models	840 to 870 kg
Vans:	
1.1 litre models	775 to 790 kg
1.3 litre models	795 kg
Courier	920 kg
Payload:	
1.1 litre Van models	310 to 335 kg
1.3 litre Van models	330 kg
Courier	515 to 525 kg

Capacities

Engine oil – with filter:	
1.6 litre CVH EFi (turbocharged) engine	3.60 litres (6.3 pints)
DOHC 16-valve engines	4.25 litres (7.5 pints)
Engine oil – without filter:	
1.6 litre CVH EFi (turbocharged) engine	3.35 litres (5.9 pints)
DOHC 16-valve engines	3.75 litres (6.6 pints)
Approximate amount required to raise engine oil level from dipstick 'MIN' to 'MAX' marks:	
DOHC 16-valve engines	0.75 litres (1.3 pints)
All other engines	0.5 to 1.0 litres (0.9 to 1.8 pints)
Cooling system (including heater) – DOHC 16-valve engines	7.0 litres (12.3 pints)

OHV (HCS) engines

General

Engine code:	
1.0 litre engine	TLB
1.1 litre carburettor engine – to 15:05 emission standard	GUD
1.1 litre carburettor engine – to 15:04 emission standard	GUE
1.1 litre CFi engine	G6A
1.3 litre carburettor engine	JBC
1.3 litre CFi engine	J6A
Power output:	
1.0 litre engine	33 kW/45 PS @ 5000 rpm
1.1 litre carburettor engines	40 kW/54 PS @ 5200 rpm
1.1 litre CFi engine	37 kW/50 PS @ 5200 rpm
1.3 litre engines	44 kW/60 PS @ 5000 rpm
Capacity – 1.3 litre engines	1297 cc
Bore – 1.3 litre engines	73.94 mm
Stroke – 1.3 litre engines	75.48 mm
Compression ratio – 1.3 litre CFi engine	8.8:1
Compression pressure – all HCS engines, at starter motor speed, engine fully warmed-up	13 to 16 bars

Cylinder block – 1.3 litre engines
Cylinder bore diameter:
    Standard class 1 .... 73.940 to 73.950 mm
    Standard class 2 .... 73.950 to 73.960 mm
    Standard class 3 .... 73.960 to 73.970 mm
    Oversize 0.5 mm .... 74.500 to 74.510 mm
    Oversize 1.0 mm .... 75.000 to 75.010 mm

Pistons and piston rings – 1.3 litre engines
Piston diameter:
    Standard class 1 .... 73.910 to 73.920 mm
    Standard class 2 .... 73.920 to 73.930 mm
    Standard class 3 .... 73.930 to 73.940 mm
    Standard (service) .... 73.930 to 73.955 mm
    Oversize 0.5 mm .... 74.460 to 74.485 mm
    Oversize 1.0 mm .... 74.960 to 74.985 mm

Gudgeon pin – 1.3 litre engines
Length .... 63.3 to 64.6 mm

Camshaft – 1.3 litre engines
Cam lift:
    Inlet valves .... 5.70 mm
    Exhaust valves .... 5.76 mm
Cam length:
    Inlet valves .... 32.586 to 32.814 mm
    Exhaust valves .... 32.646 to 33.874 mm

Cylinder head – 1.3 litre engines
Combustion chamber capacity .... 31.79 to 33.79 cc

Valves – general
Valve timing – 1.3 litre engines:

	Carburettor engine	CFi engine
Inlet valve opens	16° BTDC	12° BTDC
Inlet valve closes	44° ABDC	48° ABDC
Exhaust valve opens	51° BBDC	47° BBDC
Exhaust valve closes	9° ATDC	13° ATDC

Valve clearances – engine cold:
    Inlet .... 0.20 mm (0.008 in)
    Exhaust .... 0.30 mm (0.012 in)

Note: Revised recommendation applies to all engines built from May 1990, and to any fitted with modified rocker cover – see text.

Inlet valve – 1.3 litre engines
Head diameter .... 34.40 to 34.60 mm
Valve lift (less clearance) .... 8.86 to 9.26 mm

Exhaust valve – 1.3 litre engines
Valve lift (less clearance) .... 8.96 to 9.36 mm

Lubrication
Approximate amount required to raise engine oil level from dipstick 'MIN' to 'MAX' marks .... 0.5 to 1.0 litres

Torque wrench settings

	Nm	lbf ft
Rocker cover (modified type) bolts	5	3.5

OHC (CVH) engines

General
Engine code:
    1.4 litre carburettor engine – to 15:04 emission standard .... FUF
    1.4 litre carburettor engine – to 15:05 emission standard .... FUG
    1.4 litre CFi engine F6E
    1.6 litre carburettor engine .... LUH
    1.6 litre EFi (non-turbo) engine .... LJC
    1.6 litre EFi (turbocharged) engine .... LHA
Power output:
    1.4 litre carburettor engines:
        Engine code FUF .... 55 kW/75 PS @ 5600 rpm
        Engine code FUG .... 54 kW/73 PS @ 5700 rpm
    1.4 litre CFi engine .... 52 kW/71 PS @ 5500 rpm
    1.6 litre carburettor engine .... 66 kW/90 PS @ 5800 rpm
    1.6 litre EFi (non-turbo) engine .... 81 kW/110 PS @ 6000 rpm
    1.6 litre EFi (turbocharged) engine .... 96 kW/130 PS @ 6000 rpm
Compression ratio – 1.6 litre EFi (turbocharged) engine 8.0:1
Compression pressure – all CVH engines, at starter motor speed, engine fully warmed-up .... 12 to 14 bars

Chapter 13 Supplement: Revisions and information on later models

Piston and piston rings – 1.6 litre EFi (turbocharged) engine
All dimensions as 1.6 litre CVH (not EFi) engine – see Chapter 1

Gudgeon pin – 1.6 litre EFi (turbocharged) engine
Length .... 63.600 to 64.400 mm

Camshaft – 1.6 litre EFi (turbocharged) engine
Cam lift (inlet and exhaust)... 6.90 mm
Inlet and exhaust cam length .... As 1.6 litre CVH (not EFi) engine

Cylinder head – 1.6 litre EFi (turbocharged) engine
Combustion chamber volume .... Not available
Minimum combustion chamber depth after skimming........................ Not available

Valves (general) – 1.6 litre EFi (turbocharged) engine
Valve timing – at 1 mm of cam lift:
    Inlet valve opens... 8° ATDC
    Inlet valve closes... 36° ABDC
    Exhaust valve opens .... 34° BBDC
    Exhaust valve closes .... 6° BTDC
Valve lift – inlet and exhaust .... 10.50 mm
Valve spring free length .... Not available
Fitted valve spring height.. Not available

Lubrication
Approximate amount required to raise engine oil level from dipstick 'MIN' to 'MAX' marks .... 0.5 to 1.0 litres
Minimum oil pressure @ 2000 rpm – 1.6 litre EFi (turbocharged) engine .... 2.8 bars @ 80°C (175°F)

Torque wrench settings – 1.6 litre EFi (turbocharged) engine

	Nm	lbf ft
Alternator mounting bracket-to-cylinder block bolts	41 to 58	30 to 43
Oil pipe bracket-to-engine lifting eye	8.5 to 12	6 to 8
Air intake crossover duct-to-rocker cover nuts	2.0 to 2.5	1.5 to 2
Oil cooler mounting adaptor-to-cylinder block	55 to 60	40 to 44
Turbocharger oil feed/oil pressure switch adaptor-to-cylinder block	15 to 41	11 to 30
Oil pressure switch-to-adaptor	25 to 29	18 to 21

DOHC 16-valve engines

General
Engine type .... Four-cylinder, in-line, double overhead camshafts
Engine code:
    XR2i .... RDB
    RS 1800 .... RQC
Power output:
    XR2i .... 77 kW/105 PS @ 5500 rpm
    RS 1800 .... 96 kW/130 PS @ 6250 rpm
Capacity .... 1796 cc
Bore... 80.60 mm
Stroke.. 88.00 mm
Compression ratio 10:1
Compression pressure – at starter motor speed, engine fully warmed-up ...... Not available
Firing order .... 1-3-4-2 (No 1 cylinder at timing belt end)
Direction of crankshaft rotation .... Clockwise (seen from right-hand side of vehicle)

Cylinder block
Cylinder bore diameter:
    Class 1 .... 80.600 to 80.610 mm
    Class 2 .... 80.610 to 80.620 mm
    Class 3 .... 80.620 to 80.630 mm
    Oversizes .... None available

Crankshaft and main bearings
Main bearing shell standard inside diameter – installed .... 58.011 to 58.038 mm
Main bearing journal standard diameter... 57.980 to 58.000 mm
Main bearing journal-to-shell running clearance .... 0.011 to 0.058 mm
Main bearing shell undersizes available .... 0.02 mm, 0.25 mm
Big-end bearing shell standard inside diameter – installed .... 46.926 to 46.960 mm
Big-end bearing journal standard diameter .... 46.890 to 46.910 mm
Big-end bearing journal-to-shell running clearance .... 0.016 to 0.070 mm
Big-end bearing shell undersizes available .... 0.02 mm, 0.25 mm
Crankshaft endfloat... 0.090 to 0.260 mm

Gudgeon pin
Diameter:
- White colour code/piston crown marked 'A' .... 20.622 to 20.625 mm
- Red colour code/piston crown marked 'B' .... 20.625 to 20.628 mm
- Blue colour code/piston crown marked 'C' .... 20.628 to 20.631 mm

Clearance in piston .... 0.012 to 0.018 mm
Connecting rod small-end eye internal diameter .... 20.570 to 20.610 mm
Interference fit in connecting rod .... 0.012 to 0.061 mm

Pistons and piston rings
Piston diameter:
- Class 1 .... 80.570 to 80.580 mm
- Class 2 .... 80.580 to 80.590 mm
- Class 3 .... 80.590 to 80.600 mm
- Oversizes .... None available

Piston-to-cylinder bore clearance .... No information available at time of writing
Piston ring end gaps – installed:
- Top and second compression rings .... 0.30 to 0.50 mm
- Oil control ring .... 0.38 to 1.14 mm

Camshafts
Camshaft bearing journal diameter .... 25.960 to 25.980 mm
Camshaft bearing journal-to-cylinder head running clearance ... 0.020 to 0.070 mm
Camshaft endfloat .... 0.080 to 0.220 mm

Cylinder head
Maximum permissible gasket surface distortion .... 0.10 mm
Valve seat angle .... 45°
Valve guide bore – inlet and exhaust .... 6.060 to 6.091 mm
Hydraulic tappet/cam follower bore inside diameter ... 28.395 to 28.425 mm

Valves – general

	Inlet	Exhaust
Valve lift	7.500 to 7.685 mm	7.610 to 7.765 mm
Valve length	96.870 to 97.330 mm	96.470 to 96.930 mm
Valve head diameter	32.0 mm	28.0 mm
Valve stem diameter	6.028 to 6.043 mm	6.010 to 6.025 mm
Valve stem-to-guide clearance	0.017 to 0.064 mm	0.035 to 0.081 mm

Lubrication
Engine oil type/specification .... Multigrade engine oil, viscosity range SAE 10W/30 to 20W/50, to API SG/CD or better (Duckhams QS, QXR, Hypergrade Plus, or Hypergrade)

Engine oil capacity:
- Initial fill – engine new or rebuilt, with filter .... 4.50 litres
- At oil and filter change .... 4.25 litres
- At oil change only .... 3.75 litres

Approximate amount required to raise oil level from dipstick 'MIN' to 'MAX' marks ... 0.75 litres
Oil pressure .... No information available at time of writing
Oil pump clearances .... No information available at time of writing
Oil filter .... Champion C148

Torque wrench settings

	Nm	lbf ft
Flywheel bolts	105 to 115	77.5 to 85
Crankshaft rear oil seal carrier bolts	14 to 18	10 to 13
Main bearing cap bolts and nuts	70 to 90	51.5 to 66
Oil baffle nuts	17 to 21	12.5 to 15.5
Big-end bearing cap bolts:		
Stage 1	15 to 20	11 to 15
Stage 2	Angle-tighten a further 90°	
Piston-cooling oil jet/blanking plug bolts	8 to 11	6 to 8
Oil pick-up pipe-to-pump bolts	8 to 11	6 to 8
Oil filter adaptor-to-pump	18 to 25	13 to 18.5
Oil pump-to-cylinder block bolts	8 to 11.5	6 to 8.5
Oil pressure switch	25 to 29	18.5 to 21.5
Cylinder block and head oilway blanking plugs:		
M6 x 10	8 to 11	6 to 8
M10 x 11.5 – in block	20 to 27	15 to 20
1/4 PTF plug – in block	22 to 28	16 to 21
Sump bolts	20 to 24	15 to 17.5
Engine oil drain plug	21 to 28	15.5 to 21
Cylinder head bolts:		
Stage 1	20 to 30	15 to 22
Stage 2	40 to 50	29.5 to 37
Stage 3	Angle-tighten a further 90° to 120°	

Chapter 13 Supplement: Revisions and information on later models

Torque wrench settings (continued)

	Nm	lbf ft
Crankshaft pulley bolt	107 to 117	79 to 86
Timing belt tensioner bolt	35 to 40	26 to 29.5
Timing belt tensioner backplate locating peg	8 to 11	6 to 8
Timing belt tensioner spring retaining pin	8 to 11	6 to 8
Timing belt guide pulley bolts	35 to 40	26 to 29.5
Water pump bolts	16 to 20	12 to 15
Water pump pulley bolts	8 to 12	6 to 9
Camshaft toothed pulley bolts	64 to 72	47 to 53
Camshaft bearing cap bolts:		
Stage 1	10	7.5
Stage 2	17 to 20	12.5 to 15
Cylinder head cover bolts:		
Stage 1	1 to 3	0.5 to 2
Stage 2	6 to 8	4.5 to 6
Timing belt (inner and outer) cover fasteners:		
Cover-to-cylinder head or block bolts	6 to 8	4.5 to 6
Cover studs-to-cylinder head or block	9 to 11	6.5 to 8
Upper-to-middle (outer) cover bolts	3 to 5	2 to 3.5
Engine lifting eye bolts	23 to 28	17 to 21
Crankcase breather system:		
Oil separator-to-cylinder block bolts	8 to 11	6 to 8
Pipe-to-cylinder head bolt	20 to 25	15 to 18.5
Transmission-to-engine flange bolts	35 to 45	26 to 33
Engine front right-hand mounting:		
Alternator mounting bracket-to-cylinder block bolts	41 to 58	30 to 43
Mounting bracket-to-alternator mounting bracket bolts	Not available	Not available
Mounting through-bolt	Not available	Not available
Outer bracket-to-mounting bolts	58 to 79	43 to 58
Inner bracket-to-body bolts	58 to 79	43 to 58
Outer bracket-to-body bolts	58 to 79	43 to 58
Engine rear right-hand mounting:		
Bracket-to-cylinder block bolts	76 to 104	56 to 77
Mounting-to-(cylinder block) bracket bolts	70 to 97	51.5 to 71.5
Mounting-to-body bolt and nut	102 to 138	75 to 102
Transmission mounting fasteners	Refer to Chapter 6	

Cooling, heating and ventilation systems

Coolant mixture

Recommended coolant mixture	Clean, soft water, mixed in recommended concentration with antifreeze of specified quality
Recommended concentration – by volume:	
Standard	40% antifreeze
At topping-up – for convenience only	50% antifreeze
Protection at 40% concentration:	
Slush point	-25°C (-13°F)
Solidifying point	-30°C (-22°F)
Specific gravity at 40% concentration and 15°C/59°F (without other additives)	1.061

Antifreeze

Type/specification	Ethylene glycol-based antifreeze to Ford specification ESD-M97B49-A (Duckhams Universal Antifreeze and Summer Coolant)

Total cooling system capacity

DOHC 16-valve engines	7.0 litres

Torque wrench settings

	Nm	lbf ft
1.6 litre EFi (turbocharged) engine components:		
Intercooler-to-radiator bolts	4 to 6	3 to 4.5
Radiator cooling fan shroud fasteners:		
Shroud-to-mounting bracket	4 to 6	3 to 4.5
Shroud mounting bracket-to-radiator screws	4 to 6	3 to 4.5
Turbocharger cooling pipe banjo union bolts	21.6 to 29.3	16 to 21.5
DOHC 16-valve engine components:		
Thermostat housing-to-cylinder head bolts	18 to 22	13 to 16
Water outlet connection-to-thermostat housing bolts	8 to 11	6 to 8
Water pump bolts	16 to 20	12 to 15
Water pump pulley bolts	8 to 12	6 to 9
Temperature gauge sender unit	6 to 10	4.5 to 7.5
Engine Coolant Temperature (ECT) sensor	12 to 18	9 to 13

Fuel, exhaust and emission control systems – (HCS and CVH) carburettor engines

General

	Type	Part Number
Carburettor application and type:		
1.0 litre engine	Weber 32 TLM (1V)	89BF-9510-AA
1.1 litre engine – engine code GUE:		
Manual transmission	Weber 26/28 TLDM (2V)	89BF-9510-BA
Automatic transmission	Weber 26/28 TLDM (2V)	89BF-9510-CA
1.1 litre engine – engine code GUD	Weber 26/28 TLDM (2V)	89BF-9510-EA
1.3 litre engine	Weber 28/30 TLDM (2V)	92BF-9510-AA
1.4 litre engine – engine code FUF:		
Manual transmission	Weber 28/30 DFTM (2V)	89SF-9510-CC
Automatic transmission	Weber 28/30 DFTM (2V)	89SF-9510-DA
1.4 litre engine – engine code FUG	Weber 28/30 TLDM (2V)	91SF-9510-BA/BB
1.6 litre engine	Weber 28/32 TLDM (2V)	89SF-9510-AA
Air cleaner element – 1.3 litre engine	Champion W225	

Weber 26/28 TLDM (2V) carburettor

Throttle kicker speed – 1.1 litre engine, code GUD, manual transmission .... 1250 to 1350 rpm

Weber 28/30 TLDM (2V) carburettor – 1.3 litre engine

Note: *Refer to entries for 1.1 litre HCS engines in Chapter 3 Specifications, except for those below.*

Fast idle speed .... 2500 rpm
Throttle kicker speed .... 1900 ± 100 rpm

	Primary	Secondary
Venturi diameter	19	20
Main jet	90	122
Air correction jet	185	130

Weber 28/30 TLDM (2V) carburettor – 1.4 litre engine

Note: *Refer to entries for 1.4 litre CVH engines in Chapter 3 Specifications, except for those below.*

Fast idle speed .... 3300 rpm
Choke pull-down .... 2.80 ± 0.25 mm
Float height .... 31.00 ± 0.25 mm
Throttle kicker speed .... 1400 ± 100 rpm

	Primary	Secondary
Venturi diameter	20	22
Main jet	107	140
Air correction jet	195	170
Emulsion tube	F105	F75
Idle jet	47	60

Torque wrench settings

	Nm	lbf ft
Carburettor-to-inlet manifold Torx-type screws – 1.0 litre HCS engine	8 to 10.6	6 to 8
Exhaust manifold heatshield bolts – all HCS engines	4	3

Fuel, exhaust and emission control systems – (HCS and CVH) CFi engines

Note: *For information on 1.1 and 1.3 (HCS) CFi engines, refer to that given in Chapter 3 for the 1.4 litre CVH engine, except for the following.*

System application – 1.1 and 1.3 litre HCS CFi engines .... As described for 1.4 litre CVH CFi engine in Chapter 3, including three-way regulated catalytic converter and evaporative emissions control system (carbon canister). Pulse-air system on 1.3 litre engines only.

Fuel octane rating .... 95 RON (Premium), **unleaded only**

Fuel filter element – applies to all fuel-injected engines:
  Filter with 14 mm inlet, 12 mm outlet threaded unions .... Champion L204
  All other filters .... Champion types not available

Air cleaner element:
  1.1 litre engine .... Champion W153
  1.3 litre engine .... Champion type not available

Idle speed – 1.1 litre engine .... 750 ± 50 rpm nominal (see Chapter 3, Section 55)

Torque wrench settings

	Nm	lbf ft
Pulse-air union sleeve nuts-to-cylinder head – 1.3 litre HCS engines	32	23.5

Fuel, exhaust and emission control systems – 1.6 litre EFi (turbocharged) engine

Note: *Refer to that given in Chapter 3 for the 1.6 litre EFi engine, except for the following.*

Fuel filter element .... See recommendation above for CFi engines
Idle mixture CO content .... 1.50 ± 0.25% (radiator cooling fan on)
Base idle speed .... 650 to 700 rpm
Turbocharger type .... Garrett AiResearch T02
Boost pressure .... 0.47 to 0.51 bars

Chapter 13 Supplement: Revisions and information on later models

Torque wrench settings	Nm	lbf ft
Idle speed control valve nuts	3.5 to 5	2.5 to 3.5
Intercooler-to-radiator bolts	4 to 6	3 to 4.5
Boost control valve screws	2.2 to 2.7	1.5 to 2
Exhaust manifold heatshield bolts	21 to 26	15.5 to 19
Exhaust manifold-to-engine nuts	28 to 31	21 to 23
Exhaust manifold-to-turbocharger bolts	20 to 28	15 to 21
Turbocharger-to-exhaust downpipe nuts	35 to 47	26 to 34.5
Turbocharger cooling pipe banjo union bolts	21.6 to 29.3	16 to 21.5
Turbocharger oil feed and return line couplings	14.9 to 20.1	11 to 15
Air intake crossover duct-to-rocker cover nuts	2.0 to 2.5	1.5 to 2

Fuel, exhaust and emission control systems – DOHC 16-valve engines

General

Fuel octane rating	95 RON (Premium), **unleaded only**
Fuel filter element	Champion type not available
Air cleaner element	Champion type not available
System type	Multi-point sequential fuel injection system controlled (with ignition system) by EEC-IV engine management module. Regulated three-way catalytic converter, pulse-air system and evaporative emissions control system.

Idle speed:
- Nominal: 900 ± 50 rpm
- Base: 750 ± 50 rpm

Idle mixture CO content: 0.5% **maximum** (radiator cooling fan on)

Regulated fuel pressure:
- With regulator vacuum hose disconnected: 2.7 ± 0.2 bars
- With engine running and regulator vacuum hose connected: 2.1 ± 0.2 bars

Note: *The system should hold 1.8 bars for 5 minutes when the ignition is switched off.*

Torque wrench settings

	Nm	lbf ft
Inlet and exhaust manifold studs-to-cylinder head	10 maximum	7.5 maximum
Exhaust manifold nuts	14 to 17	10 to 12.5
Pulse-air union sleeve nuts-to-exhaust manifold	32	23.5
HEGO sensor	50 to 70	37 to 51
Inlet manifold and intermediate flange nuts and bolts	16 to 20	12 to 15
Fuel distributor rail-to-intermediate flange bolts	16 to 20	12 to 15
Idle speed control valve bolts	5 to 8	3.5 to 6
Throttle housing-to-inlet manifold screws	8 to 11	6 to 8
Fuel pressure regulator bolts	5 to 7	3.5 to 5
Fuel injector bolts	8.5 to 12	6 to 9
Fuel feed and return line threaded couplings at fuel distributor rail	24 to 30	17.5 to 22
Fuel feed and return line support bracket bolts	8.5 to 12	6 to 9

Ignition and engine management systems – (HCS and CVH) carburettor engines

General
System type:
1.0 litre engine and 1.1 litre engine – engine code GUE	Distributorless, with UESC module
1.1 litre engine – engine code GUD	Distributorless, with ESC-H2 or ESC-P1 module
1.3 litre engine, 1.4 litre engine – engine code FUG	Distributorless, with ESC-P1 module
1.4 litre engine – engine code FUF, 1.6 litre engine	Transistorised

HT leads
Type:
1.3 litre engine	Champion LS-28 (boxed set)
1.4 litre engine – engine code FUG	Champion LS-27 (boxed set)

Torque wrench settings

	Nm	lbf ft
Engine speed/Crankshaft Position Sensor (CPS)	3 to 4.5	2 to 3
Ignition coil-to-bracket fasteners	5 to 7	3.5 to 5
Ignition coil bracket-to-cylinder block fasteners	8.5 to 12	6 to 9

Ignition and engine management systems – (HCS and CVH) CFi engines

General
System type – 1.4 litre CVH engines from September 1990 onwards, all 1.1 and 1.3 litre HCS engines: Distributorless, under control of EEC-IV engine management module

Firing order:
- HCS engines: 1-2-4-3
- CVH engine: 1-3-4-2

Location of No 1 cylinder: At crankshaft pulley/timing belt end

Ignition timing
At idle speed: 10° ± 2° BTDC nominal

Coil
Type... High-output distributorless
Output... 37 kilovolts (minimum)
Primary resistances (measured at coil tower)...................................... 0.50 ± 0.05 ohms

Spark plugs (see Note at end of Specifications)
Type:
    1.1 and 1.3 litre engines... Champion RS9YCC or RS9YC
    1.4 litre engines.. Champion RC7YCC or RC7YC
Electrode gap.. 1.0 mm

HT leads
Type:
    1.1 and 1.3 litre engines... Champion LS-28 (boxed set)
    1.4 litre engine... Champion LS-27 (boxed set)
Maximum resistance per lead.. 30 000 ohms

Torque wrench settings

	Nm	lbf ft
Air Charge Temperature (ACT) sensor	20 to 25	15 to 18
Engine Coolant Temperature (ECT) sensor	20 to 25	15 to 18
Engine speed/Crankshaft Position Sensor (CPS)	3 to 4.5	2 to 3
Ignition coil-to-bracket fasteners	5 to 7	3.5 to 5
Ignition coil bracket-to-cylinder block fasteners	8.5 to 12	6 to 9

Ignition and engine management systems – 1.6 litre EFi (turbocharged) engine
Note: *Refer to that given in Chapter 4 for the 1.6 litre EFi engine, except for the following.*

Spark plugs (see Note at end of Specifications)
Type... Champion C61YC
Electrode gap.. 0.7 mm

HT leads
Type... Champion LS-26 (boxed set)
Maximum resistance per lead.. 30 000 ohms

Torque wrench settings

	Nm	lbf ft
Air Charge Temperature (ACT) sensor	12 to 18	9 to 13
Ignition coil-to-bracket fasteners	4.5 to 6.5	3 to 5
Ignition coil bracket-to-cylinder head fasteners	8.5 to 12	6 to 9

Ignition and engine management systems – DOHC 16-valve engines

General
System type... Distributorless, under control of EEC-IV engine management module.
Firing order.. 1-3-4-2 (No 1 cylinder at timing belt end)
Direction of crankshaft rotation.. Clockwise (seen from right-hand side of vehicle)

Ignition timing... 10° BTDC nominal
Note: *Ignition timing is under control of EEC-IV engine management module – it may vary constantly at idle speed, and is not adjustable.*

Coil
Type... High-output distributorless
Output... 37 kilovolts (minimum)
Primary resistances (measured at coil tower)................................... 0.50 ± 0.05 ohms

Spark plugs
Type... Champion type not available
Electrode gap.. 1.3 mm

HT leads
Type... Champion type not available
Maximum resistance per lead.. 30 000 ohms

Torque wrench settings

	Nm	lbf ft
Engine speed/Crankshaft Position Sensor (CPS)-to-bracket	6 to 9	4.5 to 6.5
Engine speed/Crankshaft Position Sensor (CPS) bracket-to-cylinder block	18 to 23	13 to 17
Air Charge Temperature (ACT) sensor	12 to 18	9 to 13
Engine Coolant Temperature (ECT) sensor	12 to 18	9 to 13
Spark plugs	14 to 20	10 to 15
Camshaft position sensor-to-cylinder head	18 to 23	13 to 17
Ignition coil bracket-to-cylinder head	18 to 23	13 to 17

Chapter 13 Supplement: Revisions and information on later models

Clutch

General
Note: *For information on DOHC 16-valve engines, refer to that given in Chapter 5 for the 1.6 litre CVH engine; note also the following.*
Type – all 1.6 and 1.8 litre engines .... Single dry plate, 'low-lift' diaphragm spring

Manual transmission
Note: *For information on DOHC 16-valve engines, refer to that given in Chapter 6 for the 1.6 litre CVH engine; however, note the following.*

Ratios
1.1 HCS CFi engine, 1.3 litre engines, 1.6 litre EFi (turbocharged)
engine, 1.8 litre DOHC 16-valve engines .... Information not available at time of writing

Driveshafts

Lubrication
Quantity – inner tripod-type joints .... 95 g (3.4 oz)

Torque wrench settings
	Nm	lbf ft
Vibration damper Torx bolts	30	21

Steering

Steering/front suspension angles
Note: *The following settings are for a vehicle at kerb weight - ie, unladen but with all lubricants and fluids. On Fiesta models before 1990 (model year) and all RS Turbo models, the nominal settings are with 3.0 litres of fuel in the tank; in all other cases, settings are for a full fuel tank. See a Ford dealer for a precise identification of the model year of the vehicle in question.*

Toe setting – RS Turbo:
  Service check .... −4.0 mm (toe-out) to 0.0 mm
  Setting if outside service check .... −2.0 mm (toe-out) ± 1.0 mm

Suspension angles – non-adjustable, given for checking purposes only:

	Castor	Camber
XR2i – 1990 to early 1991 model years	0°56'	−0°04'
XR2i – early 1991 to early 1992 model years, and RS Turbo	0°14'	−1°04'
Van models up to early 1992 model year	0°51'	−0°03'
All other models up to early 1992 model year	0°53'	−0°08'
XR2i – 1.8 litre model, RS 1800	0°22'	−0°11'
Van models, early 1992 model year onwards	0°53'	−0°06'
All other models, early 1992 model year onwards	0°52'	−0°09'

Suspension

General
Rear suspension:
  Later XR2i models, RS Turbo, RS 1800 .... 20 mm anti-roll bar fitted
  Courier .... Twist beam rear axle with torsion bar springing and anti-roll, damping by separate damper units.
Wheelbase and track dimensions .... Refer to *'General dimensions, weights and capacities'*

Rear suspension angles
Note: *The following settings are for a vehicle at kerb weight – ie, unladen but with all lubricants and fluids.*

	Toe setting	Camber
1989 model year vehicles:		
1.6 S	0.50 mm toe-in	−1°01'
XR2i	0.70 mm toe-in	−1°01'
All other models	0.20 mm toe-in	−1°01'
All vehicles, April 1990 onwards:		
Courier	1.35 mm toe-in	−0°24.5'
All other models	2.30 mm toe-in	−1°00'
Maximum variation – side-to-side	–	−1°15'

Roadwheels and tyres
Roadwheel size and type:
  XR2i – 1.8 litre model .... 14 x 5.5J pressed-steel
  RS Turbo, RS 1800, (option on XR2i – 1.8 litre model) .... 14 x 5.5J light alloy
  All other models:
    Standard fitting .... 13 x 4.5J, 13 x 5.0J or 13 x 5.5J pressed-steel, according to model
    Option on some models .... 13 x 5.5J light alloy
Tyre size:
  RS Turbo .... 185/55 R 14 78V
  RS 1800, XR2i – 1.8 litre model .... 185/55 R 14 79H
  Courier .... 145 R 13 78R or 165/70 R 13 79T
  All other models .... 135 R 13-S, 145 R 13-S, 155/70 R 13-S, 165/65 R 13-S, 165/65 R 13-H, 175/60 R 13, 175/60 R 13-H or 185/60 R 13 80H

Chapter 13 Supplement: Revisions and information on later models

Roadwheels and tyres (continued)

Tyre pressures (tyres cold) – bars(lbf/in^2):	Front	Rear
135 R 13-S, 145 R 13-S, 155/70 R 13-S, 165/65 R 13-S and 185/60 R 13 80H	Refer to Chapter 10	
145 R 13 78R, 165/70 R 13 79T (Courier):		
Normal loading – driver plus 170 kg load	2.3 (33)	2.5 (36)
Fully-laden – over normal load, up to maximum (see VIN plate)	2.3 (33)	2.8 (40)
175/60 R 13, 175/60 R 13-H:		
Normal loading – up to three persons	1.8 (26)	1.8 (26)
Fully-laden – over three persons	2.3 (33)	2.8 (40)
165/65 R 13-H, 185/55 R 14 79H:		
Normal loading – up to three persons	2.0 (29)	2.0 (29)
Fully-laden – over three persons	2.3 (33)	2.8 (40)
185/55 R 14 78V:		
Normal loading – up to three persons	2.0 (29)	1.8 (26)
Fully-laden – over three persons	2.1 (30)	2.3 (33)

Torque wrench settings

	Nm	lbf ft
Rear suspension anti-roll bar fasteners:		
10 mm (thread size) clamp bolts	41 to 58	30 to 43
12 mm (thread size) clamp bolts	88 to 113	65 to 83
Rear suspension fasteners – Courier:		
Pivot bracket-to-underbody bolts	70 to 97	51.5 to 71.5
Rear hub spindle-to-axle beam:		
8 mm (thread size) bolts	Not available	
10 mm (thread size) bolts	Not available	
Rear suspension damper mountings:		
Upper bolt and nut	102 to 138	75 to 102
Lower bolt	70 to 97	51.5 to 71.5

Electrical system

Alternator – DOHC 16-valve engines

Type:	Model	Rated output
Bosch unit	KC-70A or KC-90A	70 A or 90A
Mitsubishi unit	A002T	70 A or 90A
Minimum brush length – all types	5.0 mm	
Regulated voltage @ 4000 (engine) rpm and 3 to 7 amp load – all types	14.0 to 14.6 volts	

Starter motor – DOHC 16-valve engines

Type:	Model	Rated output
Bosch	DW	1.1 or 1.4 kW
Lucas/Magneti Marelli	M80R	1.0 kW
Nippondenso	Reduction-gear type	0.8 or 1.0 kW
Minimum brush length:		
Bosch and Lucas/Magneti Marelli	8.0 mm	
Nippondenso	9.0 mm	
Commutator minimum diameter:		
Bosch	32.8 mm	
Lucas/Magneti Marelli	Not available	
Nippondenso	27.2 mm	
Armature endfloat:		
Bosch	0.30 mm	
Lucas/Magneti Marelli	0.25 mm	
Nippondenso	0.40 to 1.20 mm	

Fuses – all models, 1992 onwards

Circuits	Fuse number (see text)	Fuse rating (amps)
All fuses in main panel	Unchanged – refer to Chapter 12	
Unused	23	–
Unused	24	–
Tailgate remote release	25	15
Heated windscreen	26	30
Heated windscreen	27	30
Full-length sunroof	28	20

Relays (on fusebox) – all models, 1992 onwards

Relay location (see text)	Circuit
I	Heated rear window
II	Windscreen wiper delay
III	Power delay (CFi)/Fuel injection (EFi)
IV	Not used on UK models
V	Ignition
VI	Automatic transmission inhibitor
VII	Headlamp main beam

Chapter 13 Supplement: Revisions and information on later models

Relays (on fusebox) – all models, 1992 onwards (continued)

Relay location (see text)	Circuit
VIII	Dim-dip
IX	Not used on UK models
X	Not used on UK models
XI	Heated windscreen
XII	Anti-lock braking system
A	Idle speed control (automatic transmission) or two-tone horn (where fitted)
B	Fuel pump
C	Front foglamps (XR2i, RS Turbo and RS 1800)
D	Headlamp dipped beam
E	Dim-dip
F	Not used on UK models

Torque wrench settings

	Nm	lbf ft
DOHC 16-valve engine alternator fasteners:		
Alternator pivot bolts and nuts	20.4 to 27.2	15 to 20
Adjuster link-to-mounting bracket bolt	10.9 to 13.5	8 to 10
Adjuster nut – to tension drivebelt	10 to 15	7.5 to 8
Adjuster centre lockbolt	18 to 26	13 to 19
Temperature gauge sender unit – DOHC 16-valve engine	6 to 10	4.5 to 7.5

Note: *Information on spark plug type and electrode gaps is as recommended by Champion Spark Plug. Where alternative types are used, refer to the manufacturers' specifications.*

3 Buying spare parts and vehicle identification numbers

DOHC 16-valve engine number location

1 The DOHC 16-valve engine number (consisting of two letters and five digits, with the three-letter engine code nearby) is stamped into a flat-machined surface on the cylinder block's forward-facing flange, between the pulse-air filter housing and the transmission (photo). It may be necessary to disconnect the flexible rubber hose from the (black plastic) air intake trunking, so that the hose and air mass meter can be unclipped and moved aside sufficiently to see the flange.

2 If the number cannot be seen in this location, possible alternative sites are on a lower flange on the cylinder block's forward face, immediately behind the starter motor (it may be necessary to remove the motor to be able to see the number), or on the left-hand end of the cylinder head, between the oil filler cap and ignition coil.

VIN plate emission standard codes

3 Vehicles built from the 1990 model year onwards have a VIN plate which has an additional code, indicating the emission standard regulations met by that particular vehicle. This code is stamped in the 15th box (see third illustration in 'Buying spare parts and vehicle identification numbers' at the start of this manual), the entries used on Fiesta models being as follows:

Blank	Vehicle complies with EEC 15:04 (also known as 'Euronorm 15:04') emission standard. This is the basic standard which, apart from the closed-circuit crankcase ventilation system, requires little in the way of specific emission control systems.
8	Vehicle complies with New EEC or 5th amendment, Reduced Severity (often abbreviated to '15:05R') emission standard. This is the interim level standard between 15:04 and (full) 15:05.
X	Vehicle complies with New EEC or 5th amendment (often abbreviated to '15:05' and also known as 'Euronorm 88/76') emission standard. While most of this standard's requirements are not relevant to Fiesta models, if the vehicle's engine is fitted with a carburettor, it will have devices such as a vacuum-controlled throttle kicker and an air charge temperature sensor, and the ignition will be controlled by a more sophisticated module (ESC-H2 or ESC-P1).
U	Vehicle complies with US '83 emission standard. This is the most stringent currently in force in the EEC – to comply with it, all engines must be fuel-injected (whether CFi or EFi) with three-way regulated catalytic converters and a full engine management system. An evaporative emissions control system (carbon canister) must be fitted, and a pulse-air system may also be fitted.

4 Routine maintenance

Vehicles built up to (and including) 1991 model year, and all RS Turbo models

1 The manufacturer's maintenance schedule for these models has been slightly revised from that given at the front of this manual; the current recommendation is laid out below. Note that the schedule starts from the vehicle's date of registration, and is based, as previously, on a service being carried out every 6000 miles or 12 months.

2 Note also that the first free service (carried out by the selling dealer 1500 miles or 3 months after delivery) although an important check for a new vehicle, is not part of the regular maintenance schedule, and is therefore not mentioned.

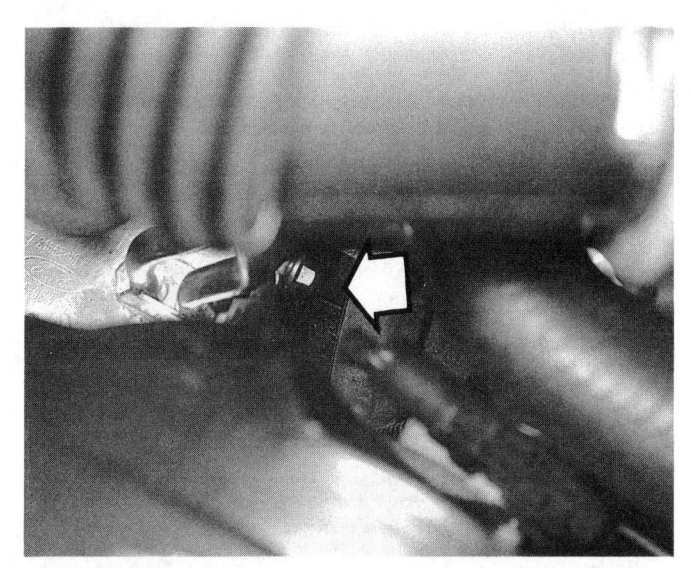

3.1 Location of DOHC 16-valve engine number and code (arrowed) – between pulse-air filter housing and transmission

Chapter 13 Supplement: Revisions and information on later models

Daily, when refuelling, or before any long journey

Check the engine oil level, and top-up if necessary (Chapter 1)
Check the brake fluid level, and top-up if necessary. If repeated topping-up is required, check the system for leaks or damage at the earliest possible opportunity (Chapter 8)
Check the windscreen/tailgate washer fluid level, and top-up if necessary (Chapter 12)
Check the tyre pressures, including the spare (Chapter 10)
Visually check the tyres for wear or damage (Chapter 10)
Check the operation of all (exterior and interior) lamps and the horn, wipers and windscreen/tailgate washer system. Renew any blown bulbs, and clean the lenses of all exterior lamps (Chapter 12)

Monthly

Repeat all the daily checks, then check the following:

Check the coolant level, and top-up if necessary (Chapters 2 and 13)
Visually check all reservoirs, hoses and pipes for leakage
Check the operation of the handbrake (Chapter 8)
Check the aim of the windscreen/tailgate washer jets, correcting them if required (Chapter 12)
Check the condition of the wiper blades, renewing them if worn or no longer effective – note that the manufacturer recommends renewing the blades as a safety precaution, irrespective of their apparent condition, at least once or twice a year (Chapter 12)

Interim service – at first 6000 miles (10 000 km) or 12 months, whichever occurs first, then every 6000 miles (10 000 km) or 12 months, whichever occurs first, from previous Standard/Extended Service

Repeat all the monthly checks, then check the following:

Engine (Chapter 1)
Change engine oil and renew oil filter – see also paragraph 3

Fuel, exhaust and emission control systems (Chapter 3)
Check condition of fuel lines

Ignition and engine management systems (Chapters 4 and 13)
Renew spark plugs – RS Turbo only

Braking system (Chapter 8)
Check pads and shoes for wear
Check condition of hydraulic pipes and hoses – see also paragraph 3

Steering (Chapter 9)
Check condition of steering components and gaiters – see also paragraph 3

Suspension (Chapter 10)
Check and adjust all tyre pressures, check condition of all tyres
Check tightness of roadwheel nuts
Check security/installation of front suspension lower arm balljoints

Bodywork and fittings (Chapter 11)
Check seat belt webbing and retractor operation
Lubricate door check straps
Lubricate bonnet lock/safety catch and check operation
Make a general check of underbody

Electrical system (Chapter 12)
Check battery electrolyte level
Check condition, routing and security of visible wiring

Standard service – at first 12 000 miles (20 000 km) or 24 months, whichever occurs first, then every 6000 miles (10 000 km) or 12 months, whichever occurs first, from previous Interim Service

Carry out all Interim Service operations, plus the following:

Engine (Chapter 1)
Check valve clearances, and adjust if necessary – HCS engines only
Renew timing belt (camshaft drivebelt) – CVH engines only, at relevant interval – see also paragraph 3
Cooling, heating and ventilation systems (Chapter 2)
Coolant renewal, at relevant interval – see also paragraph 3

Fuel, exhaust and emission control systems (Chapters 3 and 13)
Check tightness of exhaust manifold-to-turbocharger bolts and turbocharger-to-exhaust downpipe nuts – RS Turbo only
Check condition and security of exhaust system – see also paragraph 3
Check idle speed – carburettor engines only

Ignition and engine management systems (Chapter 4)
Clean distributor cap (where applicable), HT leads and ignition coil, check condition (where applicable) of leads, cap and rotor arm, and ensure that HT lead connectors are secure
Renew spark plugs

Transmission (Chapter 6)
Check manual transmission oil level
Check CTX transmission fluid level – only with engine still hot, after road test

Driveshafts (Chapter 7)
Check gaiters for security, condition and leaks

Suspension (Chapter 10)
Check balljoint covers/gaiters for security and condition

Electrical system (Chapter 12)
Check condition and tension of (water pump/) alternator drivebelt

Extended service – at first 24 000 miles (40 000 km) or 48 months, whichever occurs first, then every 6000 miles (10 000 km) or 12 months, whichever occurs first, from every second Interim Service

Carry out all Standard Service operations, plus the following:

Cooling, heating and ventilation systems (Chapter 2)
Renew coolant mixture, at relevant interval – see also paragraph 3

Fuel, exhaust and emission control systems (Chapter 3)
Renew air filter element and (where fitted) crankcase ventilation system filter
Check operation of air filter intake-air temperature control system (where fitted)
Check oil filler cap filter mesh – clean, or renew cap – HCS engines only
Renew fuel filter – CFi and EFi engines only, at relevant interval – see also paragraph 3

Transmission (Chapter 6)
Renew CTX transmission fluid

3 In addition to the regular schedule laid out above, there are some additional items which must be attended to as follows:

(a) If the vehicle is used regularly for very short (less than 10 miles) stop/start journeys, the oil and filter should be renewed between services (ie, every 3000 miles/6 months) – seek the advice of a local Ford dealer if in doubt on this point.
(b) On CVH engines, the timing belt is to be renewed every 36 000 miles.

Chapter 13 Supplement: Revisions and information on later models

(c) Depending on the quality of antifreeze used, the coolant may need to be renewed at intervals of either 36 000 miles/two years, every four years, or less often – refer to Section 6 of this Chapter.
(d) The fuel filter – CFi and EFi engines only – is to be renewed every 48 000 miles (ie, at every second Extended Service).
(e) The exhaust system should be checked for leaks, security and condition every 36 000 miles or three years, whichever occurs first.
(f) The brake hydraulic fluid must be renewed every three years – at this time, or whenever the pads/shoes are renewed, the system's rubber components must be checked carefully, and the whole system must be overhauled if necessary.
(g) The front wheel alignment must be checked every 36 000 miles or three years, whichever occurs first.

Vehicles built from 1992 model year onwards, except RS Turbo

4 The manufacturer's recommended maintenance schedule for these vehicles, including all XR2i models with the DOHC 16-valve engine and all RS 1800 models, is as described below. Note that the schedule starts from the vehicle's date of registration.

5 Note also that the first free service (carried out by the selling dealer 1500 miles or 3 months after delivery), although an important check for a new vehicle, is not part of the regular maintenance schedule, and is therefore not mentioned.

Daily, when refuelling, or before any long journey

Check the engine oil level, and top-up if necessary (Chapters 1 and 13)
Check the brake fluid level, and top-up if necessary. If repeated topping-up is required, check the system for leaks or damage at the earliest possible opportunity (Chapter 8)
Check the windscreen/tailgate washer fluid level, and top-up if necessary (Chapter 12)
Check the tyre pressures, including the spare (Chapter 10)
Visually check the tyres for wear or damage (Chapter 10)
Check the operation of all (exterior and interior) lamps and the horn, wipers and windscreen/tailgate washer system. Renew any blown bulbs, and clean the lenses of all exterior lamps (Chapter 12)

Monthly

Repeat all the daily checks, then check the following:

Check the coolant level, and top-up if necessary (Chapters 2 and 13)
Visually check all reservoirs, hoses and pipes for leakage
Check the operation of the handbrake (Chapter 8)
Check the aim of the windscreen/tailgate washer jets, correcting them if required (Chapter 12)
Check the condition of the wiper blades, renewing them if worn or no longer effective – note that the manufacturer recommends renewing the blades as a safety precaution, irrespective of their apparent condition, at least once or twice a year (Chapter 12)

Standard service – every 10 000 miles (16 000 km) or 12 months, whichever occurs first

Repeat all the monthly checks, then check the following:

Engine (Chapters 1 and 13)
Change engine oil and renew oil filter – see also paragraph 6
Check valve clearances, and adjust if necessary – HCS engines only
Renew timing belt (camshaft drivebelt) – CVH engines only, at relevant interval – see also paragraph 6

Cooling, heating and ventilation systems (Chapter 2)
Renew coolant mixture, at relevant interval – see also paragraph 6

Fuel, exhaust and emission control systems (Chapters 3 and 13)
Check idle speed – carburettor engines only – see also paragraph 6
Check base idle speed – CVH EFi engines only – see also paragraph 6
Check condition of fuel lines
Check operation of throttle/choke controls (as applicable) during road test

Ignition and engine management systems (Chapters 4 and 13)
Renew spark plugs – all except DOHC 16-valve engines, at relevant interval – see also paragraph 6

Transmission (Chapter 6)
Check manual transmission oil level
Check operation of transmission (including clutch) during road test
Check CTX transmission fluid level – only with engine still hot, after road test

Driveshafts (Chapter 7)
Check gaiters for security, condition and leaks

Braking system (Chapter 8)
Check operation of handbrake and footbrake during road test
Check pads and shoes for wear
Check condition of hydraulic pipes and hoses – see also paragraph 6

Steering (Chapter 9)
Check condition of steering components and gaiters – see also paragraph 6
Check operation of steering during road test

Suspension (Chapter 10)
Check and adjust all tyre pressures, check condition of all tyres
Check tightness of roadwheel nuts
Check security/installation of front suspension lower arm balljoints
Check operation of suspension during road test

Bodywork and fittings (Chapter 11)
Check seat belt webbing and retractor operation
Lubricate door check straps
Lubricate bonnet lock/safety catch, and check operation
Make a general check of underbody
Check for unusual noises during road test

Electrical system (Chapter 12)
Check battery electrolyte level
Check condition and tension of (water pump/) alternator drivebelt
Check condition, routing and security of visible wiring
During road test, check operation of all instruments, lights and controls

Extended service – at first 30 000 miles (48 000 km) or 36 months, whichever occurs first, then every 10 000 miles (16 000 km) or 12 months, whichever occurs first, from every second Standard Service

Carry out all Standard Service operations, plus the following:

Engine (Chapters 1 and 13)
Renew timing belt – DOHC 16-valve engines only, at relevant interval – see also paragraph 6

Fuel, exhaust and emission control systems (Chapters 3 and 13)
Renew air filter element and (where fitted) crankcase ventilation system filter
Check operation of air filter intake-air temperature control system (where fitted)
Check oil filler cap filter mesh – clean, or renew cap – HCS engines only
Check crankcase ventilation system for blockages
Renew fuel filter – CFi and EFi engines only, at relevant interval – see also paragraph 6
Check condition and security of exhaust system

Ignition and engine management systems (Chapters 4 and 13)
Renew spark plugs – DOHC 16-valve engines only

Transmission (Chapter 6)
Renew CTX transmission fluid

Steering (Chapter 9)
Check the front wheel alignment

6 In addition to the regular schedule laid out above, there are some additional items which must be attended to as follows:

(a) *If the vehicle is used regularly for very short (less than 10 miles) stop/start journeys, the oil and filter should be renewed between services (ie, every 5000 miles/6 months) – seek the advice of a local Ford dealer if in doubt on this point.*

(b) *On CVH engines, the timing belt is to be renewed every 40 000 miles (ie, every fourth Standard/Extended Service).*

(c) *On DOHC 16-valve engines, the timing belt is to be renewed every 60 000 miles (ie, every second Extended Service).*

(d) *Depending on the quality of antifreeze used, the coolant may need to be renewed at intervals of every four years, or less often – refer to Section 6 of this Chapter.*

(e) *On carburettor-engined vehicles, the idle mixture need only be checked at the first 10 000 mile service interval.*

(f) *On CVH EFi engines, the base idle speed check need only be made at the first 10 000 mile service interval, and applies only to certain engines fitted with the Hitachi-built idle speed control valve – seek the advice of a local Ford dealer if in doubt on this point.*

(g) *The fuel filter – CFi and EFi engines only – is to be renewed every 60 000 miles (ie, every second Extended Service).*

(h) *On all except DOHC 16-valve engines, the spark plugs are to be renewed every 20 000 miles (ie, every second Standard/Extended Service).*

(i) *The brake hydraulic fluid must be renewed every three years – at this time, or whenever the pads/shoes are renewed, the system's rubber components must be checked carefully, and the whole system must be overhauled if necessary.*

5 Engine

PART A: ALL ENGINES

Compression test – description and interpretation

1 When engine performance is down, or if misfiring occurs which cannot be attributed to the ignition or fuel systems, a compression test can provide diagnostic clues as to the engine's condition. If the test is performed regularly, it can give warning of trouble before any other symptoms become apparent.

2 The engine must be fully warmed-up to normal operating temperature, the oil level must be correct (particularly important on engines with hydraulic cam followers/tappets), and the battery must be fully charged. The aid of an assistant will be required.

3 Remove the spark plugs, referring to Chapter 4, Section 3 if necessary. Disable the ignition system by disconnecting the ignition HT coil wiring – where a distributor is fitted, disconnect the low tension wire from the coil's terminal 15, if a distributorless system is fitted, unplug the DIS coil's wiring multi-plug.

4 Fit a compression tester to the No 1 cylinder spark plug hole – the type of tester which screws into the plug thread is to be preferred.

5 Have the assistant hold the throttle wide open and crank the engine on the starter motor; after one or two revolutions, the compression pressure should build up to a maximum figure, and then stabilise. Record the highest reading obtained.

6 Repeat the test on the remaining cylinders, recording the pressure developed in each, and where applicable, comparing the pressures obtained with the specified values.

7 All cylinders should produce very similar pressures; any difference greater than 10% indicates the existence of a fault. Note that the compression should build up quickly in a healthy engine; low compression on the first stroke, followed by gradually-increasing pressure on successive strokes, indicates worn piston rings. A low compression reading on the first stroke, which does not build up during successive strokes, indicates leaking valves or a blown head gasket (a cracked head could also be the cause). Deposits on the undersides of the valve heads can also cause low compression.

8 If the pressure in any cylinder is reduced to the specified minimum or less, introduce a teaspoonful of clean oil into that cylinder through its spark plug hole, and repeat the test.

9 If the addition of oil temporarily improves the compression pressure, this indicates that bore or piston wear is responsible for the pressure loss. No improvement suggests that leaking or burnt valves, or a blown head gasket, may be to blame.

10 A low reading from two adjacent cylinders is almost certainly due to the head gasket having blown between them; the presence of coolant in the engine oil will confirm this.

11 If one cylinder is about 20 percent lower than the others, and the engine has a slightly rough idle, a worn camshaft lobe or faulty hydraulic cam follower/tappet could be the cause.

12 If the compression is unusually high, the combustion chambers are probably coated with carbon deposits. If this is the case, the cylinder head should be removed and decarbonised.

13 On completion of the test, refit the spark plugs and reconnect the ignition system.

PART B: HCS AND CVH ENGINES

1.0 and 1.1 HCS engines – modifications
Modified rocker cover

1 To further reduce engine noise, a modified rocker cover was introduced on vehicles built from April 1990 onwards. The cover, identifiable by the part number 91BM-6582-GA embossed in its top, is constructed of two sheets of steel, double-skinned, with a 0.1 mm thick layer of plastic sandwiched between.

2 The cover may be fitted to earlier models, if required, to help quieten a noisy engine.

3 Whenever one of these covers is disturbed, the gasket **must** be renewed to prevent oil leaks. Tighten the cover retaining bolts to the specified torque wrench setting; if the bolts are overtightened beyond this figure, the cover flange will be distorted, and oil leaks will result.

Excessive valve gear noise

4 If the top end or valve gear of an HCS engine is thought to be unduly noisy, even when fully warmed-up and with correct valve clearances, take the vehicle to a Ford dealer for expert advice. Depending on the mileage covered, and the use to which the vehicle has been put, some engines may be noisier than others; only a good mechanic experienced in these engines can tell if the noise level is typical, or whether a genuine fault exists. If further attention is required, proceed as follows.

5 First unbolt and check the exhaust manifold heat shield. The correct item should have part number 89BF-9K632-BA embossed in its underside; it should also be free from cracks, with all five spot welds that secure the plate to the moulding intact. If the heat shield is faulty, it must be renewed. On refitting, tighten the bolts to the specified torque wrench setting.

6 If the engine is an early example, some reduction in noise level may be effected by fitting the modified-type rocker cover – see paragraphs 1 to 3 above – and setting the valve clearances (at all service intervals) to precisely the value given in the Specifications Section of this Chapter.

7 If the excessive noise persists, remove and dismantle the rocker shaft assembly as described in the relevant Sections of Chapter 1, Part A. Check for signs of wear or damage at all points of contact, but particularly check the contact faces of the valve clearance adjuster bolts with the pushrods. Normally-worn items will show a smooth unbroken ring; if any adjuster bolt's contact face has worn unevenly, or if its edges are broken (see Fig. 13.1), the bolt must be renewed as a matched set with its rocker arm and pushrod. Renew all other worn or damaged components, reassemble the rocker shaft, and refit it. Set the valve clearances to precisely the specified value (note that the tolerance originally specified is no longer allowed) and check the noise level on restarting the engine.

8 Where adjuster bolts have been renewed, the valve clearances must be checked, and if necessary readjusted, after the vehicle has covered approximately 500 to 1000 miles.

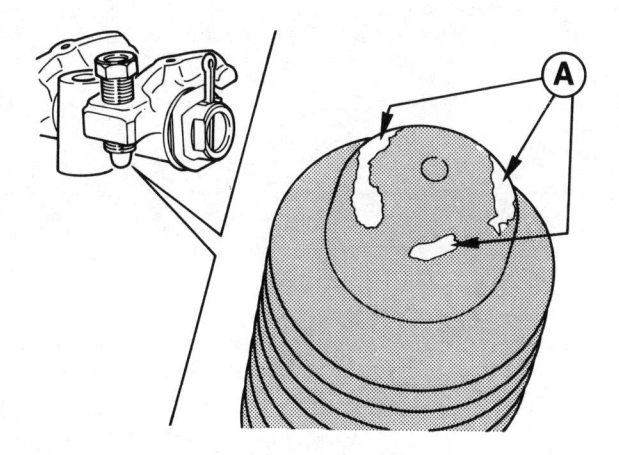

Chapter 13 Supplement: Revisions and information on later models

Fig. 13.1 Check contact faces of valve clearance adjuster bolts – if wear (as shown at 'A') is found, bolt must be renewed with rocker arm and pushrod – HCS engines (Sec 5B)

9   As noted in Section 8 of Chapter 1, the HCS engine's valve clearance adjuster bolts are secured by the stiff-thread locking method. If, on any engine, any of the bolts should lose this locking action (shown by the valve clearance increasing after only a short period), the bolt should be renewed as soon as possible.

1.1 HCS CFi engine

10   This engine is mechanically identical to the carburettor unit covered in Chapter 1, the only differences being in the components of the fuel injection system attached to the inlet and exhaust manifolds.
11   When working on one of these engines, therefore, follow the procedure given in Chapter 1, with reference to the relevant Sections of Chapter 3 (Part B), Chapter 4 (Part C) and of this Chapter, for further information on the components of the CFi system, when encountered.
12   Obviously, all references to the carburettor and the mechanical fuel pump should be ignored.

1.3 HCS engines

13   Apart from the revised bore and stroke dimensions which give the increased capacity, these engines are mechanically identical to the 1.0/1.1 litre units covered in Chapter 1. All working procedures are therefore as described in that Chapter, noting the additional information given above and in the Specifications Section of this Chapter.

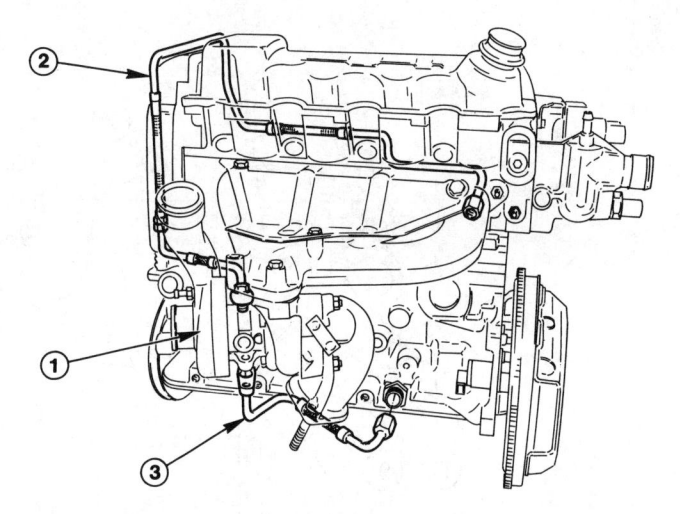

Fig. 13.2 Turbocharger lubrication system – 1.6 CVH EFi (turbocharged) engine (Sec 5B)

1   Turbocharger
2   Oil feed pipe
3   Oil return pipe

14   The 1.3 CFi engine is mechanically identical to the carburettor units – refer to paragraphs 10 to 12 above.

1.4 and 1.6 (non-turbo) CVH engines – modifications

15   The early 1.4 litre engines covered in Chapter 1, Part B are fitted with distributor ignition systems. When working on the later units with distributorless systems, therefore, ignore all references to the distributor, which is replaced on the left-hand end of the cylinder head by the DIS-type of ignition coil; the opening in the end of the cylinder head will be sealed by a blanking plug.
16   If the hydraulic tappets/cam followers are to be renewed, note that improved versions are available, identifiable by the yellow (formerly silver) clip at the top of each. These **must** be fitted to 1.6 litre EFi engines; the earlier type may still be used in 1.4 litre engines and in 1.6 litre carburettor units.
17   Always renew all the hydraulic tappets/cam followers as a set, even if only one appears to be faulty, and always renew the oil filter whenever the tappets are renewed.

Fig. 13.3 Oil cooler and turbocharger oil feed/oil pressure switch adaptor details – 1.6 CVH EFi (turbocharged) engine (Sec 5B)

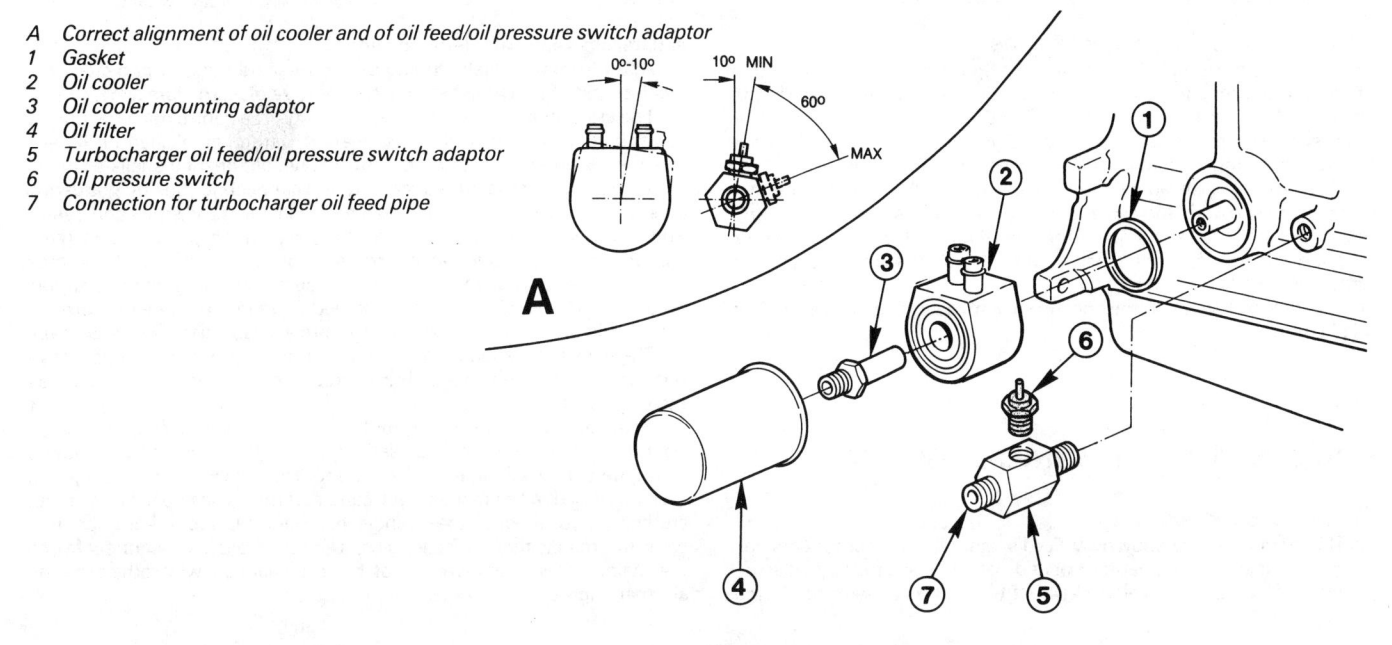

A   Correct alignment of oil cooler and of oil feed/oil pressure switch adaptor
1   Gasket
2   Oil cooler
3   Oil cooler mounting adaptor
4   Oil filter
5   Turbocharger oil feed/oil pressure switch adaptor
6   Oil pressure switch
7   Connection for turbocharger oil feed pipe

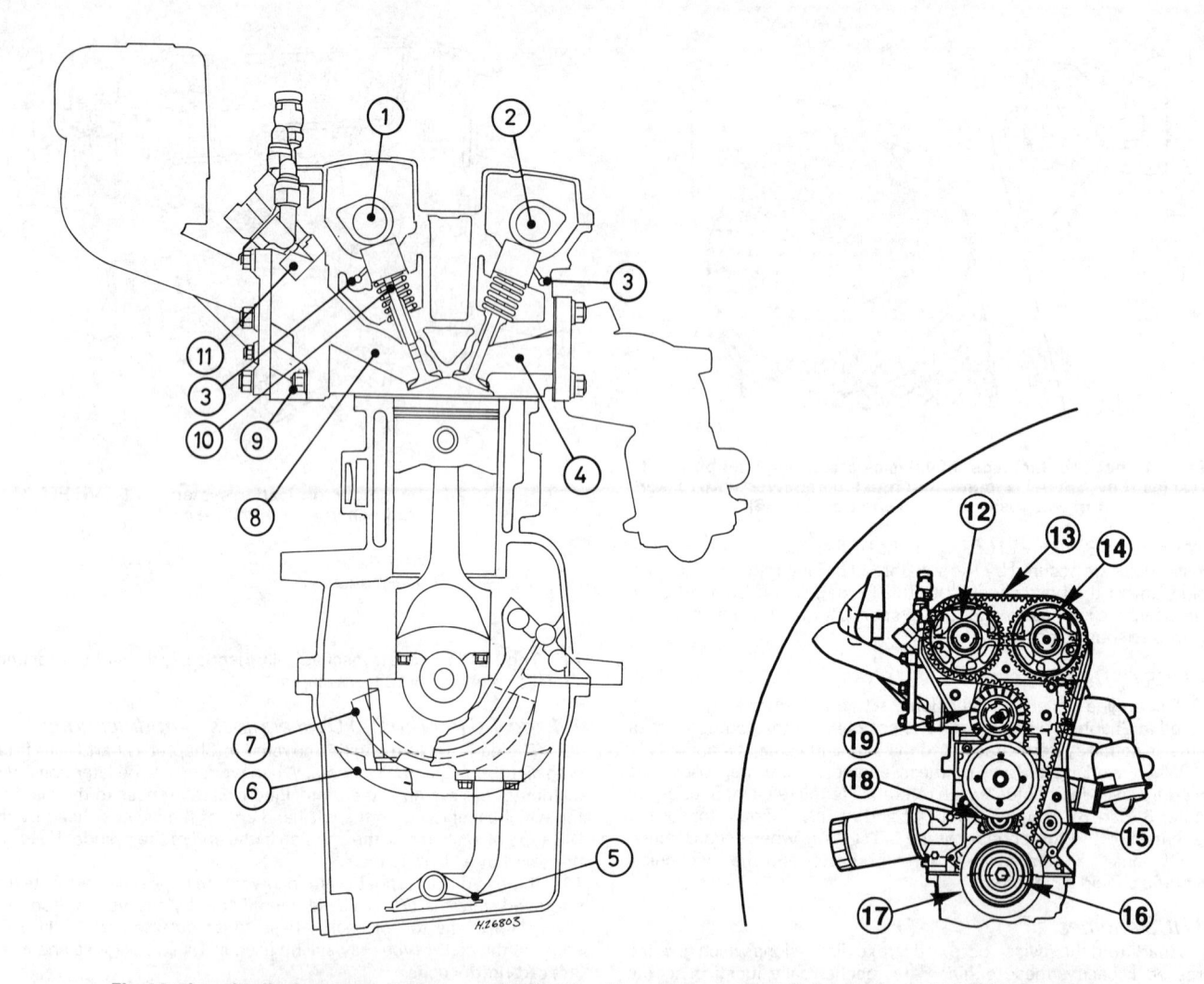

312 Chapter 13 Supplement: Revisions and information on later models

Fig. 13.4 Longitudinal cross-section through DOHC 16-valve engine – inset showing timing belt details (Sec 5C)

1 Inlet camshaft	6 Oil baffle	11 Fuel injector	15 Timing belt (front) guide pulley
2 Exhaust camshaft	7 Crankshaft	12 Inlet camshaft toothed pulley	16 Crankshaft pulley
3 Oil galleries	8 Inlet port	13 Timing belt	17 Sump
4 Exhaust port	9 Intermediate flange	14 Exhaust camshaft toothed pulley	18 Timing belt (rear) guide pulley
5 Oil strainer and pick-up pipe	10 Inlet valve		19 Timing belt tensioner

1.6 CVH EFi (turbocharged) engine

18 Apart from the changes listed in the Specifications Section of this Chapter, the turbocharged 1.6 litre CVH EFi engine is mechanically the same as the (non-turbo) 1.6 litre CVH EFi unit covered in Part B of Chapter 1. Refer to Section 7 of this Chapter for details of changes to the air intake trunking, manifolds and turbocharger installation.

19 The most significant change in the engine is the modified lubrication system shown in the accompanying illustrations. Note particularly the oil cooler, the turbocharger oil feed and return pipes, and the adaptor screwed into the cylinder block to take the oil pressure switch and turbocharger oil feed.

20 Note the changes to the cooling system, described in Section 6 of this Chapter.

PART C: DOHC 16-VALVE ENGINE – GENERAL

General description

This engine is also known by Ford's internal code name 'Zeta'; as with those described in Chapter 1 of this manual, it is of four-cylinder, in-line type, mounted transversely at the front of the car, with the clutch and transmission on its left-hand end.

Apart from the plastic timing belt covers and the cast-iron cylinder block/crankcase, all major engine castings are of aluminium alloy.

The crankshaft runs in five main bearings, the centre main bearing's upper half incorporating thrustwashers to control crankshaft endfloat. The connecting rods rotate on horizontally-split bearing shells at their big-ends. The pistons are attached to the connecting rods by gudgeon pins, which are an interference fit in the connecting rod small-end eyes. The aluminium alloy pistons are fitted with three piston rings; two compression rings and an oil control ring. After manufacture, the cylinder bores and piston skirts are measured and classified into three grades, which must be carefully matched together to ensure the correct piston/cylinder clearance; no oversizes are available to permit reboring.

The inlet and exhaust valves are each closed by coil springs; they operate in guides which are shrink-fitted into the cylinder head, as are the valve seat inserts.

Both camshafts are driven by the same toothed timing belt, each operating eight valves via self-adjusting hydraulic cam followers/tappets, thus eliminating the need for routine checking and adjustment of the valve clearances. Each camshaft rotates in five bearings that are line-bored direct in the cylinder head and the (bolted-on) bearing caps; this means that the bearing caps are not available separately from the cylinder head, and must not be interchanged with others from another engine.

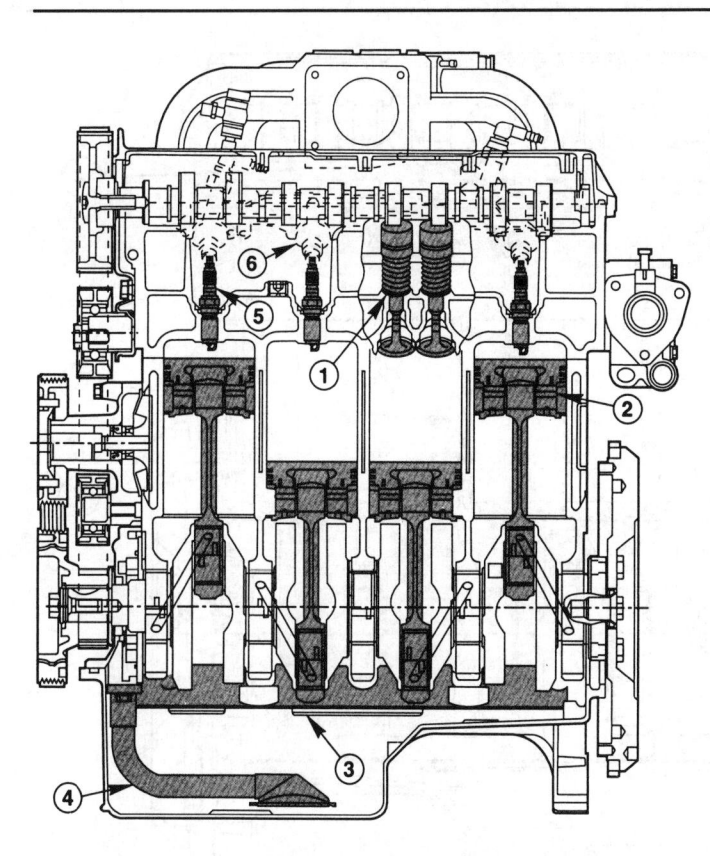

Chapter 13 Supplement: Revisions and information on later models

Fig. 13.5 Lateral cross-section through DOHC 16-valve engine (Sec 5C)

1. Exhaust valve
2. Piston
3. Oil baffle
4. Oil strainer and pick-up pipe
5. Spark plug
6. Fuel injector

The water pump is bolted to the right-hand end of the cylinder block, inboard of the timing belt, and is driven with the alternator by a flat 'polyvee'-type drivebelt from the crankshaft pulley.

When overhauling this engine, it is essential to establish first exactly what replacement parts are available; at the time of writing, components such as the piston rings are not available separately from the piston/connecting rod assemblies, valve guides are not available separately, and very few under- or oversized components are available for engine reconditioning. In most cases, it would appear that the easiest and most economically-sensible course of action is to replace a worn or damaged engine with an exchange unit. Note also that Torx-type (both male and female heads) and hexagon socket (Allen head) fasteners are widely used on this engine; a good selection of bits, with the necessary adaptors, will be required so that these can be unscrewed without damage and, on reassembly, tightened to the torque wrench settings specified.

Lubrication system – description

Lubrication is by means of an eccentric-rotor trochoidal pump, which is mounted on the crankshaft right-hand end, and draws oil through a strainer located in the sump. The pump forces the oil through an externally-mounted full-flow cartridge-type filter into a gallery in the cylinder block/crankcase, from where it is distributed to the crankshaft (main bearings) and cylinder head.

The big-end bearings are supplied with oil via internal drillings in the crankshaft. On the RS 1800 (130 PS) version of the engine, each piston crown is cooled by a spray of oil directed at its underside by a jet. These jets are fed by passages off the crankshaft oil supply galleries, with spring-loaded valves to ensure that the jets open only when there is sufficient pressure to guarantee a good oil supply to the rest of the engine components; where the jets are not fitted, separate blanking plugs are provided so that the passages are sealed, but can be cleaned at overhaul.

The cylinder head is provided with two oil galleries, one on the inlet side and one on the exhaust, to ensure constant oil supply to the camshaft bearings and hydraulic cam followers/tappets. An oil-retaining valve (inserted into the cylinder head's top surface, in the middle, on the inlet side) prevents these galleries from being drained when the engine is switched off, and incorporates a ventilation hole in its upper end to allow air bubbles to escape from the system when the engine is restarted.

While the crankshaft and camshaft bearings and the hydraulic tappets receive a pressurised supply, the camshaft lobes and valves are lubricated by splash, as are all other engine components.

Maintenance and inspection

Engine oil level check

1 Proceed as described in Chapter 1, Part A, Section 2, noting the following:

(a) The dipstick, which can be identified by its yellow grip, is located at the front right-hand end of the engine. Its maximum and minimum levels are indicated by notches; the approximate amount required to raise the oil level from the minimum to the maximum is given in the Specifications Section of this Chapter.

(b) The yellow plastic oil filler cap is screwed into the left-hand front end of the cylinder head cover (photo).

Engine oil and filter change

2 The oil should preferably be changed when the engine is still fully warmed-up to normal operating temperature, just after a run.

3 With the vehicle parked on firm level ground, apply the handbrake securely and raise its front end, supporting it securely on axle stands. Remove the front right-hand roadwheel to provide access to the oil filter.

4 Remove the dipstick and unscrew the oil filler cap, then unscrew the engine oil drain plug, which is located at the rear of the sump (photo). Allow the oil to drain into a container, and check the condition of the plug's sealing washer; renew it if worn or damaged.

5 Using a suitable filter removal tool if necessary, unscrew the oil filter from the right-hand rear of the cylinder block; be prepared for some oil spillage (photo). Check the old filter to make sure that the rubber sealing ring hasn't stuck to the engine; if it has, carefully remove it.

6 Using a clean, lint-free rag, wipe clean the cylinder block around the filter mounting. When the oil has completely drained, wipe clean the

5C.1 Topping-up DOHC 16-valve engine oil

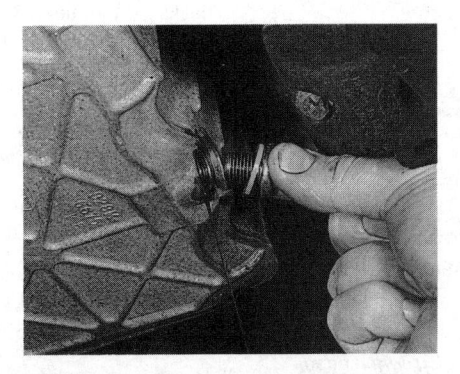

5C.4 Engine oil drain plug is located at rear of sump

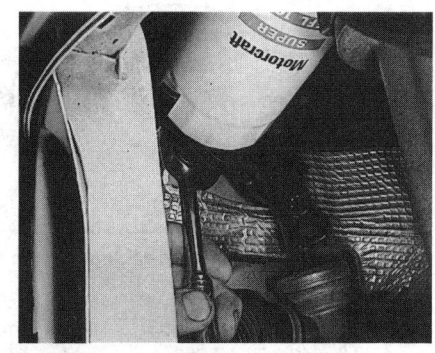

5C.5 Engine oil filter is accessible through right-hand wheel arch

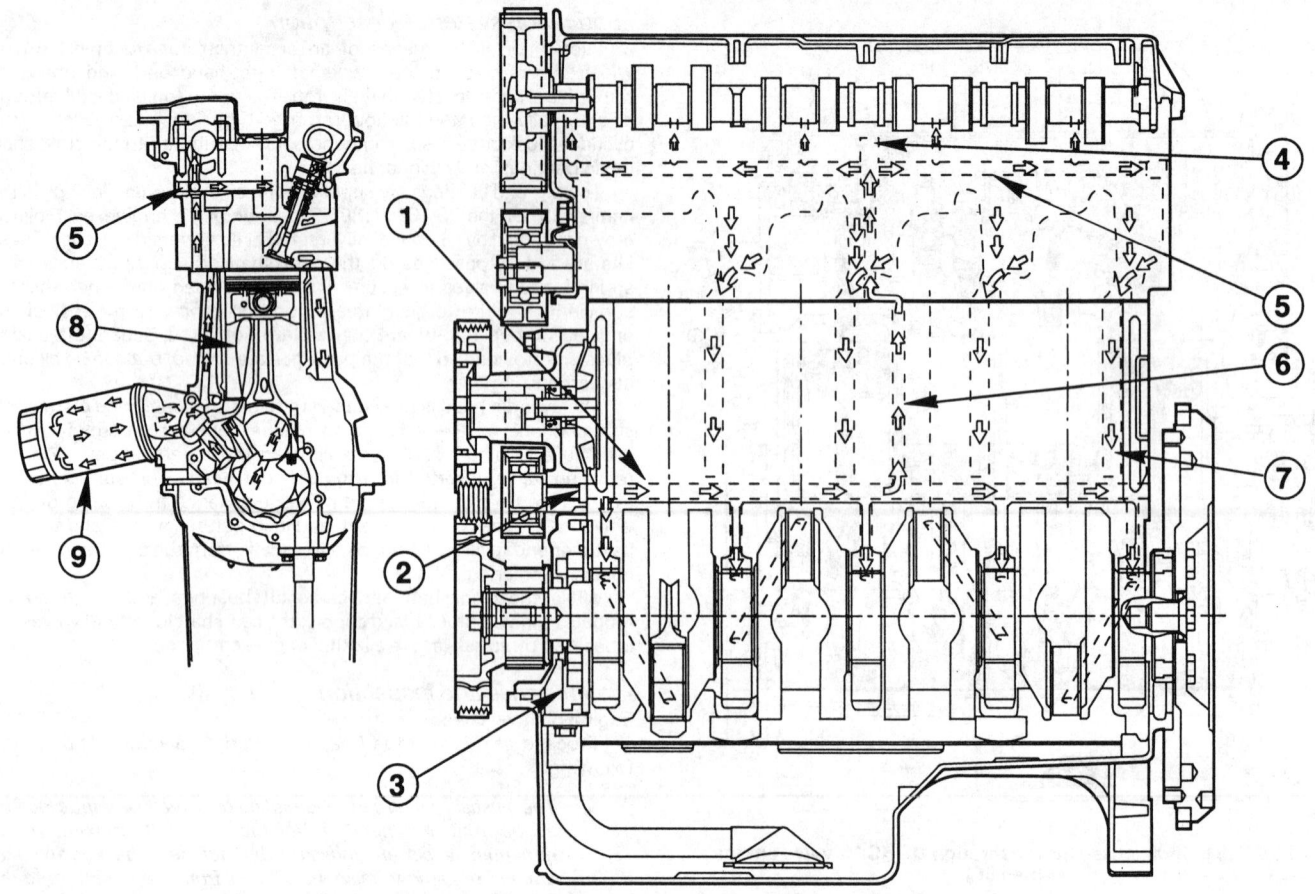

Chapter 13 Supplement: Revisions and information on later models

Fig. 13.6 DOHC 16-valve engine lubrication circuit (Sec 5C)

1 Main oil gallery
2 From oil filter
3 Oil pump
4 Cylinder head oil-retaining valve
5 Cylinder head oil gallery
6 Cylinder head oil supply
7 Oil return
8 Piston-cooling oil spray (where fitted)
9 Oil filter

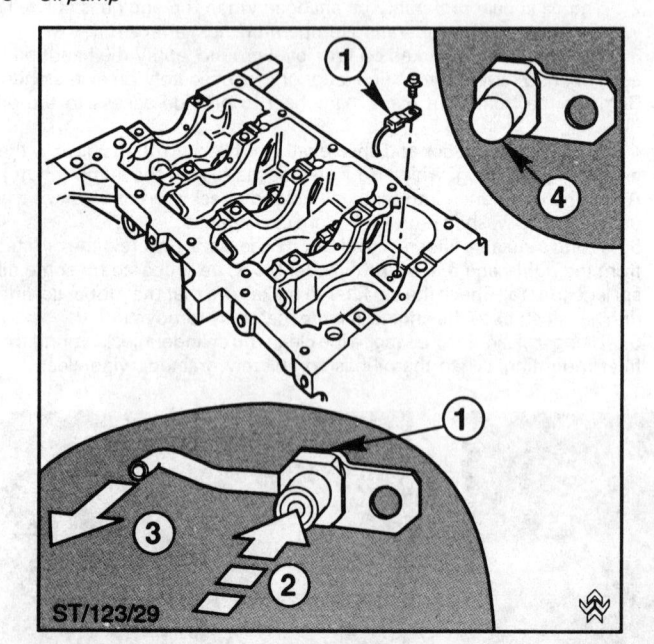

Fig. 13.7 DOHC 16-valve engine piston-cooling oil jet details (Sec 5C)

1 Oil jets (when fitted)
2 Oil flow – only when valve opens at set pressure
3 Oil spray
4 Blanking plug (when fitted)

drain plug and its threads in the sump, and refit the plug, tightening it to the torque wrench setting specified.

7 Apply a light coating of clean engine oil to the new filter's sealing ring, screw the filter into position on the engine until it seats, then tighten it through a further half-turn **only**; tighten the filter by hand only – do not use any tools.

8 Remove the old oil and all tools from under the vehicle, refit the roadwheel, and lower the vehicle to the ground.

9 Refill the engine with oil, using the correct grade and type of oil, as described earlier in this Section. Pour in half the specified quantity of oil first, then wait a few minutes for the oil to fall to the sump. Continue adding oil a small quantity at a time until the level is up to the lower mark on the dipstick. From this point, adding approximately the specified amount will bring the level up to the upper mark on the dipstick.

10 Start the engine and run it for a few minutes, while checking for leaks around the oil filter seal and the drain plug.

11 Switch off the engine, and wait a few minutes for the oil to settle in the sump once more. With the new oil circulated and the filter now completely full, recheck the level on the dipstick, and add more oil as necessary.

12 Dispose of the used engine oil safely, in accordance with local environmental health regulations.

Timing belt renewal

13 Refer to Part D of this Section.

PART D: OPERATIONS POSSIBLE WITH DOHC 16-VALVE ENGINE IN VEHICLE

Major operations possible with the engine in the vehicle

1 The following work can be carried out with the engine in the vehicle.

Chapter 13 Supplement: Revisions and information on later models

Owners should note that any operation involving the removal of the sump requires careful forethought, depending on the levels of skill and the tools and facilities available; refer to the relevant text for details.

(a) Compression pressure - testing.
(b) Cylinder head cover - removal and refitting.
(c) Timing belt covers - removal and refitting.
(d) Timing belt - renewal.
(e) Timing belt tensioner and toothed pulleys - removal and refitting.
(f) Camshaft oil seals - renewal.
(g) Camshafts and hydraulic tappets - removal, examination and refitting.
(h) Cylinder head - removal, overhaul and refitting.
(i) Cylinder head and pistons - decarbonising.
(j) Sump - removal and refitting.
(k) Crankshaft oil seals - renewal.
(l) Oil pump - removal and refitting.
(m) Piston/connecting rod assemblies - removal and refitting.
(n) Flywheel - removal and refitting.
(o) Engine/transmission mountings - removal and refitting.

Compression test - description and interpretation
2 Refer to Part A of this Section.

Cylinder head cover - removal and refitting
3 Referring to the relevant paragraphs of Section 7, Part F of this Chapter, disconnect the accelerator cable at the throttle housing, then undo the screws securing the cable abutment bracket to the throttle housing and withdraw the bracket (photo).
4 Unscrew the retaining nuts, and slacken the clamp screws so that the (black plastic) air intake trunking can be disconnected from the air mass meter's flexible rubber hose and from the throttle housing; withdraw the trunking (photo).
5 Disconnect the crankcase breather hose from the cylinder head cover union, and slacken the bolt securing the engine lifting eye on the rear right-hand side of the cylinder head (photos). Unbolt the timing belt upper cover, as described in paragraph 14.
6 Check that the HT leads are marked for position (Ford leads are normally numbered 1 to 4, by cylinder number), and label them 1 to 4 if necessary. Unplug the HT leads from the spark plugs and withdraw them, unclipping the leads from the cover.
7 Working progressively, unscrew the cylinder head cover retaining bolts, noting the spacer sleeve and rubber seal at each, then withdraw the cover (photo).
8 Discard the cover gasket; this **must** be renewed whenever it is disturbed. Check that the sealing faces are undamaged, and that the rubber seal at each retaining bolt is serviceable; renew any worn or damaged seals.
9 On refitting, clean the cover and cylinder head gasket faces carefully, then fit a new gasket to the cover, ensuring that it locates correctly in the cover grooves (photo).
10 Refit the cover to the cylinder head, then insert the rubber seal and spacer sleeve at each bolt location (photo). Tighten all bolts finger-tight to begin with, ensuring that the gasket remains seated in its groove.
11 Working in a diagonal sequence from the centre outwards, and in two stages (see Specifications), tighten the cover bolts to the specified torque wrench setting.
12 Refit the HT leads, clipping them into place so that they are correctly routed; each is numbered and can also be identified by the numbering on its respective coil terminal. Refit the timing belt upper cover, as described in paragraph 15.
13 Reconnect the crankcase breather hose and tighten the engine lifting eye's bolt, then refit the air intake trunking, the accelerator cable and its bracket.

Timing belt covers - removal and refitting
Upper cover
14 Unscrew the cover's two mounting bolts and withdraw it (photo).
15 Refitting is the reverse of the removal procedure; ensure the cover edges engage correctly with each other, and note the torque wrench setting specified for the bolts.

Middle cover
16 Slacken the water pump pulley bolts (photo). Remove the timing belt upper cover (see paragraph 14 above).
17 Remove the alternator/water pump drivebelt (Section 14 of this Chapter).
18 Unbolt and remove the water pump pulley (photo).
19 Unscrew the middle cover fasteners (one bolt at the front and one at the lower rear, one stud at the top rear) and withdraw the cover (photo).

5D.3 Disconnect accelerator cable and remove screws (arrowed) to release cable abutment bracket

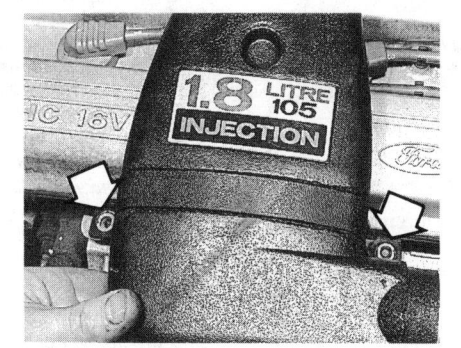

5D.4 Unscrew nuts (arrowed) to release air intake trunking

5D.5A Disconnect crankcase breather hose (arrowed)

5D.5B Slacken right-hand engine lifting eye bolt (arrowed)

5D.7 Removing cylinder head cover

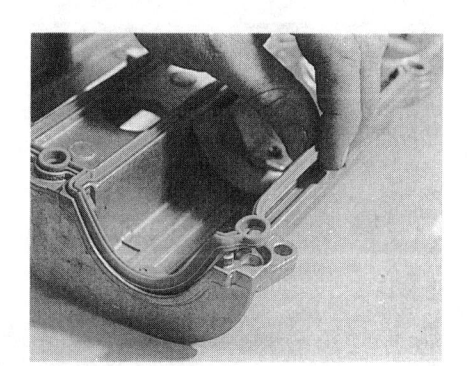

5D.9 Ensure gasket is located correctly in cover groove

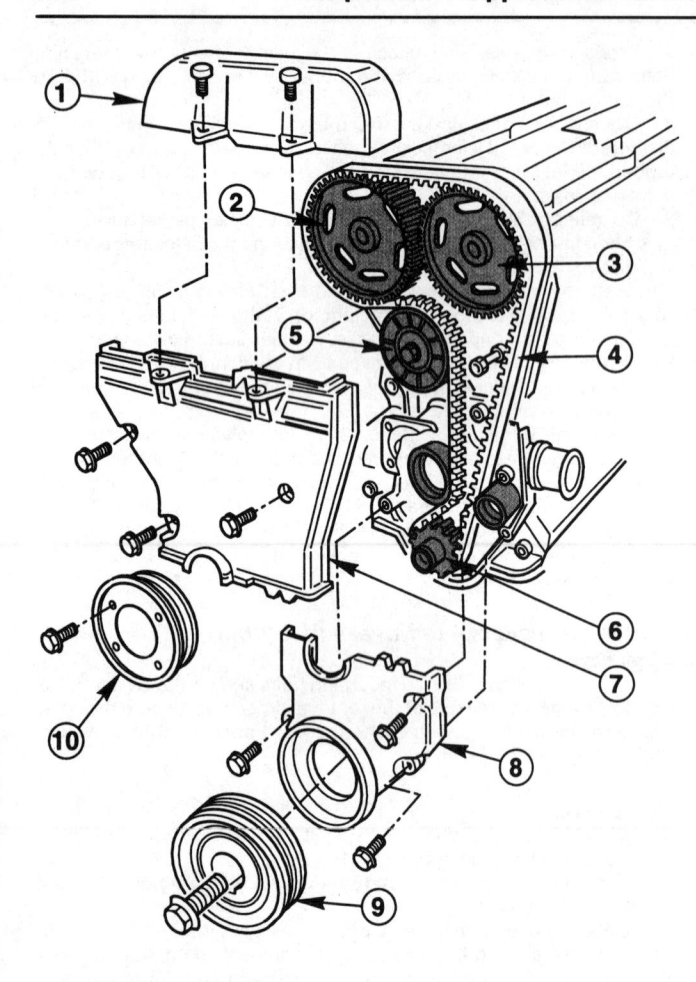

Chapter 13 Supplement: Revisions and information on later models

(negative) lead, noting the comments made in Sections 5 and 17 of Chapter 12.

22 Select top gear and apply the handbrake securely, then raise the vehicle's front end, supporting it securely on axle stands. Turn the steering to full right lock, or remove the front right-hand roadwheel.

23 Remove the alternator/water pump drivebelt (Section 14 of this Chapter). Unbolt and remove the water pump pulley.

24 Support the weight of the engine/transmission using a trolley jack, with a wooden spacer to prevent damage to the sump (photo).

25 Working in the engine compartment, unscrew the three bolts securing the engine's front right-hand (Y-shaped) mounting bracket to the alternator mounting bracket. Unfasten the engine's rear right-hand mounting from the body by unscrewing first the single nut (and washer) immediately to the rear of the timing belt cover, then the bolt in the wheel arch (photos).

26 With the engine's right-hand mountings unfastened from the body, lower the engine/transmission on the jack until a socket spanner can be fitted to the crankshaft pulley bolt.

27 With an assistant preventing crankshaft rotation by applying the footbrake as hard as possible, unscrew the crankshaft pulley bolt and withdraw the pulley (photo). Note the pulley-locating Woodruff key; if this is loose, it should be removed for safe storage with the pulley.

28 Unscrew the cover's three securing bolts and withdraw it (photo).

29 Refitting is the reverse of the removal procedure, noting the following points:

(a) Ensure the cover edges engage correctly with each other.
(b) Note the torque wrench settings specified for the various fasteners.
(c) Align the crankshaft pulley on its locating key, use the method employed on dismantling to lock the crankshaft, and tighten the pulley bolt to its specified torque wrench setting (photo).
(d) When reassembling the engine right-hand mountings, use the jack to adjust the height of the engine/transmission until the bolts (and nut, with washer) can be refitted and screwed home by hand, then tighten them securely, to the specified torque wrench settings, where given.
(e) Tension the alternator/water pump drivebelt as described in Section 14 of this Chapter.

Inner shield

30 Remove the timing belt, its tensioner components and the camshaft toothed pulleys, as described in the relevant paragraphs below. The shield is secured to the cylinder head by two bolts at the top and by two studs lower down; unscrew these and withdraw the shield (photo).

31 Refitting is the reverse of the removal procedure; note the torque wrench settings specified for the various fasteners.

Timing belt – renewal

Note: *To carry out this operation, a new timing belt, cylinder head cover gasket, and various tools (some special – see text) and other facilities will be required. If the timing belt is being removed for the first time since the vehicle left the factory, a tensioner spring and retaining pin must be obtained for fitting on reassembly.*

32 With the vehicle parked on firm level ground, open the bonnet and slacken the water pump pulley bolts. Disconnect the battery earth (negative) lead, noting the comments made in Sections 5 and 17 of Chapter 12.

Fig. 13.8 DOHC 16-valve engine timing belt and cover details (Sec 5D)

1 Timing belt upper cover
2 Inlet camshaft toothed pulley
3 Exhaust camshaft toothed pulley
4 Timing belt
5 Timing belt tensioner
6 Crankshaft toothed pulley
7 Timing belt middle cover
8 Timing belt lower cover
9 Crankshaft pulley
10 Water pump pulley

20 Refitting is the reverse of the removal procedure; ensure the cover edges engage correctly with each other, and note the torque wrench settings specified for the various fasteners. Tension the alternator/water pump drivebelt as described in Section 14 of this Chapter.

Lower cover

21 With the vehicle parked on firm level ground, open the bonnet and slacken the water pump pulley bolts. Disconnect the battery earth

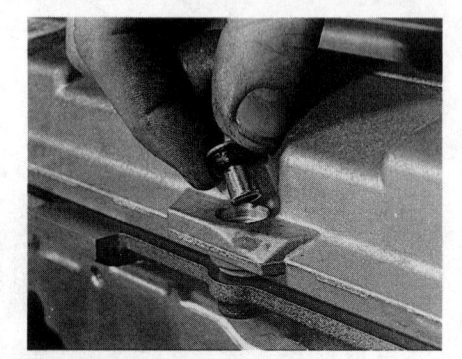

5D.10 Ensure rubber seal is fitted to each cover bolt spacer, as shown

5D.14 Remove bolts (arrowed) to release timing belt top and middle covers

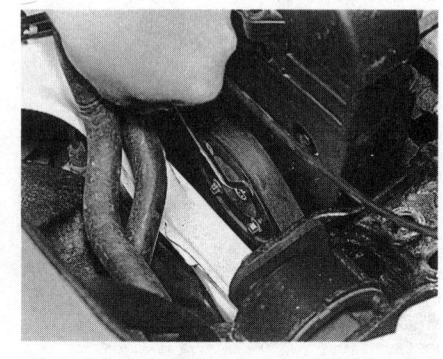

5D.16 Slacken water pump pulley bolts ...

Chapter 13 Supplement: Revisions and information on later models

5D.18 ... and remove pulley when drivebelt has been withdrawn

5D.19 Removing timing belt middle cover

5D.24 Trolley jack supporting engine, with wooden block to prevent damage to sump

5D.25A Unbolt engine's right-hand front mounting ...

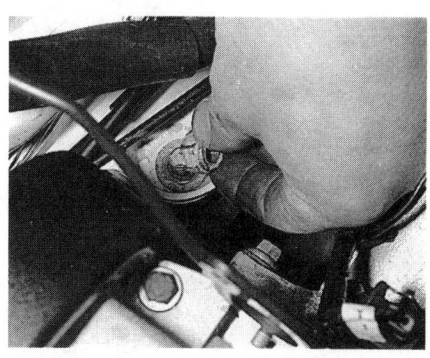

5D.25B ... and right-hand rear mounting's nut ...

5D.25C ... and bolt

5D.27 Removing crankshaft pulley

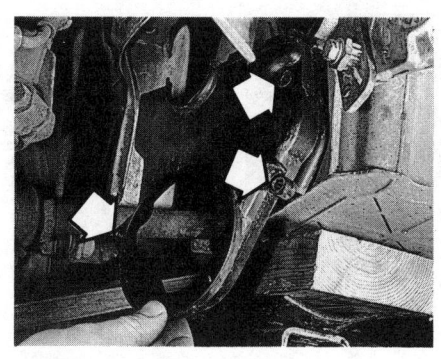

5D.28 Removing timing belt lower cover – bolt locations arrowed

5D.29 Tightening crankshaft pulley bolt

33 Remove the cylinder head cover (paragraphs 3 to 8 above) and the timing belt upper cover (paragraph 14 above). Remove the spark plugs, covering their holes with clean rag to prevent dirt or other foreign bodies from dropping in.
34 Select top gear and apply the handbrake securely, then raise the vehicle's front end, supporting it securely on axle stands. Turn the steering to full right lock, or remove the front right-hand roadwheel.
35 Remove the alternator/water pump drivebelt (Section 14 of this Chapter). Unbolt and remove the water pump pulley.
36 Support the weight of the engine/transmission using a trolley jack, with a wooden spacer to prevent damage to the sump.
37 Working in the engine compartment, unscrew the three bolts securing the engine's front right-hand (Y-shaped) mounting bracket to the alternator mounting bracket. Unfasten the engine's rear right-hand mounting from the body by unscrewing first the single nut (and washer) immediately to the rear of the timing belt cover, then the bolt in the wheel arch.
38 With the engine's right-hand mountings unfastened from the body, lower the engine/transmission on the jack until a socket spanner

can be fitted to the crankshaft pulley bolt. Rotate the crankshaft clockwise until the second pair of notches (in direction of crankshaft rotation) in the pulley rims align with the edge of the raised rib on the sump (photo).
39 Obtain Ford service tool 21-162, or fabricate a substitute from a strip of metal 5 mm thick (while the strip's thickness is critical, its length and width are not, but should be approximately 180 to 230 mm by 20 to 30 mm). Check that Nos 1 and 4 cylinders are at Top Dead Centre (TDC) – No 1 on the compression stroke – by resting this tool on the cylinder head mating surface and sliding it into the slot in the left-hand end of both camshafts (photo). The tool should slip snugly into both slots while resting on the cylinder head mating surface; if one camshaft is only slightly out of alignment, it is permissible to use an open-ended spanner to rotate the camshaft gently and carefully until the tool will fit.
40 If both camshaft slots (they are machined significantly off-centre) are below the level of the cylinder head mating surface, rotate the crankshaft through one full turn clockwise, and fit the tool again; it should now fit as described in the previous paragraph.
41 With the camshaft aligning tool remaining in place, remove the

318   Chapter 13   Supplement: Revisions and information on later models

5D.30 Timing belt inner shield fasteners (arrowed)

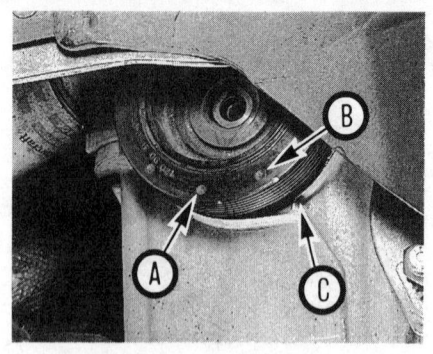

5D.38 Do not use pulley's first pair of notches 'A' – align second pair of notches 'B' with raised rib on sump 'C'

5D.39 Fit camshaft aligning tool to ensure engine is locked with Nos 1 and 4 cylinders at TDC

5D.42 Slacken tensioner bolt, and use Allen key to rotate tensioner away from timing belt ...

5D.43 ... then withdraw timing belt

5D.47A Fitting tensioner spring retaining pin

5D.47B Hook spring on to tensioner, and refit as shown – engage tensioner backplate on locating peg (arrowed)

5D.50 Slacken tensioner bolt to give initial belt tension

crankshaft pulley and timing belt lower cover as described in paragraphs 27 and 28 above, then remove the timing belt middle cover (paragraph 19 above).
42 With all covers removed and the engine locked by the camshaft aligning tool so that Nos 1 and 4 cylinders are at TDC, slacken the tensioner bolt and use an Allen key inserted into its centre to rotate the tensioner clockwise as far as possible away from the timing belt; retighten the bolt to secure the tensioner clear of the timing belt (photo).
43 If the timing belt is to be re-used, use white paint or chalk to mark its direction of rotation, and note from the manufacturer's markings which way round it is fitted. Withdraw the belt (photo). **Caution:** *Do not rotate the crankshaft until the timing belt is refitted.*
44 If the belt is being removed for reasons other than routine renewal, check it carefully for any signs of uneven wear, splitting, cracks (especially at the roots of the belt teeth) or oil contamination, and renew it if there is the slightest doubt about its condition. As a precaution, the belt must be renewed as a matter of course at the intervals given in Section 4 of this Chapter; if its history is unknown, the belt should be renewed irrespective of its apparent condition whenever the engine is over-

hauled. Similarly, check the tensioner spring (where fitted), renewing it if there is the slightest doubt about its condition. Check also the toothed pulleys for signs of wear or damage, and ensure that the tensioner and guide pulleys rotate smoothly on their bearings; renew any worn or damaged component. If signs of oil contamination are found, trace the source of the oil leak and rectify it, then wash down the engine timing belt area and all related components to remove all traces of oil.
45 On reassembly, temporarily refit the crankshaft pulley to check that the pulley notches and sump rib are aligned as described in paragraph 38 above, then ensure that both camshafts are aligned at TDC by the special tool (paragraph 39). If the engine is being reassembled after major dismantling, both camshaft toothed pulleys should be free to rotate on their respective camshafts; if the timing belt alone is being renewed, both pulleys should still be securely fastened.
46 A holding tool will be required to prevent the camshaft toothed pulleys from rotating while their bolts are slackened and retightened; either obtain Ford service tool 15-030A, or fabricate a substitute as follows. Find two lengths of steel strip, one approximately 600 mm long and the other about 200 mm, and three bolts with nuts and washers.

Chapter 13 Supplement: Revisions and information on later models

5D.53 Using forked holding tool while camshaft toothed pulley bolt is tightened

5D.54 When setting is correct, tighten tensioner bolt to specified torque wrench setting

5D.58 Removing timing belt tensioner

One nut and bolt form the pivot of a forked tool, with the remaining nuts and bolts at the tips of the 'forks' to engage with the pulley spokes, as shown in the accompanying photographs. **Warning:** *Do not use the camshaft aligning tool described in paragraph 39 (whether genuine Ford or not) to prevent rotation while the camshaft toothed pulley bolts are slackened or tightened; the risk of damage to the camshaft concerned and to the cylinder head is far too great. Use only a forked holding tool applied directly to the pulleys, as described.*

47 If it is being fitted for the first time, screw the timing belt tensioner spring retaining pin into the cylinder head, tightening it to the specified torque wrench setting. Unbolt the tensioner, hook the spring on to the pin and the tensioner backplate, then refit the tensioner, engaging its backplate on the locating peg (photos).

48 In all cases, slacken the tensioner bolt (if necessary) and use an Allen key inserted into its centre to rotate the tensioner clockwise as far as possible against spring tension, then retighten the bolt to secure the tensioner.

49 Fit the timing belt; if the original is being refitted, ensure that the marks and notes made on removal are followed, so that the belt is refitted the same way round and to run in the same direction. Starting at the crankshaft toothed pulley, work anti-clockwise around the camshaft toothed pulleys and tensioner, finishing off at the rear guide pulley. The front run, between the crankshaft and the exhaust camshaft toothed pulleys, **must** be kept taut, without altering the position either of the crankshaft or of the camshaft(s) – if necessary, the position of the camshaft toothed pulleys can be altered by rotating each on its camshaft (which remains fixed by the aligning tool). Where the pulley is still fastened, use the holding tool described in paragraph 46 above to prevent the pulley from rotating while its retaining bolt is slackened – the pulley can then be rotated on the camshaft until the belt will slip into place; retighten the pulley bolt.

50 When the belt is in place, slacken the tensioner bolt gently until the spring pulls the tensioner against the belt; the tensioner should be retained correctly against the timing belt inner shield and cylinder head, but must be just free to respond to changes in belt tension (photo).

51 Tighten both camshaft toothed pulley bolts (or check that they are tight, as applicable) and remove the camshaft aligning tool. Temporarily refit the crankshaft pulley, and rotate the crankshaft through two full turns clockwise to settle and tension the timing belt, returning the crankshaft (pulley notches) to the position described in paragraph 38 above. Refit the camshaft aligning tool; it should slip into place as described in paragraph 39. If all is well, proceed to paragraph 54 below.

52 If one camshaft is only just out of line, fit the forked holding tool to its toothed pulley, and adjust its position as required. Check that any slack created has been taken up by the tensioner, then rotate the crankshaft through two further turns clockwise, and refit the camshaft aligning tool to check that it now fits as it should. If all is well, proceed to paragraph 54 below.

53 If either camshaft is significantly out of line, use the holding tool described in paragraph 46 above to prevent its pulley from rotating while its retaining bolt is slackened – the camshaft can then be rotated (gently and carefully, using an open-ended spanner) until the camshaft aligning tool will slip into place; take care not to disturb the relationship of the pulley to the timing belt. Without disturbing the pulley's new position on the camshaft, tighten the pulley bolt to its specified torque wrench setting (photo). Remove the camshaft aligning tool, rotate the crankshaft through two further turns clockwise and refit the tool to check that it now fits as it should.

54 When the timing belt has been settled at its correct tension, and the camshaft aligning tool fits correctly when the crankshaft pulley notches are exactly aligned, tighten the tensioner bolt to its specified torque wrench setting (photo). Fitting the forked holding tool to the spokes of each pulley in turn, check that the pulley bolts are tightened to their specified torque wrench setting. Remove the camshaft aligning tool, rotate the crankshaft through two further turns clockwise, and refit the tool to make a final check that it fits as it should.

55 The remainder of the reassembly procedure is the reverse of removal, noting the following points:

  (a) Tighten all fasteners to the torque wrench settings specified.
  (b) Refit the timing belt covers as described in the relevant paragraphs above.
  (c) Align the crankshaft pulley on its locating key, use the method employed on dismantling to lock the crankshaft, and tighten the pulley bolt to its specified torque wrench setting.
  (d) When reassembling the engine right-hand mountings, use the jack to adjust the height of the engine/transmission until the bolts (and nut, with washer) can be refitted and screwed home by hand, then tighten them securely, to the specified torque wrench settings, where given.
  (e) Tension the alternator/water pump drivebelt as described in Section 14 of this Chapter.

Timing belt tensioner – removal and refitting

Note: *If the tensioner is being removed for the first time since the vehicle left the factory, a tensioner spring and retaining pin must be obtained for fitting on reassembly.*

56 While it is possible to reach the tensioner once the timing belt upper and middle covers only have been removed (paragraphs 14, and 16 to 19, above), the whole procedure outlined below must be followed to ensure that the valve timing is correctly reset once the belt's tension has been disturbed.

57 Proceed as described in paragraphs 32 to 42 above.

58 Unscrew the tensioner bolt and withdraw the tensioner, unhooking the spring, if fitted (photo).

59 On reassembly, if it is being fitted for the first time, screw the timing belt tensioner spring retaining pin into the cylinder head, tightening it to the specified torque wrench setting. Hook the spring on to the pin and the tensioner backplate, then refit the tensioner, engaging its locating peg on the backplate.

60 Use an Allen key inserted into its centre to rotate the tensioner clockwise as far as possible against spring tension, then tighten the bolt to secure the tensioner.

61 Reassemble, checking the camshaft alignment (valve timing) and setting the timing belt tension, as described in paragraphs 50 to 55 above.

Camshaft oil seals – renewal

Note: *While it is possible to reach either oil seal once the timing belt upper and middle covers have been removed (paragraphs 14, and 16 to 19, above), the timing belt tension has been relaxed (paragraph 42) and the appropriate camshaft pulley bolt has been unscrewed (paragraph 49) so*

320   Chapter 13 Supplement: Revisions and information on later models

5D.65 Using socket and camshaft toothed pulley bolt to install camshaft oil seal

5D.66 Alternatively, seal can be inserted when camshaft bearing cap is unbolted

5D.71 Using forked holding tool while camshaft toothed pulley bolt is slackened

that the pulley can be withdrawn to allow the seal to be prised out, this procedure is not recommended. Not only are the seals very soft, making this difficult to do without risk of damage to the seal housing, but it would be very difficult to ensure that the valve timing and the timing belt's tension, once disturbed, are correctly reset. Owners are advised to follow the whole procedure outlined below.

62 Proceed as described in paragraphs 32 to 42 above; if the timing belt is found to be contaminated by oil, remove it completely as described above, then renew the oil seal (see below). Wash down the engine timing belt area and all related components to remove all traces of oil. Fit a new belt on reassembly.
63 If the timing belt is still clean, slip it off the toothed pulley, taking care not to twist it too sharply; use the fingers only to handle the belt. **Warning:** *Do not rotate the crankshaft until the timing belt is refitted.* Cover the belt, and secure it so that it is clear of the working area and cannot slip off the remaining toothed pulley.
64 Unfasten the pulley bolt as described in paragraph 49 and withdraw the pulley, then unbolt the camshaft right-hand bearing cap and withdraw the defective oil seal. Clean the seal housing, and polish off any burrs or raised edges which may have caused the seal to fail in the first place.
65 To fit a new seal, Ford recommend the use of their service tool 21-009B, with a bolt (10 mm thread size, 70 mm long) and a washer, to draw the seal into place when the camshaft bearing cap is bolted down; a substitute can be made using a suitable socket (photo). Grease the seal lips and periphery to ease installation, and draw the seal into place until it is flush with the housing/bearing cap outer edge. Refit the bearing cap, using sealant and tightening the cap bolts as described in paragraphs 84 to 87 below.
66 For most owners, the simplest answer will be to grease the seal lips and to slide it on to the camshaft (until it is flush with the housing's outer edge). Refit the bearing cap, using sealant and tightening the cap bolts as described in paragraphs 84 to 87 below (photo). Take care to ensure that the seal remains absolutely square in its housing, and is not distorted as the cap is tightened down.
67 Refit the pulley to the camshaft, tightening the retaining bolt loosely, then slip the timing belt back on to the pulley (refer to paragraph 49 above) and tighten the bolt securely.
68 The remainder of the reassembly procedure, including checking the camshaft alignment (valve timing) and setting the timing belt tension, is as described in paragraphs 45 to 55 above.

Camshafts and hydraulic tappets – removal, examination and refitting

Removal

69 Proceed as described in paragraphs 32 to 42 above.
70 Either remove the timing belt completely (paragraphs 43 and 44), or slip it off the camshaft toothed pulleys, taking care not to twist it too sharply; use the fingers only to handle the belt. Cover the belt, and secure it so that it is clear of the working area. **Warning:** *Do not rotate the crankshaft until the timing belt is refitted.*
71 Unfasten the pulley bolts as described in paragraph 49 and withdraw the pulleys; while both are the same and could be interchanged, it is good working practice to mark them so that each is refitted only to its original location (photo).
72 Working in the sequence shown in Fig. 13.9, slacken the camshaft bearing cap bolts progressively, by half a turn at a time. Work only as described, to release the pressure of the valve springs on the caps gradually and evenly.
73 Withdraw the caps, noting their markings and the presence of the locating dowels, then remove the camshafts and withdraw their oil seals. The inlet camshaft can be identified by the reference lobe for the camshaft position sensor; therefore, there is no need to mark the camshafts (photos).
74 Make up sixteen small, clean containers, and number them 1 to 16. Using a rubber sucker, withdraw each hydraulic tappet in turn, invert it to prevent oil loss, and place it in its respective container, which should then be filled with clean engine oil (photos). Do not interchange the hydraulic tappets, or the rate of wear will be much increased; do not allow them to lose oil, or they will take a long time to refill with oil on restarting the engine, resulting in incorrect valve clearances.

Examination

75 With the camshafts and hydraulic tappets removed, check each for signs of obvious wear (scoring, pitting, etc) and for ovality, and renew, if necessary; refer to the relevant Sections of Chapter 1.
76 If the engine's valvegear has sounded noisy, particularly if the noise persists after initial start-up from cold, there is reason to suspect a faulty hydraulic tappet. Only a good mechanic experienced in these engines can tell whether the noise level is typical, or if renewal is warranted of one or more of the tappets. If faulty tappets are diagnosed and the engine's service history is unknown, it is always worth trying the effect of first renewing the engine oil and filter (Part C of this Section, paragraphs 2 to 12). Use **only** good-quality engine oil of the recommended viscosity and specification. If any tappet's operation is faulty, it must be renewed.
77 Examine the camshaft bearing journals and the cylinder head bearing surfaces for signs of obvious wear or pitting. If any such signs are evident, renew the component concerned.

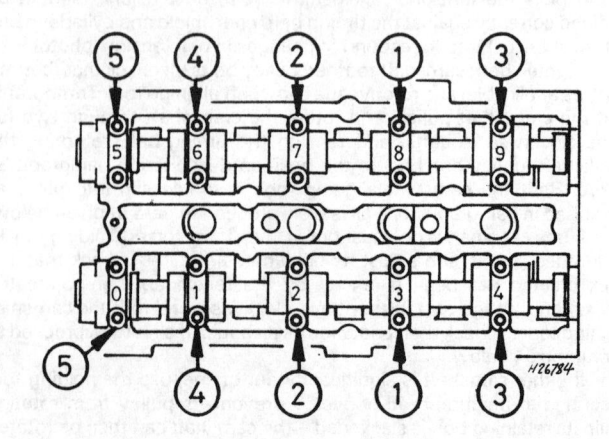

Fig. 13.9 Camshaft bearing cap slackening sequence – DOHC 16-valve engine (Sec 5D)

Note: *View from front of vehicle, showing bearing cap numbers*

Chapter 13 Supplement: Revisions and information on later models

5D.73A Note locating dowels and etched markings (arrowed – on ends of caps) when removing camshaft bearing caps. See also photo 5D.85

5D.73B Inlet camshaft has lobe (arrowed) for camshaft position sensor

5D.74A Removing hydraulic tappets

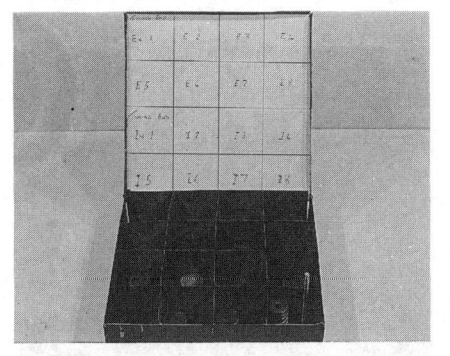

5D.74B Hydraulic tappets must be stored as described in text

5D.82 Oil liberally when refitting hydraulic tappets

5D.84 Apply sealant to mating surface of camshaft right-hand bearing caps

78 To check the bearing journal running clearance, remove the hydraulic tappets, carefully clean the bearing surfaces, and refit the camshafts and bearing caps with a strand of Plastigage across each journal. Tighten the bearing cap bolts to the specified torque wrench setting (do not rotate the camshafts), then remove the bearing caps and use the scale provided with the Plastigage to measure the width of the compressed strands.

79 If the running clearance of any bearing is found to be worn to beyond the specified service limits, fit a new camshaft and repeat the check; if the clearance is still excessive, the cylinder head must be renewed.

80 To check camshaft endfloat, remove the hydraulic tappets, clean the bearing surfaces carefully, and refit the camshafts and bearing caps. Tighten the bearing cap bolts to the specified torque wrench setting, then measure the endfloat using a DTI (Dial Test Indicator, or dial gauge) mounted on the cylinder head so that its tip bears on the camshaft right-hand end.

81 Tap the camshaft fully towards the gauge, and zero the gauge. Now tap the camshaft fully away from the gauge and note the gauge reading. If the endfloat measured is found to be at the specified service limit or beyond, fit a new camshaft and repeat the check; if the clearance is still excessive, the cylinder head must be renewed.

Refitting

82 On reassembly, liberally oil the cylinder head hydraulic tappet bores and the tappets (photo). Note that if new tappets are being fitted, they must be charged with clean engine oil before installation. Carefully refit the tappets to the cylinder head, ensuring that each tappet is refitted to its original bore and is the correct way up. Some care will be required to enter the tappets squarely into their bores.

83 Liberally oil the camshaft bearings and lobes. If necessary, refer to paragraph 73 and identify the camshafts, to ensure that they are refitted to their original locations. Refit the camshafts, locating each so that the slot in its left-hand end is approximately parallel to, and just above, the cylinder head mating surface.

84 Ensure that the locating dowels are pressed firmly into their recesses, and check that all mating surfaces are completely clean, unmarked and free from oil. Apply a thin film of suitable sealant (Ford recommend Hylosil 102) to the mating surfaces of each camshaft's right-hand bearing cap (photo). Referring to paragraph 66 above, some owners may wish to fit the new camshaft oil seals at this stage.

85 All camshaft bearing caps have a single-digit identifying number etched on them (photo). The exhaust camshaft's bearing caps are numbered in sequence 0 (right-hand cap) to 4 (left-hand cap), the inlet's 5 (right-hand cap) to 9 (left-hand cap); see Fig 13.10 for details. Each cap is to be fitted so that its numbered side faces outwards, to the front (exhaust) or to the rear (inlet).

86 Ensuring that each cap is kept square to the cylinder head as it is tightened down, and working in the sequence shown in Fig. 13.10, tighten the camshaft bearing cap bolts slowly and by one turn at a time until each cap touches the cylinder head (photo). Next, go round again in the same sequence, tightening the bolts to the first stage torque wrench setting specified, then once more, tightening them to the second stage

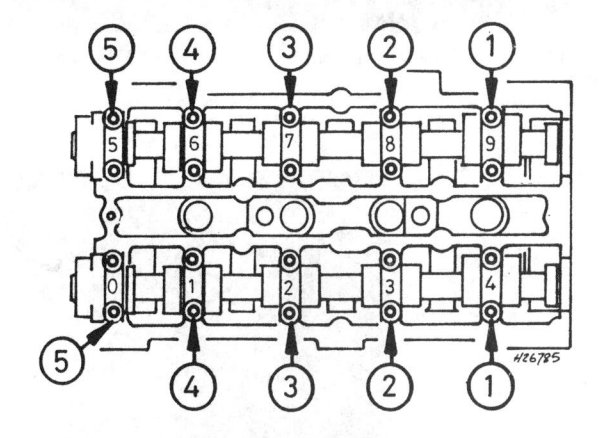

Fig. 13.10 Camshaft bearing cap tightening sequence – DOHC 16-valve engine (Sec 5D)

View from front of vehicle – locate bearing caps according to etched numbers, aligned as described in text

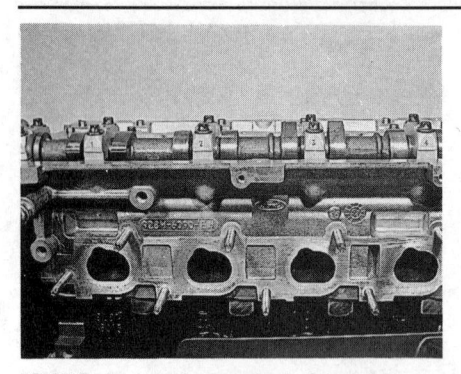

5D.85 Etched marks on camshaft bearing caps must be arranged as shown and face outwards

5D.86 Keep caps square to cylinder head at all times when tightening down

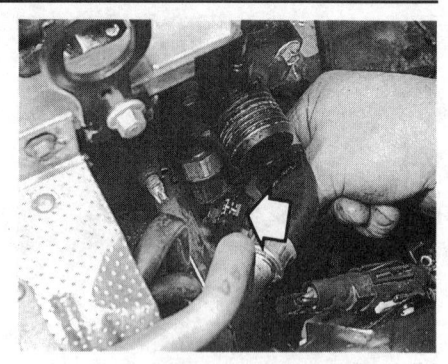

5D.94 Disconnecting hoses and wiring (temperature gauge sender unit arrowed) from thermostat housing

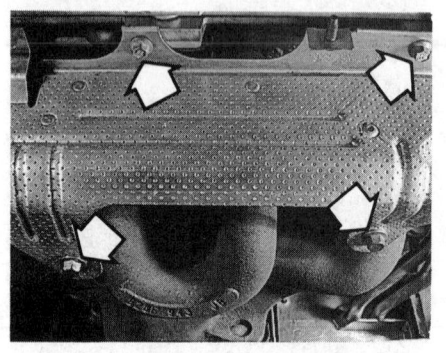

5D.95A Exhaust manifold heat shield bolts (arrowed)

5D.95B Disconnecting coolant hose (arrowed) to release assembly

5D.96 Unfastening dipstick tube from cylinder head

5D.97A Disconnect cooling fan motor wiring ...

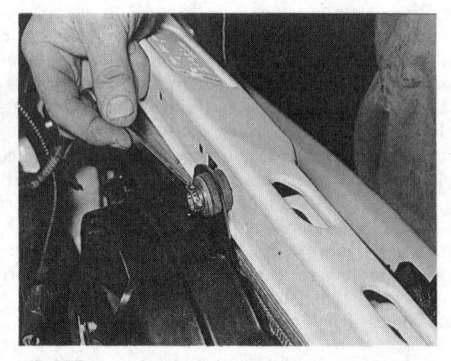

5D.97B ... and unbolt assembly to remove

5D.98A Unscrew union sleeve nuts (arrowed) ...

5D.98B ... and unbolt filter housing to remove pulse-air assembly

5D.99 Disconnecting fuel feed and return lines (arrowed)

Chapter 13 Supplement: Revisions and information on later models

5D.100A Uncoupling multi-plugs to disconnect engine wiring ...

5D.100B ... from main wiring harness (arrowed)

5D.101A Disconnect vacuum hoses (arrowed) from inlet manifold – seen from below ...

5D.101B ... disconnecting brake servo vacuum hose

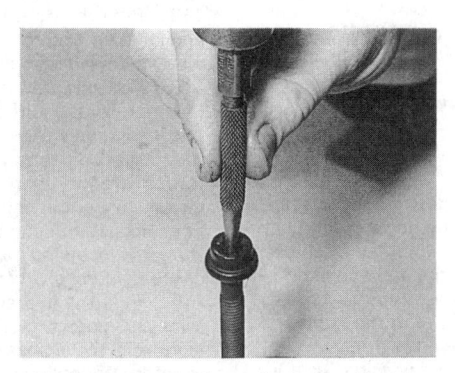

5D.102 Punch-marking cylinder head bolt to show it has been used

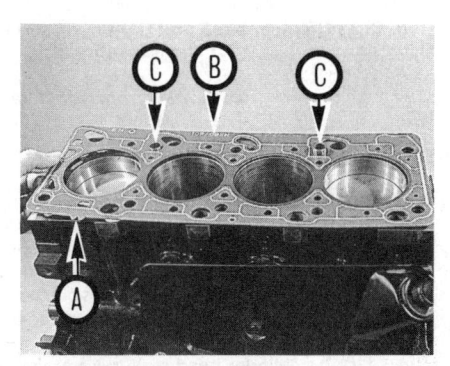

5D.109 Ensuring protruding teeth 'A' are at front and marking 'B' is upwards, locate new cylinder head gasket on dowels 'C'

setting. Work only as described, to impose gradually and evenly the pressure of the valve springs on the caps. Refit the camshaft aligning tool; it should slip into place as described in paragraph 39.
87 Wipe off all surplus sealant, so that none is left to find its way into any oilways. Follow the sealant manufacturer's instructions as to the time needed for curing; usually, at least an hour must be allowed between application of the sealant and starting the engine.
88 If using Ford's recommended procedure, fit new oil seals to the camshafts as described in paragraph 65 above.
89 Using the marks and notes made on dismantling to ensure that each is refitted to its original camshaft, refit the toothed pulleys to the camshafts, tightening the retaining bolts loosely, then slip the timing belt back on to the pulleys (refer to paragraph 49 above) and tighten the bolts securely.
90 The remainder of the reassembly procedure, including checking the camshaft alignment (valve timing) and setting the timing belt tension, is as described in paragraphs 45 to 55 above.

Cylinder head – removal and refitting
Removal
Note: *The following text assumes that the cylinder head will be removed with both inlet and exhaust manifolds attached; this simplifies the procedure, but makes it a bulky and heavy assembly to handle. If it is wished first to remove the manifolds, proceed as described in the relevant paragraphs of Sections 7 and 8 of this Chapter.*

91 Disconnect the battery earth (negative) lead, noting the comments made in Sections 5 and 17 of Chapter 12. Drain the cooling system (Chapter 2).
92 Unclip the air filter housing lid and withdraw the air filter element. Separate the flexible rubber hose from the (black plastic) air intake trunking, and unclip the air mass meter, then move the assembly clear of the working area (Section 7 of this Chapter, Part F).
93 Remove both camshafts (paragraphs 32 to 42, and 70 to 73, above); if the cylinder head is to be dismantled, withdraw the hydraulic tappets as described in paragraph 74. Remove the timing belt inner shield as described in paragraph 30, and unbolt the engine earth lead from the cylinder head lifting eye.

94 Disconnect the coolant hoses from the thermostat housing. Unplug the HT leads and wiring multi-plug from the ignition coil, and the wiring connectors from the engine coolant temperature sensor and the temperature gauge sender unit (photo).
95 Unbolt the exhaust manifold heat shield; either disconnect the coolant pipe and hose so that the heat shield assembly can be removed, or secure the assembly clear of the working area (photos).
96 Unscrew the nut securing the dipstick tube to the cylinder head (photo), and remove the tube and dipstick, or swing it out of the way.
97 Remove the radiator cooling fan and motor as described in Chapter 2 (photos).
98 Remove the pulse-air piping assembly as described in Section 7 of this Chapter, Part H (photos). Unfasten the exhaust downpipe from the manifold (Chapter 3); disconnect the HEGO sensor wiring so that it is not strained by the weight of the exhaust. Disconnect the multi-plug from the engine speed/crankshaft position sensor.
99 Depressurise the fuel system, then uncouple the fuel feed and return pipes as described in the relevant paragraphs of Section 7 of this Chapter (photo).
100 Disconnect the two multi-plugs connecting the engine's wiring to the main wiring loom (photos).
101 Unplug the two vacuum hoses from the rear of the inlet manifold, then disconnect the brake servo vacuum hose from the top of the manifold, as described in Section 16 of Chapter 8. Pull the crankcase breather hose off the stub on the left-hand end of the manifold's intermediate flange (photos).
102 Working in the *reverse* of the sequence shown in Fig. 13.11, slacken the ten cylinder head bolts progressively and by one turn at a time; a Torx key (TX 55 size) will be required. Remove each bolt in turn, punch-mark its head to show that it has been used (photo), and store it in its correct fitted order by pushing it through a clearly-marked cardboard template; if any bolt already has two punch marks, this means it has already been re-used twice after the original fitting, and must now be discarded; renew all bolts as a set.
103 Check the condition of the cylinder head bolts, particularly their threads, whenever they are removed. Keeping all the bolts in their correct fitted order, wash them and wipe dry, then check each for any sign of visible wear or damage, and renew it if necessary. Each bolt may

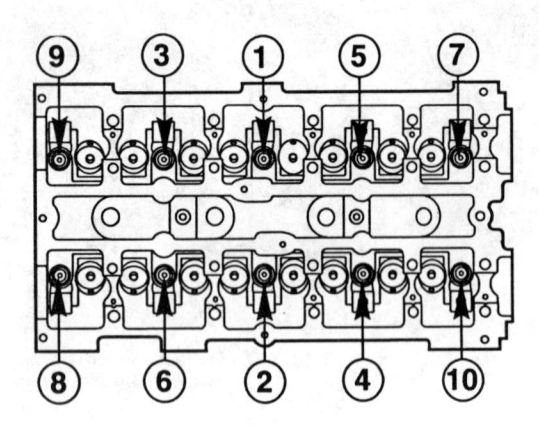

Fig. 13.11 Cylinder head bolt tightening sequence – DOHC 16-valve engine (Sec 5D)

Note: View from rear of vehicle

be used (ie, subjected to the stress of tightening-down to the specified loading) a maximum of three times, shown by the number of punch-marks on its head (see previous paragraph); once a bolt has been used three times, it must be renewed, regardless of its apparent condition. If any single bolt is to be renewed on this precautionary basis, or if there is any doubt about the condition of any of the bolts or of the number of times any has been used, they should all be renewed as a set.

104 Lift the cylinder head away; use assistance if possible, as it is a heavy assembly. Remove the gasket, noting the two dowels, and discard it.

Refitting

105 The mating faces of the cylinder head and cylinder block must be perfectly clean before refitting the head. Use a hard plastic or wood scraper to remove all traces of gasket and carbon; also clean the piston crowns. Take particular care, as the soft aluminium alloy is damaged easily. Also, make sure that the carbon is not allowed to enter the oil and water passages – this is particularly important for the lubrication system, as carbon could block the oil supply to any of the engine's components. Using adhesive tape and paper, seal the water, oil and bolt holes in the cylinder block. To prevent carbon entering the gap between the pistons and bores, smear a little grease in the gap. After cleaning each piston, use a small brush to remove all traces of grease and carbon from the gap, then wipe away the remainder with a clean rag. Clean all the pistons in the same way.

106 Check the mating surfaces of the cylinder block and the cylinder head for nicks, deep scratches and other damage. If slight, they may be removed carefully with a file, but if excessive, machining may be the only alternative to renewal.

107 If warpage of the cylinder head gasket surface is suspected, use a straight-edge to check it for distortion. Refer to Chapter 1 if necessary.

108 Wipe clean the mating surfaces of the cylinder head and cylinder block. Check that the two locating dowels are in position in the cylinder block, and that all cylinder head bolt holes are free from oil.

109 Position a new gasket over the dowels on the cylinder block surface so that the 'TOP/OBEN' mark is uppermost and the teeth protruding from one edge point to the front of the vehicle (photo).

110 Temporarily refit the crankshaft pulley, and rotate the crankshaft anti-clockwise so that No 1 cylinder's piston is lowered to approximately 20 mm before TDC, thus avoiding any risk of valve/piston contact and damage during reassembly.

111 As the cylinder head is such a heavy and awkward assembly to refit with manifolds, it is helpful to make up a pair of guide studs from two 10 mm (thread size) studs approximately 90 mm long, with a screwdriver slot cut in one end – two old cylinder head bolts with their heads cut off would make a good starting point. Screw these guide studs, screwdriver slot upwards to permit removal, into the bolt holes at diagonally-opposite corners of the cylinder block surface; ensure that approximately 70 mm of stud protrudes above the gasket.

112 Refit the cylinder head, sliding it down the guide studs (if used) and locating it on the dowels (photo). Unscrew the guide studs (if used) when the head is in place.

113 Keeping the cylinder head bolts in their correct fitted order (where relevant) and dry (**do not oil** their threads), carefully enter each into its original hole and screw it in, by hand only, until finger-tight.

114 Working progressively and in the sequence shown, use first a torque wrench, then an ordinary socket extension bar and an angle gauge, to tighten the cylinder head bolts in the stages given in the Specifications Section of this Chapter (photos). **Note:** *Once tightened correctly, following this procedure, the cylinder head bolts do not require check-tightening, and must not be re-torqued.*

115 Refit the hydraulic tappets (if removed), the camshafts, their oil seals and pulleys as described in paragraphs 82 to 89 above. Temporarily refit the crankshaft pulley, and rotate the crankshaft clockwise to return the pulley notches to the position described in paragraph 38 above.

116 Refit the timing belt and covers, including checking the camshaft alignment (valve timing) and setting the timing belt tension, as described in paragraphs 45 to 55 above.

117 The remainder of reassembly is the reverse of the removal procedure, noting the following points:

(a) Tighten all fasteners to the torque wrench settings specified.
(b) Refill the cooling system, and check all disturbed joints for signs of leakage once the engine has been restarted and warmed-up to normal operating temperature.
(c) Top-up the engine oil, and check for signs of oil leaks once the engine has been restarted and warmed-up to normal operating temperature.

Cylinder head and pistons – decarbonising

118 Refer to Chapter 1, Sections 6 and 30; note also the (relevant) information given in the following sub-Section.

Cylinder head – dismantling, overhaul and reassembly

119 Refer to the relevant Sections of Chapter 1, noting the information given in the Specifications Section of this Chapter and the procedures outlined in paragraphs 75 to 81 above.

5D.112 Refitting cylinder head (shown without manifolds)

5D.114A Tightening cylinder head bolts (to first and second stages) using torque wrench ...

5D.114B ... and to third stage using angle gauge

Chapter 13 Supplement: Revisions and information on later models

Fig. 13.12 Cylinder head components – DOHC 16-valve engine (Sec 5D)

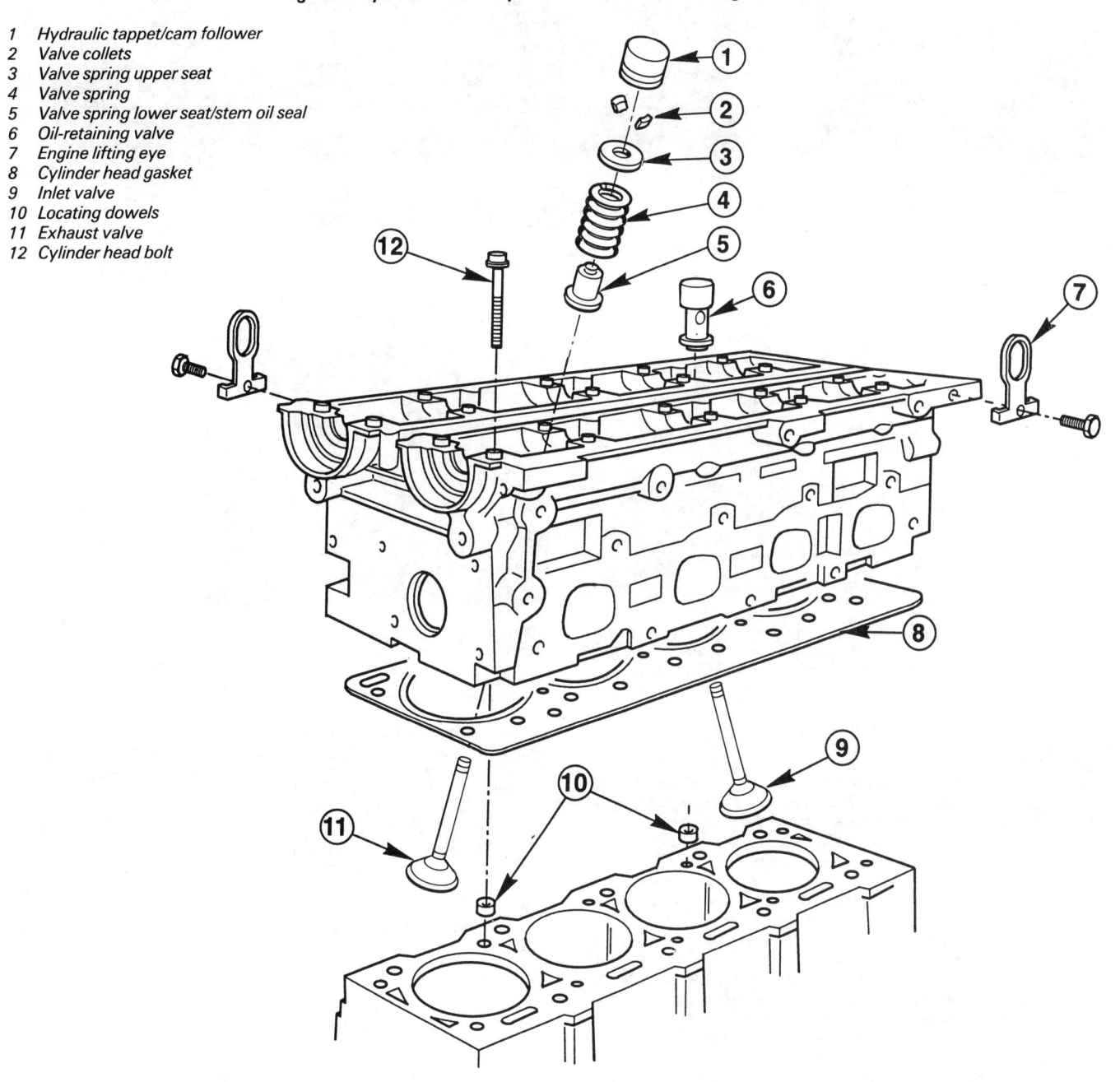

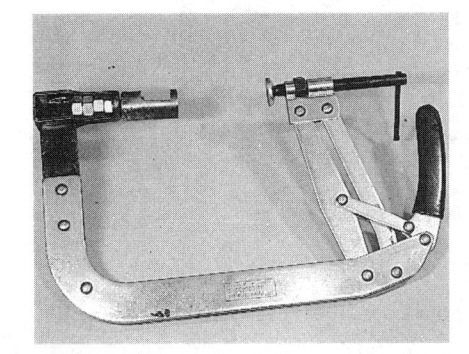

1 Hydraulic tappet/cam follower
2 Valve collets
3 Valve spring upper seat
4 Valve spring
5 Valve spring lower seat/stem oil seal
6 Oil-retaining valve
7 Engine lifting eye
8 Cylinder head gasket
9 Inlet valve
10 Locating dowels
11 Exhaust valve
12 Cylinder head bolt

5D.120A Standard valve spring compressor modified as shown ...

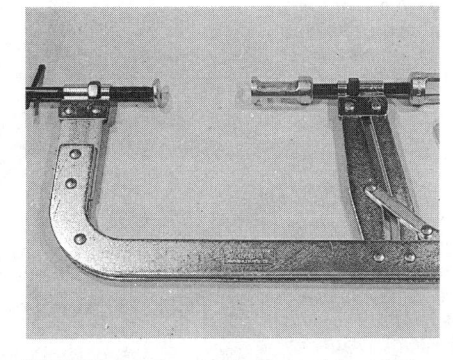

5D.120B ... or purpose-built special version ...

5D.120C ... is required to compress valve springs without damaging cylinder head

Chapter 13 Supplement: Revisions and information on later models

5D.120D Ford service tool in use to remove valve spring lower seat/stem oil seals ...

5D.120E ... can be replaced by home-made tool, if suitable spring can be found

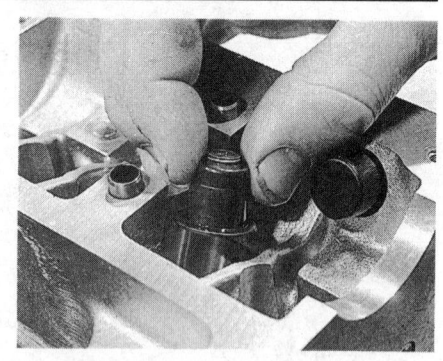

5D.120F Valve spring pressure is sufficient to seat lower seat/stem oil seals on reassembly

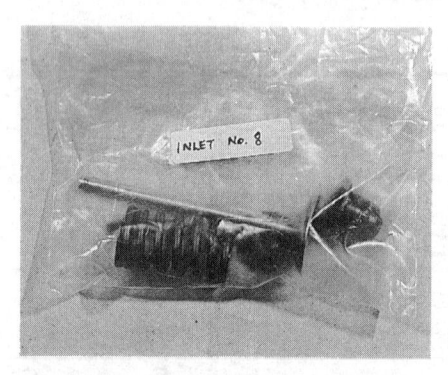

5D.120G Use clearly-marked containers to identify components, and to keep matched assemblies together

5D.120H Cylinder head oil-retaining valve (arrowed)

120 Note also the following points (photos):

(a) A special valve spring compressor will be required, to reach into the deep wells in the cylinder head without risk of damaging the hydraulic tappet bores; such compressors are now widely available from most good motor accessory shops.

(b) Ford recommend the use of their service tool 21-160 to extract the valve spring lower seat/stem oil seals; while this is almost indispensable if the seals are to be removed without risk of (extremely expensive) damage to the cylinder head, we found that a serviceable substitute can be made from a strong spring of suitable size. Screw on the tool or spring so that it bites into the seal, then draw the seal off the valve guide. On refitting, the new seal is simply placed squarely on top of the guide; the action of refitting the valve spring presses the seal into place.

(c) If the oil-retaining valve is to be removed to flush out the cylinder head oil galleries thoroughly, seek the advice of a Ford dealer as to how it can be extracted; it may be that the only course of action involves destroying the valve as follows. Screw a self-tapping screw into its ventilation hole, and use the screw to provide purchase with which the valve can be drawn out; a new valve must be purchased and pressed into place on reassembly.

(d) Note the comments at the beginning of Part C of this Section – always establish precisely what replacement parts are available before undertaking any overhaul operation.

Sump – removal and refitting

Removal

Note: *While this task appears straightforward at first glance, the full procedure outlined below must be followed, so that the mating surfaces can be cleaned and prepared to achieve an oil-tight joint on reassembly, and so that the sump can be aligned correctly. Depending on your skill and experience, and the tools and facilities available, it may be that this task can be carried out only with the engine removed from the vehicle. Note that the sump gasket must be renewed whenever it is disturbed.*

121 Drain the engine oil (Part C of this Section), then clean and refit the engine oil drain plug, tightening it to the specified torque wrench setting. Although this is not necessary as part of the dismantling procedure, owners are advised to remove and discard the oil filter, so that it can be renewed with the oil. Remove the alternator/water pump drivebelt shield (Section 14 of this Chapter).

122 Remove the exhaust downpipe and catalytic converter, disconnecting the HEGO sensor wiring (Chapter 3).

123 Remove the transmission (Chapter 6).

124 Remove the clutch (Chapter 5) and unbolt the flywheel (paragraph 168 below), then withdraw the engine/transmission adaptor plate.

125 Progressively unscrew the sump retaining bolts. Break the joint by striking the sump with the palm of your hand, then lower the sump

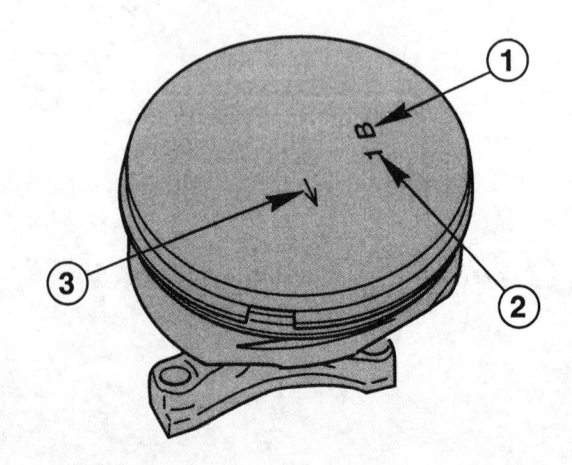

Fig. 13.13 Piston crown markings – DOHC 16-valve engine (Sec 5D)

1 Gudgeon pin diameter grade – when used
2 Piston skirt diameter grade
3 Arrow mark – pointing to timing belt end of engine

Chapter 13 Supplement: Revisions and information on later models

5D.127 Ensure gasket is located correctly in sump groove

5D.128 Apply sealant (arrowed) as directed when refitting sump

5D.129 Checking alignment of sump with cylinder block/crankcase

and withdraw it. Remove and discard the gasket; this must be renewed as a matter of course whenever it is disturbed.

126 While the sump is removed, take the opportunity to unbolt the oil pump pick-up/strainer pipe and to clean it – see paragraph 145 below.

Refitting

127 On reassembly, thoroughly clean and degrease the mating surfaces of the cylinder block/crankcase and sump, then use a clean rag to wipe out the sump and the engine's interior. If the oil pump pick-up/strainer pipe was removed, fit a new gasket and refit the pipe, tightening its bolts to the specified torque wrench setting. Fit the new gasket to the sump mating surface so that the gasket fits into the sump groove (photo).

128 Apply a thin film of suitable sealant (Ford recommend Hylosil 102) to the junctions of the cylinder block/crankcase with the oil pump and the crankshaft rear oil seal carrier (photo). Working quickly – the sump bolts must be fully tightened within 10 to 20 minutes of applying the sealant – offer up the sump to the cylinder block/crankcase and insert the bolts, tightening them lightly at first.

129 Using a suitable straight-edge to check alignment across the flat-machined faces of each, move the sump as necessary so that its left-hand face is flush with that of the cylinder block/crankcase (photo). Without disturbing the position of the sump, and working in a diagonal sequence from the centre outwards, tighten the sump bolts to the specified torque wrench setting.

130 Check again that both faces are flush before proceeding; if necessary, unbolt the sump again, clean the mating surfaces and repeat the full procedure (paragraphs 128 and 129) to ensure that the sump is correctly aligned.

131 The remainder of reassembly is the reverse of the removal procedure, noting the following points:

(a) Tighten all fasteners to the torque wrench settings specified.
(b) Refill the engine with oil, and check for signs of oil leaks once the engine has been restarted and warmed-up to normal operating temperature.

Crankshaft oil seals – renewal

Front (right-hand) seal

132 Remove the timing belt as described in paragraphs 32 to 44 above.

133 Withdraw the crankshaft toothed pulley and the thrustwasher behind it, noting which way round the thrustwasher is fitted.

134 Renew the seal as described in Sections 12 and 31 of Chapter 1 (photos).

135 Refitting is the reverse of the removal procedure.

Rear (left-hand) seal

136 Remove the transmission (Chapter 6).

137 Remove the clutch (Chapter 5) and unbolt the flywheel (paragraph 168 below).

138 Extract the seal as described in Sections 12 and 31 of Chapter 1, or punch or drill two small holes opposite each other in the seal. Screw a self-tapping screw into each, and pull on the screws with pliers to extract the seal. Clean the seal housing, and polish off any burrs or raised edges which may have caused the seal to fail in the first place.

139 Ford's recommended method of seal fitting is to use service tool 21-141 with two flywheel bolts to draw the seal into place. If this is not available, first make up a guide from a thin sheet of plastic or similar.

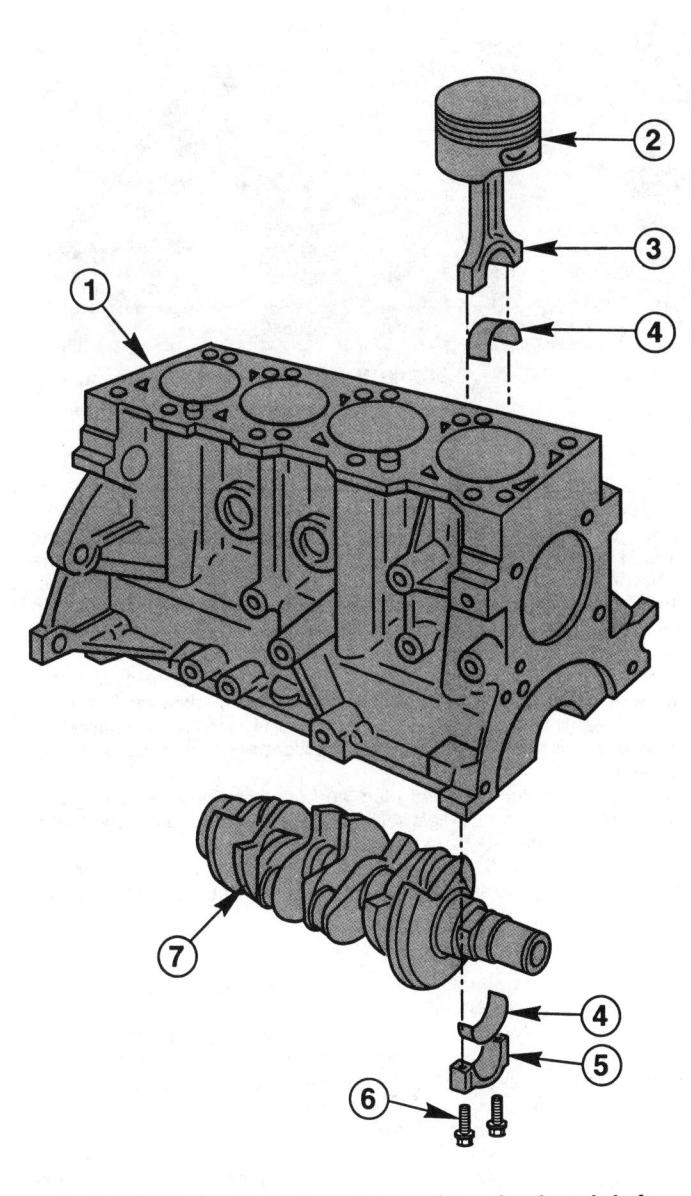

Fig. 13.14 Cylinder block, piston/connecting rod and crankshaft details – DOHC 16-valve engine (Sec 5D)

1 Cylinder block/crankcase
2 Piston
3 Connecting rod
4 Big-end bearing shell
5 Big-end bearing cap
6 Big-end bearing cap bolts
7 Crankshaft

328 Chapter 13 Supplement: Revisions and information on later models

5D.134A Using home-made guide (arrowed) to slide oil seal lips over crankshaft shoulder ...

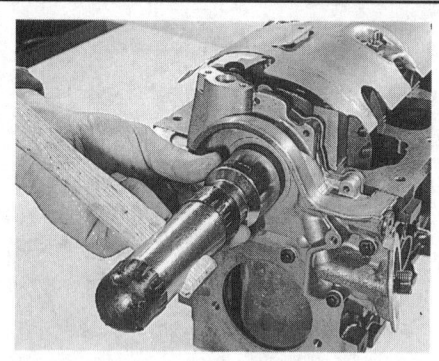

5D.134B ... install crankshaft right-hand oil seal as described

5D.139 Using home-made plastic guide to fit new crankshaft left-hand oil seal

5D.145A Unscrew bolts (arrowed) to remove oil pump ...

5D.145B ... tapping out crankshaft right-hand oil seal

5D.146 Withdrawing oil pump inner rotor

Lubricate the lips of the new seal and the crankshaft shoulder with grease, then offer up the seal, using the guide to feed the seal's lips over the crankshaft shoulder (photo). Press the seal evenly into its housing by hand only, and use a soft-faced mallet to gently tap it into place until it is flush with the surrounding housing.

140 Wipe off any surplus oil or grease; the remainder of the reassembly procedure is the reverse of dismantling.

Oil pump – removal, inspection and refitting
Removal
Note: *While this task is theoretically possible when the engine is in place in the vehicle, in practice it requires so much preliminary dismantling, and is so difficult to carry out due to the restricted access, that owners are advised to remove the engine from the vehicle first.*
In addition to the new pump gasket and other replacement parts required, an adjustable engine support bar will be needed. This should fit into the water drain channels on each side of the bonnet aperture, and have a hook which will engage the engine lifting eyes and allow the height of the engine to be adjusted; such equipment can be hired from most tool hire shops.
If the reason for removing the pump is to gain access to the pressure relief valve, note that this component can be removed without disturbing the pump, as described in paragraphs 148 to 150.

141 Remove the timing belt as described in paragraphs 32 to 44 above.
142 Withdraw the crankshaft toothed pulley and the thrustwasher behind it, noting which way round the thrustwasher is fitted.
143 Temporarily reassemble the engine right-hand mountings, and fit the engine support bar to the cylinder head's left-hand (transmission end) lifting eye so that the engine is supported securely.
144 Remove the sump (paragraphs 121 to 125 above).
145 Unbolt the pump and pick-up/strainer pipe as an assembly from the cylinder block/crankcase. Unbolt the pick-up/strainer pipe, if required, from the pump; withdraw and discard the gaskets, and remove the crankshaft front (right-hand) oil seal (photos).

Inspection
146 Unscrew the Torx screws and remove the pump cover plate; noting any identification marks on the rotors, withdraw the rotors (photo).
147 Inspect the rotors for obvious signs of wear or damage, and renew if necessary; if the pump body or cover plate are scored or damaged, the complete oil pump assembly must be renewed.
148 The oil pressure relief valve can be dismantled, if required, without disturbing the pump. With the vehicle parked on firm level ground, apply the handbrake securely and raise its front end, supporting it securely on axle stands. Remove the front right-hand roadwheel to provide access to the valve.
149 Unscrew the threaded plug, and recover the valve spring and plunger (photos). If the plug's sealing O-ring is worn or damaged a new one must be obtained, to be fitted on reassembly.
150 Reassembly is the reverse of the dismantling procedure; ensure the spring and valve are refitted the correct way round, and tighten the threaded plug securely.

Refitting
151 The oil pump must be primed on installation by pouring clean engine oil into it, and rotating its inner rotor a few turns.
152 Using grease to stick the gasket in place on the cylinder block/crankcase, and rotating the pump's inner rotor to align with the flats on the crankshaft, refit the pump and insert the bolts, tightening them lightly at first.
153 Using a suitable straight-edge and feeler gauges (photo), check that the pump is both centred around the crankshaft, and aligned squarely, so that its (sump) mating surface on each side of the crankshaft is 0.3 to 0.8 mm below that of the cylinder block/crankcase. Being careful not to disturb the gasket, move the pump into the correct position, and tighten its bolts to the specified torque wrench setting.
154 Check again that the pump's position is correct **on both sides** before proceeding; if necessary, unbolt the pump again and repeat the full procedure (paragraphs 152 and 153) to ensure that the pump is correctly aligned.

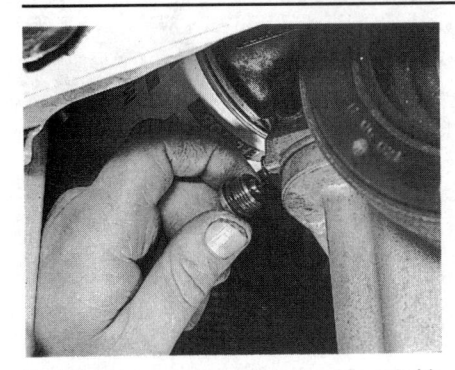

Chapter 13 Supplement: Revisions and information on later models

5D.149A Unscrew threaded plug – seen (with pump installed) through right-hand wheel arch ...

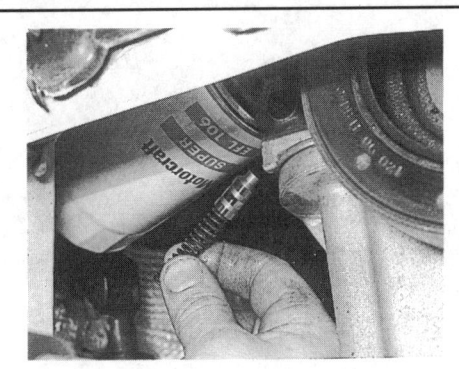

5D.149B ... to withdraw oil pressure relief valve spring and plunger

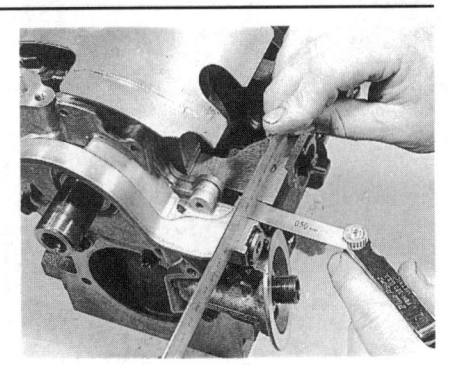

5D.153 Checking alignment of (sump) mating surfaces of oil pump and cylinder block/crankcase

5D.162 Removing oil baffle

5D.163A Check for identifying marks (arrowed) before unbolting big-end bearing caps

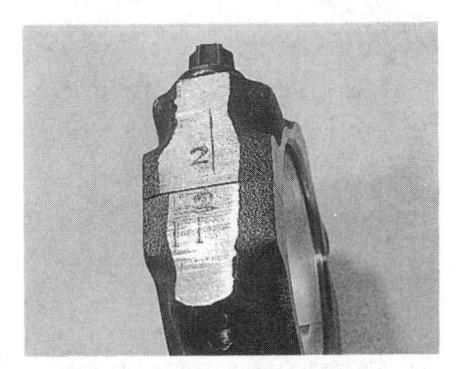

5D.163B Connecting rod and cap both have cylinder number etched into front face

155 Fit a new crankshaft front (right-hand) oil seal – see paragraph 134 above.
156 Using grease to stick the gasket in place on the pump, refit the pick-up/strainer pipe, tightening its bolts to the specified torque wrench setting.
157 The remainder of reassembly is the reverse of the removal procedure, referring to the relevant text for details where required.

Piston/connecting rod assemblies – removal and refitting

Note: While this task is theoretically possible when the engine is in place in the vehicle, in practice it requires so much preliminary dismantling, and is so difficult to carry out due to the restricted access, that owners are advised to remove the engine from the vehicle first.

In addition to the new gaskets and other replacement parts required, an adjustable engine support bar will be needed. This should fit into the water drain channels on each side of the bonnet aperture, and have a hook which will engage the engine lifting eyes and allow the height of the engine to be adjusted; such equipment can be hired from most tool hire shops.

Note the comments at the beginning of Part C of this Section – always establish precisely what replacement parts are available before undertaking any overhaul operation.

158 Remove the cylinder head as described in paragraphs 91 to 104 above.
159 Temporarily reassemble the engine right-hand mountings, and fit the engine support bar to a lifting eye bolted to a suitable point on the engine's left-hand (transmission) end, so that the engine is supported securely.
160 Remove the sump (paragraphs 121 to 125 above).
161 Unbolt the oil pump pick-up/strainer pipe from the pump.
162 Unscrew the four nuts and withdraw the oil baffle (photo).
163 Temporarily refitting the crankshaft pulley so that the crankshaft can be rotated, note that each piston/connecting rod assembly can be identified by its cylinder number (counting from the timing belt end of the engine) being etched into the flat-machined surface of both the connecting rod and its cap. These numbers are visible from the front (alternator/starter motor side) of the engine. Furthermore, each piston has an arrow stamped into its crown, pointing towards the timing belt end of the engine – see the accompanying photographs for details. If no marks can be seen, make your own before disturbing any of the components, so that you can be certain of refitting each piston/connecting rod assembly the right way round to its correct (original) bore, with the cap also the right way round (photos).

164 Removal and refitting procedures are as described in Section 33 of Chapter 1, noting the following points (photos):

(a) The caps are not located by roll pins.
(b) At the time of writing, pistons, piston rings, gudgeon pins and connecting rods are only available in assemblies – see a Ford dealer for details if any of these require renewal.
(c) As noted in Section 19 of Chapter 1, removal of the pistons from the connecting rods (even if non-genuine replacements are found elsewhere) is a task for a Ford dealer.
(d) On refitting, ensure that the etched numbers are used, so that the caps are refitted the correct way round.
(e) Note the torque wrench settings and other information given in the Specifications Section of this Chapter

165 The remainder of reassembly is the reverse of the removal procedure, referring to the relevant text for details where required.

Flywheel – removal and refitting

166 Remove the transmission (Chapter 6).
167 Remove the clutch (Chapter 5).
168 Prevent the flywheel from turning by locking the ring gear teeth (Chapter 1, photo 16.2), or by bolting a strap between the flywheel and the cylinder block/crankcase. Unscrew the bolts and withdraw the flywheel; it is very heavy – do not drop it.
169 Check the condition of the bolts, particularly their threads, whenever they are removed. Clean the bolts and wipe dry, then check each for any sign of wear or damage, and renew if necessary; if there is any doubt about the condition of any of the bolts, they should all be renewed as a set. Note that, while there is no specific recommendation to this effect, it is normal working practice to renew highly-stressed bolts such as these whenever they are disturbed.

Chapter 13 Supplement: Revisions and information on later models

5D.164A Piston ring top surfaces are marked (arrowed)

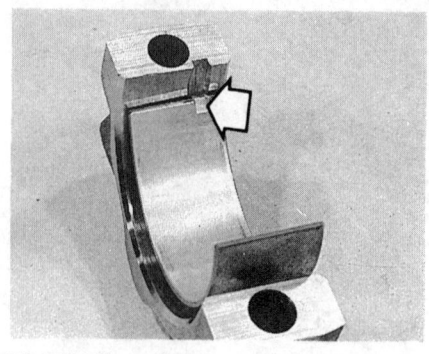

5D.164B Big-end bearing shell locating tabs (arrowed) fit into rod or cap slot

5D.164C Refitting a piston/connecting rod assembly

5D.170A Ensure engine/transmission adaptor plate is in position before refitting flywheel

5D.170B Tightening flywheel bolts

5D.174 Engine front right-hand mounting

170 On refitting, ensure that the engine/transmission adaptor plate is in place, then fit the flywheel to the crankshaft so that all bolt holes align – it will fit only one way. Lock the flywheel by the method used on dismantling, and tighten the bolts to the specified torque wrench setting (photos).
171 Refit the clutch (Chapter 5), then remove the flywheel locking tool.
172 Refit the transmission (Chapter 6).

Engine/transmission mountings – removal and refitting
General
173 Proceed as described in Sections 13 and 34 of Chapter 1, noting the following points.

Front right-hand mounting
174 This mounting consists of a two-piece bracket bolted to the inner wing panel, connected by the bonded-rubber mounting itself to a (Y-shaped) bracket, bolted (via the alternator mounting bracket) to the cylinder block (photo). Refer to paragraph 25 above.

Rear right-hand mounting
175 This mounting consists of the bonded-rubber mounting secured to the inner wing panel by a (horizontal) bolt, accessible from within the wheel arch, and a (vertical) stud, the retaining nut of which is accessible from the engine compartment. The mounting is bolted to a bracket, which is in turn bolted to the cylinder block. Refer to paragraph 25 above.

Transmission bearer and mountings
176 This mounting is as described in Chapter 1.

PART E: DOHC 16-VALVE ENGINE AND TRANSMISSION REMOVAL

Major operations requiring engine removal
1 The engine must be removed from the vehicle for the following work to be carried out:

   (a) Crankshaft – removal and refitting.
   (b) Crankshaft main bearings – renewal.

2 As noted in Part D of this Section, paragraph 1, tasks involving the removal of the sump (removal and refitting of the sump itself, of the oil pump and of the piston/connecting rod assemblies) are much easier when carried out with the engine removed from the vehicle.

5E.3 Removing engine/transmission from vehicle

5E.6A Unbolt air filter housing mounting bracket (bolts arrowed) ...

5E.6B ... and air intake duct

Chapter 13 Supplement: Revisions and information on later models

5E.7A Disconnect all coolant hoses and wiring from thermostat housing

5E.7B Disconnecting radiator bottom hose from water pump ...

5E.7C ... and from radiator

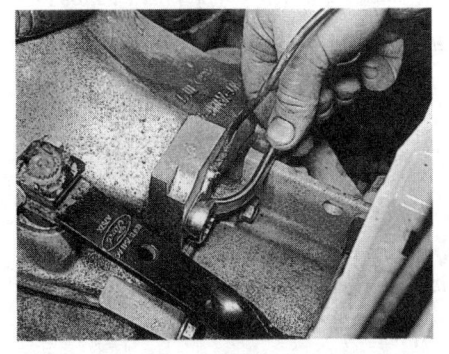

5E.9A Unbolting transmission upper ...

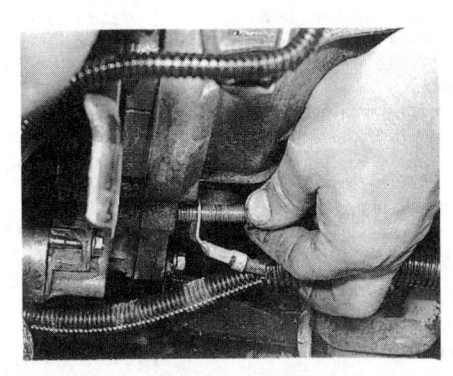

5E.9B ... and lower earth leads

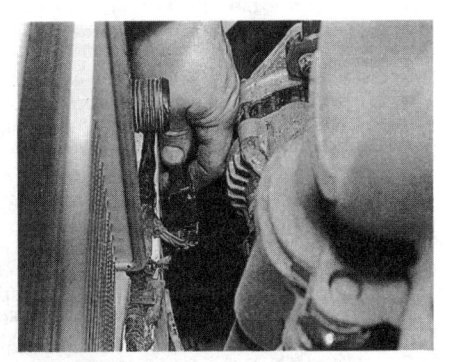

5E.9C Disconnecting wiring from HEGO sensor ...

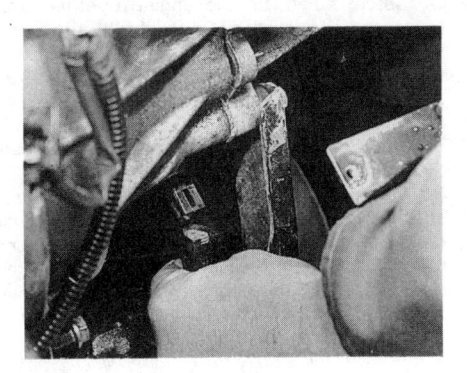

5E.9D ... reversing light switch ...

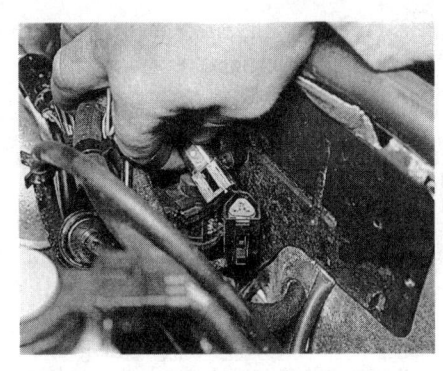

5E.9E ... and road speed sensor unit

Engine – method of removal

3 Refer to Part B of Chapter 1, Section 36 (photo).

Engine/transmission assembly – removal

4 Disconnect the battery earth (negative) lead, noting the comments made in Sections 5 and 17 of Chapter 12. Drain the cooling system (Chapter 2).

5 While the engine oil need not be drained unless the engine is to be dismantled, the oil filter must be removed to clear the driveshaft and suspension components as the engine/transmission is lowered and raised. An oil and filter change is therefore recommended as part of the operation.

6 Proceed as follows, referring to the relevant paragraphs of Section 7, Part F of this Chapter (photos):

(a) Disconnect the accelerator cable at the throttle housing.
(b) Unscrew the retaining nuts, and slacken the clamp screw so that the (black plastic) air intake trunking can be disconnected from the throttle housing. Disconnect the air mass meter's wiring, and unclip the air filter housing lid; withdraw the complete assembly.
(c) Remove the filter element and the air filter housing.
(d) Unbolt the air filter housing's mounting bracket from the inner wing panel, and remove the intake duct.

7 Removing the radiator top hose completely, disconnect all coolant hoses from the thermostat housing. Disconnecting it from the water pump, from the radiator and from the coolant pipe, remove the radiator bottom hose clear of the working area (photos).

8 Remove the radiator cooling fan and motor (Chapter 2).

9 Uncoupling the multi-plugs or unscrewing retaining bolts as necessary, disconnect the following wiring (photos).

(a) Disconnect the two multi-plugs connecting the engine's wiring to the main wiring loom.
(b) Disconnect both earth leads from the transmission.
(c) Disconnect the earth lead from the engine right-hand lifting eye.
(d) Disconnect the starter motor wiring.
(e) Disconnect the alternator wiring.
(f) Disconnect the HEGO (heated exhaust gas oxygen) sensor wiring.
(g) Disconnect the reversing lamp switch wiring.
(h) Disconnect the road speed sensor unit wiring.

10 Unplug the two vacuum hoses from the rear of the inlet manifold, then disconnect the brake servo vacuum hose from the top of the manifold, as described in Section 16 of Chapter 8. Disconnect the vacuum hose from the pulse-air filter housing.

11 Disconnect the clutch cable (photo).

5E.11 Disconnecting clutch cable

5E.13 Disconnecting speedometer cable

5E.27 Separating transmission from engine

12 Depressurise the fuel system, then uncouple the fuel feed and return pipes as described in the relevant paragraphs of Section 7 of this Chapter.
13 Disconnect the speedometer drive cable (photo).
14 On vehicles equipped with the anti-lock braking system (referring to Sections 25 and 26 of Chapter 8), disconnect the modulator belt-break switch wiring, and remove the left-hand switch. Unbolt and withdraw the left-hand modulator drivebelt cover, then unbolt the modulator and suspend it from the bodywork, unclipping (but not disconnecting) the brake pipes to permit this.
15 Raise and support the front of the vehicle, then disconnect the track rod ends from the steering arms (Chapter 9).
16 Disconnect the anti-roll bar from the suspension struts, and the lower suspension arm balljoints from the spindle carriers (Chapter 10).
17 Disconnect the left-hand driveshaft from the transmission (Chapter 7).
18 Unhook the exhaust system rubber mountings from the front suspension crossmember, unbolt the crossmember, then remove the exhaust system front downpipe.
19 On vehicles equipped with the anti-lock braking system (referring to Sections 25 and 26 of Chapter 8), unclip the right-hand modulator belt-break switch, and remove the switch assembly. Unbolt and withdraw the right-hand modulator drivebelt cover. Unbolt the modulator and suspend it from the bodywork, then unbolt its mounting plate.
20 Disconnect the gearchange mechanism shift rod and stabiliser bar from the transmission (Chapter 6).
21 Disconnect the right-hand driveshaft from the transmission (Chapter 7).
22 Remove the alternator/water pump drivebelt cover (Section 14 of this Chapter).
23 Support the weight of the engine/transmission using any of the methods outlined in Sections 13 and 34 of Chapter 1.
24 Unbolt the engine rear right-hand mounting from the body (one bolt in the wheel arch, one nut in the engine compartment), then unbolt the engine front right-hand mounting from the alternator mounting bracket. Unbolt the transmission bearer from the underbody.
25 Lower the engine/transmission to the ground, and withdraw it from under the vehicle (see also Section 37 of Chapter 1).

Engine and transmission - separation

26 Unbolt the starter motor (Chapter 12 and Section 14 of this Chapter).
27 Unbolt the transmission from the engine and withdraw it (see also Section 37 of Chapter 1) (photo).
28 Whenever the engine and transmission are separated, take the opportunity to examine the clutch components, and to renew any that are damaged or significantly worn. Considering the amount of work required to reach these components, there is a strong argument for renewing them as a matter of course (Chapter 5).

Engine - refitting to transmission

29 Refer to Section 37 of Chapter 1; reassemble the clutch (if disturbed) as described in Chapter 5, and refit the starter motor as described in Chapter 12 and Section 14 of this Chapter.

Engine/transmission assembly - refitting

30 Refitting is the reverse of the removal procedure, noting the following points:
  (a) *Check the condition of the engine mountings, renewing them if required.*
  (b) *Renew all gaskets, circlips and self-locking nuts disturbed on removal.*
  (c) *Tighten all fasteners to the specified torque wrench settings (where given).*
  (d) *Refer also to Sections 37 and 38 of Chapter 1.*
  (e) *Fit a new oil filter, and refill the engine with clean oil.*
  (f) *Check all other oil and fluid levels, refilling and topping-up as required.*

PART F: DOHC 16-VALVE ENGINE DISMANTLING, OVERHAUL AND REASSEMBLY

Engine dismantling - general

1 Refer to the introductory comments made in the first paragraphs of Chapter 1, Sections 18 and 19, 39 and 40.
2 Before beginning the engine overhaul, read through the entire procedure, to familiarize yourself with the scope and requirements of the job. Overhauling an engine is not difficult if you follow all of the instructions carefully, have the necessary tools and equipment, and pay close attention to all specifications; however, it can be time-consuming. Plan on the vehicle being off the road for a minimum of two weeks, especially if parts must be taken to an engineering works for repair or reconditioning.
3 Check on the availability of parts, and make sure that any necessary special tools and equipment are obtained in advance. Most work can be done with typical hand tools, although a number of precision measuring tools are required, for inspecting parts to determine if they must be renewed. Often the engineering works will handle the inspection of parts, and offer advice concerning reconditioning and renewal. **Note:** *Always wait until the engine has been completely dismantled, and all components, especially the cylinder block/crankcase and the crankshaft, have been inspected, before deciding what service and repair operations must be performed by an engineering works. Since the condition of these components will be the major factor to consider when determining whether to overhaul the original engine or buy a reconditioned unit, do not purchase parts or have overhaul work done on other components until they have been thoroughly inspected.*
4 As a general rule, time is the primary cost of an overhaul, so it does not pay to fit worn or sub-standard parts.
5 As a final note, to ensure maximum life and minimum trouble from a reconditioned engine, everything must be assembled with care in a spotlessly-clean environment.

Engine - complete dismantling

6 With the engine removed from the vehicle and separated from the transmission as described in Part E of this Section, remove and inspect the clutch (Chapter 5), then wash down the engine's exterior using a suitable solvent, and support it as described in Chapter 1, Section 39, paragraph 2. Unless the engine is to be completely dismantled for major overhaul and complete cleaning, try to keep it upright until the sump is removed, so that any dirt, sludge and particles of foreign matter do not fall into the engine's working components.
7 Drain the engine oil and unscrew the oil filter (Part C of this Section). Unscrew the oil pressure switch.
8 Slacken the water pump pulley bolts. Remove the alternator/water pump drivebelt (Section 14 of this Chapter), then unbolt the alternator and the water pump pulley.

Chapter 13 Supplement: Revisions and information on later models

9   Unscrew its retaining nut, then remove the engine oil level dipstick and tube.
10   Disconnect the coolant hoses from the pipe (see photo 5D.95B), then unbolt the exhaust manifold heat shield and remove the heat shield assembly.
11   Unbolt the alternator and engine mounting brackets from the cylinder block (photo).
12   Unscrewing the union sleeve nuts and remaining filter housing retaining bolts, remove the pulse-air piping assembly. If required, unscrew the exhaust manifold nuts and withdraw the manifold; discard its gasket, noting that a new one will be required, with a plastic guide sleeve, on reassembly.
13   If required, unbolt the inlet manifold, with the fuel distributor rail and intermediate flange.
14   Prevent the flywheel from turning by locking the ring gear teeth (Chapter 1, photo 16.2), or by bolting a strap between the flywheel and the cylinder block/crankcase. Slacken their bolts and withdraw the flywheel and the crankshaft pulley; refer to the relevant paragraphs of Part D of this Section for further information, if required.
15   Disconnect the HT leads from the spark plugs, and unclip them from the cylinder head cover. Disconnect the wiring multi-plug from the ignition coil, and unbolt it from the cylinder head.
16   Disconnect the engine coolant temperature sensor and temperature gauge sender unit wiring multi-plugs, then unbolt the thermostat housing from the cylinder head.
17   Working as described in the relevant paragraphs of Part D of this Section, remove the following (photos):

   (a) *Timing belt covers, timing belt, tensioner and toothed pulleys.*
   (b) *Camshafts and hydraulic tappets/cam followers.*
   (c) *Cylinder head.*
   (d) *Sump.*
   (e) *Oil pump.*

18   Unbolt the oil separator from the front of the cylinder block, unscrew the breather pipe retaining bolt from the end of the cylinder block, and withdraw the crankcase ventilation system components (photos). Check them all carefully for blockages, splits, kinks or damage, and renew them if necessary. If the separator is thought to be blocked,

5F.11 Unbolt alternator mounting bracket from cylinder block/crankcase

soak it in a bath of a suitable solvent to break down any solid matter, flush it thoroughly, and check it is clear before reassembly.
19   Remove the water pump (Section 6 of this Chapter) (photo).
20   Unbolt the engine speed/crankshaft position sensor bracket from the cylinder block (photo).
21   Unbolt the crankshaft rear (left-hand) oil seal carrier and extract the seal; refer to the relevant paragraphs of Part D of this Section for further information if required (photo).
22   Working as described in the relevant paragraphs of Part D of this Section, remove the oil baffle and all four piston/connecting rod assemblies.

5F.17A Removing timing belt

5F.17B Unbolt timing belt guide pulleys

5F.17C Note 'FRONT' marking on outside face of crankshaft toothed pulley – note which way round thrustwasher behind is fitted

5F.18A Unbolting oil separator

5F.18B Removing crankcase breather pipe with PCV valve

5.19 Unbolting the water pump

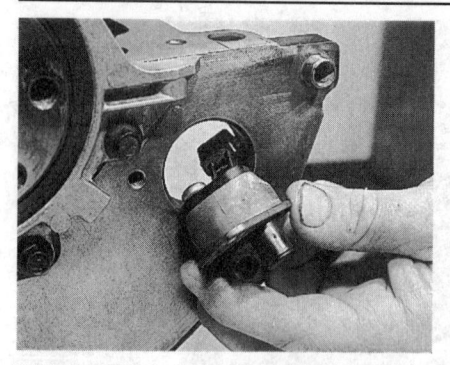

5F.20 Unbolting engine speed/crankshaft position sensor

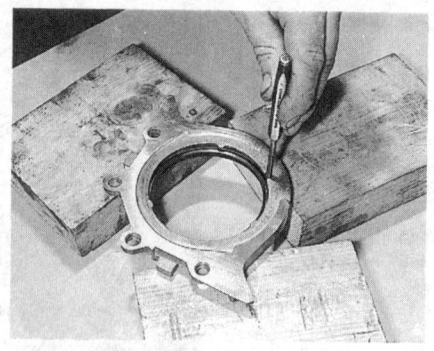

5F.21 Driving out used crankshaft left-hand oil seal

5F.23 Check for identifying marks before unbolting main bearing caps

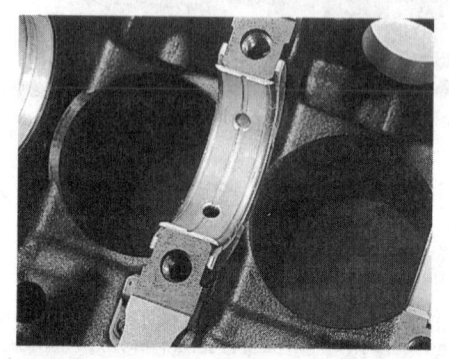

5F.24A Centre main bearing upper half has integral thrustwashers to control crankshaft endfloat ...

5F.24B ... mark or label remaining shells to ensure correct reassembly

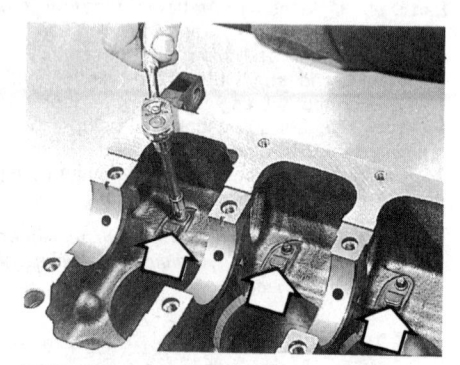

5F.25 Unbolt piston-cooling oil jets (or, as shown here) blanking plugs – three of four arrowed – to clean oilways

23 Note that each crankshaft main bearing cap can be identified by its bearing number (counting from the timing belt end of the engine) being embossed on it. Furthermore, each has an arrow embossed on it, pointing towards the timing belt end of the engine (photo). If no marks can be seen, make your own before disturbing any of the components, so that you can be certain of refitting each cap the right way round to its correct (original) location.

24 Noting the different fasteners (for the oil baffle nuts) used on caps 2 and 4, unbolt the caps and withdraw the crankshaft. Noting the thrustwashers integral with the centre main bearing upper shell, mark and/or store together (in clearly-marked containers) each pair of bearing shells so that they can be refitted only to their original locations, if they are to be re-used (photos).

25 Unbolt the piston-cooling oil jets/blanking plugs (as applicable) (photo).

Examination and renovation – general

26 Refer to the introductory comments made in the first paragraphs of Chapter 1, Sections 18 and 39, also to Sections 19 and 40. Refer also to the relevant paragraphs of Part D of this Section for further information. **Note:** *Before deciding on any course of action, establish exactly what replacement parts are available, and their cost; it may be that some components can be renewed only as part of a complete assembly, or that the only economically-viable course of action is to purchase an exchange engine unit.*

Engine – component examination and renovation
Cylinder bores

27 Note that the cylinder bores must be measured with all the crankshaft main bearing caps bolted in place (without the crankshaft and bearing shells), to the specified torque wrench settings.

Crankshaft and bearings

28 Ford's recommended method of determining main and big-end bearing running clearances is to use an American product called 'Plastigage' (type PG-1); this should be obtainable from specialist tool suppliers or engineering supply outlets.

29 To check a bearing journal's running clearance, clean the bearing surfaces carefully, and refit the crankshaft and bearing caps (at the Top- or Bottom-Dead-Centre positions) with a strand of Plastigage across each journal. Tighten the bearing cap bolts or nuts to the specified torque wrench setting (do not rotate the crankshaft), then remove the bearing caps and use the scale provided with the Plastigage to measure the width of the compressed strands (photos).

30 If the running clearance of any bearing is found to be at, or beyond, the specified service limits, have all related components measured and checked by a Ford dealer or an engine reconditioning specialist, who will then be able to decide on the most appropriate course of action.

31 To check crankshaft endfloat, clean the bearing surfaces carefully, and refit the crankshaft and main bearing caps. Tighten the bearing cap bolts/nuts to the specified torque wrench setting, then measure the endfloat using a DTI (Dial Test Indicator, or dial gauge) mounted on the cylinder block so that its tip bears on the crankshaft left-hand end (photo).

32 Tap the crankshaft fully towards the gauge, zero the gauge, then tap the crankshaft fully away from the gauge and note the gauge reading. If the endfloat measured is found to be at the specified service limit or beyond, fit a new centre main bearing upper shell and repeat the check; if the clearance is still excessive, the crankshaft must be renewed.

Engine reassembly – general

33 Before reassembly begins, ensure that all new parts have been obtained, all gaskets and oil seals have been renewed, and that all necessary tools are available. Read through the entire procedure, to familiarise yourself with the work involved, and to ensure that all items necessary for reassembly of the engine are at hand. In addition to all normal tools and materials, a suitable sealant (Ford recommend Hylosil 102) will be required for a few joint faces. In all other cases, provided the relevant mating surfaces are clean and flat, new gaskets will be sufficient to ensure joints are oil-tight. **Do not** use any kind of silicone-based sealant on any part of the fuel system or inlet manifold, and **never** use exhaust sealants upstream of the catalytic converter.

34 At this stage, all engine components should be absolutely clean and dry, with all faults repaired, and should be laid out (or placed in individual containers) on a completely-clean work surface.

Chapter 13 Supplement: Revisions and information on later models

5F.29A Using Plastigage to measure bearing running clearance – place strand as shown ...

5F.29B ... fit and tighten down bearing cap to normal setting ...

5F.29C ... unbolt, and use scale on pack to determine clearance

5F.31 Checking crankshaft endfloat

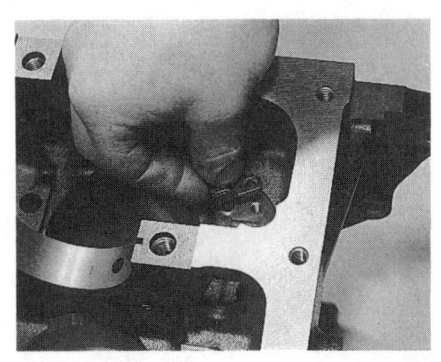

5F.35 Tighten piston-cooling oil jets/blanking plugs securely

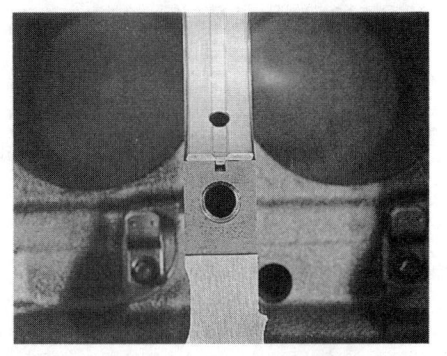

5F.36 Bearing shell locating tabs fit into housing slot

5F.37A Oil bearing surfaces liberally ...

5F.37B ... before refitting crankshaft

5F.39 Tightening oil baffle nuts

Engine – complete reassembly

35 Refit the piston-cooling oil jets/blanking plugs (as applicable), tightening their bolts to the specified torque wrench setting (photo).
36 Use the embossed marks (paragraph 23 above) to ensure that the correct cap is refitted the right way round, and that the correct shells are fitted to each bearing. Locate the shells in place so that their projecting tabs engage in the appropriate slot (photo).
37 Lubricate the bearing surfaces liberally with clean engine oil, and refit the crankshaft (photos).
38 Refit the main bearing caps, tightening the bolts/nuts to the specified torque wrench setting. Check that the crankshaft rotates smoothly and with reasonable ease.
39 Refit the piston/connecting rod assemblies as described in the relevant paragraphs of Part D of this Section, then refit the oil baffle, tightening the nuts to the specified torque wrench setting (photo).
40 Prime and refit the oil pump and pick-up pipe, using new gaskets (photos), then fit a new crankshaft right-hand oil seal as described in the relevant paragraphs of Part D of this Section.

41 Refit the water pump (Section 6 of this Chapter).
42 Using grease to stick the gasket in place on the cylinder block/crankcase, refit the crankshaft rear (left-hand) oil seal carrier, and insert the bolts, tightening them lightly at first (photo).
43 Using a suitable straight-edge and feeler gauges, check that the carrier is both centred around the crankshaft and aligned squarely, so that its (sump) mating surface on each side of the crankshaft is 0.3 to 0.8 mm below that of the cylinder block/crankcase (photos). Being careful not to disturb the gasket, move the carrier into the correct position, and tighten its bolts to the specified torque wrench setting.
44 Check again that the carrier's position is correct **on both sides** before proceeding; if necessary, unbolt the carrier again and repeat the full procedure (paragraphs 42 and 43) to ensure that the carrier is correctly aligned. Fit a new crankshaft left-hand oil seal as described in the relevant paragraphs of Part D of this Section.
45 Refit the engine speed/crankshaft position sensor bracket to the cylinder block.

Chapter 13 Supplement: Revisions and information on later models

5F.40A Use new gaskets when refitting oil pump ...

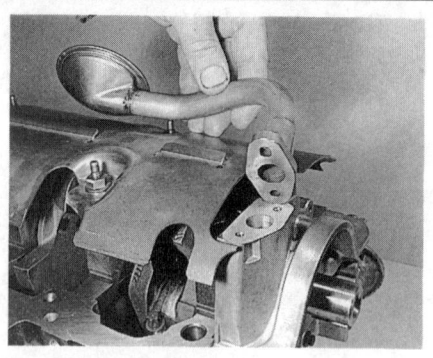

5F.40B ... and pick-up pipe

5F.42 Use new gasket when refitting left-hand oil seal carrier ...

5F.43A Centre carrier on crankshaft, and tighten bolts (arrowed) lightly ...

5F.43B ... until alignment of (sump) mating surfaces of carrier and cylinder block/crankcase has been set

5F.49A Refitting inlet camshaft

5F.49B Fit camshaft aligning tool to lock in TDC position ...

5F.49C ... while camshaft toothed pulleys are refitted (use forked holding tool described in Part D, paragraph 46) ...

5F.49D ... and timing belt is tensioned

46 Refit the sump as described in the relevant paragraphs of Part D of this Section.
47 Refit the flywheel (Part D of this Section) and the clutch (Chapter 5).
48 Using a new gasket, refit the crankcase ventilation system components.
49 Refit the cylinder head, hydraulic tappets, camshafts and timing belt components as described in the relevant paragraphs of Part D of this Section (photos).
50 Refit the ignition coil and thermostat housing, tightening their fasteners to the torque wrench settings specified, and connecting their wiring multi-plugs. Clip the HT leads into place, to prevent dirt or debris from falling into the spark plug holes; do not refit the spark plugs at this stage.
51 Refit the inlet manifold, if removed.
52 Using a new gasket and fitting a plastic guide sleeve to the stud shown in Fig. 13.30, refit the exhaust manifold (where removed), tightening its nuts to the specified torque wrench setting. Refit the pulse-air piping assembly.
53 Refit the engine and alternator mounting brackets, the exhaust manifold heat shield assembly, and the engine oil level dipstick and tube. Tighten all fasteners to the torque wrench settings specified.
54 Refit the oil pressure switch, tightening it to the specified torque wrench setting.
55 Refit the alternator and the water pump pulley, then refit and tension the alternator/water pump drivebelt (Section 14 of this Chapter). Tighten the water pump pulley bolts to the specified torque wrench setting.
56 The engine should now be fully rebuilt; check that all components have been installed and are securely fastened.

Engine – initial start-up after overhaul

57 With the engine refitted in the vehicle, double-check the engine oil and coolant levels. Make a final check that everything has been reconnected, and that there are no tools or rags left in the engine compartment.
58 With the spark plugs removed and the ignition system disabled by disconnecting the ignition coil's wiring multi-plug, turn the engine over on the starter until the oil pressure warning lamp goes out.

Chapter 13 Supplement: Revisions and information on later models

59 Refit the spark plugs and connect all the spark plug HT leads.
60 Start the engine, noting that this may take a little longer than usual, due to the fuel system components being empty.
61 While the engine is idling, check for fuel, water and oil leaks. Don't be alarmed if there are some odd smells and smoke from parts getting hot and burning off oil deposits. If the hydraulic tappets have been disturbed, some valvegear noise may be heard at first; this should disappear as the oil circulates fully around the engine and normal pressure is restored in the tappets.
62 Keep the engine idling until hot water is felt circulating through the top hose, then switch it off.
63 After a few minutes, recheck the oil and coolant levels, and top-up as necessary.
64 If they were tightened as described, there is no need to re-tighten the cylinder head bolts once the engine has first run after reassembly.
65 If new pistons, rings or crankshaft bearings have been fitted, the engine must be run-in for the first 500 miles (800 km). Do not operate the engine at full throttle, or allow it to labour in any gear, during this period. It is recommended that the oil and filter be changed at the end of this period.

Fault diagnosis – engine
66 Refer to Sections 20 and 41 of Chapter 1.

6 Cooling, heating and ventilation systems

Coolant renewal – general
1 The varying coolant renewal intervals given in this manual are the manufacturer's own, the changes being due to the better-quality antifreeze introduced at the factory during 1990, and following extensive testing after that. If owners of earlier models are prepared to pay for the better-quality antifreeze used in later models, they can extend the coolant renewal interval accordingly; check with your local Ford dealer for current recommendations.
2 All maintenance schedules assume the use of a mixture (in exactly the specified concentration) of clean, soft water and of antifreeze to Ford's specification or equivalent. It is also assumed that the cooling system is maintained in a scrupulously-clean condition, by ensuring that only clean coolant is added on topping-up, and by thorough reverse-flushing whenever the coolant is drained (Chapter 2, Section 3).
3 If antifreeze to Ford specification SSM-97B-9103-A ('Motorcraft Super Plus Antifreeze', identifiable by its pink colour) or equivalent is used, the coolant must be renewed every two years or 36 000 miles, whichever comes first.
4 If antifreeze to Ford specification ESD-M97B49-A ('Motorcraft Super Plus 4' antifreeze, identifiable by its blue colour) or equivalent is used, the original recommendation was to renew the coolant every four years. At that time, the system must be drained and thoroughly reverse-flushed before the fresh coolant mixture is poured in.
5 From late in 1992, Ford now state that where antifreeze to specification ESD-M97B49-A is used, **if** it is used in the recommended concentration, unmixed with any other type of antifreeze or additive, and topped-up when necessary using only that antifreeze mixed 50/50 with clean water, it will last the lifetime of the vehicle. If any other type of antifreeze is added, the lifetime guarantee no longer applies; to restore the lifetime protection, the system must be drained and thoroughly reverse-flushed before fresh coolant mixture is poured in.
6 If the vehicle's history (and therefore the quality of the antifreeze in it) is unknown, owners who wish to follow Ford's recommendations are advised to drain and thoroughly reverse-flush the system as outlined above before refilling with fresh coolant mixture. The coolant can then be renewed every two or four years, or left for the life of the vehicle, as appropriate to the quality of antifreeze used – see paragraphs 3, 4 and 5 above.
7 If any antifreeze other than Ford's is to be used, the renewal intervals must be made much more frequent than those recommended, to provide an equivalent degree of protection; the conventional recommendation is to renew the coolant every two years.

Cooling system modifications – 1.6 litre EFi (turbocharged) engine
8 Referring to Fig. 13.15, note the turbocharger and oil cooler lines in the cooling system of this engine.

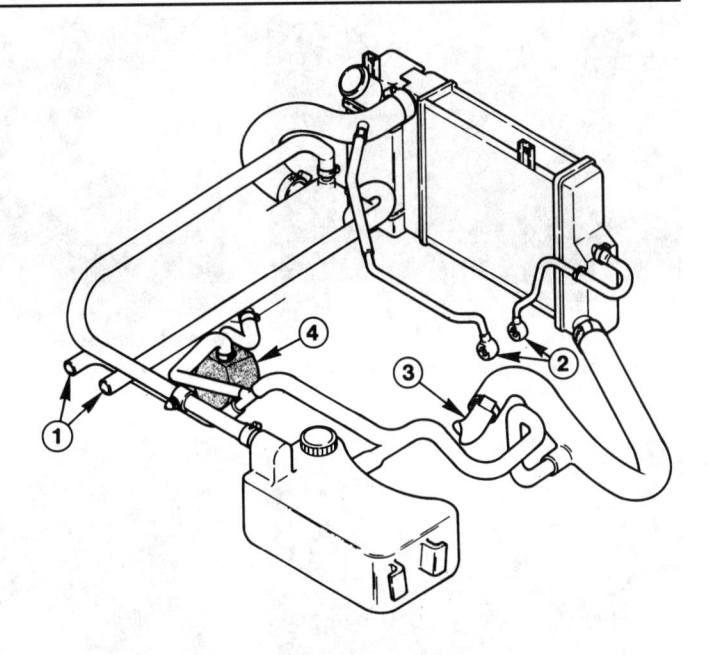

Fig. 13.15 Cooling system – 1.6 CVH EFi (turbocharged) engine (Sec 6)

1 To heater
2 To turbocharger
3 To water pump
4 Oil cooler

9 In addition to the statement on coolant renewal above, and the information given in the Specifications Section of this Chapter, note also the following modifications to the procedures given in Chapter 2.

Draining the cooling system
10 The system must be drained by disconnecting the radiator bottom hose.

Radiator cooling fan and motor – removal and refitting
11 Proceed as described in Section 5 of Chapter 2, noting the following:

(a) *Remove its two retaining screws and move the HT lead bracket clear of the working area, disconnecting the HT leads as required.*
(b) *Disconnect the auxiliary lamp wiring (Chapter 12).*
(c) *Remove the front bumper (Chapter 11).*
(d) *Referring to Fig. 2.3, the cooling fan shroud is secured by two bolts at the bottom (corresponding to tags 'B') and by a tongue at the top (corresponding to bolt 'A').*
(e) *To dismantle the assembly, pull off the fan guard from the shroud, flatten back the raised lockwasher tab, and unscrew clockwise (a left-hand thread is employed) the nut securing the fan to the motor shaft. Separate the motor from the shroud as described in Chapter 2.*
(f) *Fit a new lockwasher on refitting the fan.*

Radiator – removal and refitting
12 Noting the comments made in Section 5 (and 17) of Chapter 12, disconnect the battery earth terminal.
13 Remove the radiator cooling fan and shroud assembly, as outlined above.
14 Remove the intercooler, as described in Section 7 of this Chapter, Part E.
15 Drain the cooling system – by disconnecting the bottom hose, as noted above. Disconnect the radiator top hose.
16 Disconnect the turbocharger coolant feed by slackening its clamp and pulling the hose (at the radiator rear right-hand side) off the turbocharger's metal pipe.
17 Remove its three retaining screws, and withdraw the exhaust manifold heat shield.
18 Withdraw the radiator.
19 Refitting is the reverse of the removal procedure; check the condition of the radiator rubber mountings, and renew them if necessary. Refill the system and check the level as described in Chapter 2.

338 Chapter 13 Supplement: Revisions and information on later models

6.26 Disconnect coolant hoses from water outlet, then unbolt ...

6.27 ... to withdraw thermostat

6.30 Draining the cooling system at radiator drain plug (arrowed)

6.31 Disconnecting radiator bottom hose (arrowed) from water pump

6.32A Unscrew bolts (arrowed) ...

6.32B ... to release water pump – always renew pump gasket

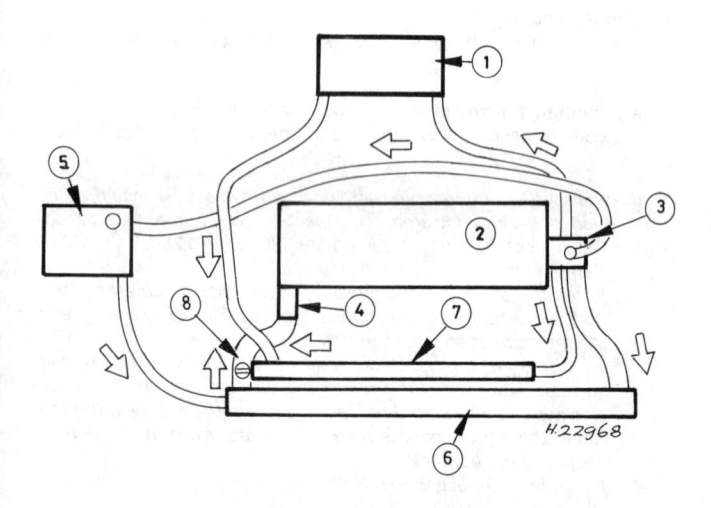

Fig. 13.16 Cooling system – DOHC 16-valve engine (Sec 6)

1 Heater matrix
2 Engine
3 Thermostat housing
4 Water pump
5 Expansion tank
6 Radiator
7 Coolant pipe
8 Radiator cooling fan thermal switch

Cooling system – DOHC 16-valve engine

20 Referring to Fig. 13.16, note the differences in the cooling system of this engine.
21 In addition to the statement on coolant renewal above, and the information given in the Specifications Section of this Chapter, note also the following modifications to the procedures given in Chapter 2.

Radiator – removal and refitting

22 Noting that the cooling fan thermal switch wiring must be disconnected, proceed as described in Section 6 of Chapter 2.

Radiator cooling fan thermal switch – removal and refitting

23 Noting that the switch is located on the right-hand end of the radiator, next to the bottom hose (therefore requiring that the cooling system be completely drained before it is disturbed), proceed as described in Section 7 of Chapter 2.

Thermostat – removal and refitting

24 Noting the comments made in Section 5 (and 17) of Chapter 12, disconnect the battery earth terminal.
25 Unclip the air filter housing lid, and disconnect the flexible rubber hose from the (black plastic) air intake trunking. Unclip the air mass meter, then move the assembly clear of the working area.
26 Drain the cooling system to just below the level of the thermostat as described in Chapter 2, then disconnect the expansion tank hose and the radiator top hose from the thermostat water outlet (photo).
27 Unbolt the water outlet and withdraw the thermostat (photo).
28 Refitting is the reverse of the removal procedure; renew the thermostat's sealing ring if it is worn or damaged, and refit the thermostat with its air bleed valve uppermost. Tighten the water outlet bolts to the specified torque wrench setting.

Water pump – removal and refitting

29 Remove the timing belt and tensioner (Section 5 of this Chapter, Part D).
30 Drain the cooling system (Chapter 2) (photo).
31 Disconnect the radiator bottom hose from the pump union (photo).
32 Unbolt and remove the water pump (photos). If the pump is to be renewed, unbolt the timing belt guide pulleys, and transfer them to the new pump.
33 Clean the mating surfaces carefully; the gasket must be renewed whenever it is disturbed.
34 On refitting, use grease to stick the new gasket in place, refit the pump, and tighten the pump bolts to the torque wrench setting specified.
35 The remainder of the reassembly procedure is the reverse of dismantling; note that a new tensioner spring and retaining pin must be fitted if the timing belt has been removed for the first time. Tighten all fasteners to the specified torque wrench settings, and refill the system with coolant as described in Chapter 2.

Chapter 13 Supplement: Revisions and information on later models

7 Fuel, exhaust and emission control systems

PART A: CARBURETTOR FUEL SYSTEM – MODIFICATIONS

Tamperproof mixture adjusting screws

Later carburettors are fitted with tamperproof mixture adjusting screws, consisting of a hexagon-shaped socket with a pin in the centre. Such screws require the use of Ford service tool 23-032 (see Fig. 13.17) to alter their settings; if this is not available, the CO level will have to be checked, and any necessary adjustment will have to be made, by a Ford dealer.

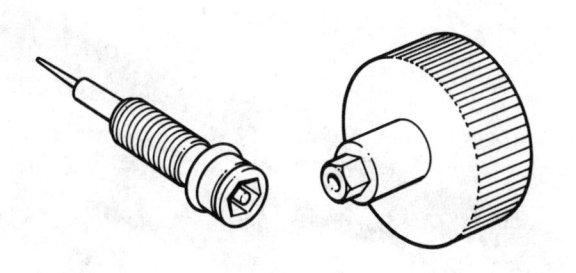

Fig. 13.17 Carburettor tamperproof mixture adjusting screw and Ford service tool 23-032 (Sec 7A)

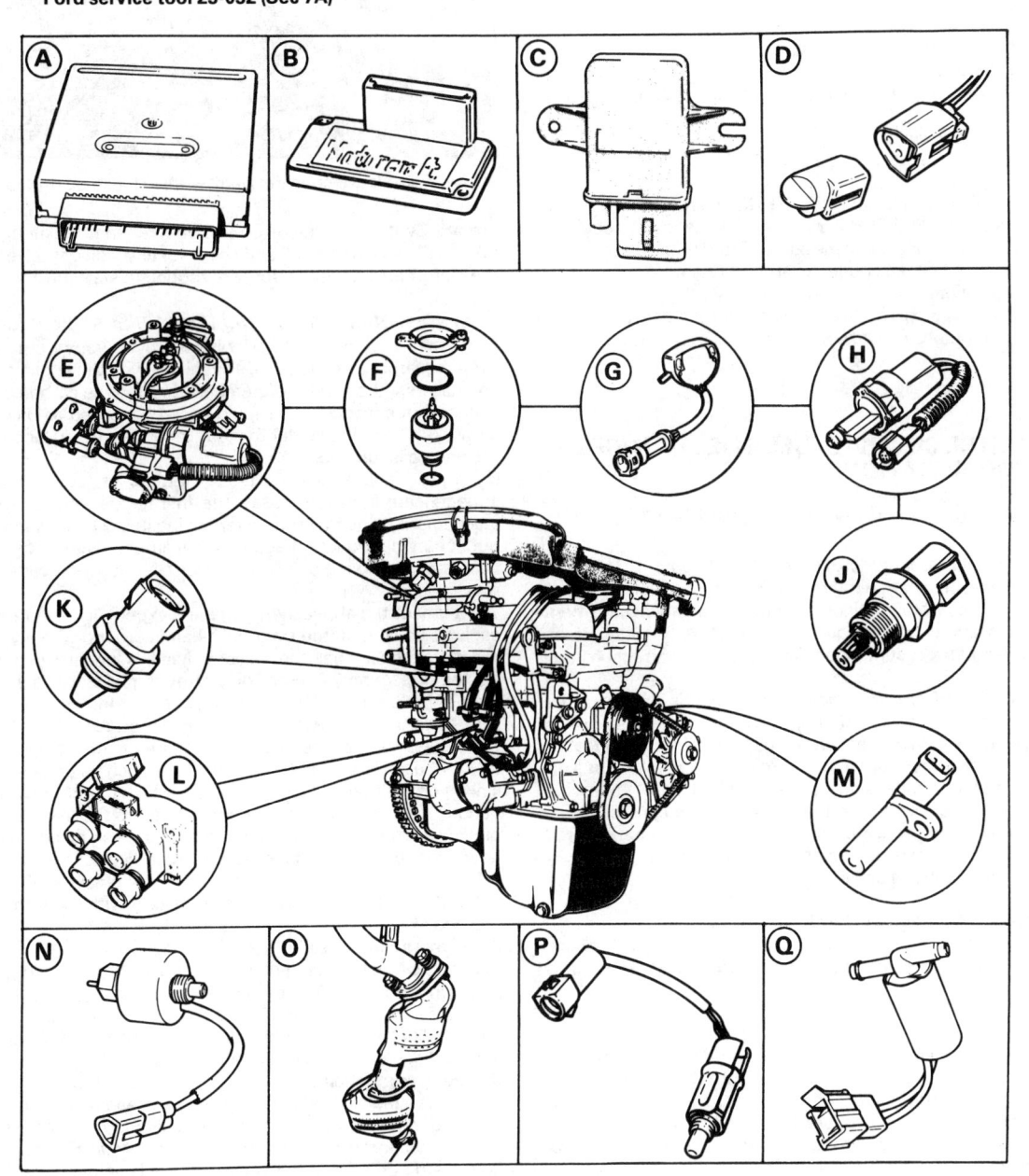

Fig. 13.18 Components of Central Fuel injection (CFi) and ignition systems – 1.1 and 1.3 litre HCS engines (Sec 7B)

A EEC-IV engine management module
B EDIS ignition module
C Manifold Absolute Pressure (MAP) sensor
D Diagnostic socket
E CFi unit
F Fuel injector
G Throttle Position Sensor (TPS)
H Throttle plate control motor
J Air Charge Temperature (ACT) sensor
K Engine Coolant Temperature (ECT) sensor
L Ignition coil
M Engine speed/Crankshaft Position Sensor (CPS)
N Vehicle road speed sensor unit
O Catalytic converter
P Heated Exhaust Gas Oxygen (HEGO) sensor
Q Carbon canister-purge solenoid valve

PART B: 1.1 AND 1.3 LITRE CENTRAL FUEL INJECTION (CFI) SYSTEM

General description

Apart from minor differences in location and mounting resulting from the different engine installation (refer to Fig. 13.18 for details), the CFi system fitted to the HCS engines is the same as that covered in Part B of Chapter 3 for the 1.4 litre CVH engine. The only significant difference is that, with the fitting of the distributorless ignition system, the EEC-IV engine management module gets its information on engine speed and crankshaft position from the cylinder block-mounted engine speed/crankshaft position sensor.

When working on the fuel system components of an HCS CFi engine, therefore, refer to Parts B and C of Chapter 3, but note also the additional information given in Part C of this Section concerning the following points:

(a) Quick-release fuel line couplings.
(b) Fuel starvation problems.
(c) Fuel filter (located in a clamp under the battery platform on these models).
(d) Throttle position sensor wiring.
(e) Injector ballast resistor.

The inlet manifold heater is not fitted to HCS engines.

Note that, like the 1.4 litre CVH engine, the 1.1/1.3 litre CFi engine's air filter assembly is basically the same as that fitted to the corresponding carburettor-engine unit described in Part A of Chapter 3.

On later models, if problems are experienced which are not covered by the relevant fault diagnosis section of this manual, refer to Section 14 of this Chapter, paragraphs 64 and 65.

PART C: 1.4 LITRE CENTRAL FUEL INJECTION (CFI) SYSTEM – MODIFICATIONS

General description – engines from September 1990 onwards

The system remains essentially as described in Chapter 3. The most significant difference is that, with the fitting of the distributorless ignition system to these later engines, the EEC-IV engine management module gets its information on engine speed and crankshaft position from the cylinder block-mounted engine speed/crankshaft position sensor.

Some engines built from November 1990 onwards may be fitted with a heater in the inlet manifold. This is controlled by the EEC-IV engine management module to ensure that, even before the effect of the coolant heating becomes apparent, the manifold is warmed-up. This prevents fuel droplets condensing in the manifold, thus improving driveability and reducing exhaust emissions when the engine is cold.

On later models, if problems are experienced which are not covered by the relevant fault diagnosis section of this manual, refer to Section 14 of this Chapter, paragraphs 64 and 65.

Quick-release fuel line couplings – general

1 All later vehicles with fuel-injected engines employ quick-release couplings at many, if not all, unions in the fuel feed and return lines (photo).
2 Before disconnecting any fuel system component, relieve any residual pressure in the system by removing the fuel pump fuse (No 19) and starting the engine; allow the engine to idle until it dies. Turn the engine over once or twice on the starter to ensure that all pressure is released, then switch off the ignition; do not forget to refit the fuse when work is completed. **Warning:** *This procedure will merely relieve the increased pressure necessary for the engine to run – remember that fuel will still be present in the system components, and take precautions accordingly before disconnecting any of them.*
3 To release, squeeze together the protruding locking lugs on each union and carefully pull the coupling apart; use rag to soak up any spilt fuel. Where the unions are colour-coded, the pipes cannot be confused; where both unions are the same colour, note carefully which pipe is connected to which, and ensure that they are correctly reconnected on refitting.
4 To reconnect, press together until the locking lugs snap into their

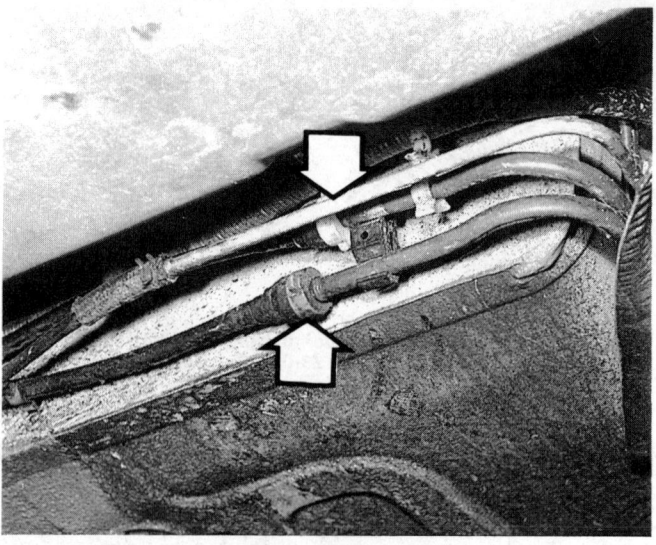

7C.1 Quick-release couplings (arrowed) in fuel feed and return lines

groove. Switch the ignition on and off five times (ensuring that fuse No 19 is in place, if removed) and check for any sign of fuel leakage around the disturbed coupling before attempting to start the engine.

Fuel starvation and/or fuel tank collapse problems

5 Where a vehicle is fitted with the evaporative emissions control system (carbon canister) – ie, all vehicles with CFi engines and catalytic converters – the fuel tank 'breathes' through the carbon canister. If the canister's fuel vapour line should become blocked or kinked, therefore, air will not be able to enter the tank to replace the fuel consumed while the engine is running.
6 If this fault should occur, symptoms of fuel starvation will be experienced; in severe cases, the fuel gauge may be thought to be inaccurate (as the fuel shown remaining in the tank cannot be pumped out, it may be thought that the tank is empty, when in fact the gauge is accurate in showing plenty of fuel left). In the worst case, the fuel tank may collapse.
7 If any of the above symptoms are experienced, start at the carbon canister (located in the front right-hand corner of the engine compartment) to check that the canister itself is undamaged and clear of anything (accumulations of dirt, or even pieces of rag!) which might restrict its 'breathing'. Work back along the fuel vapour line (which is clipped to the inner wing panel, then along the bulkhead at the rear of the engine compartment, and from there under the vehicle to the fuel tank). Check for signs of visible damage such as crushed metal pipes, kinked or trapped flexible hoses, etc., which might prevent the flow of vapour. If a blue plastic cap is found fitted to the metal pipe above the braking system servo operating link (on the engine compartment bulkhead), the cap must be removed and discarded. If any part of the system is found to be damaged or faulty, it must be renewed; if a kink is found, cut out the deformed section, and replace it with a length of 6 mm bore fuel hose.
8 If there is no visible sign of damage, disconnect the sections of metal pipe one by one (by removing the connecting lengths of flexible hose) and blow (or suck) through each to ensure that it is clear. **Do not** suck through the carbon canister; like the fuel tank, it will be full of unleaded petrol vapour, which is a significant hazard to health if inhaled. If a blockage is found which cannot be cleared by other means, renew the faulty component.
9 While a fault in that part of the system leading to the canister-purge solenoid valve and the inlet manifold should not produce these symptoms, this part can be checked as described above; remember that a flow restrictor is fitted between the canister-purge solenoid valve and the inlet manifold. To check the solenoid valve itself, use an ohmmeter to check the resistance of its windings; a reading of between 50 and 120 ohms should be obtained. To check the solenoid's operation, disconnect its wires and connect it directly to the battery. It should energise the solenoid to open the valve so that vapour can pass into the engine; if the solenoid is faulty, it must be renewed.
10 If the fuel tank has collapsed as a result of this fault, the fuel pump and fuel gauge sender unit must be renewed with the tank.

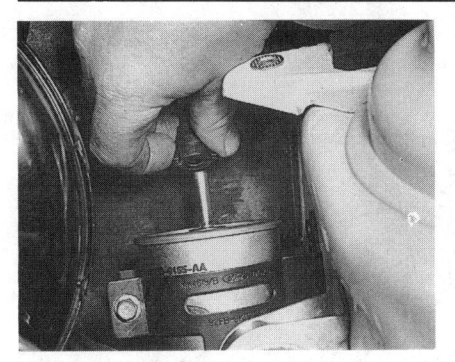

Chapter 13 Supplement: Revisions and information on later models

7C.14 Depress locking lugs to release later-type fuel filter's quick-release unions

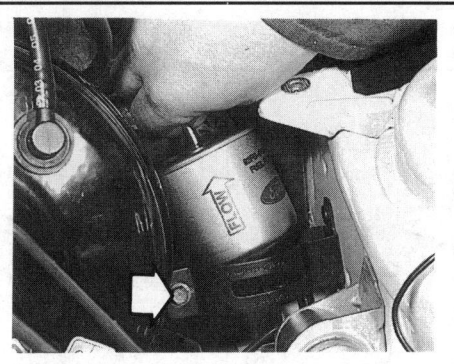

7C.15 Loosen filter clamp screw (arrowed), then withdraw filter, noting direction of fuel flow arrow/markings to ensure correct reassembly

7C.25 Injector ballast resistor

Fuel filter – removal and refitting

11 All later vehicles with fuel-injected engines are fitted with fuel filters that employ quick-release unions. Filter removal and refitting is therefore as follows.
12 Relieve any residual pressure in the system by removing the fuel pump fuse (No 19) and starting the engine; allow the engine to idle until it dies. Turn the engine over once or twice on the starter to ensure that all pressure is released, then switch off the ignition. **Warning:** *This procedure will merely relieve the increased pressure necessary for the engine to run – remember that fuel will still be present in the system components, and take precautions accordingly before disconnecting any of them.*
13 Noting the comments made in Section 5 (and 17) of Chapter 12, disconnect the battery earth terminal.
14 Using rag to soak up any spilt fuel, release the fuel feed and outlet pipe unions from the filter by squeezing together the protruding locking lugs on each union and carefully pulling the union off the filter stub (photo). Where the unions are colour-coded, the feed and outlet pipes cannot be confused; where both unions are the same colour, note carefully which pipe is connected to which filter stub, and ensure that they are correctly reconnected on refitting.
15 Noting the arrows and/or other markings on the filter showing the direction of fuel flow, slacken the filter clamp screw and withdraw the filter (photo). Note that the filter will still contain fuel; care should be taken to avoid spillage and to minimise the risk of fire.
16 On installation, slide the filter into its clamp so that the arrow marked on it faces the correct way, then slide each pipe union on to its (correct) respective filter stub, and press it down until the locking lugs click into their groove. Tighten the clamp screw carefully, until the filter is just prevented from moving; do not overtighten the clamp screw, or the filter casing may be crushed.
17 Refit the fuel pump fuse and reconnect the battery earth terminal, then switch the ignition on and off five times. Check for any sign of fuel leakage around the filter unions before attempting to start the engine.

Inlet manifold heater – removal and refitting

18 The heater is located in a recess in the manifold, directly underneath the fuel injection unit. While access is possible from underneath, owners may prefer, depending on the tools available, to remove the complete manifold (Chapter 3) to reach the heater.
19 Noting the comments made in Section 5 (and 17) of Chapter 12, disconnect the battery earth terminal. Raise and support the front of the vehicle on axle stands or similar.
20 Disconnect the heater wiring, and extract the circlip retaining the heater. Withdraw the heater.
21 Refitting is the reverse of the removal procedure. Ensure that both the heater and its circlip are correctly located in the manifold.

Throttle position sensor wiring

22 Owners of some vehicles may find that the throttle position sensor's wiring connector (shown in Chapter 3, Section 56, photo 56.3) has been cut out and the wires soldered together, the join being covered by insulating tape. Where this has been done, it is as a result of a modification carried out by a Ford dealer to prevent occasional increases in idle speed.
23 Where this modification is found, as its wires can no longer be

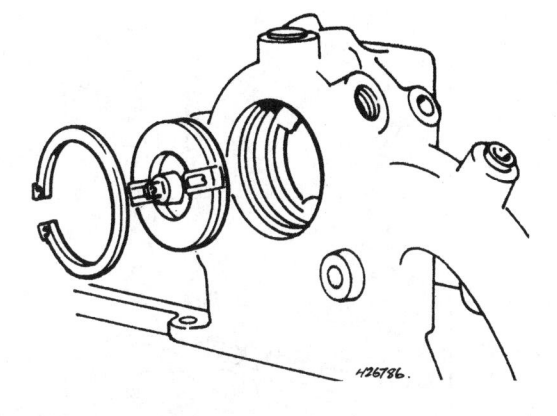

Fig. 13.19 Removing inlet manifold heater (where fitted) – later 1.4 litre (CVH) CFi engines (Sec 7C)

disconnected when required on removal and refitting of various components, the sensor will have to be removed from the CFi unit instead, and secured to the bulkhead with the main wiring harness.
24 If an unmodified vehicle is thought to have a fault producing occasional irregular and unexplained increases in idle speed, usually when braking or changing gear, it should be taken to a Ford dealer for a thorough check of the entire system, from the accelerator pedal and cable onwards. Only if no other fault is found will a modification be required to effect a solution.

Injector ballast resistor – removal and refitting

25 This component is located on the engine compartment bulkhead, next to the manifold absolute pressure sensor (photo).
26 To remove the resistor, disconnect the battery earth terminal (noting the comments made in Sections 5 and 17 of Chapter 12), then disconnect the resistor wiring at its multi-plug, remove the retaining screw and withdraw the resistor.
27 Refitting is the reverse of the removal procedure.

PART D: 1.6 LITRE (NON-TURBO) ELECTRONIC FUEL INJECTION (EFI) SYSTEM – MODIFICATIONS

Quick-release fuel line couplings – general
1 Refer to Part C of this Section.

Fuel filter – removal and refitting
2 Refer to Part C of this Section.

CO potentiometer – adjustment, removal and refitting
Adjustment
3 With reference to the procedure given in Section 60 of Chapter 3,

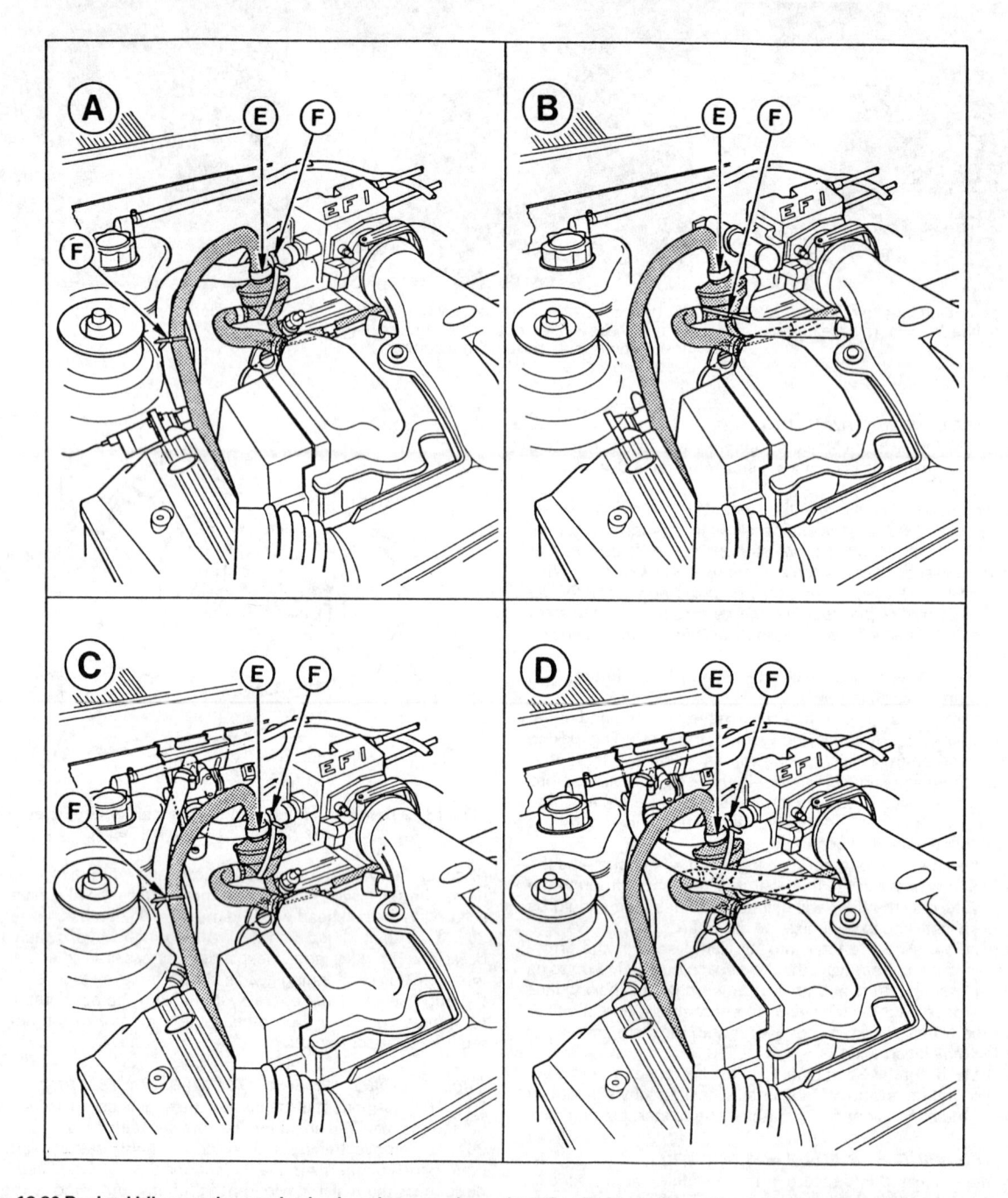

342 Chapter 13 Supplement: Revisions and information on later models

Fig. 13.20 Revised idle speed control valve locations, crankcase breather hose routing and ventilation system filter mounting – 1.6 litre (non-turbo) EFi engines (Sec 7D)

A Original installation – Weber-built valve on air filter housing
B Final version – Hitachi-built valve on inlet manifold upper section
C Hitachi-built valve on engine compartment bulkhead, clean air feed from air filter
D Hitachi-built valve on engine compartment bulkhead, clean air feed from air filter trunking
E Hose clamp
F Nylon strap

turning the fuel mixture adjusting screw anti-clockwise richens the mixture, turning clockwise weakens it. While the potentiometer influences the fuel/air mixture over the whole of the engine speed range, its most significant effect is at idle speed.

4   For the potentiometer to provide the EEC-IV engine management module with the clearest signals by which it can control the injectors (and therefore the fuel/air mixture), the potentiometer must be as close as possible to the centre of its adjustment range; this is where its output voltage is greatest.

5   If difficulty is experienced in achieving the correct fuel/air mixture (CO level) at idle and the potentiometer is thought to be faulty, first return it to the true centre point of its adjustment range by turning the fuel mixture adjusting screw 20 turns in one (either) direction, then 10 turns in the opposite direction. Since the screw has no stop at either end of its adjusting range, this is the only way of establishing the true centre of the range.

6   Repeat the adjustment procedure described in Chapter 3; note that all electrical loads (including the radiator cooling fan) **must** be switched off for an accurate result to be obtained.

7   If the potentiometer is still thought to be faulty, use an ohmmeter to

Chapter 13 Supplement: Revisions and information on later models

measure the resistance between one of the outer pins and the centre pin; a reading of between 448 and 5993 ohms should be obtained. If the reading obtained is significantly more or less, the potentiometer is probably at fault. Ideally, the vehicle should be taken to a Ford dealer for the entire system to be tested thoroughly on the correct test equipment before any component is condemned. If this is not possible, a new potentiometer must be fitted to check whether any improvement is obtained.

Removal and refitting
8 Refer to Part D, Section 43, of Chapter 4.

Base idle speed – checking and adjustment

9 When carrying out the procedure given in Section 61 of Chapter 3, while the *base* idle speed is 750 ± 50 rpm, a closer tolerance (750 ± 25 rpm) is required when resetting (paragraph 4). Note also that the *normal* (engine fully warmed-up) operational idle speed is maintained by the EEC-IV engine management module at approximately 900 rpm. **Note:** *All electrical loads (including the radiator cooling fan) must be switched off for an accurate result to be obtained.*

10 If any problem at all is encountered in getting one of these engines to idle smoothly, steadily and reliably, or in obtaining good driveability at low engine speeds and light throttle openings, take the vehicle to a Ford dealer for the engine and the entire fuel/ignition system to be tested thoroughly on the correct diagnostic equipment. In some cases, some form of modification may be required, but **only** after the system has been found to be free from faults; a Ford dealer will have details.

11 The procedure in Chapter 3 applies to vehicles with Weber-built idle speed control valves, and to those with Hitachi-built valves mounted on the engine compartment bulkhead. Owners of models with Hitachi-built valves mounted directly on the upper section of the inlet manifold (see below) who wish to attempt this check themselves rather than have it carried out (as is advisable) by a Ford dealer, should proceed as follows.

12 With the engine fully warmed-up to normal operating temperature, directly after a 2-mile road test, allow the engine to idle. Disconnect the idle speed control valve wiring at its multi-plug. If the engine continues to run, no adjustment is required, and the multi-plug should be reconnected immediately, but the engine must be left to idle for **at least** 5 minutes, to allow the EEC-IV engine management module to re-learn its idle values.

13 If the engine stalls, reset the base idle speed (using the adjustment screw shown in Fig. 3.62) to no more than 700 rpm, with the radiator cooling fan **not** operating; if the cooling fan motor starts up while adjustment is in progress, wait for it to stop before proceeding. The engine may run unevenly at such low speeds, particularly if it is new; providing it continues to run, this is acceptable. **Do not** set the base idle speed to more than 700 rpm, or throttle 'hang-ups' may occur while driving. Reconnect the valve multi-plug, and leave the engine to idle for **at least** 5 minutes to allow the EEC-IV engine management module to re-learn its idle values.

Idle speed control valve – general

14 The idle speed control valve described in Chapter 3 is a Weber-built unit mounted directly on the air filter housing. These valves require regular dismantling and cleaning, as noted in *'Routine maintenance'* and described in Chapter 3. Base idle speed check and adjustment is as described in Chapter 3, with reference to the general notes in paragraphs 9 and 10 above.

15 Later models are fitted with Hitachi-built valves, mounted either on the engine compartment bulkhead, or directly on the upper section of the inlet manifold, secured by four nuts; while these can be dismantled for the passages to be cleaned, if desired, this is not required on a regular basis. Base idle speed check and adjustment is as described in Chapter 3, with reference to the general notes in paragraphs 9 and 10 above, for bulkhead-mounted valves; for inlet manifold-mounted valves, refer to paragraphs 9 to 13 above.

16 Valve removal and refitting is essentially as described in Chapter 3 for the Weber-built valves; precise details will obviously vary with the model and installation. Refer to Fig. 13.20 for more information.

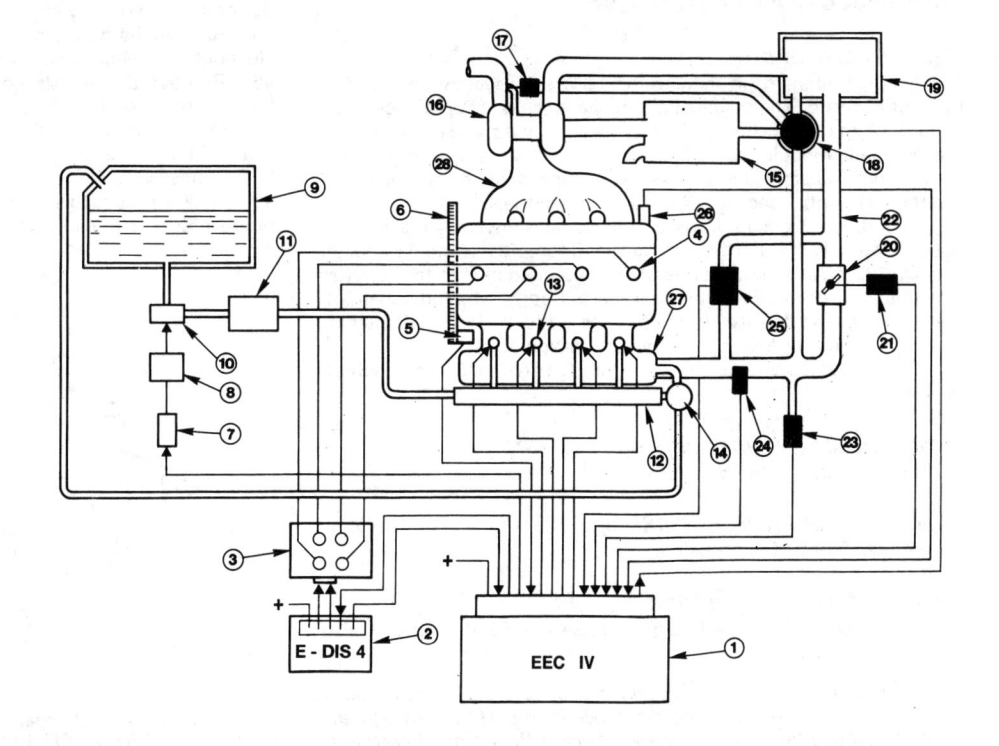

Fig. 13.21 Electronic Fuel injection (EFi) and ignition systems – 1.6 CVH EFi (turbocharged) engine (Sec 7E)

1 EEC-IV engine management module
2 EDIS ignition module
3 Ignition coil
4 Spark plugs
5 Engine speed/Crankshaft Position Sensor (CPS)
6 Flywheel
7 Fuel pump relay
8 Inertia switch (fuel cut-off)
9 Fuel tank
10 Fuel pump
11 Fuel filter
12 Fuel distributor rail
13 Fuel injectors
14 Fuel pressure regulator
15 Air filter
16 Turbocharger
17 Turbocharger wastegate control
18 Boost control valve
19 Intercooler
20 Throttle
21 Throttle Position Sensor (TPS)
22 Air intake crossover duct
23 Manifold Absolute Pressure (MAP) sensor
24 Air Charge Temperature (ACT) sensor
25 Idle speed control valve
26 Engine Coolant Temperature (ECT) sensor
27 Inlet manifold
28 Exhaust manifold

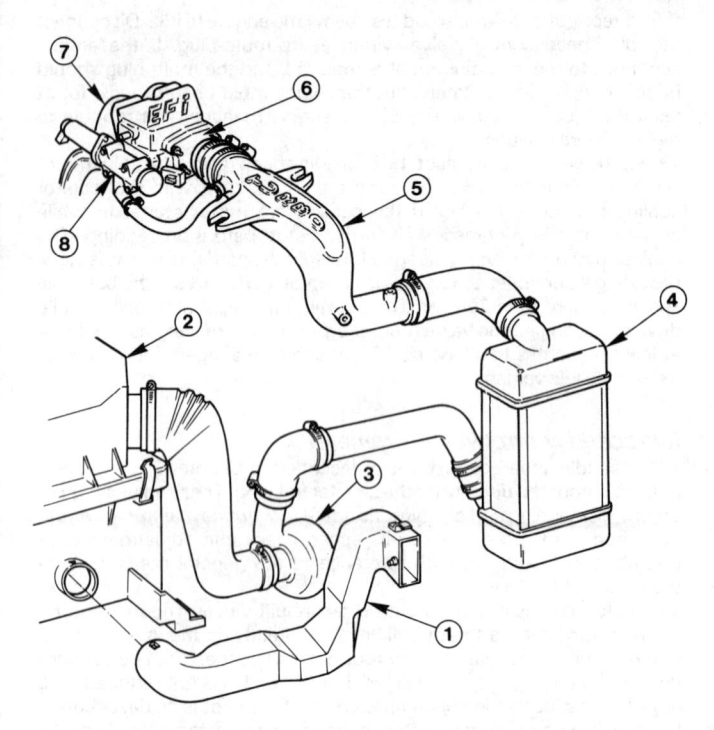

344   Chapter 13   Supplement: Revisions and information on later models

Fig. 13.22 Air intake, turbocharger and intercooler details – 1.6 CVH EFi (turbocharged) engine (Sec 7E)

1 Air intake duct	5 Air intake crossover duct
2 Air filter housing	6 Throttle housing
3 Turbocharger	7 Inlet manifold
4 Intercooler	8 (Hitachi-built) idle speed control valve

PART E: 1.6 LITRE (TURBOCHARGED) ELECTRONIC FUEL INJECTION (EFi) SYSTEM

General description

The fuel system is the same as the 1.6 litre EFi system described in Chapter 3, Part B. The principal difference is the addition of electronic control of the turbocharger boost pressure by the EEC-IV engine management module, acting through the boost control valve. This allows inlet manifold depression to be applied to the turbocharger wastegate control – see Fig. 13.21 for details of the system.

The turbocharger consists of a turbine that is driven by the exhaust gases, to suck air through the air filter and to compress it into the engine; see Fig. 13.22. An air-cooled intercooler, mounted next to the radiator, cools the intake air (heated by its passage through the turbocharger); this increases the density of the compressed fuel/air mixture entering the engine, thus improving the engine's power output.

All procedures are as described in Part B of Chapter 3, except where outlined below.

Quick-release fuel line couplings – general
1   Refer to Part C of this Section.

Fuel filter – removal and refitting
2   Refer to Part C of this Section.

Fuel injectors – removal and refitting
3   In addition to the procedure given in Section 59 of Chapter 3, also remove the following:

(a) Remove its two retaining screws and move the HT lead bracket clear of the working area, disconnecting the HT leads as required.
(b) Remove the retaining screws, slacken the clamp at each end, and withdraw the air intake crossover duct. Disconnect it from the intercooler, and move it clear of the working area as far as the idle speed control valve's air bypass hose will allow.

CO potentiometer – adjustment, removal and refitting
4   Refer to Part D of this Section.

Base idle speed – checking and adjustment
5   Refer to Part D of this Section, and also to Section 61 of Chapter 3. The check is made as described in Chapter 3, noting the following:

(a) All electrical loads (including the radiator cooling fan) must be switched off for an accurate result to be obtained.
(b) Note the different CO level and base idle speed given in the Specifications Section of this Chapter.
(c) If the cooling fan motor starts up while adjustment is in progress, wait for it to stop before proceeding.
(d) The engine may run unevenly at such low speeds, particularly if it is new; providing it continues to run, this is acceptable. Do not set the base idle speed to more than 700 rpm, or throttle 'hang-ups' may occur while driving.

Idle speed control valve – removal and refitting
6   All turbocharged engines are fitted with Hitachi-built valves mounted directly on the upper section of the inlet manifold, secured by four nuts; refer to the relevant paragraphs of Part D of this Section.

Intercooler – removal and refitting
7   Noting the comments made in Section 5 (and 17) of Chapter 12, disconnect the battery earth terminal.
8   Remove the flexible hose connecting the intercooler to the air intake crossover duct, then the pipe and flexible hose connecting the turbocharger to the intercooler. Use adhesive tape to seal the turbocharger ports against the entry of dirt.
9   Unbolt the horn nearest the intercooler.
10   Unbolt the radiator/intercooler assembly from the bonnet closure panel and the body crossmember.
11   Move the assembly as far as possible towards the engine, and unbolt the intercooler from the radiator; withdraw the intercooler.
12   Refitting is the reverse of the removal procedure; note the specified torque wrench settings, and do not forget to unseal the turbocharger openings before reconnecting the intercooler pipe and hose.

Turbocharger – removal, examination and refitting
Removal (with exhaust manifold)
13   Noting the comments made in Section 5 (and 17) of Chapter 12, disconnect the battery earth terminal.
14   Remove its two retaining screws and move the HT lead bracket clear of the working area, disconnecting the HT leads as required.
15   Remove the flexible hose connecting the intercooler to the air intake crossover duct. Remove the retaining screws and slacken the remaining clamp to withdraw the air intake crossover duct, moving it clear of the working area as far as the idle speed control valve's air bypass hose will allow.
16   Drain the cooling system – by disconnecting the bottom hose, as noted in Section 6 of this Chapter.

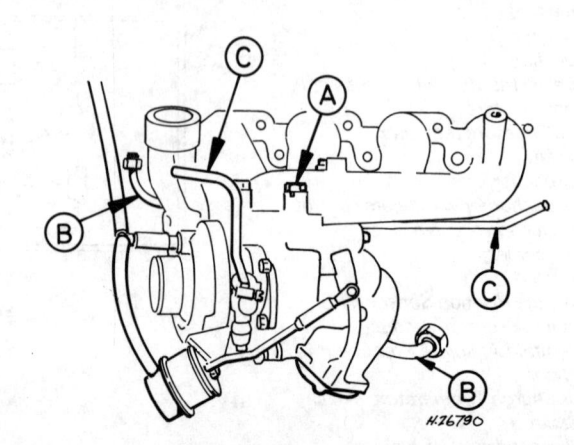

Fig. 13.23 Turbocharger/exhaust manifold assembly details – 1.6 CVH EFi (turbocharged) engine (Sec 7E)

A Exhaust manifold-to-turbocharger bolts	B Oil pipes
	C Coolant pipes (metal)

17 Disconnect the turbocharger coolant feed and return hoses by slackening the clamps and pulling the hoses off the turbocharger's metal pipes.
18 Remove the pipe and flexible hoses connecting the turbocharger to the air filter housing and intercooler. Use adhesive tape to seal the turbocharger ports against the entry of dirt.
19 Disconnect the turbocharger oil feed and return pipes by unscrewing the couplings.
20 Remove its three retaining screws and withdraw the exhaust manifold heat shield.
21 Disconnect the exhaust system downpipe from the turbocharger, then disconnect the hose from the boost control valve to the turbocharger wastegate control actuator.
22 Unscrew the exhaust manifold nuts, and withdraw the manifold and turbocharger as an assembly, protecting the radiator with cardboard or similar.
23 To separate the turbocharger from the manifold, flatten back the raised lockwasher tabs and unscrew the three retaining bolts. Disconnect the turbocharger oil and coolant pipes if required. This is as far as the unit can be dismantled; **do not** disturb the wastegate or its actuator and linkage.

Examination
24 With the turbocharger in the vehicle, check that there are no air leaks around any part of the intake trunking or crossover duct, and that the boost control valve hoses are intact and securely fastened.
25 With the turbocharger removed, check the turbine wheels and blades (as far as possible) for signs of wear or damage. Spin the turbine and check that it rotates smoothly and easily, with no sign of roughness, free play or abnormal noise. If possible, check for axial play (endfloat) of the shaft. Check that the wastegate, its actuator and linkage show no visible signs of wear, damage or stiffness due to dirt and corrosion.
26 If any sign of wear or damage is found, the turbocharger must be renewed.

Refitting
27 Refitting is the reverse of the removal procedure, noting the following points:

(a) *Refit the oil and coolant pipes, tightening the unions to the torque wrench settings specified.*
(b) *Always use new lockwashers when refitting the turbocharger to the manifold; again, tighten the bolts to their specified torque wrench setting.*
(c) *Protect the radiator when refitting the assembly, and always fit a new exhaust manifold gasket.*
(d) *Ensure that the oil feed and return pipes are absolutely clean before reconnecting them and tightening them to the specified torque wrench setting.*
(e) *Owners are well advised to change the engine oil and filter whenever the turbocharger is disturbed (Chapter 1).*
(f) *When reassembly is complete and the cooling system refilled (Chapter 2), disable the ignition system by disconnecting the ignition coil wiring multi-plug, then turn the engine over on the starter motor until the oil pressure light goes out. This is essential to ensure that the turbocharger oil supply is established BEFORE the engine is started; do not forget to reconnect the coil before attempting to start the engine.*

Turbocharged engines – notes and precautions
28 Owners of RS Turbo models should note the following points, safety precautions and general servicing notes:

(a) *Being driven by the exhaust gases, the turbocharger routinely operates at extremely high temperatures; its castings will retain heat for a very long time, and must not be touched for a period of at least three hours after the engine was last run. Even at this time, it is advisable to wear heavy gloves when handling it or when working near it, to prevent any risk of burns.*
(b) *Do not poke fingers or anything else into the turbocharger while it is still spinning – always wait for the blades to stop completely and for the unit to cool down before starting work.*
(c) *A turbocharger is a ducted fan, and like any turbine, its greatest enemy is dirt and foreign objects – a stray nut, metal chip or stone passing through the unit while it is spinning can cause extreme damage. The best defence against such damage is to work on the engine and turbocharger only when both have been cleaned, to seal (using adhesive tape) all turbocharger ports as soon as they are opened, and to account for every nut, bolt, screw and washer after reassembly is complete, before starting the engine.*
(d) *Never run the engine with the air intake duct, air filter housing and housing-to-turbocharger flexible hose missing or disconnected – see (c) above. Any foreign object entering the unit under these circumstances will not only damage the turbocharger, but may be ejected from the unit with sufficient force to risk personal injury.*
(e) *The turbocharger's high operating temperatures and speeds mean that it demands a great deal of the engine oil, and depends for long life on a constant flow of clean, top-quality engine oil to its shaft's plain bearings; the shaft actually floats on a thick film of oil.*
(f) *Careful attention should be paid to the condition and security of the turbocharger oil feed and return lines and their unions. Always fasten all couplings and banjo union bolts to the specified torque wrench settings; if they are overtightened, they will be distorted, and this may restrict the flow.*
(g) *Do not rev the engine immediately after starting-up – wait for a few seconds to allow oil pressure to become established at the turbocharger shaft bearings.*
(h) *When switching off the engine, allow it to idle for at least 10 seconds – switching off immediately after a run can leave the turbocharger rotating at high speeds and temperatures, with no oil pressure available at the shaft bearings. Similarly, do not 'blip' the throttle immediately before switching off – this spins the unit up to high speed, followed by a sudden loss of oil pressure, and is to be avoided.*
(i) *Always change the engine oil and filter at the recommended intervals (noting the more frequent interval that may be required – Section 4, paragraph 3(a)). Further, change the oil and filter whenever the turbocharger is removed, and flush the system carefully if a bearing fails in the engine, or if severe oil contamination is found.*
(j) *Use only the very best engine oil you can find – see (e) above – at the time of writing, oils classed as 'synthetic' or 'semi-synthetic' are the ones to aim for. While this type of oil may appear expensive, it will be cheaper (by far) than a new turbocharger.*

Boost control valve – removal and refitting
29 Noting the comments made in Section 5 (and 17) of Chapter 12, disconnect the battery earth terminal.
30 Disconnect the control valve wiring at its multi-plug.
31 Marking or labelling the valve hoses so that each can be reconnected to its original union, disconnect the hoses from the valve.
32 Remove the securing screws and withdraw the valve.
33 Refitting is the reverse of the removal procedure; tighten the screws to the specified torque wrench setting.

Boost pressure – check and adjustment
34 This task requires experience, skill and a considerable amount of special test equipment; the vehicle must be taken to a Ford dealer for any such work to be carried out.

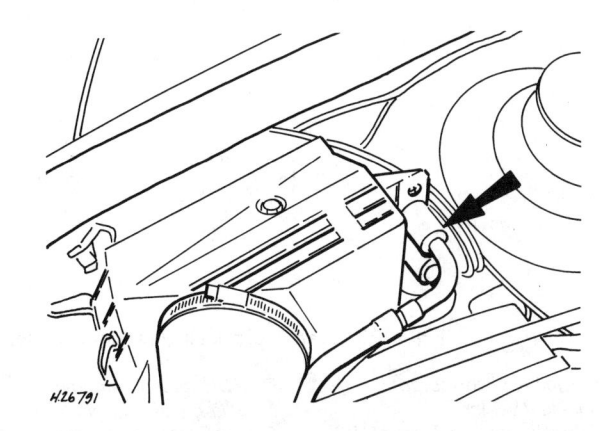

Fig. 13.24 Location of boost control valve – 1.6 CVH EFi (turbocharged) engine (Sec 7E)

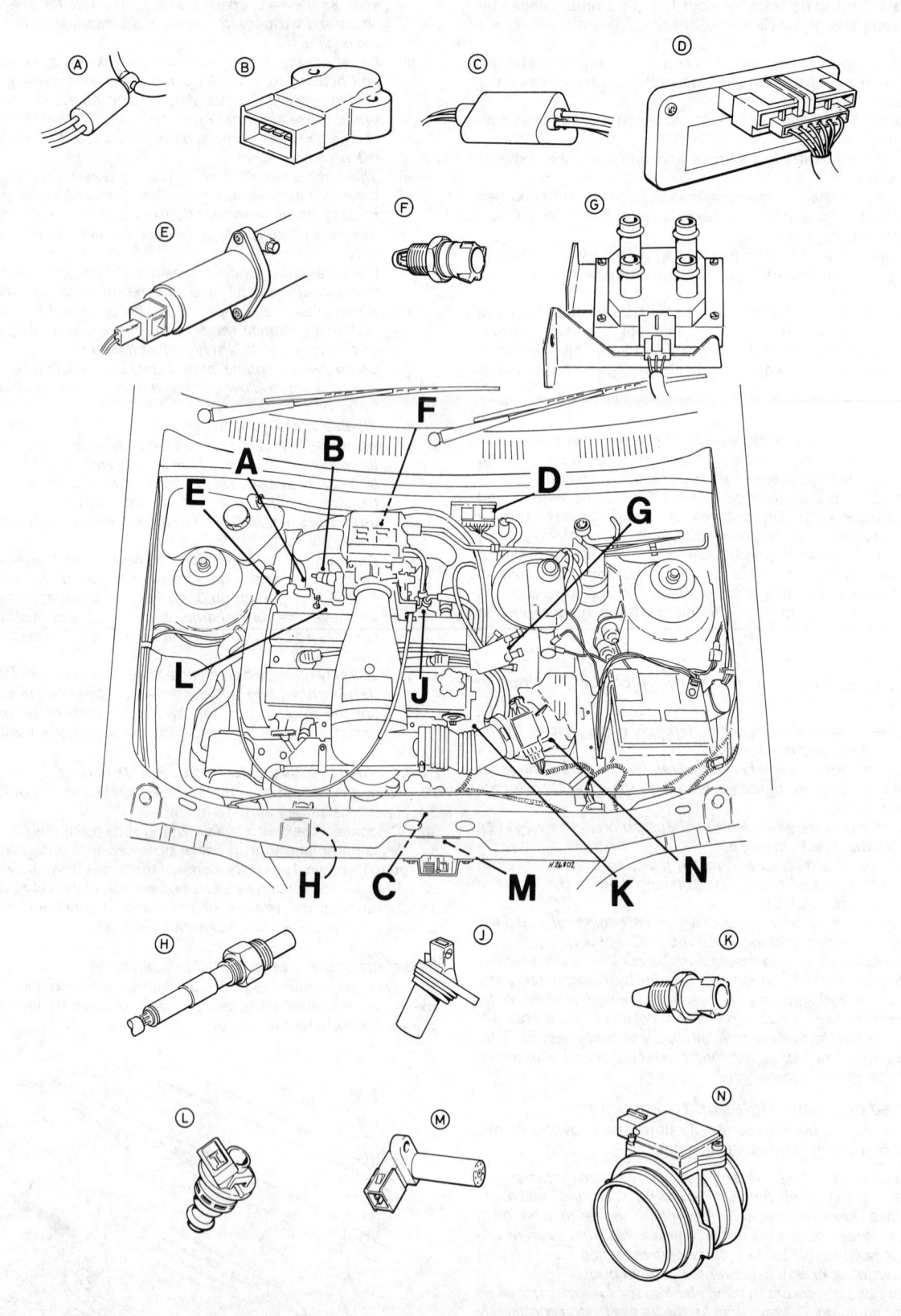

346 Chapter 13 Supplement: Revisions and information on later models

Fig. 13.25 Location of principal fuel and ignition system components – DOHC 16-valve engine (Sec 7F)

A   Carbon canister-purge solenoid valve
B   Throttle Position Sensor (TPS)
C   Vehicle road speed sensor unit
D   EDIS ignition module
E   Idle speed control valve
F   Air Charge Temperature (ACT) sensor
G   Ignition coil
H   Heated Exhaust Gas Oxygen (HEGO) sensor
J   Camshaft position sensor
K   Engine Coolant Temperature (ECT) sensor
L   Fuel distributor rail/injectors
M   Engine speed/Crankshaft Position Sensor (CPS)
N   Air mass meter

Chapter 13 Supplement: Revisions and information on later models

PART F: DOHC 16-VALVE ENGINE EEC-IV ENGINE MANAGEMENT SYSTEM – FUEL SYSTEM COMPONENTS

General description

Refer to Part B of Chapter 3, Section 39, and to Figs. 13.25 and 13.26. Although this system is very much more powerful and sophisticated than that fitted to CVH-engined models, it is essentially as described in Chapter 3, the principal differences being as follows:

(a) The EEC-IV engine management module uses the signals derived from the engine speed/crankshaft position sensor and the camshaft position sensor, to trigger each injector separately in cylinder firing order ('sequential' injection), which has obvious benefits in terms of better fuel economy and lower exhaust emissions.

(b) Instead of the Manifold Absolute Pressure (MAP) sensor being used to determine engine load, a 'hot-wire' type air mass meter measures the mass (ie, volume and density) of air entering the engine. Combined with signals from the other sensors, this allows the engine management module to control the injectors more accurately for all engine operating conditions.

(c) The system also includes features such as the flushing of fresh (ie, cold) fuel around each injector on start-up, thus improving hot starts.

Note that all these engines are fitted with full three-way regulated catalytic converters, in which the HEGO sensor is used to provide feedback related to the amount of oxygen remaining in the exhaust gases. This information is used by the engine management module to adjust the fuel/air mixture ratio, pulse-air and evaporative emissions control systems. With the ignition, all these are integrated under the control of the one EEC-IV module into a complete engine management system. Refer to the relevant Part of this Section (and of Section 8 of this Chapter) for further information.

Fuel system – precautions

1 Refer to Chapter 3, Part A, Section 2 and Part B, Section 40.
2 Depressurise the fuel system as described in Part C of this Section, paragraph 2, before disconnecting any part of the fuel system.

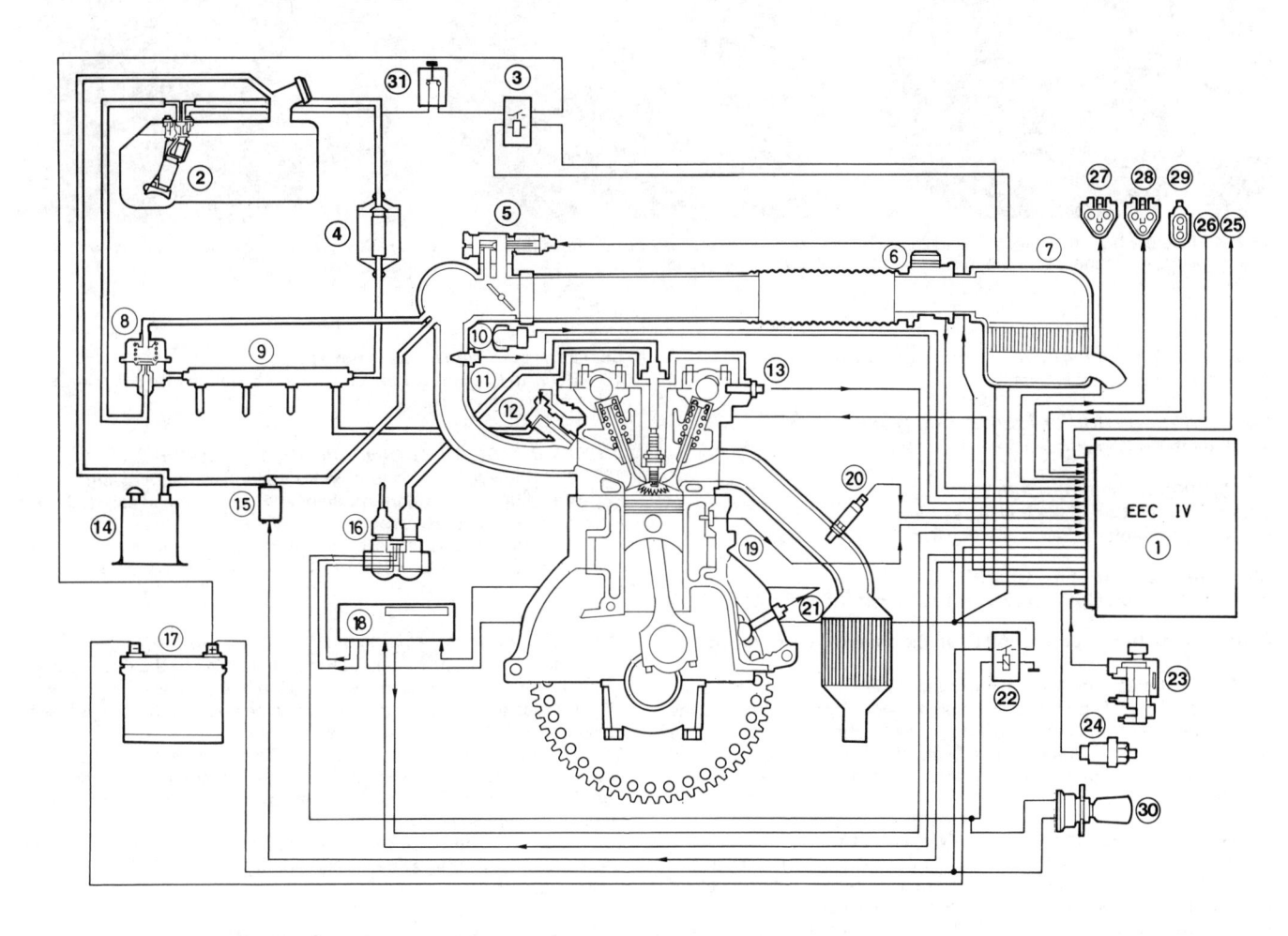

Fig. 13.26 Fuel injection and ignition systems – DOHC 16-valve engine (Sec 7F)

1 EEC-IV engine management module
2 Fuel pump
3 Fuel pump relay
4 Fuel filter
5 Idle speed control valve
6 Air mass meter
7 Air filter
8 Fuel pressure regulator
9 Fuel distributor rail
10 Throttle Position Sensor (TPS)
11 Air Charge Temperature (ACT) sensor
12 Fuel injector
13 Camshaft position sensor
14 Carbon canister
15 Carbon canister-purge solenoid valve
16 Ignition coil
17 Battery
18 EDIS ignition module
19 Engine Coolant Temperature (ECT) sensor
20 Heated Exhaust Gas Oxygen (HEGO) sensor
21 Engine speed/Crankshaft Position Sensor (CPS)
22 Fuel injection relay
23 Pulse-air solenoid valve
24 Not applicable
25 Not applicable
26 Not applicable
27 Connector for Ford diagnostic equipment (FDS 2000)
28 Diagnostic socket
29 Octane-adjust socket
30 Ignition switch
31 Inertia switch (fuel cut-off)

Chapter 13 Supplement: Revisions and information on later models

7F.15 Disconnecting air mass meter flexible hose from air intake trunking ...

7F.16 ... withdrawing air filter housing/air mass meter assembly ...

7F.17 ... and removing filter element

7F.20A Unscrew front retaining nut and lift air filter housing, disconnect breather hose (arrowed) ...

7F.20B ... and release housing from rear mountings (arrowed)

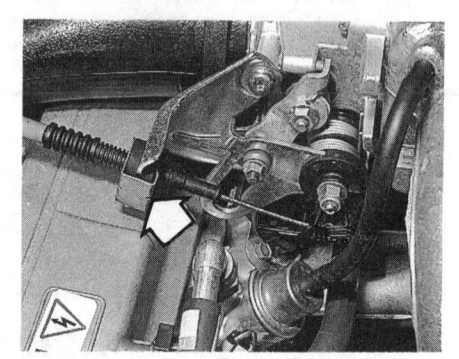

7F.22 Accelerator cable is secured by clip (arrowed)

Quick-release fuel line couplings – general
3   Refer to Part C of this Section.

Fuel starvation and/or fuel tank collapse problems
4   Refer to Part C of this Section.

Maintenance and inspection
Check the condition of the fuel lines
5   Refer to Chapter 3, Part A, Section 3 and Part B, Section 41.

Check the operation of the throttle
6   Refer to Chapter 3, Part A, Section 3 and Part B, Section 41.

Renew the air filter element and clean the crankcase ventilation system filter
7   Refer to paragraphs 14 to 17 below for air filter element renewal.
8   While the filter is removed, wipe out the housing and withdraw the small foam filter from its location in the front right-hand corner of the housing. If the foam is badly clogged with dirt or oil, it must be cleaned by soaking it in a suitable solvent, and allowed to dry before being refitted.

Check the crankcase ventilation system for blockages
9   Referring to Part H of this Section for details, check that all components of the system are securely fastened, correctly routed (with no kinks or sharp bends to restrict flow) and in sound condition; renew any worn or damaged components. If oil leakage is noted, disconnect the various hoses and pipes to check that all are clear and unblocked. The small foam filter in the air filter housing must be cleaned whenever the air filter element is renewed (see above).
10   If either the oil separator or the PCV valve are thought to be blocked – remember the PCV valve is designed to allow gases to flow out of the crankcase only, so that a depression is created in the crankcase under most operating conditions, particularly at idle – they must be renewed. In such a case, however, there is nothing to be lost by attempting to flush out the blockage using a suitable solvent.

Renew the fuel filter
11   Refer to Part C of this Section.

Check the condition of the exhaust system
12   Refer to Chapter 3, Part C, Section 65, noting also the comments concerning catalytic converters.

Idle speed and mixture adjustment – general
13   Both the idle speed and mixture are under the full control of the EEC-IV engine management module, and cannot be adjusted. **Do not** attempt to 'adjust' these settings in any way.
14   If either are thought to be incorrect, take the vehicle to a Ford dealer for the complete system to be tested using special diagnostic equipment.

Air filter element – renewal
15   Slacken the clamp securing the flexible hose to the (black plastic) air intake trunking, and disconnect the air mass meter's wiring (photo).
16   Release the four clips and withdraw the housing lid, complete with the air mass meter and flexible hose (photo).
17   Lift out the element and wipe out the housing (photo). If carrying out a routine service, note that the small foam filter in the front right-hand corner of the air filter housing must be cleaned whenever the air filter element is renewed.
18   Refitting is the reverse of the removal procedure; ensure that the element and housing lid are securely seated, so that unfiltered air cannot enter the engine.

Air filter housing – removal and refitting
19   Unclip the air mass meter and withdraw it, then unclip the filter housing lid and withdraw the element.
20   The remainder of the procedure is as described in Part B of Chapter 3, Section 43, noting that the housing is retained by a single nut at the front, and that the housing must be partially withdrawn before the crankcase breather hose can be disconnected (photos).
21   Refitting is the reverse of the removal procedure.

Accelerator cable – removal and refitting
22   Refer to Chapter 3, Part B, Section 44, and Part A, Section 7, noting that no preliminary dismantling is required in this case (photo).

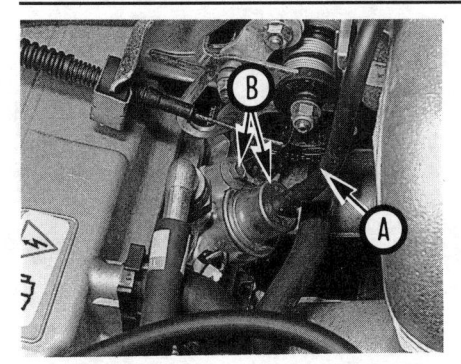

Chapter 13 Supplement: Revisions and information on later models

7F.33 Fuel pressure regulator vacuum hose 'A', retaining bolts 'B'

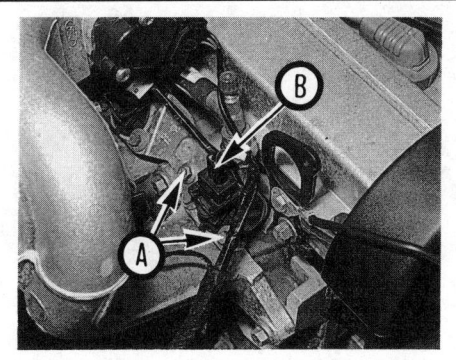

7F.35 Fuel injector retaining bolts 'A', wiring connector 'B'

7F.40A Unscrewing fuel distributor rail mounting bolts (arrowed)

7F.40B Note nose seals (arrowed) between rail and intermediate flange

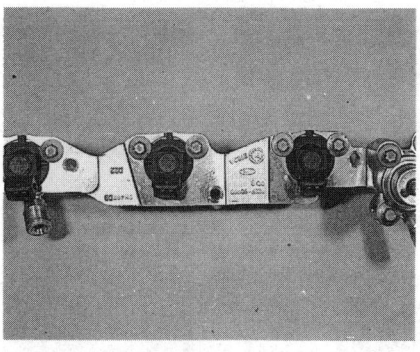

7F.41 Fuel distributor rail removed, showing injector and fuel pressure regulator mountings

Accelerator pedal – removal and refitting
23 Proceed as described in Chapter 3, Part B, Section 45.

Fuel tank – removal and refitting
24 Proceed as described in Chapter 3, Part B, Section 47.

Fuel tank filler pipe – removal and refitting
25 Proceed as described in Chapter 3, Part B, Section 48.

Fuel pump – testing
26 Proceed as described in Chapter 3, Part B, Section 49.

Fuel pump/fuel gauge sender unit – removal and refitting
27 Proceed as described in Chapter 3, Part B, Section 50.

Inertia switch (fuel cut-off) – removal and refitting
28 Proceed as described in Chapter 3, Part B, Section 51.

Fuel filter – removal and refitting
29 Proceed as described in Part C of this Section.

Fuel pressure regulator – removal and refitting
30 Relieve any residual pressure in the system by removing the fuel pump fuse (No 19) and starting the engine; allow the engine to idle until it dies. Turn the engine over once or twice on the starter to ensure that all pressure is released, then switch off the ignition. **Warning:** *This procedure will merely relieve the increased pressure necessary for the engine to run – remember that fuel will still be present in the system components, and take precautions accordingly before disconnecting any of them.*
31 Noting the comments made in Section 5 (and 17) of Chapter 12, disconnect the battery earth terminal.
32 Disconnect the vacuum hose from the regulator.
33 Unscrew the two regulator retaining bolts, place a wad of clean rag to soak up any spilt fuel, and withdraw the regulator (photo).
34 Refitting is the reverse of the removal procedure, noting the following points:

(a) *Renew the regulator sealing O-ring whenever the regulator is disturbed – lubricate the O-ring with clean engine oil on installation.*

(b) *Locate the regulator carefully in the fuel distributor rail recess, and tighten the bolts to the torque wrench setting specified.*
(c) *On completion, refit fuse No 19, then switch the ignition on and off five times to activate the fuel pump and pressurise the system without cranking the engine – check for signs of fuel leaks around all disturbed unions and joints before attempting to start the engine.*

Fuel distributor rail and injectors – removal and refitting
Note: *For simplicity, and to ensure the necessary absolute cleanliness on reassembly, the following procedure describes the removal of the fuel distributor rail assembly, complete with the injectors and pressure regulator, so that the injectors can be serviced individually on a clean work surface.*
Obviously, it is possible to remove and refit an individual injector alone (photo), once the fuel system has been depressurised and the battery has been disconnected. If this approach is followed, read through the complete procedure, and work as described in the relevant paragraphs, depending on the amount of preliminary dismantling required. Be careful not to allow any dirt to enter the system.

35 Relieve any residual pressure in the system by removing the fuel pump fuse (No 19) and starting the engine; allow the engine to idle until it dies. Turn the engine over once or twice on the starter to ensure that all pressure is released, then switch off the ignition. **Warning:** *This procedure will merely relieve the increased pressure necessary for the engine to run – remember that fuel will still be present in the system components, and take precautions accordingly before disconnecting any of them.*
36 Remove the throttle housing as described in paragraphs 52 to 56 below, moving it clear of the working area; there is no need to disconnect the accelerator cable.
37 Disconnect the crankcase breather hose from the cylinder head cover union, and the fuel pressure regulator vacuum hose from the inlet manifold.
38 Releasing the wire clips, unplug the multi-plug connectors to disconnect the wiring from all four fuel injectors, and from the air charge temperature sensor.
39 Disconnect the fuel feed and return lines at the quick-release couplings just below the braking system servo. Release by squeezing

7F.42A Always renew upper and lower (O-ring) seals (arrowed) whenever an injector is disturbed – lubricate seals with clean engine oil on reassembly

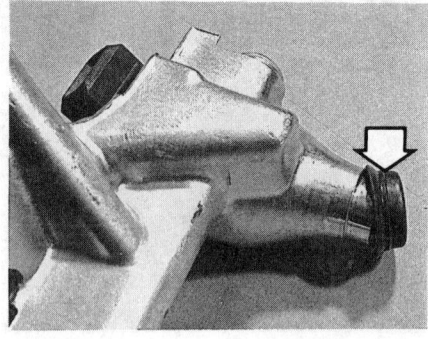

7F.42B Fit new nose seals (arrowed) before refitting fuel distributor rail to intermediate flange

7F.45 Throttle position sensor mounting screws (arrowed)

7F.48 Idle speed control valve wiring connector 'A', mounting bolts 'B' (two of three visible)

7F.49 Idle speed control valve mounting bolts (arrowed) shown with inlet manifold removed, for clarity

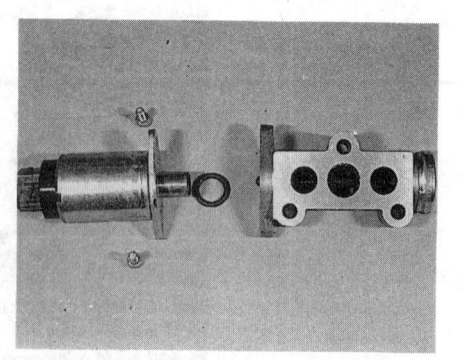

7F.50 Idle speed control valve shown dismantled for cleaning

together the protruding locking lugs on each union, carefully pulling the coupling apart, noting the colour-coding to ensure correct reconnection; use rag to soak up any spilt fuel. **Note:** *Do not disturb the threaded couplings at the fuel distributor rail unions; these are sealed at the factory. The quick-release couplings will suffice for all normal service operations.*

40 Unscrew the three bolts securing the fuel distributor rail, withdraw the rail, carefully prising it out of the intermediate flange, and draining any remaining fuel into a suitable clean container. Note the seals between the rail noses and the intermediate flange; these must be renewed whenever the distributor rail is removed (photos).

41 Clamping the distributor rail carefully in a vice fitted with soft jaws, unscrew the two bolts securing each injector, and withdraw the injectors (photo).

42 Refitting is the reverse of the removal procedure, noting the following points (photos):

(a) Renew the upper and lower (O-ring) seals whenever any injector is disturbed – lubricate each seal with clean engine oil on installation.

(b) Locate each injector carefully in the fuel distributor rail recess, ensuring that the locating tab on the injector head fits into the slot provided in the rail, and tighten the bolts to the torque wrench setting specified.

(c) Fit a new seal to each fuel distributor rail nose, and ensure that none are displaced as the rail is refitted to the intermediate flange. Ensure that the fuel distributor rail is settled fully in the intermediate flange before tightening the three bolts evenly and to the torque wrench setting specified.

(d) Fasten the fuel feed and return quick-release couplings as described in Part C of this Section.

(e) Ensure that the breather hose, vacuum hose and wiring are routed correctly, and secured on reconnection by any clips or ties provided.

(f) On completion, refit fuse No 19, then switch the ignition on and off five times to activate the fuel pump and pressurise the system without cranking the engine – check for signs of fuel leaks around all disturbed unions and joints before attempting to start the engine.

Throttle position sensor – removal and refitting

43 Noting the comments made in Section 5 (and 17) of Chapter 12, disconnect the battery earth terminal.

44 Releasing its wire clip, unplug the multi-plug connector to disconnect the throttle position sensor wiring.

45 Unscrew the two screws and withdraw the sensor from the throttle housing (photo). **Do not** force the sensor's centre to rotate past its normal operating sweep; the unit will be seriously damaged.

46 Refitting is the reverse of the removal procedure, noting the following points:

(a) Ensure that the sensor is correctly orientated by locating its centre on the D-shaped throttle shaft (throttle closed) and aligning the sensor body so that the bolts pass easily into the throttle housing.

(b) Tighten the screws evenly and securely (but do not overtighten them, or the sensor body will be cracked).

Idle speed control valve – removal and refitting

47 Noting the comments made in Section 5 (and 17) of Chapter 12, disconnect the battery earth terminal.

48 Releasing its wire clip, unplug the multi-plug connector to disconnect the valve's wiring (photo).

49 Unscrew the three retaining bolts and withdraw the valve from the inlet manifold – if required, the throttle position sensor can be removed first to improve access (paragraphs 44 and 45 above) (photo).

50 The valve can be dismantled for the passages to be cleaned, if desired; this is not required on a regular basis (photo).

51 Refitting is the reverse of the removal procedure, noting the following points:

(a) Clean the mating surfaces carefully, and always fit a new gasket whenever the valve is disturbed (photo).

(b) Tighten the bolts evenly, to the specified torque wrench setting.

(c) Once the wiring and battery are reconnected, start the engine and allow it to idle – when it has reached normal operating temperature, check that the idle speed is stable, and that no induction (air) leaks are evident. Switch on all electrical loads and check that the idle speed is still correct – connect a tachometer, if required, to make an accurate check.

Chapter 13 Supplement: Revisions and information on later models

7F.51 Always renew gasket when refitting idle speed control valve

7F.56 Throttle housing screws (arrowed)

7F.59 Disconnecting air mass meter wiring ...

7F.60A ... unclipping meter from air filter housing ...

7F.60B ... to withdraw meter

Throttle housing – removal and refitting

52 Noting the comments made in Section 5 (and 17) of Chapter 12, disconnect the battery earth terminal.
53 Unscrew the retaining nuts and slacken the clamp screws, so that the (black plastic) air intake trunking can be disconnected from the air mass meter's flexible rubber hose and from the throttle housing; withdraw the trunking.
54 Disconnect the accelerator cable – see paragraph 22 above.
55 Releasing its wire clip, unplug the multi-plug connector to disconnect the throttle position sensor wiring.
56 Undo the four retaining (Torx-type) screws (photo), and withdraw the throttle housing; remove and discard the gasket.
57 Refitting is the reverse of the removal procedure, noting the following points:

(a) Clean the mating surfaces carefully, and always fit a new gasket whenever the housing is disturbed.
(b) Tighten the screws evenly, to the specified torque wrench setting.
(c) Adjust the accelerator cable and check the throttle operation.

Fig. 13.27 Location of pulse-air valve – 1.3 litre HCS CFi engines (Sec 7G)

Air mass meter – removal and refitting

58 Noting the comments made in Section 5 (and 17) of Chapter 12, disconnect the battery earth terminal.
59 Releasing its wire clip, unplug the multi-plug connector to disconnect the meter wiring (photo).
60 Slackening the securing clamp screw, disconnect the flexible hose from the meter. Release the two clips, and withdraw the meter from the air filter housing lid (photos).
61 Refitting is the reverse of the removal procedure.

Fault diagnosis – general

62 Refer to Chapter 3, Part D, Section 72. Also see Section 14 of this Chapter, paragraphs 64 and 65.

PART G: MANIFOLDS, EXHAUST AND EMISSION CONTROL SYSTEMS (EXCEPT DOHC 16-VALVE ENGINE)

1.1, 1.3 and 1.4 litre CFi engines – general

Note that all CFi engines (HCS or CVH) are fitted with three-way regulated catalytic converters (and therefore require **only** unleaded petrol) and evaporative emissions control systems (carbon canisters). Referring to Part C of this Section, note the problems that can result if the carbon canister's fuel vapour line is kinked or blocked, and the method of checking the system.
In addition, 1.3 litre HCS CFi engines are fitted with a pulse-air system.

Pulse-air system – 1.3 litre HCS CFi engines
General description

This system, consisting of the pulse-air solenoid valve, the pulse-air valve itself and the piping, injects filtered air directly into the exhaust ports. This is achieved using the pressure variations in the exhaust gases (there is no need for a separate air pump) to draw air through from the filter housing; air will flow into the exhaust only when its pressure is below atmospheric. The pulse-air valve can allow gases to flow only one way, so there is no risk of hot exhaust gases flowing back into the filter.

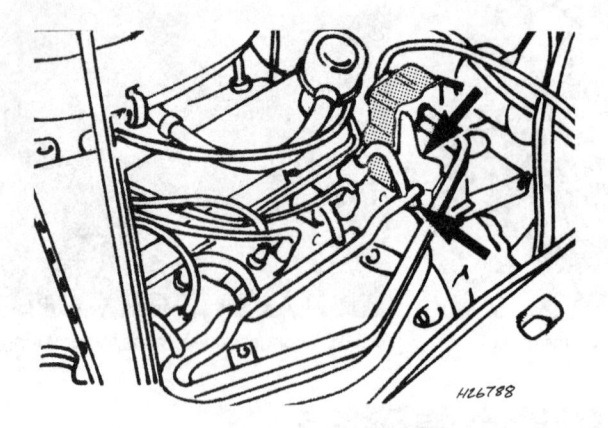

Fig. 13.28 Pulse-air valve-to-bracket retaining screws (arrows) – 1.3 litre HCS CFi engines (Sec 7G)

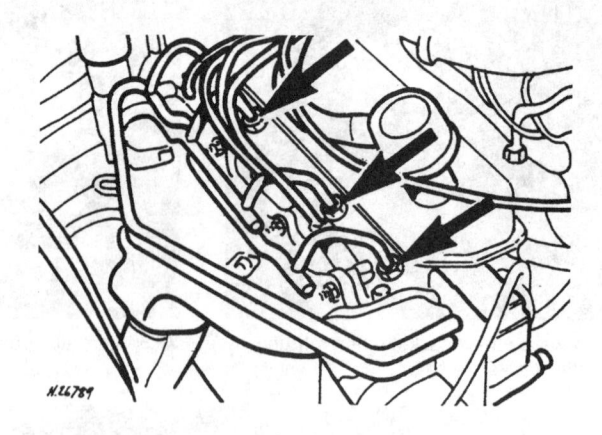

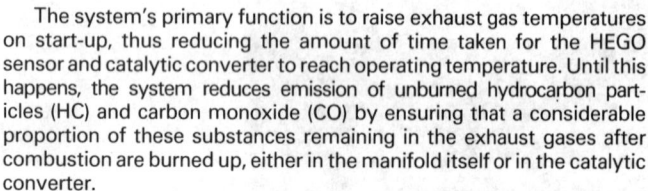

Fig. 13.29 Pulse-air piping-to-cylinder head sleeve nuts (3 of 4 arrowed) – 1.3 litre HCS CFi engines (Sec 7G)

The system's primary function is to raise exhaust gas temperatures on start-up, thus reducing the amount of time taken for the HEGO sensor and catalytic converter to reach operating temperature. Until this happens, the system reduces emission of unburned hydrocarbon particles (HC) and carbon monoxide (CO) by ensuring that a considerable proportion of these substances remaining in the exhaust gases after combustion are burned up, either in the manifold itself or in the catalytic converter.

To ensure that the system does not upset the smooth running of the engine under normal driving conditions, it is linked by the pulse-air solenoid valve to the EEC-IV engine management module, so that it only functions when the HEGO sensor is influencing the fuel/air mixture ratio.

Pulse-air solenoid valve – removal and refitting

1 This component is mounted on the engine compartment bulkhead, between the EDIS module and Manifold Absolute Pressure sensor – see Chapter 4 – and can be identified by the two vacuum hoses connected to it.
2 To remove the valve, first disconnect the battery earth terminal (noting the comments made in Sections 5 and 17 of Chapter 12).
3 Disconnect the valve wiring at its multi-plug.
4 Marking or labelling the valve vacuum hoses so that each can be reconnected to its original union, disconnect the hoses from the valve.
5 Remove the securing screws and withdraw the valve.
6 Refitting is the reverse of the removal procedure.

Pulse-air valve – removal and refitting

7 Noting the comments made in Section 5 (and 17) of Chapter 12, disconnect the battery earth terminal. Allow the engine to cool down completely, or wear heavy gloves to prevent the possibility of burns.
8 Disconnect the vacuum hose from the rear of the valve, then remove the screws securing the valve to its bracket, and withdraw the valve.
9 Refitting is the reverse of the removal procedure.

Pulse-air piping – removal and refitting

10 Noting the comments made in Section 5 (and 17) of Chapter 12, disconnect the battery earth terminal. Allow the engine to cool down completely, or wear heavy gloves to prevent the possibility of burns.
11 Disconnect the vacuum hose from the rear of the pulse-air valve.
12 Unscrew the three bolts securing the piping bracket to the sides of the exhaust manifold and cylinder head.
13 Slacken all four sleeve nuts securing the pipes into the cylinder head, then remove the piping as an assembly, taking great care not to distort any of them.
14 Carefully clean the piping, particularly its threads and those of the cylinder head; remove all traces of corrosion, which might prevent them seating properly, causing air leaks when the engine is restarted.
15 On refitting, insert the piping carefully into the cylinder head ports, taking great care not to bend or distort any of them. Apply anti-seize compound to their threads, and carefully tighten the retaining sleeve nuts while holding each pipe firmly into its port. If a suitable open-socket spanner is available, tighten the sleeve nuts to the specified torque wrench setting.
16 The remainder of the refitting procedure is the reverse of removal.

1.6 litre EFi (non-turbo) engines – general

17 With reference to Fig. 13.20, note the revised crankcase breather hose routing and ventilation system filter mounting that may be found on some models.

1.6 litre EFi (turbocharged) engine – general

18 With the obvious exception of the exhaust manifold/turbocharger assembly removal and refitting procedure given in Part E of this Section, all procedures are as described in Part C of Chapter 3.
19 With reference to Fig. 13.20, note the revised crankcase breather hose routing and ventilation system filter mounting that may be found on some models.

7H.1A Unfastening intermediate flange securing nuts (tool arrowed) ...

7H.1B ... to release flange

7H.1C Always renew heat insulating plate/gasket when refitting flange

Chapter 13 Supplement: Revisions and information on later models

PART H: MANIFOLDS, EXHAUST AND EMISSION CONTROL SYSTEMS (DOHC 16-VALVE ENGINE)

Inlet manifold – general

1 Refer to Figs. 13.30 and 13.31 for full details of the layout of the inlet manifold; note that the intermediate flange is secured by two nuts, one at each end, as well as by the inlet manifold bolts (photos).
2 The inlet manifold can be unbolted separately, leaving the intermediate flange and fuel distributor rail assembly in place on the engine; the manifold bolts can be reached from below (photos).
3 Always clean the mating surfaces carefully on reassembly, and fit new gaskets (photo). **Do not** attempt to re-use the original gaskets, and **never** apply silicone-based sealing compounds to any part of the fuel system.
4 Tighten all fasteners to the torque wrench settings specified.

Exhaust manifold – general

5 The pulse-air system piping assembly must be removed as described below before the exhaust manifold can be withdrawn.
6 When removing the manifold with the engine in the vehicle, additional clearance can be obtained by unscrewing the studs from the cylinder head; a female Torx-type socket will be required (photo).
7 Always clean the mating surfaces carefully on reassembly, and fit a new gasket. **Do not** attempt to re-use the original gasket, and **never** apply sealing compounds to any part of the exhaust system upstream of the catalytic converter.
8 A plastic guide sleeve must be obtained with the replacement gasket, to be fitted to the stud nearest the thermostat so that the manifold is correctly located (photo). **Do not** refit the manifold without this sleeve.
9 Tighten all fasteners to the torque wrench settings specified (photo).

Exhaust system – renewal

10 Refer to Chapter 3, Part C, Sections 68 and 70.
11 Note that the downpipe is secured to the manifold by two bolts, with a coil spring, spring seat and self-locking nut on each. On refitting, tighten the nuts until they stop on the bolt shoulders; the pressure of the springs will then suffice to make a leakproof joint (photos).

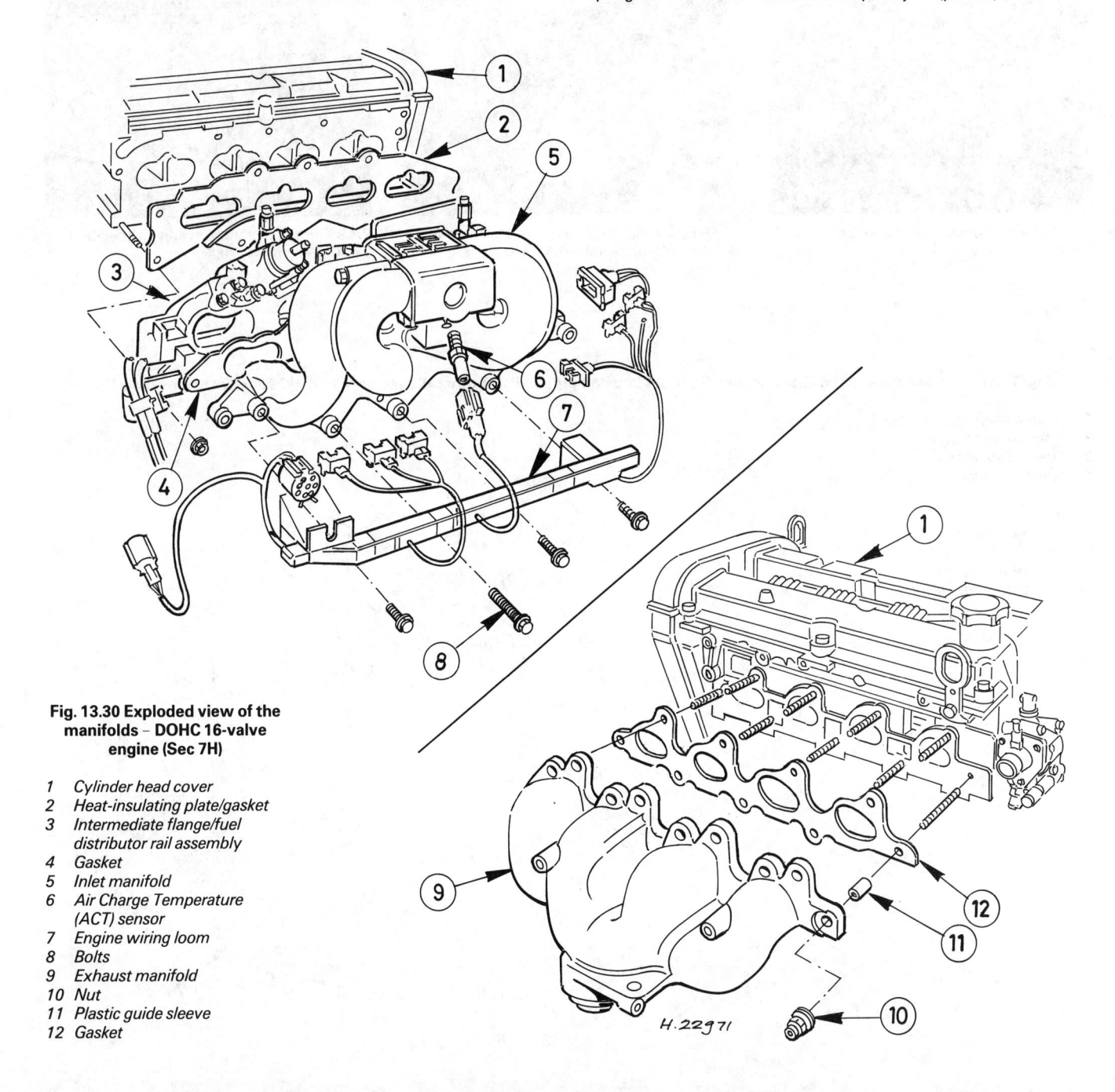

Fig. 13.30 Exploded view of the manifolds – DOHC 16-valve engine (Sec 7H)

1 Cylinder head cover
2 Heat-insulating plate/gasket
3 Intermediate flange/fuel distributor rail assembly
4 Gasket
5 Inlet manifold
6 Air Charge Temperature (ACT) sensor
7 Engine wiring loom
8 Bolts
9 Exhaust manifold
10 Nut
11 Plastic guide sleeve
12 Gasket

Chapter 13 Supplement: Revisions and information on later models

7H.2A Unscrewing inlet manifold bolts – seen from below ...

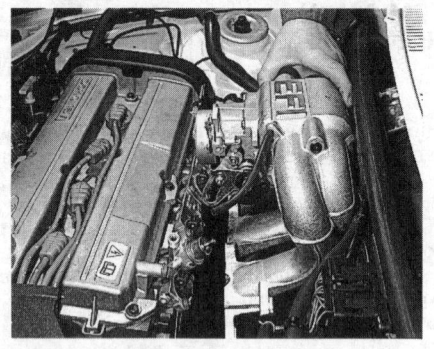

7H.2B ... to release manifold from intermediate flange

7H.3 Always renew gaskets to prevent leaks

7H.6 Exhaust manifold studs can be unscrewed, if required

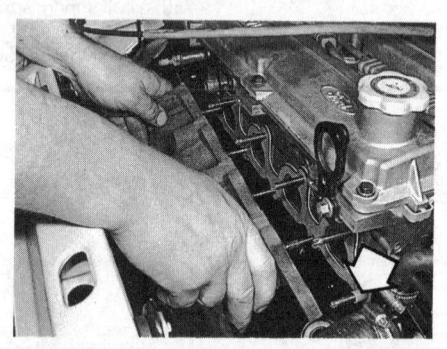

7H.8 Fit plastic guide sleeve to stud arrowed when refitting exhaust manifold

7H.9 Tighten fasteners evenly to specified torque wrench settings

Fig. 13.31 Exploded view of the inlet manifold intermediate flange/fuel distributor rail assembly – DOHC 16-valve engine (Sec 7H)

1　Heat-insulating plate/gasket
2　Intermediate flange
3　Fuel distributor rail
4　Fuel feed union
5　Fuel injectors
6　Bolts
7　Fuel pressure regulator
8　Injector (O-ring) seals
9　Fuel return union
10　Fuel distributor rail/intermediate flange seals

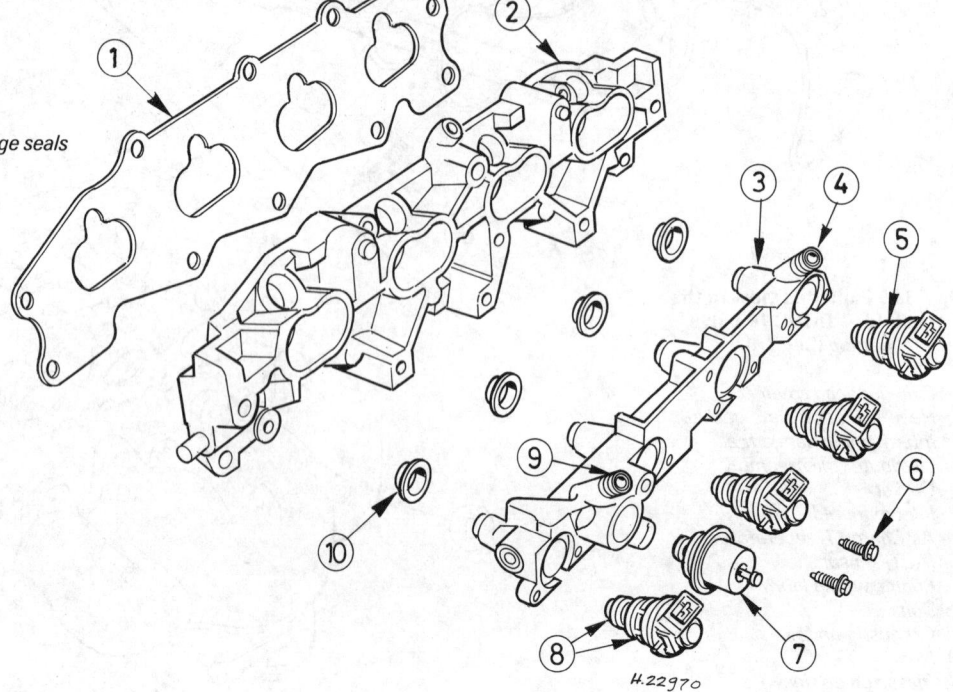

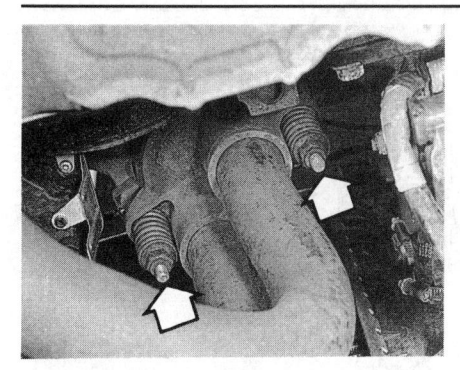

Chapter 13 Supplement: Revisions and information on later models

7H.11A Exhaust system downpipe-to-manifold securing nuts (arrowed)

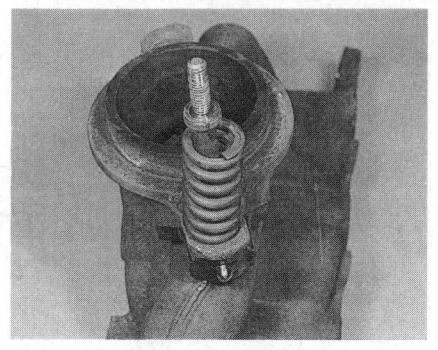

7H.11B Showing securing bolts – note coil spring, and shoulder on bolt

7H.12A Renew exhaust system downpipe-to-manifold gasket to prevent leaks

7H.12B Release spring clip to extract securing bolt from manifold, when required

7H.19 Location of carbon canister – vapour line connection arrowed

7H.20 Location of canister-purge solenoid valve (arrowed)

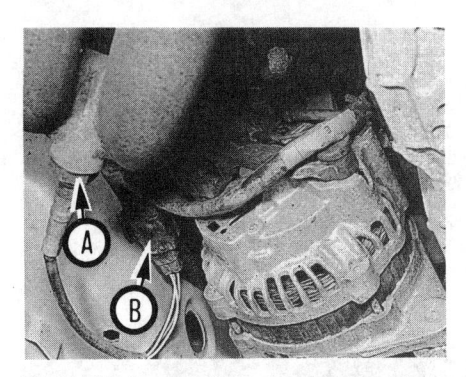

7H.21 HEGO sensor (A) and wiring plug (B)

7H.22 Catalytic converter and mountings

7H.23 Location (under EDIS module, arrowed) of pulse-air solenoid valve

12 Do not overtighten the nuts to cure a leak – the bolts will shear; renew the gasket and the springs if a leak is found. The bolts themselves are secured by spring clips to the manifold, and can be renewed easily if damaged (photos).

Emissions control systems – general description
Evaporative Emissions Control system
13 The system is as described in Chapter 3, Part C.
14 Referring to Part C of this Section, note the problems that can result if the carbon canister's fuel vapour line is kinked or blocked.

Crankcase Emissions Control system
15 The crankcase ventilation system consists of the oil separator, mounted on the front (radiator) side of the cylinder block, and the Positive Crankcase Ventilation (PCV) valve, set in a rubber grommet in the separator's left-hand upper end. A crankcase breather pipe and two flexible hoses connect the PCV valve to a stub on the left-hand end of the inlet manifold's intermediate flange. A crankcase breather hose connects the cylinder head cover to the air filter housing, where a small foam filter prevents dirt from being drawn directly into the engine.
16 The function of these components is to ensure that a depression is created in the crankcase under most operating conditions, particularly at idle; this is to achieve the burning-up of oil vapours and 'blow-by' gases collected in the crankcase, and to minimise the formation of oil sludge by positively inducing fresh air into the system.

Exhaust Emissions Control
17 Refer to Chapter 3, Part C; this engine is fitted with a full three-way regulated catalytic converter, in which the HEGO sensor is used to provide feedback related to the amount of oxygen remaining in the exhaust gases. This is used by the EEC-IV engine management module to adjust the fuel/air mixture ratio by altering the amount of time that each injector remains open.

Pulse-air system
18 Refer to Part G of this Section – the system fitted to 1.3 litre HCS CFi engines is identical in operation.

356   Chapter 13   Supplement: Revisions and information on later models

Fig. 13.32 Crankcase Emissions Control system – DOHC 16-valve engine (Sec 7H)

1   Oil separator
2   Gasket
3   Positive Crankcase Ventilation (PCV) valve
4   Cylinder block/crankcase opening
5   Crankcase breather pipe and flexible hoses

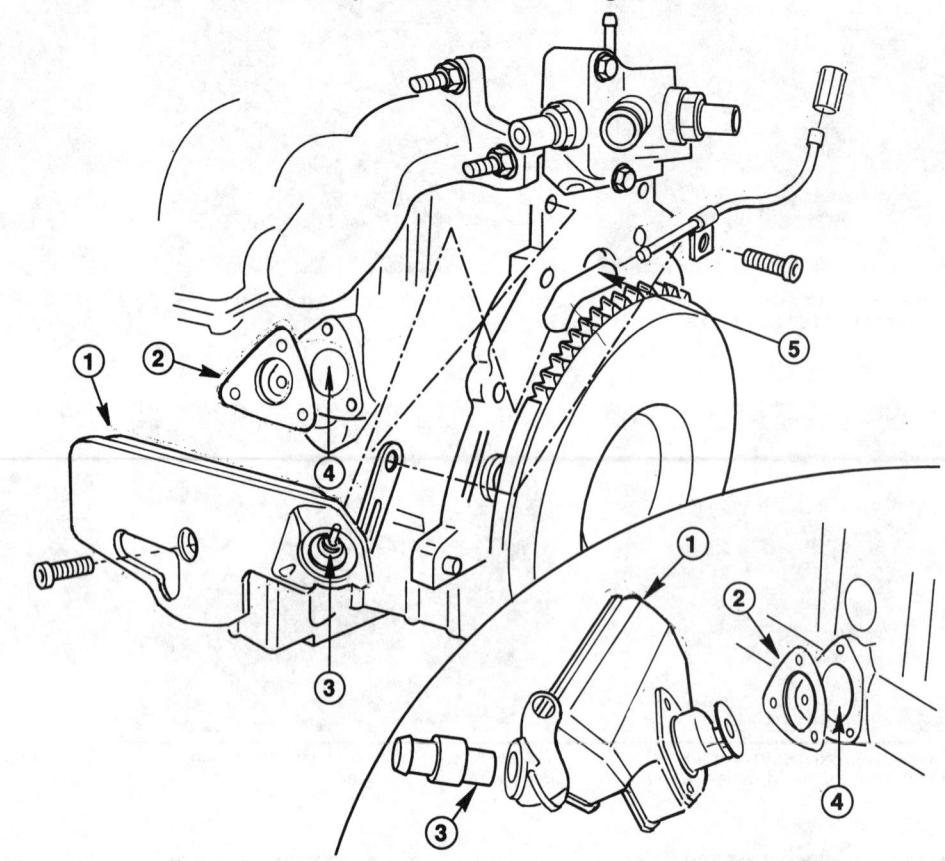

Emissions control system components – removal and refitting

Carbon canister
19   Proceed as described in Chapter 3, Part C, Section 70 (photo).

Canister-purge solenoid valve
20   Proceed as described in Chapter 3, Part C, Section 70 (photo).

HEGO sensor
21   Proceed as described in Chapter 3, Part C, Section 70 (photo).

Catalytic converter
22   Proceed as described in Chapter 3, Part C, Section 70 (photo).

Pulse-air solenoid valve
23   This component is mounted on the engine compartment bulkhead, just beneath the EDIS module, and can be identified by the two vacuum hoses connected to it (photo).
24   To remove the valve, first disconnect the battery earth terminal (noting the comments made in Sections 5 and 17 of Chapter 12).
25   Disconnect the valve wiring at its multi-plug.
26   Marking or labelling the valve vacuum hoses so that each can be reconnected to its original union, disconnect the hoses from the valve.
27   Release the locking tab by pressing it away from the valve, then slide the valve downwards to withdraw it.
28   Refitting is the reverse of the removal procedure.

Pulse-air piping assembly
29   Noting the comments made in Section 5 (and 17) of Chapter 12, disconnect the battery earth terminal. Allow the engine to cool down completely, or wear heavy gloves to prevent the possibility of burns.
30   Unscrew the retaining nuts and slacken the clamp screws, so that the (black plastic) air intake trunking and the air mass meter's flexible rubber hose can be disconnected from the meter and from the throttle housing; withdraw the trunking and hose. Secure the accelerator cable clear of the working area.
31   Remove the radiator cooling fan and motor (Chapter 2).
32   Remove the screws securing the coolant pipe to the exhaust manifold heat shield, secure the coolant pipe as far out of the way as the coolant hoses will permit, then unbolt and withdraw the heat shield.
33   Disconnect the vacuum hose from the rear of the pulse-air filter housing (photo).
34   Unscrew the filter housing retaining bolts and all four union sleeve nuts securing the pipes into the manifold, then remove the piping as an assembly, taking great care not to distort any pipes.

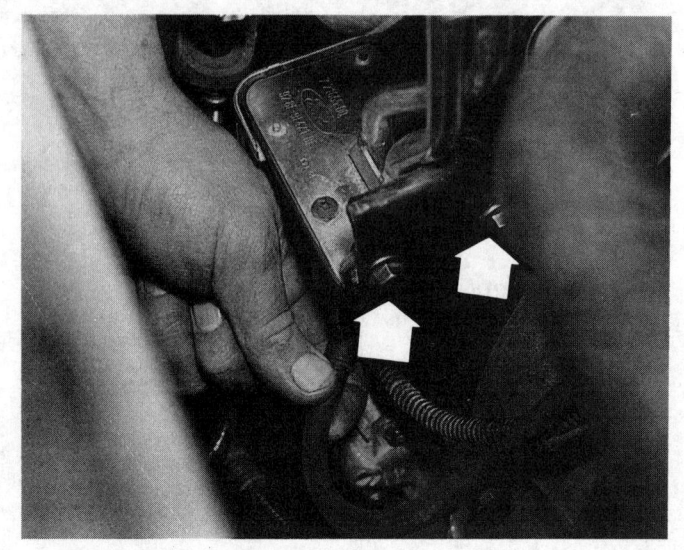

7H.33 Disconnecting vacuum hose from pulse-air filter housing – note housing mounting bolts (arrowed)

Chapter 13 Supplement: Revisions and information on later models

35 Carefully clean the piping, particularly its threads and those of the manifold. Remove all traces of corrosion, which might prevent them seating properly, causing air leaks when the engine is restarted.
36 On refitting, insert the piping carefully into the manifold, taking great care not to bend or distort any pipes. Apply anti-seize compound to their threads, and carefully tighten the retaining sleeve nuts while holding each pipe firmly into its port; if a suitable open-socket spanner is available, tighten the sleeve nuts to the specified torque wrench setting.
37 The remainder of the refitting procedure is the reverse of removal.

8 Ignition and engine management systems

PART A: HCS AND CVH CARBURETTOR ENGINES

1.0 and 1.1 litre HCS engines – modifications
1 Apart from the introduction of the more sophisticated ESC-P1 module on some engines, which then uses an air charge temperature sensor as described for the earlier ESC-H2 system, procedures are unchanged from those described in Chapter 4.
2 The ESC-P1 module itself is removed and refitted as described in Section 8 of Chapter 4 for the ESC-H2 module.

1.3 litre engine – general
3 This engine uses the ESC-P1 module (with an air charge temperature sensor); as noted in paragraphs 1 and 2 above, all procedures are therefore the same as those described in Chapter 4.

1.4 litre CVH engine – general
4 Later vehicles fitted with the 1.4 litre CVH carburettor engine employ a distributorless ignition system.
5 The system functions as described in Section 1 of Chapter 4, noting that the more sophisticated ESC-P1 module uses an air charge temperature sensor in the same way as the earlier ESC-H2 system.
6 All procedures are as described in Part A of Chapter 4, except where outlined below.

Engine coolant temperature sensor – removal and refitting
7 Refer to Chapter 4, Part C, Section 25.

Ignition coil – removal and refitting
8 Refer to Chapter 4, Part D, Section 42.

Engine speed/crankshaft position sensor – removal and refitting
9 Refer to Chapter 4, Part D, Section 40.

ESC-P1 module – removal and refitting
10 Proceed as described in Part A, Section 8 of Chapter 4 for the ESC-H2 module.

PART B: 1.1 AND 1.3 HCS CFI ENGINES

General description
These engines use the distributorless ignition system fitted to the carburettor HCS engines (refer to Section 1 of Chapter 4 for details), but under the overall control of the EEC-IV engine management module, and using an EDIS module as described in Part D, Section 35 of Chapter 4.
All procedures are as described in Part A of Chapter 4, except where outlined below. Refer also to Part D of Chapter 4, Section 47 (and to Section 14 of this Chapter, paragraphs 64 and 65) for fault diagnosis procedures.

Component removal and refitting
Air charge temperature sensor
1 Refer to Chapter 4, Part C, Section 24.

Road speed sensor
2 Refer to Chapter 4, Part C, Section 26.

Manifold Absolute Pressure sensor
3 Refer to Chapter 4, Part C, Section 31.

4 On 1.3 litre engines, note that a fuel trap is fitted in the sensor vacuum hose; if disturbed, this must be refitted as described in Chapter 3, Part C, Section 70, paragraphs 26 to 28.

EDIS module
5 Refer to Chapter 4, Part D, Section 45.

EEC-IV engine management module
6 Refer to Chapter 4, Part C, Section 33.

PART C: 1.4 CFI AND 1.6 EFI (TURBOCHARGED) ENGINES

1.4 CFi engines – modifications (September 1990-on)
These engines use the distributorless ignition system, under the overall control of the EEC-IV engine management module, and using an EDIS module as described in Part D, Section 35 of Chapter 4; refer also to Section 1 of Chapter 4 for details.
All procedures are as described in Part C of Chapter 4, except where outlined below; if the ignition timing is to be checked, refer to Part A, Section 9, noting that the fixed timing marks remain as described in Part C, Section 30. Refer also to Part D, Section 47 (and to Section 14 of this Chapter, paragraphs 64 and 65) for fault diagnosis procedures.

1.4 CFi engines (September 1990-on) – component removal and refitting
Ignition coil – removal and refitting
Refer to Chapter 4, Part D, Section 42.

Engine speed/crankshaft position sensor – removal and refitting
Refer to Chapter 4, Part D, Section 40.

EDIS module
Refer to Chapter 4, Part D, Section 45.

1.6 EFi (turbocharged) engine – general
As far as the ignition system is concerned, all procedures are exactly as described in Part D of Chapter 4.

PART D: DOHC 16-VALVE ENGINE EEC-IV ENGINE MANAGEMENT SYSTEM – IGNITION SYSTEM COMPONENTS

General description
Refer to Part D of Chapter 4, Section 35, and to Figs. 13.25 and 13.26.
Although this system is very much more powerful and sophisticated than that fitted to CVH-engined models, it is essentially as described in Chapter 4.
As with the system(s) described in Parts A and D of Chapter 4, the ignition timing is totally under the control of the EEC-IV engine management module, and cannot be adjusted. If attempting to check the timing, remember that it may vary constantly at idle speed – the setting quoted in the Specifications Section of this Chapter is a purely nominal value, given for reference only.

Maintenance, inspection and precautions
Maintenance and inspection
1 Maintenance is restricted to spark plug renewal and a check of the condition of the HT leads – refer to paragraph 3 below. Note that the ignition timing cannot be adjusted (see note above).

Precautions
2 Refer to Chapter 4, Section 2.

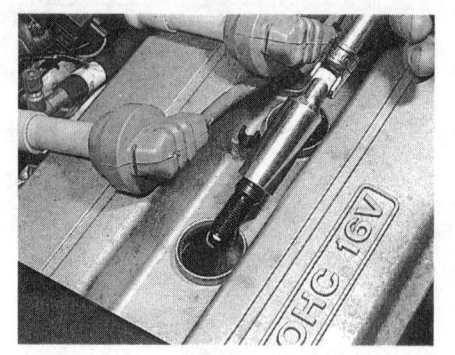

358 Chapter 13 Supplement: Revisions and information on later models

8D.3 Spark plug renewal – DOHC 16-valve engine

8D.4 Air Charge Temperature (ACT) sensor screwed into rear of inlet manifold – note tapered thread

8D.6 Location of Engine Coolant Temperature (ECT) sensor

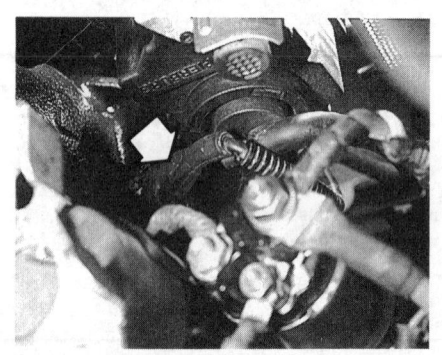

8D.7A Remove exhaust system downpipe to reach engine speed/Crankshaft Position Sensor (CPS) – wiring connector arrowed ...

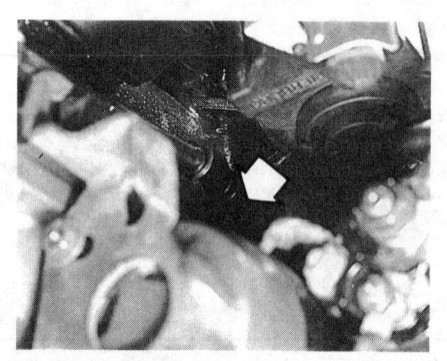

8D.7B ... unbolting sensor from mounting bracket

8D.10 Disconnecting camshaft position sensor

8D.16A Removing DOHC 16-valve engine ignition coil – unscrew Torx-type screws ...

8D.16B ... disconnect wiring ...

8D.16C ... unplug spark plug HT leads – note lead and coil terminal numbering for correct reconnection ...

8D.16D ... and invert mounting bracket to release wiring

8D.17 Location of EDIS ignition module

Chapter 13 Supplement: Revisions and information on later models

Spark plugs and HT leads – inspection and renewal

3 Refer to Section 3 of Chapter 4, noting the following differences (photo):

(a) Note the spark plug electrode gap and torque wrench setting given in the Specifications Section of this Chapter.
(b) Each HT lead can be identified by the cylinder number marked on it and on the ignition coil terminals.

Air Charge Temperature (ACT) sensor – removal and refitting

4 The sensor is screwed into the rear of the inlet manifold's upper section (photo). Removal and refitting is as described in Chapter 4, Part D, Section 38; a deep 25 mm socket will be required.
5 On refitting, tighten the sensor to the torque wrench setting given in the Specifications Section of this Chapter; if it is overtightened, its tapered thread may crack the inlet manifold.

Engine Coolant Temperature (ECT) sensor – removal and refitting

6 The sensor is screwed into the top of the thermostat housing (photo). Removal and refitting is as described in Chapter 4, Part A, Section 5; tighten the sensor to the torque wrench setting given in the Specifications Section of this Chapter.

Engine speed/crankshaft position sensor – removal and refitting

7 The sensor is mounted on a separate bracket, that is itself secured by a screw to the transmission end of the cylinder block/crankcase (photos).
8 Access is extremely restricted; depending on the tools available, owners may remove either the exhaust system downpipe (Chapter 3) or the starter motor (Chapter 12, and Section 14 of this Chapter).
9 Once the sensor is accessible, proceed as described in Chapter 4, Part D, Section 40; note the torque wrench settings given in the Specifications Section of this Chapter.

Camshaft position sensor – removal and refitting

10 The sensor is secured by a screw to the rear transmission end of the cylinder head (photo).
11 Noting the comments made in Section 5 (and 17) of Chapter 12, disconnect the battery earth terminal.
12 Releasing its wire clip, unplug the multi-plug connector to disconnect the sensor wiring.
13 Unscrew the (Torx-type) retaining screw and withdraw the sensor.
14 Refitting is the reverse of the removal procedure, noting the following points:

(a) Apply petroleum jelly or clean engine oil to the sensor's sealing O-ring.
(b) Locate the sensor fully in the cylinder head, and wipe off any surplus lubricant before securing it.
(c) Tighten the screw to the torque wrench setting specified.

9.1 Note 'Low-lift' markings (arrowed) on clutch components – 1.6 and 1.8 litre engines

Road speed sensor unit – removal and refitting

15 Proceed as described in Chapter 4, Part C, Section 26.

Ignition coil – removal and refitting

16 Proceed as described in Chapter 4, Part D, Section 42 (photos).

EDIS ignition module – removal and refitting

17 Proceed as described in Chapter 4, Part D, Section 45 (photo).

EEC-IV engine management module – removal and refitting

18 Proceed as described in Chapter 4, Part C, Section 33.

Fault diagnosis

19 Refer to Chapter 4, Part D, Section 47. Also see Section 14 of this Chapter, paragraphs 64 and 65.

9 Clutch

'Low-lift' clutch components – all 1.6 and 1.8 litre engines

1 The clutch cover assemblies and driven plates fitted to these engines are marked 'Low-lift' (photo). This is to identify them as being of a modified design which, by altering the pressure plate's internal ratio, increases the clutch's torque-handling capabilities without increasing pedal travel/effort; this results in reduced pressure plate lift, hence the designation.
2 Clutch cover assemblies and driven plates marked in this way must never be mixed with unmarked components, or the clutch take-up may be faulty; it is possible that a mixture may even result in a clutch that cannot release. **Always** use only a cover assembly and driven plate that are correctly marked when renewing these components.

Clutch release bearing guide sleeve

3 Some vehicles will be found to have a steel sleeve fitted over the (aluminium alloy) transmission guide sleeve. In such cases, only a modified release bearing with the necessary larger inside diameter can be fitted.

10 Driveshafts

Tripod-type inner CV joint

1 The ball-type inner CV joints shown in Chapter 7 have been replaced progressively in production by tripod-type joints.

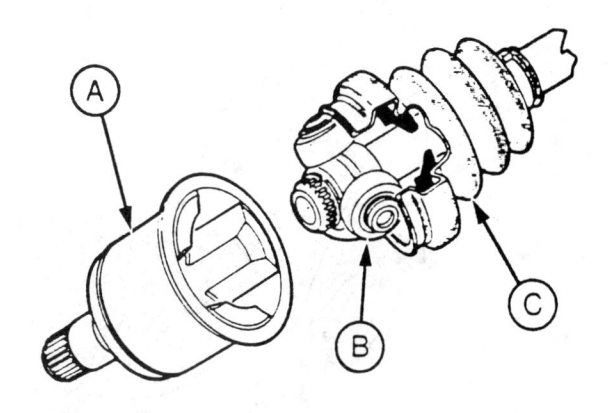

Fig. 13.33 Tripod-type inner CV joint fitted to some later models (Sec 10)

A Joint body
B Driveshaft and inner joint assembly
C Joint gaiter

Chapter 13 Supplement: Revisions and information on later models

4 Refitting is the reverse of the removal procedure; refill the system and bleed it as described in Chapter 8.

Rear brake shoe shims
5 On vehicles equipped with 180 mm rear brakes (see Chapter 8 Specifications), anti-squeal shims may be found fitted between each shoe and the lower pivot – see Fig. 13.35. Whenever the shoes are disturbed, ensure that the shims are securely located on each side of the lower pivot.

Anti-lock braking system – modulator drivebelt renewal
6 In addition to the regular checks carried out at the intervals specified in 'Routine maintenance' at the front of this manual, the modulator drivebelts should be checked for wear as follows, and if necessary renewed, whenever the driveshafts are disconnected from the transmission.
7 Remove the drivebelt cover (Chapter 8, Section 25), then use a magnifying glass to examine closely the condition of the belt over its entire length. Renew the belt if any cracks are noticed in the fabric at the roots of the teeth, if there is any abrasion of the fabric facing material, or if there are any tears starting from the edge of the belt.
8 If, since the drivebelts were last renewed, a vehicle has covered more than 30 000 miles (48 000 km) or a period of more than two years has elapsed, the drivebelts should be renewed as a matter of course whenever the driveshafts are disconnected from the transmission.

Light-laden valve (Courier)
Adjustment
Note: *To adjust the valve accurately, the vehicle must be at a known rear axle loading – owners who cannot determine the loading with sufficient accuracy must have this check made by a Ford dealer or similar expert.*

9 If the vehicle is to be raised for the check to be made, ensure that all its weight remains on its wheels. Measure the distance between the inner radius of the linkage's hooked end and the first shoulder (dimension 'X', Fig. 13.36). If the dimension is not as specified for the axle loading, adjustment is required.
10 To adjust the setting, slacken the locknut on the valve linkage, move the rod until the setting is correct, then tighten the locknut securely (photo). Check the operation of the brakes before taking the vehicle out on the road.

Removal and refitting
11 If the vehicle is to be raised for the work to be carried out, ensure that all its weight remains on its wheels.
12 To remove the valve, first place a container underneath it to catch any spilt fluid, then disconnect the four brake pipes. Plug or cap the pipe ends and valve openings, to keep dirt out and fluid in.

10.3 Driveshaft vibration damper, showing later Torx-type bolts

2 Driveshaft removal and refitting procedures remain unchanged, but note the increased amount of lubricant required when renewing the joint gaiter or the joint itself.

Vibration damper Torx-type bolts
3 On later models, Torx-type bolts have been fitted to clamp the vibration damper (where fitted) to the right-hand driveshaft (photo); note the torque wrench setting given in the Specifications Section of this Chapter.
4 These later bolts can be fitted to earlier vehicles, if the damper is found to have worked loose at any time.

11 Braking system

Master cylinder pressure-conscious reducing valves – removal and refitting
1 All later models without anti-lock braking are fitted with pressure-conscious reducing valves in the master cylinder – see Fig. 13.34.
2 To remove a valve, first place a container under the master cylinder to catch any spilt fluid, then disconnect the relevant brake pipe. Plug or cap the pipe and valve adaptor ends, to keep dirt out and fluid in.
3 Unscrew the valve; the threaded adaptor may be unscrewed if required.

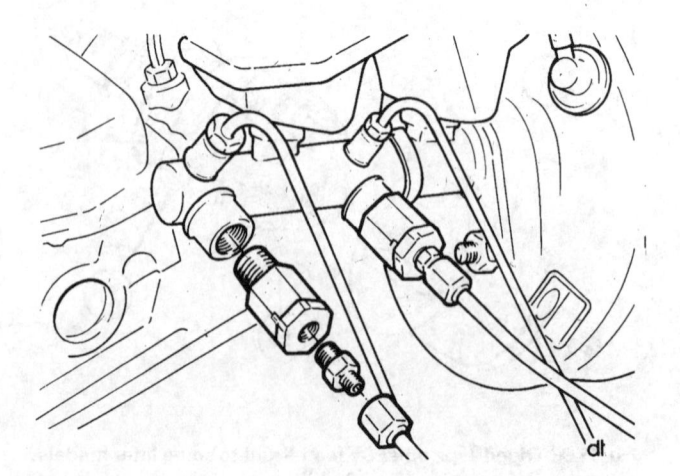

Fig. 13.34 Braking system pressure-conscious reducing valves – later models only, at master cylinder (Sec 11)

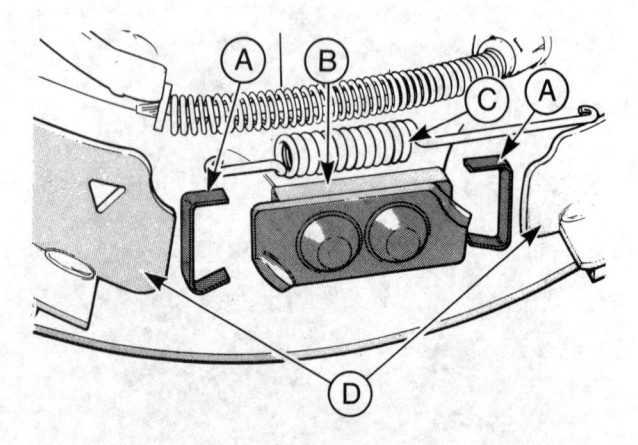

Fig. 13.35 Correct installation of rear brake shoe shims – where fitted (Sec 11)

A Shims
B Lower pivot
C Lower pull-off spring
D Brake shoes

Chapter 13 Supplement: Revisions and information on later models

11.10 Courier braking system light-laden valve – slacken locknut arrowed to adjust valve setting

11.13A Removing light-laden valve – Courier – disconnect pipes and linkage ...

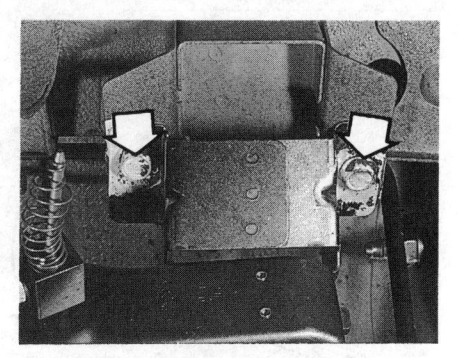

11.13B ... then unscrew bolts (arrowed) to release valve assembly

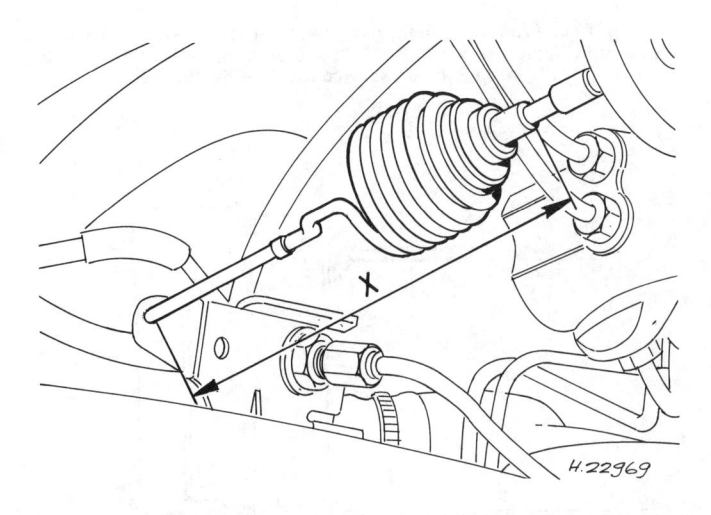

Fig. 13.36 Checking adjustment of Courier light-laden valve (Sec 11)

With rear axle load at 400 kg, 'X' should be 147 mm
With rear axle load at 850 kg, 'X' should be 166 mm

13 Unbolt the valve from its mounting bracket, unhook the linkage from the rear axle, then withdraw the valve (photos). The intermediate bracket may be unbolted if required.
14 Refitting is the reverse of the removal procedure; adjust the valve as described above, then refill the system and bleed it as described in Chapter 8. Check the operation of the brakes before taking the vehicle out on the road.

12 Suspension and steering

Front hub bearings – renewal (all models, June 1990-on)
1 On these later models, the bearing inner race is a press-fit on each front hub.
2 An hydraulic press must be used to push the hub out of the (removed) spindle carrier; the outer race can be drifted out as described in Section 4 of Chapter 10.
3 To draw the inner race off the hub without risk of damage requires the use of Ford service tools 14-038 (screwed firmly on to tool 15-050), with adaptor 14-038-01 – see Fig. 13.37.
4 On refitting, it is essential that the special tools listed in Section 4 of Chapter 10 are used, first to draw the outer race into the spindle carrier (Fig. 13.38), followed by the inner race (Fig. 13.39) and then to draw the hub through the bearings (Chapter 10, Fig. 10.9). Finally support the hub end of the assembly on service tool 15-064 (to protect the studs) and use service tool 15-036 with the hydraulic press to push the spindle carrier fully into place on the hub (Fig. 13.40).

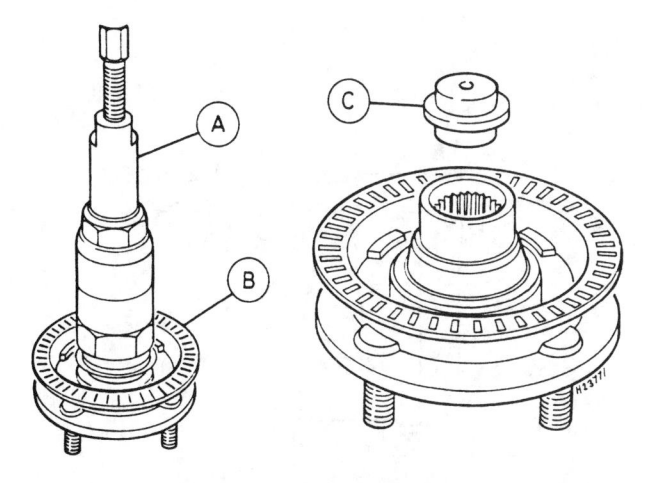

Fig. 13.37 Front hub bearing renewal – all models, June 1990 onwards – drawing inner race off hub (Sec 12)

A Service tool 15-050
B Service tool 14-038
C Service tool 14-038-01

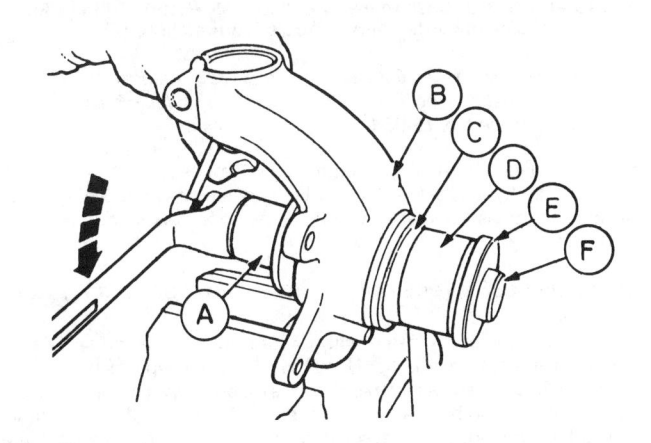

Fig. 13.38 Front hub bearing renewal – all models, June 1990 onwards – drawing bearing's outer race into spindle carrier (Sec 12)

A Service tool 15-033-01
B Spindle carrier
C Bearing
D Service tool 14-034
E Service tool 15-068
F Service tool 15-034

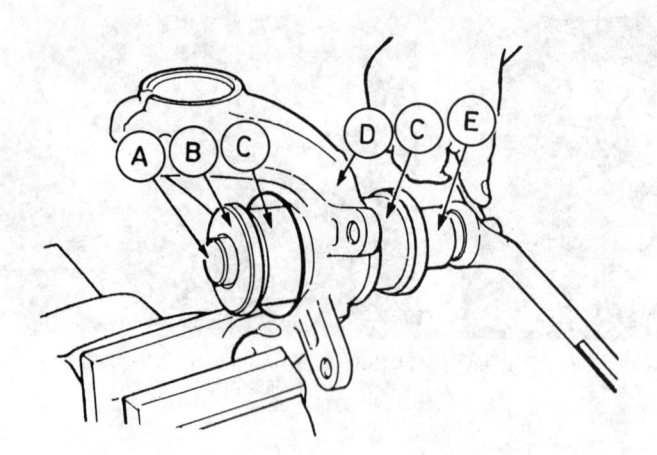

362   Chapter 13 Supplement: Revisions and information on later models

Fig. 13.39 Front hub bearing renewal – all models, June 1990 onwards – drawing bearing's inner race into spindle carrier (Sec 12)

- A Service tool 15-034
- B Service tool 15-068
- C Service tool 14-034 (2 required)
- D Spindle carrier
- E Service tool 15-033-01

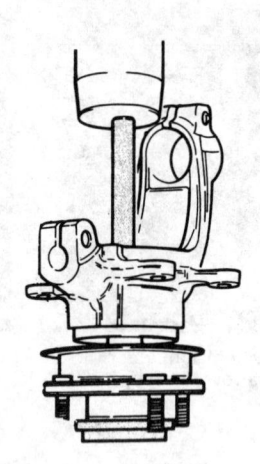

Fig. 13.40 Front hub bearing renewal – all models, June 1990 onwards – using service tool 15-036 to press spindle carrier on to hub, supported on service tool 15-064 (Sec 12)

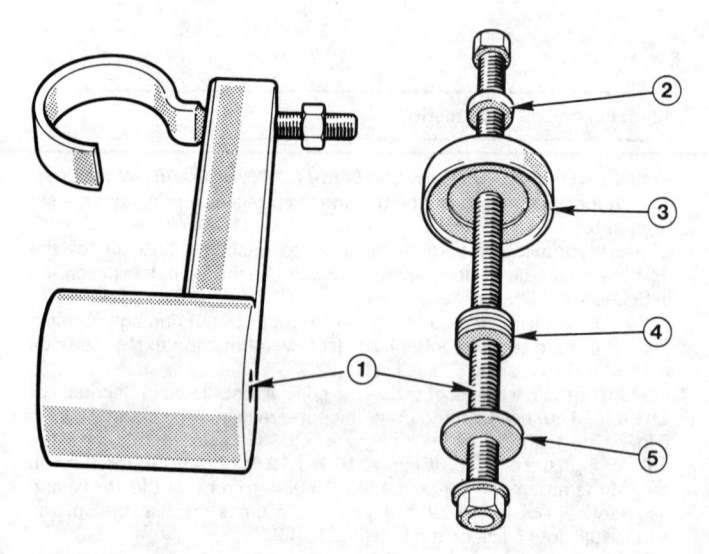

Fig. 13.41 Rear axle bush renewal – all models except Courier, April 1990 onwards – service tools required (Sec 12)

- 1 Service tool 15-086
- 2 Thrust bearing
- 3 Service tool 15-084
- 4 13 mm spacer
- 5 30 mm (diameter) washer

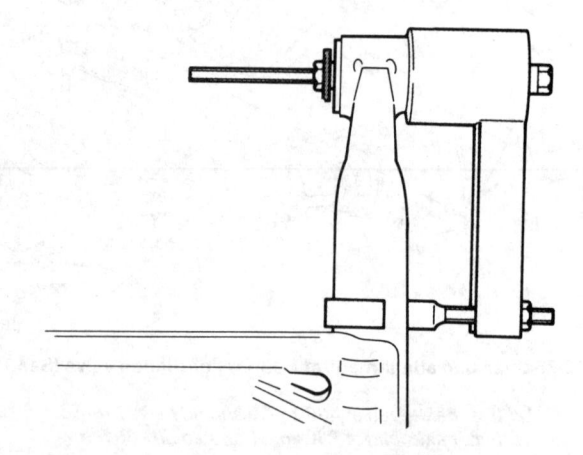

Fig. 13.42 Rear axle bush renewal – all models except Courier, April 1990 onwards – drawing old bush from rear axle (Sec 12)

5 Any variation from this procedure risks damage to the components, and the threat of greatly-reduced bearing life; owners without the facilities and tools listed are advised to have the work carried out by a Ford dealer.

Rear axle bush – renewal (all models except Courier, April 1990-on)

6 All later models are fitted with revised bushes which use 12 mm (thread size) void bush bolts. Note that the early type of bush is no longer available; if an early vehicle's rear axle bushes are to be renewed, two corresponding bolts and nuts must be obtained at the same time, and either the body mounting brackets must be replaced by the corresponding modified items, or the holes in the original brackets must be opened out (to 12.5 mm) to suit the new bolts.

7 Any attempt to fit these bushes without the special tools listed risks damage to the components, and the threat of greatly-reduced bush life; owners without the facilities and tools listed are advised to have the work carried out by a Ford dealer.

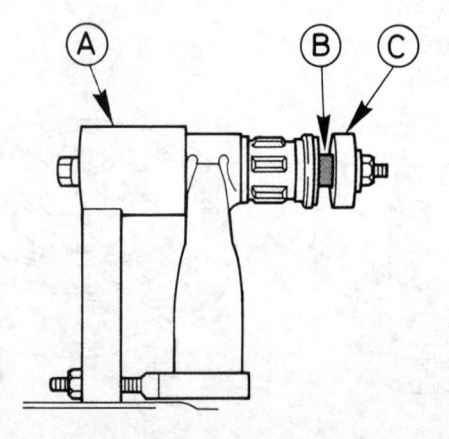

Fig. 13.43 Rear axle bush renewal – all models except Courier, April 1990 onwards – drawing new bush into rear axle (Sec 12)

- A Service tool 15-086
- B Spacer
- C Service tool 15-084

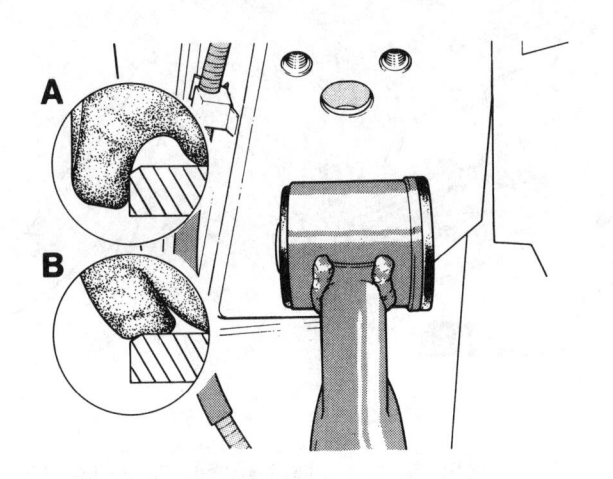

Chapter 13 Supplement: Revisions and information on later models

Fig. 13.44 Rear axle bush renewal – all models except Courier, April 1990 onwards – correct engagement of bush lip on rear axle (Sec 12)

- A New bush, installed correctly
- B New bush, installed incorrectly

8   When renewing these bushes, the old-type bushes (10 mm bolts) can be driven out as described in Chapter 10, Section 10, but new-type bushes can only be extracted using service tool 15-084, which must be drilled out to 12.5 mm (from 10.5 mm) to accept tool 15-086. Using the tools as shown in the accompanying illustrations, draw the old bushes into the tool.

9   On installation, clean the rear axle bores carefully, then lubricate the bushes with soap solution. Draw the new bushes into the axle so that their flanges are on the outside; use a spacer to ensure that each bush is pulled through far enough for its lip to engage correctly.

10   The remainder of the procedure is as described in Chapter 10, Section 10. Do not forget to refit only the 12 mm void bush bolts to suit the new bushes; the torque wrench setting remains as given in Chapter 10.

Rear suspension anti-roll bar – removal and refitting

11   Raise the rear of the vehicle, and support it securely on axle stands on each side, so that all weight is off the rear suspension.

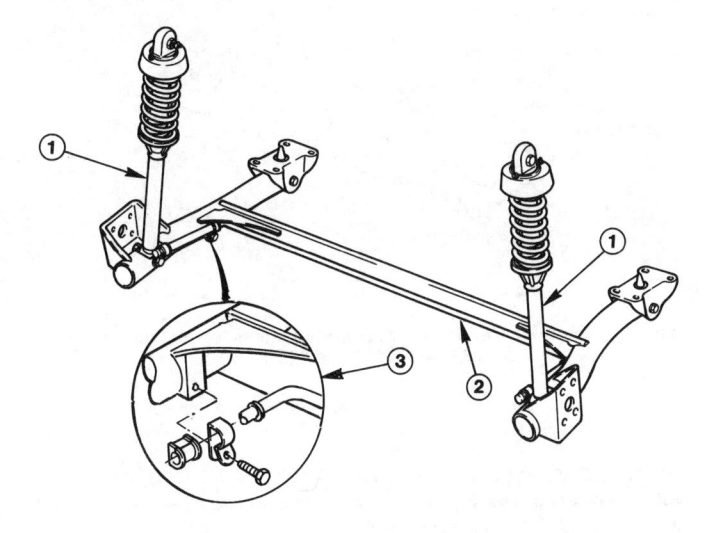

Fig. 13.45 Rear suspension anti-roll bar (Sec 12)

- 1 Rear suspension strut
- 2 Anti-roll bar
- 3 Inset showing (front) mounting clamp

12   Undo the front clamp bolt from each end of the bar, then support the bar and undo the rear bolt at each end; withdraw the bar.

13   Prise open the clamps to release the rubber bushes, if required. Fit the new bushes using soapy water as a lubricant.

14   On refitting, offer up the bar and align first the front clamp on each side, refitting the bolts loosely.

15   Align and refit the rear clamp on each side; again, tighten the bolts only loosely.

16   Lower the vehicle to the ground, rock it to settle the suspension, then tighten the clamp bolts to the torque wrench settings specified.

Rear suspension (Courier) – general description

17   The Courier model is fitted with a modified version of the twist beam rear suspension, using linked torsion bars as springs and to provide anti-roll stabilisation – see Fig. 13.46. Separate damper units are fitted to control suspension movement.

Fig. 13.46 Rear suspension – Courier (Sec 12)

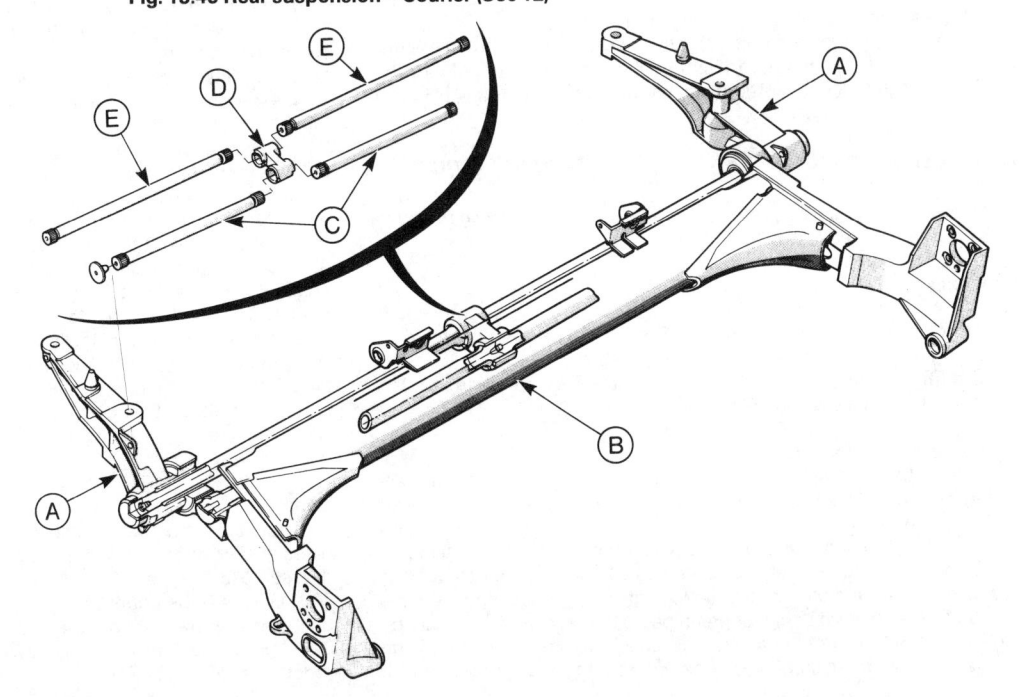

- A Pivot brackets
- B Rear axle
- C Rear (anti-roll) torsion bars
- D Connecting link
- E Front (spring) torsion bars

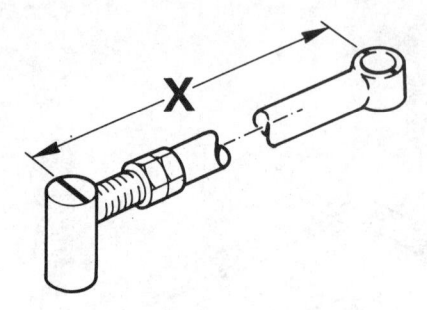

364 Chapter 13 Supplement: Revisions and information on later models

Fig. 13.47 Service tool 15-087 required to check/reset Courier rear suspension ride height (Sec 12)

Set dimension 'X' as specified in text

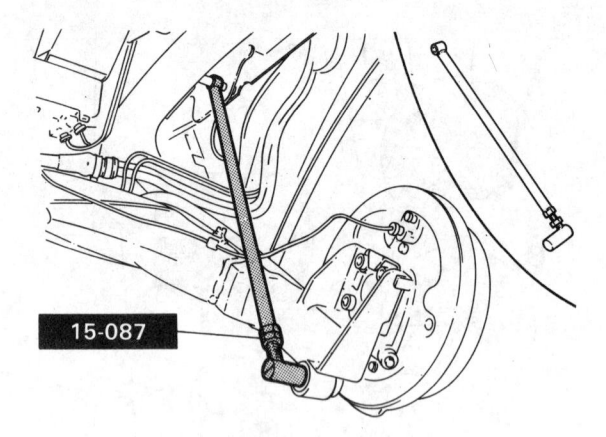

Fig. 13.48 Service tool 15-087 installed – Courier (Sec 12)

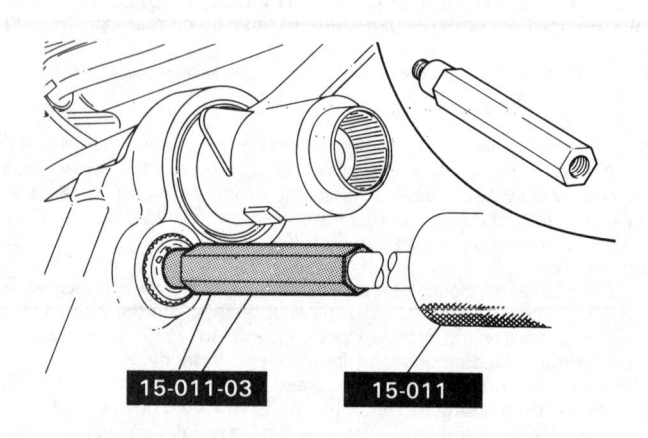

Fig. 13.49 Using slide hammer 15-011 and adaptor 15-011-03 to withdraw torsion bar – Courier (Sec 12)

Rear suspension (Courier) – ride height adjustment
Note: *Ford service tool 15-087 is required to enable the rear suspension torsion bars to be set; owners without this tool must have any rear suspension overhaul work carried out by a Ford dealer. New torsion bar spring clips will be required whenever they are disturbed.*

18 If the rear suspension ride height is thought to require adjustment, raise the rear of the vehicle, and support it securely on axle stands on each side, so that all weight is off the rear suspension. Support the rear axle and remove the rear roadwheels.
19 Remove the damper on one side, and loosely refit its upper mounting bolt. Set the service tool to a length of 457 mm between the reference lines marked at each end (dimension 'X', Fig. 13.47), then fit it to the damper's upper bolt and lower mounting point as shown in Fig. 13.48.
20 Remove the access plug from the vehicle's bodywork, followed by the dust cap from the end of the front (spring) torsion bar, then extract the spring clip securing the torsion bar in the rear axle.
21 Using a slide hammer (service tool 15-011) with adaptor 15-011-03, draw out the torsion bar until its splines are felt to disengage.
22 The ride height can only be adjusted in 3 mm increments. To raise the ride height by 3 mm, reset the service tool to a length of 459 mm between the reference lines; to lower the ride height, reset it to 455 mm.
23 With the rear axle held at the desired height by the tool, insert the torsion bar until the splines are felt to engage in the connecting link. Push gently to relocate the bar; if any greater force is required, the bar is in the wrong position, and must be relocated. When the bar is fully in place, secure it with a new spring clip, then refit the dust cap and access plug.
24 Repeat the procedure on the remaining side.
25 When both sides are correct, refit the dampers and roadwheels, lower the vehicle to the ground, and tighten the nuts and bolts to the specified torque wrench settings. Check the setting of the light-laden valve, as described in Section 11 of this Chapter.

Rear suspension (Courier) – component removal and refitting
Note: *Before starting work, refer to the note above concerning the need for service tool 15-087.*

Rear axle
26 Proceed as described in Section 12 of Chapter 10, disconnecting the light-laden valve as described in Section 11 of this Chapter, and removing the dampers as described below (photo).
27 Refitting is the reverse of the removal procedure, noting that the pivot brackets locate on dowels in the underbody (photo).

Torsion bars (front and rear)
28 Note that the front bars provide the springing, with the rear bars acting as anti-roll stabilisers. Left-hand bars (front or rear) are marked with either two notches or the letter 'G' stamped in their outboard end face, while right-hand bars are similarly stamped with three notches or the letter 'D'. In addition, left-hand bars may be marked with a band of red paint, right-hand bars with yellow paint. The front (spring) bars are each identified by two bands of green paint, the rear (stabiliser) bars by blue paint. Before removing any bar, check carefully for identifying marks, and note them carefully if different from those just described – if none can be found, make your own. **Do not** interchange left- and right-hand bars.
29 To remove a bar, raise the rear of the vehicle, and support it securely on axle stands on each side, so that all weight is off the rear suspension. Support the rear axle and remove the rear roadwheel.
30 Remove the appropriate damper, and loosely refit its upper mounting bolt. Set the service tool to a length of 457 mm between the reference lines marked at each end (dimension 'X', Fig. 13.47), then fit it to the damper's upper bolt and lower mounting point as shown in Fig. 13.48.
31 Remove the access plug from the vehicle's bodywork (if applicable), followed by the dust cap from the end of the torsion bar, then extract the spring clip securing the torsion bar in the rear axle.
32 Using a slide hammer (service tool 15-011) with adaptor 15-011-03, draw out the torsion bar.
33 On refitting, insert the torsion bar until the splines are felt to engage in the connecting link. Push gently to relocate the bar; if any greater force is required, the bar is in the wrong position, and must be relocated. When the bar is fully in place, secure it with a new spring clip, then refit the dust cap and access plug (photos).
34 Refit the damper and roadwheel, lower the vehicle to the ground, and tighten the nuts and bolts to the specified torque wrench settings. Check the setting of the light-laden valve, as described in Section 11 of this Chapter.

Damper
35 Raise the rear of the vehicle, and support it securely on axle stands on each side, so that all weight is off the rear suspension, then support the rear axle.
36 Unscrew the upper mounting bolt and nut, followed by the lower bolt, and withdraw the damper (photos).
37 If either damper is to be renewed, it is good practice to renew both together as a matched pair.

Chapter 13 Supplement: Revisions and information on later models 365

12.26 Removing rear axle – Courier – note braking system union attached to axle ...

12.27 ... rear axle pivot brackets locate on dowels in underbody

12.33A Ensure torsion bars are located correctly into connecting link

12.33B Torsion bar spring clip correctly engaged

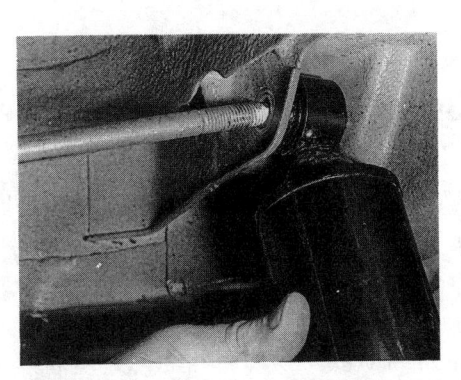

12.36A Removing Courier rear suspension damper upper ...

12.36B ... and lower mounting

38 On refitting, loosely refit the bolts and nut, lower the vehicle to the ground and rock it to settle the suspension, then tighten the bolts and nut to the torque wrench settings specified.

Steering wheel (September 1993-on models) – removal and refitting

39 Fiesta models from September 1993 onwards may be fitted with a driver's airbag, which is mounted within the steering wheel. No further details were available at the time of writing. *No attempt should be made to remove a steering wheel equipped with an airbag – there is a risk of setting off the device accidentally if the correct procedure is not followed. If you are in any doubt on this point, seek the advice of a Ford dealer.*

13 Bodywork and fittings

Bonnet vents (RS Turbo) – removal and refitting

1 Open the bonnet and remove the three retaining screws to release each vent.
2 Refitting is the reverse of the removal procedure; ensure that the vents are aligned correctly, and do not overtighten the screws.

Roof mouldings (Courier) – removal and refitting

3 These are released by lifting and gripping the moulding's inboard edge, then by rotating the whole length of the moulding towards the outside of the vehicle to release it from its outboard lip.
4 Refitting is the reverse of the removal procedure; ensure that the moulding's outboard edge is seated securely in its lip before pressing the inboard edge firmly into place.

Rear quarter mouldings (Courier) – removal and refitting

5 Remove the rear bumper (see paragraph 8 below).

6 Each moulding is secured by two screws at its forward edge, in the wheel arch, and by a single screw at the rear (photos). Remove the screws and withdraw the moulding, unclipping it at its front end.
7 Refitting is the reverse of the removal procedure.

Rear bumper (Courier) – removal and refitting

8 Unscrew the two screws securing each bumper end moulding to its respective wheel arch, then remove the two bolts securing each bumper mounting bracket to the rear underside of the vehicle (photo). Withdraw the bumper.
9 The mounting brackets can be unbolted, if required, from the bumper. The end mouldings can be unclipped for renewal separately (photos).
10 Refitting is the reverse of the removal procedure.

Front door trim panel (1992-on models) – removal and refitting

11 Removal of this panel is as described in Chapter 11, noting that the wiring for the electric front window switches (where fitted) will also have to be disconnected.

Opening rear side windows (three-door Hatchbacks) – removal and refitting

12 With an assistant standing outside ready to take the window, unscrew the single screw securing each hinge, then open the window and unscrew the two screws securing the catch to the body. Withdraw the window.
13 The weatherstrip can be removed and refitted, if required, as described in the relevant Sections of Chapter 11.
14 To refit the window, have your assistant stand outside and offer up the window so that both hinge leaves enter their sockets, then loosely refit the catch securing screws. Refit the hinge securing screws, then adjust the catch position to ensure that the window is clamped evenly on to the weatherstrip over the whole mating surface when the catch is fastened. Tighten all screws securely.

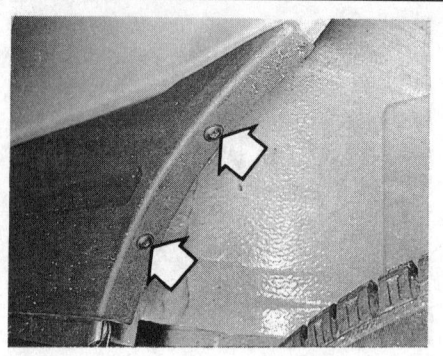

366 Chapter 13 Supplement: Revisions and information on later models

13.6A Courier rear quarter moulding mounting screws (arrowed) in wheel arch ...

13.6B ... and at rear of vehicle

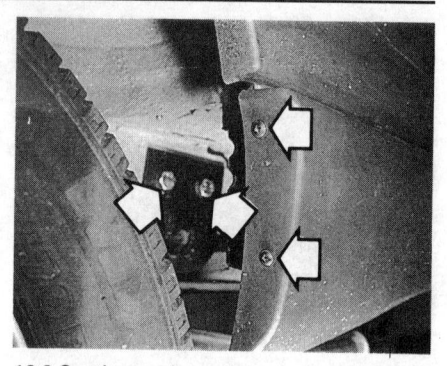

13.8 Courier rear bumper mounting screws (arrowed) at wheel arch – note also bumper mounting bracket bolts (arrowed) visible under rear of vehicle

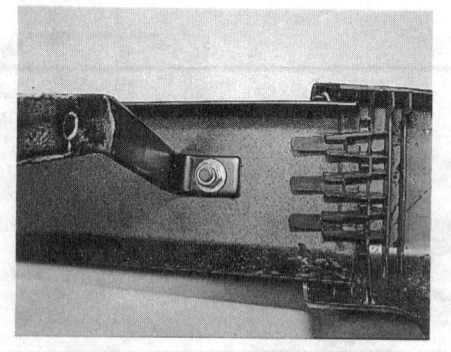

13.9A Courier rear bumper removed, showing mounting bracket retaining nut and end moulding clips ...

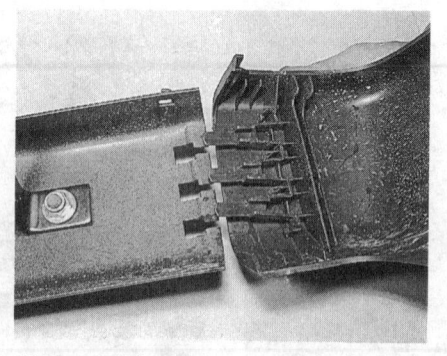

13.9B ... Courier rear bumper end moulding unclipped

Rear sliding windows (Courier Kombi) – removal and refitting

15 To remove the complete sliding window assembly requires special cutting equipment and supplies of suitable adhesive; it is recommended that this task be left to a Ford dealer or similar expert.
16 To remove the window glass, remove their retaining screws and withdraw the lock assemblies, slide the front glass as far to the rear as possible, then prise up and withdraw the front glass guide; it will be necessary to pull the last part of the guide from under the glass.
17 Withdraw the vertical seal and two support blocks from between the two glasses, slide the seal to the front of the frame, and withdraw it. Slide each glass in turn to the front of the frame and withdraw it, then prise up and withdraw the rear glass guide.
18 On refitting, check that the glass guides are undamaged; renew them if necessary. Ensure that the lock assemblies are removed from both glasses.
19 Press the rear glass guide into place. Refit first the rear glass, fitting it to the front of the frame and sliding it to the rear, then the front glass, in the same way.
20 To refit the vertical seal, sandwich it between two strips of thin sheet metal and install it, with its support blocks, first in the front part of the frame, then push it into place. This will require some effort as the glass is reached, due to the distortion caused by the location notch on the block. When the seal is fully in place, withdraw the metal strips, and check that the seal's lips have not folded back on themselves.
21 When refitting the front glass guide, locate it first around the frame's detent catch, then slide it under the glass using soapy water as a lubricant.
22 Refit the lock assemblies, clean the glasses, and check the operation of the windows and locks.

Rear door window glass (Courier) – removal and refitting

23 These are removed and refitted using the technique described for other fixed glasses in Chapter 12, Sections 31, 32 and 33.

Rear door check assembly (Courier) – removal and refitting

24 These are bolted to the door and frame, and must be renewed complete if faulty; mark around each mounting position using a felt tip marker pen or similar before unbolting them.
25 Refitting is the reverse of the removal procedure; tighten the bolts securely.

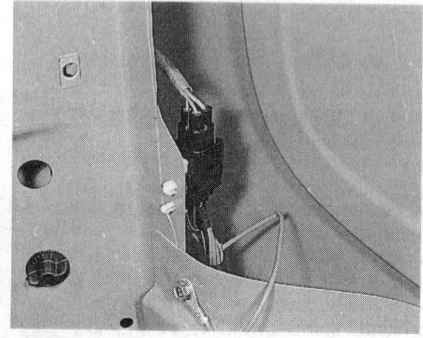

13.26 Courier rear door removal – disconnect wiring at connector inside door frame ...

13.27 ... unscrew bottom hinge's cap securing bolt ...

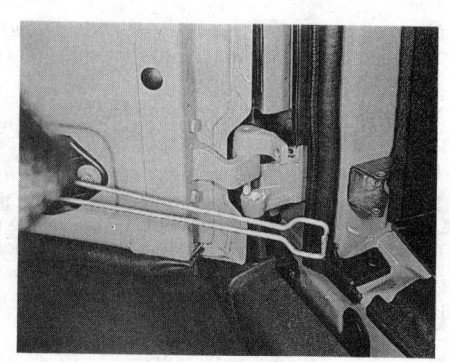

13.28 ... and lift door off hinges

Chapter 13 Supplement: Revisions and information on later models 367

13.30 Unscrew nuts inside door frame to remove door hinges

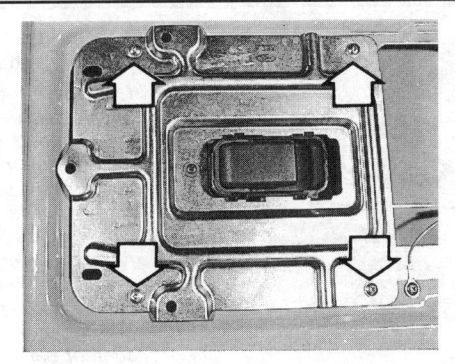

13.32A Courier rear door latch assembly retaining screws (arrowed)

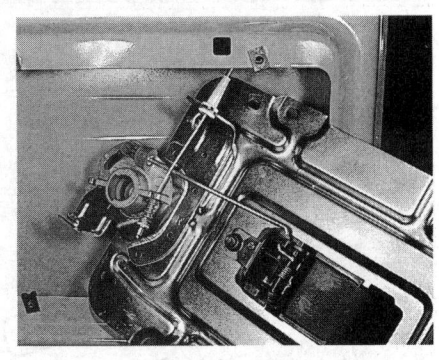

13.32B Latch assembly released – disconnect cables to withdraw

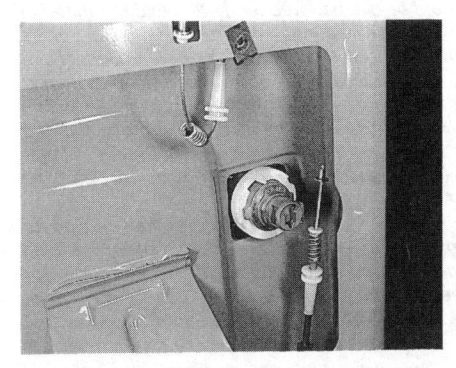

13.37A Release (white) locking ring and withdraw gaskets ...

13.37B ... to release Courier rear door exterior handle

13.42 Courier rear door lower catch removed ...

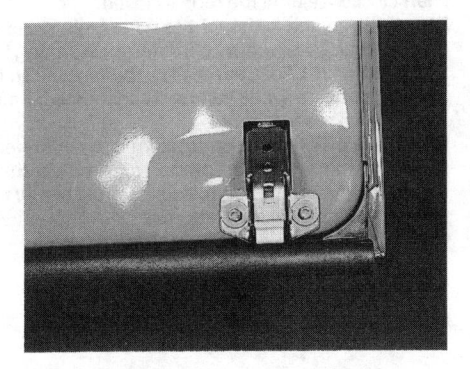

13.43 ... catch position can be adjusted in slots, on refitting

13.62 Removing Courier rear door interior handle bezel

Rear doors (Courier) – removal and refitting

26 Open the door and disconnect the wiring leading to it; all connectors are accessible from the openings inside the door frame (photo).
27 Unclip the door check arm, and unscrew the bottom hinge's cap securing bolt (photo).
28 Lift the door off its hinges and withdraw it (photo).
29 Refitting is the reverse of the removal procedure; apply grease to the hinge pins, and tighten the cap bolt securely.
30 The door hinges can be unbolted from the door and frame; either remove the door first, or support it carefully and remove the hinge with it in place (photo).

Rear door interior handle (Courier) – removal and refitting

31 Remove the interior handle bezel and the trim panel, as described in paragraphs 62 and 63 below.
32 Remove its four retaining screws, and withdraw the complete latch assembly (photos).
33 Disconnect the link rod and remove the single retaining screw to withdraw the interior handle.
34 Refitting is the reverse of the removal procedure.

Rear door exterior handle and lock assembly (Courier) – removal and refitting

35 Remove the interior handle bezel and the trim panel, as described in paragraphs 62 and 63 below.
36 Remove its four retaining screws, and withdraw the complete latch assembly.
37 Using pliers, rotate the exterior handle locking ring anti-clockwise to release it, then withdraw the handle with the inner and outer gaskets (photos).
38 The lock assembly is secured by a circlip to the handle.
39 Refitting is the reverse of the removal procedure.

Rear door upper and lower catches (Courier) – removal and refitting

40 Remove the interior handle bezel and the trim panel, as described in paragraphs 62 and 63 below.
41 Remove its four retaining screws, and withdraw the complete latch assembly. Disconnect the upper and lower catch cables from the latch assembly operating mechanism.

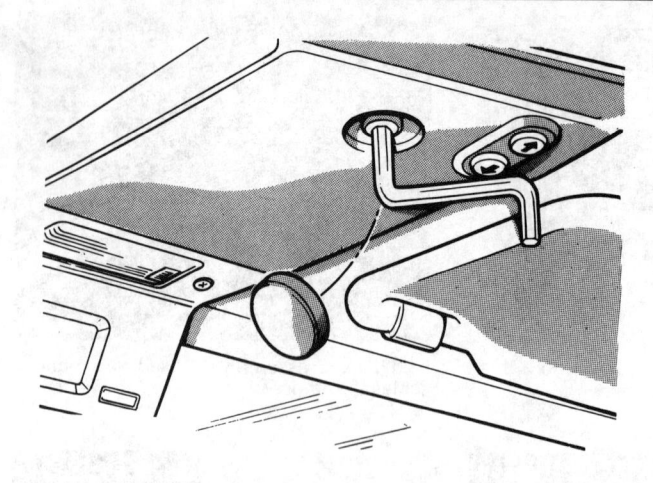

Chapter 13 Supplement: Revisions and information on later models

Fig. 13.50 Using crank provided to manually close full-length sunroof (Sec 13)

42 Unbolt the catch and withdraw it from the door (photo). If the cables are available separately (seek the advice of a Ford dealer) they can be disconnected from each catch by releasing the retaining clip.
43 Refitting is the reverse of the removal procedure; check the adjustment of the catch before finally tightening the bolts (photo).
44 The striker plates can be adjusted on their slotted mounting bolt locations, if required.

Full-length sunroof ('Calypso') – general

45 The full-length sunroof is electrically-operated, the operating system consisting of the motor and the roof-mounted switch. The circuit is fed via fuses 18 and 28 from the ignition switch; switch illumination is via fuse 6.
46 All components are mounted in the roof of the passenger compartment, above the headlining at the front of the vehicle.
47 To open the sunroof, press the switch lightly on the upper side; to close it, check first that the opening is completely unobstructed, and press the switch on the lower side.
48 Maintenance is confined to checking for freedom of action and a snug fit when shut. Check that the seals are in good clean condition and not scratched or damaged.

49 Owners must note the following to ensure the maximum trouble-free life from this feature:
(a) *It is normal for the motor to slow down as the sunroof approaches full opening.*
(b) *If the sunroof stops before it is fully opened in cold weather, this may be due to the material being too hard to fold correctly; do not force the sunroof open if this is suspected.*
(c) *If the material does not fold correctly on opening at any time, close the sunroof again, correct the folds by hand, and try again.*
(d) *Never open or close the sunroof with the vehicle travelling at more than 70 mph (120 km/h), and never allow passengers to travel standing up or with any part of their bodies in the opening.*
(e) *Ensure that any collected water, snow or ice is removed from the sunroof before opening it. Check that the deflector is clear of water, particularly after washing the vehicle; sponge it dry if necessary.*
(f) *Never place heavy objects on the sunroof or its surrounds.*
(g) *The sunroof should be cleaned frequently to avoid the material being stained by dirt. Use a sponge, soft brush or soft cloth and a neutral detergent, rinsing with a gentle flow of clean water from directly above until all traces of dirt and detergent are removed.* **Never** *use a high-pressure jet, pressure washer or similar, and do not aim the jet from a hose at the joints of the sunroof with the body (or water will enter the passenger compartment).* **Never** *use alcohol, petrol, thinners or similar products to clean the material.*
(h) *If the vehicle is parked in heavy rain, or if it is parked outside for long periods, a proprietary car cover or tarpaulin should be used to protect the roof and body. Do not leave the sunroof open for long periods; the material will stiffen in its folds, with a consequent risk of tearing when the sunroof is eventually operated again.*

50 If the system fails with the sunroof open, it can be closed in emergency by switching off the ignition, prising out the access plug in front of the switch, and using the crank provided to rotate the motor shaft clockwise until the roof is closed.
51 If the switch is thought to be at fault, it can be removed after first disconnecting the battery earth terminal (noting the comments made in Sections 5 and 17 of Chapter 12); the switch can then be eased from its housing until the wires can be disconnected. Refitting is the reverse of the removal procedure.
52 If any other failure or problem is encountered, the general inaccessibility of the system's components means that servicing and fault-finding is beyond the capabilities of most owners; the vehicle should be taken to a Ford dealer for attention.

Fig. 13.51 Full-length sunroof – 'Calypso' models (Sec 13)

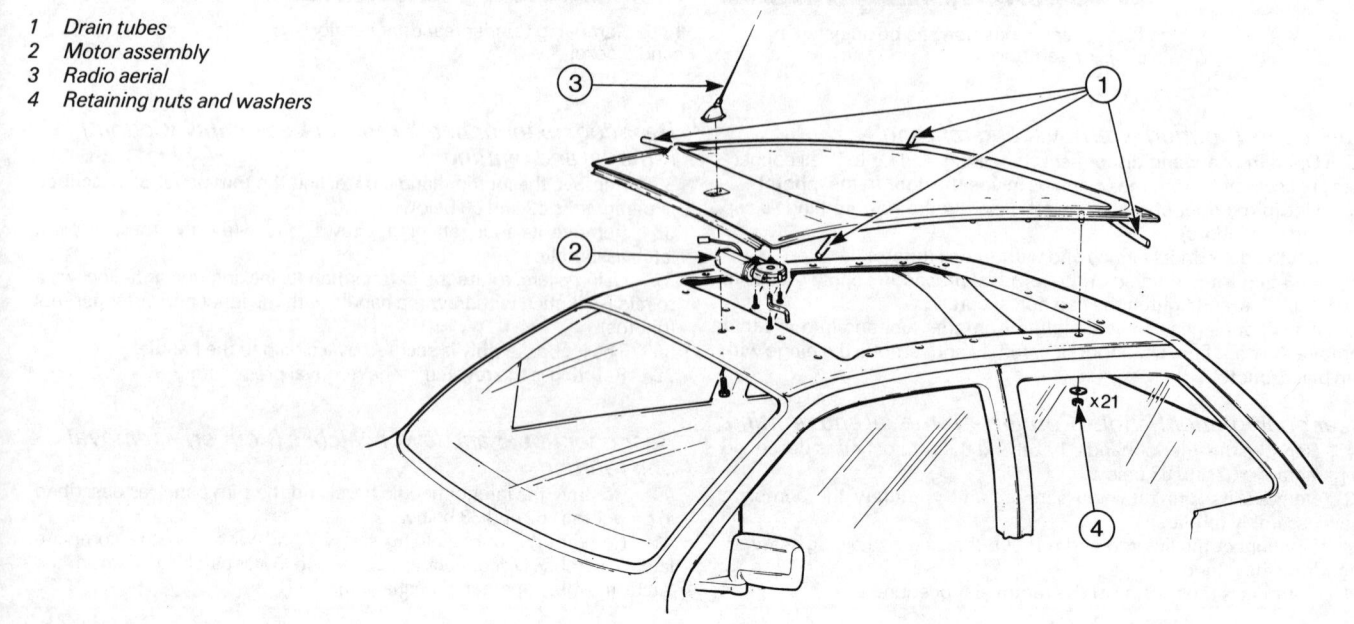

1 Drain tubes
2 Motor assembly
3 Radio aerial
4 Retaining nuts and washers

Chapter 13 Supplement: Revisions and information on later models

Rear seats (Courier Kombi) – removal and refitting

53 The seats are secured by two bolts at each forward hinge, visible once the seat has been folded forwards. Unbolt the hinges and withdraw the seat assembly.
54 Refitting is the reverse of the removal procedure.

Headlining (Courier) – removal and refitting

55 Using either a trim clip releasing tool or a screwdriver with a broad flat blade, and protecting the paintwork and trim with a layer of rag, extract the clips securing the headlining rear section.
56 Remove the two screws and withdraw the map tray/storage box (where fitted).
57 Remove the sun visors, interior (courtesy) lamp and passenger handle.
58 Remove the six screws and release the catch to withdraw the stowage shelf.
59 Remove the A-pillar trim (Chapter 11).
60 Remove the remaining trim clips (three at one side of the front section, one at the opposite side) and withdraw the headlining.
61 Refitting is the reverse of the removal procedure.

Load compartment interior trim panels (Courier) – removal and refitting

62 If removing one of the rear door trim panels, first remove the interior handle bezel by pushing it away from the door's hinges until it can be withdrawn (photo).
63 Using either a trim clip releasing tool or a screwdriver with a broad flat blade, and protecting the paintwork and trim with a layer of rag, extract the clips securing the panel and withdraw it.
64 Refitting is the reverse of the removal procedure.

14 Electrical system

Alternator drivebelt (DOHC 16-valve engine) – check, renewal and adjustment

General

1 The alternator, located in the front right-hand corner of the engine compartment, is driven (with the water pump) from the crankshaft pulley by a flat 'polyvee'-type of drivebelt. With the alternator pivoting on a mounting bracket bolted to the cylinder block, the drivebelt is tensioned by a rack-and-pinion system.
2 Before any drivebelt servicing operations can be undertaken, adequate access must first be gained by jacking up the front right-hand side of the vehicle and supporting it on axle stands. Disconnect the battery earth (negative) lead, noting the comments made in Sections 5 and 17 of Chapter 12. Remove the shield protecting the drivebelt; it is secured by three plastic quarter-turn fasteners screwed vertically into the body side member (photo).

Drivebelt check and renewal

3 Once the drivebelt and adjuster components are fully accessible, rotate the crankshaft as necessary to inspect the full length of the belt. Check for signs of cracks, splitting and fraying, or for signs of wear or damage, such as glazing (shiny patches) or separation of the belt plies. Renew the belt if worn or damaged, and check the condition and security of the pulleys, the mounting bracket and adjuster components, and their fasteners.
4 To renew the belt, slacken the two pivot bolts and nuts at the top of the alternator, then the adjuster link bolt and the adjuster's centre lockbolt (photos).
5 Rotating the adjuster nut anti-clockwise, move the alternator towards the engine until the drivebelt can be slipped off the pulleys (photo).
6 Fit the new drivebelt, ensuring that it is routed correctly and settled in the pulley grooves, then tension the drivebelt as described below.

Setting/adjusting drivebelt tension

Note: *This procedure requires the use of a torque wrench (preferably TWO) and a 22 mm crowfoot adaptor – if such equipment is not available, the work must be carried out by a Ford dealer.*
Note that there is no 'check' as such of the drivebelt's tension – the tension checking and adjusting procedures are one and the same, as described below.

7 Slacken the two pivot bolts and nuts at the top of the alternator, then the adjuster link-to-mounting bracket bolt and nut, and finally the adjuster's centre lockbolt.

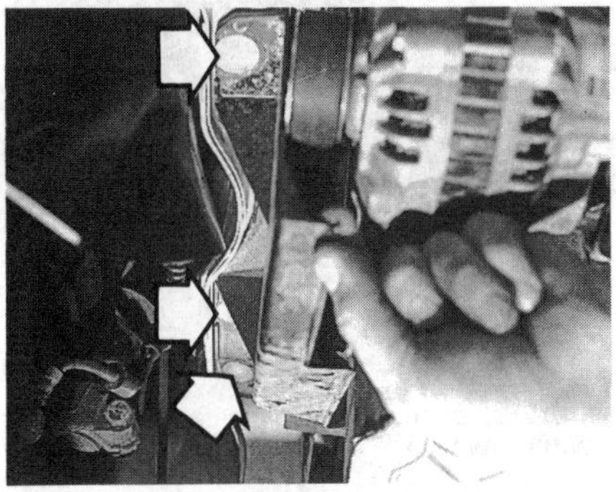

14.2 Alternator/water pump drivebelt shield fasteners (arrowed) – DOHC 16-valve engine

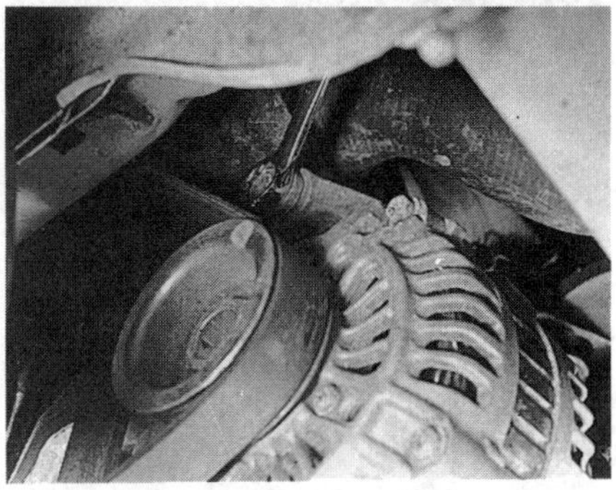

14.4A Unscrewing alternator right-hand ...

14.4B ... and left-hand pivot bolts and nuts ...

14.5 ... to release drivebelt from pulleys

14.9 Tightening drivebelt adjuster's centre lockbolt after setting alternator/water pump drivebelt tension – DOHC 16-valve engine

14.12 Showing additional starter motor support bracket and fasteners (arrowed)

370 Chapter 13 Supplement: Revisions and information on later models

14.15A Dismantling Nippondenso reduction-gear starter motor – disconnect brush link lead from solenoid ...

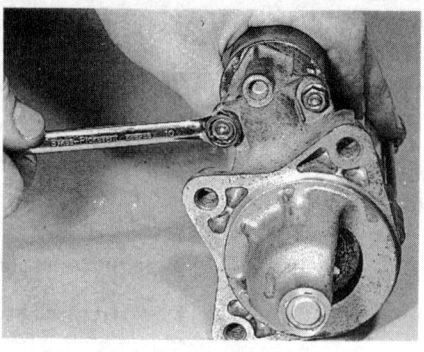

14.15B ... unscrew retaining nuts ...

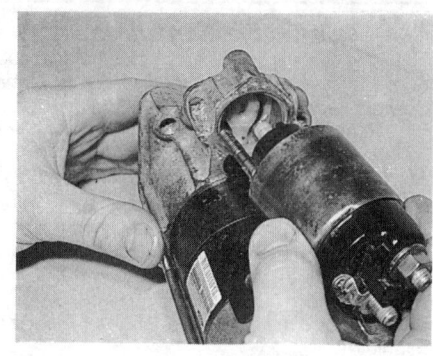

14.15C ... and withdraw solenoid

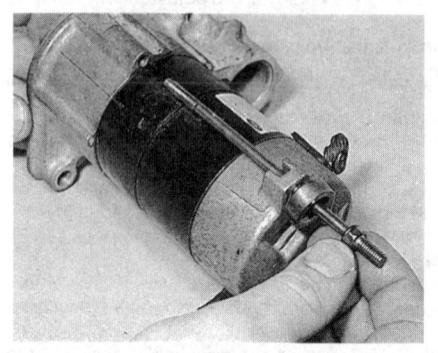

14.16A Unscrew two studs, noting spacers ...

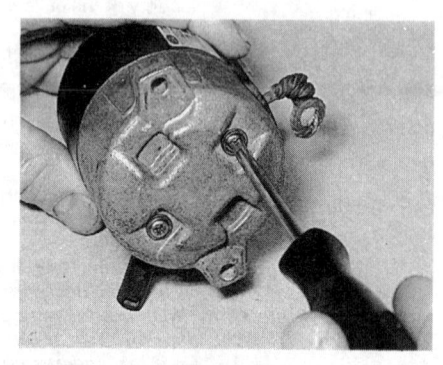

14.16B ... and two screws ...

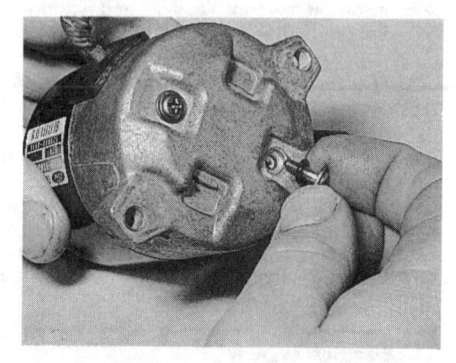

14.16C ... noting sealing O-ring under each ...

14.16D ... remove commutator end cover, noting sealing O-ring

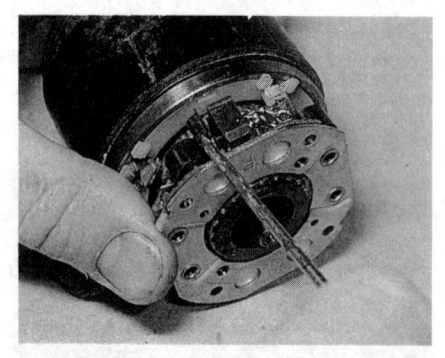

14.17A Retract springs and wedge them clear of brushes ...

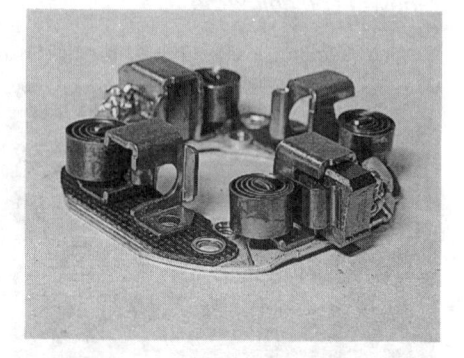

14.17B ... so that brush plate can be withdrawn

8 Using a torque wrench with a crowfoot adaptor, apply the specified torque setting to the adjuster nut; this will cause the adjuster nut to rotate clockwise. As the nut's pinion teeth are engaged with the adjuster link's rack teeth, the alternator will automatically be moved to the appropriate position for correct drivebelt tension.
9 While maintaining the pressure on the nut, tighten the adjuster's centre lockbolt securely (photo) – preferably using a second torque wrench to tighten the bolt to the specified setting.
10 Tighten to their specified torque wrench settings first the adjuster link-to-mounting bracket bolt, then the alternator pivot bolts and nuts (drivebelt end first).
11 Refit all other components removed, and reconnect the battery.

Starter motor – removal and refitting

12 With reference to the procedure in Chapter 12, on some models, the starter motor is supported by an additional bracket, which is secured by two nuts to its commutator end, and bolted to the cylinder block (photo).
13 On refitting the starter motor (support bracket fasteners loose at first), tighten first the main three mounting bolts to their specified torque wrench setting. Align the bracket on the starter motor and cylinder block, and tighten its fasteners so that there is no strain on the motor.

Nippondenso reduction-gear starter motor – overhaul

Note: *Always check first exactly what replacement component parts are available and their cost, before deciding whether to overhaul the existing unit, or to replace it with a new or reconditioned one.*

14 With the starter motor removed from the vehicle (see note above) and cleaned as described in Chapter 12, undo the two nuts and withdraw the support bracket (where fitted).
15 Unscrew the retaining nut, and disconnect the brush link lead from the solenoid's terminal stud. Undo the two nuts and withdraw the solenoid, disengaging it from the actuating arm (photos).
16 Unscrew the two studs and the two screws (noting the sealing O-ring under each of the screws) and remove the commutator end cover, noting the sealing O-ring (photos).
17 Use a small screwdriver to retract the springs, wedging them against the brushes so that the brushes are held clear of the commutator; remove the two field brushes. Withdraw the brush plate (photos).

Chapter 13 Supplement: Revisions and information on later models

14.18 Removing armature from yoke

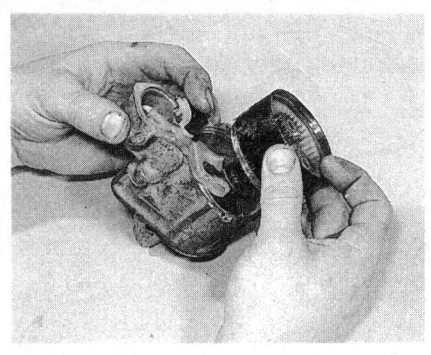

14.19 Withdrawing reduction gear housing/actuating arm assembly from drive end housing

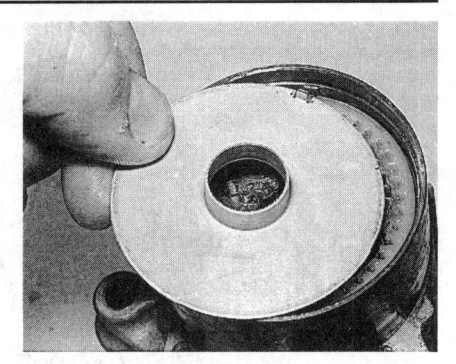

14.20A Note tab locating plate in reduction gear housing

14.20B Reduction planet gears are punch-marked on drive end face (ie on underside of gears, in this case)

14.21A Thrust collar must be driven down off C-clip to release drive pinion assembly from planet gear shaft

14.21B Extract circlip (arrowed) to release planet gear shaft from reduction gear housing

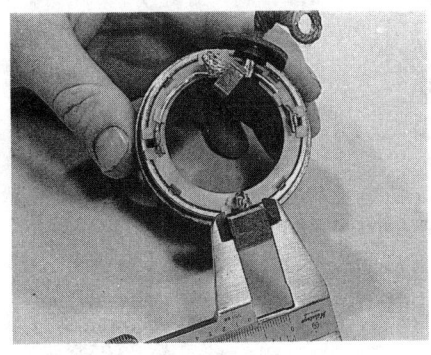

14.22 Measuring brush length – renew set if any are worn to specified minimum or less

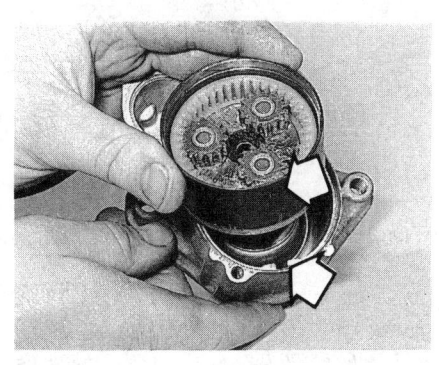

14.25 Protrusions (arrowed) on reduction gear housing engage with slot in drive end housing ...

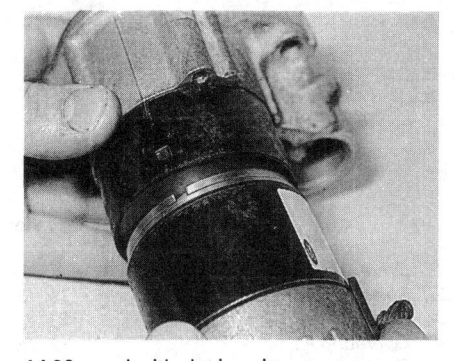

14.26 ... and with slot in yoke

18 Mark the relationship of the yoke to the reduction gear housing, then slide off the yoke and armature, noting the sealing O-ring between the yoke and drive housing. When extracting the armature from the yoke against the pull of the magnets, take care not to allow either to be damaged (photo).

19 Mark the relationship of the reduction gear housing to the drive end housing, then withdraw the reduction gear housing and the actuating arm as an assembly from the drive end housing (photo).

20 Remove the plate from the reduction gear housing, noting how it is located; if the planet gears are removed, note the punch mark on each identifying its drive end face (photos).

21 Proceed as described in Chapter 12, Section 13, paragraph 9 to remove the drive pinion assembly from the planet gear shaft. Remove the circlip, noting the thrustwasher behind it, to release the planet gear shaft from the reduction gear housing (photos).

22 Clean and check all components, renewing as necessary any that are worn or damaged; refer to Chapter 12, Section 13, paragraphs 11 to 14 (photo). Check that the one-way clutch only allows the pinion to rotate in one direction.

23 Refitting is the reverse of the dismantling procedure, but lubricate all moving parts as they are assembled, using lithium-based grease.

24 Refit the planet gear shaft to the reduction gear housing, lubricating all bearing surfaces, and securing it with the thrustwasher and circlip. Refit the drive pinion assembly, and secure it as described in Chapter 12, Section 13, paragraph 15. Check that the planet gears and plate are correctly located in the reduction gear housing.

25 Fit the rubber block to the actuating arm, match the arm to the drive pinion assembly, and locate the whole unit in the drive end housing. Note that the two will fit together only one way, as a protrusion on the reduction gear housing locates in a slot in the drive end housing (photo).

26 When refitting the yoke and armature, note that these will also only fit one way, as a slot in the yoke engages a protrusion on the reduction gear housing (photo). Do not omit the sealing O-ring.

27 Fit the brush plate to the armature, check that all four brushes slide freely in their holders and bear correctly against the commutator, then retain each brush with its spring.

28 Refit the commutator end cover, tightening the screws and studs securely; do not omit the sealing O-rings.

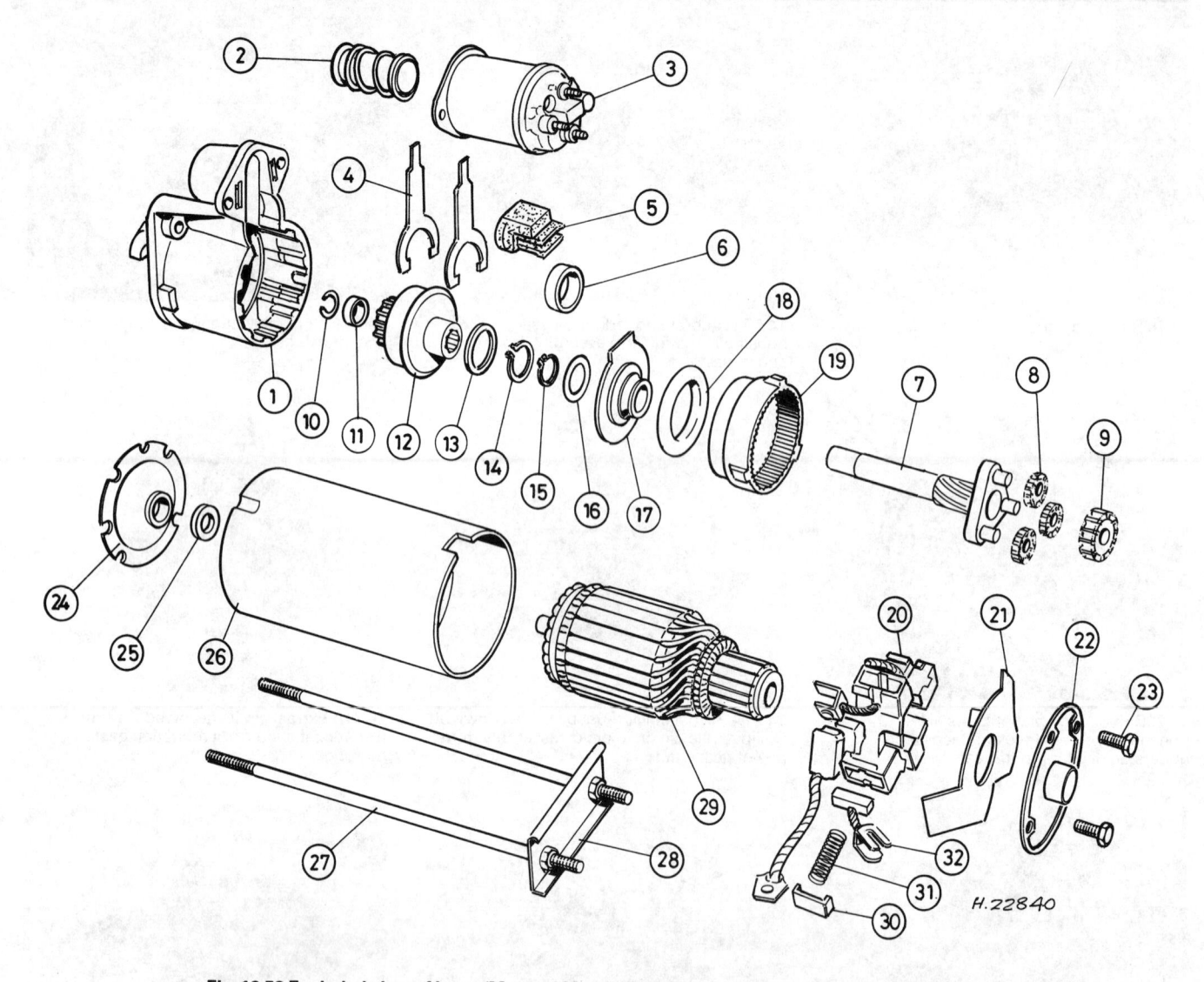

372　Chapter 13 Supplement: Revisions and information on later models

Fig. 13.52 Exploded view of Lucas/Magneti Marelli (M80R) reduction-gear starter motor (Sec 14)

1 Drive end housing	9 Sun gear	17 Bearing support plate	25 Spacer
2 Solenoid plunger with spring	10 Spring clip	18 Seal	26 Field winding yoke
3 Solenoid yoke	11 Thrust collar	19 Plastic support cup	27 Through-studs
4 Operating arm	12 Pinion and clutch assembly	20 Brushplate	28 Support bracket
5 Rubber plug	13 Washer	21 Insulator	29 Armature
6 Carrier	14 Circlip	22 Commutator end cover	30 Brush clip
7 Planet gear shaft	15 Circlip	23 Screws	31 Brush spring
8 Planet gears	16 Washer	24 Armature retaining plate	32 Brush

29   Refit the solenoid, tightening its nuts securely, and connecting the brush link lead to the solenoid's terminal stud before securing it with the nut. Finally, refit the support bracket (where fitted), tightening its retaining nuts only lightly at first.

Lucas/Magneti Marelli M80R starter motor – overhaul

Note: *Always check first exactly what replacement component parts are available, and their cost, before deciding whether to overhaul the existing unit or to replace it with a new or reconditioned one.*

30   With the starter motor removed from the vehicle (see note above) and cleaned as described in Chapter 12, undo the two nuts and withdraw the support brackets (as applicable), then undo the two screws and remove the commutator end cover and the plastic insulator.
31   Unscrew the retaining nut, and disconnect the brush link lead from the solenoid's terminal stud.
32   Withdraw the brushplate assembly (taking care to release the spring pressure from each brush before disturbing the assembly, so as not to damage the brushes), then release the brushes from their holders in the brushplate.
33   Unscrew the mounting screws and withdraw the solenoid yoke, then unhook and remove the solenoid plunger and spring.
34   Carefully withdraw the complete field winding yoke and armature assembly from the drive end housing, then remove the armature retaining plate (Fig. 13.53) and withdraw the spacer. When extracting the armature from the yoke against the pull of the magnets, take care not to allow either to be damaged.
35   Withdraw the complete planet gear shaft assembly, with the operating arm and rubber plug and the pinion and clutch assembly, from the drive end housing.
36   Proceed as described in Chapter 12, Section 13, paragraph 9, to remove the pinion and clutch assembly from the planet gear shaft. Remove the circlip and washer (Fig. 13.54) to separate the operating arm and carrier from the pinion and clutch assembly. Withdraw the operating arm members from the carrier, and remove the rubber plug.
37   Remove the circlip and washer (Fig. 13.55) from the planet gear

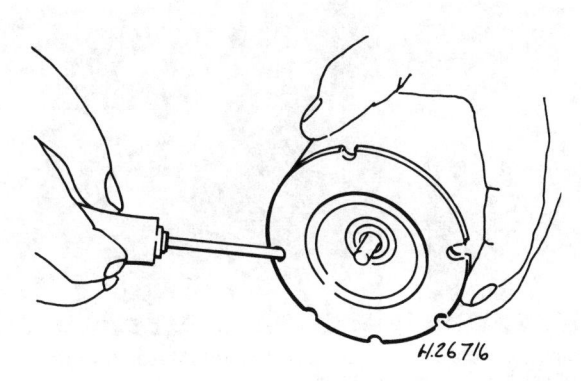

Chapter 13 Supplement: Revisions and information on later models

Fig. 13.53 Removing armature retaining plate – Lucas/Magneti Marelli (M80R) reduction-gear starter motor (Sec 14)

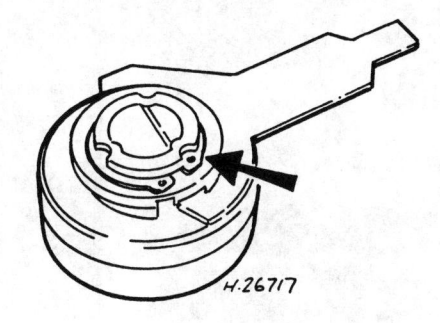

Fig. 13.54 Remove circlip (arrowed) to release washer, operating arm and carrier from pinion and clutch assembly – Lucas/Magneti Marelli (M80R) reduction-gear starter motor (Sec 14)

shaft, and withdraw the bearing support plate, the seal and the plastic support cup.
38 Clean and check all components; refer to paragraph 22 above.
39 Refitting is the reverse of the dismantling procedure, but lubricate all moving parts as they are assembled, using lithium-based grease.
40 On reassembly, ensure that the seal is correctly located in the support cup.
41 Fit both operating arm members to the carrier, then refit the carrier to the pinion and clutch assembly, securing it with the washer and circlip. Refit the pinion and clutch assembly, securing it as described in Chapter 12, Section 13, paragraph 15.
42 Refit the rubber plug to the operating arm. Refit the complete planet gear shaft assembly, with the operating arm and rubber plug and the pinion and clutch assembly, to the drive end housing. Ensure that the rubber plug is correctly located, and that the reduction gears are properly lubricated.
43 Insert the armature into the field winding yoke, ensuring that the through-studs are correctly located, then refit the spacer and retaining plate. Refit the assembly to the drive end housing, ensuring that it is correctly aligned.
44 Apply a smear of lithium-based grease to the solenoid plunger 'hook', engage it on the operating arm, then refit the solenoid yoke; do not forget the spring.
45 Fit the new brushes using a reversal of the dismantling procedure. Make sure that the brushes move freely in their holders.
46 Refit the plastic insulator, the commutator end cover and the support bracket, securing them with the screws and nuts. Connect the brush link lead to the solenoid's terminal stud, and secure it with the nut. Finally, refit the support bracket (where fitted), tightening its retaining nuts only lightly at first.

Fuses and relays – revisions

47 On 1992 and later models, the location and function of some fuses and relays has been changed from that given in Chapter 12. Refer to the Specifications Section of this Chapter, and to Fig. 13.57, for details.

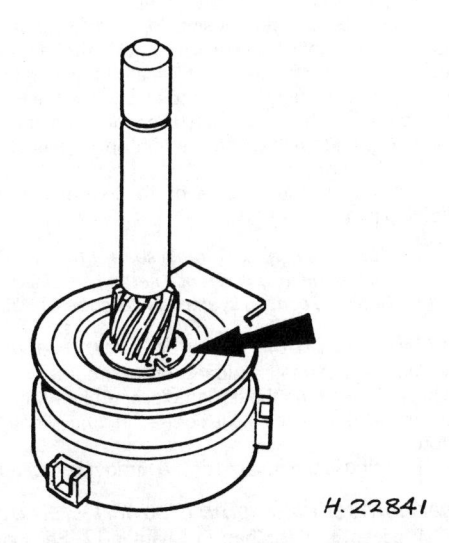

Fig. 13.55 Remove circlip (arrowed) to release washer, bearing support plate, seal and plastic support cup from planet gear shaft – Lucas/Magneti Marelli (M80R) reduction-gear starter motor (Sec 14)

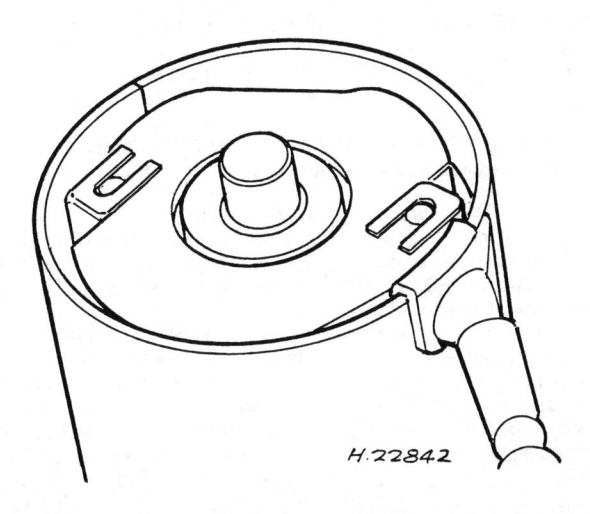

Fig. 13.56 Correct location of insulator and brush connections – Lucas/Magneti Marelli (M80R) reduction-gear starter motor (Sec 14)

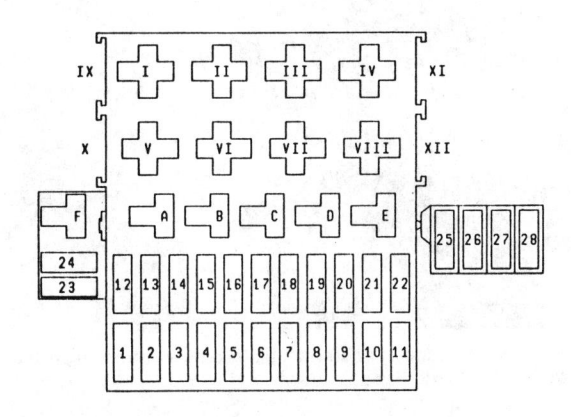

Fig. 13.57 Fusebox layout – 1992-on models (see Specifications) (Sec 14)

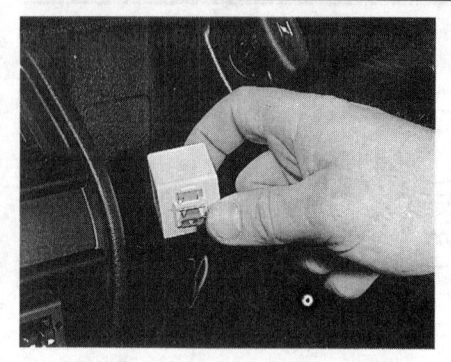

374 Chapter 13 Supplement: Revisions and information on later models

14.48 'Lights-on' warning module – 1993-on models – shown removed

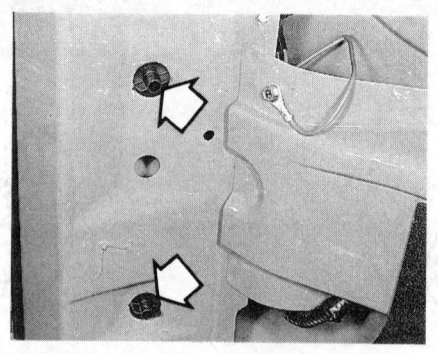

14.57A Courier tail lamp assembly – unscrew black plastic nuts (arrowed) from inside luggage compartment ...

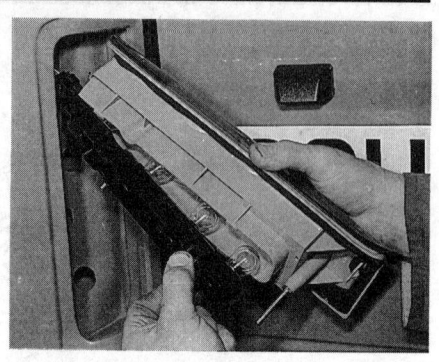

14.57B ... to withdraw assembly from outside of vehicle

'Lights-on' warning module (1993-on models) – removal and refitting

48 The module, which resembles an ordinary relay, is located behind the facia, between the steering column and radio/cassette player. It can be removed by reaching up behind the facia and unclipping it from its mounting (photo).

Electric front window switch (1992-on models) – removal and refitting

49 These switches are removed by prising them out of the door stowage pocket upper surface, using the technique described for centre console switches in Chapter 12, Section 15.
50 If the switch illumination fails, the switch must be renewed; however, there is nothing to be lost by attempting to open the switch and to find a replacement bulb at an auto-electrical specialist.

Full-length sunroof switch ('Calypso') – removal and refitting

51 The switch is removed and refitted using the technique described for centre console switches in Chapter 12, Section 15.

Temperature gauge sender unit (DOHC 16-valve engine) – removal and refitting

52 The sender unit is screwed into the front of the thermostat housing (see photo 5D.94). To remove and refit the unit, proceed as follows.

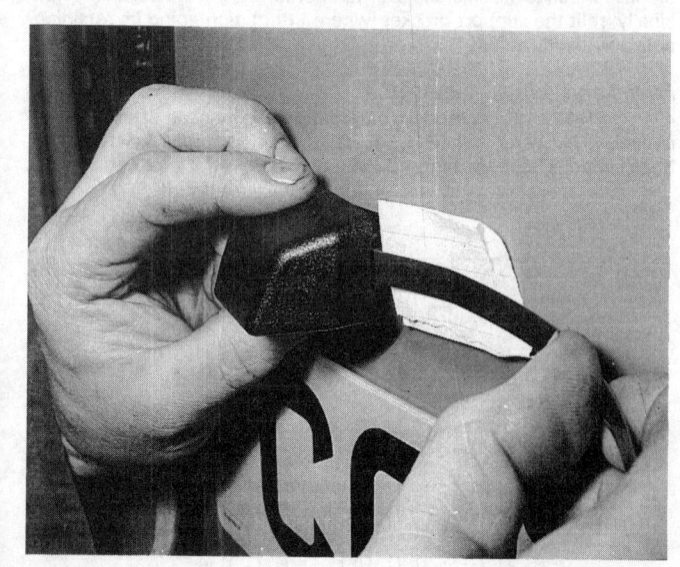

53 Noting the comments made in Section 5 (and 17) of Chapter 12, disconnect the battery earth terminal.
54 Unclip the air filter housing lid, then disconnect the flexible rubber hose from the (black plastic) air intake trunking. Unclip the air mass meter, then move the assembly clear of the working area.
55 Drain the cooling system to just below the level of the thermostat as described in Chapter 2, then disconnect the unit's wiring by releasing its wire clip and unplugging the multi-pin connector. Unscrew the unit and withdraw it.
56 Refitting is the reverse of the removal procedure, noting the following points:

(a) Apply a smear of suitable sealant to its threads, and tighten the unit to the torque wrench setting specified.
(b) Refill the cooling system as described in Chapter 2.

Tail lamp assembly (Courier) – removal and refitting

57 Working inside the load compartment, unscrew the two black plastic nuts securing the tail lamp assembly. Withdraw the assembly to the outside of the vehicle, disconnecting the wiring from the bulbholder (photos).
58 Refitting is the reverse of the removal procedure.

Rear number plate lamp (Courier) – removal and refitting

59 Proceed as described in Chapter 12, Section 27, paragraphs 16 to 18 (photo).

Tail lamp (Courier) – bulb renewal

60 With the tail lamp assembly removed (see paragraph 57 above), bulb renewal is as described in Chapter 12, Section 30.

Rear number plate lamp (Courier) – bulb renewal

61 The procedure is as described in Chapter 12, Section 30.

Load compartment lamp (Van and Courier) – bulb renewal

62 Proceed as described in Chapter 12, Section 31, paragraphs 18 to 20.

Heated rear door window glass (Courier) – general

63 All procedures are as described in Chapter 12, Sections 47 and 48.

Anti-theft immobilisation system – general

64 Models from August 1993 onwards may be fitted with an anti-theft immobilisation system, operated by a remote-control unit, and incorporating an LED tell-tale light mounted at the base of the windscreen pillar on the passenger side.
65 The system is capable of isolating the engine management system, the fuel system and the starting system – it follows that, when attempting to diagnose any starting problems, the immobilisation system should not be overlooked. Few details were available at the time of writing – if a fault is suspected with the system (which cannot be cured by checking for poor connections), refer to a Ford dealer.

14.59 Removing Courier rear number plate lamp – note card protecting paintwork

NOTES:

1. All diagrams are divided into numbered circuits depending on function e.g. Diagram 2: Exterior lighting.
2. Items are arranged in relation to a plan view of the vehicle.
3. Wires may interconnect between diagrams and are located by using a grid reference e.g. 2/A1 denotes a position on diagram 2 grid location A1.
4. Complex items appear on the diagrams as blocks and are expanded on the internal connections page.
5. Brackets show how the circuit may be connected in more than one way.
6. Not all items are fitted to all models.

INTERNAL CONNECTION DETAILS

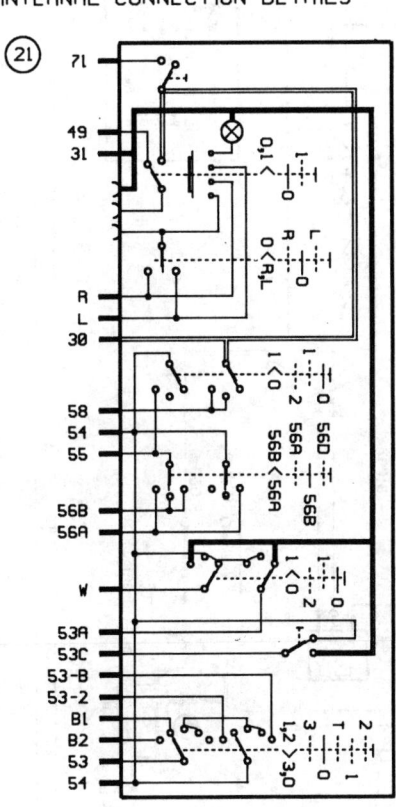

FUSEBOX

FUSE	RATING	CIRCUIT
1	3A	Electric Engine Control System
2	15A	Interior Lamp, Cigar Lighter, Clock And Radio Memory
3	20A	Central Locking
4	30A	Heated Rear Window
5	10A	Dim Dip Lighting
6	10A	LH Side Lamp And Rear Foglamp
7	10A	RH Side Lamp
8	10A	LH Dipped Beam
9	10A	RH Dipped Beam
10	15A	LH Main Beam And RH Spot Lamp
11	15A	RH Main Beam And LH Spot Lamp
12	20A	Heater Blower And Reversing Lamp
13	30A	Radiator Cooling Fan
14	15A	Front Foglamps (XR2i Only)
15	15A	Horn
16	20A	Wash/wipe
17	10A	Brake Lights And Instruments
18	30A	Electric Windows
19	20A	Electric Fuel Pump
20	10A	HEGO Sensor
21	10A	LH Direction Indicators
22	10A	RH Direction Indicators
23		Free
24		Free
25		Free
26	15A	Tailgate Release
27	30A	Heated Windscreen
28	30A	Heated Windscreen

FUSEBOX (CHANGES FROM 1992)

FUSE	RATING	CIRCUIT
25	15A	Tailgate Release
26	30A	Heated Windscreen
27	30A	Heated Windscreen
28	20A	Power Sunroof

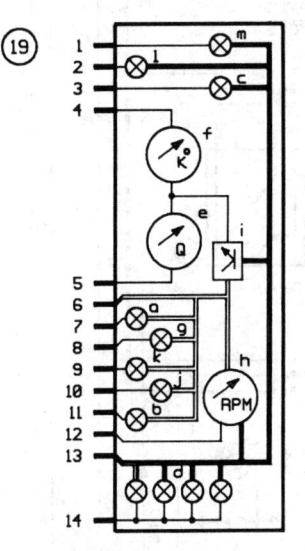

a = No Charge Warning Lamp
b = Handbrake Warning Lamp
c = Main Beam Warning Lamp
d = Instrument Illumination
e = Fuel Gauge
f = Temperature Gauge
g = Oil Pressure Lamp
h = Tachometer
i = Voltage Stabilizer
j = ABS Warning Lamp
k = Choke Warning Lamp
l = Dir. Ind. Warning Lamp LH
m = Dir. Ind. Warning Lamp RH

KEY TO INSTRUMENT CLUSTER

KEY TO SYMBOLS

PLUG-IN CONNECTOR	→
EARTH	G1004
BULB	⊗
DIODE	▷⊢
SOLDERED JOINT	S1003
FUSE/FUSIBLE LINK	—⊏▭⊐—

Notes, internal connection details, and key to symbols

H24500
T.M.MARKE

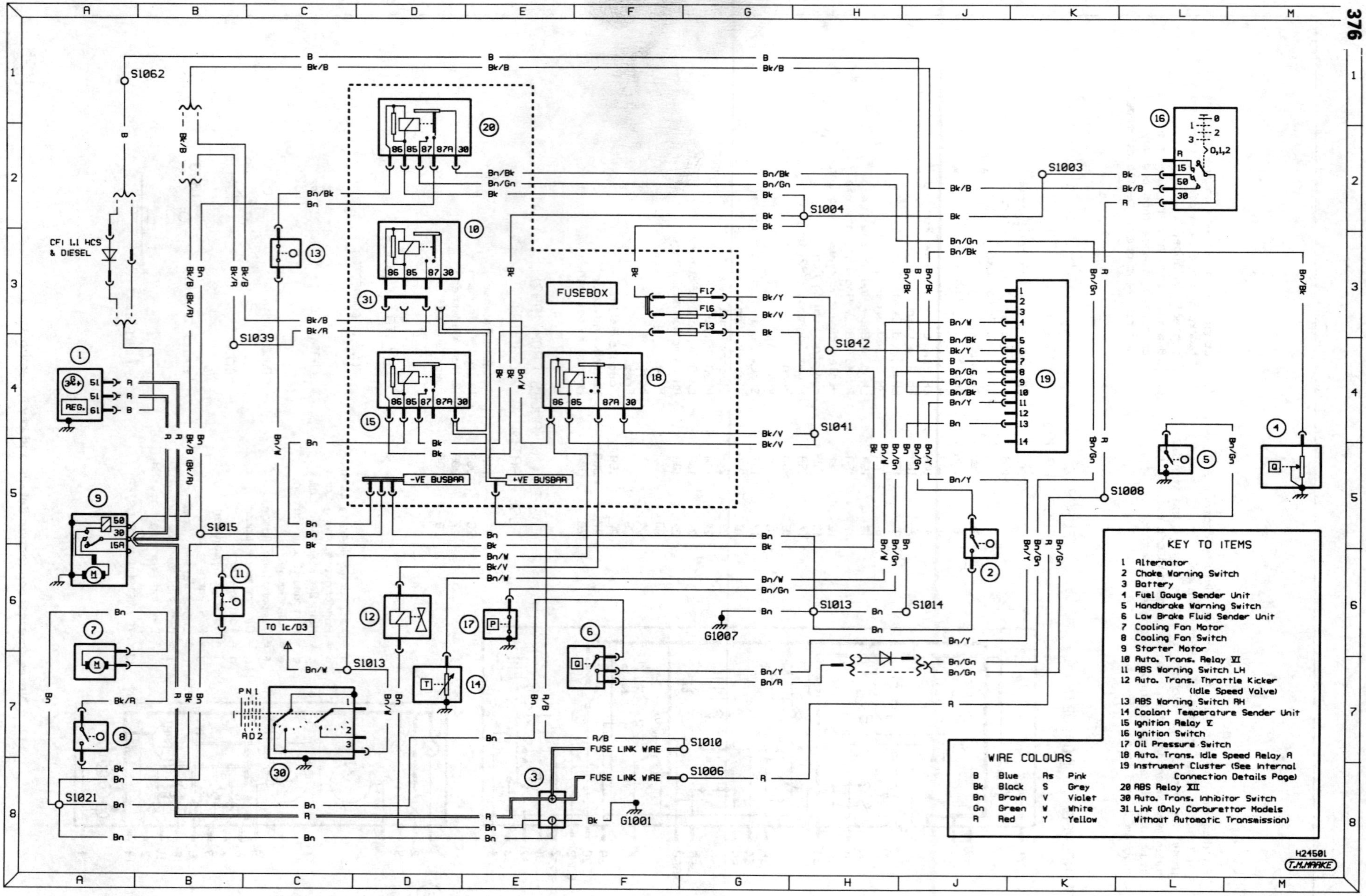

Diagram 1: Starting, charging, cooling fan, ABS, warning lamps and gauges

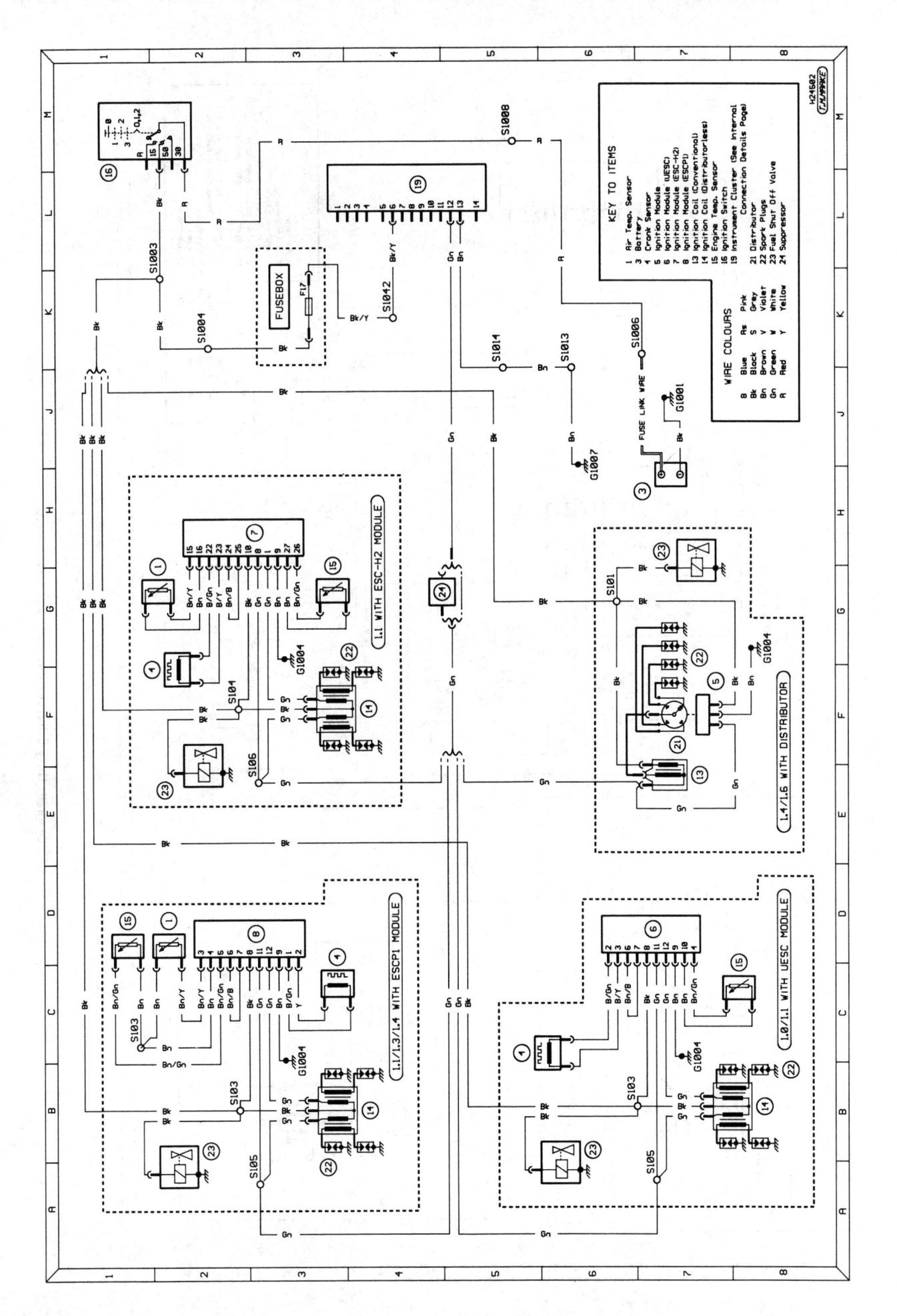

Diagram 1a: Ignition system – carburettor models

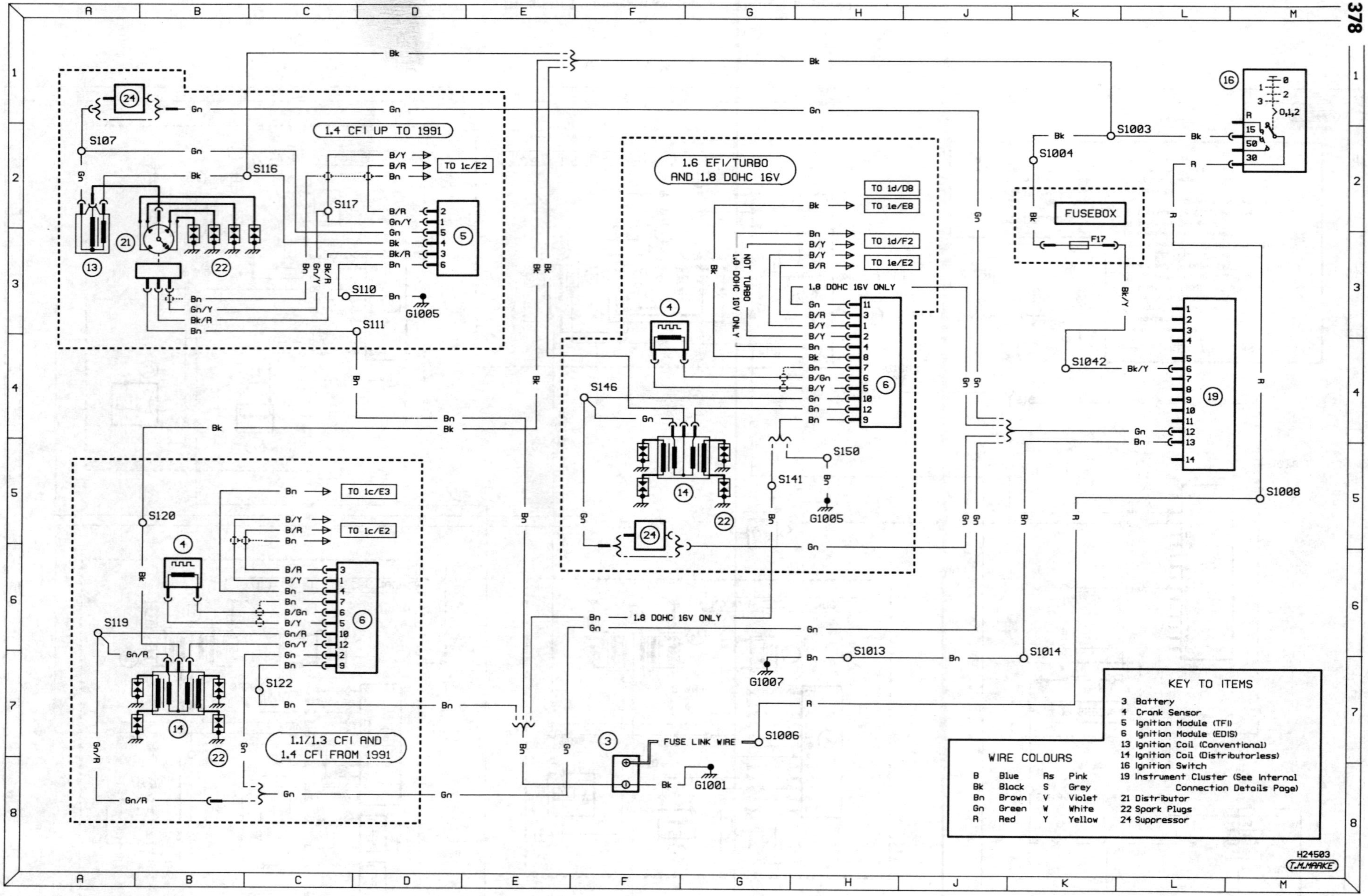

Diagram 1b: Ignition system – fuel-injected models

Diagram 1c: Typical 1.1/1.3/1.4 CFi fuel injection

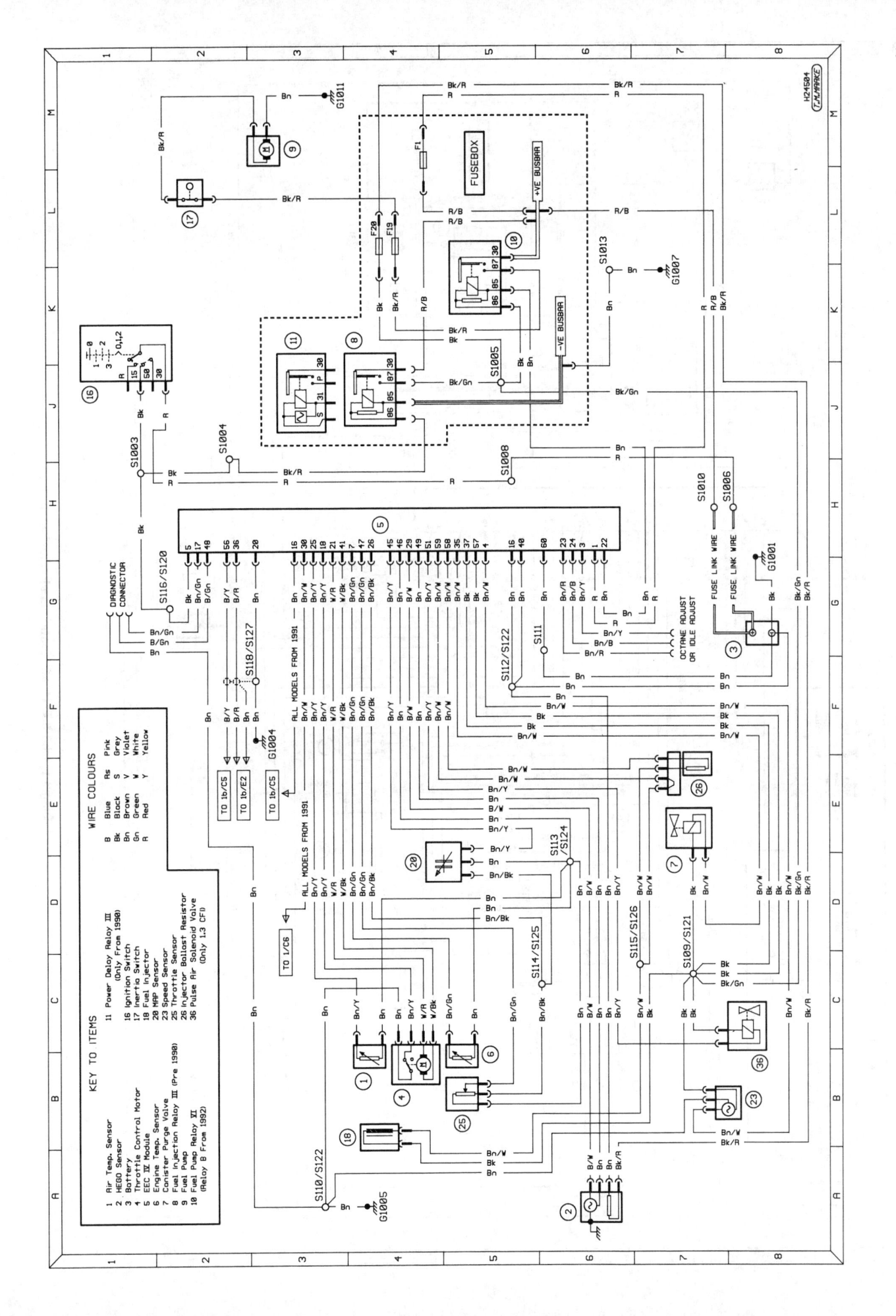

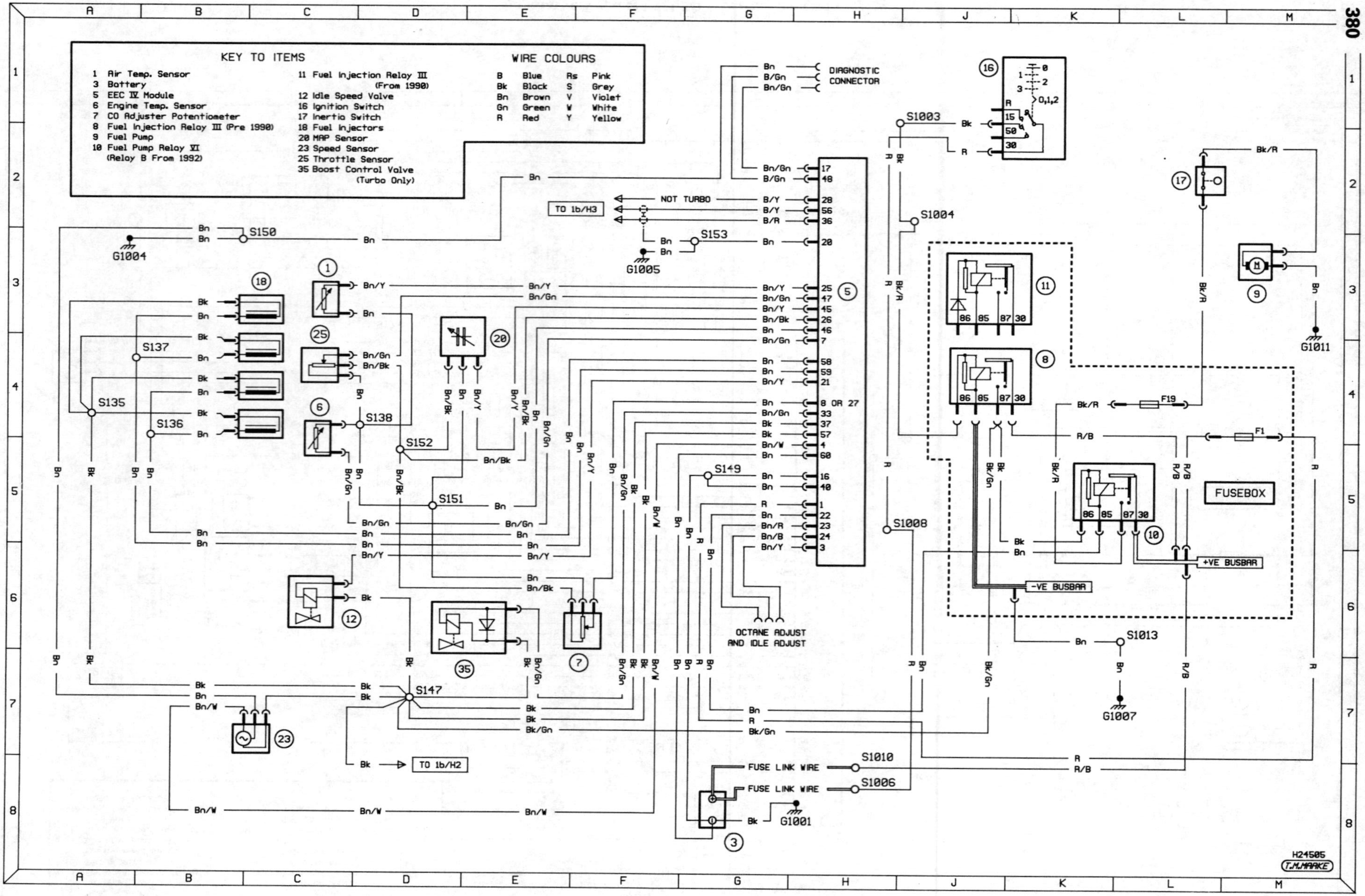

Diagram 1d: 1.6 EFi fuel injection (including RS Turbo)

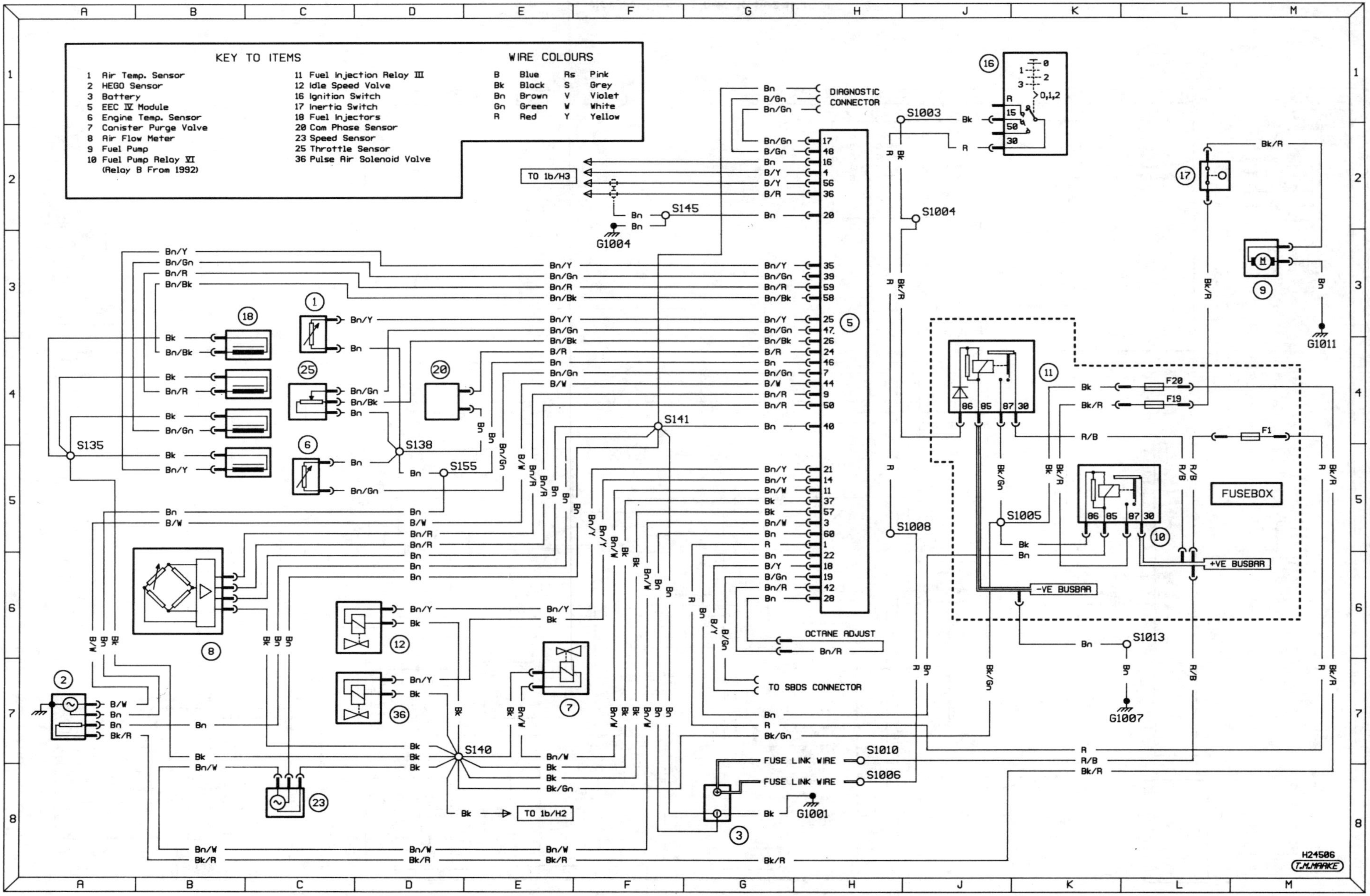

Diagram 1e: DOHC 16-valve fuel injection

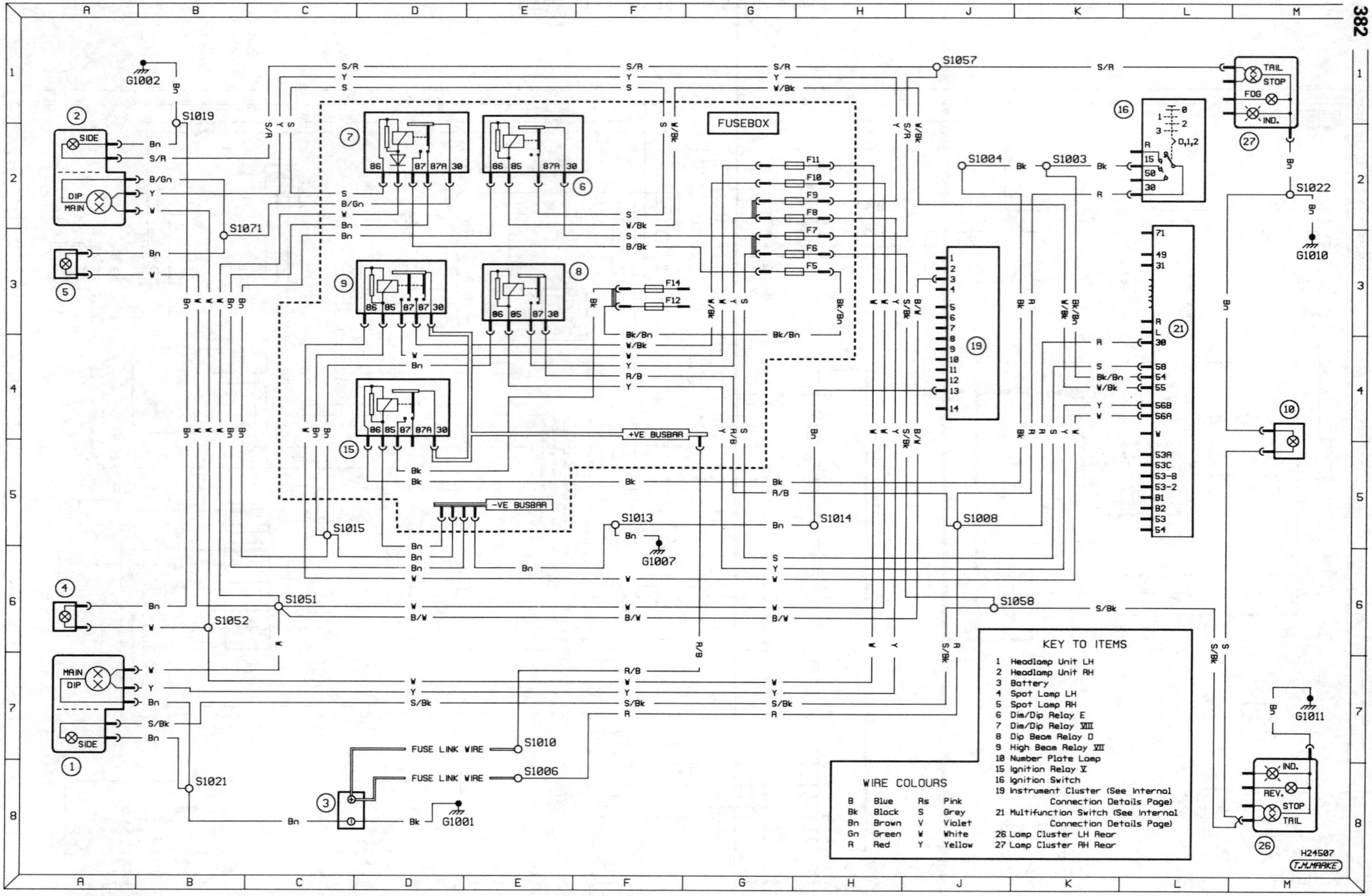

Diagram 2: Exterior lighting – side and headlamps

Diagram 2a: Exterior lighting – stop, reversing, fog and direction indicator lamps

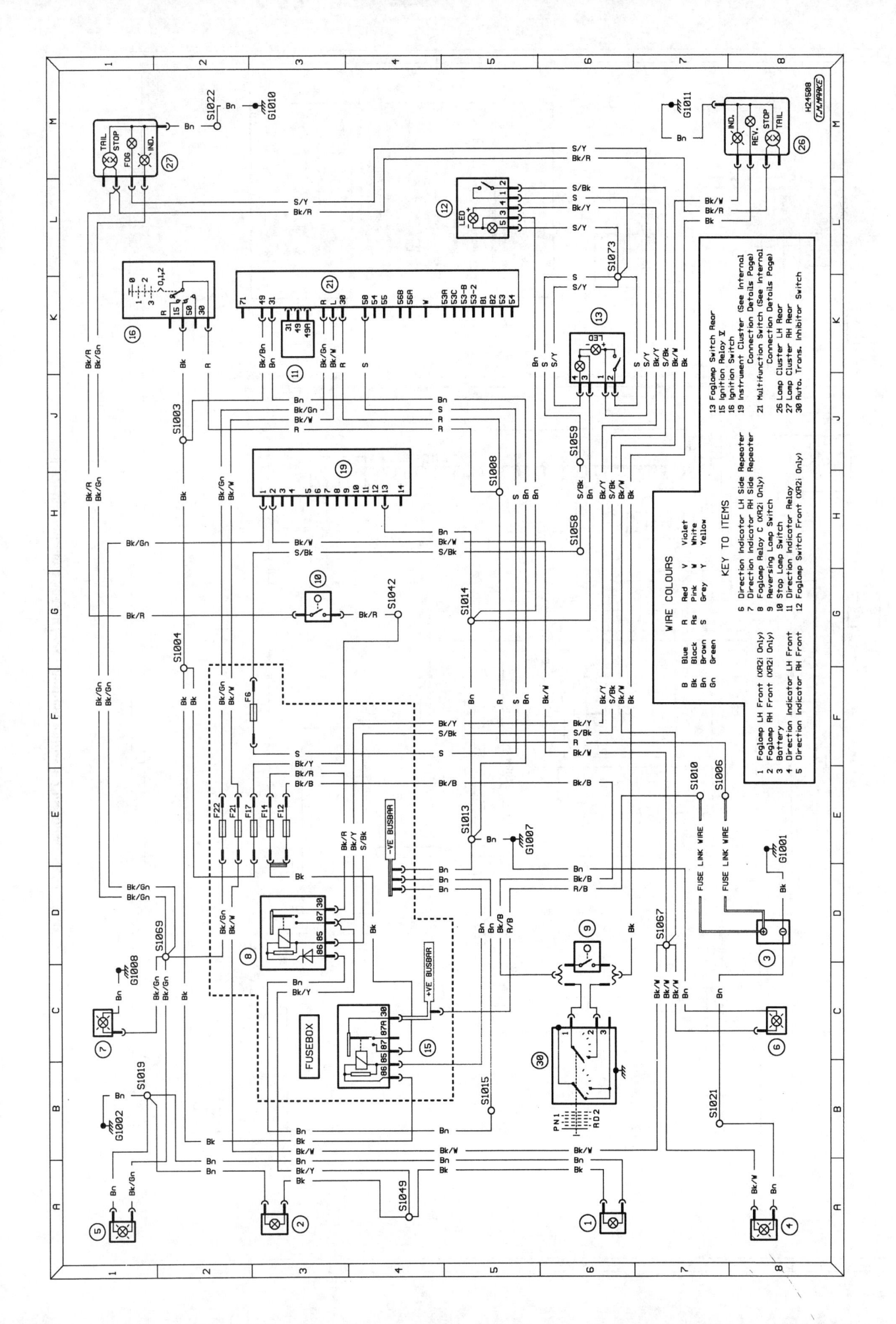

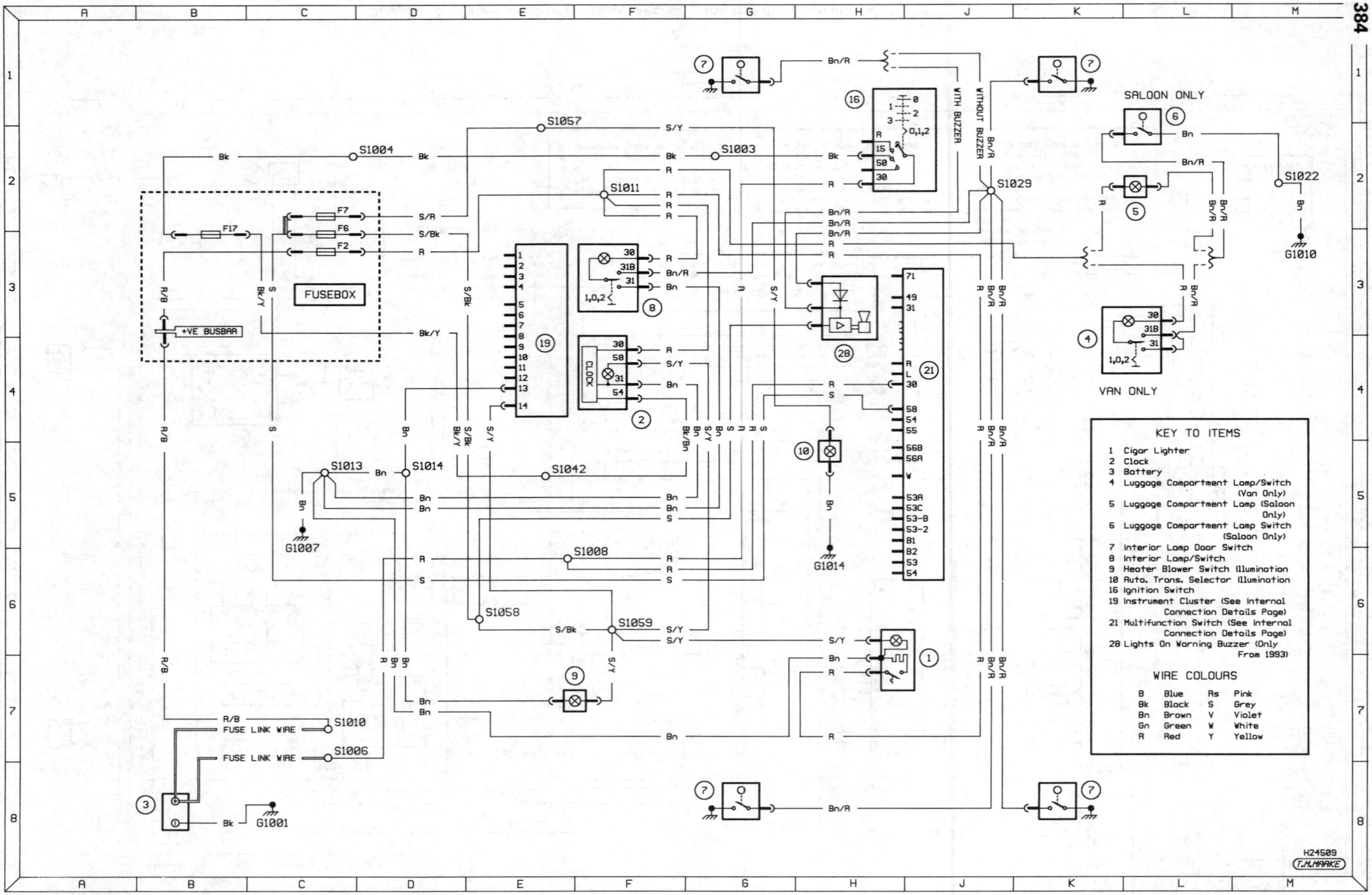

Diagram 2b: Interior lighting, lights-on warning buzzer, clock and cigar lighter

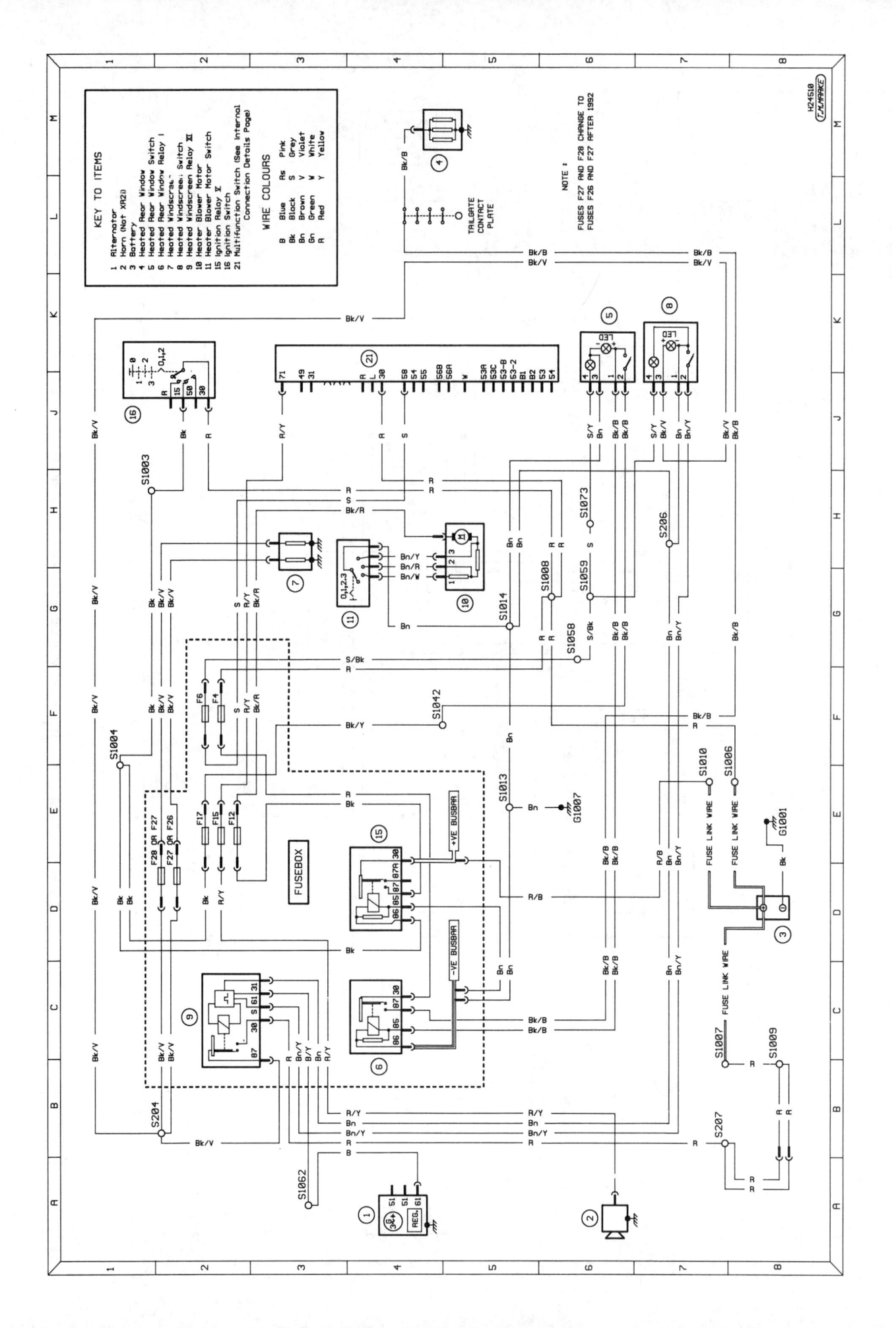

Diagram 3: Ancillary circuits – horn (except XR2i/RS models), heater blower and heated front/rear screens

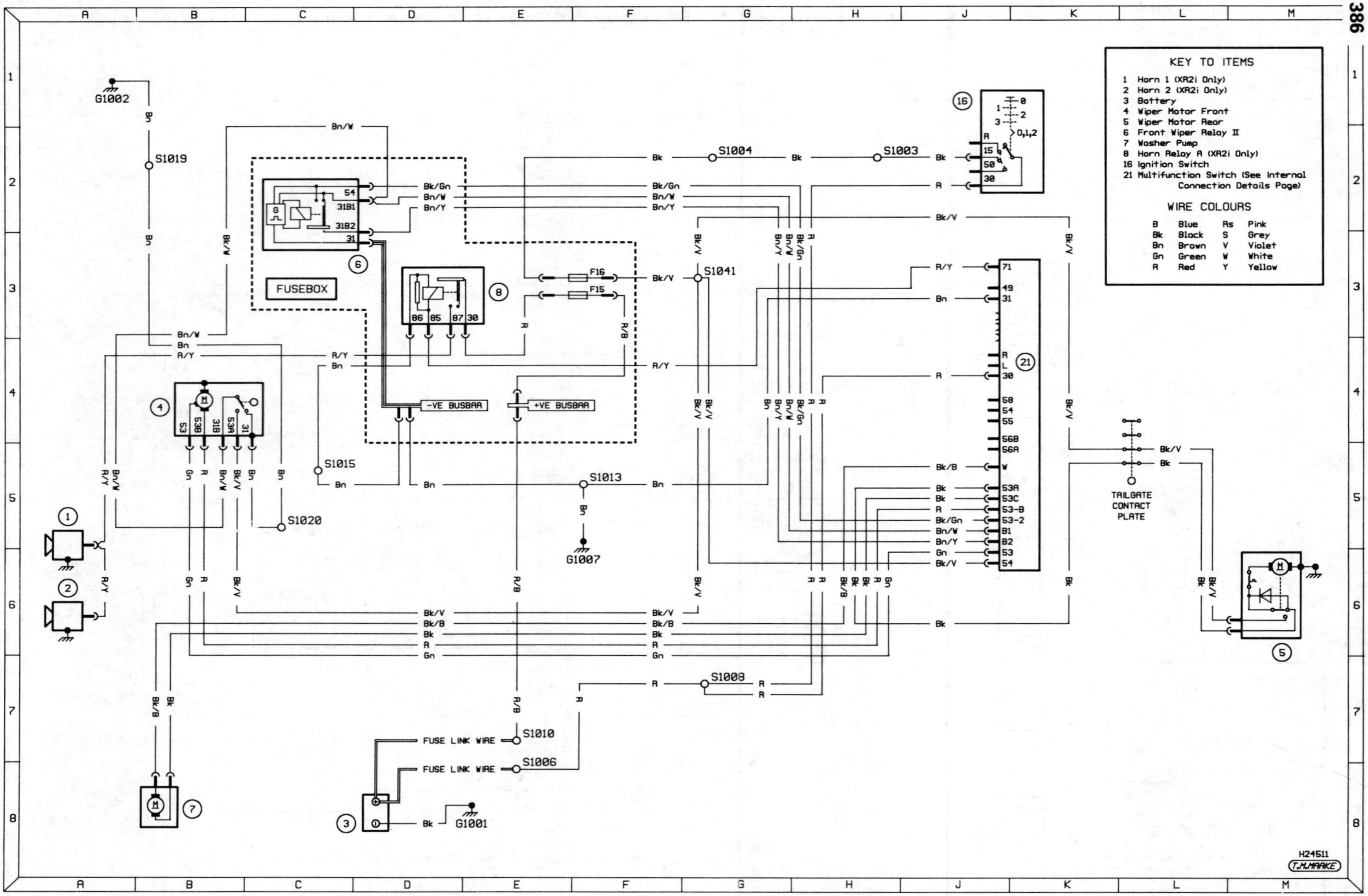

Diagram 3a: Ancillary circuits – horn (XR2i and RS models) and wash/wipe

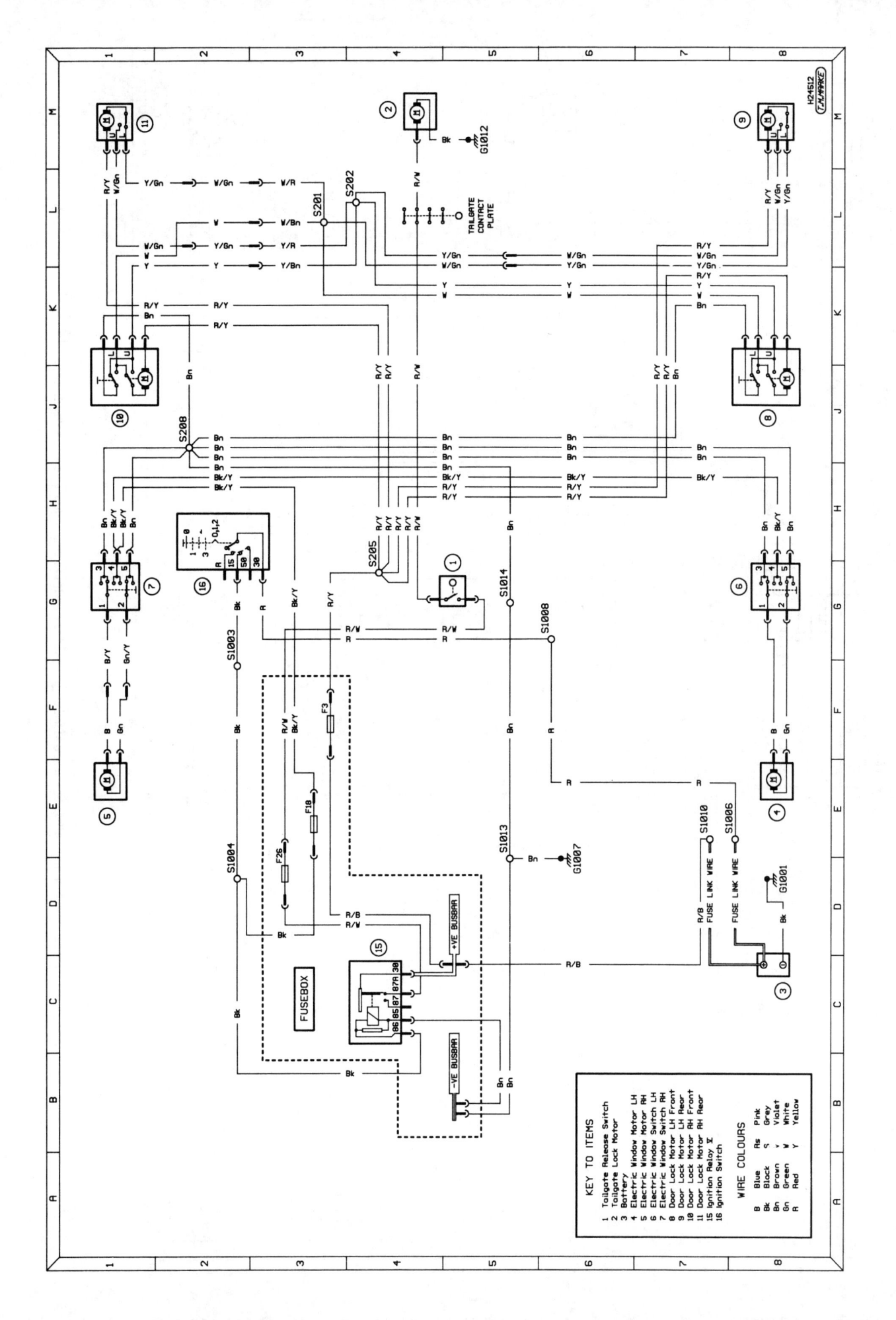

Diagram 3b: Ancillary circuits – central locking, electric windows and tailgate release (up to 1992)

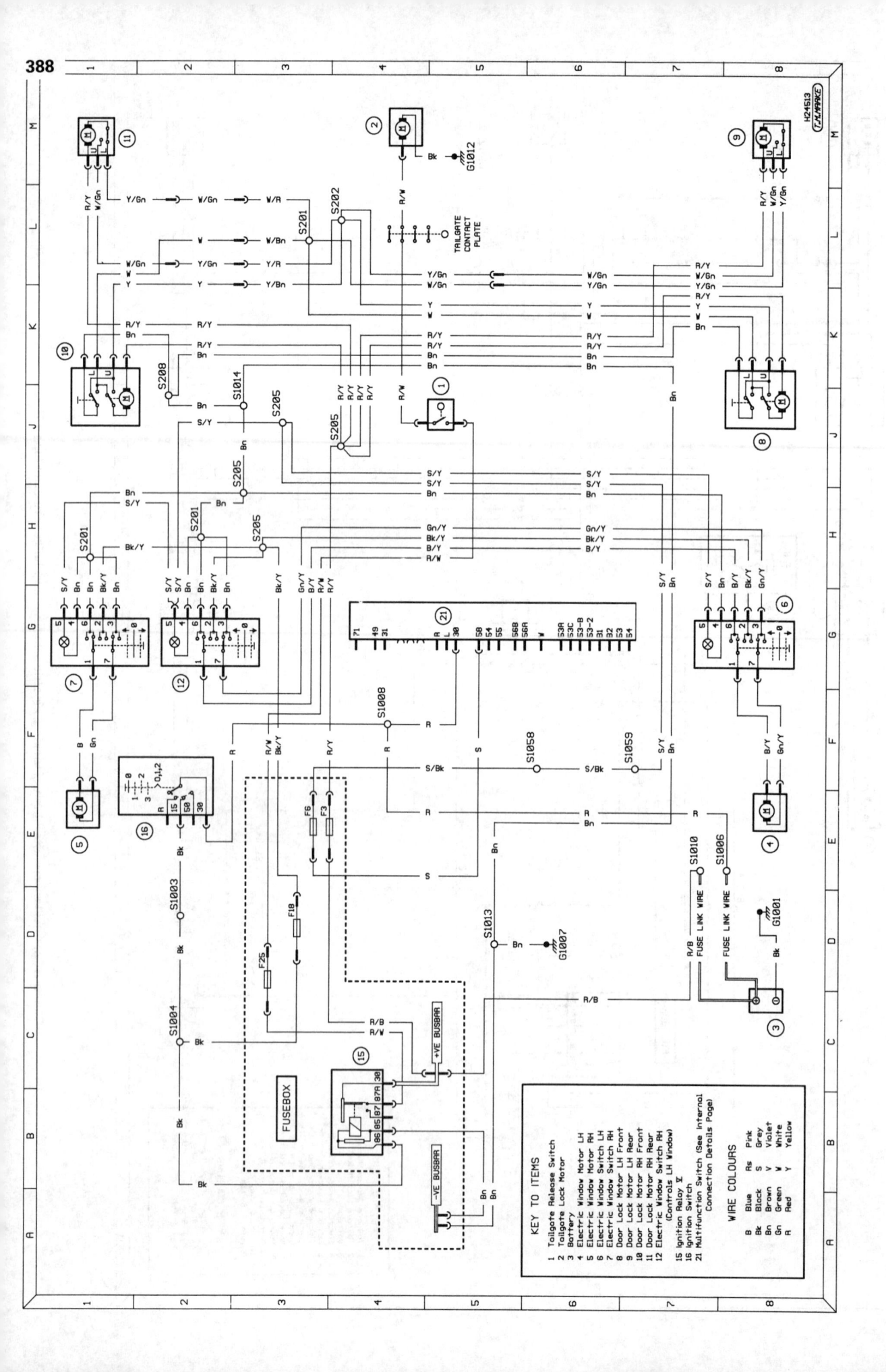

Diagram 3c: Ancillary circuits – central locking, electric windows and tailgate release (1992-on)

Diagram 4: In-car entertainment

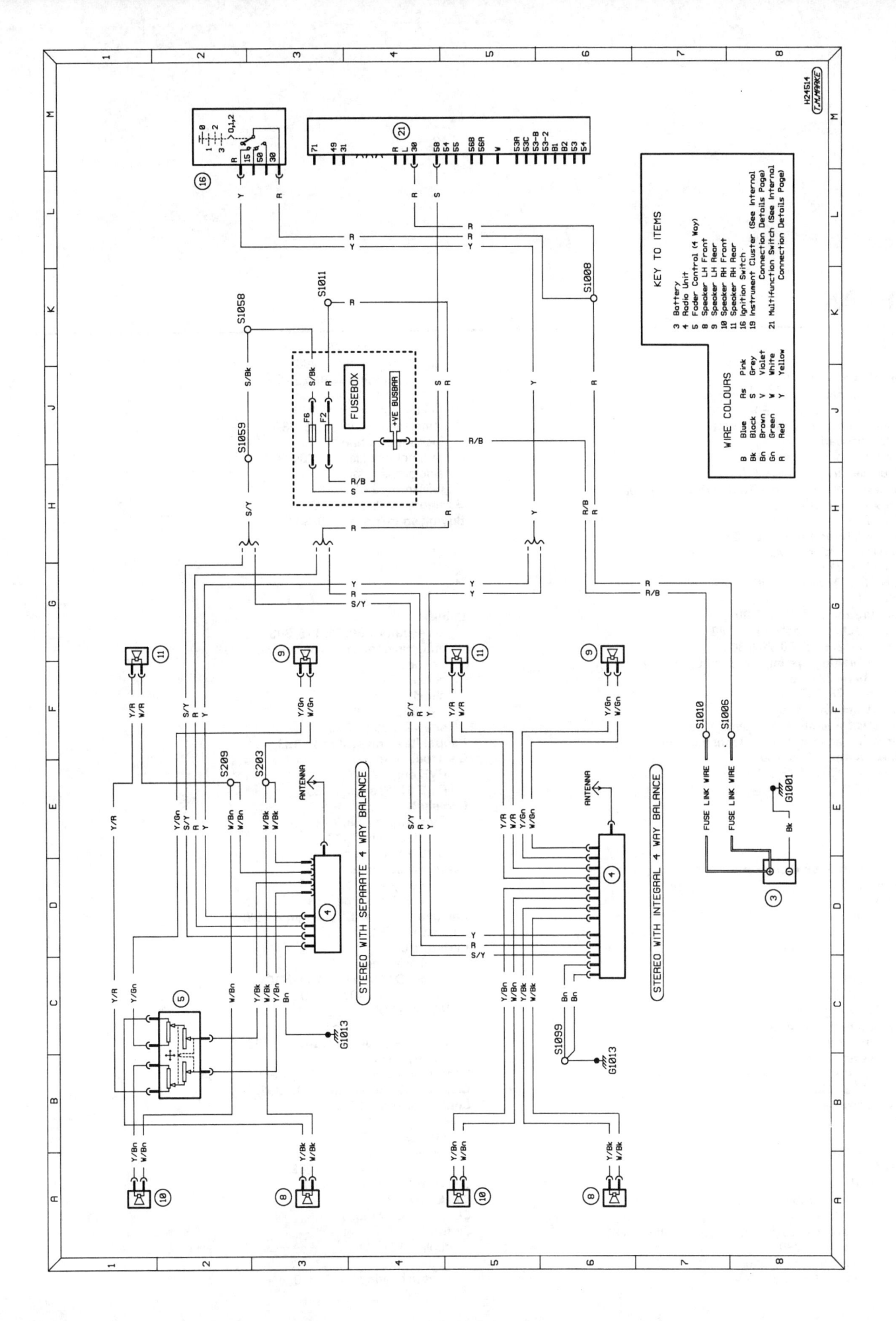

Index

A

About this manual – 5
Accelerator cable
   carburettor models – 90
   fuel injection models – 115, 348
Accelerator pedal
   carburettor models – 91
   fuel injection models – 115
Accelerator pump (Weber TLM carburettor) – 98
Acknowledgements – 2
Aerial – 275
Air change temperature sensor
   CVH CFi engine – 143
   CVH EFi engine – 147
   OHV (HCS) engine – 136
Air filter
   carburettor models – 89, 90
   fuel injection models – 114, 348
Alternator – 258 to 260, 262, 369
Anti-lock braking system – 203 to 207, 272, 360
Anti-roll bar – 221, 363
Antifreeze – 78, 337
Automatic choke – 108
Automatic transmission oil seal – 183
Automatic transmission see **Transmission**
Auxiliary lamps – 279, 280
Axle (rear) – 225 to 227, 362

B

Balance control (loudspeakers) – 274
Battery – 258
Bearings
   big-end – 51, 72
   camshaft – 53, 73
   clutch release – 150
   crankshaft – 51, 72
   hub – 220, 224, 361
   main – 51, 72
Big-end bearings – 51, 72
Bleeding (braking system) – 202, 207
Body damage repair – 231, 233
Bodywork and fittings – 230 et seq, 365 et seq
Bodywork repair – see colour pages between pages 32 and 33
Bonnet – 233, 365
Boost control valve – 345
Boost pressure – 345
Bores
   CVH engine – 72
   OHV (HCS) engine – 51
Brake pedal – 203
Braking system – 190 et seq, 271, 272, 360 et seq
Bulbs
   automatic transmission selector illumination – 281
   auxiliary lamps – 280
   cigarette lighter illumination – 281
   clock illumination – 281
   courtesy lamp – 281
   direction indicator – 280
   hazard warning switch – 281
   headlamps – 280
   heater illumination – 281
   instrument panel – 276, 281
   luggage compartment lamp – 281
   number plate lamp – 280, 374
   sidelamps – 280
   tail lamp – 280, 374
Bumpers – 235, 236, 365
Buying spare parts – 10, 307

C

Cables
   accelerator – 90, 91, 115, 348
   automatic transmission selector – 179
   bonnet – 233
   choke – 91
   clutch – 151
   handbrake – 198
   speedometer – 276
Caliper (braking system) – 193
Cam followers
   CVH engine – 74
   OHV (HCS) engine – 53
Camshaft
   CVH engine – 58, 73
   DOHC engine – 320
   OHV (HCS) engine – 53
Camshaft bearings
   CVH engine – 73
   OHV (HCS) engine – 53
Camshaft oil seal (CVH engine) – 58
Capacities – 7, 297
Carburettor
   modifications – 339
   Weber DTFM – 100, 103 to 105
   Weber TLD – 107 to 110
   Weber TLDM – 98 to 101
   Weber TLM – 96 to 98
Carrier plate (braking system) – 196
Cassette player – 273
Central fuel injection (CFi) system – 111, 340
Central locking system – 240, 282, 283
Centre console – 250, 271
Choke
   automatic – 108
   cable – 91
   warning light switch – 289
Cigarette lighter – 275, 281
Clock – 275, 281
Clutch – 149 et seq, 359
Clutch
   cable – 151
   pedal – 152
   release bearing – 150, 359

Index

Coil (ignition)
   CVH carburettor engine – 142
   CVH CFi engine – 144
   CVH EFi engine – 147
   OHV (HCS) engine – 37
Connecting rods
   CVH engine – 63
   DOHC engine – 329
   OHV (HCS) engine – 46, 51
Console – 250, 271
Conversion factors – 24
Coolant mixture – 78, 337
Coolant temperature sensor
   CVH CFi engine – 144
   CVH EFi engine – 147
   OHV (HCS) engine – 137
Cooling fan – 78, 337
Cooling, heating and ventilation systems – 76 et seq, 277, 337 et seq, 338
Courtesy lamp – 272, 281
Crankcase
   CVH engine – 73
   OHV (HCS) engine – 53
Crankshaft
   CVH engine – 62, 72
   OHV (HCS) engine – 42, 46, 51
Crankshaft bearings – 51, 72
Crankshaft oil seal
   CVH engine – 62
   DOHC engine – 327
Crankshaft position sensor
   CVH EFi engine – 147
   OHV (HCS) engine – 137
Crossmember (front suspension) – 222
CV joint (driveshaft) gaiter – 186
Cylinder bores
   CVH engine – 72
   OHV (HCS) engine – 51
Cylinder head
   CVH engine – 59, 61, 62, 74
   DOHC engine – 315, 323, 324
   OHV (HCS) engine – 37, 39, 53

D

Decarbonising – 39
Differential unit (manual transmission) – 168
Dimensions – 7, 297
Direction indicators – 278, 280
Disc (braking system) – 194
Distributor
   CVH carburettor engine – 140
   CVH CFi engine – 144, 145
Doors – 237 to 244, 282, 365, 366, 367
Draining cooling system – 77, 337
Drivebelt
   alternator – 258, 369
   modulator (anti-lock braking system) – 205
   water pump – 80
Driveshaft oil seal – 185
Driveshafts – 184 et seq, 359
Drum (braking system) – 194

E

Electrical system – 255 et seq, 369 et seq
Electric windows – 282, 374
Emblems – 234
Emission control systems – 125, 128, 355, 356

Engine – 28 et seq, 310 et seq
   CVH engine – 55, 310, 311, 312
   DOHC engine – 312, 314, 330, 331, 332
   OHV (HCS) engine – 36, 310, 311
Engine management module (EEC-IV)
   CVH CFi engine – 146
   CVH EFi engine – 148
   DOHC EFi engine – 347
Engine speed/crankshaft position sensor
   CVH EFi engine – 147
   OHV (HCS) engine – 137
Exhaust manifold – 123, 353
Exhaust system – 123, 353
Expansion tank (cooling system) – 82

F

Facia – 251, 271
Fan – 78
Fast-idle adjustment
   Weber DFTM carburettor – 104
   Weber TLD carburettor – 108
   Weber TLDM carburettor – 99
   Weber TLM carburettor – 96
Fault diagnosis
   braking system – 207
   clutch – 153
   cooling system – 85
   driveshafts – 189
   electrical system – 290
   engine – 54, 75
   fuel, exhaust and emission control systems – 129, 130
   general – 25
   heating and ventilation system – 85
   ignition and engine management systems – 138, 142, 146, 148
   suspension – 228
   steering – 215
   transmission – 178, 183
Filling cooling system – 77
Flushing cooling system – 77
Flywheel
   CVH engine – 73
   DOHC engine – 329
   OHV (HCS) engine – 52
Front brakes – 192, 193, 194
Fuel cut-off switch (fuel injection models) – 116, 289
Fuel, exhaust and emission control systems – 86 et seq, 277, 289, 339 et seq
Fuel filler pipe
   carburettor models – 93
   fuel injection models – 115
Fuel filter (fuel injection models) – 116, 341
Fuel gauge – 277
Fuel injection systems
   1.1 and 1.3 CFi – 340
   1.4 CFi – 111, 340
   1.6 EFi (Non-Turbo) – 111, 113, 341, 347
   1.6 EFi (Turbocharged) – 344
Fuel injection unit
   1.4 litre CFi – 118, 340
   1.1 and 1.3 litre CFi – 340
Fuel injectors
   1.4 litre CFi – 117
   1.6 litre EFi – 119, 344
Fuel level sensor unit
   carburettor models – 94
   fuel injection models – 116
Fuel pressure regulator
   1.4 litre CFi – 16
   1.6 litre EFi – 118, 349
Fuel pump
   carburettor models – 94
   fuel injection models – 115, 116

Index

Fuel starvation – 340
Fuel tank
    carburettor models – 92, 93, 94
    fuel injection models – 115, 340
Fuses – 270, 373

G

Gaiter
    driveshaft – 186
    steering rack – 210
Gaskets – 52, 73
Gearbox *see* **Transmission**
Gearchange mechanism (manual transmission) – 158, 162, 171
Gearchange selector (automatic transmission) – 179, 181, 281
Glove compartment – 251
Grab handle – 252
Grille – 254

H

Handbrake – 197, 198, 271
Hazard warning lamps – 281
Headlamps – 277, 279, 280
Heated rear window – 288
Heated windscreen – 288
Heater – 82 to 84, 271, 281, 288
Horn – 289
HT leads – 136, 139, 143, 147
Hubs
    front – 220, 361
    rear – 194, 224
Hydraulic pipes and hoses (braking system) – 192

I

Idle speed adjustment
    1.6 litre EFi – 120, 343, 344, 348
    Weber DTFM carburettor – 104
    Weber TLD carburettor – 107
    Weber TLDM carburettor – 99
    Weber TLM carburettor – 96
Idle speed control valve (1.6 litre EFi) – 121, 343, 350
Ignition amplifier module (CVH carburettor engine) – 141
Ignition and engine management systems – 132 *et seq*, 357 *et seq*
Ignition coil
    CVH carburettor engine – 142
    CVH CFi engine – 144
    CVH EFi engine – 147
    OHV (HCS) engine – 137
Ignition control module
    CVH CFi engine – 146
    CVH EFi engine – 148
    OHV (HCS) engine – 137
Ignition switch – 272
Ignition timing
    CVH carburettor engine – 140
    CVH CFi engine – 145
    OHV (HCS) engine – 138
Indicators – 278, 280
Inertia (fuel cut-off) switch (fuel injection models) – 116, 289
Injectors
    1.4 litre CFi – 117
    1.6 litre EFi – 119
Inlet manifold – 122, 341, 353
Inner CV joint (driveshaft) gaiter – 186
Input shaft (manual transmission) – 168
Instrument panel – 276, 281
Intake-air temperature control (carburettor models) – 90
Intercooler – 344
Introduction to the Ford Fiesta – 5

J

Jacking – 8

L

Light-laden valve (Courier) – 360
Load compartment dividers (Van) – 254
Load-apportioning valve (anti-lock braking system) – 206
Locks
    central locking – 240, 282, 283
    door – 240
    tailgate – 245
Loudspeakers – 252, 273, 274
Lower arm (front suspension) – 222
Lubricants and fluids – 23
Lubrication
    CVH engine – 65
    OHV (HCS) engine – 41
Luggage compartment lamp – 281

M

Main bearings – 51, 72
Mainshaft (manual transmission) – 164
Maintenance *see* **Routine maintenance**
Manifold absolute pressure sensor
    CVH CFi engine – 145
    CVH EFi engine – 148
Manifolds – 122, 123, 353
Manual gearbox *see* **Transmission**
Manual transmission oil seal – 177
Master cylinder (braking system) – 199
Mirrors – 239, 252
Mixture adjustment
    1.6 litre EFi – 120, 348
    tamperproof caps – 339
    Weber DFTM carburettor – 104
    Weber TLD carburettor – 107
    Weber TLDM carburettor – 99
    Weber TLM carburettor – 96
Mixture/CO adjustment potentiometer (CVH EFi engine) – 148
Modulator (anti-lock braking system) – 204, 205
Mountings
    automatic transmission – 183
    CVH engine – 65

N

Number plate lamp – 278, 280, 374

O

Octane adjust (OHV (HCS) engine) – 138
Oil pressure switch – 273
Oil pump
    CVH engine – 73
    DOHC engine – 328
    OHV (HCS) engine – 42, 52
Oil seals
    automatic transmission – 183
    camshaft – 58
    crankshaft – 42, 46, 62, 327
    driveshaft – 185
    general – 52, 73
    manual transmission – 177
Outer CV joint (driveshaft) gaiter – 186
Overriders (bumper) – 236

Index

P

Pads (braking system) – 192
Parcel shelf – 252
Pedals
    accelerator – 91, 115
    brake – 203
    clutch – 152
Pistons
    CVH engine – 62, 63, 72
    DOHC engine – 329
    OHV (HCS) engine – 39, 46, 51
Pressure control valve (braking system) – 202
Pressure regulator
    1.4 litre CFi – 116
    1.6 litre EFi – 118
Printed circuit – 276
Pulse-air system – 351

Q

Quick-release fuel line couplings – 340

R

Radiator – 79, 337, 338
Radiator cooling fan – 78, 79, 273, 337, 338
Radiator grille slat – 254
Radio – 273, 275, 290
Rear axle – 225 to 227, 362
Rear brakes – 194, 196, 360
Rear suspension (Courier) – 363, 364
Relays – 270, 373
Repair procedures – 12
Reversing light switch – 272
Road speed sensor unit
    CVH CFi engine – 144
    CVH EFi engine – 147
Roadwheel – 228, 253
Rocker arms (CVH engine) – 74
Rocker gear (OHV (HCS) engine) – 53
Rocker shaft (OHC (HCS) engine) – 40
Rotor arm
    CVH carburettor engine – 140
    CVH CFi engine – 144
Routine maintenance – 16 *et seq*, 307 *et seq*
    bodywork and fittings – 16, 231, 308, 309
    braking system – 16, 191, 308, 309
    cooling system – 16, 77, 308, 309
    driveshafts – 16, 185, 308, 309
    electrical system – 16, 257, 308, 309
    engine – 16, 36, 56, 308, 309
    fuel, exhaust and emission control systems – 16, 89, 114, 122, 308, 309
    ignition and engine management systems – 16, 135, 139, 143, 147, 308, 309
    steering – 16, 210, 308, 309, 310
    suspension – 16, 218, 308, 309
    transmission – 16, 158, 179, 308, 309, 310

S

Safety first!
    carburettor fuel system – 89
    electrical system – 257
    fuel injection system – 114
    general – 15
    ignition system – 135, 139, 143, 147
    transmission – 179
Seat belts – 248, 249, 250
Seats – 246, 369
Servo (braking system) – 200, 201
Shoes (braking system) – 194

Sidelamps – 280
Spare parts – 10
Spare wheel carrier – 253
Spark plugs – 136, 139, 143, 147
Spark plug condition – see colour pages between pages 32 and 33
Speedometer – 171, 276, 277
Spindle carrier (front suspension) – 219
Spoiler – 245
Starter inhibitor switch (automatic transmission) – 182, 272
Starter motor – 264, 265, 370, 372
Steering – 209 *et seq*, 361 *et seq*
Steering column – 212, 281
Steering rack – 210, 213
Steering wheel – 211, 365
Stop-lamp switch – 271
Strut
    front suspension – 223
    rear suspension – 226
    tailgate – 244
Sump
    CVH engine – 63
    DOHC engine – 326
    OHV (HCS) engine – 41
Sunroof – 246, 368
Sun visor – 252
Supplement: Revisions and information on later models – 292 *et seq*
Suspension – 215, 216 *et seq*, 361 *et seq*
Suspension angles – 215
Switches
    anti-lock braking system modulator belt-break – 272
    brake fluid level warning – 272
    centre console – 271
    choke warning light – 289
    cooling fan – 79
    courtesy light – 272
    facia panel – 271
    fuel cut-off (fuel injection models) – 116
    handbrake-on – 271
    heater – 271
    ignition – 272
    oil pressure – 273
    radiator fan – 273
    reversing light – 272
    starter inhibitor (automatic transmission) – 182, 272
    steering column – 281
    stop-lamp – 271
    sunroof – 374
    window – 374

T

Tachometer – 277
Tail lamp – 278, 280, 374
Tailgate – 244, 245, 283, 284, 286, 288
Temperature gauge – 277
Thermostat – 80
    DOHC engine – 338
Throttle kicker
    Weber DFTM carburettor – 100, 105
    Weber TLDM carburettor – 99, 100
Throttle position sensor
    1.4 litre CFi – 118, 341
    1.6 litre EFi – 121, 350
Throttle-plate control monitor (1.4 litre CFi) – 117
Timing belt
    CVH engine – 57
    DOHC engine – 315, 316
Timing belt tensioner (DOHC engine) – 319
Timing sprockets and belt (CVH engine) – 72
Timing sprockets and chain (OHV (HCS) engine) – 52
Tools – 13
Torsional vibration damper (automatic transmission) – 183
Towing – 8
Track rod ends – 211
Transmission – 43, 65, 155 *et seq*, 272, 281

Index

Trim panels – 234, 236, 238, 245, 247
Turbocharger – 344, 345
Tyre pressures – 216, 306
Tyres – 228

U

Unleaded petrol (OHV (HCS) engine) – 138

V

Vacuum diaphragm unit (CVH carburettor engine) – 141
Vacuum servo (braking system) – 200, 201
Valve clearances (OHV (HCS) engine) – 40
Vehicle identification numbers – 10, 307
Ventilation see **Cooling, heating and ventilation systems**
Voltage regulator – 262

W

Washer jets – 286, 287
Washer pump – 287
Washer reservoir – 287
Water pump
   CVH engine – 81
   DOHC engine – 338
   OHV (HCS) engine – 80
Weatherseal
   door – 237
   sunroof – 246
Weights – 7, 297
Wheel alignment – 215
Wheelarch liner – 253
Wheelarch moulding – 253, 254
Wheel changing – 8
Wheel cylinder (braking system) – 196
Wheels – 228, 253
Wind deflector – 254
Window regulator – 241, 242, 243
Windows – 244, 282, 365, 366
Windscreen – 244, 284, 286, 287, 288
Wiper blades and arms – 284
Wiper motor – 284, 286
Wiring diagrams – 375
Working facilities – 13